Paul Boudreau

Rees

Mold Engineering

SPE Books from Hanser Publishers

Belofsky, Plastics: Product Design and Process Engineering
Bernhardt, Computer Aided Engineering for Injection Molding
Brostow/Corneliussen, Failure of Plastics
Chan, Polymer Surface Modification and Characterization
Charrier, Polymeric Materials and Processing—Plastics, Elastomers and Composites
Ehrig, Plastics Recycling
Gordon, Total Quality Process Control for Injection Molding
Gordon/Shaw, Computer Programs for Rheologists
Gruenwald, Plastics: How Structure Determines Properties
Macosko, Fundamentals of Reaction Injection Molding
Malloy, Plastic Part Design for Injection Molding
Manzione, Applications of Computer Aided Engineering in Injection Molding
Matsuoka, Relaxation Phenomena in Polymers
Menges/Mohren, How to Make Injection Molds
Michaeli, Extrusion Dies for Plastics and Rubber
Michaeli, Training in Plastics Technology
O'Brien, Applications of Computer Modeling for Extrusion and Other Continuous Polymer Processes
Progelhof/Throne, Polymer Engineering Principles
Rauwendaal, Polymer Extrusion
Rees, Mold Engineering
Saechtling, International Plastics Handbook for the Technologist, Engineer and User
Stoeckhert, Mold-Making Handbook for the Plastics Engineer
Throne, Thermoforming
Tucker, Fundamentals of Computer Modeling for Polymer Processing
Ulrich, Introduction to Industrial Polymers
Wright, Injection/Transfer Molding of Thermosetting Plastics
Wright, Molded Thermosets: A Handbook for Plastics Engineers, Molders and Designers

Herbert Rees

Mold Engineering

With 552 Figures

Hanser Publishers, Munich Vienna New York

Hanser/Gardner Publications, Inc., Cincinnati

The Author:
Herbert Rees, 248386-5 Side Road (Mono), RR #5, Orangeville, Ontario, Canada L9W 2Z2

Distributed in the USA and in Canada by
Hanser/Gardner Publications, Inc.
6600 Clough Pike, Cincinnati, Ohio 45244-4090, USA
Fax: (513) 527-8950 or 1-800-950-8977

Distributed in all other countries by
Carl Hanser Verlag
Postfach 86 04 20, 81631 München, Germany
Fax: +49 (89) 98 12 64

Library of Congress Cataloging-in-Publication Data
Rees, Herbert, 1915–
Mold engineering / Herbert Rees.
p. cm.
Includes index.
ISBN 1-56990-131-7
1. Plastics--Molds. 2. Plastics--Molding. I. Title.
TP1150.R447 1995
668.4'12--dc20 95-2590

Die Deutsche Bibliothek - CIP-Einheitsaufnahme
Rees, Herbert:
Mold engineering / Herbert Rees. - Munich ; Vienna ; New York : Hanser ; Cincinnati : Hanser/Gardner, 1995
(SPE books)
ISBN 3-446-17729-9

Typeset in the USA by pageAbility, Watertown
Printed and bound in Germany by Ludwig Auer GmbH, Donauwörth

Foreword

The Society of Plastics Engineers is pleased to sponsor *Mold Engineering* by Herbert Rees. This presentation is intended to provide an uncomplicated, yet in-depth understanding of the design, operation and life expectancy of an injection mold.

This publication has been developed not only as a teaching guide, but for anyone concerned with making, buying or operating injection molds.

SPE, through its Technical Volumes Committee, has long sponsored books on various aspects of plastics. Its involvement has ranged from identification of needed volumes and recruitment of authors to peer review and approval and publication of new books.

Technical competence pervades all SPE activities, not only in the publication of books but also in other areas such as sponsorship of technical conferences and educational programs. In addition, the Society publishes periodicals—including *Plastics Engineering, Polymer Engineering and Science, Polymer Processing and Rheology, Journal of Vinyl Technology* and *Polymer Composites*—as well as conference proceedings and other publications, all of which are subject to rigorous technical review procedures.

The resource of some 38,000 practicing plastics engineers, scientists, and technologists has made SPE the largest organization of its type worldwide. Further information is available from the Society at 14 Fairfield Drive, Brookfield, Connecticut 06804, U.S.A.

Michael R. Cappelletti
Interim Executive Director
Society of Plastics Engineers

Technical Volumes Committee
Claire Bluestein, Chairperson
Captan Associates, Inc.

Preface

Mold Engineering deals with injection molds for thermoplastic molding materials and their performance. However, many of the subjects treated here also apply to other molds such as blow molds and molds for thermosets. It presents in an easy to understand language, with a minimum of mathematics or difficult theories, a practical approach for the design of injection molds.

The subjects are broken down into parts as follows:

Part I: Mold Engineering
- Section 1: Basics About Molds, Machines, Plastics, and Products
- Section 2: General Mold Design Guidelines
- Section 3: Specific Subjects for the Mold Designer

Part II: Mold Performance

A significant portion of the book is devoted to explaining the relationships affecting mold performance, productivity, and mold life. Understanding these relationships is not only important for mold designers but also for any person who is involved with planning (and cost estimating), buying, and operating injection molds.

The book gives step-by-step guidelines for the design of molds, from product drawing to complete mold assembly drawing. The designer is shown how to study any product drawing before starting to design a mold.

There is an infinite number of molded products and possible arrangements of molds for them; therefore, rather than showing complete mold assemblies that were built for some selected products, molds are broken down into basic elements. The various possibilities the designer could follow are then discussed.

Subjects such as shrinkage, venting, cooling, ejection, heat expansion, gating, hot and cold runner systems, balancing of cavity layouts, and tolerancing are covered, as well as mold materials selection, heat treat, and finishing of molded parts. There is also a multitude of reference and conversion charts, and information about mold handling. In addition, there is also a wealth of information for technicians working in areas such as molding machine set-up, servicing, and sales of molds and molding machines, for moldmakers, designers, estimators, and quality control personnel at molders and moldmakers.

This book is intended primarily as a text for students in polymer engineering and a reference book for mold designers, particularly for designers of injection molds, although there are many subjects also applicable to die casting molds, compression molds, as well as any other type of permanent mold into which materials are injected to create a product.

We hope this approach will be of value and help for all those trying to sort out the puzzles they are confronted with during their work with injection molds.

Herbert Rees

Contents

Part I Mold Engineering
Section 1 Basics About Molds, Machines, Plastics, and Products

Part I Mold Engineering

Section 1 Basics About Molds, Machines, Plastics, and Products

1 Introduction to Mold Engineering

Mold Engineering deals with injection molds for thermoplastic molding materials. However, much of the subjects treated herein apply also to other molds, such as blow molds, and to some types of thermoset molds.

The purpose of Part I: Mold Engineering is to familiarize not only beginners in mold design but also more experienced mold designers with an engineering approach to mold design. *Mold Engineering* is also recommended literature to any individuals, such as sales and service personnel, in the plastics industry who need to understand molds in the course of their activities. Part I: Mold Engineering is divided into three sections:

Section 1: Basics About Molds, Machines, Plastics and Products,
Section 2: General Mold Design Guidelines, and
Section 3: Specific Subjects for the Mold Designer.

There are several other good books on mold design. Some of them are listed at the end of this chapter [1–5]. These books contain mostly designs and illustrations of molds built for very specific applications and plastics. They show many complete mold assembly drawings having many different features, without explaining in detail why these features were selected. In many cases, the molding technology shown is about 20–30 years behind current standards. This is not to say that these books are without value. On the contrary, a designer may find many interesting solutions to problems that may arise when designing a mold, provided one does not just copy them but uses ideas from these designs in combination with today's technology.

Part I: Mold Engineering, rather than showing complete assemblies, breaks the mold design down into its various features (elements) and treats them separately, from an engineering point of view. It also explains why and when certain features are to be selected, and when not.

1.1 What is an Injection Mold?

An injection mold is a arrangement, in one assembly, of one (or a number of) hollow *cavity spaces* built to the shape of the desired product, with the purpose of producing (usually large numbers of) plastic parts, or *products*.

The cavity space is generated by a female mold part, called the *cavity*, and a male mold part, called the *core*. To fill the cavity spaces, the mold is mounted in an injection molding machine that is timed (usually automatically) to:

- *Close* the mold,
- *inject* the (hot, more or less fluid) plastic into the cavity spaces,
- *keep the mold closed* until the plastic is cooled and ready for ejection,
- *open* the mold, and

- *eject* the finished products. Also,
- if necessary, the machine may stay open an additional *mold open (MO) time* to ensure that the mold is ready for the next (injection) cycle, before closing.

The *molding cycle* (in seconds) in a fully automatic (FA) operation is defined as the time from the moment the mold is closed for one injection, or *shot*, until it is closed again for the following shot. Usually, number of shots per minute (or shots per hour) are given to indicate productivity of a mold, rather than the molding cycle in seconds.

1.2 What is an Injection Molding Machine?

It is important that the designer first understands the action and terminology of the injection molding process. An injection molding machine consists essentially of four (4) distinct elements:

1. Clamping mechanism,
2. plasticizing unit,
3. injection unit, and
4. all necessary controls.

1.2.1 Clamping Mechanism

The clamping mechanism opens and closes the mold (preferably rapidly) as required during the cycle. It must also supply the necessary *clamping force* to keep the mold closed during injection, because the injection pressure acting on the internal, or *projected*, surface of the cavity space tends to open the mold at the *split-* or *parting-plane*, also called *parting line (P/L)*.

1.2.2 Plasticizing Unit

Today's plasticizing unit is almost exclusively an *extruder* that heats the cold plastic material to the required temperature to make it fluid for injection, or *melt*. The heating is generated mostly by the mechanical energy (created by the screw motor) as the extruder screw rotates in the barrel and works the plastic. This screw action also advances the plastic toward the tip of the screw.

Heaters around the barrel, usually in three or more heating zones, provide additional heating, which is mainly required during start-up of the machine but also where the mechanical working of the screw alone would not plasticize the amount of plastic required for each shot.

1.2.3 Injection Unit

An injection unit forces the melt (under pressure) into the mold. The level of pressure required to fill a mold depends largely on the wall thickness of the product.

Heavy-walled products require relatively low pressure (500–1,000 kg/cm^2, or 7,000–14,000 psi). There are even cases where the extruder pressure alone is sufficient to fill a cavity (flow molding). Thin-walled products, especially if the *L/t* ratio (see definition on page 7) is greater than 200, require very high injection pressures (1,400–2,000 kg/cm^2, or 20,000–28,000 psi) to ensure that the cavities will fill before the plastic freezes.

There are two injection methods used, either single-stage or two-stage, and the types of machine used are discussed below.

1.2.3.1 RS Machines

Today, in most injection molding machines, the extruder and the injection unit are combined into one unit. The extruder screw is stopped when enough melt is prepared for the next injection, then the screw is pushed forward to inject the melt accumulated in front of the screw tip. These units are called by various descriptions: in-line, ram screw or reciprocating screw (RS), or single-stage injection units.

Manufacturers rate screws by the amount of plastic they can plasticize per hour. However, due to limits imposed by the thrust bearing size and strength, the extruder can only plasticize during that portion of the molding cycle when it is *not* injecting; therefore, it is actually used at less than its "rated" plasticizing capacity (indicated in the machine specifications). With large shots, and with slow injection, the time required to inject takes a relatively large portion of the molding cycle, and the screw is often able to plasticize , or *recover*, only 60–80% of its rated capacity.

The amount of plastic pushed into the mold depends also to a large extent on the efficiency (tightness) of the *check valve* at the tip of the screw. If this valve is poorly designed or worn, some plastic will leak past or through the check valve during the high pressure injection cycle. This will affect the amount of plastic entering the cavity spaces, and thereby the productivity of the mold, by causing *short shots* or *packing* (over-filling); this affects the weight (density), size (due to varying shrinkage) and, in general, the quality of the product.

To avoid some of the effects of varying shot size, molders using RS machines usually prepare a shot size greater than required for the shot, so that the screw will never come completely forward, or *bottom out*, but create a *cushion* of melt of about 5–10 mm at the tip of the screw.

1.2.3.2 P Machines

A preplasticizing machine system separates the functions of the extruder and the injection unit. The extruder plasticizes the material and fills an injection cylinder, or *pot*, of the injection unit. These machines may be called preplasticizing, two-stage, or simply P machines.

Advantages of the two-stage system are:

- The screw can run continuously, therefore plasticizing 100% of the available time (and its rating). This may permit the use of a smaller extruder than for an equivalent size RS machine.
- By running continuously, the P machine produces melt that is better mixed and can be held at lower temperatures than with an RS machine. This may be very desirable with certain heat-sensitive materials.

- There is no check valve at the screw. Also, the shot volume in the injection pot is mechanically measured, and the repeatability and accuracy of the shot size is greater than with an RS machine. No cushion is required, and the volume of plastic injected can be matched perfectly to the volume of the cavity space(s).
- Because the transfer from extruder to pot takes place under very low pressure, it is easily possible to place an effective *filter* in the path of the plastic to remove any dirt in the plastic. This will not affect (reduce) the injection pressure from the pot to the mold. Such filtering is not practical with RS machines,where too much pressure would be lost in a filter.

Disadvantages of this system are:

- Higher cost of the machine, because more hardware and more controls are required, and
- this system is not suitable for some very heat-sensitive materials, such as PVC.

1.2.4 Controls

Controls make the molding machine operate. There are four basic elements of molding machine controls:

- The *command module* is located near the safety gate of the clamp, where the operator can observe the mold. There, the operator has easy access to the pushbuttons to operate all functions manually. In some machines, the operating push buttons and the controls for the machine settings are in the panel near the safety gate.
- The *control logic* executes the machine settings and manipulates the signals from positional sensors, timers, etc., to make the machine perform as specified. Today, the machine logic operates almost exclusively using electronic switching or uses a microcomputer. [Note that mechanical relays, limits switches, and timers have a much shorter service life than electronic switches or timers, and are much less reliable and repetitive. However, they are easier to understand and to service than electronic switching, which requires better qualified service personnel and better electrical measuring instruments for checking and servicing. On the other hand, electronics are more sensitive to elevated temperatures found in some hot countries and may require provisions for cooling the control cabinet.]
- The *power supply* and distribution to the motors and heaters, and
- The *heat controls* for the machine and mold.

There are other features of an injection molding machine, mostly for the convenience of the user. However, for the purpose of understanding the injection process, the above described basic elements are sufficient.

1.3 Mold Timing and Terminology

Dry-cycle: The *total* time required for the (machine) clamp to close and open, or the sum of the *mold opening* and the *mold closing time*. Today's fast machines have dry-cycles in the order of 1–3 seconds. A short dry-cycle is of particular importance with fast-cycling molds. The dry-cycle also depends on the length of the clamp stroke.

Opening time: Usually quite fast. The *ejection* preferably should take place during this time, to reduce (or to omit completely) any *mold open time*. Occasionally, mold opening speed may need to be slowed down to suit the ejection method.

Closing time: Usually quite fast, except for the final approach before the mold is fully clamped up, to permit the mold protection system to operate in time before serious damage is caused to the mold.

Mold protection: A system which senses (at the moment of final closing the mold) whether there is foreign material (dirt, plastic pieces, products which failed to eject, etc.) between the mold halves which could cause damage to the mold. A signal from the mold protection system will cause the mold closing to stop before damage occurs and sound an alarm. It usually automatically reopens the clamp so the foreign material can be removed.

There are many types of mold protection systems, such as electric, optical, or pressure activated. Some are more sensitive than others and may not always save a mold from damage.

Mold open (MO) time: This is lost time. This time should be as little as possible. It can be zero.

Mold closed (MC) time: Time from the moment the mold has closed until it reopens. It is the sum of the following times:

Injection time: the time to fill the mold with plastic (usually with high injection pressure).

Hold time: the plastic in the cavities is held under pressure lower than injection to add plastic volume as the plastic shrinks within the cavity.

Cooling time: the time from the moment the injection (or hold) pressure is off until the mold starts opening. (This term is actually a misnomer, since the cooling is always on and starts to remove heat from the plastic as soon as the plastic enters the mold.)

Ejection time: the time required to eject the products from the molding area so that the mold can re-close without catching an ejected piece. Preferably, this should take place during the opening time so as to eliminate the need for additional MO time. In some molds, it is not possible or practical to eject during the mold opening, and the ejection takes place partly or solely during the MO time.

The above-defined terms can be shown on a graph describing a complete molding cycle. Figure 1.1 is a very simple graph. All motions in a mold can and should be described in such a graph. This is particularly useful where there are several motions or timed air functions

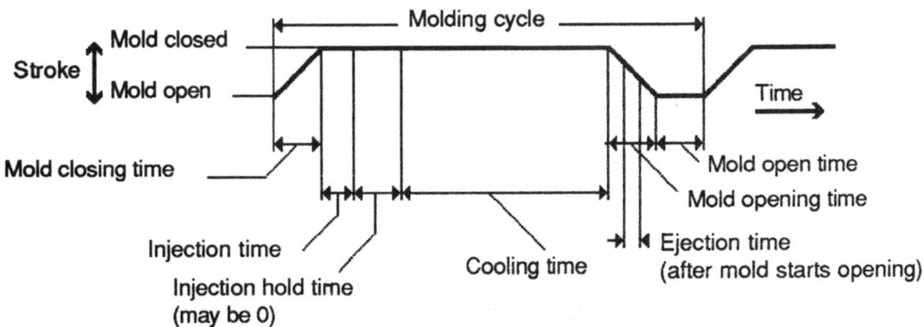

Figure 1.1 Timing diagram showing complete molding cycle.

within the mold and wherever there are auxiliary mechanisms, such as product removal systems (robots, chutes, guide rails, etc.). For more discussion of this type of graph, see Chapter 7, Operation Sequences.

Following are detailed explanations of the times shown in the above schematic, for better understanding of the operation of a mold and of the various features influencing the times.

1.3.1 Mold Closing and Opening Times (Dry-Cycle)

Some times are machine related. Some machines are faster than others, and speeds can be varied by settings of the machine. It is very important to understand that both the closing and opening times are more or less "wasted" times; that is, the longer they are, the smaller the productivity. Note, however, that during the mold opening stroke, the ejection can take place. The shorter the stroke, the less time is required, both for opening and for closing.

In Fig. 1.1, the entire ejection takes place during mold opening, and there should be no need for the additional MO time shown. If, however, the ejection time becomes larger and is not finished by the time the mold arrives in the mold open position, some MO time will be required.

In some cases, it may be unavoidable to delay the ejection until the mold is fully open; for example, when unloading the mold with a robot or other mechanism which is *not* driven and interlocked (synchronized) mechanically with the clamp motion. But as a rule, the mold should be designed to eject during the opening stroke without the need for MO time.

Elimination of MO time is of particular importance with thin-walled products and any other type of product which can run at very short molding cycles.

Example: A container could run at a total mold closed time (injection + cooling) of 4 seconds. With a dry-cycle of 2 seconds, the total cycle would be 6 seconds, or 3,600 seconds/hour ÷ 6 seconds = 600 shots/hour. With a longer dry-cycle of 3 seconds, the cycle would be 7 seconds, or 3,600 seconds/hour ÷ 7 seconds = 514 shots/hour; obviously a large loss of production.

There is a limit of how fast a machine can cycle. A larger machine, with much larger masses to be moved, will usually run slower than a small machine. As a rule, the better a machine, the shorter the dry-cycle.

In some cases, the opening of a mold must be slowed down at the start of opening for mechanical reasons. For example, there may be a large mold separating force required, or the vacuum in the mold must be allowed to dissipate without damaging the product.

1.3.2 Ejection Time and Mold Opening Stroke

Ejection time is usually mold related. For example, in free fall ejection, the mold must open sufficiently far (long opening stroke) to allow all products to clear the molding area. (The larger the opening stroke, the more time is required for opening and closing, thus increasing the dry-cycle).

A long, vertical mold requires more time for the products to fall and requires more ejection time than a similar mold with a long, horizontal layout. For more information on this, see Chapter 6, Mold Layout, "Arrangement of Cavities".

1.3.3 Mold Open Time

Mold open (MO) time is *always* mold design related. MO time should be avoided whenever possible by trying to complete the ejection before the end of the open time.

1.3.4 Injection Time

The injection time depends on three factors: the machine, the mold design, and the plastic.

1.3.4.1 Injection Time and the Machine

The faster the machine can inject the required amount of plastic, the shorter the injection time. All machines are rated in the number of cm^3 PS (polystyrene) per second. This rate usually depends on the size of the hydraulic pump and motor of the machine, and the availability of accumulators to assist the injection by supplying stored high-pressure oil when the pump cannot deliver enough hydraulic oil to the injection cylinder in the required time.

1.3.4.2 Injection Time and Mold Design

Fast filling (and short injection time) depends on:

- The *pressure drop* in the runner system from machine nozzle to, and including, the gate in the cavity. The smaller the pressure drop, the faster the mold will fill. But there are several restrictions. Large runners will reduce the pressure drop, but with cold runners, they will take a long time to cool, and with hot runners, the plastic inventory in the manifold becomes too large, which may affect the quality of the melt. Large gates in cold runners are unsightly and may need cutting after ejection. In hot runners, valve gates can provide large passages.
- *Number of cavities*. The more cavities, the longer it takes to fill the mold. Also, as the length of cold runners increases, they require additional plastic at every cycle. In hot runners, there is no need for more plastic once the manifold is filled, but the pressure drop increases with the runner length.
- *Product shape*. There is little the mold designer can do about product shape, but one must understand that the length of flow from gate to rim, and the wall thickness of the product are greatly influencing the speed of filling. A short flow length and a heavy wall thickness offer little resistance to filling, but a long flow and thin walls resist the filling severely, and require much higher injection pressures to push the plastic from the gate toward the rim of the product. In addition, as the hot plastic enters the cold mold, it will immediately start to freeze and further restrict the passage between the cavity and core walls.
- *L/t Ratio* (pronounced "*L* over *t* ratio"). For thin-walled products, any ratio between flow length (*L*) and wall thickness (*t*) of greater than 200 must be treated with special attention. High injection pressures will require stronger cavity walls; also, the higher mold clamping force required to counteract the high injection pressure will affect the specific pressures on the parting line of the mold.

1.3.4.3 *Plastic*

Some plastics cannot be injected at very high speeds, to avoid degrading or burning. This must be established before starting the design by checking records of molds using similar materials or, for new plastics, with the materials suppliers.

1.3.5 Injection Hold Time

The injection hold time may be required by the product design. The method of gating selected by the designer must be suitable for injection hold pressure to be effective.

Heavy-walled products shrink considerably after injecting and will show unsightly shrink marks as the plastic solidifies where it touches the cooled cavity and core surfaces while the plastic between these colder layers is still hot and continues to shrink. This is especially noticeable at thick sections and at surfaces under ribs and hubs in the product.

The purpose of injection hold time is to maintain pressure on the plastic at the gates so that it continues filling the cavity as the product shrinks inside the cavity space. This requires relatively large, open gates which will not freeze too soon. Cold runner molds require larger than usually required edge or pin-point gates, and hot runner molds must have large, open gates or valve gates.

Thin-walled products usually have very small gates which freeze off as soon as the injection is completed; therefore, injection hold time or pressure would not be useful.

1.3.6 Cooling Time

The cooling time depends on many factors. Some are outside the influence of the mold designer, such as the cooling water supply of the molding plant; others are the direct result of the mold design and construction.

1.3.6.1 *Cooling Water Supply*

A designer should be interested in these features:

- *Water temperature*. Should be in the order of 5–10 °C, although certain plastics may require higher cooling water temperatures, up to 60 °C. In injection blow molds, cooling of the core may require hot oil in the range of 100–150 °C.
- *Water flow* and *water pressure*. There is no sense in having cooling water unless it is available at the mold in sufficient quantity (flow) and with sufficient pressure to maintain good circulation through the mold.
- *Clean, filtered water* which will not clog or corrode the cooling channels.
- *Connections* to the mold must be of adequate size, and laid out and installed properly, to avoid kinks causing flow restrictions.
- *Fittings* and *hoses* installed by the set-up crew must not restrict the water flow.

1.3.6.2 *Cooling Layout*

Proper design of the cooling layout within the mold is one of the main objectives of the mold designer. See Chapter 13, Mold Cooling.

Recommended Reading

1. Menges, G., Mohren, P. (1983). *How to Make Injection Molds*, 2nd ed. Hanser Publishers, Munich (English translation).
2. Rubin, I. I. (1972). *Injection Molding: Theory and Practice.* John Wiley & Sons, New York Toronto.
3. Gastrow, H. (1992). *Injection Molds—108 proven Designs.* Hanser Publishers, Munich Vienna New York.
4. Stoekert, K. (1983). *Moldmaking Handbook for the Plastics Engineer.* Hanser Publishers, Munich New York.

2 Basic Mold Functions

2.1 Shaping the Product

Product shape is provided by the *stack*; that is, all mold parts which are in contact with plastic. For specific discussion, see Chapter 6, Mold Layout, which gives in steps 1 through 12 a procedure suggested for designing the stack. To the beginner, stack design may appear one of the more difficult tasks, but actually, in most molds, it is not difficult.

Round, square or rectangular shapes, such as containers and lids, require usually fairly simple cores and cavities that can be machined from solid steel, with a minimum of inserts. This is especially so when all the latest methods of machining, using Computer Numerical Controls (CNC) or Electric Discharge Machining (EDM) are employed, and the use of inserts can be avoided. There may be complications in cases where moving cavities are required to release the product from deep undercuts in the cavity or where the product requires two-stage ejection and/or side cores, but, in general, these molds are fairly easy to design. Usually, a greater problem is the layout of the cooling channels for best productivity.

Complicated technical products with narrow slots, small holes, and unavoidable sharp outside corners (in the product) will usually require inserts in a core or cavity base—partly for ease of manufacture, partly to enable easy repairs in case small projections (pins, etc.) in the core or cavity break. However, the older methods of mold making, where most cavities and cores were made from many (hardened and ground) inserts, based on a technology similar to metal punching dies, is rarely used now. The disadvantages of such construction are that it is usually impossible to provide good cooling and that all mold parts require high precision to be interchangeable. (It also required individual fitting of the inserts, which then had to be clearly identified as to location in the mold. Such fitted inserts are not interchangeable.) With EDM, including WIRE-EDM, it is now possible to produce many cavity shapes and slots for inserts in the solid core or cavity base, thus simplifying the mold making considerably.

The mold designer must realize that the product designer is not necessarily familiar with the intricacies of mold making or molding. Certain simple features in the product design may present difficulties in mold making or molding; often, a small design change that may be insignificant to the product could greatly simplify the mold or improve the moldability.

It is up to the designer to point out any possibilities to simplify the mold. Changes are often suggested based on past experience in molding similar products that required unexpected long molding cycles, had difficulties in ejection, warped due to poor cooling and uneven shrinkage after ejection, or had other problems. The more experienced designer can more productively criticize the proposed design and help the product designer to arrive at a moldable product. It is of great advantage for the mold designer to be familiar with the type of product and understand the purpose of the submitted design.

2.1.1 Product Shape

The product shape is usually one or a combination of the following three shapes:

1. Plain utility shape,
2. artistic shape, or
3. engineering (functional) shape.

2.1.1.1 Plain Utility Shape

The plain utility shaped group includes, for example, plain containers, household wares, toys, and any utility article where the shape itself does not mean too much as long as the product does its intended job, whether alone or in an assembly. Design changes required to simplify the mold are usually easy to obtain.

Typical design changes may apply to draft angles, location and/or size of ribs, wall thicknesses, location of gate(s), shape and size of pockets (for adequate cooling), and the need for and the size of undercuts. The shape of holes and openings in the sides of the product may specify whether the mold requires side cores (costly) or just shut-off areas that usually are included at considerably lower cost.

Some undercuts may be required not for the functioning of the product but to hold the product on the desired mold half for proper ejection. If they affect the appearance or the functioning of the product, such undercuts require the approval of the product designer.

The designer also must point out where shrinkage may distort the product, unless special provisions for contraction of the plastic are provided. Typically, a polyethylene (PE) snap-on lid with a perfectly flat top surface is more difficult to mold to the proper size than if there are some concentric steps or waves in the top surface. The same applies to the bottom shape of a container. Flat surfaces are always difficult to achieve.

Another area of concern is the cost of the finishing of the product. For example, *texturing* may be specified where inexpensive sand blasting could yield the same effect. The finishing operations (polishing, texturing) are often done manually or must be subcontracted to specialists in this field and are, therefore, expensive.

Engraving is often shown *depressed* in the product, which is expensive to produce in the mold. It is much easier to produce *raised* letters or logos, which are simply engraved into the mold steel.

2.1.1.2 Artistic Shape

For products with artistic shapes, changes to the product design (for improved mold design) are about the same as in utility shapes, but approval is often more difficult to obtain because these artistic shapes may be required for their sales appeal. If there can be no concessions to facilitate the mold building, the mold may have to be more expensive and/or may have to cycle slower than anticipated.

It is desirable to be able to define shapes geometrically, combining portions of circles, ellipses and straight lines. At this time, "free-hand" artwork is very difficult to translate for use in CNC machines used for cavity work; free-hand translation may require the use of expensive models, using duplicating machines.

Note that many utility products are a combination of utility and artistic shapes.

2.1.1.3 Engineering (Functional) Shape

For products with an engineering shape, the function of the product is the key requirement. Accuracy of dimensions is necessary as demanded by the product designer, and the finish may also be important if the product is visible in use, for example, as an enclosure (computer hardware, cassette box, etc.) or a structural member (automotive) and, therefore, must have sales appeal and artistic value.

The mold designer will look out for features such as gate location and uniformity of wall thickness, and must be especially aware of any ribbing and hubs or projections required for assembly of other components. Often, shifting a rib or reducing its size, or reducing the size of a hub, will reduce or eliminate the risk of having bad sink marks in the outside surface of the product. If such changes are not possible, the alternative is to use larger gates (or valve gates) to provide effective holding pressure, but this will almost certainly increase the molding cycle.

The finish on the inside of technical products (which is not normally seen by the end user) needs to be only good enough to ensure good release from the core. It is not necessary to spend much time on polishing.

2.2 Ducting the Plastic from Machine to Cavities

The following discussion applies to the whole range of runner systems, from the simplest to the most complicated. Cold and hot runner systems will be explained, and several examples of each system will be described. For detailed design information, refer to Chapter 10, Gates and Runners.

2.2.1 Cold Runners

All cold runner systems have the same features in common:

- The *cost* of cold runner systems is usually less, often considerably, than a hot runner system for the same cavity layout. There are no controls, heaters, or sensors needed, and there are fewer mold components.
- *Shot volume.* More material is injected into the mold than is required for the products alone, and this, in effect, reduces the injection capacity of the machine. This may be critical if the shot size is close to the capacity of the extruder or the injection unit.
- *Regrinding.* The runner material can usually be reground and used again, but it may have some physical properties reduced from the original, virgin material. In such cases, the amount of regrind used and mixed with virgin material may have to be limited to a percentage acceptable for the required quality of the product. A percentage of regrind is always lost, which adds to the material cost of the product. With certain products, the use of regrind is not at all permissible.
- *Labor.* Additional labor is required for handling of runners and for grinding, proportioning, and mixing the regrind with virgin material. Such labor adds directly to the piece cost

or may appear as an increase in overhead. Also, additional labor and/or sorting equipment may be required for separating the runners from the product.

- *Power.* The excess mass of material required to mold the runner must first be plasticized, then cooled; this is wasted energy, which adds to the piece cost. Note that in some cold runner molds, for small products, the mass of runners may be as much as 80% of the mass of the total shot.

Typical examples of cold runner systems include:

- *Sprue gate.* The *sprue* feeds directly into a single cavity. It needs sprue cutting or machining after molding. This is (historically) the oldest method of molding medium- and large-size products.
- *Cold runners* are used for molding small- and medium-size products in two or more cavities. The sprue feeds into channels (runners) that duct the plastic to the gates. To reduce the amount of plastic in the runners, they should be small (i.e., as short as possible) and of a small cross section.

  Too thick a runner often requires more time to cool than the product itself and thereby lengthens the molding cycle. Too thin a runner may cause too much pressure drop before the plastic arrives at the gates. The plates on both sides of the runner must be well cooled to reduce the cycle time.
- *Two-plate mold.* The runner is in the parting plane of the mold and is ejected together with the products. This type of mold is simple and is selected where the mold cost is a prime consideration, usually where the production requirements are low. Note: The product must be suitable for edge gating.

  Edge, fan or *diaphragm gates* require cutting, breaking off, or, rarely, machining (milling) of the products from the runners after ejection. (Labor, and possibly also fixtures and other equipment, is required.)

  Tunnel or *submarine gates* are self-degating. In this case, the runners are separated from the product as the mold opens, but both runners and products fall together out of the mold.
- *Three-plate mold.* The runner is in a different plane than the parting plane of the mold, and it can be ejected separately from the products as the mold opens. The product is usually *outside center gated* (OSCG); that is, gating is provided anywhere into the product except into the rim or the side wall. Note: The product must be suitable for center gating.

Three-plate molds are inherently self-degating; the products and runners fall out from different mold planes and can be easily conveyed separately away from the mold. They also permit regrinding of runners with small in-press or beside-the-press grinders, with automatic return of the reground material to the machine hopper. This process requires more equipment but eliminates the need for runner handling and reduces risk of contamination of the plastic. Because of the added plates and the required mechanism to separate and eject the runners, three-plate molds are much more expensive to build than a two-plate mold for the same product and the same number of cavities.

Inside center gated (ISCG) three-plate molds are sometimes used for cup-shaped products which must not have any external gate marks, as for cosmetic products (face powder boxes, etc.) or for "fancy" (artistically designed screw caps perfume bottles, etc.). ISCG molds may also be required where the top surface of the product will be printed or where a fancy label will be applied, since any gate mark would be either ugly or could perforate the label.

As a rule, ISCG molds are slower in production because of the difficulty of cooling the core next to the sprue and because, often, the ejection mechanism must be on the same side as the injection. In view of the above, it is important that the mold and product designers discuss the need for inside center gating a product, and the effect it will have on productivity (less) and mold cost (higher).

Gates for three-plate molds are usually very small (pinpoint) and, if properly designed, are almost invisible and can easily be hidden inside an ornament or a lettering on the surface of the product.

2.2.2 Hot Runners

All hot runner systems have the following in common:

- The *cost* of hot runner systems with the same number of cavities, for a similar product, is usually considerably higher than that of a cold runner, two-plate mold, but (for a small number of cavities) it is not much higher than the cost of a three-plate mold.
- *Shot volume.* The entire amount of plastic injected is converted into products. The injection unit can therefore be used to greater capacity.
- *Regrinding.* There are no runners to be reground, thus, savings in material, labor, and/or overhead are realized.
- *Power* requirements are also lower, since no power is wasted to mold and cool runners. Only at start-up is more power required to heat the plastic within the distribution *manifold*, but, in a well-designed and efficient hot runner system, once the mold is on cycle, the power input into the hot runner manifold and the nozzle heaters is negligible.
- *Labor.* No direct labor is required to mold, remove, or degate the products.
- *Sensitivity to dirt.* In two- or three-plate molds, any dirt large enough to plug a gate remains "molded-in" in the runner and will be ejected together with the runner so that the next cycle will find a clean gate for uninterrupted production. However, with hot runner molds and open gates, any dirt lodged in the gate will remain there and block subsequent injections so that the affected cavity will not produce until the dirt is removed, which requires shutting down the mold. Therefore, good housekeeping and use of clean plastic is of special importance when using hot runner molds. Note: valve gated molds are less sensitive to small dirt.
- *Start-up from cold* should take place within a reasonable time, in the order of 15–25 minutes.
- *Color changes* must be as easy and fast as possible. The target is a change-over time of 10–15 shots, from dark to light color.

Typical examples of hot runner molds include:

- *Hot sprue.* This is, in the hot runner mold, the equivalent of a cold sprue for gating directly into the cavity of a single-cavity mold.
- *Multi-cavity* hot runner molds. With today's technology, there are hot runner molds that range from 2 to 96 cavities.

- *Insulated runner molds* are a different type of hot runner technology. For single-cavity molds, a "through-shooting" method is similar to a hot sprue. For multi-cavity molds, a maximum of 12–16 cavities can be used.

  This method is used with commodity plastics (PE, polypropylene (PP), and PS); it has the disadvantage of a difficult start-up procedure (and related safety problems) but the advantage of extremely simple construction, low cost, and high efficiency, provided the system can be left running undisturbed for long periods. The molds for PE do not need any heaters, but those for PP and PS may require internally heated *torpedoes* to prevent the gates from freezing. Note that the faster the mold cycles, the better this system works. The longest possible cycle is about 25 seconds for PE and PP, and about 10 seconds for PS.

  Note that some commercial hot runner systems are derived from the insulated runner molds by having heaters not only near the gates but also in the distribution channels. The disadvantages of such derivations are longer start-up times, long color change times, and the presence of many "dead" pockets where plastic can remain stagnant and decompose, thus contaminating the products with decomposed plastic.

2.3 Removing Air from the Cavities During Filling

Detailed discussion of air removal from mold cavities can be found in Chapter 11, Venting.

2.3.1 Single Gate

In a well-built mold, with good matching fit at the parting plane, the air within the cavity space is trapped. As the plastic is injected into this space, the air cannot escape and will be compressed into any pockets within the cavity and in the space near the parting plane. As the air is compressed, it heats rapidly to a temperature above that permissible for the plastic, with these results:

- The plastic in contact with the hot air will char or burn, and the product will be ruined.
- The plastic cannot completely fill the cavity space where there are air pockets; this also results in scrap products.

There is rarely a perfectly fitting parting plane, and there are also sometimes natural vents such as the clearance of ejector pins and sleeves in the core. Also, the effect of air compression is not so noticeable with slow injection. However, especially with an always desired rapid injection, vents must be provided to ensure that the air can easily escape from the cavity space, at the rim and *especially* at all deep ribs and recesses.

2.3.2 Uneven Walls

Plastic flows faster along a heavier section, or through a rib, than in the narrower sections of the cavity space and will double back toward the gate, thereby trapping air in the wall space. The location where the air will be trapped can usually be predicted, and venting can be provided there with vent pins, ejector pins, or vent slots.

2.3.3 Multiple Gates

When gating from two or more gates into one cavity, as the plastic enters the cavity from two or more spots, portions of the plastic streams will converge toward each other and trap air, with the result that a burnt spot will appear in the product, or the product will have a hole where the air was trapped. Vent pins in such locations will let the air escape and ensure a good product.

2.4 Cooling the Plastic

The mold is a heat exchanger—most of the energy (heat) which has been added to the plastic in the extruder, to make it soft and suitable for injection, must be removed before the mold can be opened to eject the product, which must be stiff enough for ejection.

The plastic enters the *hopper* at room temperature, T_R. It is then heated in the extruder to the operating temperature T_O and is cooled in the mold to an ejection temperature, T_E. T_E is usually much higher than T_R. For example, in many cases, T_R is about 20–25 °C, but TE may be anywhere between 50 and 80 °C, just cool enough that the product will not be damaged during or after ejection. This depends, of course, on the type of plastic molded.

In general, ejecting as hot as possible saves cycle time. However, there are certain disadvantages when ejecting very hot products:

1. More heat is dissipated into the molding room which, in winter may be acceptable. In summer, it would be better to remove as much heat as possible to the outside; that is, to the outside cooling towers, rather than adding more heat to the already hot plant.
2. Because of the effect of shrinkage, the hotter the product is at ejection, the smaller it will be after having cooled completely. With certain products, and with materials with high shrinkage (e.g., PE and PP), the temperature at ejection has great influence on the final size of the product.

The key to a good mold is the *quality* and *efficiency* of the cooling layout in the mold stack and in the plates, and the ability to remove as many calories per second as is practical without making the mold too complicated and thereby too costly. The more efficient the cooling layout, the faster the heat is removed and the shorter is the cycle time.

This is where the designer often finds the greatest challenge: to select suitable mold materials and to arrange the cooling channels so that the plastic in the cavity space is quickly and evenly cooled. It is often difficult—and therefore expensive—to provide cooling to some small areas near large accumulations of plastic, but every second saved from the cooling time works out to a proportional increase in productivity and may be well worth the (reasonable) additional design effort and any higher manufacturing costs.

We are mostly interested in the effects of temperature differences (ΔT) between 1) incoming plastic and ejected product and 2) between the plastic and the cooling water. Also important is the heat conductivity of the mold materials used.

The efficiency of the cooling can be seen in the difference of the cooling water temperatures between IN (entering the mold) and OUT (leaving the mold). If there is no rise in the cooling water temperature, there is too much water flowing through the mold (oversized channels and/or excessive laminar flow) without carrying away heat. This is a sign of inefficiency. Smaller channels would provide more turbulence and better, faster cooling, with less energy required to pump the water. A temperature rise of about 3–5 °C is often adequate.

The actual temperature of the cooling water is not of too great importance. However, the flow (volume) of cooling water through the mold is very important, and if the flow is properly distributed and turbulent, good cooling can be achieved even with cooling water at a higher temperature. In most plants, cooling water is supplied at about 4–10 °C and at a pressure of about 3–5 kg/cm^2.

It is very important that the main cooling supply pipes to the mold are of sufficient capacity to supply the required flow of coolant and that there are no restrictions caused by hoses or fittings of the wrong size.

Chapter 13, Mold Cooling, provides more details regarding cooling layout. It also includes instructions for calculating the amount of water required for a mold under various conditions.

2.5 Ejecting the Product

The key to all ejection is reliability. In any automatic molding operation, there must be no possibility that the product cannot eject from the mold, at every shot, without fail. The mold must function 100% of the time. Even if a product fails to eject only once out of 10,000 products, for example, with a 16-cavity mold running at a 10-second cycle, this amounts to a possible stoppage of once every 1.74 hours, which of course is not acceptable. This can be simply calculated as follows:

Example: 3,600 seconds/hour ÷ 10 seconds/shot = 360 shots/hour. With 16 cavities: 360 shots/hour × 16 pieces/shot = 5,760 pieces/hour. For a failure rate of 1 in 10,000: 10,000 pieces ÷ 5,760 pieces/hr = 1.74 hour.

The basic consideration when deciding on the method of ejection is the recognition of Murphy's law: "*If it can happen, it will.*" A 100% ejection system need not necessarily be a complicated one, but the designer must envision all possibilities that can cause the ejection to fail.

There are catastrophic failures which can cause improper ejection, for example, breakage of mechanical components of the machine or the mold. There is little that can be done about machine failures, except to select a machine of appropriate size for the type of mold. Also, there is little that can be done about total failure of the compressed air to the mold or power supply to machine.

The designer, however, can do much to design reliable and sufficiently strong mold parts. A typical, frequent failure is the breakage of under-designed (too small) ejector pins, where larger pins could have been used. Proper guiding of ejector and stripper plates will also remove a common source of frequent ejection failure—misalignment, which causes bending of pins and wearing of ejector pin holes.

Within the mold, with properly designed air circuits that will not be blocked or leak, the designer can avoid some of the foreseeable problems in air and vacuum systems. Also, selection of adequate hardware (controls, etc.) is important for repetitive, accurate operation. Springs, when improperly applied or poorly sized, are also a common source of mold failures.

When ejecting by air pressure acting directly on the product, it is important to distribute and size the channels so that there is enough pressure available at all points required, even after some products have already ejected from some cavities and the air is freely escaping there. If a number of air cylinders are acting on a mechanical ejection system, the air flow must be evenly distributed so that all actuators operate simultaneously, without cocking that may stop or slow down the ejector plate and cause ejection failure. Another common trouble occurs when products are hanging up in the mold because they are only partially ejected, which may be caused by the following:

- Improper air ejection method (poor design and/or poor set-up),
- insufficient stroke of mechanical ejectors (poor design),
- insufficient mold stroke (poor set-up), or
- any marginal condition which does not properly define where the product will stay for ejection (core or cavity). This is often corrected with undercuts, after tryout of the mold, but this is not considered proper engineering.

For an in-depth discussion of this topic, refer to Chapter 12, Ejection.

2.6 Producing Economically and Suiting Requirements

2.6.1 Mold Type and Number of Cavities

The designer should always be aware of the actual requirements of the mold. It is very important to establish the size of production expected from a mold before starting to design.

Often, the type of mold and the number of cavities is included in the order and is not left to the designer to decide. However, it is good to understand some of the reasoning for selecting certain designs and sizes of molds:

- Experimentation before launching a new product: The shape would have to be right, but many features of a production mold may not be necessary. For example, high polish may be omitted, holes could be drilled rather than molded, engraving could be omitted, and side cores could be avoided by machining the molded piece after ejection. Also, mold cooling may not be required, and any provision for ejection could be either completely omitted or simplified. Such a mold should not be treated as an efficient production mold but should be inexpensive, even if the products will be more costly than when coming from a production mold.

Example: The mold is intended for *experimental* purposes only (usually a single cavity only) to make few (one to a few hundred) *prototypes*. [Note that the time and cost factors are important; such a mold will usually be required as soon as possible, and should cost little.]

- Study of the behavior of a new material or determining shrinkage for a difficult shape: No large production is expected from the mold, but the shape must be right, and the cooling and ejection should be the same as planned for a production mold. However, savings in the mold construction can be made by selection of lower cost mold materials and by omitting some costly features that will be in the final mold where such omission will not affect the test results. Such omissions could be: polishing, side cores, special finish, engraving, etc., unless it is important for the test to have any or all of these features provided. Using mold materials such as mild steel, aluminum, brass, etc., for such molds yields little saving in machining time over using proper mold materials (except by the elimination of heat treat and grinding after heat treat). Where experiments are important to establish cooling, the heat conductivity of the selected mold materials must be considered to obtain good test results.
- Test molds for engineering purposes: These molds are usually for very limited production of *test specimens* but must be constructed very accurately, from the best materials, hardened, and with proper cooling and election, to produce accurate test specimens to established standards.

Example: The mold is for a limited production run, in the order of a few thousand or even ten thousand pieces.

This may be an initial production run of a new product, for example, to test its customer acceptance before committing to a very high production. It is quite possible that, after an initial run, changes to the product may be required, or it may be abandoned.

Such a mold must produce the proper shape of product, but shortcuts may be advisable if the savings in mold cost by eliminating some features are much greater than the labor cost required to add the omitted features after molding. Consider also the cost of fixtures which may be required for such after-molding operations. Typically, this applies to drilling or milling side holes or slots instead of molding them with side cores, or by using a simpler 2-plate mold with sprue gate or edge gates rather than a 3-plate mold or a hot runner mold. However, in some cases different runners may create significant differences in the appearance of the product, and it may be a mistake to use a cheaper runner system or gating.

For limited production, a single-cavity mold will usually be sufficient; however, 2-, 3- or 4-cavity molds with lower cost mold features could be of advantage because of better usage of the molding machine and better approximation of the mold to the future production mold.

- The mold is intended for high production: It does not really matter if a mold is planned to produce 200,000 shots per year over, for example, five years, for a total mold life of 1,000,000 shots, or whether the molder requires 1,000,000 shots in the first year and then it becomes obsolete. The accuracy of the shape of the cavity space and the productivity of the mold must be the prime concern.

The molder is interested in the following: 1) delivery of the products when required, 2) production of the products at the lowest possible piece cost, and 3) keeping the investment in the mold low. These concerns affect the selection of the number of cavities required.

Example: 4,000,000 pieces of a product are required in one year and will be required for several years to come. The product can be molded in a 10-second cycle.

One cavity will produce:

$$\frac{3{,}600 \text{ seconds/hour}}{10 \text{ seconds/piece}} = 360 \text{ pieces/hour},$$

and production of 4,000,000 pieces will require:

$$\frac{400{,}000 \text{ pieces}}{360 \text{ pieces/hour}} = 11{,}110 \text{ hours.}$$

For convenience, assume that the year has only 5,400 hours:

$$24 \text{ hours/day} \times 5 \text{ days/week} \times 50 \text{ weeks/year} = 6{,}000 \text{ hours.}$$

Assuming that the machine is used only 90% of the time, or 5,400 hours/year, at least two cavities will be required to produce all 4,000,000 pieces in one year (actually 5,555 hours are required, which is slightly more than 5,400 but possible). It does, however, leave little allowance for contingencies and down time. For practical purposes, at least three or, better, four cavities will be needed to suit the stipulated production while keeping the number of cavities low (for lowest mold cost).

2.6.1.1 Machine Hour Time

If the size (and weight) of the product is such that in an available machine, for example, an eight-cavity mold could be used, the piece cost will decrease further because less machine time per piece will be added to the cost of the product. The incremental decrease in piece cost by adding more cavities is usually quite large, with few cavities, and can be best shown with an example:

Example: A machine is rated at $100 per hour. With a 4-cavity mold at a 10-second cycle, the mold produces:

$$\frac{3{,}600 \text{ seconds/hour}}{10 \text{ seconds/shot}} \times 4 \text{ pieces/shot} = 1{,}440 \text{ pieces/hour.}$$

for a machine cost chargeable to the product of

$$\frac{\$100\text{/hour}}{1{,}440 \text{ pieces/hour}} = \$0.0694\text{/piece.}$$

For eight cavities, a mold could produce 2,880 pieces/hour, and the machine cost would be $0.0347/piece, or half the above. For 16 cavities at the same molding cycle, the cost would be half of the eight-cavity mold above, or $0.0174/piece. As the number of cavities increase, the incremental savings become smaller and will slowly approach zero.

Note: As long as the same machine can be used, the above reasoning is correct. But if a larger machine becomes necessary for the larger mold, the machine hour cost may rise. This depends on accounting procedures. Some plants use an average machine hour cost for a number of various size machines; others consider the actual cost of the selected machine.

Another advantage of having more cavities in a mold is that fewer machines are required for large requirements, using less floor space and fewer services.

A disadvantage of having more cavities is that if production depends on only one or a few molds with large numbers of cavities, the reliability of the mold becomes more important. Any breakdown can have a severe effect on the production.

The mold cost per piece will increase as the number of cavities increases, but an eight-cavity mold is usually only little less than twice the cost of a four-cavity mold, and the mold cost *per piece* will therefore be little affected. See Chapter 3, Mold Requirements, for further discussion of costs.

2.6.1.2 High Productivity Within a Short Time

Consider the example below, in which the problem is high productivity:

Example: The same product as in the example above is required as a promotional item, for a "one-shot deal" and should be available as soon as possible.

A four-cavity mold may not produce the required quantities in the desired time, and a larger number of cavities must be provided, either in the form of an eight-cavity mold (in one machine) or two four-cavity molds requiring two machines, to produce the 4,000,000 pieces in half a year, approximately. If this timing is still unsatisfactory, more four-cavity and/or eight-cavity molds (and machines) must be added. In such cases, shortcuts in productivity to save mold costs are not recommended, even though the molds may be scrapped at the completion of the deal.

Note that in many, if not all, plastic products, the piece cost is mostly affected by the weight or cost of the plastic (often between 50–70% of the total piece cost), and that "light-weighting" the product, for example, with smaller wall thickness will considerably reduce the materials cost and have a far greater effect on the piece cost than increasing the number of cavities.

2.6.2 Molding Different Shapes in One Mold

As a rule, a mold should be made for only one type of product, the main reason being that different shapes may have different molding characteristics, and the slower running shape will control the molding cycle and the production output. However, products such as toys and some technical (engineering) products—which require the same or multiple quantities of a number of different shapes of the same plastic and the same color, but in relatively small total quantities—may be produced in a "family mold."

One problem with such family molds is that if one cavity produces bad pieces without being found out soon enough, the mold will have to be run longer just to make up for the missing pieces. This may also be the case if, for example, one technical product coming from a family mold in an assembly is required in larger quantities for service, due to wear or breakage in operation, and needs replacement while the other pieces are still working. In general, it is usually difficult to match the molded quantities with the requirements while keeping the inventories low.

Another problem is that all cavities must use the same type of plastic. Thus, it is possible that one or more of the pieces produced in the family mold uses a plastic which is not necessarily the best type for its use. However, if the planned production is low and the mold cost becomes an important part of the product pricing, a family mold may be very economical, even if the piece cost is higher, due to the need to select a higher priced material and/or because the mold must run slower than is possible with a mold designed for one product and one material only.

High productivity family molds are of advantage if:

1. The molding cycle for all shapes can be about the same, i.e. the molding cycle will not be sacrificed.
2. The products will be assembled immediately after molding, either within the take-off system or immediately after removal from the machine. Products in this category include Petri dishes (bases and lids), jewel boxes (two matching halves), cassette enclosures for video and audio tapes, some toys, etc. Also included is plastic cutlery, such as forks and knives to be used in matching quantities that are separated (with or without a robot) at the machine.
3. It is imperative that the identical color be used for both halves (e.g., for face powder boxes) or, with some technical products, if there is a possibility that color matching cannot be achieved when producing the requirements in two (or more) separate molds. Of course this is only of value if the matched color products are packaged with labels clearly identifying them as coming from the same mold and color batch.

2.6.3 Multiple Color or Material Molds

In this molding specialty, the mold consists of two (or more) sets of cavities and cores which are fed from two (or more) injection units. The partly finished product from one set of cavities is shifted, usually together with the core, to the other (or next) set of cavities, where the final injection completes the product or where the next, intermediate step takes place.

In some molds, (only rarely) the partly made pieces are transported to the next step with a carrier rather than with the cores. In other molds, for example, for some designs of typewriter keys, the cavities travel with the products, rather than the cores. The transporting mechanism is usually part of the mold, and very rarely part of the machine.

Typical products include typewriter keys, escutcheons, car tail light lenses, etc. This technology is very specialized and will not be further described here.

3 Mold Requirements

3.1 Accuracy and Finish

The first requirement of a mold is that its products are as the customer expects them: dimensionally accurate, within the tolerances given, and with the appearance (finish) as specified. This must be so whether the mold has only one cavity or any number of cavities.

3.1.1 Accuracy

The main problem affecting mold accuracy is the shrinkage of the plastic used. (For further discussion of product shrinkage, refer to Chapter 8, Shrinkage, and Chapter 9, Molding Surface Tolerances.)

Tolerances shown on the product drawing must be reasonable and achievable. If there is disagreement about the finished sizes of a product, the mere acceptance of the product drawing "as is" places the responsibility of producing the correct sizes with the mold maker. If a tolerance is impossible to keep, the mold designer should immediately bring this to the attention of the product designer. It is usually possible to make the steel (mold) to extremely close tolerances, but this does not guarantee that the molded products will be within the requested (but unreasonable) limits.

Some plastics, notably PS in the commodity plastics and some engineering plastics, have a relatively low shrinkage factor which is fairly consistent from batch to batch, even when supplied by different plastics manufacturers. Other plastics, such as PP and PE in the commodity plastics and also some engineering plastics, can have large and variable shrinkage factors; they can also vary greatly from batch to batch or between the various manufacturers, even when the specifications read the same. Note: Glass or other fillers in the plastic reduce the shrinkage factor considerably.

In Chapter 8, Shrinkage, it is shown that many different factors affect the shrinkage dimensions of a molded product (e.g., temperature, pressure, cycle time, etc.). With "low-shrinkage" plastics (less than 0.6%), there is usually no problem, and the mold (steel) dimensions can be easily calculated to give the accurate finished product dimension. With "high-shrinkage" plastics (more than 0.6%), it is very important to make sure that the product dimensions make sense, or, in other words, that they are feasible for molding.

In some cases, the dimensions shown on the product drawing are only an indication for the requirement to fit another product with matching dimensions, while in itself they are not very important (e.g., a container of one material and a matching lid of the same or a different material). In some containers, in addition to the proper fit of a lid, the contents (volume) is closely defined, and the accuracy of dimensions becomes very important. In such cases, experimenting with the container and lid steel sizes (building experimental cavities or molds) may be the only method to ensure the proper steel (and product) sizes.

3.1.1.1 *Steel Sizes*

A common practice is that the customer has experience with similar molds, and both the container and lid dimensions given are the actual steel dimensions, rather than product sizes. A drawback of this method is that the older dimensions may have been arrived at by molding with a slower cycle than could be possible with a new, better designed mold. This could still result in the need for dimensional changes to the steel to get the best mold performance; if not, the mold may have to run slower to produce acceptable products.

3.1.2 Finish

The customer expects the appearance of the product as specified on the drawing. Unless the product drawing is properly done, however, some clarifications may be necessary. Does the whole product require the specified finish, or only the areas visible to the end user? Or only areas which have special functions (such as "optical" finish, which is very expensive)?

Often, the core needs only a finish good enough to reliably release the product at ejection. NOTE: With some plastics, a rougher finish may be needed to ensure that the product stays on the side desired for proper ejection.

3.1.2.1 *Texturing*

There are different methods of texturing available to the mold maker:

1. *Real texturing*, with photo (chemical) etching, etc., requires expensive work by specialists, often involving transportation costs and long delays.
2. *Texturing with EDM*, for more or less rough stipple finish, can be controlled by different current intensities. It requires electrodes to match the shape of the areas desired to be stippled. This method could be fairly expensive but can usually be done in-house. Specially shaped electrodes can also provide a finish comparable to real texturing, but are quite expensive to produce.
3. *Sand blasting* can often produce an acceptable finish at a very low cost. It usually requires some masking to protect areas which must stay polished. In molds for PE products, a finely sand blasted (vapor honed) and then buffed surface facilitates ejection of the product. A highly polished surface makes the plastic adhere to the steel. Vapor honing or sand blasting, or even some light roughing of the surface by hand with sand paper, must be done occasionally after the mold has been in operation for some time and proper ejection becomes affected.

In all cases, the designer must understand what the customer really wants. If it is different from what the drawing specifies, any proposed change must be approved by the customer. The use of SPI standards, or indicating the finish in "micro inches" or "microns" is recommended when specifying finishes.

3.2 Productivity

Productivity of a mold depends on a number of factors, which will be discussed below:

1. Number of cavities,
2. quality of cooling,
3. speed and timing of ejection,
4. strength and durability of the mold, and
5. ease of installation and start-up.

3.2.1 Number of Cavities

The portion of the mold cost included in the cost of the molded product (piece cost) is in many cases very small. Review the example below:

Example: A container is required in a quantity of 5,000,000 pieces over its product life. A single cavity mold will run at a 10-second cycle, and will cost \$50,000. The *mold* cost per piece is therefore:

$$\frac{\$50{,}000}{5{,}000{,}000 \text{ pieces}} = \$0.0100/\text{piece}.$$

By producing this product in a two-cavity mold, at a cost of \$80,000, the mold cost per piece becomes higher, by a relatively small amount:

$$\frac{\$80{,}000}{5{,}000{,}000 \text{ pieces}} = \$0.0160/\text{piece}.$$

For 5,000,000 pieces, the total cost will be $5{,}000{,}000 \times 0.0060 = \$30{,}000$ higher than with a single cavity mold. The productivity of the mold has doubled, now producing 720 pieces/hour, and the piece cost is significantly reduced. If a machine hour cost of \$100 is assumed, the *machine* cost per piece attributed to the product in a single-cavity mold is:

$$\frac{\$100/\text{machine hour}}{360 \text{ pieces/machine hour}} = \$0.2778/\text{piece},$$

but for a two-cavity mold, the machine cost is only one half of this:

$$\frac{\$100/\text{machine hour}}{720 \text{ pieces/machine hour}} = \$0.1289/\text{piece}.$$

Multiplying the savings of \$0.1289/piece (reduction in cost if produced in a two-cavity mold rather than a single-cavity mold), the savings with 5,000,000 pieces are:

$$5{,}000{,}000 \times \$0.1289 = \$644{,}500.$$

This means the added cost of \$30,000 for the larger mold is paid many times over by the savings of \$644,500 in machine hour cost due to higher productivity of the mold.

These simple calculations can be made for any mold and make a good case for multicavity molds. It is also important to recognize that fewer machines and less plant space will be required. Note that there are other considerations regarding the number of cavities, and they may be different for larger or smaller products, as explained in the following paragraphs.

Theoretically, for the same product, molds with more cavities could be made to run as fast as molds with fewer cavities. However, this is not always the case, and often the cycle time will increase as the number of cavities increases, for the following reasons:

1. *Recovery time*: Time required for the extruder to prepare the plastic for the next cycle. The more cavities, the more time is required, in direct proportion to the number of cavities. This increase in time may be insignificant for small shots but can become very significant for large shots.
2. *Injection time*: The injected volume increases in proportion to the number of cavities; therefore, the oil volume required to push the ram screw or the injection piston forward will increase proportionally. This increase in time may be insignificant for small shots, but can become very significant for large shots.
3. *Pressure drop*: The more cavities, the longer will be the distance travelled by the plastic and the higher can be the pressure drop between the machine nozzle and the gates. With large, thin-walled products, this pressure drop may limit the number of cavities possible to be injected with the necessary high pressure and fast speed.

The following points are not directly related to the mold design but are intended to point out that the productivity of multicavity molds is more sensitive to poor mold operation.

1. *Operation of plant* applies especially to the productivity of hot runner molds. The mold designer has little or no influence at all, except by better education of a customer who may not be very familiar with multicavity and hot runner molds. As the number of cavities increases, the possibility of a mold stoppage increases unless care is taken by ensuring that the conditions for operating the mold are made as reliable as possible.
2. *Housekeeping* is very important. The molder must use proper preventive maintenance of the machine and provide clean cooling water. Molding only virgin material is best; the molder must be extremely careful when using reground plastic. (Dirt in the plastic accounts for most stoppages of hot runner molds.) Note that the lack of reliability caused by poor housekeeping is the main reason why it took so many years for hot runners and, in particular, multicavity hot runners to be accepted by the molders.
3. *Power supply* can be another problem and does cause production stoppages. In the worst case, there can be frequent stoppages due to overload of the electric power network, usually (but not only) in developing countries. Also, large fluctuations in voltage can affect the controls and cause burned out heaters and/or malfunction of controls. In areas with very unreliable power supply, simple, cold runner molds may be preferable to the more complicated multicavity and/or hot runner molds.

3.2.2 Quality of Cooling

The need for good cooling and its influence on productivity was explained in detail in Section 2.4. Productivity and cycle time are inversely proportional: The better the heat exchange within the mold, the shorter the cycle and the higher the productivity of the mold.

3.2.3 Speed and Timing of Ejection

As shown in the Introduction and in Section 2.5 of this book, the time required for ejection is actually a lost time, similar to the mold opening, mold closing, and mold open times, which means that during these times, nothing is produced (injected or cooled).

Obviously, we need mold opening and mold closing (times) to remove the products from the mold, but, to increase productivity, we try to limit the ejection time to take place, if possible, during mold opening time. If this is not possible, time must be added to keep the mold open until ejection is completed (mold open time). It is important to consider all practical alternatives to increase the speed of ejection and to consider its timing; that is, the moment when the ejection can start and when it must be finished.

As an example, random (free fall) ejection of most products is inexpensive, but requires more mold stroke to ensure that all pieces have cleared the molding area before re-closing the mold. A positive removal system such as the HUSKY "guide chutes", while more expensive, will require less stroke and therefore less mold opening and closing times. The added cost of this system can pay for itself by higher productivity.

3.2.4 Strength and Durability of the Mold

Productivity is dependent on the quality of the mold (strength and durability), as shown in detail in Section 3.3. Production interruptions due to failure of an improperly designed or built mold are costly for the customer, but just as costly for the reputation of the mold maker. A properly designed and built mold, barring accidents, should last forever.

3.2.5 Ease of Installation and Start-Up

Some molds stay in a machine all year long, and the installation and start-up time is not of great concern to their productivity. However, many molds run in a machine only a relatively short time (days, weeks, or a few months) before being changed. In these cases, the productivity of a mold can be greatly enhanced by making mold changes easy. Also, there are still some plants where the machines are shut down every weekend and started up again on Monday. In both these cases, easy start-up is a very important consideration.

Efficiency E (in %) of a molding operation can be defined as follows:

$$E = \frac{\text{Number of good pieces produced}}{\text{Number of pieces produced in available time}} \times 100 \tag{2.1}$$

or

$$E = \frac{\text{Hours producing good pieces}}{\text{Hours machine is reserved for mold}} \times 100. \tag{2.2}$$

"Available time" and "hours machine is reserved for the mold" means the same thing—the machine time planned (or reserved) is the total time available from the moment the machine becomes available for a mold until it is free again for another mold to be installed. Examples will best explain the above formulae:

Example 1: A mold should run 40 hours, and the installation time is 6 hours. Installation time includes (in this context):

1. Mounting the mold in the machine,
2. making all connections,
3. heating up the mold,
4. running the mold until it can run "on cycle" fully automatically, and
5. shutting down, disconnecting, and removing the mold from the machine.

Assuming that all pieces produced are *good*, the efficiency E can be calculated as

$$E = \frac{(40\text{ hours} - 6\text{ hours installation})}{40\text{ hours}} \times 100 = 85\%.$$

Assuming further that only 90% of the products are good, the real efficiency E_r of this operation becomes

$$E_r = 0.85 \times 0.90 \times 100 = 76.5\%.$$

Example 2: By halving the installation time (to 3 hours), the efficiency will be quite improved.

$$E = \frac{(40\text{ hours} - 3\text{ hours installation})}{40\text{ hours}} \times 100 = 92.5\%.$$

Example 3: If the production run is much longer, the effect of the installation time becomes much less significant. If, for example, the run is planned for 240 hours (2 weeks), the efficiency will be

$$E = \frac{(240\text{ hours} - 6\text{ hours})}{240\text{ hours}} \times 100 = 97.5\%$$

for a 6-hour installation, and

$$E = \frac{(240\text{ hours} - 3\text{ hours})}{240\text{ hours}} \times 100 = 98.75\%$$

for a 3-hour installation.

This shows that the so-called *Quick Mold Changer* methods are only of significant advantage if the production runs are short and the time for installation can be significantly reduced. Note that some quick mold changers today permit changes in as little as 0.5 hour time lost between two producing molds.

For high efficiency and productivity, it is much more important that the mold produces good (acceptable) products, without interruptions caused by poor housekeeping or breakdowns caused by poor molds.

In some molds, the start-up time (to get the mold on cycle) takes a long time and can be a significant portion of the installation time, as defined above. The start-up time is directly related to the complexity of the product, the mold design and the design of the hot runner system, if such a system is used.

3.3 Physical Strength

Mold strength depends on many factors:

1. Selection of appropriate mold materials,
2. design of adequate strength into all mold components,
3. allowing for fatigue strength,
4. care taken in avoiding stress raisers,
5. proper heat treat specifications,
6. suitable direction of grain in a steel, where applicable, and
7. direction in which mold parts are ground.

Most of these factors are interrelated, and all must be considered at all times by the designer.

The designer must not go by "feel" or "intuition" but by proper engineering methods available to properly determine the adequacy of the selected materials and to calculate the strength of the mold components, and to make sure that they will not break or wear during production. The following describes some of the most important areas the designer must consider in mold design.

3.3.1 Tensile Strength

A mold must be strong enough to withstand the high *internal pressures* during injection. This applies mainly to deep products where the depth of the cavity is such that a large (projected) molding area is under internal pressure at right angles to the opening direction of the mold. This is comparable to the internal forces in a pressure vessel.

There are several problems connected with the stresses generated by these internal forces when the wall thickness of the cavity is inadequate:

1. *Bursting strength*: If the walls are too weak, the cavity will deform (stretch outward) permanently, or even burst.
2. *Stretching* (outwardly): No matter how heavy the walls are, they will stretch, somewhat. As long as the stresses are sufficiently far below the yield point of the steel, the cavity will return to its original shape after the (injection) pressure is released. If the a) cavity walls are too thin, b) draft angles of the side walls are very small (less than 3°), and c) wall thickness of the product is very small (less than about 1 mm), this outward stretch may have a very serious effect on the operation of the mold.

   During injection, the cavity walls will stretch as the cavity space becomes filled with plastic and the injection pressure acts fully on the plastic at the end of the filling period. As the injection pressure drops, the cavity will return (spring back) to its original size and clamp down on the plastic trapped between the cavity and the core. If the mold opening force of the machine is not strong enough, the trapped plastic can prevent the mold from being reopened. Extraordinary efforts may be required to pry cavity and core apart, and such a situation may lead to the destruction of the mold.

   Note that with heavier plastic wall thickness, shrinkage will reduce the thickness of the plastic in the cavity space, and the above described "locking-up" effect is not a problem. Also, with greater draft angles (over 3°), even with small wall thicknesses, there is not much risk of locking up. However, with sufficient cavity wall thickness, the cavity will expand so little as to be insignificant, and the mold will not lock. Refer to Section 19.1, Formulae Used in Mold Design, for further discussion.

3. *Fatigue*: Any mold is subjected to steadily varying loads as it is injected at every cycle. The load cycle is always from *zero to full to zero* , which is the second-worst condition when considering the effect of metal fatigue. For in-depth discussion, see Chapter 18, Metal Fatigue. Generally, when a steel has been subjected to about 2,000,000 stress cycles without failure, it is assumed that it will last forever. Such a number of cycles is not unusual in plastics molding.

   Example: A mold cycling at 4 seconds, or 900 shots/hour, will cycle 900 shots/hour x 5,400 hours/year or 4,860,000 cycles per year.

   For this reason and as also explained in Chapter 18, the yield strength of a steel cannot be used in strength calculations since it is a measure (or strength value) of just *one* load cycle. To be safe, it is practical to use only 10% of the yield strength (safety factor of 10) of a steel when determining the wall thickness of a cavity. Note that hardened mold steels are more subject to fatigue failure than mild steels or (prehardened) machinery steels such as P20, which need only a safety factor of about 5.

   This becomes often a problem for the designer. If one goes "by the book", the cavities may become too large to fit a planned layout. If one makes them thinner, they may fail soon. The designer must carefully consider whether it is worth the risk to make the cavities too thin, with a reduced mold life, or to make them properly thick, requiring possibly a larger mold (and machine). The designer must in all cases consider the effect of size, number, and location of screw holes, bores for water and air, and other weakening features such as steps and grooves in a mold part to make sure that the weakest cross section is considered when calculating a wall strength.

3.3.2 Compressive Strength

Injection pressure causes compressive stresses on the core but they are usually no problem, except when cooling channels or other passages or screw holes in the core are too close to the surface. In this case, the core will also deflect (cyclically) at these weak spots and can fail because of fatigue stress after a number of molding cycles.

Clamping pressure on the parting plane is often underestimated and can lead to early mold failure. Clamping pressure is the total clamping force acting on the mold divided by the area where cavity and core halves meet under pressure when the mold is clamped up.

Only too often in practice, the mold is installed and operated not with the *minimum* clamping force required to keep it shut during injection, but with the *available* total clamping force of the machine. This practice is especially harmful to the mold if it is installed in a machine which is larger than required for the size of the mold. There is little the designer can do about this situation except to make the molder aware of the possible damages to the mold due to overclamping, and to place warning nameplates on the mold indicating the *maximum permissible clamp force*.

It does not matter if the full tonnage is used for a few cycles, but the molder should reduce it to a permissible tonnage before starting production. The constant pounding of an excessive clamp closing force stresses the surface from zero to maximum to zero; again, a very bad condition for the fatigue life of the steel. The result of such pounding can be usually seen after a few weeks of operating the mold as small depressions in the parting plane where the steel has collapsed due to fatigue failure. Also, very shallow vents will soon be closed and disappear, resulting in poorly vented cavities and, eventually, in rejects due to poor filling or burning of the products.

If there is no possibility of increasing the actual parting plane, symmetrically located stops outside the molding area must be provided to share the load and to reduce the surface pressure on the parting plane. Area of venting channels and vent gaps in the parting plane must be taken into account when calculating the area of the parting plane.

3.3.3 Plate Deflection

There are two specific areas where molds are often too weak: backing plates and ejector plates. Deflection of backing plates can be serious when molding with more than one cavity (cavities on each side of the center line of the machine) and is particularly bad with deep, thin-walled products. The effect of weak backing plates is more critical when the platens of the molding machine are also weak, as is the case with some makes of molding machines, usually in the lower machine price range, and with toggle machines where the toggles are attached near the edges of the moving platen.

If the backing plates are too weak, they will deflect in the center as the mold is clamped up and make the cores stand at an angle converging toward the center line of the machine. This will mean that inside the cavities, the tip of the cores will be closer to the side wall of the cavity toward the center of the mold and farther away toward the outside, so that the mold, even before injection, has some core shift. This core shift will be aggravated during injection and may lead to poor or unacceptable products with uneven wall thickness. Such deflection can also lead to poorly fitting parting plane and flashing.

Any plate must always be properly backed up by an underlying plate, or with support bars and/or pillars. For more information, see also Section 17.2, Mold Plates.

Deflection of ejector plates can occur with ejector plates used in an "ejector box" for ejector pins and/or sleeves, or for stripper rings. It applies also to stripper plates without an ejector box. Usually, it is almost impossible to predict the ejection force required for a mold, but as a safe approximation, the designer should assume for calculations that the whole ejection force of the hydraulic ejector mechanism available on the machine will be used. Forces on the ejector plates are cyclical, and the maximum stresses should not exceed 20% of the yield strength of the machinery steel used for the plate.

3.4 Wear Resistance

3.4.1 Wear of Mold Parts Moving Under Pressure

There are several factors affecting the wear of moving mold parts that are under pressure:

1. *Materials selection*: Materials of mold parts which move under pressure against each other must have a different crystalline or grain and surface structure so that the grains of

one piece will not interlock with the (similar) grains of the other piece. Most steels have grain structures determined by their composition so that by selecting different steels, this problem can usually be solved. Different hardness of the same steel for the two rubbing pieces is *not* a solution to this problem.

2. *Surface finishes*: Applying special surface finishes can also avoid this type of wear. For example, by nitriding one of two pieces made from the same steel (such as H13), the surface structure becomes sufficiently different and will wear less. Other finishes such as chemical depositions are also good but may not last as long and will have to be repeated from time to time.
3. *Wear strips* of nonferrous materials (bronze, plastics, etc.) may be used where the specific pressures are low enough to ensure that the wear strips will last.
4. *Surface condition*: Taper fits are usually ground surfaces. If the finish is too rough, and if the grinding is done so that the grooves are at an angle or at right angles to the motion, they will act like a file and rapidly wear down the surface of the matching part. (This happens only too often when grinding taper seats, which will then not last very long.)
5. *Lubrication* of sliding and rubbing surfaces can be used where it is possible to do so, even with materials of the same type, provided:
   a. The lubrication can be easily maintained during operation, either manually or tied in with the automatic machine lubrication, and
   b. There is no danger that the lubricant contacts and contaminates the molded products. This may be important with food containers, medical products, etc.

3.4.2 Fretting

Fretting is a surface wear caused by moving, even under light pressure, two parts repeatedly over a very short distance (≈ 0.1–0.5 mm) back and forth against each other. Fretting is not yet fully understood but is probably some sort of fatigue failure of the metal surfaces. Lubrication and other methods do not help to avoid this type of wear and, when designing the mold, it is best to ensure that such small motions do not take place in the mold.

3.4.3 Wear from Abrasive and Corrosive Plastics

Abrasive wear can usually be quite readily controlled by the materials selection. It applies particularly to the gate where the abrasive plastic flows through at very high speed, and will soon increase the gate opening beyond its permissible size. However, even with the best materials and hardness specification, the gate will wear sooner or later, and the proper solution is to design the mold so that the gate becomes a *wear item* that can be easily replaced at minimum cost and with the least interruption in the production.

The same applies to the cavity and core, and all mold parts which touch the plastic. Any area where the plastic is expected to wear down the mold should be constructed using inserts so that worn pieces can be exchanged.

For wear due to corrosive plastics, the same comments apply as for abrasive plastics. In any case, stainless steels suitable for the specific corrosive properties of the plastic must be selected for the mold. Chrome plating is frequently used to protect the steel surfaces from corrosion, but since stainless steels can be hardened as well as tool steels, there are little if any cost savings in plating, and it is usually better to make all mold parts from stainless steels.

Wear may also be caused by corrosive gases. Some plastics release corrosive gases that will attack not only the cavities but also the mold shoe as the gases escape through the vents. The use of stainless steel plates is suggested.

3.4.4 Rust

Rust is simply the oxidizing of steel. Rust may affect the mold surfaces both externally and internally.

To prevent external rust, there is little the designer or mold maker can do about poor housekeeping at the molding facility. The molds must be properly maintained immediately after every stoppage (shut down over the weekend, etc.) or after the mold is removed from the machine. The molding surfaces must be wiped clean and sprayed with rust preventive oil, and the rest of the mold (the mold shoe) should be cleaned and kept dry to prevent rust.

Some molders paint the shoe with oil paint to keep the outside from rusting, particularly in humid climates, but the molding surfaces are still subject to manual cleaning and oiling. Rust on the molding surfaces will eat away the steel. Removing the rust to repolish the surface will affect the size of the product. This may not mean much, but there is certainly unnecessary labor cost required to fix the mold before starting production. *Electroless nickel plating* the shoe will improve the rust resistance of the plate but cannot be used on the molding surfaces because it will rub off.

Stainless steel cavity stacks and mold shoes are a solution to these problems, provided the customer wants to pay the additional cost of using these materials. For molds operating in very humid climates, such additional first cost may be paid back easily by the assurance that the mold will stay rust free during its life.

Internal rust applies essentially to the inside surfaces of the cooling channels. The problem is that the rust in the channels creates a barrier to the heat transfer and will reduce the efficiency of the cooling system. Also, the rust may start cracks in the steel. Electroless nickel plating helps to some extent but is not as good a protection as stainless steel shoes. Today, there is a trend to make all cavities and cores from stainless steel, regardless of whether it is needed for anticorrosion purposes or not, to ensure that the cooling lines inside are not subject to rust.

For additional information on wear resistance in molds, see also Chapter 16, Mold Materials.

3.5 Safety in Operation

A mold must be constructed so as to:

1. Prevent any damage to the mold itself, and
2. prevent injury to persons operating the mold or others in the vicinity of the mold, in any foreseeable circumstances.

3.5.1 Damage to the Mold

Damage usually occurs if the mold is improperly designed by being either too weak or having any mechanism within that can fail and cause a "catastrophic" failure; that is, cause other parts to break as a consequence. Some typical examples of such failures are described below:

1. *Side cores or slides fail to maintain their position when the actuating cam is disengaged.* As the mold closes, the side cam will collide with the side core or slide. Very solid pieces may be saved by mold protection, but usually cams and/or slides will break. For this reason, slides should best be moving in a horizontal direction so that gravity cannot act on them and push them out of position. However, this is not always possible.

   Ball springs locking into grooves in the slides are particularly risky and, as a rule, should not be used. Compression springs to pull the slides to the correct outside position are acceptable provided the springs are properly designed and of sufficient length so that they will not break because of fatigue. Springs should not be compressed more than 25% of their total compression if they are visible when the mold is in the machine, and not more than 20% if they are built inside the mold where they cannot be seen.

   The safest method is if the cams never leave the engagement with the slide. There is then no risk that the slides move out of position. Unfortunately, this is not always possible.
2. *Return pins must always be provided when using ejector pins or ejector sleeves,* to prevent a sticking ejector plate from causing ejector pins to damage the cavity surface as the mold closes.
3. *Ejector pins can collide with side cores*. With some products, it may be necessary to place an ejector pin or sleeve right under a hub in the product where a side core must come in, for example, to mold a hole. The motions in the mold must be so designed that the side core is sufficiently withdrawn before the ejector pin is allowed to pass without collision. During mold closing, the ejector must be out of the way before the side core is allowed to advance safely. The best way to solve this problem is to prevent the two parts from crossing their paths. Occasionally, a slight redesign of the product may be permissible to allow this condition. If this is not possible, it will be necessary to provide hydraulically operated slides with proper timing and safety interlocks, or use delayed ejection and positive ejector plate return mechanisms, to withdraw the ejector pin before the cam actuates. Poor designs in this area are quite common.
4. *Returning the stripper plate or stripper rings* should not be done with the closing mold. This is especially important when the closing speed of the clamp is very fast at the point of impact of the cavities with the strippers, as would be the case with long ejection strokes. Such "return by the machine" is not only noisy, but, in the long run, the impact will damage both the parting plane surfaces of the stripper and the cavity, and also the taper seat of stripper and core. An independent method such as springs or, preferably, air cylinders should always be used to return ejector and stripper plates.
5. *Latches, links and pivots* for actuation of plates, such as ejector plates (or third plates in a three-plate mechanism, etc.) must be strong enough for the forces required. They must be accurately aligned and secured against loosening during operation.
6. There are many cases where the designer can ensure the safe operation of the mold by *avoiding misalignment, vibration, and excessive moving masses,* which may cause early fatigue failure and/or catastrophic failure of components in these systems.

3.5.2 Personal Injury

Any poor mold design can produce failures which can result in personal injury. The mold designer must always look for what is *possible* or *foreseeable*. In other words, "If it can happen, it will."

Failure to foresee by the mold designer and/or mold maker is often quoted in civil court actions when someone gets injured because of the breaking of a mold part or the losing of a seal where hot plastic or oil escapes. Some of the areas of risk are:

1. Lack of proper communication and instruction (manuals, nameplates, drawings, etc.),
2. handling of the mold on the hoist,
3. suspension for the mold in the machine,
4. external springs breaking and flying off,
5. hot plastic escaping,
6. compressed air escaping,
7. hydraulic pressure oil escaping,
8. screws snapping and heads flying off,
9. sharp edges,
10. heated exposed mold parts,
11. unprotected electric power distribution points,
12. moving parts outside the standard safety gate area,
13. motion which needs adjustment with safety gates open, and
14. possible access to the open mold with gates closed.

Unless such risks can be completely avoided (by design), at the least, prominent nameplates must be posted in critical areas to warn of the possible dangers and their consequences.

3.6 Maintenance and Interchangeability

3.6.1 Maintenance

The mold must be so constructed that maintenance is easily performed.

1. *Cleaning of the hot runner gates*: Since dirt in the gates is an operational hazard and is difficult to avoid, particularly where reground plastic is frequently used, it should be possible to clean the gates in the machine *without* need for removing the hot runner side or the entire mold. Some hot runner systems permit easy access to the gates by pulling the cavity plate to the core plate, thereby exposing all hot runner nozzles and at the same time gaining access to the rear of the gates so that any dirt can be easily removed.
2. *Replacement of nozzle heaters*: This too should be easily possible, with the same procedure as for cleaning the gates.
3. *Readjustment of taper fits*: In most cases, this is easily done by grinding under the respective parts with the taper seats or by adding spacers, as required.
4. *Flashing* occurs in any area where it is possible that plastic can enter (flash) between or under mold parts, such as stripper rings. This can happen during start-up or during a

machine failure. Such areas must be easily accessible, preferably by providing sufficient operating stroke or easy removal of some screws.

3.6.2 Interchangeability

Hardware items (heaters, nozzles, hoses, etc.) are usually interchangeable. However, with fastening devices (screws), care must be taken to specify the proper strength class, because this is not easily seen by just looking at a screw. It is good practice to use only class 12.9 screws.

If special high-strength, high-heat screws are absolutely necessary, this fact must be clearly shown on the assembly drawing and in any mold manual or description that the mold maker may supply with the mold. This should avoid a possible future mold failure caused by the use of a regular screw during a mold maintenance.

Spare parts should also be interchangeable. This applies particularly to multicavity molds and/or where several molds are made for the same product. All mold parts should be interchangeable; however, this may not always be possible.

3.7 Ease of Installation in Molding Machine

The mold must fit the machine(s) for which it was intended. It must be easily installed and connected to the required services. Some of the critical considerations are:

1. *Platen size*: If possible, the mold should cover a large percentage (at least 50%) of the available platen area; in other words, the machine size should be in proper relation to the mold size. It is not a good practice to place a small mold in too large a machine.
2. *Platen area* (also, distance between tie bars): Preferably, the mold should fit the platen area without the need for pulling a tie bar to install the mold. In some machines, the removal of tie bars is not difficult, but it is time consuming. In some machines, it is not practical to remove a tie bar for mold installations.
3. *Tonnage*: The rated machine tonnage should be adequate for the requirement of the mold but not too large. Depending on the product (size, wall thickness, L/t ratio), the suggested tonnage required is between about 1–3 tonnes per square inch of projected molding area, with the higher value for a high L/t ratio.
4. *Ejector hole pattern*: Important when the mold must fit different machines.
5. *Mounting hole pattern*: Important when the mold must fit different machines.
6. *Force of hydraulic machine ejector* (if required): This may be critical in some technical applications requiring mechanical ejection where the initial ejection force required is very high, and may require additional actuators to assist the machine ejectors.
7. *Injection and plasticizing capacity*: Is the machine suitable to process the required amount of plastic for the estimated shot size and cycle time?
8. Are *quick mold change systems* intended?
9. *Clearance to remove the products*: How will the products clear the molding area after ejection: Free fall? Mechanical take-off? Robot? Is there enough room for any of these methods?

10. *Clearance for services*: Is there clearance for cooling water lines? Air lines? Vacuum lines and pumps? Electric cables? Lubrication hoses? Is there enough clearance, without any possibility of collision with the mold motion during opening and closing, and with the motions of any product handling equipment?

3.8 Reasonable Mold Cost

The cost of a mold must be competitive. It must be affordable for the customer but also be designed and built so that the mold maker will make his profit on the mold.

A "cheap" mold is not necessarily cheap in production. A mold can only be judged by the relation between mold cost and molded piece (product) cost. For *low production*, a simple mold, with little sophistication in cooling or ejection may be quite satisfactory. For *high production*, every possibility must be explored to provide the best cooling, fastest ejection, and longest mold life, even if the mold cost is higher.

It can be easily shown that for a product, the *mold cost per piece* molded when millions and millions of pieces are produced from a good, even if more expensive, mold is only negligibly higher than the mold cost per piece from a cheaper mold. However, the molding cycle of a cheaper mold is invariably longer then that of a good mold and requires more machine time. This can *substantially increase* the piece cost.

Also, a faster cycling mold allows faster response to orders, earlier shipment, and better usage of plant capacity. A well-designed mold should also have less down time and, therefore, higher productivity than a cheap mold.

The molder is primarily interested in the *cost of the product* and in the *productivity of the mold,* particularly if large quantities must be produced. In the examples in Section 3.2 Productivity, we have seen that the piece cost depends much more on the productivity (machine hour time) than on the mold cost per unit. If large quantities are required in a very short period, a large number of cavities may be required, even if the mold cost is high. For small quantities, the mold cost will become more important than the cost per molded piece, and it may not be possible to justify economically the investment in a large mold.

What determines the mold cost? The mold cost increases almost in direct proportion to the number of cavities. The mold cost consists essentially of eight elements:

1. Cost of mold design engineering (CE),
2. cost of production engineering (CP),
3. cost of each stack (CS),
4. cost of mold shoe (CM),
5. cost of assembling (CA),
6. cost of testing (CT),
7. cost of overhead (CO), and
8. profit P.

Provided there is no repair necessary, a properly designed and built simple mold will mount easily in the suitable molding machine and require under normal circumstances not more than a few hours for testing. Molds with takeoffs, etc., may take a few days to set up and to test. The total mold cost is the sum of CE + CP + CS + CM + CA + CT + CO + P.

3.8.1 Cost of Mold Design Engineering (CE)

Standardization of designs and use of computer assisted designing (CAD) has the potential of reducing this cost considerably, while at the same time improving the quality of designs because it makes it easy to recall earlier, good designs and standardized parts and assemblies.

The design cost consists of the design of the stack, which is essentially the same whether for one cavity ($n = 1$) or any number n of cavities, and the design of the mold shoe. For n cavities, therefore, the design cost for the stack will not increase, but the design cost for the mold shoe will be higher as the shoe becomes larger and more complicated.

3.8.2 Cost of Production Engineering (CP)

This cost can also be greatly simplified by the use of standard designs. The cost of production engineering involves determining sequences of machining operations, the choice of machine tools, the need for fixtures and gauges to make this mold, and any other costs to produce the mold from the mold drawings produced by the mold design engineering.

3.8.3 Cost of the Stack (CS)

Modern manufacturing methods, automatic turning, milling and grinding machines, as well as extensive use of EDM (including WIRE-EDM) and modern methods of finishing the molding surfaces greatly reduce the cost of producing the stack components.

To produce one stack only requires a certain time. To produce n stacks usually requires less time per stack, since the setup times for one or a multitude of identical parts are virtually the same. Also, for more than one stack, it may be advantageous to provide special methods which would not be worthwhile for one stack only.

3.8.4 Cost of the Mold Shoe (CM)

Fully automatic milling and deep hole drilling machines eliminate much of the manual operations usually associated with making mold plates and other components of the mold shoe. The cost of the shoe increases with the number of cavities, because the size of plates and the number of cooling and air channels increases in proportion.

3.8.5 Cost of Assembly (CA)

Here, too, standardization of designs and the use of standard hardware and proper assembly tooling makes the assembly fast and easy. This cost will increase proportionally to the number of cavities. Note that the cost of assembling a hot runner mold is considerably higher than that of a cold runner mold because of the number of added elements and the electric wiring required.

3.8.6 Cost of Testing (CT)

There should be little difference in cost if testing a mold with few or many cavities, unless the mold design is new and/or experimental. Note that the cost of testing is always there: at many mold makers it is added to the mold cost, and the customer has the guarantee that the mold functions as promised. Some mold makers ship the mold untested and leave it up to the customer to test and to return it for any corrections if necessary. Having the customer do the testing can be more costly than when the mold maker does the testing before shipping, not only because of the savings in shipping costs and time, but mostly because current production setups do not have to be interrupted to test a new mold.

3.8.7 Cost of Overhead and Profit (CO and P)

These last two items add to the mold cost and depend on the type of plant and the method of accounting used. They will not be further discussed here.

3.8.8 Cost of a Multicavity Mold

The cost of a multicavity mold will be less than the cost of a single-cavity mold for the same product, and is multiplied by the number of cavities. The amount by which it will be less depends on many factors, but mainly on the size and complexity of the product. The following examples are provided only for the purpose of illustrating a fact borne out by experience and must not be used indiscriminately.

Example 1: A simple large container mold for one cavity may cost $50,000, and for two it may cost $80,000, or 20% less than the cost of two single-cavity molds.

Example 2: A two-cavity mold for a very complicated technical product may cost $60,000, but a four-cavity mold for the same piece may cost $115,000, or only 5% less than two two-cavity molds, because most of the work was with the stacks rather than with the mold shoe.

Example 3: A 16-cavity mold for a simple product may cost $80,000, and a 24-cavity mold for the same product may cost $110,000, less than the $120,000 one would expect when adding 50% more cavities.

3.8.9 Cost of Stack Molds

The great advantage of a stack mold is that it can have a number of cavities in two levels, back to back, which requires only a little more clamping force than a one-level mold for half that number of cavities would require. Also, a certain number of cavities in a stack mold will require almost twice the area when placed in a single-face mold and require a much larger machine platen to accommodate the larger mold area.

The mold cost for a stack mold for a certain number of cavities is almost the same as, or only slightly less than, the cost of two single-level molds, each for half the number of cavities. The gain using the stack mold is that only one machine is required to produce the required quantities. On the other hand, the stack mold will require a larger injection capacity to be able to inject the larger number of cavities.

For more discussion of stack molds, see Chapter 15, Stack Molds.

3.8.10 Cost of Piece (Product)

As was shown earlier in Section 3.3, Productivity, the mold cost attributable to the piece cost because of an increased number of cavities varies little or only insignificantly as the number of cavities increases beyond four cavities, provided that the production requirements are very high. Further discussion of product cost is presented in Chapter 24, Frequently Asked Questions About Mold Performance, Section 24.4, How Does Performance Affect the Cost of a Molded Product?

4 General Mold Design Guidelines

4.1 Before Starting to Design a Mold

The mold designer starts with the design of a new mold when he receives a *part drawing*, a *sample* or, rarely today, a *model* of the product. Additional information includes the machine (or machines) the mold will be run in, the number of cavities required, and, if it is not shown on the drawing, the type of plastic that will be used for this product.

While the above information is important, it is incomplete. There are other concerns which should be addressed before going into the mold design:

Molding characteristics of the specified plastic?
How many parts will be molded? (approximately annually, over the life of the mold)
Anticipated molding cycle?
Where and how is the product used?
Must the product fit with other parts? (tolerances)
Shrinkage?
Draft angles?
What type of runner system is required?: Cold runner (2-plate, 3-plate)? Hot runner? Combination of both?
Gate location, flow and weld lines, ejector marks?
Finish?
Engraving?
Cavity numbering?
Spares required with mold?
Is indicated machine suitable: Tonnage? Shot size? Plasticizing capacity?
Is mechanical (automatic) product removal planned?
Timing of project? (How soon is the mold required?)

Some of these questions and concerns may have been answered with the accompanying mold order; there may be other questions which will have to be answered by dialogue with the customer.

After a mold order has been received, it is important to find out how the *quoted price* for this mold has been arrived at. Mold prices are usually estimated by an experienced mold designer or estimator, often by the owner of the mold making business. At the time when the estimator receives a request for a quotation, he or she may sketch a mold design which is believed to be appropriate for the product, based on experience and/or from records of similar molds. The designer then bases all cost figures on this sketch. The designer may, according to the complexity or novelty of the product, add a safety factor or "fudge factor" before arriving at an estimated mold price.

There is a *risk* that the quoted mold price is lower than it might have been if all the parameters had been considered at the time of the quoting. The quoted mold price must be competitive with that quoted by other mold makers; there is an added risk in times when mold orders are scarce and prices are very competitive in that, just to get the order, the quoted mold price is lower than would normally be quoted.

It seems to be impossible to find a better system unless the estimator completely designs each mold; this would take much more time than normally allotted for estimating. An exception to this is when the customer is willing to purchase some preliminary mold design time before the mold will be quoted.

Another problem is that many more molds are estimated than will eventually result in orders. It is common practice in all industries to send out requests for at least three quotations from different mold makers before a mold order is placed. Also, occasionally, a mold estimate is requested by a manufacturer for pricing purposes only, simply to find out whether a product is economically interesting, without any intention of soon or ever placing an order.

It is important to understand that the mold designer must know the type of mold that the estimator had in mind when quoting a price. If not, he or she may design a mold that greatly exceeds the quoted price.

Note that the purpose of any industry is to make money with their product. (The product of the molder is the plastic product; the product of the mold maker is the mold.)

The foregoing does not imply that one should design and build a "bad" mold just to stay within the budget of the quoted price. In mold estimating, it is impossible for the estimator to be always right; however, it is important that when averaging many molds the financial results of the "good" and the "bad" molds balance out or are in favor of the mold maker if the business is to prosper.

Also, the mold envisaged by the estimator is not necessarily the best design. There is a good chance that in the course of the design process a better design will be found. But it is necessary that the mold designer is aware of the quoted mold price at the start of a mold design project. It is quite possible that the estimator has erred, has underestimated some difficulties in molding the product, or was missing some important information which was subsequently supplied with the mold order.

If, after a preliminary study by the mold designer of the product and the final specifications, it becomes apparent that some parameters have been either *inadvertently* or *deliberately* changed, the customer must be immediately advised of any increase in the mold cost caused by such changes before more time is spent on the project. In this way, much aggravation can be avoided, and the customer may revise or eliminate some of the new or previously unspecified parameters. The customer may not have been aware that such changes between the time of the quote and the order would affect the mold price.

On the other hand, if the error was due to poor estimating, there is nothing that can be done by the designer but to try to stay within budget and to design whatever is necessary to build the mold as quoted; under no circumstances must the quality or the performance of the mold be compromised. Since the reputation of the mold maker is at stake, any loss suffered due to poor estimating must be written off as *learning experience* and *good will.* There is a possibility that the customer may agree to carry some or all of the extra costs involved, but it is a policy decision of the mold maker whether to approach the customer for an increase of the contracted mold price.

Occasionally, a mold price may be deliberately quoted low as the result of a sales policy; for example, to win a new customer or to enter a new field of products in which the molder has little or no experience. Regardless of the low price, the designer must still design the best possible mold to perform as specified, at a reasonable cost.

4.2 Molding Characteristics of the Specified Plastic

We will here consider only characteristics which directly affect the mold design:

1. *Flow characteristics.* Easy flowing materials usually present no problems, but "stiff" materials require higher injection pressure and, therefore, heavier construction of the mold. This will also affect the need for more accuracy and strength of the alignment elements.
2. *Melt (processing) temperature.* The higher this temperature, the more important becomes the cooling design and, sometimes, the method of heat insulation between hot and cooled portions of the mold.
3. *Material degradation.* Every thermoplastic is to some degree heat-sensitive (or subject to degradation) when exposed to high temperatures over a length of time. Figure 4.1 shows schematically a typical curve: T_l represents the highest temperature at which the material could be injected safely; T_2 is the lowest temperature required to keep the plastic injectable. The difference between T_l and T_2 is the operating (molding) temperature range (*OR*) of the plastic; t_l and t_2 show the elapsed time before degradation starts at the temperatures T_l or T_2.

Some materials, such as PVC, are molded at low temperatures and have a narrow *OR* and a short time span ($t_1 - t_2$) (Fig. 4.2A); they are very heat-sensitive. Some plastics are molded at much higher temperatures and may have a wider *OR* but also a fairly short time span ($t_1 - t_2$) (Fig. 4.2B); they are also heat-sensitive. Such materials include PET, all acetates, and others. Figure 4.2C represents low heat-sensitive materials, such as PE, PP, PS, etc.

The graphs shown are schematics only; curves with the appropriate values for each plastic are available from the plastic materials suppliers. The mold designer must understand this

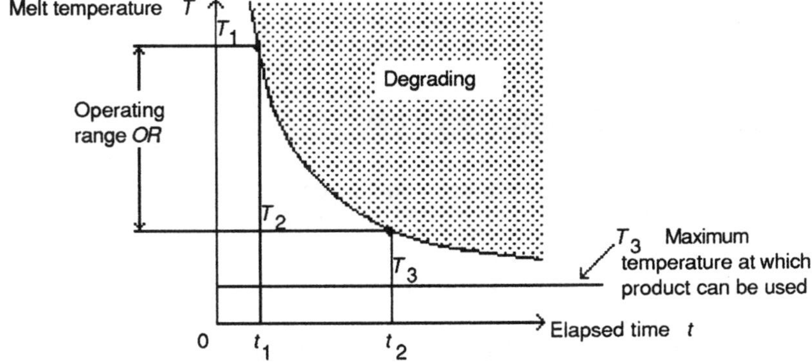

Figure 4.1 Material degradation curve.

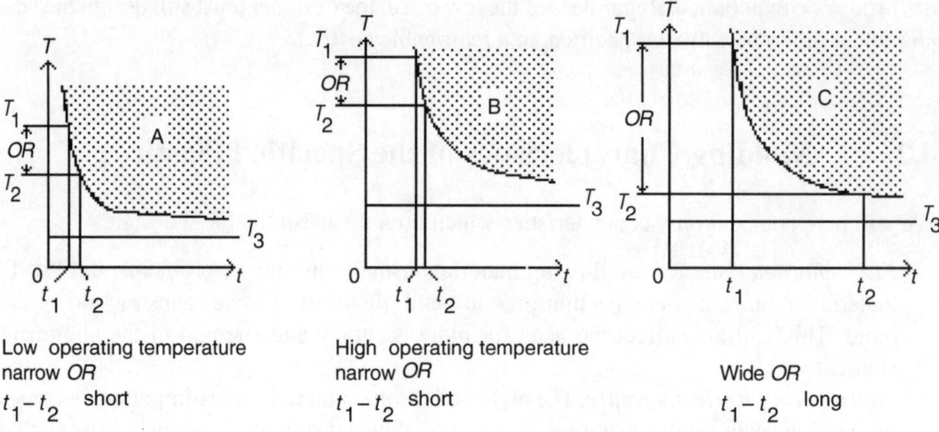

Figure 4.2 Degradation curves for various types of plastics: A. Highly heat-sensitive, B. Heat-sensitive, and C. Not very heat-sensitive.

relation between time and temperature because it affects the design of the runner system and the design and construction of the hot runners, provided that hot runners can be used at all. It will also affect the mold material (steel) selection, for example, in the presence of corrosive agents in the plastic.

Some materials, typically acetal, give off poisonous gases when heated for even a short time above their upper permissible temperature and corresponding safe time, if even the material does not have visible signs of degradation at the very first. Other materials give off highly corrosive gases and will require specially selected mold materials and/or finishes.

All materials show degradation by changing color to yellow or brown; the ultimate form of degradation is when the plastic becomes charred (carbonized) and black. All products containing even small portions of degraded material must be discarded, not only because of poor appearance of the product (streaks, etc.) but because physical properties may have been lost and the product will not perform as expected.

4.3 Anticipated Molding Cycle

There are several reasons why the estimator and the designer should know the anticipated molding cycle. It is always desirable to build a mold with the shortest possible molding cycle, but this can only be achieved at a cost. Special cooling methods, added ejection features, special mold materials, lubrication, etc., will certainly add to the cost of the mold.

The designer (and estimator) should always question the molding cycle demanded or suggested. While the customer (molder) probably has experience on which the expected cycle time is based, there is no harm in double-checking it against the experience of the mold maker and the mold designer. Seldom are two parts so similar that cycle times can be extrapolated from one to the other. Differences in the plastic, wall thicknesses, draft angles, methods of gating, and other factors may have a significant impact on the molding cycle.

This is one of the areas where the mold designer can prevent a financial disaster right at the beginning of a project by questioning the molding cycle specified with the order, particularly if there is a performance guarantee involved.

4.4 Studying the Product

The first task for the mold designer is to become familiar with the intended product. Steps for studying the product are outlined below.

4.4.1 Check the Product Drawing

4.4.1.1 Clarity

Are all necessary views, sections, and section lines shown? At this point of the job it is almost impossible to make sure that all dimensions are given and are correct. Frequently, a product designer who is very familiar with the product forgets that the mold designer may not be and, hence, may take shortcuts which can lead to errors in the interpretation of the drawing.

4.4.1.2 Projection

Is the drawing in 1st angle (European) or 3rd angle (American) projection? This is a common source of error and may not be discovered until some time into the project; it could result in production of the mirror image of the desired shape.

Some draftsmen, to save time, rather than redraw a view or section which was drawn incorrectly in relation to the main view, or to save space on the paper, will mark the direction of a view with an arrow, regardless of the system of projection used. Views indicated by such arrows should be carefully checked to avoid subsequent machining errors.

4.4.1.3 Tolerances

This is probably the most important area to be checked before the mold design should be started. The estimator should check and question the tolerances at the time when the job is quoted.

The final drawing released with the order must be compared with the product drawing used for the quotation to make sure that there were no changes made. Quite often, the drawings used for quotations are only preliminary and incomplete, and tolerances may not have been shown.

Since a large portion of the machining cost of the mold components is directly related to the tightness of the tolerances, it is important to ascertain that *critical tolerances have not been made tighter* since the quotation. If this should be the case, and if the new tolerances will affect the mold cost, the mold designer must immediately approach the customer to bring the tolerances back to what was quoted or to have the contract requoted.

Often, the product designer assigns close tolerances to the product that are not necessary for the function of the product. While in some places the tight tolerances are probably justified, many are not. Also, the general tolerance may have been specified unnecessarily tight. Although it may require a bit more work and understanding by the product designer, the proper method is to give a relatively large general tolerance and tighten up dimensions only where it is necessary for the function of the product or where it is required in the assembly with matching products.

Note that, often, tight tolerances are impossible to hold by simply molding because of the wide range of the shrinkage of plastic. If such close tolerances are absolutely necessary, it may require selection of the proper size pieces and disposal of the molded products that are outside of the tolerances; this is an extremely expensive way of producing anything. For more information regarding plastic shrinkage and tolerances, see Chapter 8, Plastic Shrinkage, and Chapter 9, Mold Surface Tolerances.

It is usually easy enough to make the mold steel to close tolerances, but this does not mean that every product will be within tolerance. Various methods are used to show that the tolerances will increase as the sizes increase.

The product tolerances can be given as percentage of the size; for example, a general tolerance may be ±0.05 mm per 10 mm (or ±0.005 in. per 1 in.), which reduces the tolerance for smaller sizes and increases the tolerance for larger sizes. But there must also be a reasonable minimum tolerance set for very small dimensions to prevent them from being overly restricted and difficult to produce.

Whichever method is specified, the designer must make sure that the tolerances shown on the product drawing make sense and can be achieved. The mold designer must not forget that the sizes also depend on the operating temperatures; in some cases, the product keeps shrinking hours and even days after molding. With some critical dimensions and certain materials, it must be established beforehand when, and under what conditions, the product will be measured.

This study of the tolerances at the beginning of the job will prevent arguments later on. If the tolerances specified are unreasonable, the mold designer must discuss them with the customer and get a *release in writing* so that the mold maker will not be held responsible for sizes of the molded product which are outside the (unreasonably) specified ones.

4.4.1.4 Product Use

Where and how is the product to be used? This question must be asked not just out of idle curiosity; it will give the mold designer some idea of the importance of certain aspects and critical areas of the product such as required fits with other products, finish, physical strength, location of gate, ejectors, etc. The designer may then suggest changes, especially in the areas of fragile mold cores or thin ribs, not only to make the mold easier to build but also to extend the life of the mold and to improve its serviceability.

Understanding the product may save on unnecessary polishing; it may indicate to the designer where looser tolerances could be in order and where to request sharp edges instead of the round edges specified, or round edges where sharp edges are shown (see typical examples in Fig. 4.3).

Some suggestions that can be pointed out right at the beginning of the project refer to improvements of the filling path of the plastic which may need only small changes in the

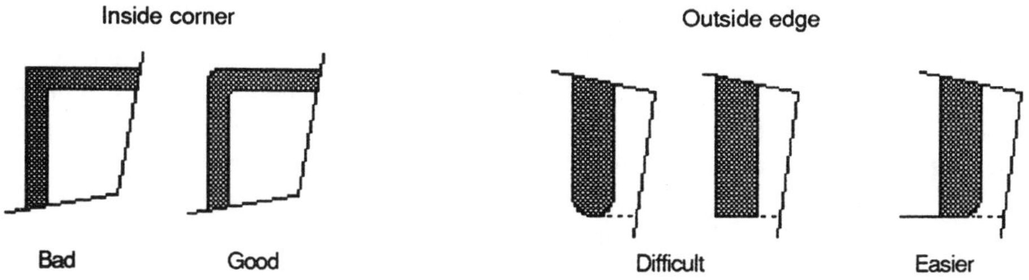

Figure 4.3 Typical examples of product wall design improvements.

Figure 4.4 Typical improvement of flow path of plastic.

product shape to give faster filling (and molding) speeds. One such example is shown above (Fig. 4.4).

Such requests for changes will benefit not only the mold maker but also the molder, and it is the duty of the mold designer, after understanding the purpose of the product, to discuss such proposed changes in the product design with the customer, who would be the ultimate beneficiary of such changes.

The customer must also be made aware of the fact that the mold maker (and designer) often has had specific experiences in the areas of discussion and does not request such changes solely to save some money for the mold maker but in a desire to give the customer some advice gained from molding similar products, with earlier molds they have built.

Other typical but highly undesirable features (for molding rather than for mold making) are severe changes in wall thicknesses and heavy sections at the end of long, thin passages. The product designer may not be aware of the fact that this is bad for molding. It affects the filling of the cavity space and may cause sink marks in the product. To overcome such poor flow conditions, higher injection temperatures and/or pressures and longer holding times will be required, resulting in longer cycles.

Heavy walls and bosses can sometimes be redesigned by coring out or by relocating them so that more uniform wall thicknesses can be created, and cooling can be provided inside heavy sections to permit faster cooling. Usually, it is the heaviest section of a product—and its cooling—which determines the molding cycle and the productivity of the mold.

In summary, there are many areas in which an experienced mold designer can pinpoint possible improvements which may have a bearing on the quality, productivity, longevity, and serviceability of the mold, and which should be discussed with the product designer (or customer).

4.4.1.5 Notes on the Drawing

Scrutinize all notes on the product drawing(s). It is a good practice to show with notes on a drawing all information which cannot be easily shown by conventional drawing techniques or which, if shown, would clutter the view(s) on the drawing. These notes may add information regarding tolerances, draft angles, finishes, concentricity, fits, material selection, etc.

Make sure the notes are clear and make sense. To overlook or misinterpret a note is as serious a mistake as to overlook or misinterpret a dimension.

Notes may show changes from earlier designs of the product; they may point out which areas of the product demanded changes by the product designer. This could have significance for the mold design by indicating that future changes are possible. It may suggest provision of inserts where they otherwise would not be necessary from the mold maker's point of view.

4.4.1.6 Draft Angles

Check the drawing for draft angles. Generally, a draft angle greater than 1° per side presents no problem for pulling the molded part out of the cavity, or for the ejection of internal ribs, provided the finish is adequate. Draft angles smaller than 1° should be carefully considered, and if they appear to be unreasonable, the customer should be approached with a request to increase the angles. While it is possible to mold sides and ribs without any draft angle (or even with negative draft) by using side cores, two-stage ejection, or collapsible cores, such features could add considerably to the complexity and cost of the mold.

Thin-walled products are usually more delicate and more affected by small draft angles. They will require better finish, draw polishing (polishing in the direction of the mold motion), and/or the use of inserts where otherwise no inserts would be required. Increase in the use of inserts, however, reduces the possibility of using maximum cooling and, therefore, may affect the molding cycle, which is of prime interest to the molder. In many cases, this argument alone will convince the customer to increase insufficient draft angles.

The stiffness and the condition of the molding machine used for the planned mold also has a bearing on the minimum draft angles. If the clamp opening motion is not perfectly straight and in line with the center line of the product, it will permit the moving mold half or the cavity plate to sag as the clamp opens; the molded product will show drag marks where the plastic (between the core and the cavity) supported the mold halves during opening stroke. If the machine is not solid enough to ensure that the mold will stay on its center line, additional supports as part of the mold will be required, thereby adding to the mold cost.

If thin, deep ribs are specified, with small draft angles per side, they will require special ejection features. Ejector pins should always be at the *bottom of the ribs* to ensure that the ribs will not break and remain in the core. To be effective, such ejectors pins can have a narrow, flat cross section where they contact the rib, which is very expensive to produce and to maintain.

Round pins (with a diameter equal to the width of the rib) are usually too small and fragile. To use a pin diameter larger than the width of the rib, the mold designer needs a concession from the product designer because the path of the ejector pin will create a circular thickening in the rib (Fig. 4.5).

For more in-depth discussion of ejection factors in draft angles for mold design, see Chapter 12, Ejection.

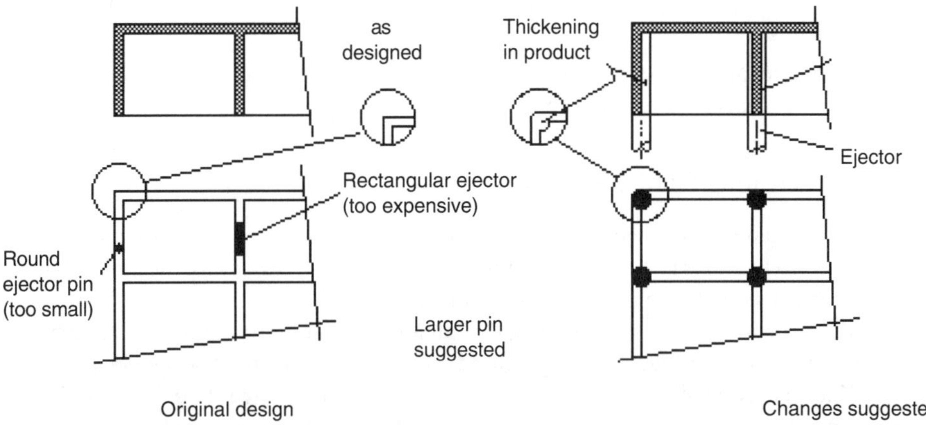

Figure 4.5 Two drawings show an original design (left) in which an ejector pin is too small and a revised design (right) depicting a correction to improve the product.

Note that the preceding is just a sample of the kind of requests by the mold designer from the product designer or customer. Each design feature of the product must be studied before starting to design.

The designer must well understand the product before making any suggestions and must not suggest anything that may affect the performance, strength, or safety of the product—even if it will mean more difficulty or higher cost in mold making or in molding.

Agreement on any changes must be in writing; at the very least, any marked-up changes on the issued product drawing should be signed and dated.

The study of drawings can usually be done very rapidly. In many cases, it may not even take an hour to scrutinize a drawing, but as in all endeavors, it is better to first spend some time to look where you are going before starting to run.

4.4.1.7 Responsibility for Shrinkage

When only a product drawing and the molding material are specified, it must be clearly defined who will assume the responsibility for selecting the proper shrinkage values. Sometimes, the customer assumes the responsibility and specifies the suggested shrinkage, or supplies in addition to the product sizes all critical mold (steel) sizes. This is usually the case if the customer has had similar cavities in use in an experimental or earlier production mold.

Another point to ascertain at this time is whether the mold will also be used with different materials, which may have different shrinkage factors. If a mold is built to use polystyrene (PS) with 0.6% shrinkage, it probably could also be used for other materials such as polypropylene (PP) with 1.5% shrinkage, but because of the higher shrinkage factors, the products will be smaller. To know the materials expected to be molded in the mold is also useful because it may affect the selected finishes of the molding surfaces.

It is important that all uncertain areas of the drawing and the specifications are clarified before starting any mold layout. Few things are as frustrating as to arrive at a "beautiful" mold design, only to find out that it would be very difficult or even impossible to change some critical dimensions or features, should that become necessary as a result of wrong shrinkage allowances.

For more relevant discussion, see Section 3.1, Accuracy and Finish, and also Chapter 8, Shrinkage, and Chapter 9, Product Tolerances.

4.4.2 Working from a Sample or Model

It is generally not possible to work from a sample or model. However, since in some cases the customer has no facilities to make proper drawings, it may be necessary for the mold designer to become involved with the product design drawing.

The designer will study the product and, with all necessary information supplied by the customer, will make a product drawing. The advantage is that the mold designer can specify all the points discussed in Section 4.4.1 to give the mold all advantages both for molding and for mold making. But it is *absolutely necessary* that the customer agrees with all these points and accepts the product drawing by *signing* it before work on the mold design is started. It is also necessary that the customer holds the mold designer and mold maker free from any liability resulting from infringing on the rights of others due to the product so designed.

4.5 Typical Check List for Estimators and Designers

The following is a summary of the questions and considerations explained earlier in this chapter. It is the basis of any formal check list used by a mold maker and may vary according to individual requirements and priorities:

4.5.1 Machine Specifications

Mold mounting data:

Tie bar spacing: vertical ______ horizontal ______

Tie bar size: ____________ Can tie bars be pulled? ____________

Platen specs: ________________ Locating ring diameter ______

Hole patterns: SPI? ______ Euromap? ______

Special mountings: ______________________________

Special ejectors: ______________________________

Machine nozzle data:

Shape: Flat ______ Round ______ Radius ______

Nozzle opening: ____________

Nozzle length: ____________

Controlled flow: Open ______ Pin ______ Other ______

Injection unit:

Shot capacity: ____________________ Injection speed: ____________________

Profiling of speed possible? ____________________

Profiling of pressure possible? ____________________

Carriage travel: ____________________ Recovery rate: ____________________

Clamp:

Clamping force: ____________________

Shut height: Minimum _______ Maximum _______

Bolster thickness: ____________________

Stroke: Maximum _______ Minimum _______

Stroke limiter: ____________________

Ejection stroke: ____________________ Ejection force: ____________________

Mold cooling (or heating) on the machine:

Available pressures: ____________________ Available flow: ____________________

Air and hydraulic functions:

Air system pressures: ______________ Hydraulic system pressures: ______________

Heat controls for mold available?: _______

Automatic? _______ Number of zones controlled? _______

Manual? _______ Number of zones controlled? _______

Ampere ratings for the various zones: ____________________

Voltage: ______________ Type of outlets: North American? _______ Other? _______

How will the products be removed?

Manually _______ Free fall _______ Chutes _______ Conveyors _______

Take-off _______ What kind? ____________________

Any other method? ____________________

4.5.2 Product, Product Drawing, and Production Specifications

Production volume: ______________________________

Estimated molding cycle: ______________________________

Is the drawing clear? ________ Projection: 1st angle ________ 3rd angle ________

Tolerances: ______________________________

Notes on drawing?____________________ Draft angles: ____________________

Where and how is the product used?______________________________________

Molding material:________________________ One type only?________________

Shrinkage:________________________

Runner system:

Cold runner:__________________

Two-plate________ Three-plate________

Tunnel gate________ Edge gate________ Other________

Hot runner: __________________

Center gate________ Edgegate________ Valve gate________

Multiple gating________ Inside gated________ Outside gated________

Combination hot and cold runner:__________________

Appearance:__________________ Location of gate:__________________

Flow path:__________________ Weld lines?:__________________

Ejection:

Pins________ Full ring________ Partial Ring________

Bars________ None________

Air only________ Air combined with mechanical?________

Manual________ Automatic________

Side cores________ Collapsible cores________

Stripping heavy undercuts?________ Unscrewing?________

Appearance:__________________ Ejector pin marks?________

Finish:

Polish specifications:__________________________________

Texturing__________________ Engraving__________________

Raised in product?________ Depressed in product?________

Art work:

Is art work supplied?________ How?______________________________

Is form of art work acceptable for the shop?______________________________

Proper magnification for work: ________________________

Cavity numbering:

Desired identification: ________________________

Size: ________ Height: ________ Location: ________

Once all the above-listed questions have been answered before work is started, the project should proceed very smoothly. Costly delays are caused when a project must be stopped because some unanswered questions appear at a later date, and valuable time may be lost in such late communications with the client before it is possible to proceed with the design.

The most aggravating condition arises when the designer has knowingly gone ahead with insufficient data or information, and as a consequence has to scratch much of the design work done. This can be the major reason for cost overruns in the design section.

5 Plastics Identification

The question of how to identify plastics is often asked when looking at a product, for example, when requested to quote on a sample with no other information available. The methods suggested are not truly scientific but often provide a quick answer with simple means.

5.1 Plastics Identification by Type

Note that the following chart and tables are reprints of an article published in *Canadian Plastics Magazine* in 1961. Except for some newer engineering materials, little has changed, and the listed methods of identification are still valid.

In case of doubt, and where the answer is important, chemical analysis and testing will be the only true method to establish the identity of the material in question. Material suppliers or independent laboratories can usually provide the answers, but it may take several weeks to get an answer.

When using the following tables, take the considerations below into account:

1. Caution should be exercised in smelling vapors; those from fluroresins and fluoroelastomers are highly toxic, and some are deadly.
2. Plastics on copper conductors should be removed before flame testing. A large volume of metal contacting the plastic can change the characteristics to self-extinguishing when it is normally not so.
3. Plastics and elastomers containing chlorine can be identified as follows: Heat a copper wire in a gas flame until no color is imparted to the flame. Then touch the unknown with the hot copper and put it back into the flame. A vivid green color indicates the presence of chlorine in the material. Numbers 8, 29, 30, 31, 35 and 37 will give the green color. Chlorinated waxes are sometimes compounded with flammable plastics to impart self-extinguishing properties. These will also give the green color.
4. As a class, thermoplastics may be distinguished from thermosets in that the former will often melt, drip, and/or bubble in the flame. Thermosets will char, crack and discolor.
5. Plastics numbers 3, 4, 5, 6, 11, 14, 21, 23, 26, 29 and 30 are available with a water-like transparency.

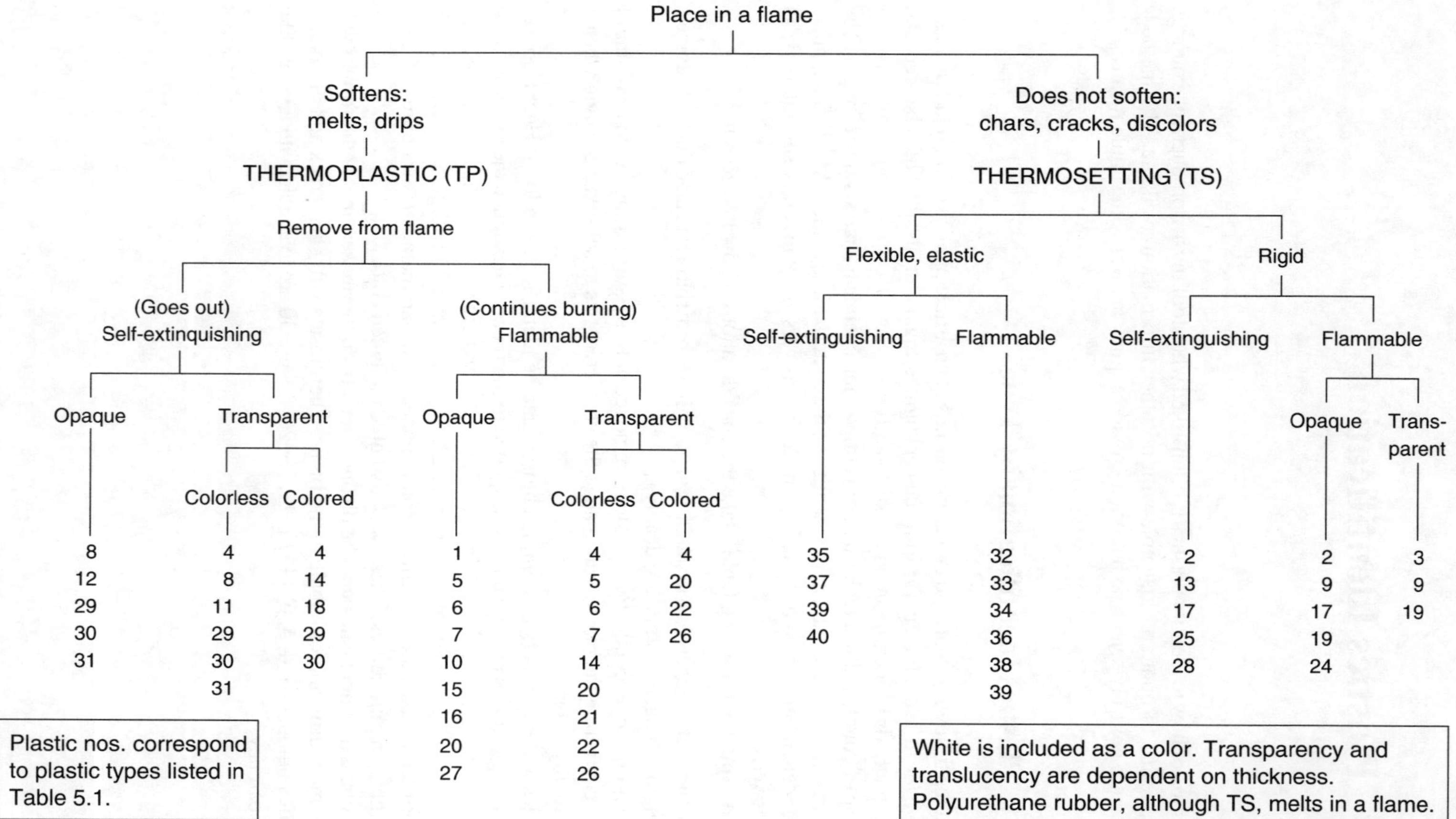

Figure 5.1 Plastics identification chart. Reprinted with permission from *Canadian Plastics Magazine,* Toronto, Ontario, Copyright © 1961.

Table 5.1 Plastics Identification by Type

I.D. no.*	Type	Ease of ignition	Self-extinguishing	Character of flame	Odor of flame	Behavior of material	Remarks
1	Acetal T.P.	Moderate	No	Pure blue, no carbon	Very sharp formal-dehyde	Melts blackens	Insoluble in all common solvents
2	Alkyd T.S.	Moderate	No (electrical moldings–yes)			Discolors, chars and cracks	Insoluble in all common solvents
3	Allyl T.S.	Moderate	No	Yellow, smokeless, sparks		Browns, where burn-ing, bubbles	Insoluble in all common solvents
4	Cellulose acetate T.P.	Readily	No (highly acetified–yes)	Dk. yellow, some black smoke	Sharp, vinegary	Melts and drips	Insoluble in carbon tet. and toluene, soluble in acetone
5	Cellulose acetate butyrate T.P.	Readily	No	Dk. yellow with blue edges, some smoke	Rancid cheese or butter	Melts and drips	Same as acetate, above
6	Cellulose propionate T.P.	Readily	No	Dk. yellow, some smoke	Sharp, of propionic acid	Melts and drips	Same as acetate, above
7	Cellulose nitrate T.P.	Very readily	No	Yellow	Camphor	Burns violently	Same as acetate, above
8	Chlorinated polyether T.P.	Moderate	Yes	Bright yellow with green core	Antiseptic, like iodoform	Melts and blackens	Insoluble in all common solvents
9	Epoxy T.S.	Readily	No	Yellow and smoky	Pleasant, like burnt flour	Blackens, softens	
10	Ethyl cellulose T.P.	Readily	No	Yellow with blue edges	Sweetish	Melts and drips	Soluble in acetone, carbon tet., and toluene

* Identification Nos.:
1. Delrin
2. Plaskon, Glaskyd, Polylite, Laminac
3. Cocor, CR 39
4. Celanese, Hercocel A, Tenite I, Plastacele
5. Tenite II, Nixon CAB, CAB
6. Forticel
7. Nitron, Pyralin, Celluloid
8. Penton
9. Araldite, Epon, Epocast, Stycast, Epoxylite
10. Ethocel, Hercocel E, Nixon E/C

Examples of trade names of plastics are listed by generic names throughout chart. No attempt is made to provide all trade names.

Table 5.1 Plastics Identification by Type *(continued)*

I.D. no.*	Type	Ease of ignition	Self-extinguishing	Character of flame	Odor of flame	Behavior of material	Remarks
11	Fluoro-carbons: Monofluoro-ethylene T.P.	Difficult	Yes		Biting, pungent, like bromine	Melts and chars	Insoluble in all solvents
12	Tetrafluoro-ethylene	Cannot be ignited T.P.	Yes	No flame		Turns transparent and jelly-like	Insoluble in all solvents
13	Melamine T.S.	Difficult		Yellow	Sharp, of formal-dehyde	Discolors, chars and cracks	Insoluble in all common solvents
14	Methyl methacrylate T.P.	Readily	No	Blue with yellow top, slight carbon	Fruity	Melts and bubbles	Soluble in ethylene dichloride
15	Methyl (α) styrene T.P.	Moderate	No	Yellow and sooty		Melts and turns brown	
16	Nylon T.P.	Moderate	Varies	Blue with yellow top	Like celery	Melts and froths	Insoluble in common solvents, soluble in phenol and conc. HCl
17	Phenolic T.S.	Moderate	Mostly yes	Yellow with black smoke	Phenol	Chars and cracks	Insoluble in common solvents
18	Poly-carbonate T.P.	Difficult	Yes	Yellow, smoky	Pleasant, faintly phenolic	Softens, bubbles and carbonizes	Soluble in methylene chloride, insoluble in aliphatics
19	Polyesters T.S.	Readily	No	Yellow with black smoke	Strongly of styrene	Blackens	

* Identification Nos.:
11. Kel-F, Fluoron, Fluorothene
12. Teflon, P.T.F.E., Chemion, Fluon
13. Melmac, Cymel, Resimens, Plaskon
14. Lucite, Plexiglas, Perspex
15. Cymec 325
16. Zytel, Polypence N
17. Bakelite, Resinez, Durite, Durex
18. Lexan
19. Laminac, Atlac, Paraplex, Plaskon, Vibrin, Selectron

Table 5.1 Plastics Identification by Type *(continued)*

I.D. no.*	Type	Ease of ignition	Self-extinguishing	Character of flame	Odor of flame	Behavior of material	Remarks
20	Poly-ethylene T.P.	Readily	No	Blue with yellow top	Like paraffin wax	Melts and bubbles	Floats on water, insoluble in common solvents
21	Poly-ethylene terephthalate T.P.	Moderate	No	Yellow, slightly smoky	Pleasant, elusive odor	Melts, opacifies, blackens	Insoluble in all common solvents
22	Poly-propylene T.P.	Readily	No	Blue with yellow top	Like paraffin wax	Melts	Floats on water, nsoluble in all common solvents
23	Poly-styrene T.P.	Readily	No	Yellow, dense black smoke much soot	Oppressive, flowery	Melts and bubbles	Soluble in benzol, carbon tet., acetone
24	Poly-urethane T.S.	Normally readily	Normally no–can be made so				
25	Silicone T.S.	Difficult	Yes	No flame, white smoke	Very pungent	Blackens	Insoluble in all common solvents
26	Styrene-acrylonitrile T.P.	Readily	No	Yellow, smoky	Weakly of styrene	Softens, bubbles and blackens	Insoluble in common solvents
27	Styrene-butadiene T.P.	Readily	No	Yellow, smoky	Mostly styrene	Melts and bubbles	
28	Urea T.S.	Difficult	Yes	Yellow with green-blue edge	Fishy	Swells, cracks, discolors	Insouble in common solvents
29	Vinyl chloride, unplasticized T.P.	Moderate	Yes	Yellow	Acrid	Softens and blackens	Soluble in tetrahydro-furan

* Identification Nos.:
20. Alkathene, Alathon, Dylan, Fortilene, Marlex, Poly-Eth
21. Mylar, Dacron, Terlene
22. Profax, Moplon
23. Styron, Lustrex, Dylene, Erinoid
24. Isoglas, Isotube
25. Silco-Flex
26 Styrex, Kralasic
27. Plio-Tuf
28. Beetle, Plaskon
29. Geon, Koroseal, Marvinol, Pliovic

Table 5.1 Plastics Identification by Type *(continued)*

I.D. no.*	Type	Ease of ignition	Self-extinguishing	Character of flame	Odor of flame	Behavior of material	Remarks
30	Vinyl-chloride, plasticized T.P.	Moderate	Yes	Yellow, smoky	Acrid paraffin wax	Melts and drips	Same as for unplasticized
31	Vinylidine chloride T.P.	Very difficult	Yes	Yellow and green		Melts	Insoluble in all common solvents
				Elastomers			
32	Butadiene acrylonitrile	Readily	No				
33	Styrene	Readily	No	Yellow, smoky	Rubbery		
34	Butyl	Readily	No	Yellow and practically smokeless	Rubbery	Chars and turns white	
35	Chloro-sulfonated polyethylene	Moderate	Yes, slowly	Sooty, yellow top, blue core		Chars	
36	Natural rubber	Readily	No	Yellow, smoky	Rubbery		
37	Poly-chloroprene	Moderate	Yes, slowly	Yellow, very sooty	Sharp, acidic	Chars	
38	Polysulfide	Readily	No	Blue, no smoke		Foul	
39	Poly-urethane polyester Polyether	Moderate Readily	Yes No	Yellow, smoky Yellow and blue, smokeless		Melts (!), drips and blackens, chars	
40	Silicone	Moderate	Slowly	Yellow, white smoke	Nonrubbery	Leaves white ash	

Identification Nos.:

30. Flamonel, Densheath, Krone, Ultron
31. Saran
32. Hycar, Paracryl, Perbunan, Krynec, NBR
33. Synpol, Ameripol, Plie-Tuf, Polysar, Krylene, SBR
34. Polysar, Super-Coronel, IIR
35. Hypalon
36. NR
37. Neoprene, Geoprene, Okoprene, Nerprene, CR
38. Thiokel
39. Vulkellan, Disogrin, Adipre ne, Vulcaprene
40. Silastic

Part I Mold Engineering

Section 2 General Mold Design Guidelines

6 Mold Layout

6.1 Mold Layout Rules

The following refers to molds in general. It is essentially an analysis of the thought process when designing a mold. An experienced designer will follow the suggested steps almost automatically and will make many decisions without even being aware that they have been made.

For the inexperienced designer, the following should be a help to organize thoughts and to make reasonable decisions.

6.1.1 Before Starting

Make sure that:

- An accepted, fully detailed and toleranced product drawing is on hand.
- Molding material(s) and shrinkages are known.
- Molding machine(s) specifications are available.
- Other data, usually given with the mold order, is known:
  Number of cavities
  Type of runner system
  Method of product removal
  Note: Occasionally, only production requirements are given, and the above data must be determined by the designer for best productivity with the specified molding equipment.
- If possible, the designer should have the sketches used for estimating and pricing the mold.

Even experienced designers may overlook some points and make wrong decisions. The purpose of the various meetings *before* a job is started is to scrutinize these decisions by other designers and by machining specialists, who may have important input to offer to arrive at the best possible layout for a mold which will not only function and produce as intended but will also be easy to manufacture and to maintain.

Any given number of designers will come up with as many different mold layouts, which may all be satisfactory. For efficient manufacturing, it is important that established in-house standards are used. This does not mean that the designer should not have new ideas or that these ideas must be stifled to comply with established designs. On the contrary, if the designer has good, new ideas, they should be brought forward, explained, and "sold" in a "concept meeting" to colleagues and others concerned with the project. This is the best way to achieve progress.

Frequently, however, the designer will have to accept established methods and must compromise between new ideas and the practical and well-proven practices of mold construction.

6.1.2 First Considerations

Is there a precedent, or is the job completely new? Here, "precedent" means a mold built earlier, for a similar product. It need not necessarily have the same number of cavities or the same layout, but the stack construction and the method of injection and ejection would be the same for the new mold as for the precedent.

1. There is a precedent. The precedent may be a mold for:

   - An identical product,
   - a similar product (only sizes are different, within a small range),
   - a product with a slightly different shape but otherwise similar, or
   - a different number of cavities but similar stack construction.

   Find out how the precedent was functioning, in test and later in production. Also, find out whether there were any problems in manufacture. If the answer is satisfactory, there is no need to redesign, and the designer can proceed in using the old design.

   The new mold may require some newer standards (and/or standard parts) which have been introduced since the precedent was built and where the older standard parts are no longer available. Also, it is necessary to check the precedent mold file to see if all drawings have been updated, to avoid copying errors which were corrected in the mold but not in the drawings.
2. There is actually no precedent, or the precedent was not good enough to be repeated. If the precedent was not good enough, find out first what was wrong with it. There is no point in redesigning a mold without knowing why!

The inexperienced mold designer will proceed step by step as outlined in the following, paying special attention to details which contributed to earlier design, mold making or molding problems, but even an experienced designer will occasionally use the step-by-step method. An experienced mold designer or estimator will go through steps 1–11 as a matter of routine, often without using a drawing board or a computer screen, as follows:

All the steps 1–11 are usually sketched in pencil (free hand or using a ruler) on a copy of the product drawing with a significant cross section. For a simple product, this would require not more than 1–2 hours; for a more elaborate one, it may take a day or even more.

6.1.3 Step-by-Step Layout: Mold Stack Design

Step 1 Product Drawing

Draw a significant (cross) section of the product. Show the section on the CAD screen, or on drawing paper, so that there will be enough room to illustrate these significant stack elements:

- Product shape,
- ejection method,
- side coring (slides) if required,
- cooling,
- gating, and
- venting.

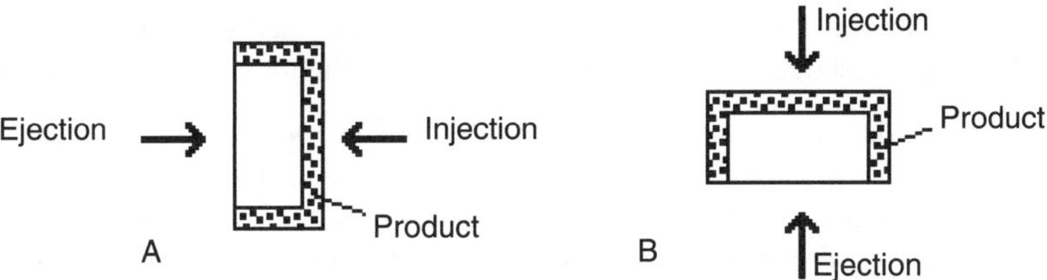

Figure 6.1 Two possible orientations of the product drawing: preferred (left) and earlier designs (right).

If the product is simple (e.g., a round container, etc.), one section can be sufficient. If the product is more complicated, additional sections may be required to bring out any possible problems, particularly with the shape, the gating, and the ejection. There are two possible orientations of the section:

a. Draw the section(s) so that the injection side is on the right and ejection is from the left of the view (Fig. 6.1A) This is the preferred method because it shows the mold as it is located in the machine.
b. In earlier designs, the injection side was shown on top and the ejection from the bottom (right sketch). This is still sometimes useful when drawing smaller molds. It may then be possible to show all views on one sheet of paper.

Step 2 Will the Product Pull out of the Cavity?

Yes: Proceed to Step 3.

No: Can the product (the section) be turned so that the product will pull out without the need for side cores (Fig. 6.2B)? This requires an "angled" parting line (P/L), but the overall mold cost may be much less than when using side cores, which not only add complications to the mold construction but also require more space and possibly a larger mold shoe. If this is possible, redraw the section and proceed to Step 3.

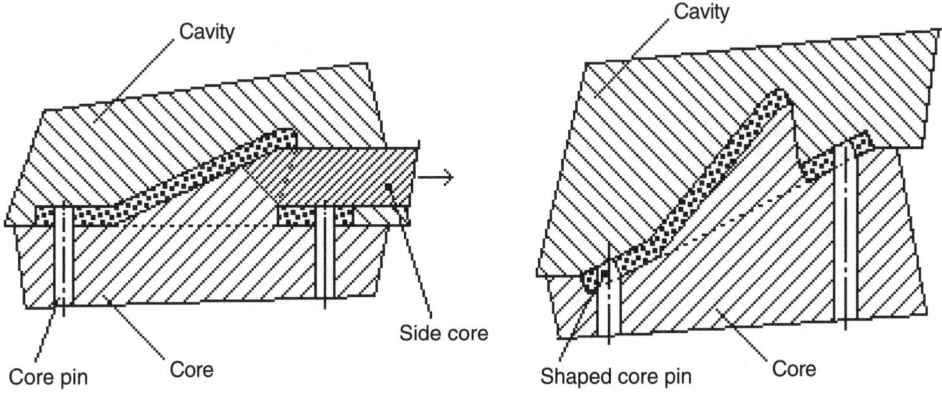

Figure 6.2 Example of a louvre mold: core pin and side core (left) and shaped core pin (right).

If turning the product as proposed above is not possible, then side core(s) or splits are required. Next, the designer must decide on which side of the mold they should be placed.

a. Side core is on the *cavity side*: The side core must pull out *before* the mold opens. This is the least desirable method because the side core must be operated independently of the mold operation with hydraulic core puller requiring piping and controls, or it requires the cavity to follow the core while the side core is withdrawn. An example would be a requirement for one (or more) holes in the side wall of the product, remote from the P/L (e.g., the holes in the sides of pails to receive a handle).

   With low productivity molds, such as prototype molds, but also in case of high productivity molds where after-molding operations will be performed on the product, there is another alternative. The designer can consider, and discuss with the customer, whether it would not be more economical to produce the holes (or openings) in the side of the product by drilling or stamping after molding, in a setup simultaneous with another operation being performed. This method may not add any labor cost to the product but could make the mold much simpler and less expensive to build and to maintain.

   The above option should be pointed out to the customer at the time of estimating. However, certain things are not always obvious at the beginning and will only be seen by the designer, who spends more time with the project. Regardless of the wording of the order, if such a suggestion makes sense and saves money for the customer, it should be discussed.
b. Side core or splits are on the *core side* (usually, on the moving mold half). During mold opening, as the core (with the product) pulls out of the cavity, the splits (or side cores) open. Typical examples include external screw threads or side holes near the P/L.
c. *Reentrant* (false bottom) of the product: If there is much side draft or a ratio of height to thickness less than 1:1, proceed to Step 3. If there is little draft (less than 5° per side) and a large ratio of height to thickness, the cavity must provide two-stage motion (moving cavity) to permit withdrawal of the thin section. For further discussion, see Chapter 12, Ejection.

Step 3: Will the Product Pull off the Core?

Yes: Proceed to Step 4.

No: Consider these options: stripping, two-stage action, unscrewing of internal threads, collapsible cores, or internal side cores.

a. (Forced) *Stripping*. This means pushing the plastic out of the grooves (or threads) which prevent the product from easily pulling off the core. This method is very common but depends on:

   - The shape (design) of the grooves, which must be designed so that the product can slip out of the groove without deforming or shearing the plastic.
   - The type of plastic: Only somewhat elastic materials can be stripped, typically PE and PP. Occasionally, even hard plastics such as PS can be stripped if the amount of stretch during stripping is within the elastic limit of the cold material, or if the product is stripped while the plastic is still warm enough to stretch easily without deforming permanently.

b. *Two-stage action.* Part of the core moves in relation to the other (first stage) to permit the product to free the undercut (groove, etc.), which would otherwise be trapped between two core parts. The product can then be ejected (second stage) by pins or strippers. For more discussion, see Chapter 12, Ejection
c. *Unscrewing of internal threads.* See Chapter 12, Ejection, Section 12.14, Unscrewing Molds.
d. *Collapsible cores.* See Chapter 12, Ejection, Section 12.13, Collapsible Cores.
e. *Internal side cores.*

Neither *c*, *d*, or *e* above will be discussed further in this section.

Step 4: Establish the P/L

With most products, the P/L is in an obvious location (i.e., at the rim of a container or at the seating surface of a technical product). However, in other products, the P/L is not so obvious and requires considerable thought.

a. *Ejection*: An important consideration in selecting the P/L is to ensure that the product will stay on the side which will have the ejection mechanism.
b. *Shut-offs, offset P/L*: If there are openings in the sides or inside of a product which do allow the product to pull off but which will require shut-off areas between core and cavity, then, to prevent sliding motion (metal to metal) of a shut-off area between cavity and core (including side cores and splits), this area must always be at an angle (Fig. 6.3). Such an angle: 1) prevents flashing, because a vertical area cannot be preloaded, and 2) reduces wear.
c. *Strength of P/L*: After establishing the (tentative) P/L, it is necessary to check the strength of the area where cavity and core will touch. If strength is insufficient, increase the area or add supports to take some of the clamping force.
d. *Bursting strength of cavity*: This must now be calculated to ensure that the cavity walls are strong enough to resist bursting. Again, see Chapter 19, Formulae and Calculations for Designers.
e. *Machining of P/L*: Consider how the P/L will be produced to provide flash-free matching of the mating surfaces. Usually, the P/L is ground on both cavity and core side. This produces the correct surface finish necessary but is not always easy to accomplish because, usually, at least some sections of the core project beyond the P/L, and these must be avoided by the grinding wheel. (On the cavity side, the cavity itself is usually below the surface of the P/L and is rather easy to grind.)

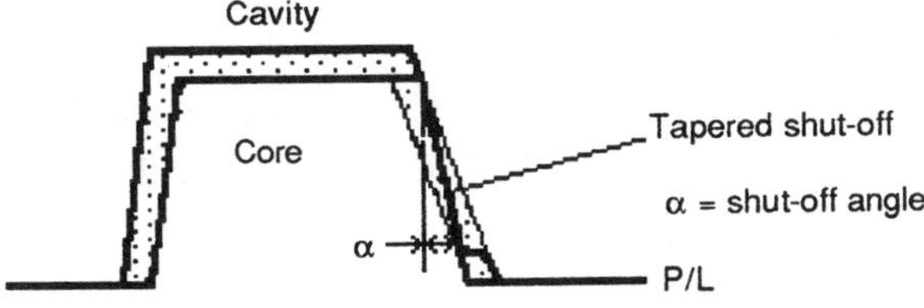

Figure 6.3 Vertical tapered shut-off.

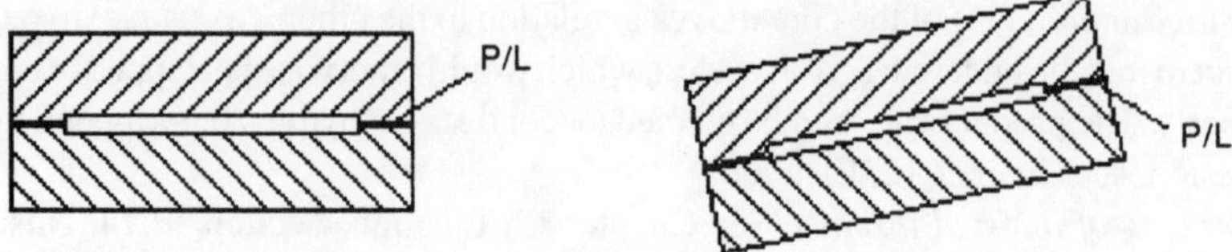

Figure 6.4 The P/L is preferred to be in one plane, whether at a level (left) or at an angle to the direction of the mold (right).

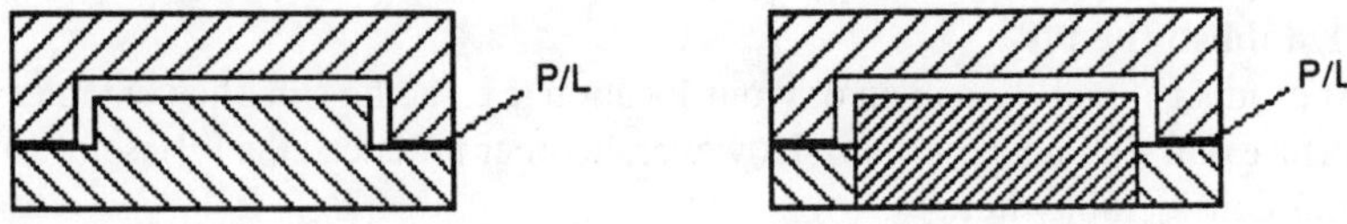

Figure 6.5 The core may project beyond the P/L, as is the case with a protruding core (left) and an inserted core (right).

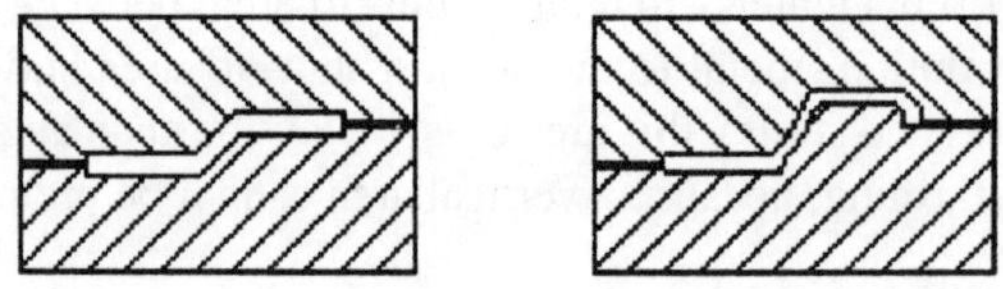

Figure 6.6 An offset P/L (left) and a protruding offset P/L.

- The preferred design is where the parting line is in *one* plane. This is easy to grind and lowest in cost. This applies also to a P/L which is at an angle to the direction of the mold motion (Fig. 6.4), as long as there is sufficient space for the run-out of the grinding wheel.
- If the core projects beyond the P/L (Fig. 6.5), it can be more difficult to grind on the core side because the wheel must avoid the core. If the whole core is inserted within the the boundaries of the P/L, the P/L can also be easily ground, but at the (often considerable) cost of having an insert.
- It is much more difficult when the P/L is offset (Fig. 6.6), and must be ground at two or more levels. The complication is worse if part of the core (or cavity) projects beyond the P/L.

Sometimes, it will be impossible to grind all of the P/L, and it then must be fitted either by hand (*hand fitting* or *bluing*) or with EDM, both quite costly methods.

Step 5 Is the Product Balanced?

This balance applies to the injection forces pushing at right angles to the center line of the mold.

Yes: Proceed to Step 6.

No: Refer to Section 6.2, Balancing of Forces in Molds. Wedges may be required to counteract these forces, or the stack may be designed so that two products are molded with side forces opposing each other.

Step 6 Gating

Usually, the method of gating (hot runner, cold runner, three-plate, edge, tunnel, etc.) has been indicated with the mold design order. If it was not specified and was left to the designer to propose a suitable method how and where to gate, any such proposal must be carefully considered before proceeding with the mold layout.

We shall assume that the specified method is suitable for the size of the product and for the material to be molded. However, under certain circumstances, the specified method may be risky, and the designer, or others, may have to point out possible problems which could be the result of using the specified type or location of the gate.

In some instances, the customer will specify gate location (or locations) where a gate vestige is acceptable, or unacceptable for cosmetic or technical reasons. This too must be carefully considered. The suggested location may be good for the appearance but bad for important molding considerations, such as filling, flow marks, weld lines, shrinking, etc.

If the designer is confronted with such a situation where the suggested gate location may be detrimental to the performance of the mold or the product, or where such location will seriously complicate the mold, the problem must be discussed with the customer before proceeding. For detailed discussion of mold gates, see Sections 6.2 and 6.3.

Step 7 Inserts and Core Pins

Inserts and core pins are often used where it is not practical to machine the cavity or core from one piece without leaving small, delicate sections of the mold material standing. This applies particularly to small, deep, narrow grooves and holes in a product.

Inserts are also required for openings in the product which are at right angles to the mold movement and which can be created by shut-off areas having metal-to-metal contact, similar to the case shown in Fig. 6.3 (openings in the outside wall of a product). The use of inserts in this case is suggested because of the probable wear of the shut-off area. Thin sections and shut-off areas will wear and/or break sooner or later, and it must possible to replace any inserts easily.

At this point in the preliminary design, the designer will consider and indicate where inserts are required and show the shape of these inserts.

Step 8 Venting

Rule 1: Vent to suit the anticipated flow of plastic at all points remote from the gate(s).

Rule 2: Vent at the closed end of all pockets, that is, any area not in the direct plastic stream, into which plastic must flow to properly fill the cavity space. If the pockets (for ribs, bosses, etc.) are created by inserts, it is easy to provide vent channels where the inserts seat in the main body of the cavity (or core). If the pocket is machined from the solid, vent pins must be provided at the deepest location. Note that ejector pins or sleeves, which should always be placed at the lowest point (of a rib, etc.), provide a natural vent and usually do not require additional vent pins.

Rule 3: Vents should be as large as possible, without causing flash. (This depends mainly on the type of plastic used.)

Rule 4: If the plastic flow from the gate(s) can circle around an area in the cavity because it flows easier through wider wall spaces or ribs, such areas must be vented. Because plastic flow always chooses the path of least resistance, it will engulf such areas and trap the air inside it.

Rule 5: Wherever possible, provide moving vents, that is, vents that will be operated every time the mold opens and closes. Typical moving vents: ejector pins (or sleeves), two-stage ejection, moving cavity. If this is not possible, provide stationary vent pins, but design the mold so that these pins can be easily removed for cleaning.

It is good practice to pressurize stationary vents pins or slots so that compressed air blows through the vent every time the mold is opening. The disadvantage is that it requires additional channelling for the compressed air. It may sometimes be acceptable to leave the air pressure ON permanently, thus not requiring valve operation for this circuit.

In some molds (for medical, food, etc., products), the use of air which could contact the product may not be acceptable for sanitary reasons.

Rule 6: Consider the injection pressure required to fill the mold and the necessary clamp force to hold the mold closed during injection. The land area between cavity and core subjected to this force may compress and permanently deform the mold material (steel, beryllium copper, etc.) and block the vents.

Rule 7: Every vent must lead to the outside. If this is not practical, vent grooves and channels must be provided to permit the air to escape into the open.

An exception to this is when the vent is pressurized. The air will escape into the channels that bring compressed air to the vent (see Rule 5, above). Even if the pressure is permanently ON, it is low compared with the pressure of the air escaping from the cavity during injection, and is acceptable as escape for the air from the vents.

For an in-depth presentation on venting in molds, refer to Chapter 11, Venting.

Step 9 Ejection

It is established in Step 3 how the product will be ejected from the core. The designer must now show, in detail, how this can be done by drawing the selected method (stripper, ejector pins, sleeves, air poppet, air gap, etc.). Refer to Chapter 12, Ejection, for further explanation of these methods.

Step 10 Cooling

The designer must show exactly how the cavity and the core will be cooled and, if possible, any other part that is in contact with the plastic such as gate pad, inserts, side cores, stripper ring, etc. Remember that the mold is a *heat exchanger*. The better and more evenly distributed the cooling is, the better and faster the mold will operate. It is often a little detail, such as an insert which is not cooled, that will considerably slow down the operation.

It is the mark of a good mold design to avoid any poorly cooled or uncooled areas where the plastic will remain hot for a long period. It is, of course, more difficult to design and more expensive to manufacture, but in the long run, the extra cost is paid back many times by the increased productivity of the mold.

For in-depth discussion of mold cooling, see Chapter 13, Cooling.

Step 11 Alignment

How will cavity and core, core and side core, etc., be aligned, to compensate for manufacturing tolerances, heat expansion, wear, etc.? The possibilities are:

a. Cavities are aligned only by the mold shoe, typically with leader pins. This is inexpensive but suitable only where the product is such that slight (unavoidable) variations are within the tolerances of the product (i.e., toys, screw caps, etc.). Taper locks may be added to improve the alignment at the P/L.
b. Cavities and cores are individually aligned with tapers, wedges, or pins. In conjunction with floating cores, this makes for excellent alignment under all circumstances, but it is more expensive to make.

Step 12 Consider All the Work Done So Far, and Ask These Questions:

- Is it really the best layout? Are there any other possibilities?
  A good designer will always find different ways to make the same product. He or she will have an open mind and not necessarily select the very first layout that comes to mind; before continuing, the designer will search to find a better method, even if it means more design time. Time "wasted" at this early stage may produce very valuable gains later when a better mold, with better productivity and/or with easier ways to manufacture, has been found.
- What materials will be used for the various parts?
  Note that the materials selection could affect the design of the stack, because of strength and heat transfer considerations.
- Are the mold proportions reasonable?
  This applies to wall thicknesses, length of taper engagements, contact areas, etc.
- In general, scrutinize what has been shown so far, before proceeding to design the next steps: actual mold layout.
- Establish minimum outside dimensions (plan view) of the stack. This is important when starting the next phase of the design, the layout of the cavity (cavities) in the mold shoe.

6.1.4 Step-by-Step Layout: Mold Shoe Design

After completing steps 1–12, we assume that we have arrived at a suitable stack layout. If there was a precedent of an identical or a similar size and shape product, but for a mold with a different number of cavities, the designer could have saved the effort of going through steps 1–12. However, even with such a precedent, in view of the rapid progress of technology it should be carefully viewed to see if the earlier design could be improved. The next steps relate to the arrangement of the stack(s) in the mold shoe.

Step 13 Draw Machine Platen Layout

Show the size and location of all tie bars, mounting holes or T-slots, ejector holes, and edges of platens. If the specified machine is not suitable, inform the customer immediately. ("Show" in this context means to display on the screen or draw on the drawing board.)

In extreme cases, a cavity layout may require that the mold overhangs the platen. This is acceptable, but all *stacks* must be located within the tie bars. In some cases, the mold may be so large that it is necessary to pull one top tie bar (or even both) to install the mold. Make sure that the machine for which the mold is intended has features to facilitate the pulling of tie bars. Some machines do not have provisions to pull the tie bars just for mold installation. If this should be necessary, discuss it with the customer before proceeding.

Step 14 Complete the Mold "Elevation" (Cross Section)
Show all the necessary plates to the right and left of the stack, with the planned (adequate) thickness. At this time, do not yet show the width or length of these plates. This will be established later. Where possible, use standard plate thicknesses.

Step 15 Show the "Shell" (Stack Outline)
The shell showing only the outlines (outside) dimensions of the stack in all directions will simplify the drawing of the stacks in the other cavity locations and save design time.

Step 16 Show Both Plan Views
Show the cavity shells and the core shells. The larger of the two outlines (in core and cavity halves of the mold) will indicate where the other stacks may be located so as to leave sufficient space between them. All views (plan of cavity, plan of core, and cross section or cross sections) should be done essentially at the same time. This prevents surprises when one view is well-advanced and then is found to be unsuitable because the other views have not been considered.

Step 17 Symmetry of Layout
Try to position all stacks (shells) in a symmetrical layout about the vertical and horizontal center lines of the mold. This will permit balanced runners and balanced clamping forces.

This may not be the final location. If there are ejector pins or moving core pins, they too must be shown to help in establishing the required ejector plate size and to prevent collision with cooling and air lines. If return pins are required, select suitable locations as far apart as possible but where cooling and air lines will not intersect with them.

Step 18 Review Meeting
Call a meeting involving anyone that can have an input on the mold design to review the work done to date, and discuss fits, clearances, operation, etc., per the check list.

Step 19 Cooling Lines
Show cooling lines and how they connect to the cooling channels in the stack. Allow space for O-rings and mounting screws for cavity and core stack parts. If possible, use cross drilling rather than connections outside the mold (except for moving stack elements, which require flexible connections).

Step 20 Alignment, Other Mold Features
Select location of leader pins. If possible, place them near the corners of the mold. If the mold shoe size has limitations (e.g., due to molding machine size), leader pins and bushings may

be placed in the path of cooling lines, which will then have to be routed around them. Make sure the leader pins are not in the way of the ejected products or a take-off, etc.

Show other features which may be required, such as spring returns, stroke limiters, taper locks, tie bar shields, leader pin shields, mold supports, linkages, etc.

Step 21 Mold Supports

Select locations of support pillars, if required. See Chapter 17, Mold Plates, Section 17.2, Deflection of Mold Plates, for more detailed information.

With ejector pins, these supports will be usually between the stacks, to allow the placement of the pins under the core. With stripper plates or rings, the supports are placed right under the stack for a better support.

Step 22 Mold Mounting

Show mounting of the mold to the machine platens. The preferred method is with screws, either from the rear of the platen through holes in the platens or from the front using threaded holes or T-slots.

The use of mold clamps is not as efficient, and should not be used unless specifically required by the customer or if the mold must be operated in a wide variety of machines.

NOTE: *At this point, all stack features may still require relocation to suit some of the elements introduced in Steps 15 to 21.*

Step 23 Recheck Machine Specifications with Current Mold Layout

Is the shut height acceptable? May be it is too large, and some plates may have to be reduced in thickness. This may require recalculation of deflection and addition of more supports. The shut height may be too small and require addition of a bolster plate under the mold or selection of thicker plates.

Step 24 Finalize the Mold Shoe Outside Plan Dimensions

Wherever possible, select a size that allows the use of existing stock plate sizes or commercial standard plate sizes without wasting material.

If the layout (to this point) is slightly larger than the closest stock size, try to rearrange the layout to be able to utilize the smaller size. This may be better than selecting the next larger stock size, making the mold unnecessarily large. It is also better than to machine off the unwanted material. But, never forget that the first requirement is to make the best possible mold, even if it results in a larger or an odd-size mold shoe and in higher costs.

Step 25 Screws

Show the placement of all screws. Note that, sometimes, mounting screws planned (in the stack design) in certain locations may have to be rearranged because of interference with cooling or air lines.This is satisfactory as long as the basic requirement of interchangeability of mold parts is maintained.

In other words, if a screw pattern must be changed, make sure it is the same pattern on all stacks and not, for example, one half with screws in a right-hand layout, the other half with a mirror symmetric, left-hand layout. Such a condition would require building right and left spares, which is undesirable from the point of view manufacturing and mold maintenance.

Step 26 Complete the Assembly Drawings
Ensure that all features are shown in all views and that each view contains all information required to detail and build the mold so that the detailer, assembler, and mold technician can follow the drawings clearly. The detailer must make all required detail drawings to make all parts for the mold. This means that all fits and clearances which are not obvious must be shown on the assembly drawing, as well as any notes that help the detailer and assembler in their tasks.

The assembler must have all specifications (screw torque, adjustments required, etc.) to complete the mold assembly.Finally, the molding technician must have the information necessary to properly install the mold. The assembly drawing must also show the sequence of the molding operation, including a timing diagram and electrical, cooling, and air circuit schematics. It is not necessary to repeat the same features all over. Once a feature (leader pin, bushing, screw, wedge, etc.) is shown clearly in one location, it is usually sufficient to show only the center lines of the other locations, and maybe the outlines, while indicating with a code letter and/or number (e.g., LP1, LPB1, S1, S2, etc.) what the center line represents. This shortcut can save much time in drawing on the board, but will not save much time with CAD; however, even with CAD, it will help by not cluttering up the drawing with many redundant details which make the reading of the drawing more difficult.

Step 27 Review Design
Review the design and, if necessary, the manufacturing process. This review is the last opportunity to scrutinize the drawing and to make changes before detailing the design and starting to build the mold.

Step 28 Cross Hatch Sections
Show cross hatch sections only where they will help to clarify the picture.

Step 29 Identify (Balloon) the Components
Each mold item (part) must be identified with letters and/or numbers, which will be also shown on a list, called the bill of materials, to be used to purchase the materials for the mold.

Step 30 Fill in Title Blocks and Finish the Drawing
Show all information on the drawing regarding the following:

- Scale;
- molding material, heat treat, and finishes;
- all other pertinent notes to clarify the drawing, if needed;
- job number;
- mold number;
- name of designer;
- name of checker [Note: It is *bad* practice to check one's own drawing. Any drawing should be checked by a person who also understands the drawing and the problems. This applies not only to the assembly drawings but more so to the detail drawings. A wrong drawing, or even one wrong dimension, can be very costly.];
- date of completion of drawing;
- date of checking; and
- date of release to the shop.

6.2 Balancing of Lateral Forces in Cavities

In most mold stacks, the forces *F* inside the cavity at right angles to the mold opening motion balance each other (Fig. 6.7).

If, however, the projected area on one side of the cavity is larger than on the opposing side, the cavity will want to move away from the core during injection, in the direction indicated by the heavy arrows shown in Fig. 6.8.

If the separating force *F* is small, it may be taken up by the leader pins. This is not recommended except in low-production molds, because it will eventually lead to excessive wear of the leader pins and bushings and to loss of dimensional accuracy of the product.

There are two recommended methods for balancing: 1) the use of wedges as part of the cavity/core stack and 2) orientation of two cavities symmetrically so that the forces oppose each other.

1. Use of wedges as part of the cavity/core stack: For practical reasons, the wedges in both cavity and core should be inserts. This will allow replacement in case of wear and adjustment of the proper relative location between cavity and core if necessary.
2. Two cavities are arranged symmetrically so that the forces oppose each other. In the simplest case, it will be as shown in Fig. 6.10.

The orientation of the cavities can also be reversed. The force *F* is taken up in the core backing plate (Fig. 6.11).

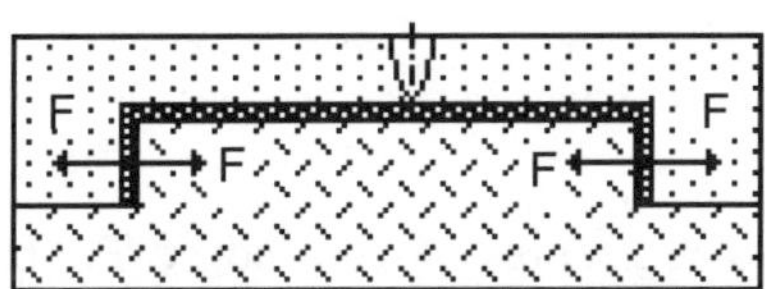

Figure 6.7 Balanced lateral forces *F* in the cavity.

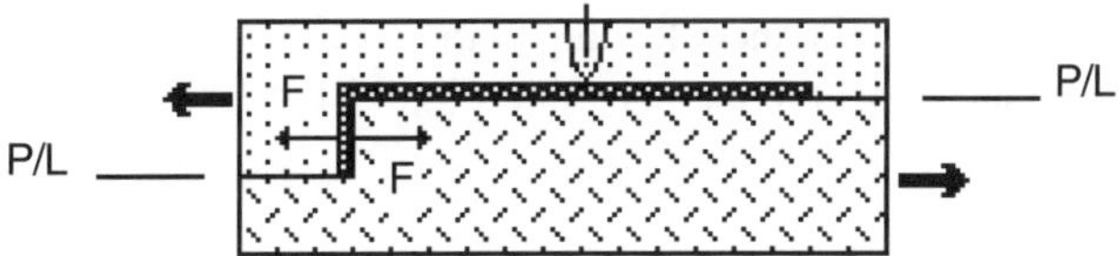

Figure 6.8 An offset P/L, with unbalanced forces *F*. Heavy arrows indicate movement of cavity away from the core during injection.

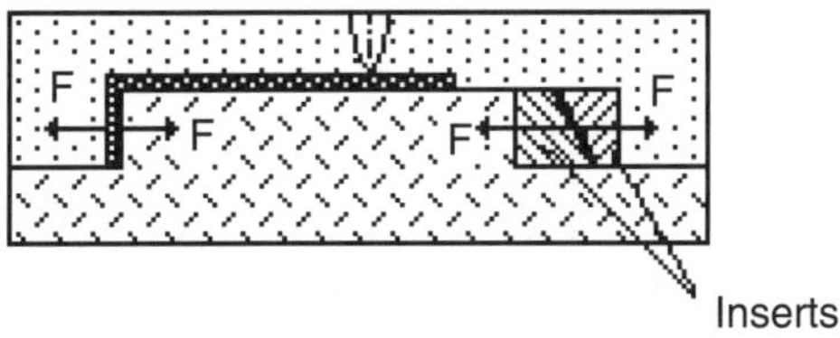

Figure 6.9 Balancing the forces *F* with wedge inserts.

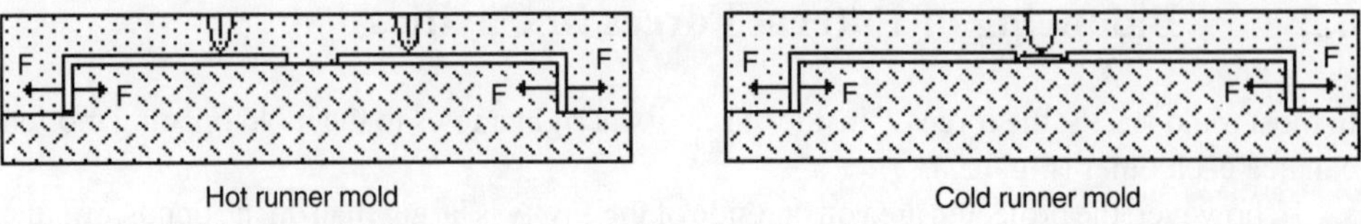

Figure 6.10 Symmetrical balancing with an opposing cavity pair in a hot runner mold (left) and a cold runner mold (right).

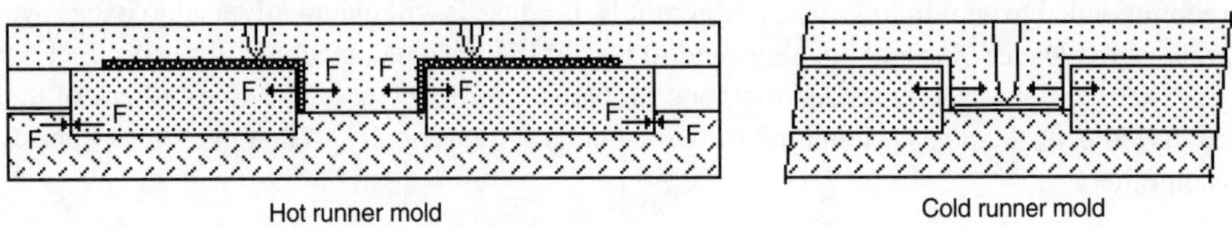

Figure 6.11 Balancing is achieved through reorientation of the cavities and use of the core backing plate in a hot runner mold (left) and a cold runner mold (right).

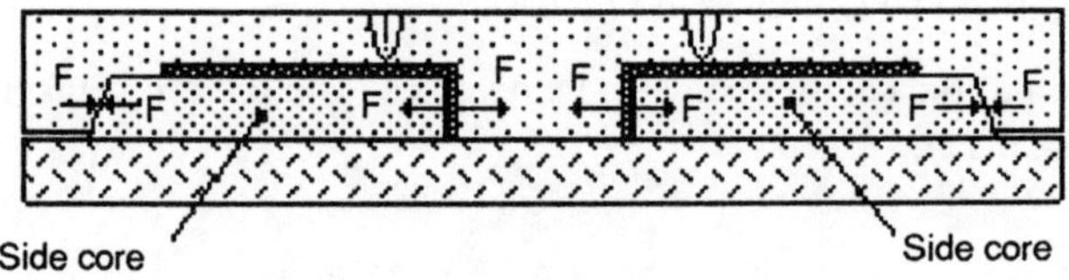

Figure 6.12 Balancing with wedged side cores.

If side cores are required, the side cores must be wedged (Fig. 6.12). Note that all side cores must be locked in position by individual wedges (*never* by their actuators) to counteract the forces *F*.

In Figs. 6.7–6.12, where center gates are shown, they could be three-plate or hot runner gates. For two-plate molds, the runners leading to the cavities would be placed in the level where the cavities are close together, as shown in Fig. 6.10 and 11 on the right.

6.3 Balancing of Runners and Arrangement of Cavities

The following lists the basic considerations in selecting the layout of cavities and the geometry of runners to distribute the plastic to all gates.

6.3.1 Balancing of Runners

Balancing ensures virtually equal flow of plastic through each gate of a multicavity mold, and/or through each gate (if there is more than one) into each cavity. The plastic should arrive at all gates with the same pressure and temperature, so that all products within one mold have uniform characteristics. "Virtually equal" in this context means within reasonable limits.

To achieve balancing, there are several lines of thought. For many years, it was preferred to arrange the runners symmetrically from the point of entry of the plastic (the sprue) all the way to each gate. While this is fairly easy with some layouts, especially with a small number of cavities, it may become difficult with a large number of cavities. On the other hand, even though the geometry is symmetrical, the desired balanced flow may not be achieved, because:

1. The flow of plastic through the channels changes at every point of division into branch channels and depends on the amount of redirection of the plastic at the point of change.
2. The flow depends on the accuracy of machining and the finish inside the channels.
3. There may be temperature differences within the runner due to uneven heating and cooling of the steel surrounding the runners.
4. There are machining tolerances of the gates.
5. Some resins are flow-sensitive to changes in direction and generally will not flow evenly around several "bends" in a runner system, making balanced runners for a multicavity mold impractical.
6. Uneven venting can affect the filling of the cavities.
7. Quality differences in molding surfaces will affect the flow.
8. The phenomenon of the so-called "plastic memory" can also affect the flow within the runners.

The speed of injection plays a major role; in cold runner molds, the influence of mold cooling gradually narrows and restricts the cross section of the channels toward the gates. This is especially noticeable with slow injection speeds, which give the plastic within the runners more time to cool.

Despite the balancing attempted with geometric symmetry, it may be necessary to correct the mold after tryout by increasing some gate sizes and/or runner sizes, to compensate for inequalities between the products coming from different cavities. In this case, the concept of "true" balancing is just ignored, and it is, in fact, a simple, nonbalanced runner system that is used. This method is often used by some mold makers, but it is not recommended because it may make a desired interchangeability of cavities impossible.

A commonly used approach today is to balance *approximately*; that is, to create a geometry which attempts to balance as much as possible but without going into complicated configurations. The theory is that at the last branches, even if not truly balanced, the differences in flow and pressure at the gates are small and within an acceptable range so that the quality of the products is not affected. This approach is often used with multicavity hot runner molds for commodity plastics (PP, PE, or PS) and with some cold runner molds when injecting with high speed and pressure.

As a general rule: *Runners should be as short as possible and as balanced as practically possible.* This applies to all runner systems, cold or hot. More will be explained about this in the following examples and discussions of various runner layouts.

Why discuss cold runner molds at all? Two-plate molds and three-plate molds are "old" technology. However, there are many occasions where a hot runner mold requires a combination with two-plate or three-plate systems. Also, two-plate molds are frequently used where it makes better economical sense. In 1994, about 90% of all molds worldwide were cold runner, two-plate molds. The designer should, therefore, be familiar with all systems and understand the similarities and differences between them.

6.3.2 Two-Plate Molds (Cold Runners)

In older molds, cold sprues were used to feed the runner from the machine nozzle. Today, most two-plate systems use "hot sprues", which are essentially heated extensions of the machine nozzle built into the mold. This eliminates the often large plastic mass in the sprue which is slow to cool, and thereby increases the molding cycle.

The general rules for two-plate systems apply to any number of cavities, and not necessarily to cold runners only. When speaking of multicavity molds, we mean usually two or more cavities of the same product. However, molds often produce pairs of matching products (e.g., a Petri dish base and cover, two matching halves of audio and video cassettes, etc.) or families of related products (family molds) such as for plastic model kits in which the products are left on the runner and shipped with the runner). The size of the different products in one kit may vary greatly; the runner layout must position all products within the molding area, possibly in a logical arrangement for the user, but the designer must still attempt to obey the following rules, which apply to all multicavity molds.

6.3.2.1 Runner Length

The runners should be as short as possible, to reduce the mass, but must be of adequate length to satisfy the other conditions.

6.3.2.2 Runner Cross Section

The cross section should be as small as possible but still compatible with the plastic flow requirements. This can be achieved in some instances by special shapes.

A circular cross section provides the largest flow area for the least mass but is more expensive to produce because the runner must be cut into two plates. A trapezoidal cross section is best because it is easier to machine. Being tangential to a circle, the flow is as good as in circular cross section; however, the mass of plastic is 35% greater.

A flat runner has reduced flow but will cool faster than either a circular or trapezoidal runner. Finally, runners shaped as crosses provide better cooling because of their larger surface. They also provide stiffening to the runner during ejection but are more expensive to manufacture; they are rarely used.

6.3.2.3 Cavity Spacing

Cavities should be spaced closely, but consideration must be given to the runner width and to the location of ejector pins, the shape of the stripper plate, and the layout of the cooling. Refer back to Section 6.1.

6.3.2.4 Cavity Supports

When planning the cavity spacing, the designer must also consider the space requirement to allow for supports under the cavities, which in turn may interfere with the location of ejector pins of products and runners.

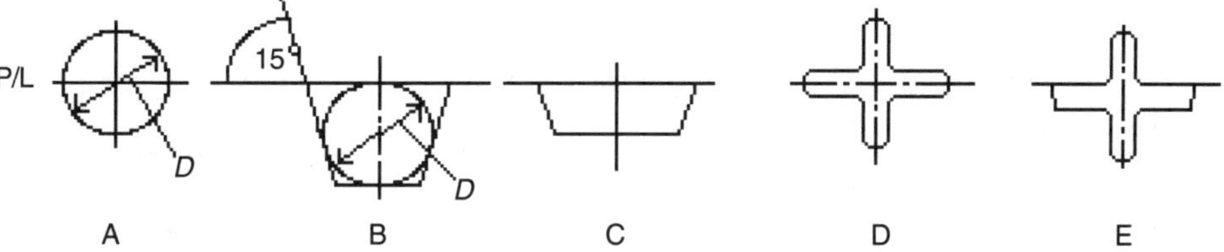

Figure 6.13 Runner cross sections may be designed in many shapes: A. circle, B. trapezoid, C. flat, D. and E. crosses.

6.3.2.5 Gate Location

Gate location depends on the product. It must be acceptable for the appearance of the product, ensure proper filling of the cavity, and suit the selected type of gate.

6.3.2.6 Distance and Time Required for the Products to Fall Free of the Mold

This is particularly important in high speed molding. The equation for calculating the time is shown below as

$$t = \sqrt{\frac{2S}{a}} \tag{6.1}$$

where S = distance in feet, a = gravity (32 ft/sec^2), and t = time of free fall in seconds. This assumes that the product, at the point of ejection, has no initial speed downward. If there is an air blast to give an initial acceleration, the time to clear the mold can be greatly reduced.

In the two-cavity layouts shown in Fig. 6.18 on page 82, the product from the upper cavity in Fig. 6.18B will take longer to clear the mold then the products from Fig. 6.18A. If the difference is large, it could require additional mold open time and result in longer cycle time.

6.3.2.7 Clearance Between Tie Bars (Free Fall or Take-Off)

The distance W (Fig. 6.18A) must be less than the clearance between the lower tie bars; if not, the products will hit the tie bars as they fall freely from the cavities and may become contaminated with tie bar lubrication grease. In some cases, deflectors can be built to guide the products down without hitting the tie bars, but this adds to the mold cost.

When using take-offs, the spacing must be such that the products held by the take-off will clear the tie bars as they pass horizontally (or vertically) between the tie bars.

6.3.2.8 Molecular Alignment

The gate location must ensure best plastic flow within the cavity for proper filling and for maximum product strength. This applies generally to elongated products. It is specially important in products utilizing a "live hinge" feature. In such cases, the plastic must flow in a direction at virtually right angles to the hinge. Also, there must never be a weld line at or near the live hinge.

Layout A in Fig. 6.14 would rarely be used because it tends to create a weak spot in the product near the gate, and the product will easily break there. Layout B is better and has a

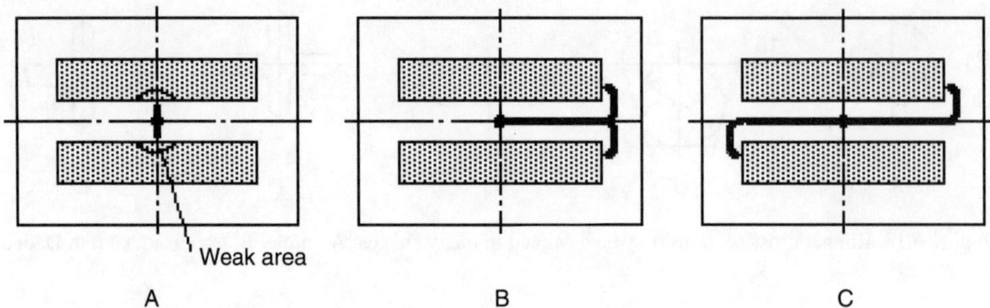

Figure 6.14 Gate location for proper filling: A. weak area near gate, B. unbalanced but shorter runner, and C. balanced but longer runners.

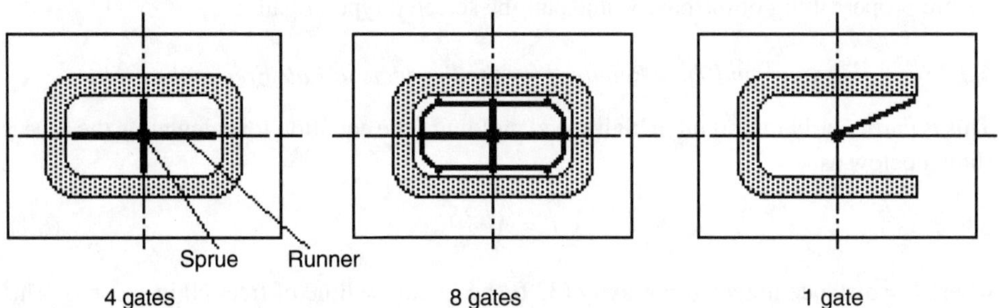

Figure 6.15 Edge gating in single-cavity molds with open center: A. 4 gates, B. 8 gates, and C. one gate.

shorter runner than C, but the clamp forces are unbalanced. In many cases, this lack of balance can be neglected; if not, then layout C is suggested.

The understanding of the above highlighted points regarding two-plate molds will help the designer greatly in selecting the best cavity layout.

6.3.2.9 Cavity Layout

6.3.2.9.1 Single-Cavity Molds If there is a large enough opening within the outline of the product, the sprue can reach the P/L and connect with runners to the gate(s) from the inside (e.g., frames, rings, U-shaped products, etc., as shown in Fig. 6.15. This permits gating into one end of a U-shape, or into any number of gates in the case of a closed or an open shape.

Visible weld lines and strength considerations may be a deciding factor in where to gate and how many gates to include. See also Chapter 10, Gates and Runners, Sections 10.3.2–10.3.4 on cold runner gates.

If the projected area of the product is uninterrupted, a sprue gate (cold or hot) can be used (Fig. 6.16). This will result in a (possibly unsightly) gate mark, usually in the center of the product. A cold sprue requires cutting after ejection; a hot sprue will leave a smaller, self-degated vestige. If either method is acceptable, this is a very simple and often preferred method (e.g., pails, cups, trays, many large products).

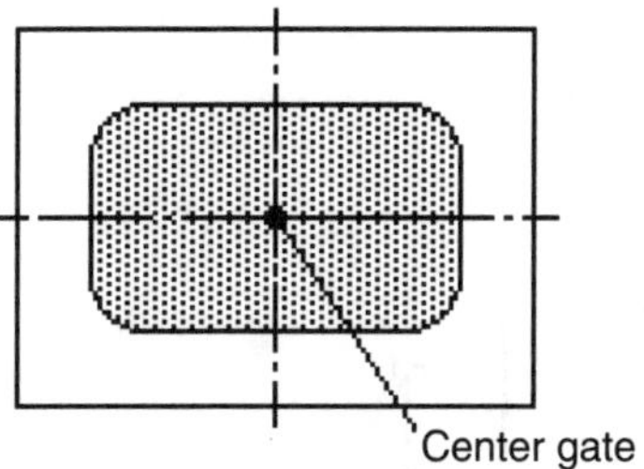

Figure 6.16 Center gate into large product.

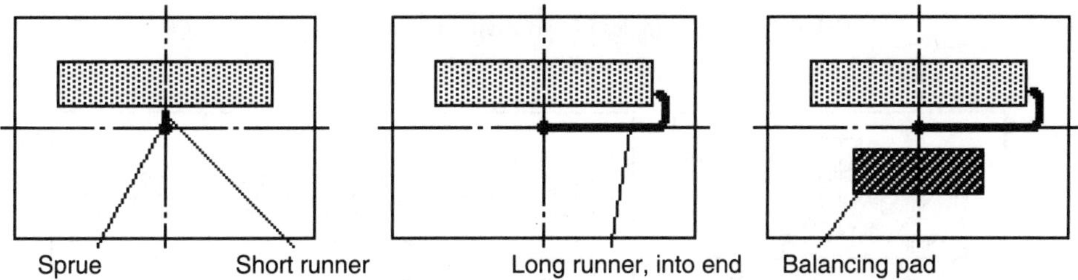

Figure 6.17 Examples of edge gating in a single-cavity mold.

If this is not acceptable, the product must be edge gated, and it becomes necessary to locate the cavity off-center in the mold (Fig. 6.17).

To place a cavity off center is generally not recommended unless the amount of the shift from the center is small. If it is absolutely necessary, a balancing load equal to the expected cavity clamping force must be provided (i.e., a pressure pad, symmetrically opposite the cavity), so that the clamp mechanism will not be loaded off center. The mold shoe will then be practically the same size as a two-cavity mold. It could then be worthwhile to add another cavity for a relatively small additional cost and gain at the same time greater productivity.

It must also be considered whether to gate the cavity from the side, which results in the shortest runner, or from the end, for best molecular alignment of the plastic and maximum strength of the product.

6.3.2.9.2 Two-Cavity Molds Illustrated in Fig. 6.18 are two basic layouts for two cavities, regardless of the shape of the product. Both layouts are practical and commonly used. Selection depends on the general rules listed above for any number of cavities. In the examples in Fig. 6.18, H2 is much larger than H1, and the more time is required for the products to clear the molding area (see equation 6.1).

Any reasonable number of cavities greater than two can be laid out in a circular layout or pattern. Rectangular layouts require multiples of two cavities. Occasionally, a combination of both circular and rectangular layouts is used (e.g., a 20-cavity mold with four 5-cavity circular patterns laid out in a square.

Shown schematically in Fig. 6.19 are two 8-cavity layouts. There are advantages and disadvantages of either layout shown in Fig. 6.19. A circular pattern allows the shortest

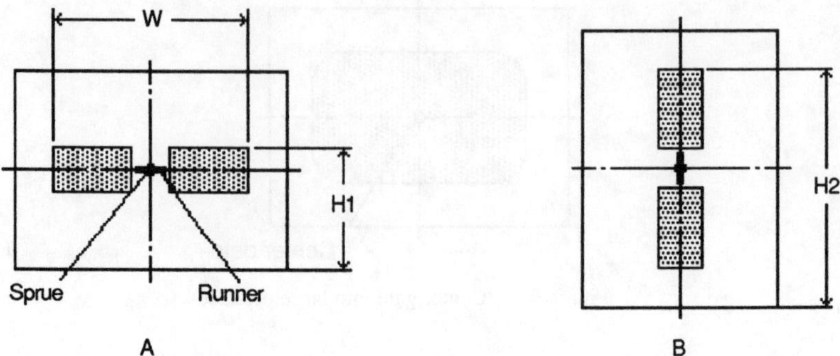

Figure 6.18 Examples of two-cavity mold layouts.

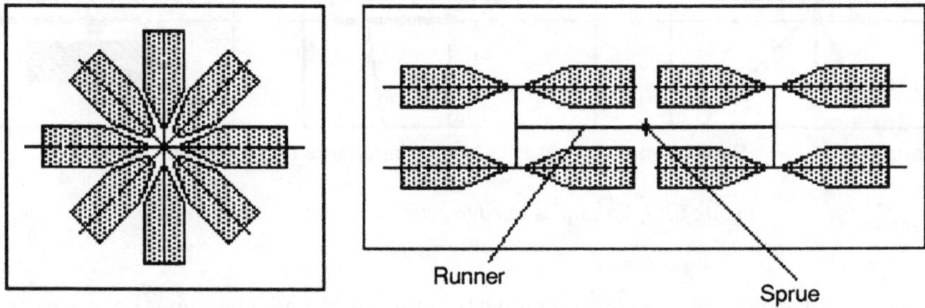

Figure 6.19 Two schematic drawings of eight-cavity layouts in a circular pattern (left) and a rectangular pattern (right).

possible runners as they radiate from the central sprue. However, the circular pattern allows only a limited number of cavities to be arranged in a circle closely around the sprue. The rectangular pattern is usually easier to produce by designing, drafting, and machining with *x*, *y* coordinates rather than with radial coordinates.

1. *Shape of product.* Small, elongated products radiating from the sprue lend themselves best to a circular layout, especially with a small number of cavities (e.g., spoons). It is sometimes used in conjunction with a three-plate or hot runner system, where each "drop" represents a sprue for a circular two-plate layout.
2. *Size of mold.* A circular layout may result in a smaller mold than is possible when using a rectangular layout. In such a case, it may be worthwhile to accept the complications of working in a radial system. It is also used where the mold size is limited due to the available molding machine size.
3. *Balancing of runners.* The circular layout has inherently balanced runners.
4. *Point of gating.* Where the shape of the cavity lends itself to a circular layout, the gate will be at the narrowest feature of the product, which is often the most desirable location (e.g., typically cutlery, measuring spoons, etc.). Note: Circular layouts can be used with hot runner edge gates.

It is not practical to show all possible arrangements of cavities within a mold, for any number of cavities. Such arrangements depend on the shape and size of the product, the proper location of the gate for best flow of the plastic, and on considerations such as the location of leader pins, ejectors, return pins, and core supports. It also depends on adequate spacing and location of cooling channels and on the positioning of side cores, if required. Also, the size of machine must be considered.

However, certain general rules based on experience should be observed and may be quite helpful; they will be listed below. Some rules have already been listed earlier in this section.

1. *Balancing of runners.* In general, balancing is recommended; however, the need for "perfect" balancing should be carefully weighed.
2. *Symmetry of layout.* Wherever possible, cavities should be arranged symmetrically around the vertical and horizontal axes of the mold to ensure even and adequate clamping force over the entire mold area. If this is not possible, pressure pads must be employed to balance any offset condition. A symmetrical layout is always preferable, for designing, manufacturing, cooling, ejection, etc.
3. *Drop height of product.* This was shown earlier for a two-cavity mold; the same rule applies for any number of cavities. Figure 6.20 shows the difference in drop heights.

6.3.2.9.3 Three-Cavity Molds The difference in drop height may seem small, as in the three-cavity mold layouts shown in Fig. 6.20, but in high-speed molding, a gain of even a fraction of a second in mold open time can be significant. For example, a savings of 0.15 second on a 3.00-second molding cycle (20 shots/minute) could permit the molding of one more shot per minute (21 shots/minute), or an increase in production of 5%.

The examples in Fig. 6.20 show a circular layout. The cavities are spaced at 120°; the clamping force is, therefore, properly balanced; that is, the areas of cavities and runners are symmetrical about the mold center lines. There is no rectangular layout for three cavities.

6.3.2.9.4 Four and More Cavities in a Mold The following are practical arrangements and will be accompanied by some comments, in addition to comments already made earlier.

4 Cavities Figure 6.21 shows circular layouts for a four-cavity mold. In layout A, the product is gated on the wider side, while in layout B, the gate is on the narrower side. The choice

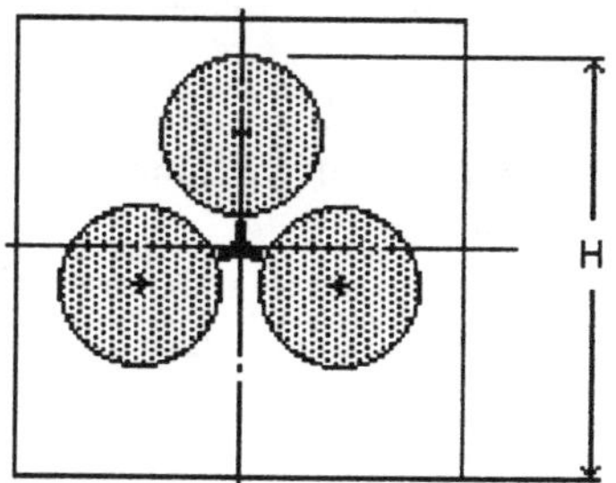

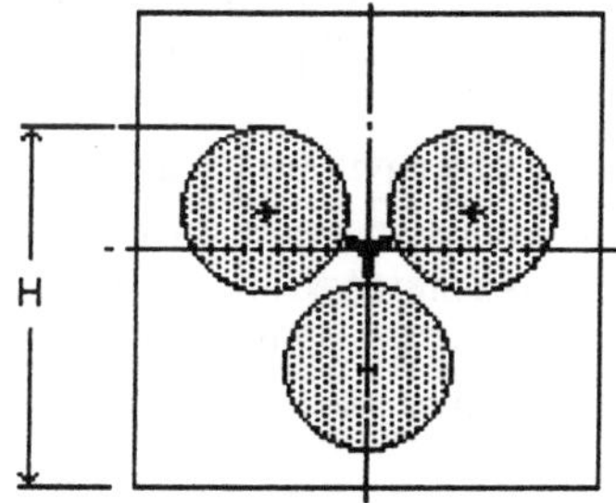

Figure 6.20 Examples of edge gated three-cavity layouts.

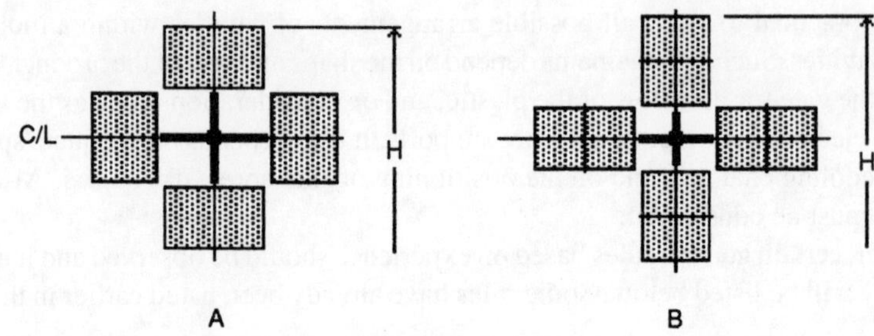

Figure 6.21 Two circular layouts of a four-cavity mold: A. gates on the wider side of the product and B. gates on the narrow side of the product.

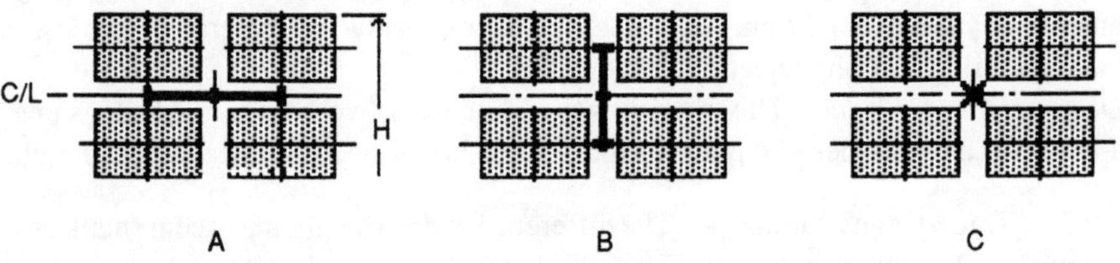

Figure 6.22 Rectangular layouts of four-cavity molds: A. H-style with gate on the wider side of the product; B. H-style with gate on the narrow side of the product, and C. X-style with the gate on a corner of the product.

depends on the location most suitable for the product. In either case, the distance *H* is larger than in a rectangular layout.

In Fig. 6.22, layouts A and B are typical H-style runners, C an X-style runner. Both styles (H or X) are used in mold design. The H-style has advantages for manufacturing by working to *x, y* coordinates. The X-style is preferred because of the shorter (and lighter) runners. Both styles may be used in combination.

5 Cavities It is only practical to design a five-cavity mold in a circular layout, with the five runners radiating from the sprue. It is occasionally used in molds with hot-cold runner combination for small products.

6 Cavities Six cavities in a mold may be in a circular pattern (not shown) or in rectangular layouts as shown in Fig. 6.23. These are widely used arrangements.
Fig. 6.23A is a preferred arrangement of two Y runner systems; Fig. 6.23B and C are typically H-style runners, although Fig. 6.23B is preferred because of the smaller number of changes in the direction of plastic flow in the runners.

7 Cavities Although it is possible to arrange seven cavities in a circular layout, it is not recommended because of the difficulties in dimensioning and in machining setup.

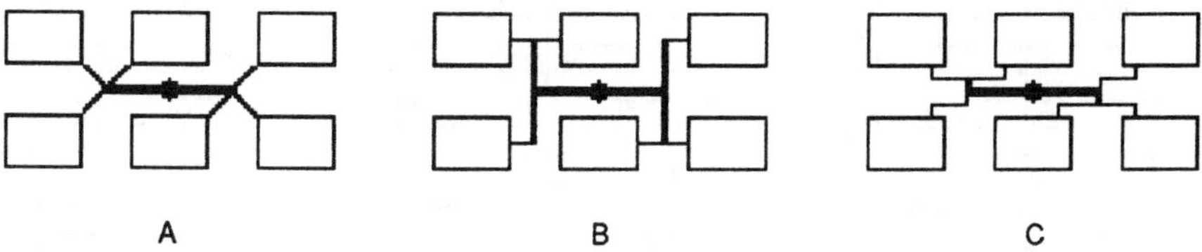

Figure 6.23 Variations of rectangular six-cavity layouts: A. Y-style runners, B. H-style runners with two changes in direction of plastic flow, and C. H-style runners with three changes in direction of plastic flow.

Figure 6.24 Two variations of an eight-cavity mold: A. two X-style runner systems in combination and B. two H-style runner systems combined.

8 Cavities The rectangular layout is a very common arrangement for eight-cavity molds, with two variations as shown in Fig. 6.24. Fig. 6.24A, with its combination of 2 X-style runners, is the preferred arrangement to Fig. 6.24B.

Alternatives are to turn the layout 90° so that the long axis of the layout is vertical. The decision of which one to use depends on the shape of the cavities, on the time required for the products to fall clear of the mold, and on the clearance between tie bars. There may be some flow differences at the final branches of the X (or Y) systems which may cause different pressure drops before the gates. With the X system, two branches continue at 135° to the main branch flow, the other two branches double back at 45°. This may have an effect in some molds but, usually, the differences will fall within acceptable limits and can be ignored. With the H system, all branching is at 90°, and flow and pressure drops are presumed to be equal in all branches.

It is impossible to show all possible layouts; shown in Fig. 6.25 are typical runner systems for eight-cavity layouts for elongated products. The arrangement of the cavities is compact, and the location of the gates is good, but the runners present some problems.

In Fig. 6.25A, we see a fully balanced layout, with alternatives A1 and A2. The distance from the sprue to each gate is the same; however, this distance is long, and the projected area of the runner system is large. In Fig. 6.25B, there is a difference in distance from the sprue to the various gates, but it can be claimed that this difference is sufficiently small and not significant. The dotted lines show an alternative layout to balance the clamping force.

In Fig. 6.25C is an unbalanced layout, resulting in the simplest layout and the smallest runner and projected runner area. The difference in distances from sprue to gates is large, and in some molds may require adjustment of the gate and runner size. The faster the injection speed, the less significant is the difference of the distances. Today, there are many molds built successfully with any of the above three layouts, with a tendency toward the simplest layout).

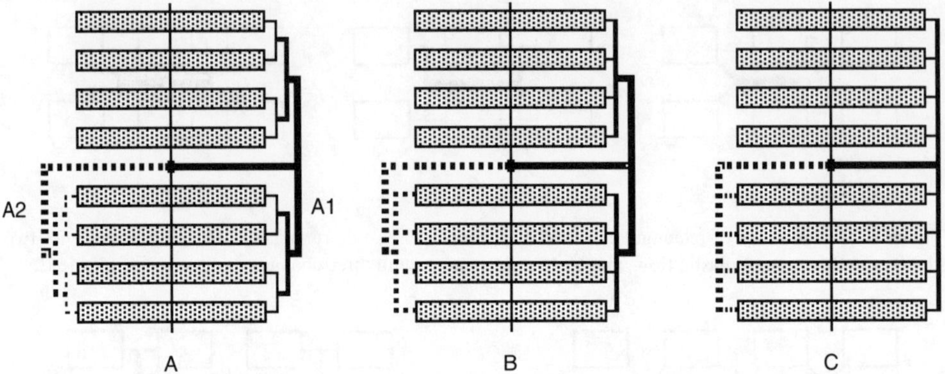

Figure 6.25 Variations of eight-cavity mold layouts for elongated products: A. fully balanced layout, B. partly balanced layout, and C. unbalanced layout of runner system.

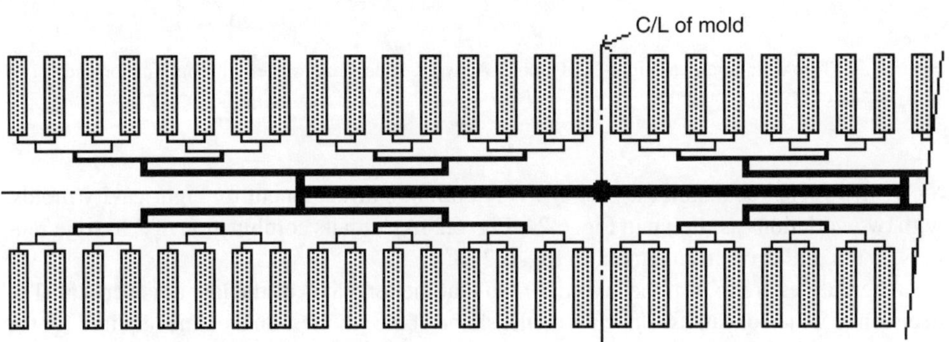

Figure 6.26 An example of a 64-cavity mold with a balanced runner system.

The same considerations apply to layouts for larger numbers of cavities. There are 64-cavity molds built (e.g., for cutlery, which have runners completely balanced with a very intricate runner shape.

From the main runner in Fig. 6.26, the flow volume leading away from the sprue to the last branch runner leading to the gates continues to diminish; in consequence, the runner cross sections start large and reduce in size up to the gates. Including the short connecting runners at each "split", there are in Fig. 6.26 ten runner sections between the sprue and the gates. For practical reasons, no more than four or five different runner sizes would be selected (i.e., from 10 mm down to maybe 3 mm). As mentioned earlier, the runners must not be too large, but they must be strong enough to stay reasonably in one plane during ejection.

Today, such a balanced layout is considered "old-fashioned"—the runner is very complicated without substantial benefits. There are similar molds, running just as well with fast injection, with a layout as shown in Fig. 6.27, without having balanced runners.

This layout is used today in many molds (e.g., cutlery) with up to 96-cavities. With fast filling speeds, there are few problems; if some cavities are not filled, the gates may have to be adjusted. The simplicity of this layout and the small projected area and space requirement for this runner makes it attractive when compared with the balanced layout.

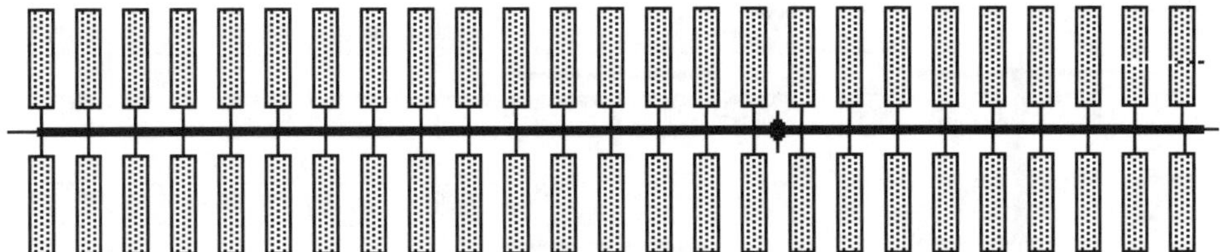

Figure 6.27 A 64-cavity mold built to current practices, without balanced runners.

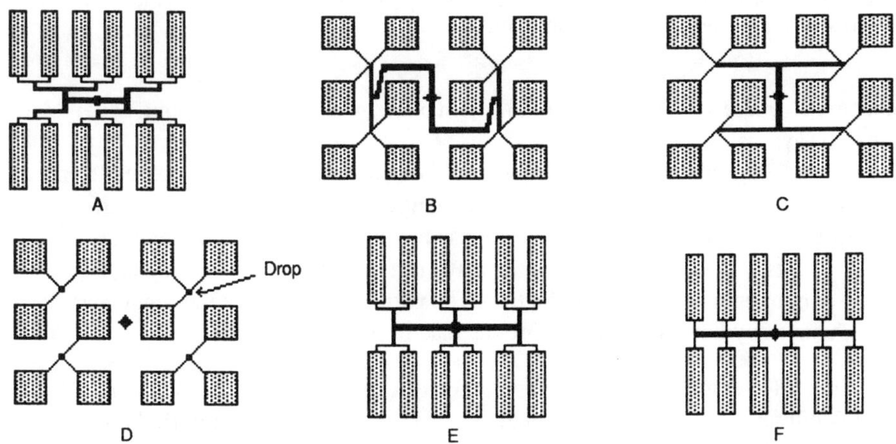

Figure 6.28 Six examples of a two-plate, twelve-cavity mold: A., B., and C. are fully balanced cold runner systems with differing runner layouts, D. is fully balanced and combined with a hot runner system, E. is partly balanced, and F. is not balanced.

Another advantage with such a layout is that the number of cavities is not be restricted to a 2^n sequence (i.e., 2, 4, 8, 16, 32, 64, etc.); any number (within reason) is feasible. However, in general it is still better to lay out the cavities and runners fully balanced or as well-balanced as possible.

9–15 Cavities (Circular Layout) In a circular layout, the cavities should be spaced at an angle that is divisible by whole number into 360°. For example, 9 cavities should be spaced at 40°, 10 cavities at 36°, (360°÷10=36°), and 12 cavities at 30°.

9-11 and 13-15 Cavities Such odd numbers of cavities would be used only in model kit sets. Ten cavities are sometimes used in rectangular, unbalanced layouts.

12 Cavities Such a mold would usually have fully balanced layouts. See Fig. 6.28 for examples of twelve-cavity molds.

Figure 6.28A, B & C show fully balanced layouts. In views B and C, note the difficulty of placing the main runner around the two central cavities. If this were a three-plate or hot

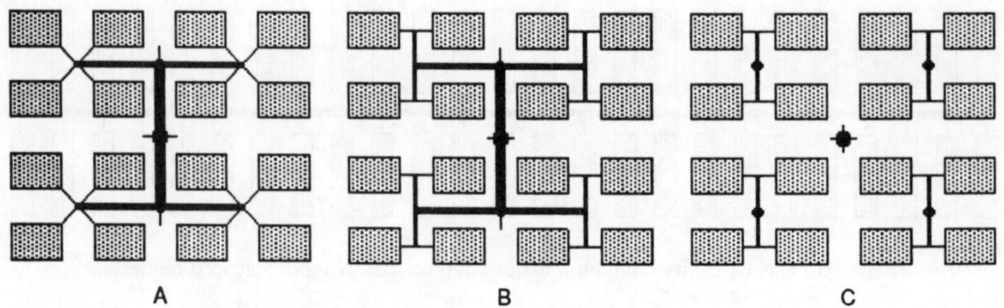

Figure 6.29 Examples of fully balanced 16-cavity molds: A. H and X style runners, B. H and H style runners, and C. H style runners fed from a hot runner drop.

runner mold, where the runners are in a different plane in the mold, they could "run right over" these two cavities.

Figure 6.28D is also fully balanced, but its four cold runner systems are fed from a four-drop hot runner system. Considering all the changes in direction of the flow, there may be some differences in pressure drop. All of these layouts (A–D) are commonly used.

Layout E is not recommended. There is little difference between this and layout F, which is much simpler and easier to manufacture. The parting plane where the runners are located must not have any breaks or gaps where the plastic could run (flash) into.

16 Cavities Refer to Fig. 6.29 for examples of 16-cavity molds. Either layout A or B can be used, in preference over a "linear", unbalanced layout, but A is the preferred layout. Layout C is also fully balanced; the cold runner system is fed from a four-drop hot runner system. Instead of feeding an H, as shown, the cold runner portion could be (preferably) an X shape, as shown in layout A.

17–23 Cavities In a circular layout, the cavities should be spaced at an angle which is divisible by whole number into 360°. For example,18 cavities should be spaced at 20° and 20 cavities at 18°, etc. Odd numbers of cavities would be used only in model kit sets.

20 Cavities A 20-cavity mold is sometimes used in rectangular, unbalanced layouts.

24 Cavities In a rectangular layout, 24 is a frequently used cavity number. Examples are shown in Fig. 6.30, including the illustration of a bridged runner system in layout A. In layout B, the cold runner system is fed from a four-drop hot runner system.

Another possible layout is shown in Fig. 6.31. This "almost" balanced layout could be used for elongated products. One should question the need for this rather intricate layout as compared to a linear, unbalanced layout.

Figure 6.31 shows a feature which is often used called *bridging*. The bridges connect the runners to stiffen them at their loose, open ends. This is important when the runners are thin and tend to bend or to curl, which may prevent the runner from falling out of the mold.

The tendency of a runner to deform out of plane is increased if the cross sections of the runners vary so much that at the time of ejection, some runner sections are still hot and continue to shrink, and thereby deform the runner. Bridges or connecting runners ($\approx 1 \times 2$ mm)

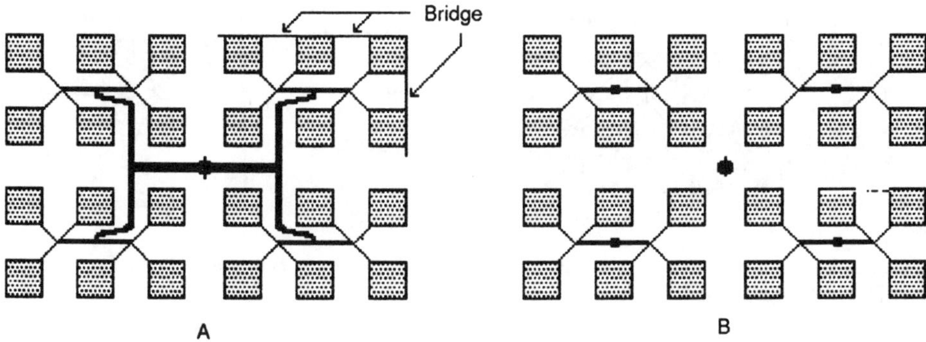

Figure 6.30 Two fully balanced 24-cavity mold layouts: A. plain cold runners, and B. combination of hot and cold runners.

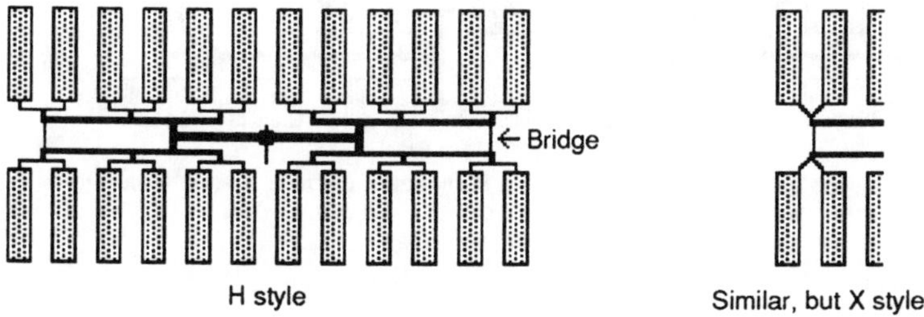

Figure 6.31 Shown with both H style and X style runners, this partly balanced layout includes bridging at the open ends.

are sometimes machined into the mold only after mold testing because of runner ejection problems. Experienced designers can foresee the need for bridges and design them into the mold.

Bridges are sometimes used with two-plate runners in molds with tunnel gates, and are often required with three-plate runners where there is usually little clearance between the mold plates to let the runner drop. Because the bridges are thin, they do not add much weight and do not affect the cooling time or have any measurable effect on the plastic flow.

Sometimes, it may become necessary to connect the products with similar bridges, i.e., if the products stay on the runners and must remain in proper relationship to each other after ejection. The bridge connects to the product with small passages similar to gates which will break off easily (see Fig. 6.30A).

The point where the sprue enters the runner (with cold sprues) is always hotter than the rest of the mold, because the runner is thickest at this point and will remain hot longer, thereby lengthening the cycle time. Because this area of the runner flexes most, it is in some molds reinforced with a rib (Fig. 6.32), which will cool and become stiffer even though the runner next to it stays hot. This does not always work, because the runner will contract more than the stiffener (rib) and will force the runners out of plane after ejection.

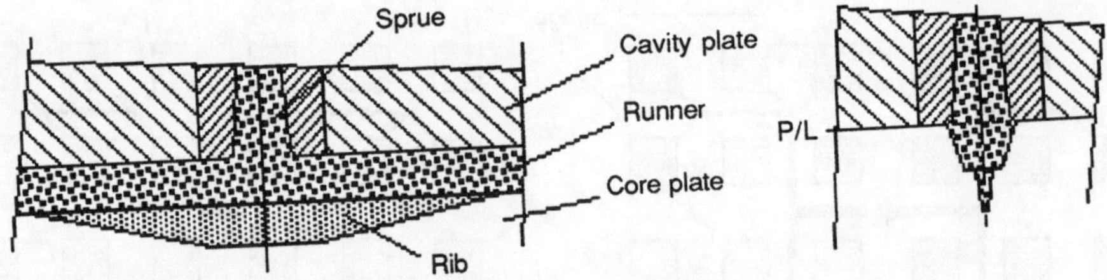

Figure 6.32 (Left) A rib is used to reinforce the mold where the sprue enters a runner. (Right) Cross section of left figure through sprue.

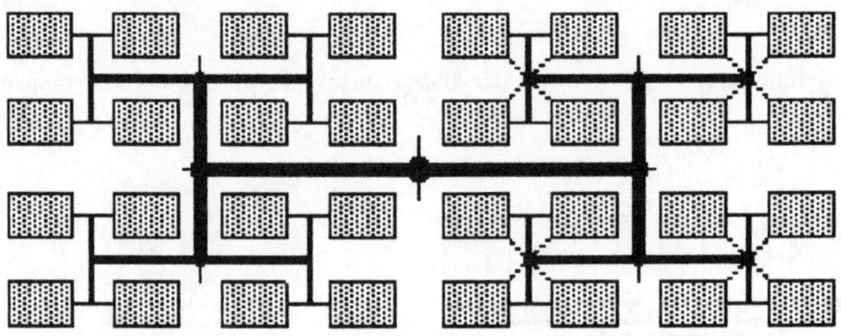

Figure 6.33 A fully balanced 32-cavity mold layout with H runners (alternate X style indicated by dotted lines).

Generally, such ribs are not recommended. The proper design method is to select runner shapes that will cool easily while remaining large enough for the necessary flow, and to provide adequate cooling in the area where the sprue meets the runner.

It should be noted that today, "cold" sprues as shown are rarely used. However, "hot" sprues, which eliminate the large mass of plastic at the junction of sprue and runner, are not without problems because the heating of the hot sprue also makes it difficult to cool this area as well as would be desirable.

32 Cavities We will only consider one, balanced layout, as shown in Fig. 6.33. The last branches (H style) could be X runners, as indicated with dotted lines. For elongated products, the choice would be between a balanced layout and a linear layout, as shown earlier for molds with fewer cavities. For the layout in Fig. 6.33, it would be better to use four-drop hot runners, each feeding eight cavities, or to use an eight-drop hot runner, each feeding four cavities, and eliminate the heavy runners.

The designer should always try to arrange the cavities in a reasonably proportioned rectangle, that is, the longer side of the total projected area should not be more than twice the length of the shorter side. This helps optimize the length of the runners, the drilling of cooling channels, and the ease of handling of the mold in manufacturing. Often, this can be achieved by simply turning the product 90°.

Other Multicavity Layouts Figure 6.34 shows a schematic layout of *any* number of cavities able to be divisible by four, or better yet, by eight; it combines an H runner with two long linear

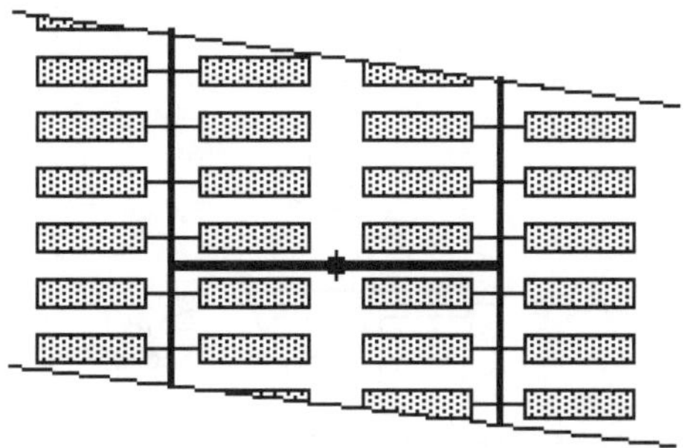

Figure 6.34 "Generic" mold layout for any number cavities divisible by 4, preferably by 8, that can be used for a large number of cavities.

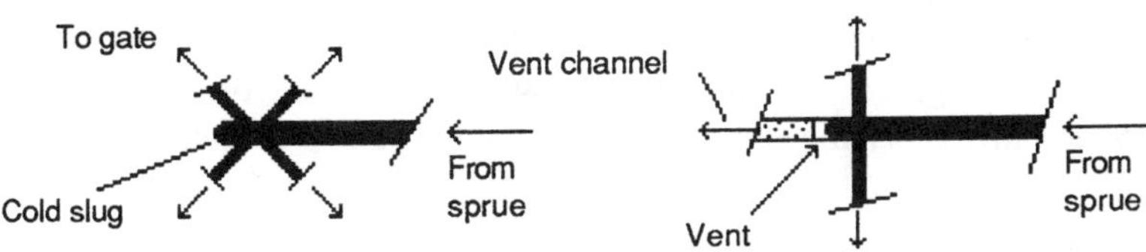

Figure 6.35 A cold slug pocket is an extension of the runner at a change in direction in both X style (left) and H style (right) runner systems.

runners. Such layouts are often used for a large number of cavities and permit very compact layouts. Balancing is often ignored.

As a reminder, note that cavity layouts not only depend on the runner system but also on other mold features such as the method of ejection, the support between the cavity stacks, and the mold alignment, to name just a few points to consider.

6.3.2.9.5 Cold Slug Pockets This feature is quite common in cold runner design, both for two-plate and three-plate runner systems. Cold slug pockets are runner extensions at every change of direction, as shown in Fig. 6.35.

The reason for cold slugs is based on the assumption that the front of the plastic advancing in the runner cools as it travels through the cold mold. When this front arrives at a branching-off point, the cold material (cold slug) is pushed into the pocket, and the hotter material following behind continues flowing toward the gate. This theory may be quite valid for molding machines with slow injection speed where such cold slugs may actually form. With high injection speeds, these pockets are not necessary.

If the runners are large, it is recommended that the cold slug pockets be provided with vents. The vent gap should be about 50% of the cavity vent gap size, because at this point, the plastic pressure is quite high. Venting applies only to runners in cold runner molds where the runners are ejected at every cycle. In a hot runner channel, any dead pocket where the plastic would gradually degrade must be avoided at all cost.

With or without pockets, it is good practice to vent the cold runner channels at every point where they severely change direction to permit the trapped air ahead of the inrushing plastic

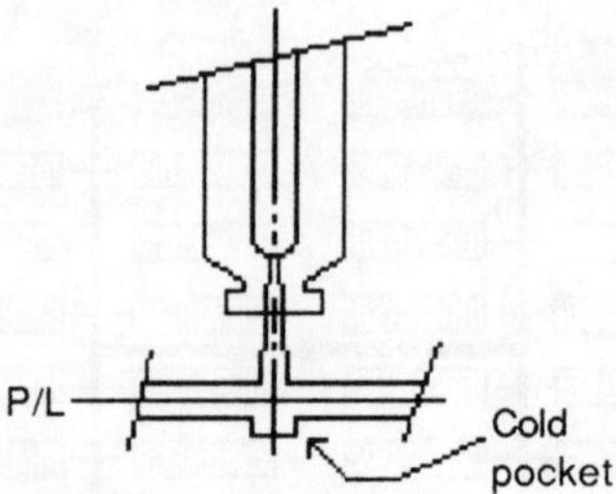

Figure 6.36 A cold pocket opposite the gate helps prevent the runner path from becoming blocked.

to escape. Permitting some of the air to escape from the runners will assist the venting of the cavity itself.

In a hot runner mold, using a gate with thermal shut-off feeding an auxiliary (cold) runner, a cold pocket opposite the gate (Fig. 6.36) may be beneficial. At the start of injection, the cold plug which keeps the gate sealed after injection will be pushed out of the gate and lodge into this pocket, and thereby is prevented from blocking the path of the plastic into the runners.

6.3.3 Three-Plate Molds

6.3.3.1 General Multicavity Layouts

The following section covers the general layout of cavities in a three-plate mold and points out the similarities and the differences from the two-plate layouts.

The main features of three-plate molds are listed below:

1. The cavities can be positioned closer to each other, because there is no need to provide space between them for sprue and runners. This does not mean that the layout of cooling lines, support for the cores, or the location of ejectors need not be considered. But, in general, three-plate layouts require less area than equivalent two-plate layouts.
2. The mold is inherently self-degating. This is an important difference from two-plate molds, which can be made either self-degating or not.
3. The gate vestige is usually very good.
4. Total runner length can be smaller, equal to, or greater than for an equivalent two-plate mold, because the length of drops must be considered as part of the runner length.

General rules for cavity layout patterns are presented below.

6.3.3.1.1 Rectangular Cavity Layouts Rectangular layouts are preferred over irregular or circular patterns for ease of design and manufacture. Occasionally, irregular patterns are advantageous. This could be the case in molds for small products with a large number of cavities, which can be "squeezed" into a layout resembling a honeycomb pattern, as illustrated in Fig. 6.37, where an equal number of cavities requires less area than a rectangular pattern. The space-saving effect of the honeycomb pattern becomes significant when used with very large numbers of cavities which must be placed in a restricted area. However, it is rarely used.

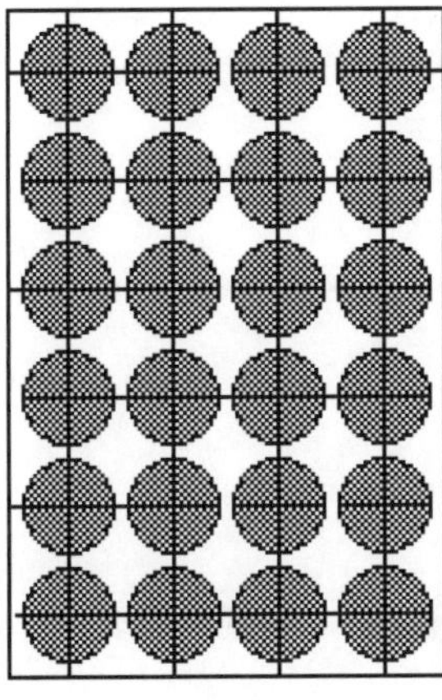

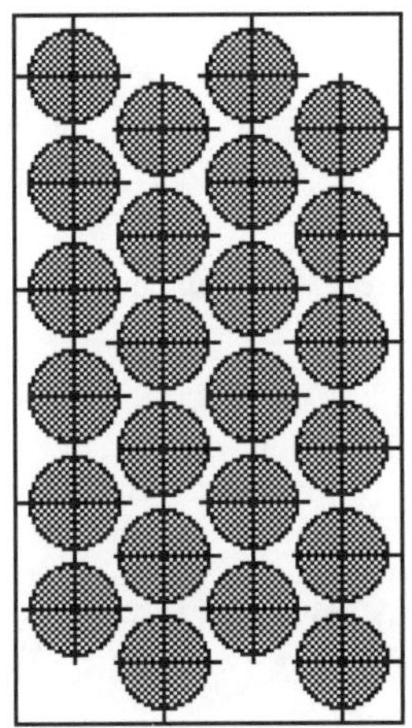

Figure 6.37 A rectangular cavity layout pattern (left) is compared with a honeycomb pattern (right).

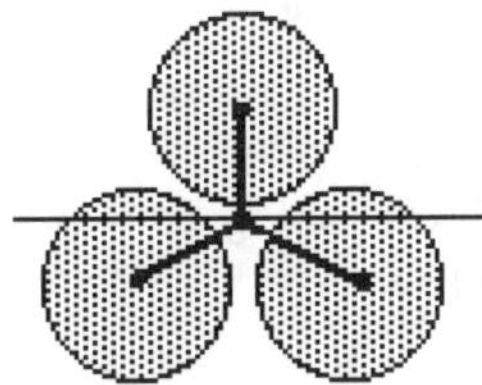

Figure 6.38 Three-cavity layout with Y style runners for a three-plate mold.

6.3.3.1.2 Runner Layout, Balancing In three-plate molds, as in two-plate molds, balancing is desired but not an absolute necessity. Many molds are "nearly" balanced, that is, the balancing has been pursued to a certain extent, and there may be a slight difference in flow and pressure at each gate, but not enough to affect the product quality or the molding speed.

6.3.3.1.3 Three Cavities Circular layout is similar to a two-plate layout, but the cavities can be placed closer (Fig. 6.38). The runners are fully balanced, the drops being equidistant from the sprue. The illustration shows a typical layout, with the runners in Y formation at 120° angles.

6.3.3.1.4 Four Cavities Cavities can be set closer together than is possible for two-plate molds. Figure 6.39 shows a typical layout, with the runners in X formation. It is also possible to use the H formation, but it is not required since the runners are in a different level from the parting line and can run over the cavities.

6.3.3.1.5 Five Cavities Such a mold is only practical in a circular layout. However, five drops could lead into one cavity, such as a large container; the runners would normally be arranged in a radial pattern.

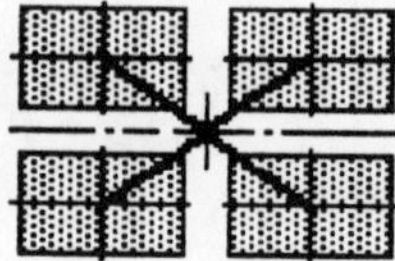

Figure 3.39 A four-cavity mold layout with X style runners for a three-plate mold.

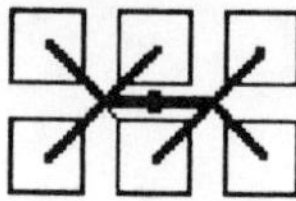

Figure 6.40 This common six-cavity layout for three-plate molds is called a "double Y" layout.

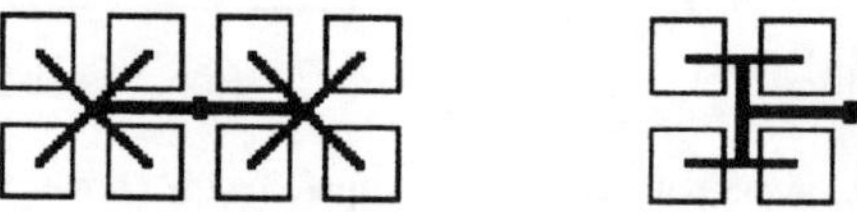

Figure 6.41 Two commonly used patterns for eight-cavity molds use a double-X (left) or a double-H (right) runner system.

6.3.3.1.6 Six Cavities The layout in Fig. 6.40 is a very common mold configuration. The runners consist of two Y's and a straight connection.

6.3.3.1.7 Eight Cavities The layouts shown in Fig. 6.41 are widely used patterns. The runners consist of two X's or H's and a straight connection.

Orientation of cavities within the mold can be an important factor in the mold layout. If the products eject at random (free fall) or their orientation is not important, the cavity stacks should be arranged to result in the shortest possible and most efficient runner. If, however, the location of the vestige is important for appearance, or if the product orientation is important and all cavities must be located in the same attitude (e.g., for transfer to an assembly fixture or to printing, etc.), then the routing of the runner may become affected.

Fig. 6.42A shows an example of an eight-cavity mold for an elongated product which will be randomly ejected. The double-H runner is short and fully balanced. Layout B shows the same product, but the orientation in all cavities must be the same. The runner is much longer and only partially balanced, although the difference in the main branch may not be serious enough to affect the product or the molding. If perfect balancing is required, an arrangement shown in B1 could be used.

Layout C shows a much longer product with similar requirements. The runner is fully balanced. It is not serious that the runner in C is completely on one side of the mold. Chapter 15, Stack Molds, explains that the forces on the plate in one level—in this case, the runner—are counteracted by the forces of the cavities on the same plate in the second level. If the projected area of the runner were larger than the area of the underlying cavities, a larger clamp force would be required, and the clamp in the mold would be unevenly loaded. This condition, however, is very unlikely to occur. The examples shown for eight cavities for

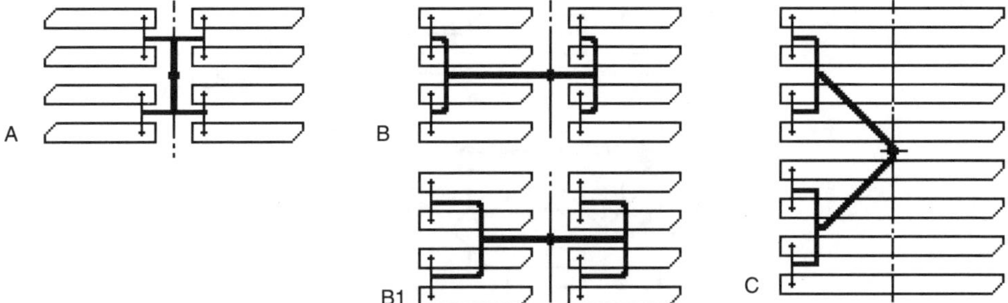

Figure 6.42 Various layouts for an eight-cavity mold for an elongated product are shown: A. layout suitable for random ejection; B. layout for the same product as in A but orientation of cavities is the same, while B1 shows an alternative to achieve balancing; and C. a balanced layout for a much longer product.

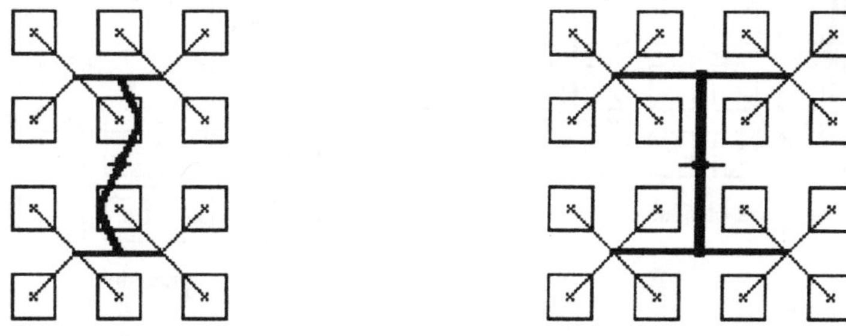

Figure 6.43 Twelve-cavity layout.

Figure 6.44 Sixteen-cavity layout.

elongated products are typical for other number and layouts of cavities, and will not be further discussed.

6.3.3.1.8 Twelve Cavities Figure 6.43 shows a commonly used, fully balanced twelve-cavity layout. The gates are shown in the center of the product, but they could be anywhere within the confines of the cavity, or they could be in a tab. For further discussion of gate location, see Chapter 10, Gates and Runners.

6.3.3.1.9 Sixteen Cavities Figure 6.44 shows a common, fully balanced 16-cavity layout layout. The same comments as those for twelve-cavity molds apply.

6.3.3.1.10 Twenty-Four Cavities A fully balanced layout as shown in Fig. 6.45 is very common. Bridges, as explained earlier, ensure that the runner hangs together, stays in its plane, and ejects easily, thus reducing the risk of getting caught on some mold components such as the leader pins. The size of the bridge is the same as for two-plate runners. Even though they connect the drops, they do not seriously affect the flow pattern of the plastic. In general, bridges are useful with any number of cavities where branch runners stick out and create "hooks"; bridges should be planned from the beginning and not put in as an afterthought.

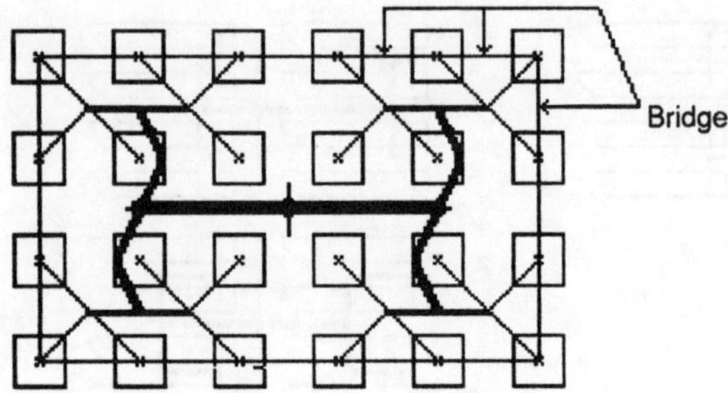

Figure 6.45 Typical 24-cavity layout, with bridges connecting the runners to keep them in the same plane.

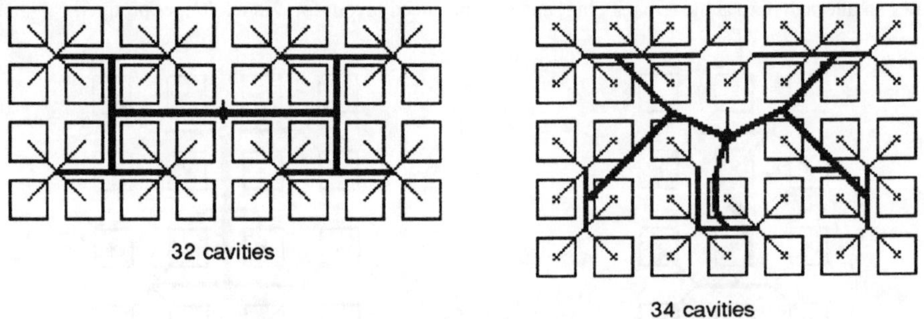

Figure 6.46 A typical 32-cavity mold layout (left), and an alternative 34-cavity mold (right).

6.3.3.1.11 More Than Twenty-Four Cavities A 32-cavity layout is shown in Fig. 6.46 (left). Such a layout tends to be disproportionately long, compared to its width. A 34-cavity mold is not normally considered, but Fig. 6.46 (right) shows how such a layout is possible. The runners are not fully balanced, but the difference is insignificant. It is an alternative if the long shape of the 32-cavity layout is not practical for space reasons. The length and the weight of runners in either case is about the same. There cannot be a cavity opposite the sprue because there is no practical way to provide a drop puller in this location. In some layouts, where the sprue is located between cavities but the drops are too close to the sprue, it may not be possible to provide drop pullers; such cavities must be then omitted also.

6.3.4 Hot Runner Layout

Cavity arrangement and layout for hot runners is quite similar to layouts for three-plate molds. Provided the cavity spacing is adequate for hot runner nozzle spacing, three-plate molds can often be fairly easily converted into hot runner molds by replacing the three-plate mechanism and the three-plate drops with a hot runner system and hot runner drops.

General rules, based on past experience, are listed below:

1. Hot runner molds should be fully balanced where mass of the product is significant and, therefore, where each cavity requires a large volume (and flow) of plastic at every cycle.

2. Unbalanced hot runner molds: Multicavity molds, particularly for lightweight products such as cutlery, can use a layout similar to two-plate molds where one or two "linear runners" feed all gates with little or no balancing.

The reasoning for rule 2 is that the product mass (or the plastic volume) required to fill each cavity is relatively small in comparison with the mass (volume) of the hot runner channel behind the gate, so that the plastic flow in the runner and the respective pressure drop to the gate for the various cavities is not much different whether the gate is near or far from the sprue. However, with "lightweighting" of products—in other words, with ever thinner wall thicknesses—it becomes very important that the pressure and the plastic flow at all gates is the same. This makes balancing of runners necessary.

7 Molding Operation Sequences

It is good practice in mold design for the mold assembly drawing to show the various elements of the molding operation and the relative time when each motion or other action (air blast, etc.) starts and ends during continuous running of the mold (i.e., in the fully automatic mode of operation). This is necessary for several reasons:

1. The designer can check to ensure that all functions required to make and release the product are covered by the design and that nothing has been left out.
2. It gives the molding technician a clear description of the planned operation of the mold and the timing of the various events within the mold and in conjunction with the peripheral equipment, such as robots, etc.
3. It provides a legal protection for the mold maker. If the mold is improperly set up, and as a consequence a breakage of mold or equipment, or worse, a personal accident occurs, the fact that the proper installation and set-up procedures and timings were shown on the assembly drawing will help in establishing the responsibility of such an occurrence.

The sequences are best shown with graphs. In Fig. 7.1, the X-axis of the graph represents time. It need not necessarily be drawn to scale, but it is important to show clearly when events start and stop in relation to each other.

Every molding operation consists of a number of events (or motions) which are initiated either from signals (push buttons, encoders, or limits switches) or from timers. The graph shows clearly when such event (or motion) starts, when it finishes, and the origin of the signal.

The valve gates described in Fig. 7.2 stay open as long as injection is forward. If earlier closing would be required, a timer would have to be shown, similar to cavity and core air. The start of core air is shown to coincide with the mechanical ejection start. However, since there is no connection between this event and the initiation of the core air, the method of initiation

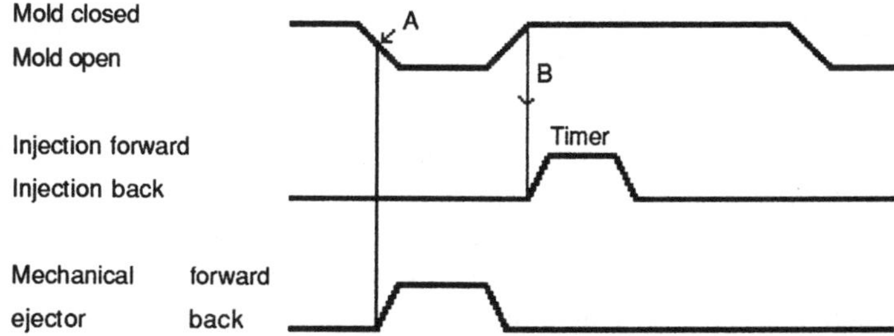

Figure 7.1 Operational sequence of a simple mold, mechanical ejection tied with clamp motion graphed over time: A. the point during the opening travel when the ejectors start, B. a signal that the clamp is closed starts the injection cycle. The timer (or T) may be shown where it is not obvious. In this example, it would not normally be shown.

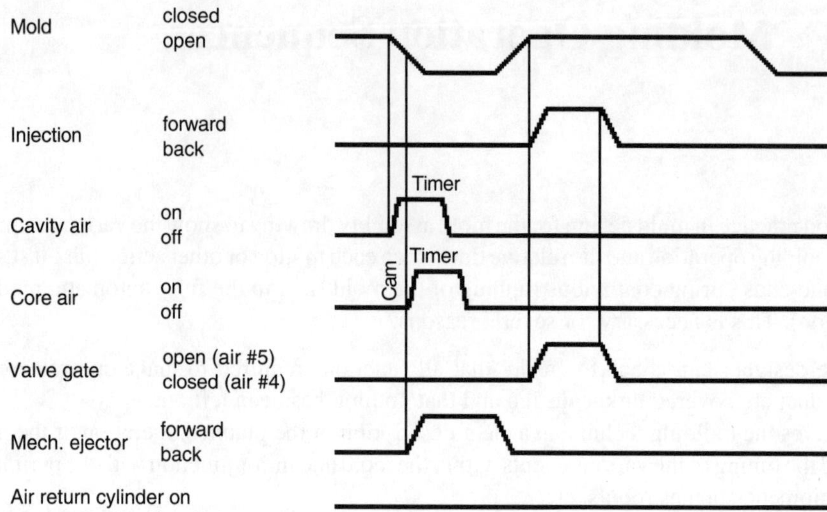

Figure 7.2 The operational sequence is shown for a container mold that is valve gated and includes cavity and core blow-off, mechanical ejection, and a central air return cylinder.

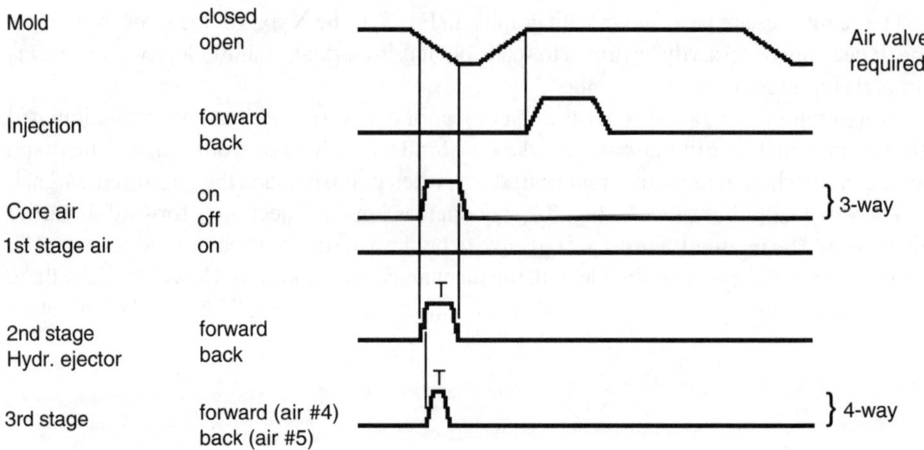

Figure 7.3 Graph of an operation sequence for a mold with three-stage ejection.

must be shown; in Fig. 7.2, "cam" has been added. This indicates that the start of air can be adjusted independently of the instant when the machine ejector starts.

Note that the three-stage sequence in Fig. 7.3 takes place during mold opening stroke to achieve optimum cycle time. The hydraulic ejector is back in place by the time the mold is fully open. Mold open time is kept to a minimum.

Notes on the right side of the graph indicate the type of control valves needed. It is sometimes useful to indicate not only the type of valve but also its capacity, to ensure that the

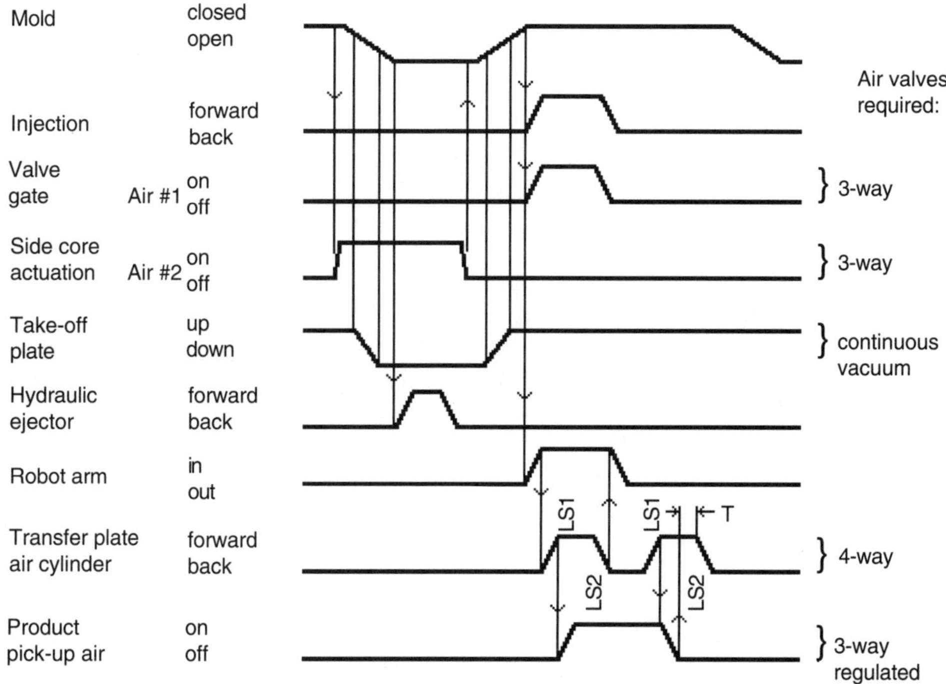

Figure 7.4 This example shows a mold with combined mechanical take-off and hydraulically operated robot.

actuator not only receives air (or hydraulic fluid) when wanted but also in sufficient volume; this may be important if very fast actuation is required.

The vertical relation lines must be shown, either with dotted or with solid lines. Arrow heads may be added to better show the sequence, as illustrated in Fig. 7.4.

The hot runner in Fig. 7.4 has a single-acting actuator (three-way valve). The take-off plate uses a vacuum system that runs continuously. The transfer (robot) plate advances, and grippers pick up the products; the plate retracts, rotates via the robot arm, advances again, and grippers release the product. The plate is ready for the next cycle. Limit switches LS1 and LS2 on the transfer plate are required to signal the completion of its stroke.

In the machine set-up described in Fig. 7.4, the motions of the take-off plate are mechanically tied with the clamp motion and need no controls. Air #2 must be ON before the clamp opens and must be OFF before the clamp closes. All transfer plate actions must be finished before the take-off plate is UP again. In this example, there is ample time.

8 Plastic Shrinkage

Plastic shrinkage will be covered from two angles: *theory*, or what affects shrinkage, and *practice*, as shown by actual examples. The designer must first understand what shrinkage is and what affects it before following blindly some formulae which may or may not apply in a given case.

8.1 Theory

Shrinkage is the amount by which a molded product is smaller than the (steel) size of the cavity space wherein it was produced by injecting plastic under high pressure injection and at high temperatures. For practical usage, shrinkage is generally given in mm/mm (in./in.) or in % [1].

Rule 1: There is a definite relationship between pressure (*P*), volume (*V*), and temperature (*T*). This relationship is different for various plastics. Any and all conditions that affect temperature, pressure, and timing will affect the shrinkage.

Rule 2: When a volume of plastic is heated, it expands. When it cools to the original temperature, it will contract to the original volume.

Rule 3: When a plastic is compressed, its volume is reduced. When the pressure is reduced to the original pressure, it will return to its original volume.

The following sketches will illustrate—in a greatly exaggerated manner—the above three rules.

1. Imagine a solid container. A volume *V* is indicated with an arrow *V*.
2. A certain amount of solid, cold plastic fills the volume *V*.
3. The plastic is heated. It will expand to the volume *VH*. The greater the temperature difference between the starting, cold temperature and the temperature at the end of the heating, the larger is the difference between *VH* and *V* (heat expansion).

Coefficient of Heat Expansion: $\frac{(VH - V)}{V} \times 100$, in % of V

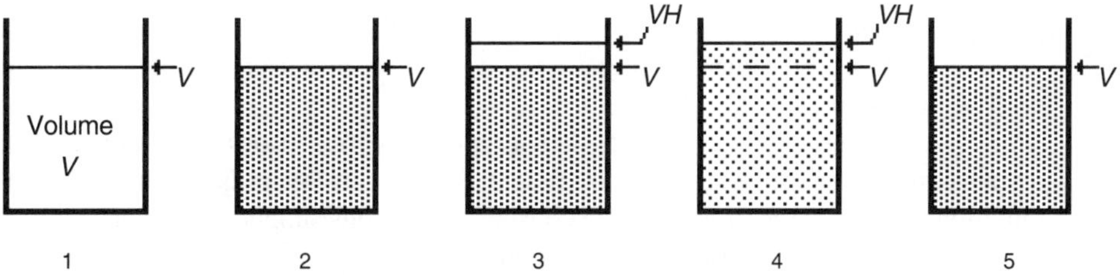

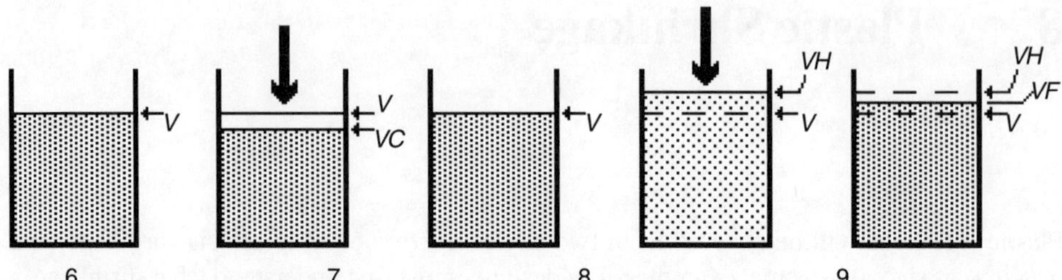

4. A volume of hot plastic is poured into this container (or cavity) to the mark *VH*, and is left there to cool.
5. The hot plastic will now contract (shrink). The amount it will shrink depends on the temperature difference between the hot plastic when poured and the cold plastic, usually at room temperature. If the cooled plastic is removed (ejected) from the container (cavity) before it has fully cooled, it will continue to shrink outside the container until it has reached room temperature.
6. The same amount of plastic is in the container as above in 1.
7. All plastic is compressible. When subjected to a compressive force, it will contract to a smaller volume *VC*. The greater the force, the greater is the contraction.
8. When the force is released, the plastic will return to its original volume *V*.
9. When the plastic is heated, it will expand to the volume *VH*. If it is now also subjected to a compressive force, it will diminish to another, not further specified volume.
10. As the plastic cools under pressure, which may be the full "injection pressure" or the lower "holding pressure", or a combination of both, it will shrink less and reach a larger final volume *VF*. (One can also say that more hot plastic is compressed into the cavity as long as the pressure is maintained on the plastic (e.g., through an open gate.) The applicable "shrinkage factor" is therefore smaller.

    If the pressure is low (or nil, as when *starve feeding*), the plastic is little or not at all compressed. It will shrink more and end in a smaller final volume VF. The applicable "shrinkage factor" is therefore larger. (Starve feeding is when the plastic volume injected into a cavity is exactly the right amount to fill the cavity, therefore not creating any pressure within the cavity.)

Summary:

1. *Temperatures*: The greater the temperature difference between that of the injected plastic and room temperature, the greater the shrinkage.
2. *Pressure*: Where the pressure on the plastic (in the cavity) is high, less shrinkage will take place. Where the pressure is low, the plastic will shrink more.
3. *Timing*: The longer the injection pressures are kept on the plastic in the cavities, the less will be the shrinkage.
4. *Plastic characteristics*: Each plastic has a typical coefficient of temperature (heat) expansion. In some plastics, the coefficient is small, in others quite large, and in addition to being large, it may not be constant for various suppliers or batches, even when supplied under the same specifications.

From the preceding discussion, it should be clear that it is in most cases *impossible to predict with certainty* the correct shrinkage of a material, since it depends on so many factors. This will be shown in more detail in the following section.

8.2 Practice and Applications

Shrinkage depends on the following variables (some of the points discussed here repeat statements made above, for added emphasis):

1. *Material (plastic)*: Different materials have different heat expansion values. But even materials with the same chemical and physical specifications may have significant differences in heat expansion and, therefore, in shrinkage.
2. *Product geometry*: This applies mainly to variations in wall thickness and the shape of surfaces, ribs, etc.
3. *Mold design*: The designer must take shrinkage into account, particularly in the cooling layout within the mold, the geometry of runners and gates (pressure drops), and the uniformity of heating in hot runners, etc.
4. *Molding conditions*: This includes the machine setup, mold cooling temperatures, humidity of plastic, cycle time elements, injection and hold pressures, etc., and the plant environment.
5. *Type of molding machine*: Injection speed, available injection pressures, accuracy of time, temperature and pressure controls, including closed loop controls, all affect shrinkage.
6. *Condition of molding machine and mold*: A neglected machine may have unreliable controls or a worn check valve, etc. A mold may have plugged or corroded cooling lines.

8.2.1 Injection Variables and Shrinkage

The hot plastic is injected into the cavity; since there is little resistance until the cavity is filled, filling happens with relatively low pressure. Cooling starts immediately, as soon as the plastic contacts the cooled walls. Since the specific volume decreases with the temperature, the solid plastic will occupy less space than the molten plastic.
After the plastic has filled the cavity, pressure builds up rapidly. This compresses the plastic in the cavity space. The pressure in the cavity is, therefore, a major factor affecting mold shrinkage. Normally, this pressure is maintained until the gate freezes, or is closed (valve gate), sealing off the material in the cavity. From this moment on, the pressure within the cavity will drop as the plastic cools and shrinks. Therefore, at constant mold and melt temperatures, variations in injection pressure and timing of the sealing of the gate are the most important factors affecting variations in shrinkage and differences in product sizes coming from the same cavity.

8.2.1.1 Heavy-Walled Products

As the plastic solidifies where it touches the cold walls, it shrinks; as long as the pressure is maintained on the plastic (while the gate is open), more plastic is packed into the cavity space. Higher temperatures result in more shrinkage when the packing pressure ceases too soon.

8.2.1.2 Ejection of the Product

The product can be ejected as soon as it is stiff enough that it will not be deformed or damaged during ejection; this is desirable to reduce the cycle time. But since the material has not yet reached room temperature, it will continue to shrink outside of the restraining shape of the cavity space.

If, however, the product is kept in the (cool) cavity beyond the time in which it could be ejected, it will cool more and may even reach the temperature of the coolant. It will then have the steel dimensions of the inside (core) and/or the steel distances of, for example, two pin locations. This would not only create unnecessarily long molding cycles but also may create excessive stresses in the product which may cause immediate or early failure of the product, depending on the type of plastic molded (i.e., a brittle plastic may crack during ejection, or may set up such severe stresses in corners that they will fail when stressed in service).

8.2.1.3 Time Limit of Shrinkage

In most materials, 90% of shrinkage is reached within a few hours after ejection, and the remaining 10% within the next 10 days. Others may take a few months before shrinking has completely stopped.

8.2.1.4 Annealing the Product

Annealing the product above room temperature will speed up the process of "relaxation" to end the shrinkage. Hygroscopic materials will also change dimensions, according to the amount of water absorbed.

8.2.1.5 Product Thickness

Thickness will affect shrinkage, particularly when molding plastics with low heat conductivity (PP, PE, etc). The frozen skin insulates the interior from losing its heat to the surrounding core and cavity walls, and the plastic within the two layers of frozen skin will stay hot long after ejection. Products can often be ejected as soon as the skin is stiff enough.

Such products will be cold to the touch as they come out of the mold, but will be much hotter after a while when the interior heats the outer skin. An advantage of this residual heat is that it relieves unwanted stresses in the product.

As a rule, a thick-walled product will shrink more than one with a smaller wall thickness. Most shrinkage data sheets acknowledge this fact by specifying a range of shrinkage or by tying the values to specific thicknesses.

Different shrinkage values caused by uneven wall thickness can cause trouble in relatively flat products (particularly those with rectangular shapes) that are surrounded by a thick-walled rim or have other heavy sections near the periphery of the product (Fig. 8.1). While the flat surface is relatively cold when ejected, the rim or heavy section will shrink after ejection and generate stresses within the product which will deform or warp the flat surface.

If this is not foreseen at the time the output of a mold is estimated, it can result in a very disagreeable surprise when the mold has to run at a much slower cycle to ensure that the product will not warp after ejection. This condition is worse with flexible than with stiff plastics.

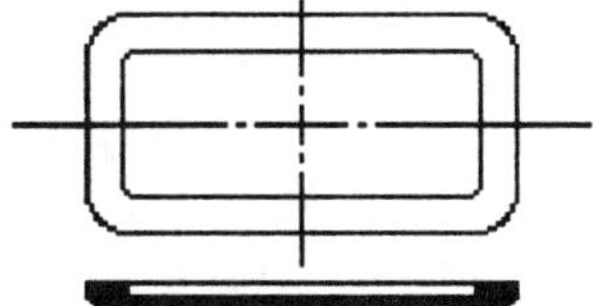

Figure 8.1 A flat product with a thick-walled rim may warp after ejection because of its uneven wall thicknesses.

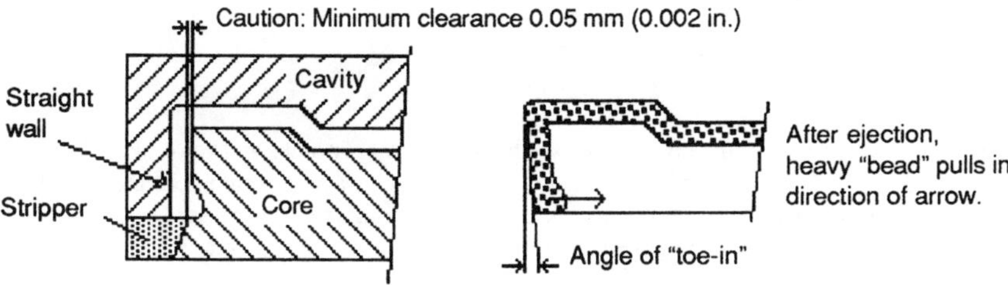

Figure 8.2 In this coffee can lid mold, shrinkage is desired so that the lid rim is pulled in the direction of the arrow to provide the proper toe-in angle.

On the other hand, shrinkage of a heavy section may be desired, as in the case of coffee can lids, which are required to have a rim that turns inward to provide the necessary holding force when pushed over the rim of the can. By increasing the thickness of the bead, the shrinkage of the rim and the resulting toe-in angle can be increased (Fig. 8.2).

As a general rule, the best condition for any product is uniform wall thickness. This must originate with the customer, but the mold designer also has a responsibility to point out to the customer where an (often minor) change in the product design could prevent uneven shrinkage, improve the cooling time, and reduce the product weight, thereby improving the productivity of the mold and reducing the cost of the product.

This rule applies also to any internal heavy section created by an intersection of ribs and by hubs, bosses, or other thick sections in the product that are required for accepting fasteners, etc. If possible, such heavy sections should be cored out to give access to cooling, even if that will increase the cost of the mold (Fig. 8.3).

Heavy sections, in addition to creating undesired stresses within the product, will shrink much more than other sections of the product and create *sinks* (on the surface) or *voids* (on the inside). Sinks may not be acceptable for appearance reasons, and voids could create weak spots in the product.

8.2.2 Basic Shrinkage Formula

The diagram and variables defined below are used in calculating S, a shrinkage factor. The equation is

$$D_c = D_p + (D_p \times S) + (D_p \times S^2), \text{ or } D_c = D_p\,(1 + S + S^2). \qquad (8.1)$$

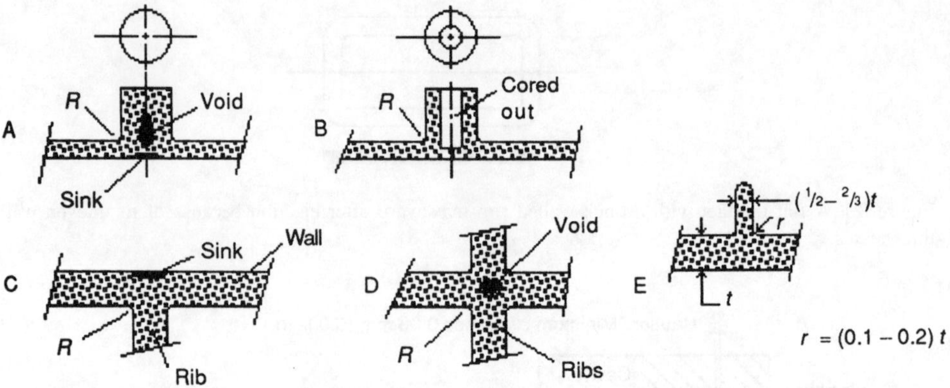

Figure 8.3 Several cross sections show uneven shrinkage, resulting in sinks and voids in the product: A. both sink and void at a hub, B. cored out hub prevents sink and void, C. sink at a rib intersection, D. void at a rib intersection, and E. rib proportions prevent sinks. Note that a radius *R* is necessary to reduce stress in the corners, but too large a radius will increase the bulk of plastic in this area and increase the risk of sinks or voids.

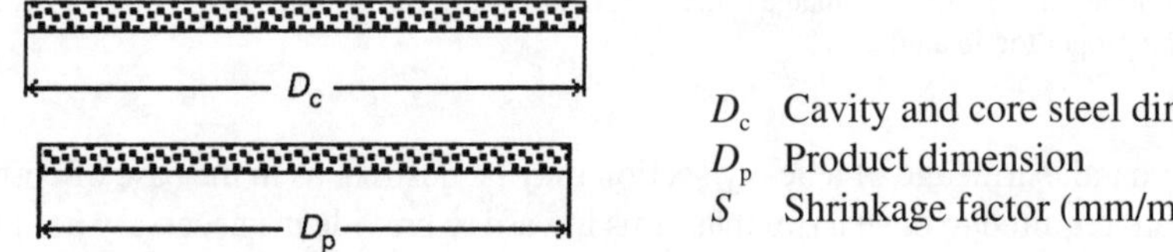

For practical purposes, because S is usually very small, S^2 can be ignored, and the formula simplified to

$$D_c = D_p\,(1 + S) \tag{8.2}$$

Occasionally, we must differentiate between *axial shrinkage*, which occurs in the direction of the flow of the plastic, and *radial shrinkage*, which is in the direction perpendicular to the flow (Fig. 8.4). This may be important when product tolerances are very tight and the plastic shows considerable differences in shrinkage in these two directions. This difference is caused by the orientation of the long, chain-like molecules of the plastic during injection.

The following are typical examples of shrinkage. The examples include suggestions as to what the designer can do about shrinkage problems.

Example 1: An LDPE strap has holes which must be accurately spaced in the product.

When designing the mold, the shrinkage was assumed to be a certain value and was used to determine D_c. When molding with the estimated cycle, the dimen-

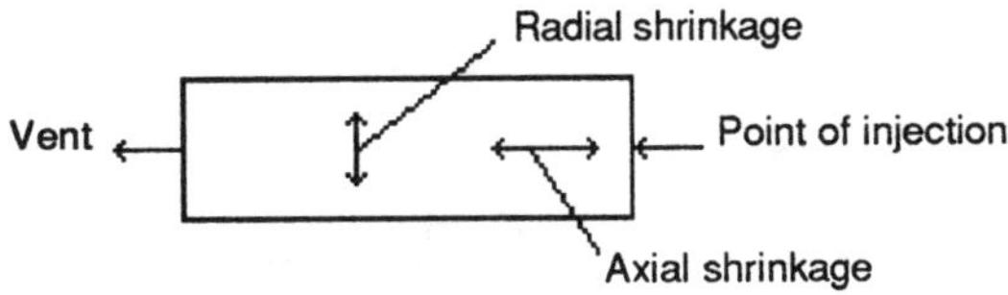

Figure 8.4 Axial and radial shrinkage are illustrated in relation to the direction of flow of the plastic.

Figure 8.5 To avoid the costly rework required to increase the pin diameter D (left), the designer can prepare for unexpected shrinkage by increasing D_p (right), so only the insert need be changed to correct pin diameter.

sion D_p was correct. However, the product was fairly cold at ejection and could have been ejected much earlier to increase production. But by doing so, as the product was hotter, it shrank more and D_p became too small. The question was then whether to run the mold slower (higher product cost), or to rebuild the mold (additional mold cost). Obviously, if production is high, the mold should be rebuilt, even if the cost is high. With low production requirements, the mold could be run "as is".

The importance of this example is to show that shrinkage and cycle time have a definite relationship. This is more serious with material with a large shrinkage factor, such as PP and PE. It is then up to the mold designer to be aware that the estimated or given shrinkage values may not be correct for product accuracy and mold productivity. This can be achieved by making allowances in critical areas so that the mold can be easily reworked.

It is always easy to remove steel if the cavity is too small. It is practically impossible to add steel if the cavity is too large. Welding or chrome plating is sometimes done but, as a rule, not suggested. If a correction is not possible to achieve by solely removing steel, a new mold part or insert should be made.

In Fig. 8.5, when D_p equals D, it is difficult to change the pin diameter D if the shrinkage selected was greater than expected, and the diameter D must be increased. The designer can foresee such a possibility and make $D_p > D$ so that only the insert (pin) needs replacement, and not the mold part where it is inserted.

Similarly, if a distance between holes in critical, stepped pins could be used (Fig. 8.6) to allow "off-set" positioning in case the distance was not correct in the first place. If the offset pin is round, it must be prevented from turning.

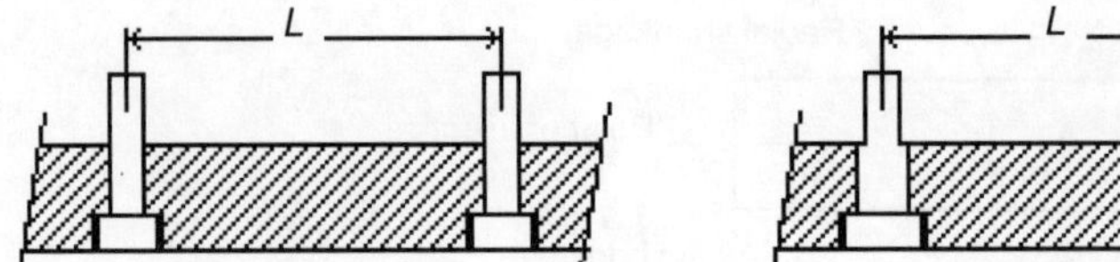

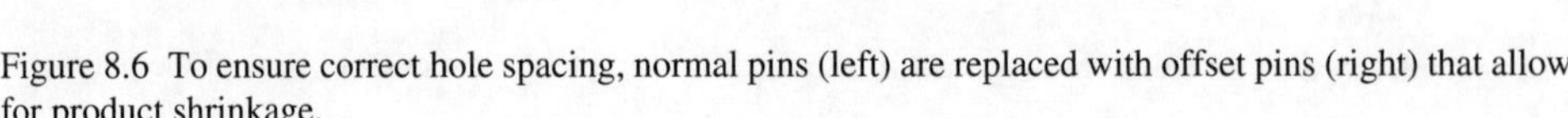

Figure 8.6 To ensure correct hole spacing, normal pins (left) are replaced with offset pins (right) that allow for product shrinkage.

8.2.3 Responsibility for Shrinkage: Selection of Shrinkage Factors

In general, the mold maker cannot be held responsible for selection of shrinkage factors used. If the purchase contract requires that the mold maker assumes this responsibility, it must be clearly specified, and as a rule, the quoted mold cost will include provision for costs necessary for changes due to shrinkage after the mold is sampled.

8.2.3.1 Steel Sizes

Occasionally, the customer will supply some or all steel dimensions, usually after having had similar molds in operation before. While this information is of great advantage, the shrinkage may still be different when a mold runs with higher injection speeds and pressures or is ejected earlier. Also, there may be a difference in shrinkage due to a different cooling layout and/or cooling conditions. Therefore, even when steel sizes are supplied by the customer, the responsibility for the product size must be clearly defined in the purchase order.

8.2.3.2 Molding Material

It is important that the mold maker uses the same plastic for testing (same specifications, same supplier) that eventually will be used in production. In some cases, the differences in shrinkage of plastic from different suppliers is significant. Occasionally, there are even significant differences from batch to batch, under identical specifications, from the same supplier.

This must be clearly understood before designing the mold, especially where close product tolerances are required. In some cases, the customer may want to run the same mold with two different materials. The effect on the product sizes due to possibly different shrinkage must be clearly established and accepted by the customer before proceeding with the mold design.

8.2.3.3 Other Rules Affecting Shrinkage

Rules that apply in selecting shrinkage factors:

1. Shrinkage is often difficult to foresee and requires close cooperation with the customer.
2. Because of shrinkage, products with flat surfaces are always more difficult to mold.
3. Wherever possible (and if tolerances permit it), if two products must match, such as a container and lid, it is best to first finish the more complex mold to produce samples, and

only then to settle on the final dimensions of the other (simpler) mold. If the second mold is then still dimensionally incorrect, it is easier and less costly to correct the simpler mold.

8.2.4 Effect of Shrinkage on Mold Opening and Stripping Force

When a product such as a cylindrical container is injected, the cavity will expand and the core will compress as a result of the forces created by the high pressure plastic, called *elastic deformation* of the steel. Therefore, as soon as the injection pressure has built up, the cavity space will be larger than at the beginning of injection.

The high-pressure injection phase and, if required (for a short time period), a (lower) hold pressure, is maintained while the plastic is cooling down. When pressure on the plastic ceases, the elastically deformed cavity and core will return to their original size, thereby reducing the cavity space. Provided the shrinkage of the plastic product is greater than the sum of the elastic deformations of cavity and core, there will be no problem.

If, however, the material shrinks less than the amount of elastic deformation of the steel, the product will be clamped (squeezed) between the surfaces of the cavity and the core. This can occur if the injection pressure is high, where the cavity walls are too weak, and where the product wall thickness is small and the actual amount of shrinkage is, therefore, very small.

This condition can be serious with side walls in thin-walled containers with little or no draft; the machine may not even have enough force to open the mold. The opening force required is the product of the coefficient of friction (between plastic and steel) and the (squeezing) force from the elastic cavity exerted on the plastic which is at right angles to the direction of the mold opening motion. Molds have had to be destroyed because it was impossible to open them after injection.

With high molding pressures, particularly in thin wall molding, cavities (solid or inserts) must be designed to expand as little as possible (heavy walls), and the core must also be stiff enough not to yield excessively.

The elastic deformation is directly proportional to the force, and inverse proportional to the modulus of elasticity E (e.g., a BeCu part (E = 15,000,000 psi) will compress or stretch about twice as much as a core made from steel (E = 30,000,000 psi) under the same injection pressure).

Shrinkage will also affect the force required to strip the product from the core. The still warm product tends to hug the core, similar to a stretched rubber band. A force which is proportional to the shrinkage and the modulus of elasticity of the plastic, multiplied by the coefficient of friction (plastic on steel), gives the force required to strip the product.

If the product is warm, the force is greater. If the product has completely cooled on the core, there is virtually no residual shrinkage, and little force is required for stripping.

However, if there is a bead or undercut (e.g., coffee can lid), the colder the plastic, the more force will be required to push the product off the core. If the cold plastic is not resilient enough, it will break.

8.2.5 Practical Suggestions for Designing with Shrinkage

So far, we have discussed in general terms what shrinkage is, how it affects molding, and what the designer must know on the subject. The following guidelines show a designer how to calculate shrinkage dimensions and how to show shrinkage on the drawings. (This may be different at various mold makers).

A CAD system can have a program which translates automatically all product dimensions with the shrinkage that has been selected. This eliminates the need to manually calculate the steel dimensions.

The product drawing(s) should show the product sizes. A second drawing, which is used to build the mold, should show the steel dimensions. Occasionally, the designer shows the steel dimensions in red on the product drawing. However, this is not recommended unless there are only very few dimensions.

8.2.5.1 *Product Tolerances and Shrinkage*

Example: A product size is 135.50 mm ± 0.30 mm. Adding shrinkage of 1.5% would result in

$$135.50 \text{ mm} \times (1 + 0.015) = 137.5325 \text{ mm and}$$
$$0.30 \text{ mm} \times (1 + 0.015) = 0.3045 \text{ mm,}$$

or a steel dimension of 137.5325 mm ± 0.3045 mm. The last two decimals of the dimension can be easily ignored, to result in 137.53 mm ± 0.30 mm.

If there is doubt concerning the accuracy of the assumed shrinkage value and the resulting product size, the steel sizes can be modified by specifying them closer to the maximum product tolerance size in case of a core dimension, and by specifying closer to the minimum product tolerance size in case of a cavity dimension. This will allow the removal of steel should rework of the mold be required because the assumed shrinkage dimensions were wrong.

8.2.5.2 *Shrinkage of Threads*

Some lathes and many grinding machines have no provision to change the tool advance, except for standard pitch sizes, and cannot produce other sizes. The designer must establish how important it is for the product to have shrinkage values added.

In most cases, a screw pitch is small (Fig. 8.7), and, for one or two threads, the effect of shrinkage will be well within the product tolerances. If the threaded section has a larger number (n) of threads, shrinkage of the pitch must not be ignored.

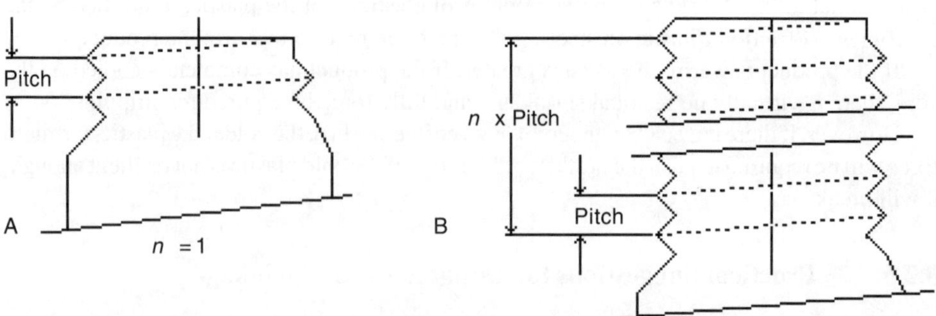

Figure 8.7 Cores for screw threads: A. core for only one thread, and B. core for *n* threads.

Table 8.1 Approximate Shrinkage Values for Various Molding Materials

Material	Shrinkage
ABS	0.005–0.007
Acetal, axially	0.021–0.026
Acetal, radially	0.018–0.020
Acrylic	0.004–0.007
EVA	0.007–0.020
Nylon 6	0.006–0.014
Nylon 66, axially	0.012–0.033
Nylon 66, radially	0.020–0.028
Polycarbonate	0.006–0.008
PE	0.015–0.050
PE, 30% Glass reinforced	0.004–0.0045
PET (bottle grade)	0.005–0.012
PP	0.012–0.022
PP, 30% Glass reinforced	0.004–0.0045
PS	0.002–0.006
PS, 30% Glass reinforced	0.0005–0.0010
PVC	0.003–0.008
PVC, 30% Glass reinforced	0.001–0.002

The mold maker may have to subcontract thread grinding; therefore, the availability of thread grinding other than standard pitches by the supplier may have to be investigated. It may affect the selection of a supplier (and the cost) for this operation.

8.2.6 Approximate Mold Shrinkage for Some Materials

Table 8.1 is provided solely to indicate approximate shrinkage values to show a novice designer the wide variations within one material, under various conditions such as product thickness, orientation, and filler. *Do not use these values for design*—always use data provided by the supplier.

8.2.7 Other Factors Affecting Shrinkage

It is important to understand that with each product design, one or more of the following conditions may be present; some are even contradictory: The shrinkage may be greater because of one condition but smaller because of another. It is impossible to predict accurately what the shrinkage will be in any specific mold unless the new mold is an exact replica of one from which the shrinkage data were taken.

8.2.7.1 Wall Thickness of Product

With thick-walled products ($L/t < 100$), there is little pressure drop between the point(s) of entry (gates) and the portions of the cavity distant from the gates. Approximately equal pressure throughout the plastic before the gate freezes will produce about equal shrinkage.

With thin-walled products ($L/t > 100$), the plastic can suffer a severe pressure drop as it advances from the gate(s) through the narrow passages within the cavity. The pressure is highest near the gate and least in remote areas. The shrinkage is inversely proportional to the pressure in the plastic after gate freeze-off; therefore, less shrinkage occurs near the gate and more shrinkage near the rim.

8.2.7.2 *Product Shape*

If the product is free to shrink (e.g., a plain disk), with center gating the plastic will shrink evenly, assuming cooling is equal on both surfaces. If the product is restricted from shrinking (e.g., by a rim surrounding it or by core pins at extreme locations), stresses will be set up in the portions between the points of resistance. In particular, sharp corners at such locations will create heavy stress raisers. This may be acceptable in the use of the product, even though it may create distortion after ejection. In the worst case, the product will crack or break as the mold opens, even before ejection.

In all cases, it is important to provide ample draft or tapers (also if possible, radii) to ensure that the product can "slip out" of its confinement while it contracts rather than break or create substandard (stressed) products. Note that such changes may require the approval by the customer.

Example 1: A plastic disk (diameter D_c) will shrink to diameter D_p. Provided cooling is equal on both sides, the disk will remain flat.

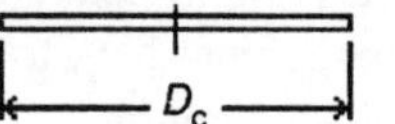

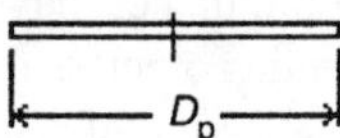

If, however, the disk is restrained from shrinking by walls, it will pull these walls and deform them, setting up heavy stresses in the product. This pulling can be greatly reduced by giving the disk a curved shape, as shown below. Since the

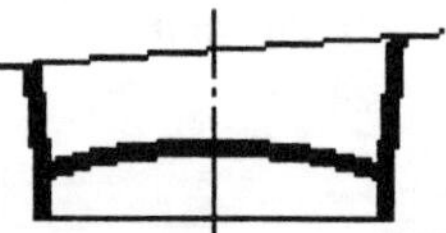

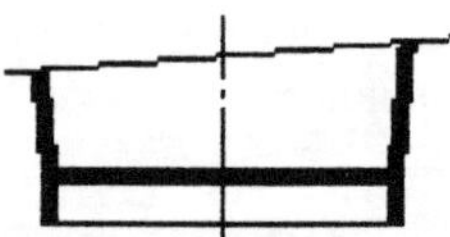

curved surface is larger than the flat one, the plastic will—while shrinking—reduce this surface without pulling on the side walls. Provided the estimated or calculated curvature was correct, the product will have a flat surface. If the product specifies a shallow curvature, a deeper curvature is required in the mold.

In the case of the preceding example, such curvature *in the mold* does not require approval of the customer, but it would be of advantage to have a slight curvature with sufficient tolerance included in the product drawing so that even in case the estimated shrinkage was different from the actual one, the product will be still acceptable. This is very important, especially with close tolerances for the volume (contents) of the container.

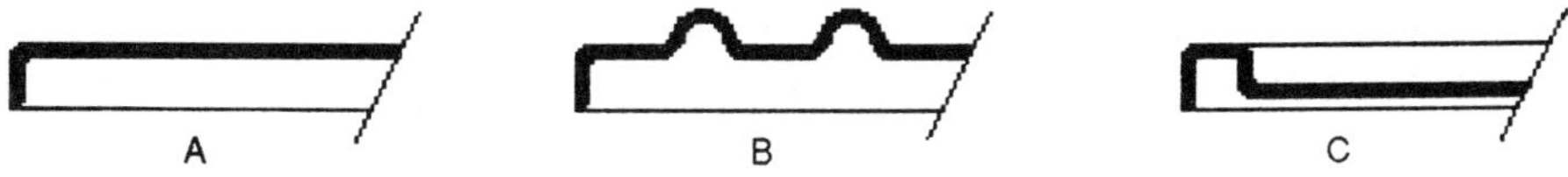

Figure 8.8 Examples of top surfaces: A. a flat top susceptible to much shrinkage, B. waves added to the surface add area to reduce shrinkage effect, and C. step added to the surface to add area.

The *principle of added area* can also be used on large areas (mainly with plastics with large shrinkage) by adding one or more waves or steps to the surface, as shown in Fig. 8.8. Otherwise, it may be impossible to arrive at a flat top surface. Needless to say, this is a severe change in product design which must be approved by the customer.

8.2.7.3 Gates

The areas closer to the gate are better packed than more remote areas, which may already have cooled down enough to prevent additional plastic to make up for volume lost through shrinkage. The result is that the areas near the gate shrink less than the areas farther away.

Small gates freeze off sooner. Once frozen, there is no plastic added during the holding (packing) phase, and the product will therefore shrink more.

Larger gates stay open longer. They either freeze slowly, naturally, leaving an unsightly vestige, or, with hot runner systems, if this is not permissible they must be valve gated, in which case the closing of the gate is controlled, and the vestige is clean. In either of these cases, plastic continues to feed under pressure through the open gate, adding more plastic as it shrinks in the cavity. Timing of the closure of the valve gate and the level of holding pressure, therefore, have much influence on the amount and location of shrinkage.

The number of gates also affects shrinkage. The effect of more than one gate is that the plastic will not need to flow as far as with only one gate, and the cavity will fill faster, with less pressure. (See above, the effect of packing on shrinkage.) Whether to provide more than one gate per cavity is an important decision. The mold cost is higher, but this may be more than offset by the potential for improvement in productivity of the mold.

The location of gates also affects shrinkage. *Center gating* is usually best for even shrinkage but is not always possible because of appearance requirements or intended use of the product. *Off-center gating* will create uneven flow lengths within the cavity, and if this is combined with large pressure drop (e.g., in thin-walled products), it can create uneven shrinkage along the flow paths. In the worst case, even though the cavity has a regular (e.g., rectangular) shape, the mold may produce an odd (e.g., trapezoidal) shape.

Off-center gating combined with multiple gating may be used to combat the above described distortion but may not be enough, and a deliberate distortion in the outline of the cavity may be required to compensate for uneven shrinking. (The distorted outline will, if properly selected, result in a shape as specified on the product drawing.) It should be noted that such compensation is usually very costly, depends often on trial and error, and should only be a last resort when product tolerances are very tight.

8.2.7.4 Equal Cooling

Flat products, such as disks (records, trays, etc), are extremely sensitive to uneven shrinkage. If the plastic is subjected to different temperatures in cavity and core, and if the rate of heat removal is unequal and the product is ejected while still warm, the side closer to the poorer cooling will be warmer than the other side and will shrink more after ejection, thereby warping the disk.

Cooling of heavy sections such as the rim (or any thick area) is more important than the cooling of a flat (or thin) area to ensure that the heavy section is cold enough before ejection. It will also ensure that the walls do not freeze too soon and starve the heavy section.

Deeper products are generally not very sensitive to variation in cooling between cavity and core. As the product shrinks, it pulls away from the cavity wall and depends then mostly on the cooling from the core.

8.2.7.5 Orientation of Plastic Within the Cavity

In general, plastic shrinks more along the flow path than across it. In the case of a container, the shrinkage in the direction from the gate toward the rim will be greater than the shrinkage at right angles to it (i.e., circumferential).

8.2.7.6 Operating Conditions

Molding conditions have a great influence on shrinkage: cycle time elements, injection pressure, holding pressure and time, cooling time, temperature and flow, etc., all affect shrinkage. A mold can produce perfect products under one set of conditions and poor ones under different molding conditions. This points out the importance of recording accurately the molding conditions when testing the mold, and making the customer aware of these conditions.

Fortunately, many products have large enough tolerances that many of the above conditions do not affect most molds. But with close tolerances, everything must be considered.

8.2.8 SPI Standard Molding Tolerances

Because of the many conditions which affect shrinkage, the Society of Plastics Industry (SPI) has published tables of recommended tolerances for plastic products (sizes, materials). (See also Chapter 9, Mold Surface Tolerances). The (published) tolerances are quite large so that all of the conditions affecting shrinkage fit easily within a large tolerance "window". This is usually satisfactory with general purpose products, such as toys, kitchen ware, electrical appliances, wiring devices, and many other commodity plastics products.

Close tolerances found frequently in engineering products, such as in electronics, computers, etc, but occasionally even in closures, etc, require the understanding of the causes contributing to and affecting shrinkage. Any anticipated problems due to shrinkage and the effect on tolerances must be clarified before mold design is started.

8.2.9 Effects of Shrinkage on Container Shape

The various factors discussed in this chapter may have combined effects on shrinkage. Several examples of shrinkage effects on container shapes are presented below.

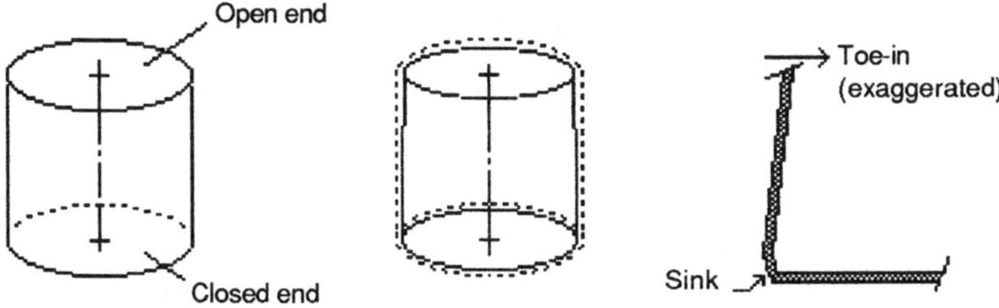

Figure 8.9 Shrinkage of a cylindrical container is higher at the open end, where tow-in occurs.

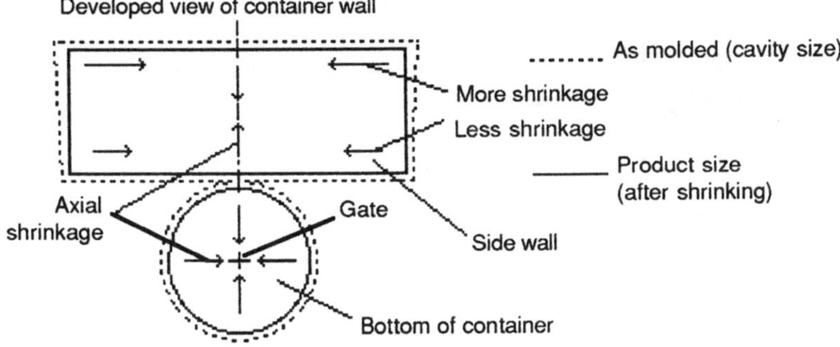

Figure 8.10 Cross section of the container shows shrinkage increasing in the less packed areas (toward the open end of the container), and axial and radial shrinkage are indicated by arrows.

8.2.9.1 Cylindrical Containers

Because the bottom of the container is better packed than at the rim, there is more shrinkage near the open end, and the container will "toe-in" as shown (exaggerated) in Fig. 8.9. This difference in shrinkage where the mold is packed more and packed less (Fig. 8.10) because of the pressure drop as the plastic travels along the wall must be taken into account when determining the steel sizes.

8.2.9.2 Conical Containers

The same shrinkage factors apply as for the cylindrical container. The wall near the top will pull in more, showing a distinct toe-in (Fig. 8.11).

Beads, lips, and stacking shoulders (Fig. 8 12) can introduce even more toe-in and other distortions in these products with possible serious effect on final size, volume, fits, stacking clearances, and other critical dimensions.

As seen earlier, molding conditions such as melt and cooling temperatures, pressures, timing, and resulting packing conditions can seriously affect the final shape and the dimensions of the product. It is often impossible to exactly predict the shrinkage of a product. The best solution is usually to find earlier, similar designs and rely on such experience for the selected shrinkage factors.

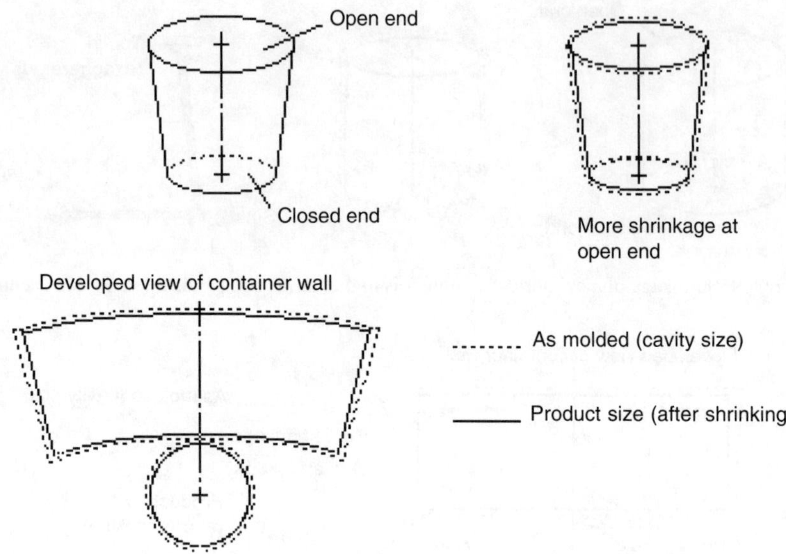

Figure 8.11 A conical container shrinks more at its open end.

Figure 8.12 Container mold features that affect shrinkage and toe-in: A. bead and B. lip.

References

1. Rubin, I.I. (1972). *Injection Molding: Theory and Practice*, John Wiley & Sons, New York Toronto, p. 270.

9 Mold Surface Tolerances

This chapter deals with the philosophy of tolerancing rather than the technique. It applies only to tolerances which affect the product, not the fit of mold parts within the mold.

9.1 Basic Tolerance Considerations

The mold designer must ask for tight tolerances only where they are really required. At the beginning, this may be more difficult and possibly take more time, but in the long run it will save much time and money. The designer will also earn the respect of the machinists by tolerancing reasonably.

A typical, unfortunately quite frequent, scenario is described in the following:

1. A number of "identical" parts have been given unnecessarily tight tolerances. (This alone is more costly because of more frequent checking and interruptions than would be necessary with looser tolerances.)
2. Despite the extra care, one or several of these parts are out of tolerance. Inspection and/or supervision gets involved for instruction on what to do. (Lost time, need for inspection reports).
3. The designer is approached about what could be done to solve the problem. (More lost time).
4. The designer may now, after a closer look (lost time), decide that a looser tolerance for this part will make some or all of the faulty parts usable. The drawing will be changed to include looser tolerances (more lost time), and the parts are returned to production.

Obviously, the machinist will wonder why the designer did not allow looser tolerances in the first place, if they were possible. All this lost time and paper work could have been avoided.

9.1.1 Misconception #1

It is a common misconception that all molded products coming from several cavities in one mold (or from several molds) must be virtually identical, provided this (much closer limited) size is within the specified tolerance.

This is not so!! The true meaning of tolerances shown on a product drawing is that the product, regardless of whether it comes from several cavities in one mold or from several molds, can have any size within the limits given on the product drawing. If there is any doubt, it must be clarified with the customer.

Example: If a product drawing shows a length $L = 100.0 \pm 0.5$ mm, it means that products coming from any cavity can have a dimension anywhere between 99.5 and

100.5 mm. It does not mean that the length must be "virtually the same" (e.g., $L \pm 0.025$ mm), as long as the established length is within the specified $L = 100 \pm 0.5$. If the customer requires such a high degree of accuracy, it should have been shown on the product drawing (e.g., $L = 100.000 \pm 0.025$ mm).

In many cases, such accuracy is impossible to achieve, and if it were, the mold cost would be considerably higher than if it were quoted with ± 0.5 mm.

9.1.2 Misconception #2

Because mold making (manufacturing) can be held to very close tolerances, it is often thought that by assigning close mold tolerances, the product is more likely to be within the desired limits. This is only partly true.

As explained in more detail in Chapter 8, Shrinkage, the influence of shrinkage on the product size is far larger than the influence of the steel dimension of the mold.

What affects the accuracy of a molded dimension?

- Assumed shrinkage factors: This alone makes it difficult to arrive at a targeted dimension. It is subjective, usually based on past experience, on experimental data, or on data given by the materials supplier or the customer.
- Actual shrinkage, which is directly related to molding conditions such as timing, temperatures, pressures, cooling, and molding material. Note that even the "same" material will vary from batch to batch. Any of these variables can alter the desired dimension.
- Mold making variations from cavity to cavity (mold tolerances).

9.1.3 Misconception #3

While the mold maker may have the equipment to make steel dimensions extremely accurate, it does not mean that working to close tolerances is as easy (and economical) as working to more open tolerances. It is not only necessary to gradually approach the tightly specified size, which means more cuts and more time, but it also means more inspection time and more rejects. This can be very expensive.

It may also mean the use of more accurate equipment, such as jig grinders or EDM, where boring or milling could be used. Thus, less expensive and more available equipment could be utilized with looser tolerances.

To utilize to the best advantage (both for machining and for the product), any tolerances given on the product drawing require careful consideration by the mold designer, and they may require some additional design time, including challenging some of the specified tolerances. This will require close cooperation with the customer.

This additional effort may take only a few hours, while the savings in machining time may amount to tens or even hundreds of hours. In other words, it is very easy and requires little effort by the designer to assign a very close tolerance (e.g., ± 0.02 mm) to every dimension, but it takes more time (and experience) to really study the drawing and assign tolerances so that the mold part can be easily made while still producing the product to the specified sizes.

This has been recognized by the Society of Plastics Engineers (SPE), in a suggested method of applying steel tolerances in relation to product tolerances. In general, they suggest

Table 9.1 Proposed Standard for Mold Making Tolerances per 1965 Data Sheet (inches)

Product Print dimension	Print tolerance (±)	Percent of tolerance allowed to mold maker	Suggested mold tolerance (±)
0 to 1	.002	20	.0004
	.005	25	.0013
	.010	25	.0025
	.020	30	.006
1 to 2	.005	20	.001
	.010	30	.003
	.020	30	.006
2 to 3	.005	20	.001
	.010	30	.003
	.020	30	.006
3 to 5	.005	20	.001
	.010	20	.002
	.020	25	.005
	.030	30	.009
5 to 8	.010	20	.002
	.020	20	.004
	.030	30	.009
8 to 12	.010	20	.002
	.020	20	.004
	.030	25	.0075
	.040	30	.012
12 to 16	.020	20	.004
	.030	30	.009
	.040	30	.012
16 to 20	.020	20	.004
	.030	25	.0075
	.040	30	.012

a certain percentage (20–30%) of the product tolerance, increasing with the size of the dimension, as shown in Table 9.1. However, the application of these rules may be affected by other factors.

9.2 Tolerances for Wall Thickness of Products

It is important for the designer to realize that wall thicknesses of the product are the result of two dimensions: the outside of the core and the inside of the cavity. For example, if a container has an overall height of 4 in. and a wall thickness of 0.05 in, and the drawing shows a general tolerance of ±0.020 in., the wall thickness must have another, tighter tolerance (e.g., ±0.002 in.). But even then, the depth of the cavity and the height of the core cannot be worked to the tolerance of ±0.020 in. as suggested in Table 9.1.

In the above example, the steel sizes for cavity depth and core height would be ±0.005 in. If both dimensions were to the extreme, the wall thickness could be 0.010 in. smaller or greater than specified. Similar considerations apply to diameters and side wall thickness.

Close tolerances for wall thicknesses and gate configurations are especially important in all multicavity molds, where the balance of filling speed affects quality and mass of product from cavity to cavity; they also affect molding cycle. This is most important, especially in the manufacture of thin-walled plastic products.

All other dimensions which will not affect the wall thickness should take advantage of the more open tolerances. This applies mainly to positional tolerances or dimensions specifying areas and depths (e.g., ribs, lettering, etc).

9.3 Responsibility for Tolerances

In general, it is up to the estimator to scrutinize the product drawing for tolerances before pricing a mold, and to bring apparently unnecessary tolerances to the attention of the customer before releasing the drawing to the mold designer. The estimator will also highlight those dimensions which are unnecessarily tight.

This practice, however, does not relieve the designer, who will be working much closer with the drawing, of the responsibility to question any (unnecessarily tight) tolerance which may, by loosening, save time in manufacture.

10 Gates and Runners

What are gates? We can define a gate as a passage through which the plastic material enters the cavity space.

The requirements for a gate are contradictory. Large gates are desirable to facilitate filling of the cavity space and to reduce stresses in the product. Large gates keep the slowly cooling and shrinking plastic in the cavity space connected for a longer period with the hot plastic supplied from the injection. This permits packing before the gate freezes. (See also Chapter 8, Shrinkage.)

Small gates freeze faster and produce higher molded-in stresses but are desirable to 1) facilitate separation of products from the runner, and 2) to make the gate mark, or *vestige*, more inconspicuous. Hot runner (HR) molds with proper heat control at the gate and proper gate configuration can use a relatively small gate, which will not freeze prematurely.

The designer must select the most suitable gate configuration for the job from existing standards or suggested gate sizes, or must calculate and design a new gate configuration.

10.1 Gate Location and Number per Cavity

As a rule, one gate per cavity should be sufficient for most products and is usually satisfactory to avoid undesirable weld lines in the product; however, there are exceptions to this rule. The following sections explain some basic gating locations using examples.

10.1.1 One Gate per Cavity

10.1.1.1 Outside Center Gating (OSCG)

Used with hot runner or three-plate molds, the OSCG should be located so that an approximately equal volume of plastic will flow about the same distance toward the outside (rim) of the product. To locate the gate *offset* from the centerline of the product (Fig. 10.1) may sometimes help in filling. This must be cleared with the customer to ensure that it is acceptable. Venting is in general no problem, since the plastic flows toward the parting line (P/L). If there are dead pockets, they must be carefully vented (Fig. 10.2).

OSCG may be used for a slender product (Fig. 10.3). We define a product as slender when its length is more than twice its diameter, as in a vial, etc.

The main problem is core shift or core deflection, as explained in the following:

1. The centerline (C/L) of the cavity may not coincide with the C/L of the core because of a design or manufacturing error, or as the result of mold and/or machine deflection during clamp-up of the mold. The result is that one side of the cavity space is filled faster than the other, thus exerting a side pressure on the core and bending it.

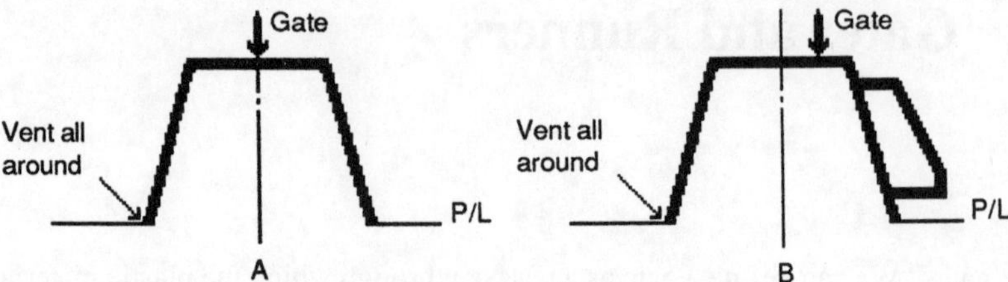

Figure 10.1 Two examples of OSCG: A. center gating, and B. offset gating to aid in filling the mold.

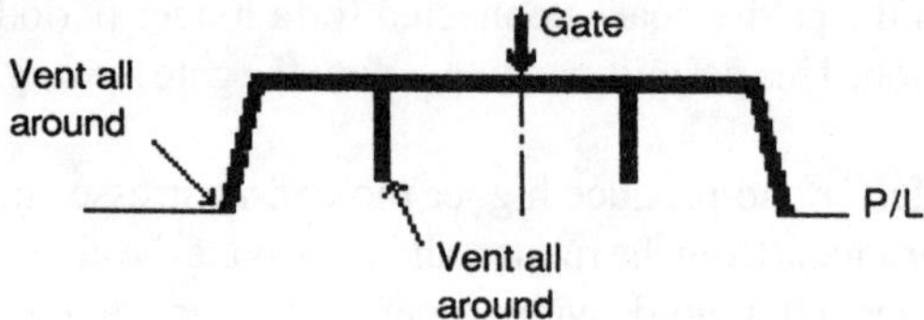

Figure 10.2 Dead pockets must be vented carefully.

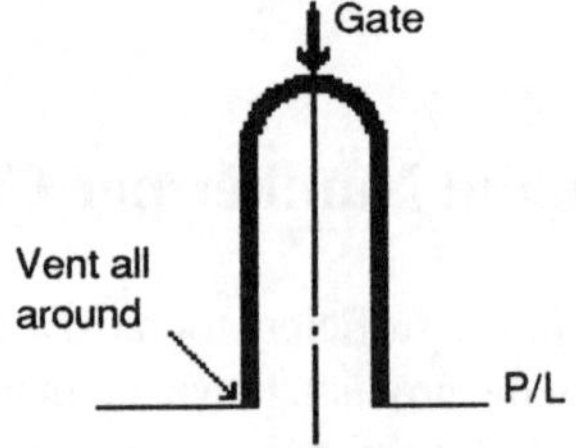

Figure 10.3 Outside center gating for a vial.

2. The C/Ls coincide, but the gate is offset just enough to produce a result similar to the above.

Note that the effect of core shift is considerably more serious with thin-walled products, where higher injection pressures are needed, and where a slight deflection may represent a substantial portion of the wall thickness. The effects of core shift are also worse with products having very little draft angle, because the product may become jammed between the cavity and the (bent) core. This prevents it from pulling out of the cavity, thus affecting mold opening, requiring more force, and causing damage (scratches) to the cavity walls. Even insignificant core shift existing before injection may become much larger during injection.

A visible and easily measurable effect of core shift is that the walls of the product are uneven, which makes the thicker side shrink more when still warm after ejection, thereby curving the part toward the heavier wall side. In the case of long, cylindrical products, the effect may be that the product looks like a banana. (This can be easily seen when rolling the product on a flat surface.) Core shift may also affect filling, as the thinner side may not fill completely.

There is a *cost limit* of how close the tolerances can be specified to ensure concentricity of all components. The proper selection of practical tolerances, well-designed (long and correctly preloaded) tapers for alignment, and selection of stiffer but very expensive core materials (e.g., tungsten-carbide cores with a modulus of elasticity about twice that of steel) must be considered. Normally, the gate selection (location, type) has the most influence on core shift. If all this does not help, the core may have to be anchored on the injection side, provided it is acceptable to the product design. Such anchoring may require a much more difficult gate design.

10.1.1.2 Inside Center Gating (ISCG)

Used in hot runner or three-plate molds, the same gate location considerations apply for ISCG (Fig. 10.4) as for OSCG. Venting problems are also the same. An added complication is ejection of product from the same side as the injection, the difficulty sometimes of holding product on the ejection side, and the difficulty of providing adequate cooling because of spatial interference with the runner.

ISCG should be used only in special applications when it is absolutely necessary to avoid a gate vestige on the visible side of the product (e.g., for dinner plates, bowls, higher priced screw caps, etc.).

10.1.1.3 Side Gating Near Top

A side gate near the top of a product (Fig. 10.5) may be hot runner edge gating (HREG) or tunnel gating (TG). A product design may not permit gate vestige in the top for cosmetic or performance reasons, such as optical requirements.

This side location is better than gating near the rim when filling speed is important, as with thin-walled products. Because the gate is near the closed end (bottom) of the product, the plastic will usually fill the bottom first and then flow toward the rim, behaving similar to OSCG.

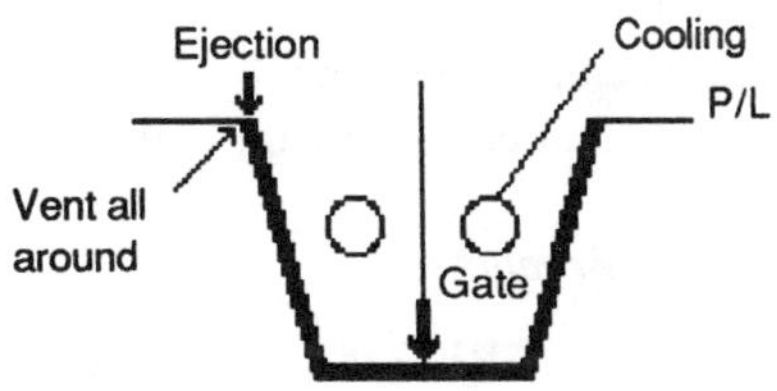

Figure 10.4 Inside center gating.

Figure 10.5 Side gating near the top of the product.

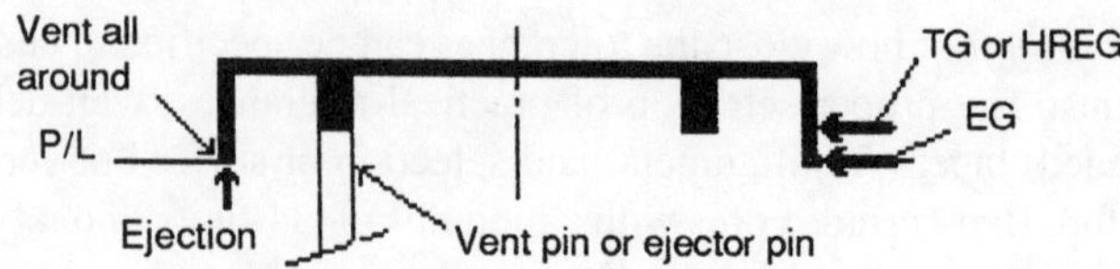

Figure 10.6 Outside side gating near the product rim; gating may be edge gating (EG), HREG, or TG.

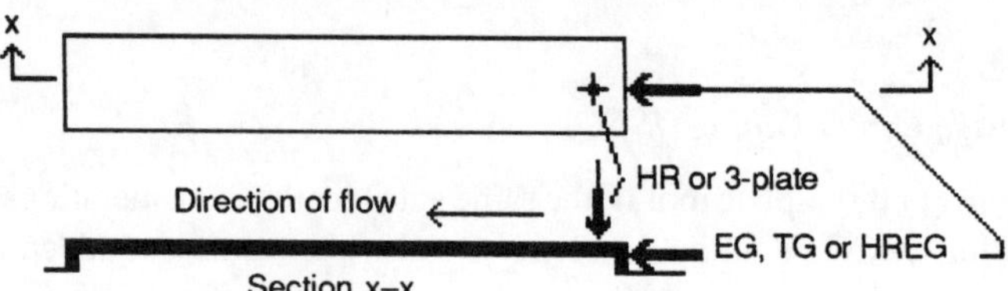

Figure 10.7 Heavy arrows show the location of various types of gating for an elongate product.

It is highly recommended that the side gate be located so that the plastic stream will not flow freely into the top surface. Instead, it should be directed to hit (at least partially) the core. The stream could also be directed against a core pin in the top surface, near the gate. This creates turbulence in the plastic flow and avoids streak marks in the top surface.

Venting is critical because the plastic may travel so that it encircles a portion of the cavity space and traps the air in the bottom. Ejector pins act as natural vents; otherwise, vent pins or vents in inserts must be used if plastic is trapped from all sides and air cannot escape at the rim. (See Chapter 11, Venting.)

10.1.1.4 Outside Side Gating Near Rim

Outside side gating near the rim (Fig. 10.6) is used in HREG, two-plate, and TG. It is used in basic, general purpose molding, usually for flat products, of any wall thickness. It is not recommended for containers, particularly thin-walled products, because of the difficulty of avoiding weld lines and air entrapment.

Venting is very important, especially if the rim is heavy. Ejector pins act as natural vents; otherwise, vent pins must be used if air cannot escape at the rim.

10.1.1.5 Gating of an Elongate Product

Elongated product molds may have HR, HREG, three-plate, two-plate edge, or TG (Fig. 10.7). The general rule is to locate the gate so that the plastic will flow the whole (or almost the whole) length of the product, to avoid formation of a weak area. Venting is usually no problem.

The area at the gate is always weaker (Fig. 10.8). The plastic stream is divided as it flows out of the gate and creates a weak area, which is similar to the weakness at a weld line.

10.1.1.6 Gating a Product with a Live Hinge

This type of gating is used in HR, three-plate, and two-plate molds. A "live hinge" (usually for PP products) is a very thin passage from one portion (e.g., box) to the matching other

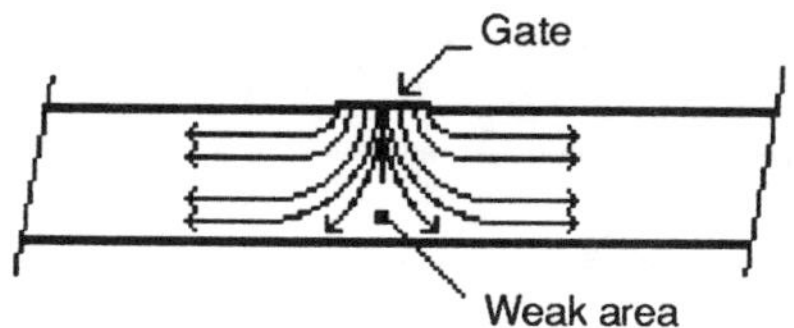

Figure 10.8 Directional arrows show division of plastic flow and the weakest area created at the gate.

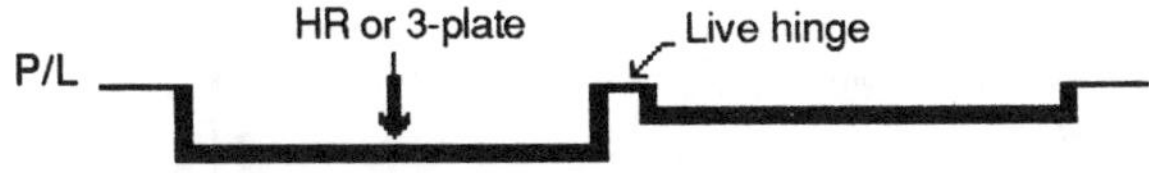

Figure 10.9 One gate is preferred for products with live hinges.

portion (e.g., cover) of the product, thereby forming a hinge (Fig. 10.9). Usually, the gate is located in the larger portion, and the smaller portion is filled through the hinge.

As a rule, gating in both parts should be avoided because it would create a weak weld line at the hinge so that, in service, it would not flex more than a few times before breaking. If for some reason two gates are necessary, the streams coming from the gates must not create a weld line at the hinge.

10.1.2 Two or More Gates per Cavity—Large Products

Two or more gates per cavity are sometimes required for very large products (e.g., automotive products, bottle crates, etc.) where the flow distances from a single gate would be too long and/or one gate could not supply all the plastic required to fill the cavity in time before freezing off. It is also used in molds for large containers where the flow through two or three gates speeds up the filling when compared with a single gate.

There are problems connected with multiple gating of a cavity:

1. *Freeze-up of gates*: With more than one gate per cavity, freeze-up is no problem for cold runner (two-plate or three-plate) gates because the plastic in the gates is ejected with the product and the gates are always open for the next cycle.

   In hot runner molds with open gates, the plastic pressure opens the gate, which has partially frozen at the end of the previous cycle. If the pressure at the gates is uneven, or if one gate is cooler than the other, the cavity will be filled from the gate that is easier to break through, and no plastic will enter from the other gate(s).

   For valved gates, this is not a problem since all gates are opened mechanically or with special heaters and are not affected by pressure or temperature differences.
2. *Minimum distance between gates*: Hot runner gates require a necessary minimum distance between gates because of the physical dimensions of the hot runner components.
3. *Weld lines and venting*: At the meeting front(s) of two or more streams of plastic, there will be a weld line, which can cause a weakness when compared to the strength in other cross sections of the product. It may also create an unsightly line or even voids on the surface of the product.

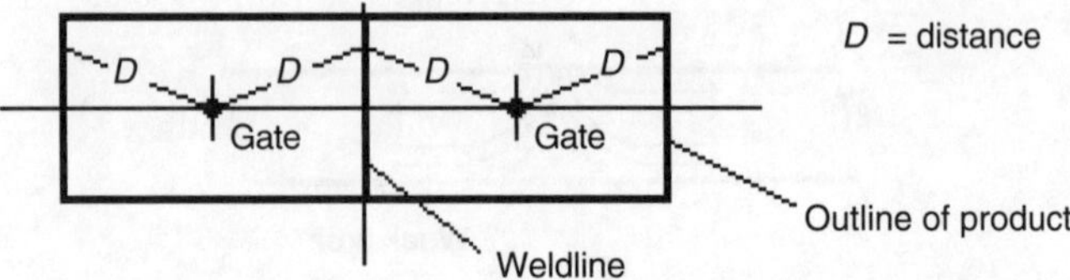

Figure 10.10 The distance *D* from the gate to the joint and to the rim should be equal.

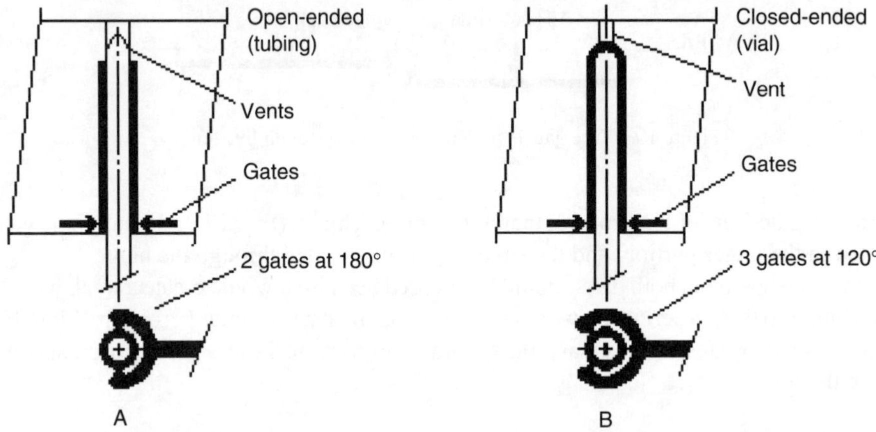

Figure 10.11 Multiple gating for slender products: A. two gates for tubing, and B. three gates for a vial.

Strength in the area of weld lines can be improved if the air trapped between the inrushing fronts of plastic is permitted to escape, thereby preventing weakening of the weld because of air entrapment and/or burning because of the rapid rise in temperature as the trapped air is compressed. It is important that the plastic arrives at the junction points (welds) hot enough to form an acceptable joint.

In large products, the flow distances should be about the same from gates to the rim and to the joints (Fig. 10.10). If the joints are weak or weld lines are pronounced, higher injection and/or mold temperatures and slower cycles are suggested.

10.1.3 Two or More Gates per Cavity—Slender Products

Conditions are similar to other slender (usually cylindrical) products, as explained earlier, except that the cavity is filled from the open end of the product from two or more gates. This system is essentially restricted to two-plate molds using edge or tunnel gates; however, such a two-plate system can be combined with a hot runner system (hot sprue) to bring the plastic to the P/L. The plastic enters on opposite sides of the cavity space and rises toward the end (Fig. 10.11). The forces on the core, starting at the end where the core is solidly held (and does not bend as easily), are balanced, and the rising plastic holds the core steady, even if it is very slender (pen barrels, etc.).

Two gates are placed at 180° and three gates at 120°. If there is concern about weld lines along the sides, a continuous gate may be selected which, however, is not self-degating. Another method is to gate into a solid ring surrounding the product which then must be removed by machining after molding.

This method also permits easy gating of a slender tubular product in a mold where the core enters the cavity (see Fig. 10.11A) and is held against deflection. To place a hot runner in the location where the core enters the cavity is usually very difficult.

The most important consideration in such molds is venting of that end which is farthest from the gates, to ensure that air will not be trapped. Even slender products (vials) with closed ends can be gated like this, provided the air can be reliably vented with stationary vent pins (see Fig. 10.11B) or, better, with moving, self-cleaning pins.

10.1.3.1 Inside Side Gating at Rim

This method is used with hot runner edge gates (HREG) for thin-walled containers, etc. (Fig. 10.12). The main problems are venting the closed end (bottom of the product), weld lines, the limited space for suitable cooling channels (which must be on the same side where the hot runner is located), and limited space for ejection from this side. The advantage of this system is faster cycling and thinner walls, because core shift is a less serious problem.

10.1.4 Gate Vestige

Vestige is the visual appearance (on the product) of the point of separation of the plastic between runner and product (break-off point). Its outline conforms to the shape of the gate. The gate shape is usually round (HR, HREG, 3-PL), occasionally elliptical (HREG, TG), half moon shaped, rectangular, or trapezoidal (EG, TG). The surface appearance of the break may be shiny when separated from hot plastic or dull when cold plastic is broken.

Protrusion is the amount (height) by which the vestige extends above the surface into which the plastic is injected. It depends on the land, the plastic injected, and on molding conditions.

The designer must keep the protrusion to a minimum by selecting (if possible, from existing standards) a gate design suitable for the plastic, for the type of product, and for the injection flow rate, and by making full use of any permissible gate size indicated by the product designer.

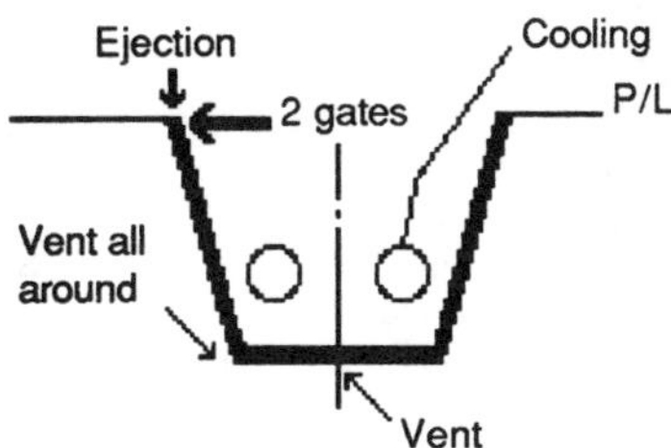

Figure 10.12 Inside side gating at the rim of a slender product.

There is no point in increasing the mold cost by introducing unnecessary features (e.g., valve gating) if the product design specifies a permissible size of protrusion. In general, properly designed gates will leave a minimal protrusion.

10.1.4.1 Hiding the Gates

In both hot and cold runner molds, the gate can be hidden in a textured, matte surface; in the enclosed portion of a letter, such as D, O, P, etc; in an engraved ornament, such as the inside of several concentric circles; or in the mouth or eye of a face, etc. This may require that the gate is located not in the exact center of the cavity but where there is a suitable hiding place. If too far off center, it may affect the flow of plastic and the filling of the cavity. If necessary, the gate can be located in a textured (rough) area, similar in roughness to the gate break, to be less noticeable.

The need for hiding the gate must be discussed with and approved by the customer before the designer decides on any gate location that may affect the appearance of the product. Experienced product designers often specify gate location on the product drawing, as well as maximum permissible protrusion.

If any gate mark on the outside of the product is objectionable:

1. For three-plate (rarely HR) molds, the designer must select inside gating, with the added complications (and costs).
2. For two-plate molds, the designer can select to eject from the injection side (complicated) or select "submarine" or "undergating".

Various HR gates will be discussed later, and the difference between open and valved gates will be shown. At this point, we will consider only the effect of the gate on the gate vestige.

10.1.4.2 Dimple

A dimple is a thickening of the product, usually in the shape of a spherical radius (Fig. 10.13B). It aids in providing an unrestricted flow of resin from the gate. If there is no dimple, the plastic must flow around a sharp corner, which creates a restriction (Fig. 10.13A) in the flow pattern and causes a considerable pressure drop, slowing down the filling speed. The dimple eliminates the effect of the restriction caused by the flow around a sharp corner.

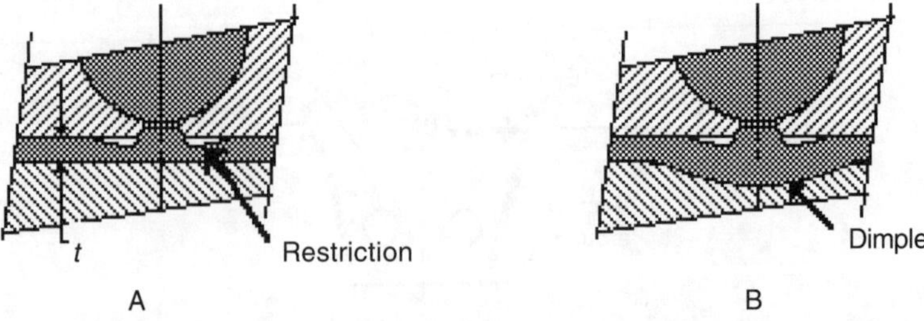

Figure 10.13 Flow of plastic from the gate: A. restricted by a sharp corner, and B. aided by a dimple in the product.

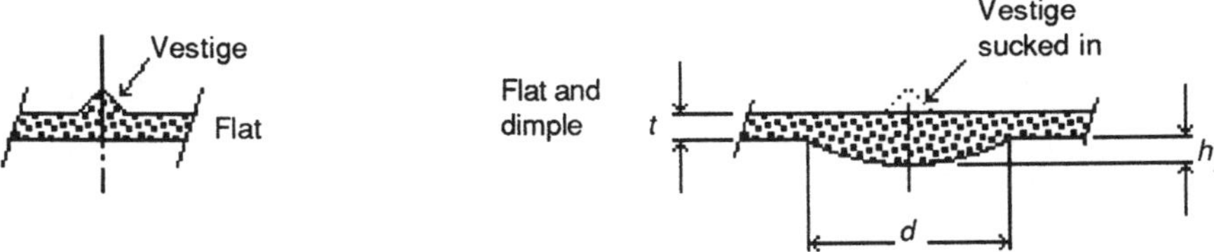

Figure 10.14 Depending on the product shape, the vestige may remain obvious (left) or be sucked in where a dimple is provided (right).

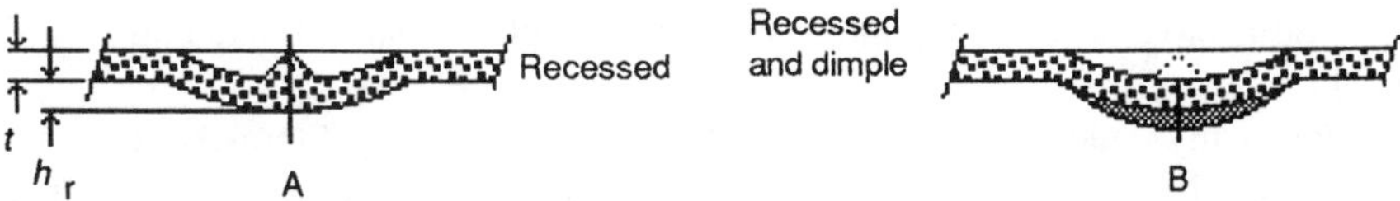

Figure 10.15 Recessed gates: A. without a dimple, and B. with a dimple.

Another benefit of the dimple, especially for thin-walled cups is that the added mass of plastic at this spot remains hot (for a short period) after ejection. As it shrinks, the vestige is "sucked in" and almost made to disappear (Fig. 10.14).

Dimples should always be used, unless the customer has serious objections. Without a dimple, particularly in thin-wall molding, the product quality may be seriously affected, and the molding cycle may increase.

The following dimple dimensions are suggested:

Diameter: $d = 4–10$ mm,
Height: $h = 0.75\ t$, for $t < 0.75$ mm, or $h = 1.0\ t$, for 0.75 mm $< t <$ 1.5 mm.

For heavier walls ($t > 1.5$ mm), there is little need for a dimple. The plastic finds enough area to flow away unrestricted from the gate.)

10.1.4.3 Recessed Gate

A small area can be recessed below the level of the product (Fig. 10.15). Even if a protrusion is left standing, it should not go beyond the level, which might later be covered with a label that could be damaged by a gate protrusion. A (sharp) protrusion may also be undesirable at the underside of a drinking cup, etc.

If the gate can be recessed, the depth of the recess h_r should be at least equal to t. With a wall thickness $t < 1.5$ mm, the recess should be combined with a dimple (Fig. 10.13B), similar to a level surface. Note that the recess shown is spherical, but it could also be any shape which suits the product.

10.1.4.4 Hot Runner Edge Gate with a Dimple

The vestige of this gate is similar to that made by a cold runner tunnel gate. It is created when the mold opens and the plastic is sheared off from the (partially frozen) gate by the relative motion between the cavity wall and the product (which moves with the core). The vestige is

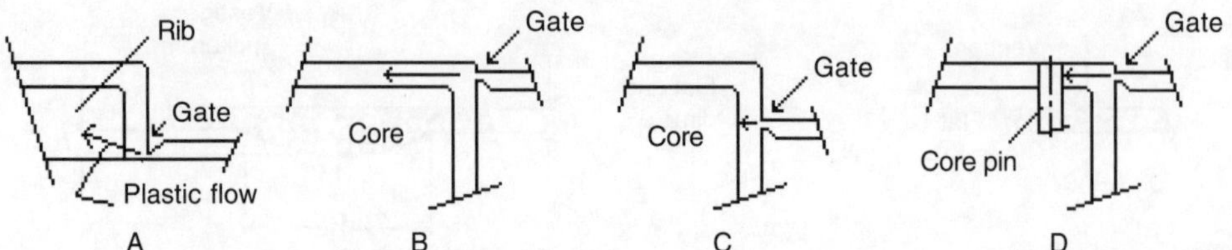

Figure 10.16 Different locations of gates affect plastic flow: avoid injecting A. into a rib or B. into open space. Instead, inject C. against the core or D. against a core pin.

flat (no protrusion) and somewhat more shiny than the plastic that is broken off or sheared when cold.

It is practically impossible to hide this gate, and the best thing is to make it as small as possible. If possible, instead of a dimple and for the same reasons, it is useful to thicken the area where the plastic enters or to gate into an area in the product which is already thicker.

Often in molds for thin-walled products, in addition to a dimple, the thickness t of the bottom of the product should be increased by 0.025–0.05 mm (0.001–0.002 in.), to facilitate flow toward the sidewalls. If the product drawing does not show a heavier bottom, it should be suggested. Such an increase may fall within the product tolerances given and not even require a change in specifications.

10.1.4.5 Valved Gates

The vestige of a valved gate is always a circle and looks similar to an ejector pin mark. The length of the protrusion can be controlled at the design stage by the design of the valve pin. Theoretically, there should be no protrusion; because of tolerance build-up, the protrusion should be planned to be negative (i.e., from zero to a slight depression in the product). This will be discussed later.

A comment that applies to all gates, including hot runner gates, is that as the plastic flow from the gate enters the cavity space, it should be broken up into "swirls" to prevent flow marks and jetting. This is best done by directing the flow against a solid obstacle in the cavity space, such as the core or a core pin opposite the gate. Injecting (gating) into an open space such as the flat bottom of a product or into ribs should be avoided if possible (Fig. 10.16).

For most hot runner and three-plate molds, this is usually no problem since the gate is opposite a solid core; but if a very deep rib is located so that the gate would inject into the open space, it would be better to move the gate to where the plastic stream would hit the solid core.

10.2 Hot Runner Gate Types and Configurations

10.2.1 Open Gate

This gate is open for the flow of plastic under pressure, but at the end of the injection (or hold) cycle, the plastic in the gate freezes sufficiently to act as a plug. This prevents the hot plastic in the runner from oozing out ("drool") into the cavity while the mold is open for ejection of the product.

As the mold opens, some of the material in the gate remains with the product and creates an unsightly (usually conical) vestige. The size and shape of the vestige depends on the shape of the gate and the operating conditions (temperatures, pressures, timing) of the machine; it depends, therefore, on the mold set-up as much as, or even more than, on the mold design.

At the next cycle, the plastic pushes the plug into the cavity, and the gate is again open for filling the cavity. The plug usually melts and mixes with the inrushing plastic. It is important that the plug is shorter than the cross section of the product opposite the gate to make sure that the plug can clear the gate area (i.e., the land must be smaller than the clearance between cavity and core). A dimple creates a favorable condition.

The goal in designing open gates is to find the geometric balance such that the plug freezes readily in the land but can then be easily pushed out into the cavity. The portion of the land facing the fresh plastic supply is tapered to create a widening of the passage which will keep the plastic "bubble" hot (fluid), ready for the next cycle. In many cases, a heated nozzle tip brings additional heat to this bubble.

To repeat, the operation of the molding machine (i.e., temperatures, pressures and timing) have an important role to play in the proper performance of an open gate. A valved gate is less critical than an open gate, from the point of view of operating a mold, but is more expensive.

Open gates are used extensively for PE and PP, and in general for any material which has little or no tendency to "string" (i.e., to pull a thread of plastic out of the hot center of gate as the mold opens).

There are three basic types of open gates:

1. Circular gate,
2. annular gate, and
3. edge gate.

10.2.1.1 Circular Gates

Figure 10.17 illustrates the disadvantage of a long, cylindrical land. The break point in the land is not determined, and the plastic can break anywhere along the length of the land *L* but will

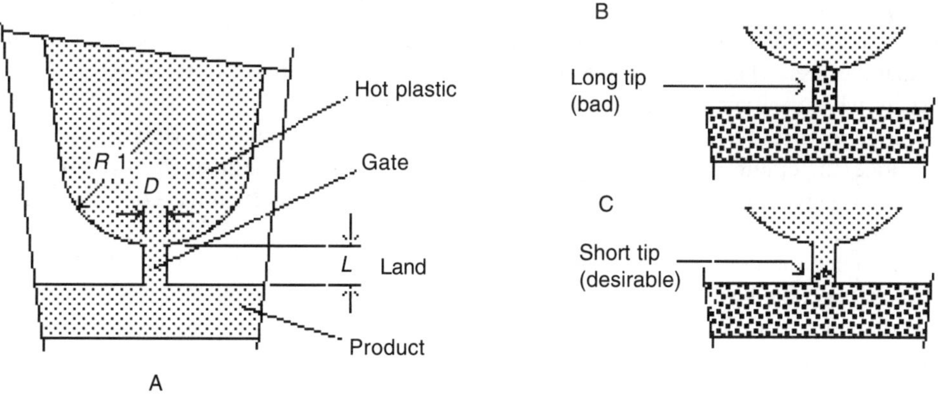

Figure 10.17 Circular gates: A. Long, cylindrical land, B. after injection, often leaves a long projection on product, and C. short tip is desirable but not determined.

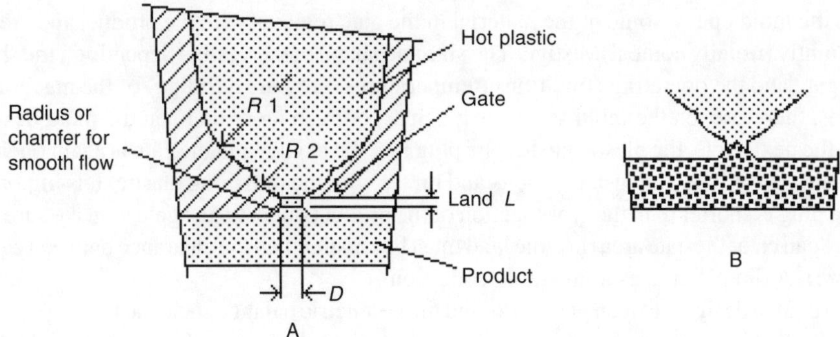

Figure 10.18 Short, cylindrical land: A. after injection and B. after cooling.

usually break near the hot plastic and leave a long projection on the product. This gate is easier to produce, but it is no good.

The gate design in Fig. 10.18 shows a short land *L*, which provides a well-defined break point above a conical extension of the product. There will still be a small conical projection of plastic above the break point where the hot plastic will adhere to the cold cone, but the total projection is quite predictable.

With a heavy wall, or when using a dimple with thin-walled products, the projection may disappear if the plastic in the product (in the gate area) is still warm enough after ejection. The actual shape of the open gate is the result of experiments and experience.

The main problems are:

1. The strength of the gate area: the plastic pressure tends to push the steel around the gate outward into the cavity space, and
2. control of the heat flow away from the gate.

The gate shown in Fig. 10.18 with a combination of two radii (*R*1 and *R*2) has a much shorter land than that shown in Fig. 10.17 (one radius *R*1 only), thereby gaining strength and the ability to conduct more heat away from the gate. Instead of radii, conical sections are sometimes used. The designer must consider the foregoing when selecting an existing gate size or when designing a new gate.

Figure 10.19 illustrates another open gate design. While the gate shown is very simple, accurate temperature control in the gate area is difficult because of the difficult-to-control heat losses through radiation and conduction. It also exposes the product to direct contact with the hot nozzle.

A better solution is the gate shown schematically in Fig. 10.20. This nozzle does not contact the cooled cavity but is insulated from it by a plastic layer; the temperature in both nozzle and cavity can be better controlled, and it is easier to create the proper freeze-off conditions for the gate.

The advantage of circular open gates is that they can be small. They are well suited for heat-sensitive materials and are less expensive to produce and easier to operate than other open gates.

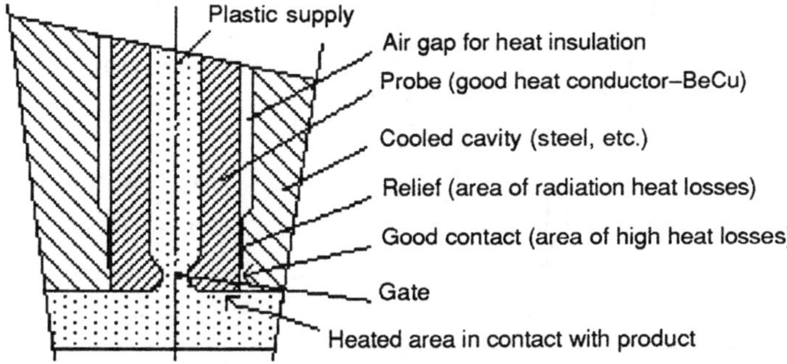

Figure 10.19 An open gate design providing poor heat loss control and direct nozzle contact with product.

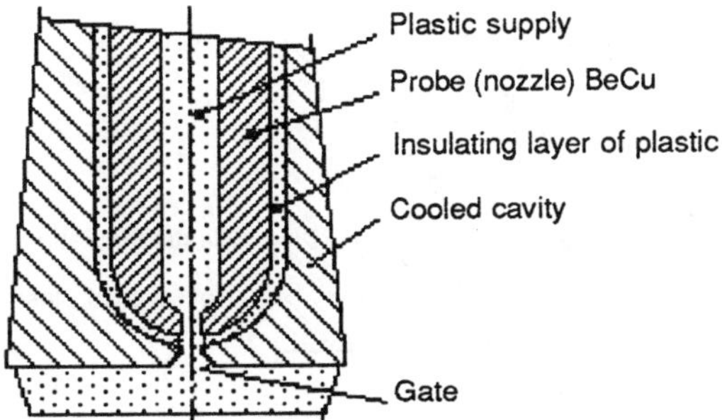

Figure 10.20 An open gate design that protects the product from direct contact with the hot nozzle and allows for easier heat loss control.

10.2.1.2 Annular Gates

A basic annular gate is essentially an open gate with a heated probe at its center to prevent premature freeze-off (Fig. 10.21). Note that the gate shape is intimately related to the shape of the nozzle tip.

In both circular and annular gates, a heated *probe* or *nozzle* or *nozzle tip* is centrally located within a well in the cavity block. At the end of the well is the gate, and the pointed nozzle tip ends near or within the gate.

Being exposed to the often abrasive, high-speed plastic stream, the tip will gradually wear and need replacement. In some hot runners, the nozzle tip is a separate unit, screwed into a *nozzle housing*, and is easily replaceable. The material of the tips is usually BeCu because of its good heat conductivity and strength. Tool steel tips are also used where severe wear is anticipated or where (from experience) the lower heat conductivity is not significant and the higher wear resistance of steel is desired. Typical shapes of tips and gates have been developed by the various manufacturers of hot runner hardware, and are shown in their catalogs.

In an annular gate, the plastic coming from the manifold enters the space upstream around the nozzle and flows toward the gate. Since the tip within the gate creates an annular opening,

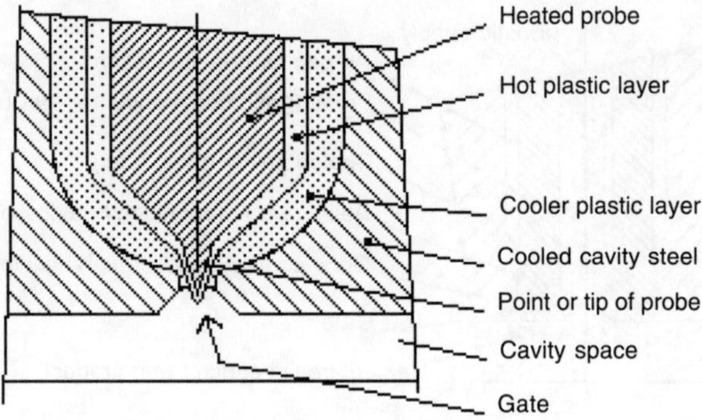

Figure 10.21 An annular gate with a heated probe at its center.

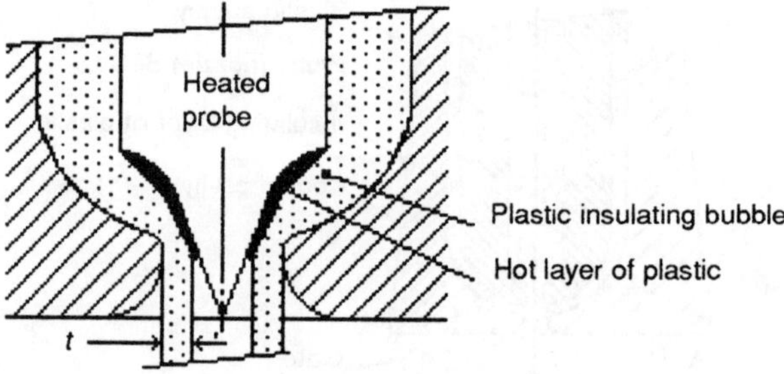

Figure 10.22 A heated probe surrounded by its plastic insulating bubble.

the plastic enters the cavity like an extruded tubing. Plastic fills the originally open space between the nozzle and the surrounding (cooled) cavity. Since plastic is a poor heat conductor, there is little heat lost through this "plastic insulating bubble" (Fig. 10.22).

In molds which process easily degradable and heat-sensitive materials, it has been found useful to fill open space between the nozzle and the cavity with a molded or machined high heat-resistant plastic, such as Vespel™. There is little heat lost through this nozzle tip insulator to the cooled cavity, but it prevents the sensitive plastic from entering the open space where it would degrade and could contaminate the products.

The advantages of the annular gate are:

1. The wall thickness *T* of the tubing can be much smaller than the diameter of a circular gate of a cross section equal to that of the tubing. Also, a much larger cross section (flow passage) can be created by only little increase in the dimension of the gate diameter. An equivalent circular gate could be too large to be controllable and would easily drool. The size of the passage (the thickness of the tubing) *T* depends much on the size and shape of the nozzle tip and the axial location and concentricity of the tip within the gate. Heat expansion of the probe has a significant influence on this passage, and must be calculated to arrive at the proper depth and diameter dimensions.

2. Frozen gates rarely occur. A thin layer of plastic close to the nozzle will remain hot (and viscous) after the rest of the gate has frozen (Fig. 10.22); during the next shot, the plastic will easily enter the cavity because the hot plastic stream will quickly melt the surrounding layer of frozen plastic.
3. Less stringing occurs so, in general, annular gates are used for plastics that string easily and are not very heat sensitive (e.g., PS). They are also used in molds requiring very high filling speeds for rapid cycling, as with disposable cups, etc.
4. Smaller vestiges occur in annular gates than circular gates.
5. A wider operating window is open in annular gates because of the ability to better control temperature within the gate.

Disadvantages of the annular gate are:

1. Since the plastic coming from the manifold enters at the centerline of the nozzle, it must be diverted from the center to the outside of the nozzle through two or more branch passages. If these are too close to the gate, and/or if the temperatures are too low, the diverted streams may not have enough time to melt into a homogeneous tubular stream. This may result in visible flow lines in the product which may be unacceptable.
2. Annular gating is not applicable to all resins because high pressures are required to overcome high flow resistance in the narrow gap.
3. The narrow gap between tip and gate can easily plug up with contamination in the plastic. Annular gates should not be used when the molder uses "dirty" plastic.
4. The plastic, after long-time exposure to heat from the nozzle, will degrade. The danger is that some of the degraded material may, from time to time, be washed out by the fresh stream of plastic and contaminate the products.

Shown in Fig. 10.23 (schematically) are three commonly used annular gates. Figure 10.23A and B were used in earlier designs; most designs used today are similar to the system shown in Fig. 10.23C.

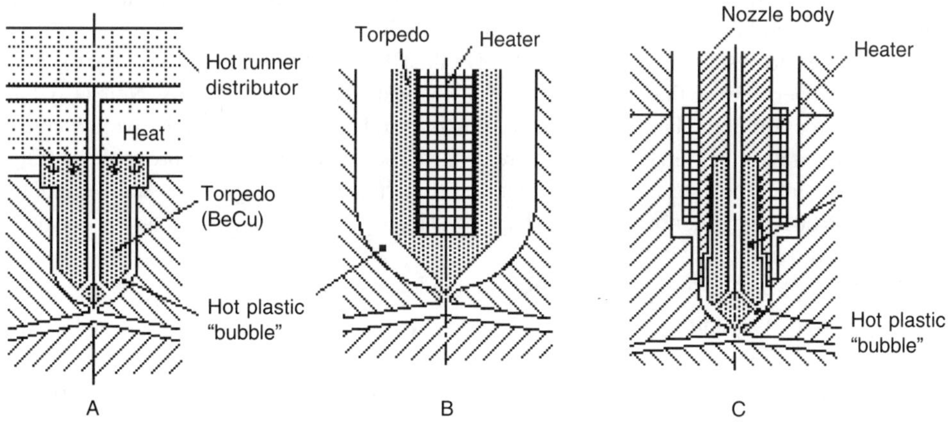

Figure 10.23 Three common annular gate designs: A. heat conducted from a hot runner distributor, B. a torpedo heated by an inside cartridge heater, and C. a nozzle tip heated by an outside band heater.

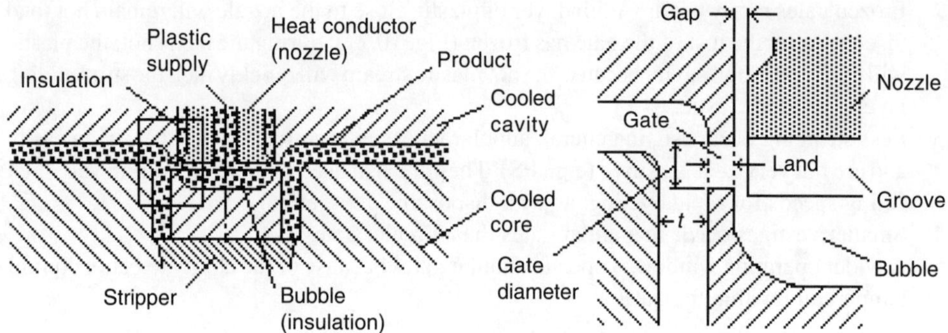

Figure 10.24 Typical hot runner edge gate design.

Circular and annular open gates can be used with any hot runner mold but can also be used in insulated runner molds, which distribute the plastic without heated manifolds. Some years ago, many molds were built with this "insulated" type of hot runner system, which is based essentially on the insulating properties of the plastic. The manifold ducts the plastic, without any heaters, to the nozzles. For certain materials and within certain size and cavity number limitations, this system is very reliable and relatively inexpensive to build and to operate; but it requires special skills by the mold and machine operators which may not be always available to guarantee safe operation. It will not be further discussed here.

10.2.1.3 Hot Runner Edge Gates (HREG)

Figure 10.24 shows a typical hot runner edge gate arrangement. The principle is the same as that for circular open gates. At the end of the injection cycle, the material in the gate freezes.

As the mold opens, the product moves with the core out of the cavity, thereby shearing the gate, leaving a plug (or "slug") between the now open cavity and the hot plastic in the runner system. The plug prevents plastic from drooling into the cavity. With the next injection, the plug is pushed into the cavity space, melts, and usually disappears.

While the above schematic shows the land of the HREG as cylindrical, it is better to taper it slightly (5° per side) so that the plug can be easily pushed out into the cavity. Note also that while the direction of the land in this illustration is at right angles to the cavity wall, it could also be at an angle (up to maybe 30°) to create a sharper shear angle to improve the cutting action of the gate; the gate vestige will then be elliptical. While the gate is usually circular, it could also be rectangular or any other shape by using EDM to produce the passage between the nozzle well and the cavity.

Suggestions for the design of the HREG include:

1. The "bubble" should be as large as possible to create a plastic pool that will not easily freeze.
2. The land L should be small, in the order of 0.5–1.0 mm. The smaller the better (Fig. 10.25), but limits are set by the strength of the steel. In Fig. 10.25A, the land L is large to ensure sufficient strength between the cavity and the bubble well. In B, because of the shape of the product, the land can be smaller, and in C smaller yet, without sacrificing strength.

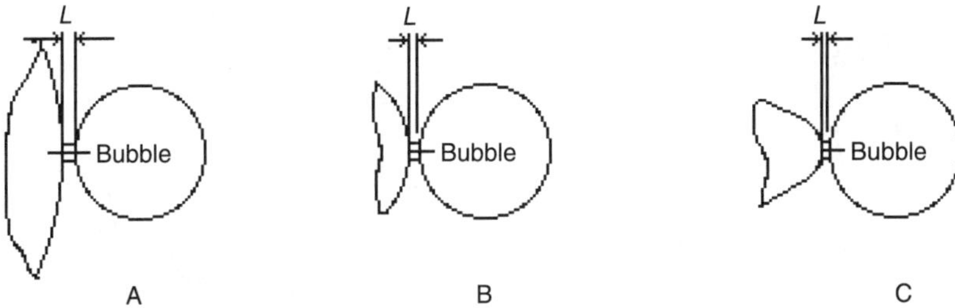

Figure 10.25 Three top views of an HREG showing different sized lands: A. land L is large, B. land is smaller, and C. land is smallest.

3. The land *L* must be smaller than the wall t of the product opposite the gate so that the slug can easily be pushed out of the gate by the inrushing plastic during the next injection.
4. The gap (Fig. 10.24) should be as small as possible (0.03–0.05 mm), to bring the heat conducting nozzle (and the heat) to the plastic close to the gate area. However, the nozzle must not touch the cooled cavity to ensure proper temperature control of the gate area.
5. The reaction force from the plastic injected through the gate must be well supported to prevent deflection of the nozzle away from the gate. Two gates can be located in the well at 180° to feed two cavities, or three small or even four very small cavities can be fed at 120° or at 90°, respectively. If there is only one cavity, there must be a mechanical support opposite to the gate, or the nozzle design must be stiff enough to withstand the deflecting force.

10.2.2 Valved Gates

The principle of valved gates is that the gate opening and/or closing is achieved independent of the injection pressure. In some systems, it opens under the injection pressure at the beginning of injection. It does not depend on the injection pressure for removal of the frozen plastic from the gate. Gates can be opened and closed independently, mechanically (using a pin), or thermally (using a special heater).

Mechanically controlled gates can have single or double acting operators:

1. Single acting: The gates are opened by the plastic pressure acting on a step in the valve pin; they may be closed by:
   a. a spring, which acts as soon as the pressure drops enough. This creates a problem: With low injection pressure required, the spring must be weaker; with high pressure it must be stronger (e.g., With high injection and lower hold pressure, a strong spring will close the gate too early.) Also, a weak spring may not be strong enough to ensure closing. Springs can anneal at the high temperatures found in some hot runners.
   b. an (in-line) air cylinder or a wedge (air or hydraulically operated) acting any time after the pressure drops.
2. Double acting: The gates are opened and closed by in-line air cylinders at any desired time. They are usually opened at the start of the injection cycle. There is no step in the valve pin.

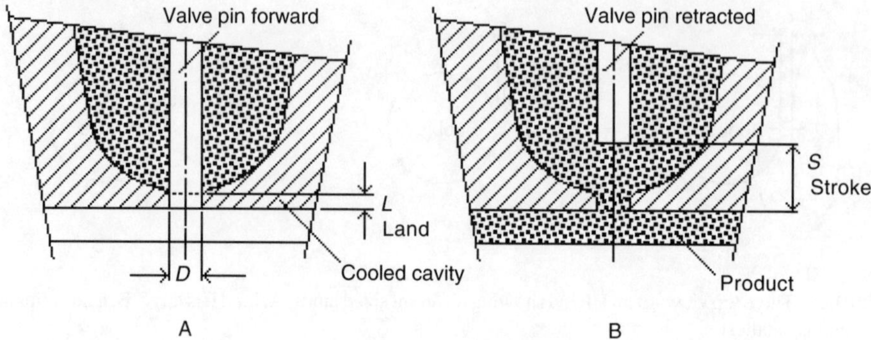

Figure 10.26 Early valve gate design: gate closed (left) with pin forward, and gate open (right) with pin retracted.

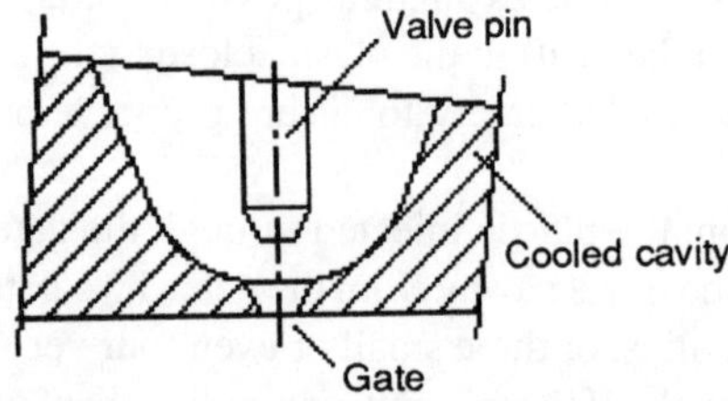

Figure 10.27 Valve gate with a tapered valve pin point and matching seat.

Some valve gate systems use hydraulically operated valve stems, but the proximity of the hot manifold poses a fire hazard and requires special, noninflammable hydraulic fluids, which are not normally available from the hydraulic system of the machine. Some systems use air cylinders that are not in line with the valve stems, but they require more space than the in-line systems.

With thermally controlled gates, an electrically heated point is within the gate. Control of the heat and temperature in the point provides opening (melting) or closing (freezing) of the gate.

10.2.2.1 Basic Valve Gate

Figure 10.26 shows an early design of a valve gate. A cylindrical pin enters a cylindrical gate. Problems are poor alignment, deflection of the pin, wear of the gate, and valve pin breakage. The stroke S must be sufficient to clear the gate and to ensure that the end of the pin which was cooled while inside the gate is heated again while immersed in hot plastic.

A tapered point of the valve pin, with a matching seat as gate (Fig. 10.27), avoids some of the problems of alignment but creates the problem of the closing forces acting on the gate.The gate must be strong enough to resist this force.

With some designs of valve actuator (pneumatic piston), the length of the valve pin is calculated so that with all tolerances "favorable", the pin will just touch the seat without

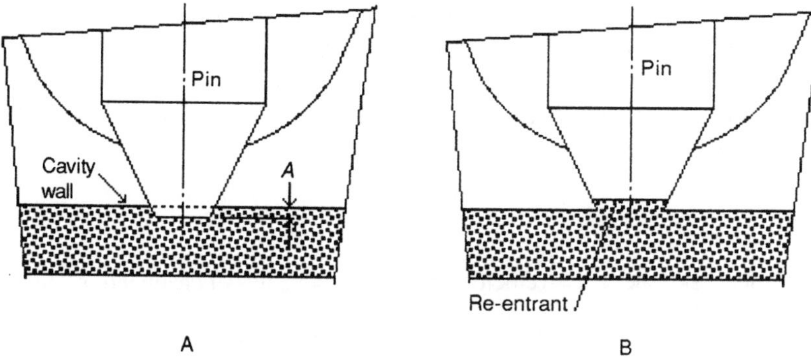

Figure 10.28 Length of the valve pin pint is illustrated: A. pin passes through the gate to create a depression in the product, B. pin is too short, creating an undesired projection on the product.

pressing down on the gate, but with "unfavorable" dimensions, there will be a slight gap at the seat. A slight plastic skin at the seat prevents metal-to-metal contact. A stop inside the valve bushing should limit the valve pin stroke to prevent excessive load at the gate. The stop also limits the pin travel when the gate is not tapered but cylindrical, and is not seating on the gate, as in certain gate configurations.

Additional length of valve pin point is important in the gate design. The point of the valve pin should always pass through the gate by the amount *A* in Fig. 10.28A to create a small depression in the product. If the pin is too short (Fig. 10.28B), the plastic will enter the gate and produce a projection in the product with a reverse taper, which may tear out and create an unsightly vestige. Dimension *A* should be between 0.03 and 0.25 mm, depending on valve stem design and product (vestige) requirements.

10.2.3 Factors Affecting Gate Size and Shape

Before discussing the actual design or sizing of gates, we will list some useful definitions and terms used:

Rheology: Science of deformation of plastics in response to an applied pressure or stress.

Melt index (MI): Expressed in numbers. These numbers can give an indication as to how well any particular polymer (plastic) will flow. A a low MI will indicate a plastic that is hard to inject.

Shear rate is the rate of change of velocity of the moving plastic with respect to the change of radius (distance from the centerline) of a round channel. It is maximum at the wall of the melt channel.

An easier way to look at the effect of shear rate on plastic is as follows: Consider shear rate to be like velocity; as the velocity increases, the resisting force preventing the plastic from flowing (shear stress) increases. This, in turn, has the effect of reducing the viscosity and makes the plastic more fluid.

All plastics are *non-Newtonian fluids* (n-Nfs). The difference between n-Nfs and *Newtonian fluids* (Nfs) (e.g., water) is that *viscosity* (resistance to flow) η of the flowing Nfs is unchanged across the flow path within a (round) channel, but viscosity of a n-Nfs changes greatly with temperature and in relation to a shear rate $\dot{\gamma}$ [$\eta = f(\dot{\gamma}, T\,°\mathrm{C})$]. This can be expressed with the following formula:

$$\dot{\gamma} = \frac{du}{dr}\ \mathrm{sec}^{-1} \tag{10.1}$$

where $\dot{\gamma}$ = shear rate (sec^{-1}), u = velocity (cm/s), and r = radius of channel (cm). Shear rate is also dependent on:

1. Type of plastic,
2. processing variables (injection pressure, speed, temperature), and
3. conditions of channel wall (finish, temperature).

A convenient approximation for the average shear rate is

$$\dot{\gamma} = \frac{4Q}{\pi r^3}, \tag{10.2}$$

where Q = average plastic flow rate (cm^3/sec):

$$Q = \frac{\text{Shot volume (cm}^3\text{)}}{\text{Injection time (sec)}}. \tag{10.3}$$

The shear rate $\dot{\gamma}$ increases with the volume (flow rate) of the plastic, as it would when higher pressure pushes the same amount of plastic (per time unit) through the channel; but $\dot{\gamma}$ increases significantly more with any reduction of the channel diameter (2*r*).

An example will illustrate two interesting points:

Example: A product has a mass of 106 g PS, which corresponds to a volume of 100 cm^3. This will be injected in 1.0 sec, so Q = 100 cm^3/sec. The selected gate size has a cross section of 1 mm^2, corresponding to a diameter (2*r*) of 1.27 mm = 0.127 cm. Therefore, r = 0.0635 cm.

The speed of injection is, therefore:

$$\frac{100{,}000\ \mathrm{mm}^3}{1\ \mathrm{mm}^2} = 100{,}000\ \mathrm{mm/sec, or}\ 100\ \mathrm{m/sec.}$$

This speed is about one-third the speed of sound (Mach 1 = 307 m/sec). If the volume Q were three times that of this example, the plastic would shoot through the gate at about the speed of sound. Considering these high velocities of plastic racing through a gate, it is no wonder that the gates will wear, particularly if the gates are small and the plastic is abrasive.

Using equation 10.2, the maximum shear rate is:

or about 500,000 sec^{-1}. This is a very high value for most plastics, and it suggests that a larger gate must be selected and/or the injection time increased to more than one second to bring the value closer to what is permitted for the plastic.

Shear stress: This is the unit pressure on the fluid that is subjected to the action of shearing.

Shear sensitivity: The various plastics respond differently to applied stress. Their response is related to the configuration of the molecules from which they are made. The term applied to such differences is called shear sensitivity.

Shear-sensitive materials respond to pressure by having their molecules readily shifted in the direction of flow.

Shear-insensitive materials: The application of a higher stress only causes the molecules to be further entangled (e.g., like a rope being pushed through too small a passage). The viscosity does not change much with any increase in the shear stress (pressure).

10.2.3.1 Factors Affecting Gate and Land Size in Open Gates

1. Part weight and size: The longer the flow length and the larger the cavity surface, the larger must be the required gate to reduce fill pressure and to ensure a sufficient flow of plastic to prevent premature freeze-off of the gate.
2. Product wall thickness: A product with a large wall thickness requires a larger gate to supply fresh plastic to the shrinking product during the holding (packing) stage of the injection cycle. An undersized gate would produce a warped or unfilled product.

   NOTE: Normally, the gate diameter is smaller than the wall thickness, because once the cavity is full and cooled, the gate must also freeze to allow ejection of the product without creating an ugly vestige. However, for thin-walled products (wall thickness $t < 1$ mm), the gate diameter is often larger than t to reduce the pressure drop at the gate.
3. Resin: If the resin is very viscous (i.e., has a low melt flow index and high molecular weight distribution), a larger gate and a shorter land are required to reduce the restriction in the gate area. See Table 10.1 regarding suggested permissible shear rates for various materials.
4. Location of cooling lines mold temperature): Improperly located cooling lines, if too close to the gate, will cause premature freeze-off. If too far from the gate, they will cause drooling. Gate diameter selection can be used to compensate. Select a larger gate for close cooling lines and a smaller gate for far away lines.
5. Injection time: Very fast injection requires a larger gate to reduce the local pressure drop at the gate and to prevent excessive shearing and degradation of the plastic. Note: A 10% increase in the diameter, with a 10% decrease of the land, results in a 47% increase of the allowable flow rate to the cavity.

   6. Melt temperature: If the melt is processed at the maximum possible temperature but the cavity still cannot be filled, a *smaller* gate can be used to transmit more heat to the resin because of the increased shear heating, which reduces the viscosity further and makes the resin easier to flow.

   7. Entrance effects: Sharp corners or restrictions impede the flow of resin and can cause shear-induced degradation. Use a generous radius on the cavity side of the gate to help provide smooth laminar flow and to prevent jetting or blushing.

8. Nozzle tip position, tip type: If the tip is closer to the gate, the gate is less likely to freeze off prematurely, and, as a result, a smaller gate can be used provided the maximum shear stress and pressure drop are not exceeded.
9. Dimple: See earlier comments for more details. If a dimple is too large, it slows down the cycle.

10.2.4 Requirements for a Correctly Designed Gate

1. Permit unrestricted flow—to the greatest possible extent—to prevent degradation of the plastic.
2. Prevent drooling or stringing.
3. Provide correct shearing to condition the resin and to reduce its viscosity to achieve the greatest flow length possible.
4. Hide or disperse the cold slug without impeding flow. This is also helped by a dimple.

Improper gating techniques can cause:

- Jetting (visible flow lines away from the gate),
- blushing (a concentric, cloud-like blemish around the gate),
- stringing (threads of resin sticking to the product),
- warping (deformation of the product),
- degradation of the resin,
- improper filling (short shots),
- premature freeze-off of gates, and
- bad vestiges (gate marks).

10.2.5 Gate Shape and Size

10.2.5.1 Land Length (Gate Height)

The land should be as short as possible to achieve lower pressure for filling and to improve degating by reducing the height of vestige. The only reason against short lands is the loss of strength of the steel at this area, which may cause the steel to break. In the author's experience, land lengths should be from 0.13 to 0.25 mm.

10.2.5.2 Gate Diameter

Too small a gate can be recognized by blemishes at the gate and surface imperfections. However, even before it affects the resin, it dramatically affects the injection pressure so that the product will not be filled, and the gate will freeze off too soon.

Too large a gate can be recognized by an unsightly vestige. It will affect the mold-closed time, and requires an increase in cycle time. While too small a gate can be easily corrected by increasing the gate size, too large a gate will require a new gate insert, or even a new cavity if there was no insert.

10.2.5.3 Role of Shearing in Gate Diameter Sizing

High shear rates can raise the local melt temperature at the small gate significantly, thus reducing the viscosity and making the resin flow easily within the cavity space. High shear

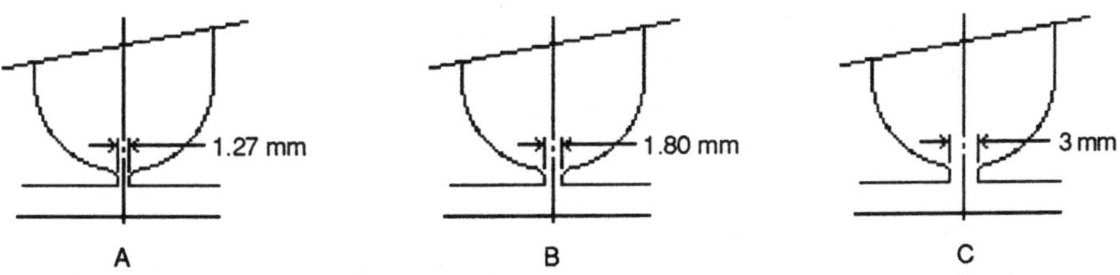

Figure 10.29 Three examples of gate diameter showing the role of shearing in sizing: A. smallest diameter shown has most shear, B. less shear than A, and C. largest diameter has relatively no shear.

rates can significantly improve the appearance of the product, especially its gloss. If the gate is too large, there is no temperature rise as a result of shear in the gate, which may result in premature gate freeze-off and sinks in the product. This can be explained with an example, as shown in Fig. 10.29.

The gate in Fig. 10.29A creates about four times as much shear as the gate in Fig. 10.29B. The gate in Fig. 10.29C is so large that the shear (and the heating of the resin) is insignificant. Note that this example applies to a specific combination of resin, product mass, injection pressure, etc. It only shows that a smaller gate may be better than a larger one, because it will not freeze off easily and will create a cleaner vestige. The actual method of calculating is shown in the following discussion.

From experience, we know that shear rate at the gate must be greater than 1,000 sec^{-1}. In thin wall molding, shear rates of 100,000–1,000,000 sec^{-1} may be needed for best results (i.e., to be able to use the lowest possible molding temperatures). This reduction of viscosity by shearing is often essential. High shear rates (caused by high velocities), which cause a reduction in viscosity, may be the most effective way to mold otherwise unprocessable plastics. However, there is a limit to the amount of shear a resin can take before it starts to degrade.

What happens at shear rates greater than 1,000,000 sec^{-1} is not fully known. There is a point where the molecules can no longer slide against each other in response to an applied shear stress. In other words, the material can only be stretched so far before it starts to be torn apart.

In a fully stretched condition, the viscosity is as low as it can be, because the molecules have aligned as much as possible in the direction of flow. If this condition occurs at 100,000 sec^{-1}, there is no point in shearing the material any further, because it will not provide any more viscosity reduction. If we could determine the limit to viscous stretching, we would obtain a physical limit to the amount of shearing.

All materials have a maximum permissible shear rate at which they degrade. The more heat sensitive they are, the lower the permissible shear rate. Note, however, that it is difficult to find the maximum shear rate, because degrading is also affected by the length of time at that shear rate.

10.2.5.4 Time of Exposure to Shear

In general, it is the longest time exposure to shear which has the most effect on the resin. This must be carefully considered.

For example, a remote disturbance (restriction) upstream will be remembered by the resin, even though the resin has subsequently passed through a viscous flow length. Also, the short time the plastic has been sheared at an extremely high rate at the gate is not as significant as the long duration of time the resin is exposed to shear in the manifold or in the nozzle tip. It is more difficult to over-shear a resin in the gate if it is subjected to the high shear stress for a short time only.

Large variations in viscosity can also lead to molding difficulties (e.g., surface imperfections, uneven fill, high stress, warp, differences in shrinkage). In *Injection Molding* [1], it is suggested that it is best to mold in the region where changes in shear rate do not significantly affect the viscosity (i.e., shear rates in the range of 500 to 2,000 sec^{-1}).

10.2.5.5 *Establishing a Proper Gate Size*

There are three methods commonly available to us for estimating gate size:

1. Use past experience from earlier designs or from information given by experienced designers or molders. This is the most common method, often practical, but it may be misleading when applied to a mold which is (erroneously) believed similar to another mold.
2. Use a sophisticated computer analysis, as described below. This (better) method is used today by many mold makers.
3. An empirical approximation, as described below, is based on experiments as described by R. W. G. Pye [2]. This method may yield a good enough gate size and can be used when no access to a computer and the necessary software is available.

10.2.5.5.1 Computer Analysis The gate is part of the hot runner and should be designed together with the rest of the hot runner system. The complete system can then be modeled, and pressure, shear stress, and temperatures can be checked for the entire system.

The following limitations apply when determining the correct channel sizes in the HR manifold:

Pressure drop (Δp)	$\Delta p < 6{,}000$ psi ($\Delta p < 5{,}000$ psi for general purpose (GP) plastics)
Melt Temp rise (ΔT)	$\Delta T < 15$ °C, but less for shear-sensitive plastics
Shear rate ($\dot{\gamma}$)	$\dot{\gamma} = 1{,}000$ sec^{-1} (for *manifold* for easy color changes; see Table 10.1 for *gate)*
Shear stress S_s	See Table 10.1 for *gate and manifold*

The critical fill rate of the cavity must not exceed the maximum permissible shear stress (or shear rate) for a material. If this value were reached, flow-induced brittleness (melt fracture) may result, causing molding difficulties and poor product quality. These values can be calculated for the entire hot runner system using Cisigraph's Procop 2-D flow analysis for hot runners.

10.2.5.5.2 Empirical Analysis Use the formula below to determine gate diameter *d*:

$$d = N\,C\,\sqrt[4]{A} \tag{10.4}$$

Table 10.1 Maximum Shear Stress and Shear Rate Values for Various Molding Materials

Material	Max. shear stress (MPa)	Max shear rate (sec^{-1})
PP	250,000	100,000
HDPE	80,000	40,000
LDPE	80,000	40,000
PS	250,000	40,000
HIPS	300,000	40,000
SAN	300,000	40,000
ABS	300,000	50,000
PPS	345,000	50,000
Nylon	500,000	60,000
PET	500,000	6,000
PUR	250,000	40,000
PBT	400,000	50,000

where A = total surface area of product (not the projected area) in mm^2, and N and C are empirical factors, as described below (In many cases this method provides a fairly good gate size approximation.):

Wall thickness of product t (mm) =	0.75	1.00	1.25	1.50	1.75	2.00
C =	0.178	0.206	0.230	0.242	0.272	0.294

N = 0.6 (PE, PS)
0.7 (PC, PP, Acetal)
0.8 (Nylon)
0.9 (PVC)

10.3 Cold Runner Gate Types and Configurations

10.3.1 General Features

The following highlights certain features associated with cold runner gates. Even though many high production molds built today are hot runner molds, the designer will find, from time to time, that certain products, for certain reasons, cannot be hot runnered, and either a cold runner or a combination hot-cold-runner must be selected. It is, therefore, useful for the designer to be familiar with cold runner as well as with hot runner gates. Also, the majority of molds built today are for low production requirements where cold runner molds are more economical.

Small gates are important for the appearance of the product. However, there is a limit. If the gate is too small, it will slow the filling of the cavity or even make it impossible to fill before it freezes.

Short lands are required to reduce the restriction to flow; however, there are limits because of the strength of the cavity material. If the runner ends too close to the cavity, the wall may break.

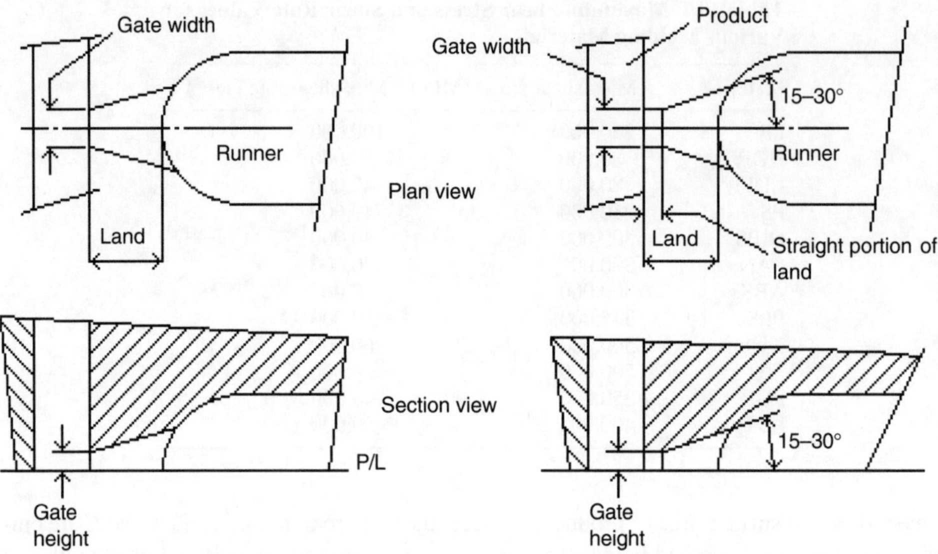

Figure 10.30 Two edge gate configurations have different lands, with the right drawing showing a short, flat portion of the land.

Tapered lands are important to ensure that the plastic will break where it joins the product. A sharp corner at the entry into the cavity will ensure that the gate breaks cleanly at that spot, leaving no or only a minimum projection (vestige). In most molds, a very small vestige projection is acceptable, and a short, straight section at the end of the gate of between 0.1–0.2 mm (0.004–0.008 in.) is suggested to provide an "almost" sharp corner, which ensures that the gates are of uniform cross section. A slight difference in the length of the straight portion of the land is insignificant for the flow of plastic.

Identical gate size is important throughout the mold. To ensure equal flow into all cavities, the manufacturing of the gates is very difficult. Close tolerances are required to ensure minimum variations in gate size, land length, and shape. A slight move of the tapered section during machining (in the direction of the land) will result in a large increase in gate size (Fig. 10.30).

For interchangeability of cavities, all gates must be virtually the same. Wherever there is a risk that pressures at the gate may vary enough to affect the quality of the products, the runners must be designed (balanced) such that all pressure drops to the gates will be the same. Some mold makers adjust gate sizes to achieve uniform filling of all cavities and to compensate for variations in pressure drops in runners. This is not a good practice.

The above statements apply to all cold runner gates.

10.3.2 Edge Gates

An edge gate is the simplest form of gate and is used wherever the product can or must be gated from the parting line and where self-degating is not required or practical. The sketches in

Fig. 10.30 illustrate the ideal, sharp, configuration (left) and the practical gate configuration with a short, flat portion (right). Note that erosion, especially from an abrasive plastic, would soon wash out the sharp corner and increase the gate size.

Suggested dimensions for general applications of edge gates are outlined in Table 10.2:

1. Gate width (*W*) and gate height (*h*): $W = 3h$. The values in Table 10.2 are only suggested sizes, they could vary according to the plastic injected. For very heavy product cross sections, larger gates will be required to maintain the back pressure on the cooling (and shrinking) plastic.
2. Length of land *L*. For small products, *L* should be 0.5–0.8 mm, and for larger products, up to 1–2 mm. It is most important, that these dimensions are the same for all cavities.
3. Taper should have an included angle of at least 30° (15° per side), but preferably 60° (30° per side). Too large an angle weakens the cavity; a smaller angle is bad for flow of plastic
4. Cross section of gate: There must be a slight draft angle on the sides of the gate to facilitate ejection of the runner (Fig. 10.31). Semi-round gate shape is sometimes used, but it is more difficult to ensure that all gates will be the same. In general, it is not recommended.

The above dimensions (1, 2, and 3) are only *general* guidelines. Gates can be calculated using flow calculations as explained in Section 10.2.5. In case of doubt, it is better to start with small gates and, if required, open them up after sampling the mold.

Table 10.2 Suggested Dimensions for Edge Gate Height, Width, and Cross Sections

Product size	Mass (gram)	Height (*h*) (mm)	Width (*W*) (mm)	Cross section (mm^2)
Very small	0–5	0.25	0.75	0.19
Small	5–40	0.50	1.50	0.75
Medium	40–200	0.75	2.25	1.69
Large	200+	1.00	3.00	3.00

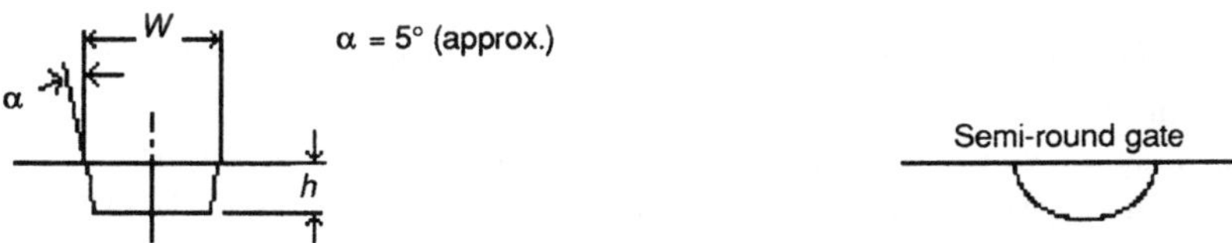

Figure 10.31 Cross section of gates showing slight draft angle for ejection of runner (left) and semi-round gate shape (right).

10.3.3 Fan Gates

Fan gates are a variation of the edge gate. Gate width *W* is much greater than $3h$. The gate height *h* may be only 0.1 mm (0.004 in.), and *W* may be 10 mm or more (Fig. 10.32).

Fan gates are used where edge gating is appropriate except that the typical gate vestige should be avoided. The fan gate vestige looks like a flash mark.

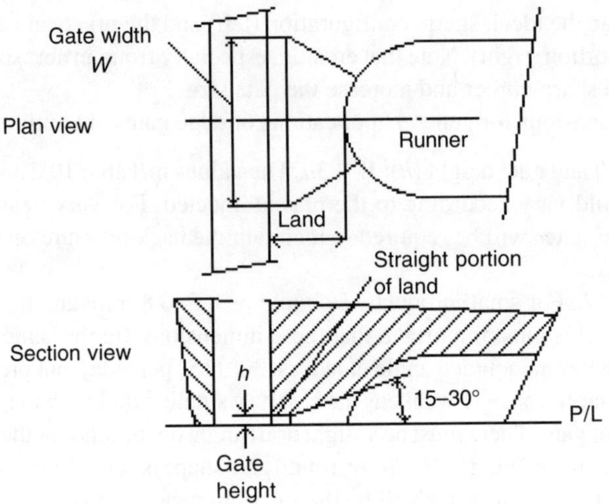

Figure 10.32 Plan view and cross sections of a typical fan gate.

Example: If a suggested edge gate size were 0.5×1.50 mm, or 0.75 mm², a comparable fan gate would need at least the same cross sectional area, or $h \times W = 0.1 \times 7.5$ mm. In fact, because of the severe restriction of the small *h*, the width *W* should be increased to an even greater length by about 25–50% to allow easy filling of the cavity. In this example, 10 mm would be a practical size.

10.3.4 Diaphragm Gate

This is a variation of the edge gate. In fact, a diaphragm gate is a circumferential fan gate. In its simplest form the gate is a disk, where the width *W* equals the length of the inside circumference. The gate height may be usually about 0.1–0.15 mm (0.004–0.006 in.), and the shape is similar to the fan gate, with an angled passage from a disk or circular runner, ending in a short straight section.

A diaphragm gate can be located on the inside (Fig. 10.33A) or on the outside of the product, in which case the plastic is fed by the runner system (Fig. 10.33B), instead of by a sprue. A great advantage of this type of gating is that the product will remain round and not become distorted as it would be with a single gate or with a small number of gates.

In the following examples of diaphragm gates, the plastic enters through a sprue, which could be a cold sprue (three-plate) or a hot runner drop. On the left of Fig. 10.34, the runner is a solid disk, which is easier to produce but has a large mass that requires additional cooling and more plastic to be reground.

On the right of Fig. 10.34, two runners bring the plastic to a distributing, circular runner, which is concentric with the product; the diaphragm gate connects it with the cavity. This design is preferred to the solid disk, particularly with larger products, because of the reduced mass of the runner system.

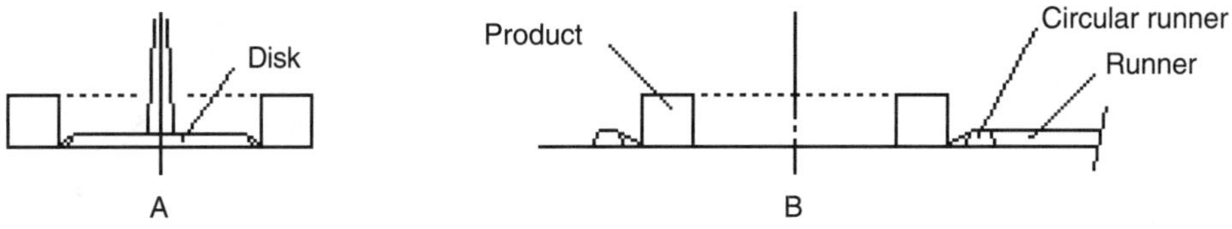

Figure 10.33 Diaphragm gates located A. on the inside of the product, or B. on the outside of the product and fed by a runner system.

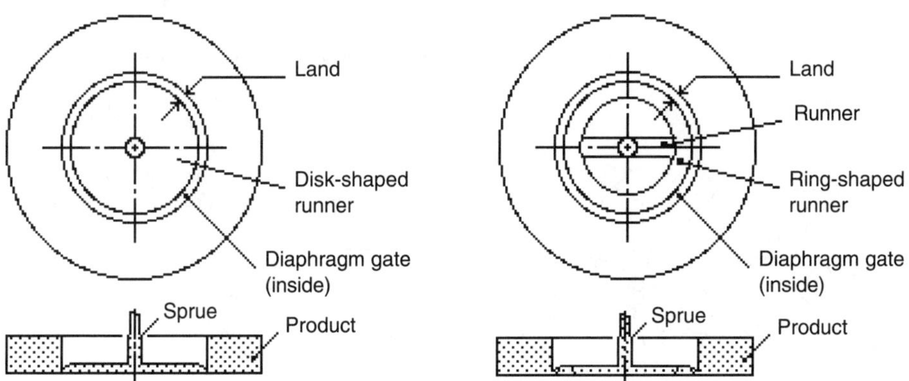

Figure 10.34 Diaphragm gates fed by a sprue via a disk-shaped runner (left) or a ring-shaped runner system with reduced mass (right).

Because the plastic enters the cavity all around through a diaphragm gate, there are no weld lines (weak spots) such as those that are unavoidable with one or several distinct edge gates. In Fig. 10.35, there are as many weld lines as there are gates. Since the plastic flow divides at each gate into two streams, this creates another weak spot per gate which must be considered.

The relationship between gate location and weld lines applies to all individual gates, whether cold runner or hot runner systems. Diaphragm gates eliminate this problem. In some cases, weld lines can be avoided by running the mold hotter to ensure that the plastic has time to fuse properly before cooling. However, this will add to the cycle time.

10.3.5 Tab Gates

Tab gates (Fig. 10.36) are sometimes used in conjunction with three-plate family molds when two or more products of different shape are produced in one mold. This allows some products to be pin point gated directly into the top (three-plate), while some others must be edge gated, either because they are very small or because they must not be pin point gated for some other reasons.

The basic runner system is three-plate; the cavities which must be gated from the edge have (usually small) *tabs* (auxiliary runners) in the P/L outside the cavity space. The three-

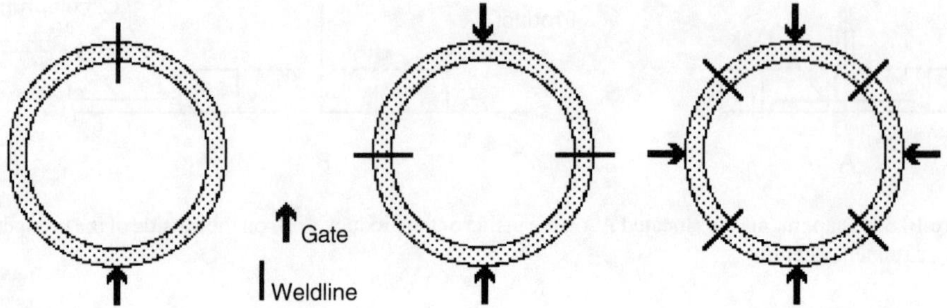

Figure 10.35 Cross sections show the effect of gating a ring from one (left), two (center), and four (right) gates.

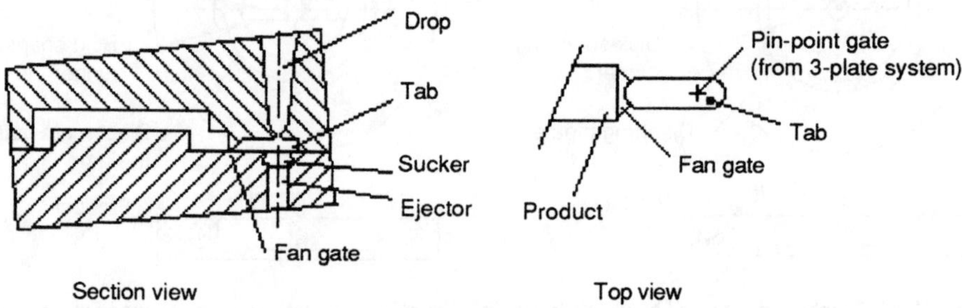

Figure 10.36 Typical tab gate in a cold runner mold.

plate drop feeds the tab, which is connected to the cavity with a gate—usually a fan gate, but it could also be an edge or tunnel gate.

In a hot runner mold, a tab gate may be required when hot runner gates in the product are not permissible or not practical. The hot runner gate enters the tab, and from the tab, the product is either edge or fan gated.

10.3.6 Tunnel Gates

Tunnel gates (Fig. 10.37) are used in two-plate molds to provide automatic, in-mold separation of the product from the runner (self-degating). In most molds, self-degating is desirable.

The cross section of the tunnel gate is similar to edge gates, except that tunnel gate passages are usually circular (see Table 10.3 for diameters). As in Table 10.2 for edge gates, the dimensions provided are only general guidelines. Gates can be calculated using flow calculations (See Section 10.2.5.)

It is important to keep the gate diameter as small as practical to facilitate the shearing-off (degating) without creating an ugly vestige or even deformation of the product. The same considerations regarding gate size and maintaining back pressure apply as for edge gates. If

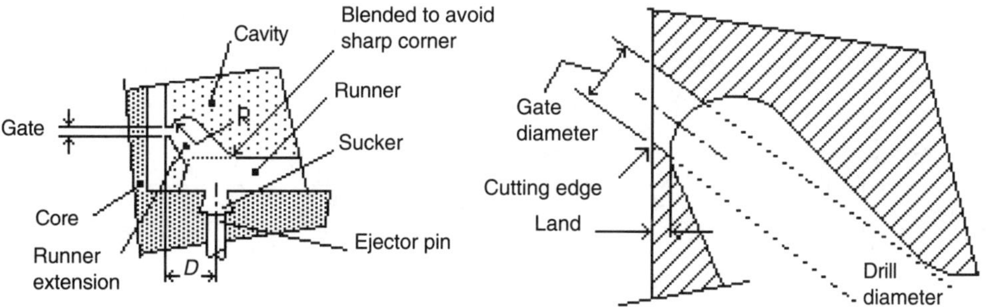

Figure 10.37 Suggested tunnel gate design: A. section through runner, and B. detail of the gate.

Table 10.3 Suggested Dimensions for Tunnel Gate Diameter and Cross Sections

Product size	Mass (gram)	Gate diameter (mm)	Cross-section (mm^2)
Very small	0–5	0.5	0.19
Small to very small	5–20	0.62	0.31
Small	10–20	0.75	0.44
Small to medium	20–40	1.00	0.78
Medium	40–100	1.25	1.23
Medium to large	100–200	1.50	1.76
Large	200+	2.00	3.14

the gates are too heavy, the plastic is more difficult to shear off and may cause the mold steel to break soon.

The land (*L*) should be as small as possible and is usually between 0.5 and 0.8 mm, although it may be 1 mm on larger gates. The intent is to make the land as short as possible while providing strength against breakage at this thin point.

The runner extension should be rounded, as shown in Fig. 10.37, to provide enough bulk in the plastic to prevent the gate from freezing too soon. Tapered extensions, even though easier to produce, are not recommended because they freeze off easily.

The product stays with the core as the mold opens (Fig. 10.38) and shears the gate at the sharp corner where the gate enters the cavity space. In the left illustration in Fig. 10.37, the gate is at right angles to the wall; the right illustration shows a gate at an acute angle to the cavity wall that creates a knife-like cutting edge. This reduces the shearing force required for cutting the gate and will enhance the gate vestige. In the first case, the vestige will be circular; in the other case, it will be elliptical.

If the passage is cylindrical, as shown in Fig. 10.37, it is possible that a frozen plastic slug will remain in the gate after the runner is pulled. It is, therefore, better to taper the gate so that it is pulled out of the cavity together with the runner extension when the mold opens.

Before specifying tunnel gates, the designer must be sure that this is what the customer wants. Tunnel gating is usually not suitable in the following cases:

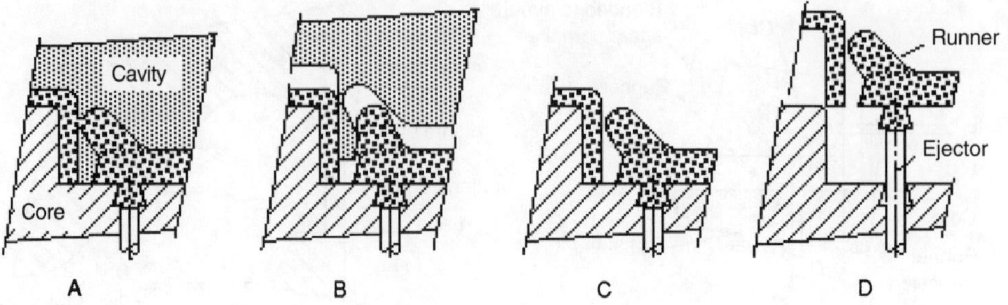

Figure 10.38 Action of the tunnel gate: A. injection, B. mold opening and gate sheared, C. mold open, and D. ejection.

1. If, in family molds, it is difficult to separate the products after ejection if they fall freely.
2. If the products must be kept on the runners for handling and/or inspection after ejection.
3. If the products are shipped together with the runners, when the end-user prefers to remove the products from the runner only when needed.

The distance *D* of the sucker from the base of the runner extension is shown in Fig. 10.37. This distance depends on the flexibility of the plastic and on the length of the extension. If in doubt, the designer should make a model to find the best location for the sucker to ensure that the extension will slide out without breaking. If available, the designer should use earlier, successful mold designs for the same plastic.

The runner extension and the runner must be as small as possible, not only to reduce waste and to permit rapid cooling but also to ensure that they are flexible enough to allow easy withdrawal of the extension as the mold opens. Note also the large, blended radius where runner and extension meets that helps to avoid breakage.

The steel between the cavity wall and the runner extension is very thin and, therefore, easily damaged, particularly if the gate is close to the P/L. Damage is often caused by operators who try to remove a frozen, broken slug from a cylindrical gate passage; this is another reason why the land should be tapered.

When molding stiff and/or reinforced, abrasive materials, the land should be longer to provide more strength, rather than what was suggested above, and is meant essentially for PE and PP.

Mold steel selection for tunnel gating may differ from that for other gates. Because the tunnel gate area is so delicate, steels selected must be tough rather than hard. H13 is a good selection when hardened up to 46–49 Rc. Carburizing steels are not recommended. During carburization, carbon enters the small steel mass at the gate area from all sides so that in the hardening process, it becomes through-hardened at 58–60 Rc and is very fragile.

A few more examples of tunnel gates are described and illustrated in the following paragraphs for different types of products:

1. Very shallow products (shallow lids, etc.): The runner can be either in the cavity side (Fig. 10.39A) or in the core side, whichever is more suitable for the design. Note how small the delicate area is under the gate (Fig. 10.39B). In this design, it may be impossible to avoid injecting into the open.

Figure 10.39 Tunnel gating for shallow products: A. in the cavity side, with B. small area under gate.

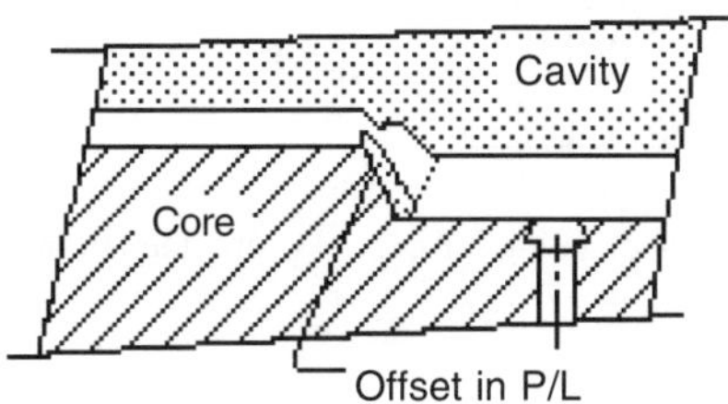

Figure 10.40 Tunnel gating for a flat product.

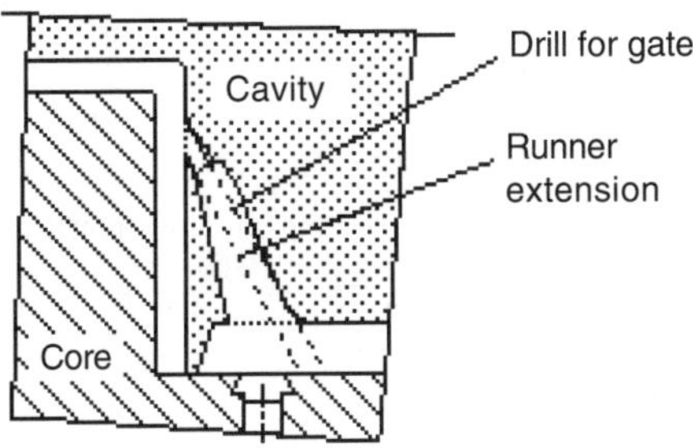

Figure 10.41 Tunnel gating for a deep product often results in a more elliptical gate opening as the angle of the runner extension approaches 90°.

2. Completely flat products (disks, etc.): An artificial offset in the P/L permits creation of space for the runner extension (Fig. 10.40).
3. Deep products (technical products, containers, etc.) (Fig. 10.41): As the gate is farther away from the parting line, the angle of the runner extension with the P/L approaches 90°, and the gate opening becomes more elliptical. At the same time it is more difficult to produce a small gate and to control the land, because the drill becomes long and slender and is difficult to control. EDM may then be the best method to produce good, identical gates in all cavities.

   Another alternative is to reduce the angle of the runner extension, but this may affect the mold layout by increasing the cavity spacing and resulting in a larger mold. With some products, this problem can be avoided, as illustrated in the gating for a spool (Fig. 10.42). The runner extension goes through a cavity insert. The gate is then created by making a rectangular slot in the insert, sized similar to an edge gate, connecting the extension with the cavity space.

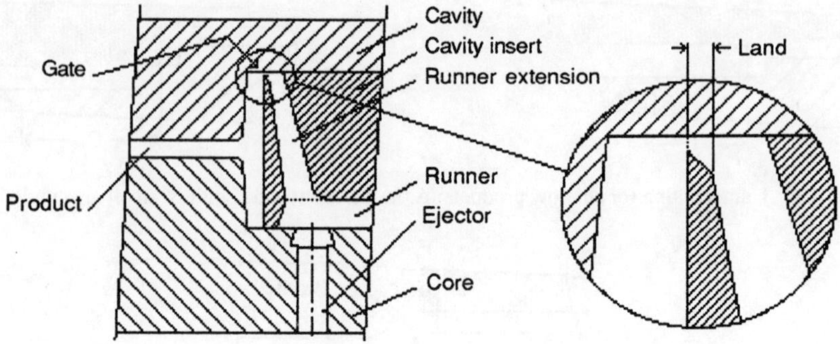

Figure 10.42 Tunnel gating for a spool is provided through a cavity insert.

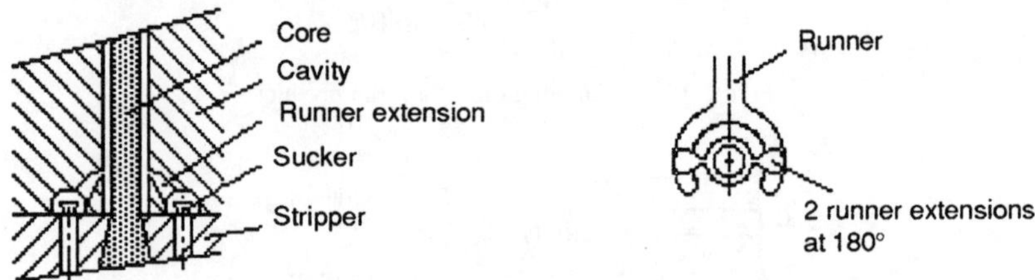

Figure 10.43 Multiple tunnel gating with two runner extensions.

10.3.7 Multiple Tunnel Gating

Typical applications of multiple tunnel gating (Fig. 10.43) are slender products such as vials or hypodermic barrels, with the end remote from the P/L closed or open (see also Section 10.1.3, Two or More Gates per Cavity. Occasionally, to provide a swirling action for the two streams of plastic entering, the axes of the two runner extensions can be skewed at an angle so that the plastic streams enter in a screw-threadlike fashion.

Note that Fig. 10.43 shows a different method of suckers holding the runner extensions, because the part is ejected from the core with a stripper. Also, it shows only two gates, but occasionally three or even four are used.

10.3.8 Curved or Submarine Tunnel Gating

This method could be selected where gating into the side of the product is not acceptable, for example, for a decorative article, such as a wall plate, where three-plate molding would be suitable for the product except that it is not practical to eject from the same side as the three-plate runner system. The runner must extend to the distance *E* within the ejector hole (Fig. 10.44) to ensure that the curved runner is fully ejected by the time the ejector is out of its hole in the core. This method is rarely used because it is essentially restricted to flexible materials. It is also very expensive because of the need for precise core inserts, which are split to allow the machining of the curved runner extension.

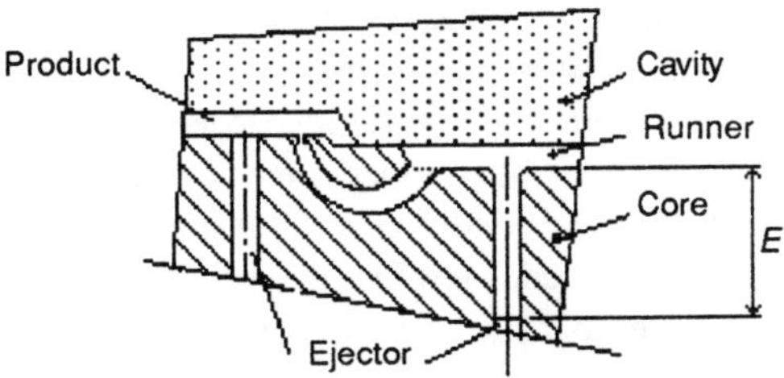

Figure 10.44 An example of curved tunnel gating.

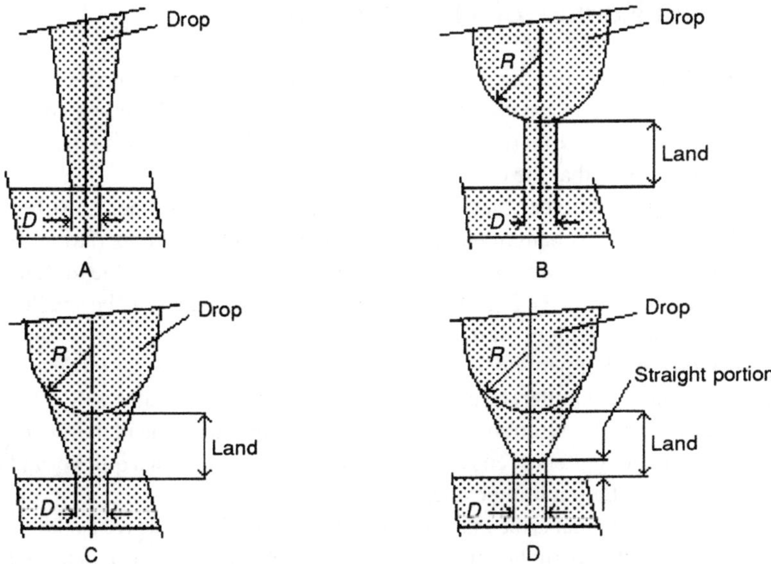

Figure 10.45 Examples of several three-plate gate designs: A. poor design with little mass near gate, B. poorly designed land length, C. impractical design although ideal, D. recommended design.

The gate is similar in shape and size to that of a three-plate, pin point gate, tapered, and with a sharp, well-defined break point.

10.3.9 Three-Plate Gates

These gates are inherently self-degating, since the gate breaks off as soon as the mold opens and the product is pulled out of the cavity. It is important to keep the gate as small as possible to minimize the vestige and to minimize any possible projection, as would be the case with an excessively large gate (Fig. 10.45).

Figure 10.45A and B illustrate poor designs of three-plate gates, although there are many molds made this way. In view A, there is not enough mass near the gate, and back pressure will be cut off early. In view B, there is no well-defined tear (or break) point. The gate is usually a fairly long projection, which must be trimmed after molding.

Figure 10.45C presents the ideal condition, but it is impractical because of the difficulty in machining and measuring involved. View D is the recommended design because it creates a good cutoff point and is easier to make. (See also Fig. 10.30, which illustrates the shape of an edge gate.)

The included taper angle should be between 60° and 90°. Recommended gate sizes are similar to sizes for tunnel gates listed in Table 10.3, or even smaller.

10.4 Cold Runner Ejection and Pullers

Cold runner pullers are required to remove the runners in cold runner molds, and also in the cold runner sections of a combination of hot and cold runner mold. There are certain standards in the design of runner ejectors (two-plate applications) and sucker pins (stripper plate and three-plate applications) that the designer will consider.

It is important to remember that for automatic molding, the runner must be positively ejected without a risk of its hanging up in the mold and requiring an operator to remove the runner. In general, the shape and the location of the ejector depends on the molding material used. The reentrant angle *A* (see Fig. 10.51 and 10.53 on later pages) can be smaller for stiffer plastics than for softer, flexible plastics.

The expected mold temperature at ejection must also be considered, as well as the volume of plastic at or around the puller. Even with the stiffest plastics, a puller will not work while the plastic is too hot and, thus, too soft. It is, therefore, important to consider the cooling layout in the area of the runner pullers to ensure that the runner will cool down quickly and not cause any undue increase in the molding cycle.

The mass of a cold runner is the main reason why the molding cycles of a cold runner mold are longer than those of a hot runner mold for a similar product. It is important to keep the cross sections (and the mass) of the runner to a minimum while still allowing proper, fast filling of the cavities. Similarly, it is important to reduce the mass around (or in) the pullers to a minimum without causing restrictions in the plastic flow.

The following illustration (Fig. 10.46) shows clearly the large bulk of plastic where the sprue enters the runner system. It will take a (relatively) long time to cool this area enough for ejection. Selection of smaller runners, a smaller sprue, and a smaller ejector can reduce this mass.

There are many other shapes of core pullers, which are not illustrated here. The purpose of this illustration (Fig. 10.46) is to show the importance of the design in this area to reduce the bulk of plastic.

Note also that with a hot sprue, the heavy junction of sprue and runner is avoided altogether. The following two examples apply where a hot sprue, or a hot runner feeding into the runner, is not possible.

The puller arrangements shown in Fig. 10.47 have the advantage of 1) reducing the mass of the plastic, 2) bringing additional cooling into the center of the junction of sprue and runner,

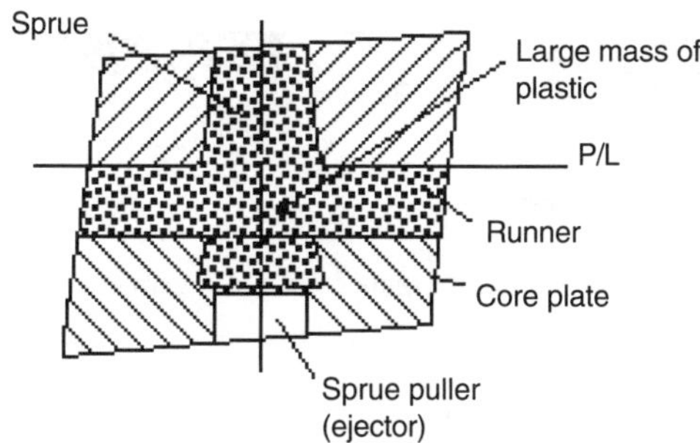

Figure 10.46 A large mass of plastic is located where the sprue enters the runner system.

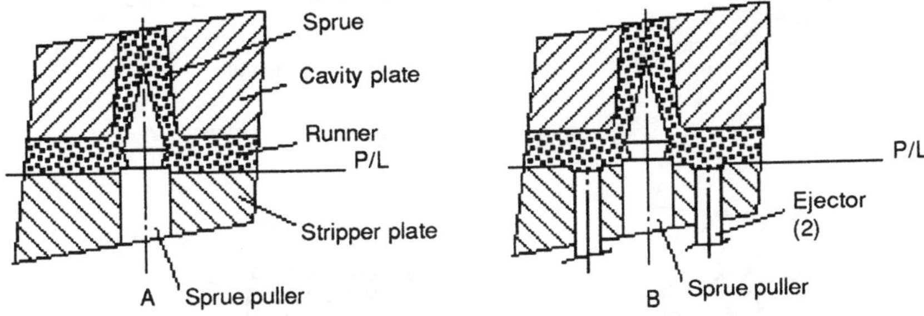

Figure 10.47 Two examples of puller arrangements: A. using a stripper plate and B. using a puller with ejector pins.

and 3) eliminating the large mass of plastic at the head of the ejector pin. This can be achieved in two ways: by using a stripper plate (Fig. 10.47A) or by using a pin together with a couple of ejector pins (Fig. 10.47B).

The above two examples are intended to show how a typical problem of reducing the plastic mass can be solved. With combination of hot and cold runners in a mold, a heavy sprue is avoided, but the design principle is the same for the runner pullers and/or suckers required in such molds.

10.4.1 Runner Pullers and Ejection of the Runner

1. Using ejector pins for runners, without pulling. Regardless of the cross section of the runner, the ejector pin must *never* extend into the runner channel but must be short by the distance *P* (Fig. 10.48). The mold parts are toleranced so that *P* will be between 0.0 and 0.2 mm. This prevents the runner from molding onto, and sticking on, the ejector pin.
2. Pulling the runner using ejector pins only. It is possible to provide the runner with undercuts on its sides near the ejector pins, as shown in Fig. 10.49, but this is would be done only if the suckers do not retain the runner as expected. Note that rough finish of the runner walls could act in a similar way as machined undercuts.
3. Pulling the runner with suckers: In many cold runner molds, the runner is retained by suckers (Fig. 10.50). A typical shape of sucker is shown in Fig. 10.51.

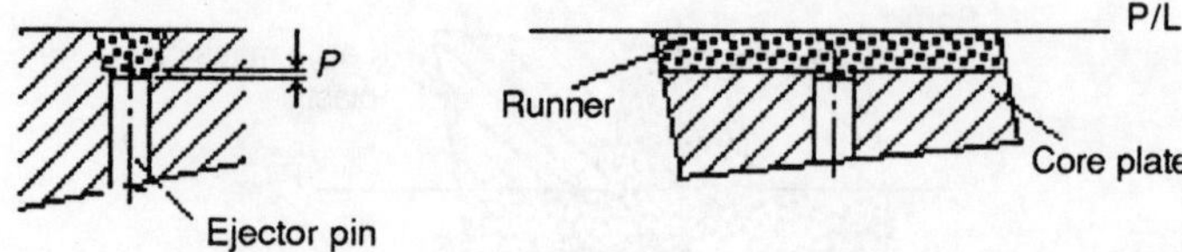

Figure 10.48 Ejector pins must never enter the runner channel, regardless of its cross section. (The runner is in the core plate.)

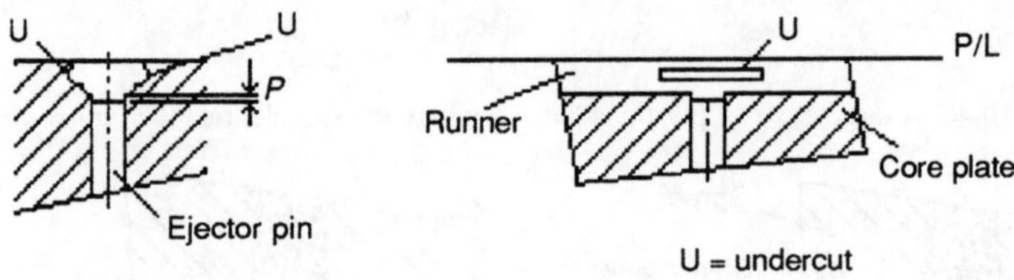

Figure 10.49 Ejector pin is used to pull the undercut runner.

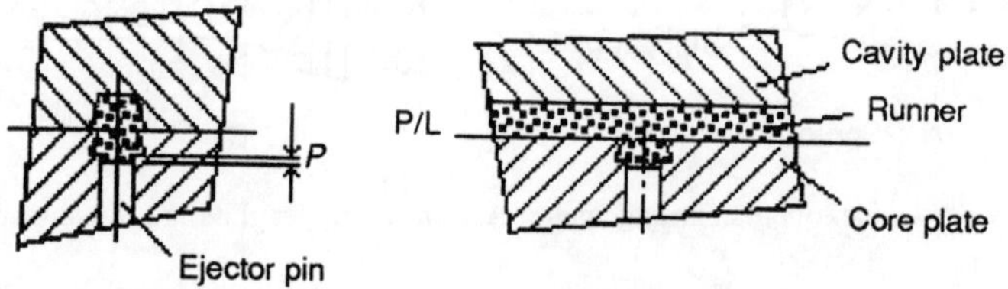

Figure 10.50 Suckers in a two-plate mold. (The runner is in the cavity plate.)

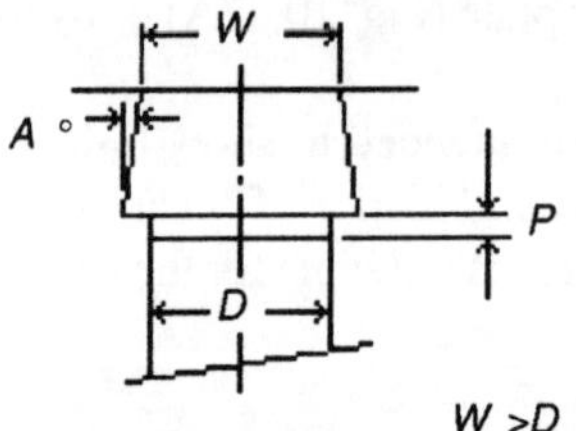

Figure 10.51 Typical detail of a sucker.

As for the variables in Fig. 10.51, D is a standard ejector pin size, preferably not less than 4 mm. W must be larger than D, to let the ejector pass. A, the reentrant angle of the sucker, depends on several factors, as discussed earlier.

4. Pulling the runner with sucker pins (three-plate molds, and two-plate molds with stripper plates): As the mold opens, the runner stays on the stripper plate. The stripper plate then strips the runner off the sucker pins. It is important that the runner not remain with the stripper plate after ejection. Therefore, the sucker pin must enter the runner by the amount P (Fig. 10.52). The mold parts are toleranced so that P will be between 0.0 and 0.2 mm.

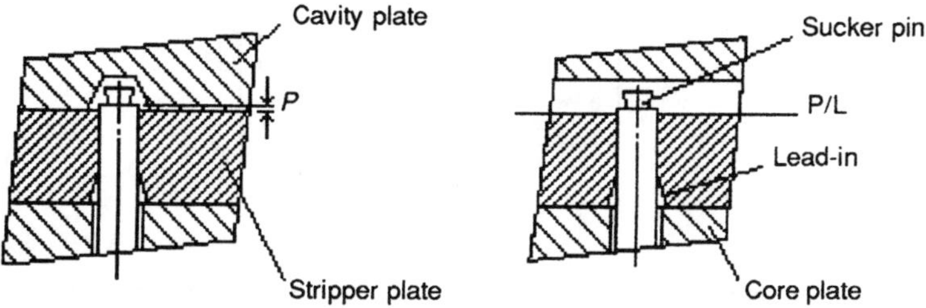

Figure 10.52 Cross sections of a sucker pin for pulling the runner. (The runner must be in the cavity plate.)

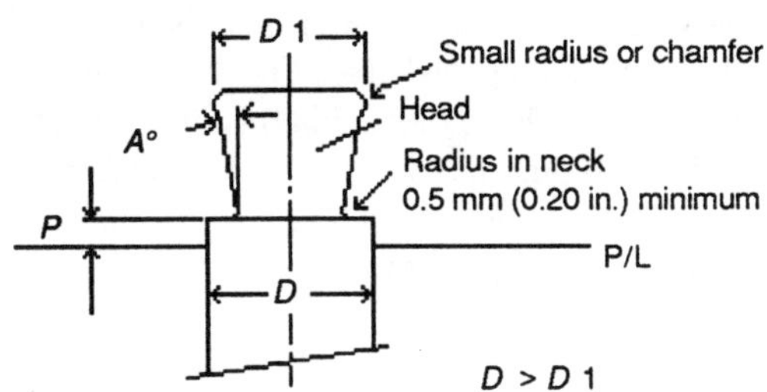

Figure 10.53 Schematic drawing of sucker pin.

The stripper plate is usually hardened steel; if prehardened steel is used, hardened bushings are recommended around the sucker pins. Occasionally, even for hardened stripper plates, commercial, very hard drill bushings are used to minimize wear. The bushings must be held by shoulders, screws, or a retainer plate to stay in place. It is not good enough to press them into the plates.

The diameter *D* of the sucker pin should be as large as possible, but should be selected from standard pin sizes (e.g., 4, 5, 6, 8 mm, or similar standard inch sizes). Note that the *pulling force* of the sucker depends on the circumference of the head, and on the angle *A*. This is especially important with three-plate molds, where the resistance against pulling a drop (the runner extension toward the gate) may be considerable, especially where a very small draft angle is selected to keep the mass of the plastic in the drop small.

If the pin diameter is large, the angle can be smaller and there is then less risk of breaking the plastic when the stripper moves forward to remove the runner. Also, the pin head is less likely to break off if the diameter of the base is larger, as is the case with a smaller angle.

Ideally, the diameter *D*1 should be only slightly less than *D*, and the angle *A* should be not more than 5° (Fig. 10.53). It is then always possible to increase the pulling force required by increasing the angle, without greatly weakening the sucker pin.

The pin is usually made from a standard ejector or core pin. The radius (chamfer) at the top (Fig. 10.53) is needed to remove the sharp and brittle nitrided surface if an ejector pin was used.

Common problems are that the sucker pin fails to pull the runner, or that the head breaks off. This can be avoided by using the sucker pin shape as suggested above. It is important to provide a good lead-in for the pins in the stripper plate (or bushing), not only for ease of

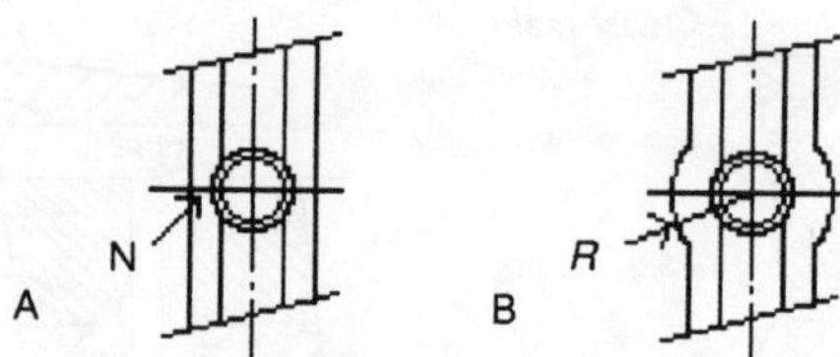

Figure 10.54 Plastic flow around sucker pin heads: A. restricted flow in the runner, B.runner widened to a radius *R* to allow unrestricted flow.

assembly but also in cases where the stroke of the stripper is so large that the pin disengages from the stripper plate at every molding cycle (see Fig. 10.52). As a rule, such stroking out of the plate during operation should be avoided, but this may not always be possible.

As with ejector pins, all sucker pins must be allowed to float freely sideways in the ejector retainer plate to ensure proper alignment without forcing the pins, which would result in rapid wear of pins and/or bushings, or breakage of the pins.

10.4.2 Flow of Plastic Around Sucker Pin Heads

Since it is important to make the sucker pins as large as possible, there is the danger that they may restrict the flow of plastic in the runner at N (Fig. 10.54A). Flow must be carefully laid out and, if necessary, the runner must be widened as indicated schematically, with a radius *R* (Fig. 10.54B).

This is especially important with sucker pins which pull right above the drops in three-plate molds. The clearance around the pin must not be too large, because it must not permit too large a hot mass of plastic, which takes longer to cool down. It is important to keep the runners small, but at the same time to use large sucker pins.

10.4.3 Location and Number of Ejectors, Suckers, or Sucker Pins

Both location and number of ejectors or suckers depend on the plastic used. The designer should keep the number as small as possible, not only to reduce the cost of the hardware and of machining but also to minimize the effect of the many drilled holes in the plates which may be in the way of a good cooling layout.

In general, the stiffer the plastic is (at the moment of ejection), the fewer ejectors are needed; also, the designer then has more freedom to place the ejectors or sucker pins.

Since the runners are usually thicker than the product itself, it is important to provide good cooling to remove the heat from the plastic. It is good practice to have cooling lines just below the runners; however, this makes it then impossible to place ejectors (or sucker pins) right under the runners and may require cooling lines on either side of the runners (Fig. 10.55).

Where cavities are fed from a network of balanced runners, it may be possible to place the ejectors (or sucker pins) under the connecting runners (bridge runners), and the cooling directly under the runners (Fig. 10.56). This may not be practical with soft, flexible plastics, which are often sticky and tend to adhere to the runner surface (Fig. 10.57).

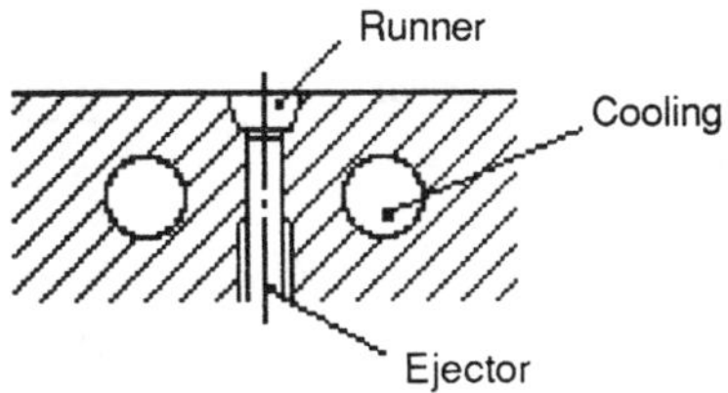

Figure 10.55 Cooling lines located on either side of the runners allow placement of ejectors under the runners.

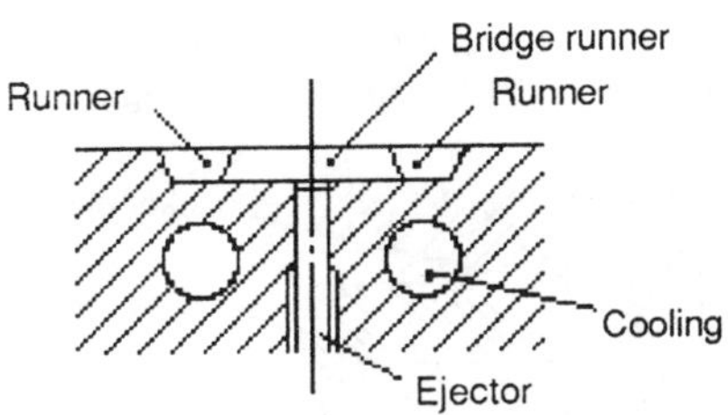

Figure 10.56 Cooling lines may be placed directly under runners in balanced runner systems where ejectors are placed under connecting runners.

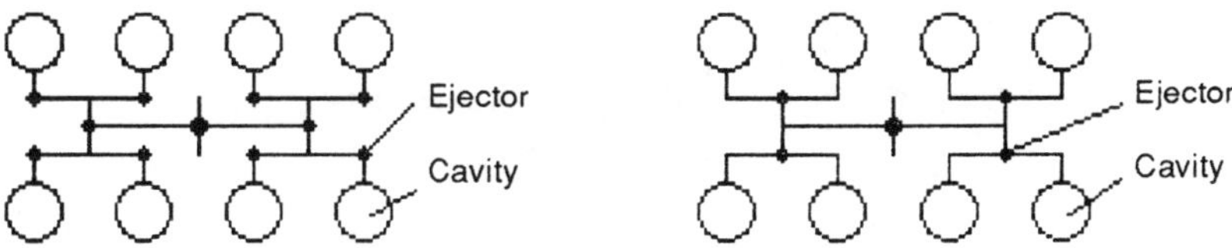

Figure 10.57 The schematic layouts depict on the left a layout for soft plastic (10+1 suckers), and on the right, a possible layout for stiff plastics (4 +1 suckers).

10.4.4 Placement of Suckers Near a Tunnel Gate

In the following examples, suckers using ejectors are illustrated. The same considerations apply to sucker pins.

When molding soft and flexible plastics, as the mold opens, the runner and the drop are held in the core plate by action of the sucker; consequently, the gate shears off. Because the plastic is soft, the runner can flex at the radius *R* so that the drop pulls out of its well. The sucker can be positioned close to the runner extension (distance *D*). Note the importance of the draft angle and a good finish in the extension (Fig. 10.58 left).

If the same layout as for soft plastics was used for stiff plastics, the plastic could break at the sucker location. It is, therefore, necessary to increase the distance *D* to a distance *D*1 (Fig. 10.58 right) to provide sufficient length for the plastic to flex as it pulls out of the extension. This greater dimension *D*1 may require greater cavity spacings and a larger mold shoe.

A larger radius (*R*1) would also help, not only because of the geometry but also because it will create a greater mass, which will keep the plastic there warmer and facilitate flexing;

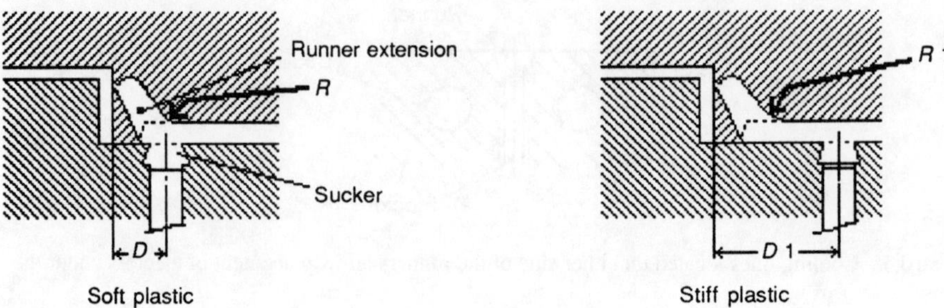

Figure 10.58 Placement of suckers near a tunnel gate differs depending on the flexibility of the plastic: *D* is small for soft plastic (left) and increases to *D*1 for stiffer plastics (right).

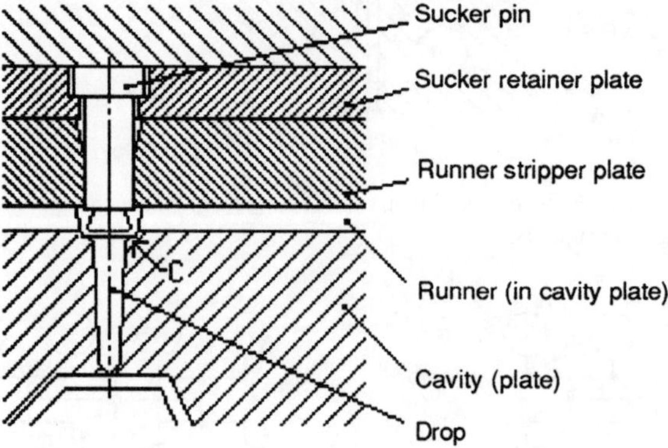

Figure 10.59 Sucker pin for use in a three-plate drop.

but this will also increase the cooling time. To provide a greater radius may be required after the mold test, when the designed distance *D*1 proves to be too short or the radius *R* too small.

10.4.5 Suckers for Three-Plate Drops

Essentially, the same rules apply as for the sucker pins in two-plate runners. The sucker should be as large as possible, and the reentrant angle as small as possible, but suitable for the stiffness of the plastic molded (Fig. 10.59).

The *drop* should be as small as possible, both in length and in diameter, to reduce the mass of plastic. A draft angle of 1–1.5° max. (per side) is recommended. The finish must be *fine draw polish* to reduce the force required to pull the drop out of the well.

The circular runner widening at the drop should be large enough for the plastic to flow to any other drops supplied by the runner, but not larger than necessary to avoid excessive mass of plastic and slow cooling.

The annular clearance *C* between the top of the sucker pin and the entry to the well for the drop must be large enough so as not to restrict the plastic flow toward the gate. For sizes and shapes of gates, refer to Section 10.2.5.

10.5 Cold Runner Molds

This section of the chapter provides more details about cold runner molds. In earlier years, all molds were made this way and were mostly two-plate molds, although occasionally three-plate molds were used. The main advantages of using cold runner molds are their lower cost as compared to hot runner molds, and their ability to produce any size product, even the smallest size.

Today, many molds with large enough production are hot runner molds. Occasionally, a combination of hot and cold runners is used where the product is too small for hot runner application, or for other, special reasons. But even today, the large majority of molds built are cold runner molds when the a hot runner mold would have no economic advantage. This would usually be the case where the number of products required over the life of the mold is so small that the piece cost of the product (which must always include the mold cost per piece) would be disproportionately high when using a hot runner mold.

The foregoing does not mean that there are not occasions where even with low quantities, a hot runner mold would be built, usually for reasons related to the quality and performance of the product or for testing and experimental purposes.

10.5.1 Two-Plate Molds

For more information on gating in two-plate molds, see Sections 10.1, 10.3 and 10.4. For simplicity, ejection, alignment, and cooling are not shown in the following illustrations.

10.5.1.1 Sprue Gates (One Cavity Only)

The cold sprue leads directly from the machine nozzle into the cavity space (Fig. 10.60). After the mold opens, the product and the sprue are ejected; the sprue must be cut off after molding, and the cutoff is usually very unsightly. If good appearance is required, the remaining vestige can be machined off.

With outside center gating (OSCG) (Fig. 10.60A), the cold sprue is short, and even though the necessary draft increases the sprue diameter near the product, the mass of the sprue is relatively small. The diameter of the sprue opening where it contacts the machine nozzle must be at least 1.0 mm larger than the hole in the nozzle to prevent a *hook* and possible failure to pull the sprue out of the sprue bushing, which is often caused by misalignment of the nozzle and the mold (sprue).

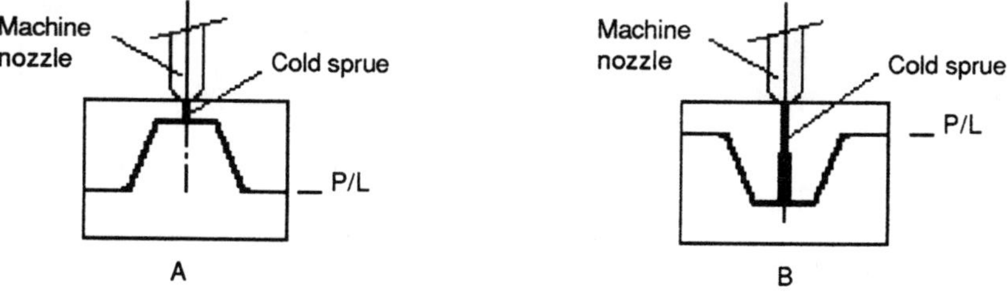

Figure 10.60 Cold sprue gates: A. outside center gated and B. inside center gated.

Since the machine nozzles usually have hole diameters of 3 mm (⅛ in.) or larger, the sprue at the nozzle end should have a minimum diameter of 4 mm, and the diameter of the sprue at the thicker end where it enters the product or the runner is greater, because of the necessary draft in the sprue bushing. Obviously, the longer the sprue, the greater the diameter becomes.

NOTE: Sprue diameters (5⁄32, 7⁄32, 9⁄32 and 11⁄32 at the small end) and draft angles (taper) of ½ inch per foot, included angle) are standardized. They can be found in the various mold maker supply house catalogs. Because the standard sprue bushings are very large, some mold makers design their own small sizes. Today, however, the cold sprue bushings are almost completely replaced with hot sprues.

Typical examples include molds for large articles, such as pails, trays, or large technical articles, where the total number of products required from the mold does not warrant a hot sprue, or where the additional labor to cut the sprue can be ignored. Cold sprues are inexpensive and require little space.

With inside center gating (ISCG), if any gate mark on the outside is undesirable, the sprue must enter through the core (Fig. 10.60B). The sprue then becomes long and very thick near the product, and its mass is relatively large. This method should be avoided if possible. The cooling of the core around the sprue may be very difficult; it is also more difficult to make the ejection mechanism on the same side as the sprue. The use of a hot sprue would avoid the large mass of plastic of the inside center gated cold sprue, and the difficulty of cutting the sprue deep inside the product, but it increases the problem of cooling inside the core.

10.5.1.2 Simple Runner (Two or More Cavities)

To reduce the mass of the sprue, the length of the sprue should be kept to a minimum (Fig. 10.61). Often, by using an extended machine nozzle, the nozzle seat can be brought closer to the P/L and the runners (Fig. 10.62).

If there is enough space, the nozzle can go right into the runner. This has been done frequently with three-plate runners for fully automatic operation so that there is no sprue attached to the runner. For two-plate molds, provided there is enough space between the cavities, it is easier to go to a short sprue length of about 15 mm. The sprue is then about the same size or slightly larger than the runner itself. As the extended nozzle gets longer, it may require a heater band, and therefore more space than the nozzle alone.

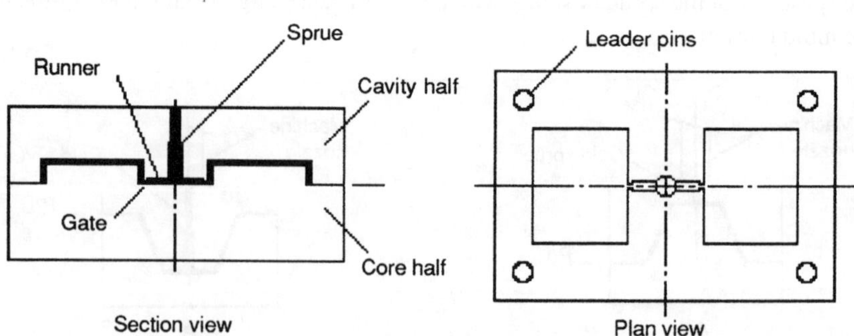

Figure 10.61 Simple runners used in a typical two-plate mold.

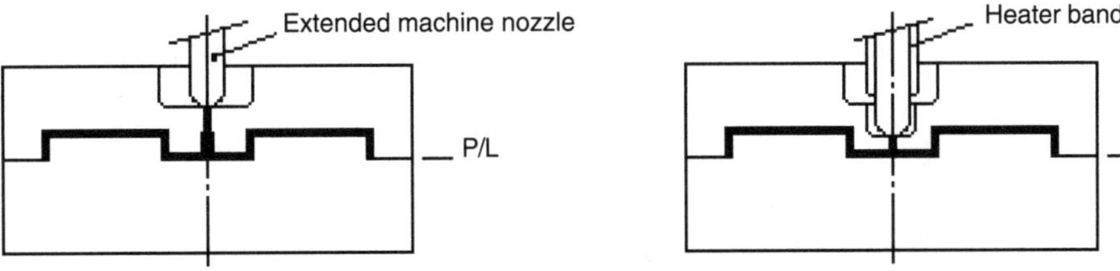

Figure 10.62 Variations on the nozzle: an extended machine nozzle (left) shortens the sprue length, but if extended long enough, the nozzle may require heating.

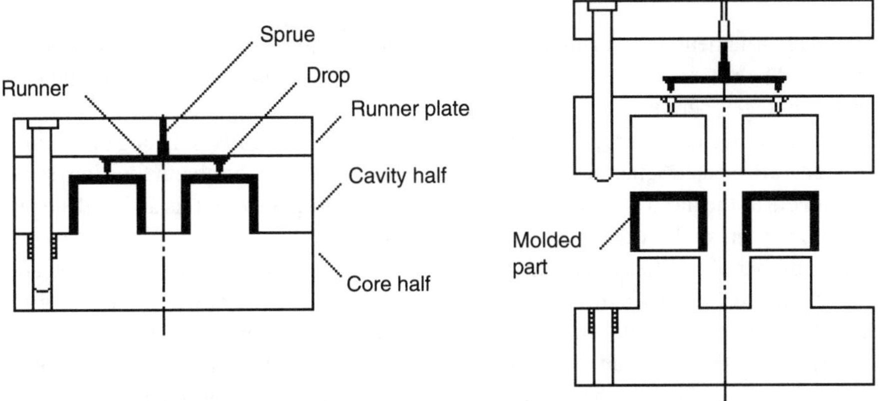

Figure 10.63 Typical three-plate mold with simple runners.

The illustrations do not show an inserted sprue bushing. Normally, a prehardened bushing is used, or the plate with the cavities must be hard enough to withstand the wear caused by the machine nozzle as it comes into contact with the mold.

10.5.2 Three-Plate Molds

A typical three-plate mold is illustrated in Fig. 10.63. A designer may select a three-plate rather than a two-plate mold arrangement

1. if the product must be gated at the top, rather than at the edge or in the side (with a tunnel gate); the reason is mainly better flow of plastic in the cavity space.
2. if the ejection is with *floating stripper rings* that are separated from the stripper plate by a gap (Fig. 10.64). Using the two-plate system, the plastic from the runner would fill this gap where it crosses over it, and would cause the runner to hang up in the mold, also preventing the rings from floating freely. Flashing into the gap would obviously not be permissible.
3. if cosmetic reasons require it. A three-plate gate (pin point gate) can often be much smaller than an edge gate of equivalent flow. Also, a pin point gate can sometimes be easily hidden

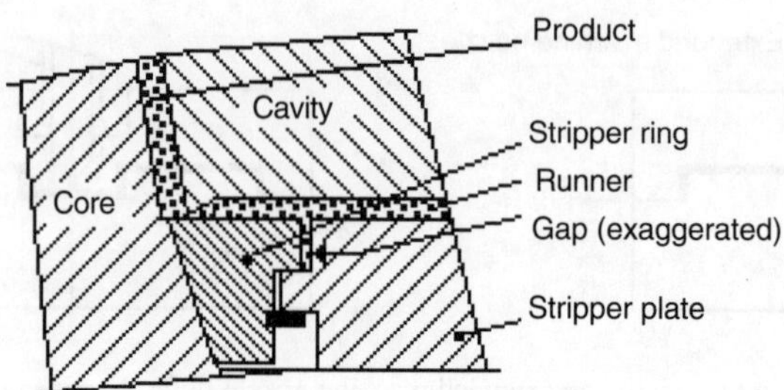

Figure 10.64 A gap separates the stripper ring from the stripper plate.

inside an ornament or lettering in the cavity, and give the impression of a "perfect" gate vestige. The restriction caused by the small gate will increase the velocity of the plastic, increase its temperature, and facilitate filling.

4. if self-degating is required. The three-plate mold is inherently self-degating, and the products are ejected separated from the runners and can easily be channelled into separate containers as the mold opens. Note: Tunnel gating (two-plate mold) is also self-degating, but the runners and products are ejected together in the same plane.
5. if the number of cavities is a critical factor. Because there is no need to provide space for runners in the P/L, the cavities can be spaced closer, and the mold will require less platen area than a two-plate mold with a similar number of cavities. This could become important if the platen size rather than the available clamping force (of the molding machine planned for the job) limits the number of cavities necessary for a certain production requirement.
6. if clamping force would be affected. The projected area of the runners in a two-plate mold *adds* to the projected area of the cavities; therefore, more clamping force is required. In a three-plate mold, the runners are in a different plane than the cavities, and the projected area of the runners does not add to the area required to hold the cavities clamped (see Chapter 15, Stack Molds). However, with very small products, the total projected area of the three-plate runners could be larger than the total projected area of all cavities; in that case, the clamp force required would depend on the projected runner area, not on the projected cavity area.

10.5.3 General Comments About Three-Plate Molds

A three-plate mold is more complicated and, therefore, more expensive to build than a two-plate mold. Except for the immediate gate area and the necessary runner ejection mechanism, a three-plate mold is very similar to a hot runner mold. Provided the cavity spacing is sufficiently large, an existing three-plate mold can quite readily be converted into a hot runner mold. This has been frequently done to improve the productivity of a mold.

10.5.3.1 Dimples

For thin-walled products, dimples opposite the gate are required similar to the dimples for hot runner gates, for the same reasons.

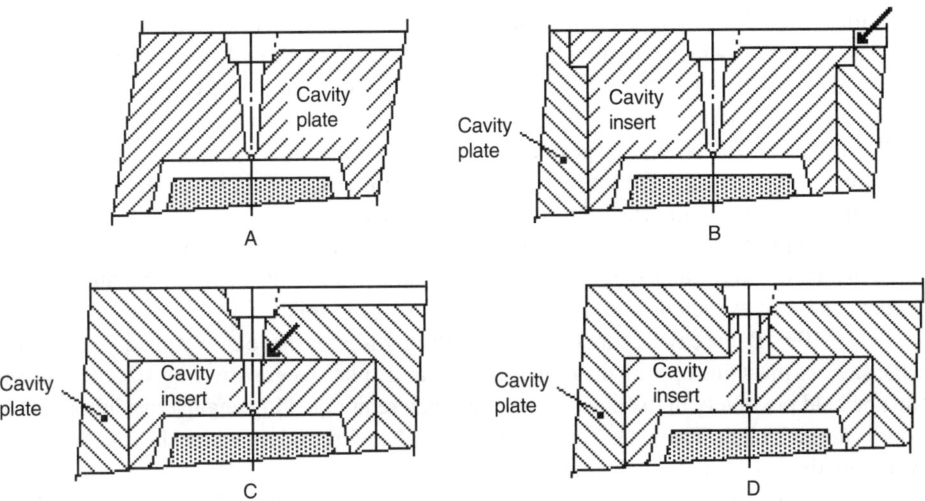

Figure 10.65 Design of drops: A. cavity and runners are cut into the cavity plate, and B. the cavity is an insert in the cavity plate, and runners must align with the plate and the insert, C. the cavity inset is pocketed in the plate, and D. the cavity insert extends into the cavity plate.

10.5.3.2 Runners

When laying out the path of the runners between the cavity plate and the runner ejector plate, the designer must ensure that, in either of these two plates, the runner does not go over openings, joints or screw holes where the plastic could enter and hang up.

The runner layout should be balanced as symmetrically as possible, and the flow path between sprue and cavities (drops) as short as possible. Usually, straight line runners in X and H formation are the best. Occasionally, curved sections may be necessary to circumvent obstacles such as screw holes, but this adds to the cost of machining.

Cross section of the runners should be a trapezoid cut into the cavity plate, and into part of the cavity insert if such an insert goes through the plate, Fig. 10.65. The runner ejector plate is always flat, with only the sucker pins projecting at the drops.

Size (cross sectional area) of main and branch runners can be determined by the using a flow analysis program, or by relying on past experience. In general, runners should be as small as possible to reduce the (wasted) percentage of injection capacity required for molding runners and to reduce the amount of regrinding.

10.5.3.3 Drops

Mass of the drops adds to the mass of the runners; therefore, the drops should be made as small as possible. The mass of drops depends on length, diameter, and draft angle. Drop length is a function of the mold layout and should kept as short as possible. Drop diameter depends on plastic flow characteristics, similar to the runners. Draft angle is a manufacturing problem. The better the finish inside the drop, the smaller can be the angle. The problem is that a poor finish will prevent the plastic from pulling out.

With a small draft angle (0.5° per side or less), the finish must be excellent (draw polish), without tool marks which can lock the plastic in place. Commercial sprues reamers have a draft angle of 1°19' per side (½ in./ft.), which creates a very large sprue; however, it is less sensitive to tool marks.

With small products and a large number of cavities, the runner system may have as much plastic or even more than the total of all products in the shot. This is usually not a design problem; however, it will affect the cost of the product.

In the simplest drop design method (Fig. 10.65A), the cavity and the runners are cut into the cavity plate. This is done if cavities are simple and replacement of cavities because of wear or changes is not expected. The sprue (drop) can be made small.

In Fig. 10.65B, the cavities are inserted into through-bores in the cavity plate. The runners in the cavities must line up with the runners in the cavity plate (see arrow, Fig. 10.65B). Inserts must be held positively in the cavity plate, and they require keying against turning. Press fit alone is not sufficient. There must be no gap and no chamfered edges where the runner crosses over from the cavity plate to the cavity insert (see arrow).

In Fig. 10.65C, the cavities are set into pockets in the cavity plate. There is no need to have right and left cavities. In Fig. 10.65D, the cavity inserts also are set into pockets, but they extend into the cavity plate so that the flat end of the insert represents the bottom of the recess for the sucker. There is no need for right and left cavities. In both Fig. 10.65C and D, the cavities are held in the plate with screws. Press fit alone is not sufficient.

It is preferable to make the sprue in the same part as the cavity itself (Fig. 10.65A, B and D). If this is not practical, the point of transition from the portion of the sprue in the cavity plate to the portion of sprue in the cavity itself must be stepped as illustrated (Fig. 10.65C); the step should be at least 0.05 mm (0.002 in.) to ensure that the plastic will pull out easily.

The decision of which method to select depends usually on the cavity cooling method used and the available space in the cavity plate. The design shown in Fig. 10.65B should be avoided if it requires right and left cavities.

10.5.3.4 Number of Gates per Cavity

As a rule, one gate is sufficient. However, with very large products, it may be necessary to provide more than one gate, if

1. the volume of plastic is so large that one small gate alone would slow down the filling of the cavity. For example, this could be the case in a heavy, large pail where three gates could be used in a triangular location at the bottom of the pail.
2. the products are large and oddly shaped product. Even more than three gates may be needed. The flow length within the cavity could be very great, while the wall thickness is relatively thin, as in large automotive panels. By placing drops at several, equidistant locations, the filling and the filling speed of the cavity can be often much improved.

However, there are two important factors to consider: weld lines and venting.

When using more than one gate in a cavity, the plastic, as it flows from each gate, will meet about halfway between the gates. This will create weld lines. If the plastic arrives too cold at this point, the product could be severely weakened.

This must be considered when deciding on the location of the gates to ensure that the weld line is in an area where it does not matter for appearance and/or strength. Thickening the

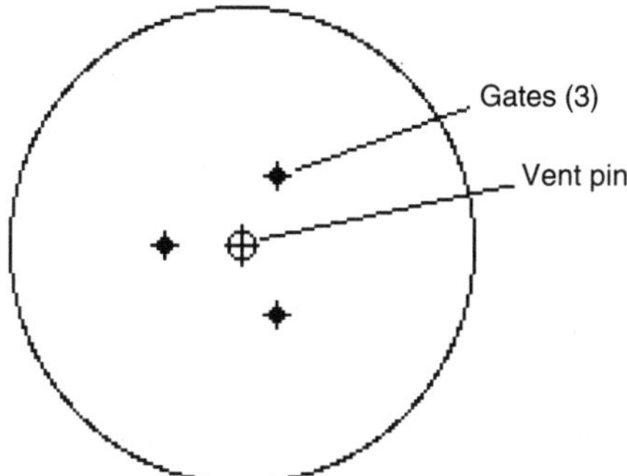

Figure 10.66 A vent pin is located centrally between three gates in the bottom of a pail mold.

product in this area may help so that the plastic has more time to weld properly. Otherwise, the melt temperature must be increased, thereby increasing the molding cycle.

In addition to the usual cavity venting, it is highly recommended to vent the areas where plastic flows meet and trap the air as the plastic advances from different directions, especially if the wall thickness of the product, or ribs in the gate area, will cause the plastic flow from the gates to join rapidly before flowing toward the center. Typically, a vent pin should be located between three gates at the bottom of a pail, as shown below (Fig. 10.66).

Today, large products would usually be made with hot runner molds, rather than with three-plate molds, but the problems of weld lines and venting are the same.

10.6 Hot Runner Molds

Basic understanding of the hot runner (HR) technology will greatly assist in design of hot runner molds. The following describes some of the more important points.

1. *Distribution.* The purpose of the hot runner system is to distribute the hot plastic coming from the machine nozzle to all gates of the mold at virtually the same temperature and pressure as it was at the machine nozzle.
2. *Heat losses.* Because the HR manifold is surrounded and supported by the cooled mold shoe, the heat losses due to radiation and direct contact (convection) must be replaced by the heaters in the HR system. The better the insulation from the surrounding cold steel, the less heat needs to be supplied to the hot runner. (With fast cycles, there may not be any heat required.)
3. *Start-up temperature.* With a cold mold, the plastic within the HR system is stiff and must be brought up as fast as practical to a temperature which permits the plastic to flow in the passages. For this reason, the heaters are of relatively large capacity.

   The higher the capacity (wattage) of the heaters, the sooner the mold can be started up. However, there are practical factors (size of heaters, rating of controls, capacity of power supply, etc.) which limit the size of heaters. The heaters should allow the mold to start up quickly, preferably from cold to operating temperature in about 15–30 minutes. It could be longer in cases where the masses of the HR manifold are very large.

4. *Sealing and heat expansion.* Temperature differences (T) between the HR system (manifold and the other HR components) and the surrounding cooled mold may be, depending on the plastic molded, in the order of $T = 100$–300 °C (or 180–540 °F). Heat expansion must be calculated to be suitable for the temperature of the plastic to be molded and for the anticipated cooling water temperature. The calculated expansion values apply to the following two areas:
   a. *The location of centerlines of HR nozzles and the gates in the cavities.* These locations in the hot (expanded) manifold and in the cold (not expanding) gates in the cavities must align properly.
   b. *Proper sealing against plastic leaks.* In many commercial HR systems, there are flat seats or cylindrical joints between various components. These joints are sealing the high pressure plastic only when the proper operating temperatures are reached. Since this is different for the various makes and models of commercial hot runners, it would go beyond the scope of this book to explore them all. The mold designer must select the right system for his application after studying the various presentations in sales literature describing each system, and make his choice accordingly. It is absolutely necessary to follow each manufacturer's instructions for the selection of the suitable model for an application, and to adhere to the given clearances and fits.
5. *Cleanliness of plastic.* "Dirt" in this context means any contamination in the plastic such as paper, cigarette butts, wood chips, floor sweepings, sand, metal chips, burnt plastic, etc. Dirt can cause serious operating trouble. In hot runner molds, the plastic remains in the runner system, and any dirt which cannot go through the gate will accumulate at the nozzle behind the gate and will eventually plug the flow of plastic. This will then require the mold operator to open the HR system at the nozzle to remove the dirt. This can cause lengthy interruption of production.

   The exclusive use of virgin material would be ideal. However, even though there are (in a hot runner mold) no runner systems to be reground and reused, every bit of material is precious; rejected products are frequently reground and reused, and some molders save old, reused plastic or buy reprocessed plastic, which is cheaper than virgin material. In these circumstances, it is more difficult to control the cleanliness of the plastic, and the molder must use every precaution to ensure that dirt is kept out of the system. Hopper magnets help only to prevent magnetic metals from reaching the extruder, not any of the other dirt.

   Two-stage machines can have a filter between the extruder and the shooting pot that makes it possible to filter out most dirt, provided the filter is frequently cleaned. In RS machines, a fine filter between the extruder and the machine nozzle is not practical because of the severe pressure drop resulting from such a filter, and it is not recommended where high injection pressures and fast filling is required.
6. *Nozzle tip insulation.* In some HR systems, the nozzle tip is separated from the surrounding cold mold by the plastic injected with the first shot. This is quite adequate for most plastics and applications. However, this plastic will gradually degrade (decompose) and can cause visible streaks in the product if some of the decomposed plastic is carried along with the fresh plastic. If this is not acceptable, a plastic (high heat) insulator (Vespel™, etc.) can be inserted to fill the space between nozzle and mold. The assembly drawing must show whether the mold is equipped with such insulator.

7. *Quality of plastic.* Even with virgin materials, there are differences from batch to batch; however, in most cases, the differences are within reasonable limits and should not affect the operation of the hot runner or the quality of the products. As a rule, the larger the lot size, the less the effect of variations will be noted in production.

   With reprocessed and especially with reground material, the mold operation and the product quality is much more difficult to control. Reground material will, in addition to losing some physical properties, lose some of its molding properties, such as ease of flow, every time it passes through the extruder. Therefore, mixing reduces the quality of the virgin material to some extent.

   As long as the proportion of the mixture can be strictly controlled, the temperature of the hot runner system and the pressure settings of the machine can be adjusted to account for the lower quality of the mix. If the ratio of mixture is not closely controlled, frequent machine setting adjustments will be required to avoid excessive scrap.
8. *Color of product.* Precolored virgin material is best but very expensive, and used only in rare cases. Most common is the use of virgin material with properly measured MasterBatch (color concentrates) added and mixed; its use is clean, and it will keep variations in color to an acceptable range. Use of reground colored material will make color matching more difficult.

   Use of powdered colorants, even though cheaper than MasterBatch, is less popular because of the fine color dust that is unavoidable when handling powder pigments, and because the measured amounts of pigments are so small and difficult to measure accurately. There are other methods of adding colors, such as liquids, etc.
9. *Method of gating.* There are two basically different systems of gating of HR molds (open and valved gates), which are discussed in Section 10.2.
10. *Power requirement.* HR systems can be quite sensitive to severe fluctuations in power supply voltage. The nozzle and manifold heaters are usually rated at 240 volts. While the manifold is always controlled by thermocouples and will compensate for voltage variations, in many HR molds the nozzle heaters are regulated by percentage timers (not by thermocouples), and the heat output will be directly affected by voltage variations.

    For example, a reduction of the voltage by only 10% will affect the output (in Watt) by about 20%, which will reduce the nozzle temperatures by the same percentage and, therefore, require adjustment. In severe cases where the stability of the power supply is known to be unreliable (frequent fluctuations), it may be advisable to provide thermocouple controls for the nozzles, also. Or, the customer may install an automatic voltage stabilizer rated large enough for the power supply to the nozzle heaters.

    But even with thermocouple-controlled nozzles, extreme low voltage may make start-up more difficult (slower or even impossible) if the nozzles, even with full power, will not reach the required temperature. Excessive high voltage can shorten the life by causing premature burn out of the heaters.

    Thermocouples must be properly fastened to the part to be controlled. False readings due to wrong or poor connections in the thermocouple circuit, or poor contact with the controlled mass will result in wrong temperatures of the system and either make molding impossible or create bad molding conditions, such as burning, flashing, drooling, etc. It may, with some plastics, create dangerous conditions such as rapid decomposing and release of poisonous gases into the air.

Mold plates surrounding the HR manifold must be properly cooled. Failure to cool these plates will affect the planned heat expansion of the hot runner components and may result in plastic leaks and/or in permanent damage to some of the HR components. The proper operating temperature window should be shown on a nameplate on the mold, as well as on the assembly drawing.

Nozzle heaters may be controlled in two ways:

a. Manual controls (percentage timer). The controller is manually set at number from 0 to 100, which indicates the percentage of time the power is ON (e.g., 60 means that the power is connected to the heater 60% of the time). Once the mold runs on cycle, the setting is usually reduced to the lowest setting achievable without freezing off any of the gates.
b. Automatic controls. Nozzle heaters are controlled by thermocouples and automatic temperature controllers. Automatic nozzle controls may be required for some engineering plastics or may be requested by the customer plagued with an unreliable voltage stability of the power supply. Manifold heaters are always controlled with automatic controls (i.e., thermocouples and automatic controllers).

References

1. Rubin, I.I. (1972). *Injection Molding: Theory and Practice*, John Wiley & Sons, New York Toronto.
2. Pye, R.W.G. (1989). *Injection Mold Design: Manual for the Thermoplastics Industry*, 4th ed., Halsted Inc., New York.

11 Venting

11.1 Background and Theory

The need for and the importance of venting has been recognized ever since materials have been poured into forms. Bells, statues, cannons, etc., could never have been cast successfully without venting of the cavity space by the piercing of the sand or clay forms before the bronze or iron was poured. In view of this, it is hard to understand why it took so long for the plastic industry to provide venting for molds, not as an after-thought but as a basic engineering requirement.

The theory of venting is simple: The air inside the cavity space must be allowed to escape so that the inrushing plastic can fill the whole space. Air, or any gases, follow a simple law which states that

$$\text{Pressure} \times \text{Volume} = \text{Constant} \tag{11.1}$$

or

$$p_n \, v_n = C. \tag{11.1a}$$

For example, in Fig. 11.1 (left) a cavity space $v_1 = 100$ cm³ at atmospheric pressure $p_1 = 1$ atm, the air inside has no route to escape. This air is pushed by the inrushing plastic into a corner; if it is thereby compressed to a volume $v_2 = 1$ cm³, the air pressure p_2 will now be 100 atm, which will offer strong resistance to the plastic to fill this corner. The smaller this compressed volume, the greater the pressure and resistance to filling the last little spaces in the cavity. This is visible in poorly filled ribs or recesses, in unfilled sharp corners, or in poor definition of details in the finished product.

As the air is compressed, its heat content is now concentrated in a small volume, resulting in a large temperature increase. The temperature can reach several hundred degrees Celsius and cause the leading edge of the inrushing plastic to burn.

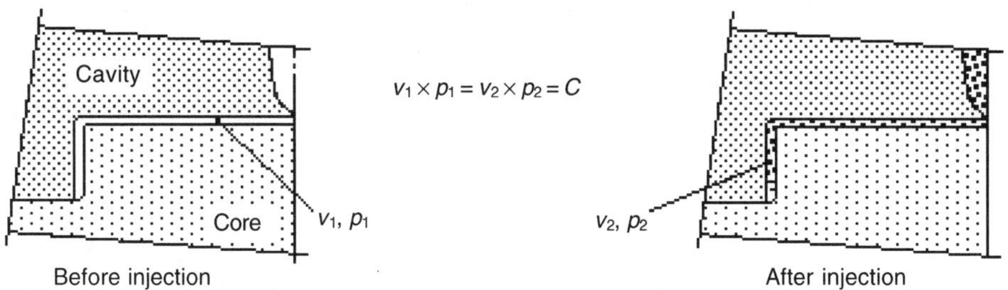

Figure 11.1 A mold before injection with air in the cavity space (left) and after injection (right), when the compressed air prevents plastic from filling the mold.

In old mold building methods, these poorly filled areas and burned edges of the product were used to indicate where the cavity required venting, after the mold was tried out. In a well-engineered mold, however, this need should be recognized *before* the mold is built, and adequate venting must be designed into it from the start. Even though the experienced designer anticipates most areas where air could be trapped, some venting may still have to be added after testing.

Vents must always be made to dimensions that can be recorded and, if damaged, repeated by machining. An earlier, still quite common practice in the industry is to provide vents with hand tools by eye or by feel; this must be strongly discouraged.

Venting in molds is not the same as venting of an extruder barrel, which provides escape for the gases and humidity extracted during heating and working of the plastic within the extruder barrel. It also must not be confused with the venting or "breathing" of compression molds for thermosets; these molds are opened about 1–2 mm for a second or so about 15 seconds after the first clamping to allow the escape of steam and gases formed as the plastic is heated by the contact with the hot mold before curing begins.

11.2 Definitions and Rules

The designer must be familiar with the terminology in venting systems:

Vent: The clearance (depth) between two surfaces through which the air can escape.
Land: Distance air travels through the vent.
Vent groove: A collector close to the cavity space into which the vents lead.
Vent channel: A channel connecting the vent groove to the outside.
Natural vents: A vent created by the clearance between mold parts (e.g., between an ejector pin and the core).

Several rules apply in the design of vents in molds:

1. Provide as large a vent as is practical for the plastic injected and for the injection pressures anticipated to create minimum resistance for the air to escape, but not large enough for the plastic to enter the gap and flashing.
2. Place a vent at the end of all plastic flow paths anticipated; additional vents (e.g., at cavity bottoms and runner vents) are also useful. If the flow originates at two or more locations, and/or if the plastic flow from a gate splits and rejoins to form a weld, the anticipated location of the point of air entrapment must be vented.
3. Cleaning of vents is important. Many plastics tend to leave minute residues on the molding surface which, over time, tend to plug the vents. Parting line vents, which open at every cycle, can be easily wiped clean from time to time to remove any residue or flashing. Moving vents, such as are provided by ejector pins will usually self-clean because of their motion. Sometimes they can also be wiped clean in the open mold if necessary.

   Stationary vents, usually vent pins, or vent slots between an insert and a base (cavity or core), tend to "gum up" after a time and require removal from the mold for cleaning. It is, therefore, important to either not use stationary vents or to design the mold so that these vents can be easily accessed without a need for major interruptions for maintenance.

It should be noted that even stationary vents can be self-cleaning if the vent is short (short land). The pressure (and flow) of the air escaping from the mold, particularly with rapid injection, will flush out residues left in the vent gap. It is also helpful to make the vent channels and vent grooves large enough so that residues can accumulate there, so less frequent cleaning of the mold is required.

Vents can also be pressurized by connecting them to an outside air pressure system. The compressed air will blow the vents clear every time the mold is open.

Vacuum venting is rarely used. It was used in some cases to facilitate the filling of the cavity. Obviously, if there is no air in the cavity space, the plastic will find no air resistance. But with good venting, the added complication for providing a vacuum system and the seals required to maintain the vacuum can usually be avoided.

11.3 Parting Line Venting

The illustrations below show a round cavity, but the remarks apply to any shape parting line. The best venting is shown in Fig. 11.2B, a *continuous vent*. It guarantees the maximum vent area through which air can escape. The disadvantage is a certain loss of support (contact area) as the mold is closed. *Spot venting* (Fig. 11.2A) may be selected for this reason and, also, in cases where a continuous vent is difficult to manufacture.

The vent groove and vent channel can be located wherever suitable for mold making. The vents are usually ground in that surface which permits easy run-out for a grinding wheel. Figure 11.3 shows examples of continuous vents; the cross section of channels should be at least the same as the cross section of the groove to facilitate the escape of air.

Spot vents can be connected directly to a vent channel leading to the outside (Fig. 11.4A). They can also lead to a vent groove, which collects the air from several vents; one or several vent channels lets the air escape to the outside (Fig. 11.4B).

The shape of vent channels and grooves should be trapezoidal or rounded to prevent plastic from catching within, should the mold flash during start-up. The trapezoidal shape has the advantage that even repeated grinding of the parting line will affect the cross section of

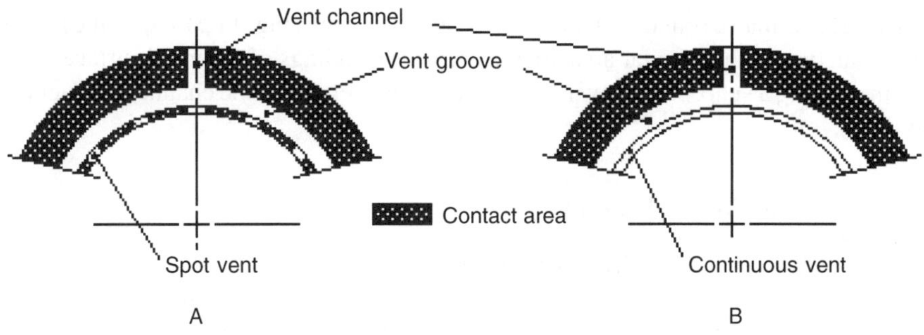

Figure 11.2 Common vents: A. spot vents spaced along the parting line, and B. a continuous vent along the parting line.

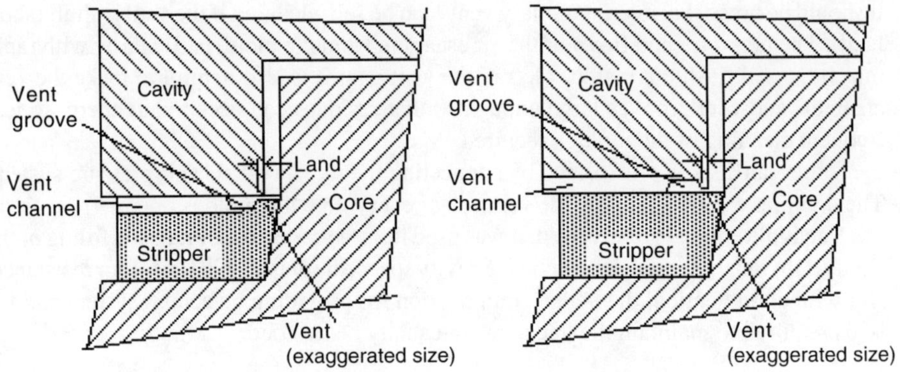

Figure 11.3 Examples of continuous vents.

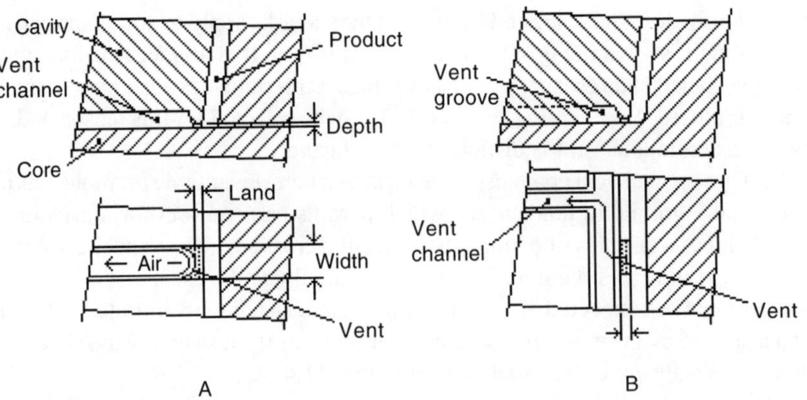

Figure 11.4 Examples of spot vents: A. a spot vent connected directly to a vent channel, and B. a spot vent is connected to a vent groove, which leads to a vent channel.

the channel less than a rounded shape. The schematic below (Fig. 11.5) shows the relatively larger reduction of the circular area after successive grindings of the mold surface.

If machining a vent channel over a taper lock, however, the round channel is easier to mill (Fig. 11.6).

11.3.1 Vent Channels and Grooves

The size of vent channels and grooves depends on the volume of plastic entering the cavity and the speed of injection. With extremely fast injection, as large a size channel as space permits must be selected. However, the vent groove must not be too deep (*DG*), or too close to the cavity (*L*) to avoid weakening the steel; also, the width (*W*) must not take away too much of the contact area (Fig. 11.7).

Figure 11.5 Cross sections of a trapezoidal (left) and rounded (right) vent channel show the effect of repeated grindings of the mold surface on area (exaggerated).

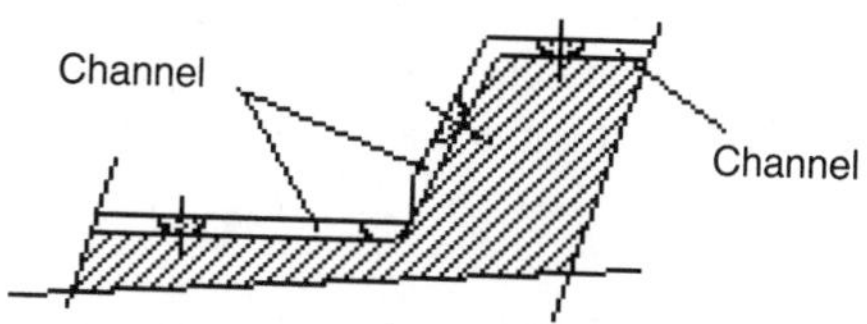

Figure 11.6 A rounded vent channel over a taper lock.

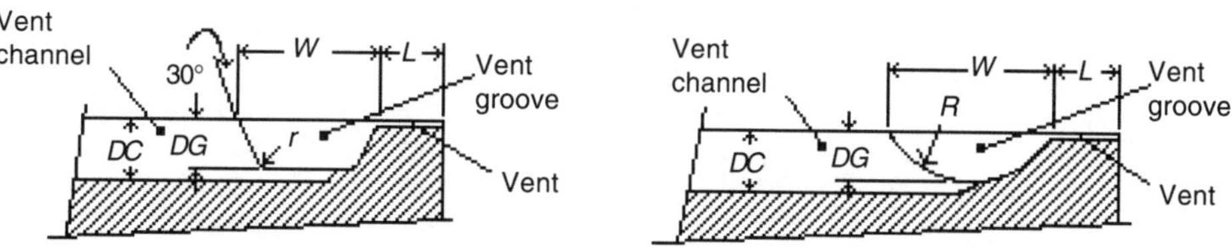

Figure 11.7 Cross sections show depth, width, and closeness to cavity of trapezoidal (left) and rounded (right) vent grooves (*DC* = depth of channel, *DG* = depth of groove, *W* = width of groove, *L* = length of groove land, *r* = radius at bottom of trapezoidal vent groove or vent channel, and *R* = radius of vent groove).

There is no formula or computer program in existence which takes all these conditions into account, and the designer is left to decide, using common sense, which of the suggested standard sizes to select. Table 11.1 lists suggested dimensions.

Channel cross section should be at least as much as the total of cross section of the vent leading into that portion of the groove from which the channel leads to the outside. For large cavities (large volume of air to be vented), the number of channels may have to be increased to provide a satisfactory total cross section. This applies also if the vent channel is created by a chamfer or by a drilled hole. If the channel is deeper than the groove, it should run out into the groove as shown in Fig. 11.7.

Dimensions for vents should fall in the ranges below:

1. Depth: 0.01–0.02 mm, depending on plastic injected,
2. Land: 1.0, 1.5, or 2.0 mm, depending on size of cavity, and
3. Width (spot vents): minimum 3 mm, preferably 5–6 mm (where possible, continuous venting should be selected). Where the product does not permit such wide vents, narrower (1–2 mm) vents are acceptable.

There is nothing wrong with selecting even larger channels if they can be accommodated, because larger tools can be used and the channels are easier to machine.

Example: Select vent dimensions for a round cavity with a 200-mm diameter, with fast injection.

Select: Vent depth: 0.01 mm.
Total vent area: 200 mm × π × 0.01 mm = 6.28 mm²
Groove: W = 5.00, DG = 0.8 mm, area = 3.64 mm²
Channel: W = 5.00, DC = 1.0 mm, area = 3.40 mm²
Number of channels required: minimum 2 (2 × 3.40 = 6.80 > 6.28)
[NOTE: It is better to have 4 or even 8 channels equally spaced, particularly with very fast injection (Fig. 11.8).]

Table 11.1 Suggested Dimensions for Land (*L*), Vent Grooves, and Channels

Vent grooves	*L* (mm)	*W* (mm)	*R* (mm)	*r* (mm)	*DG* (mm)	Area (mm²)
Round	1.00	2.00	1.50	—	0.40	0.51
	1.50	3.60	2.50	—	0.80	2.07
	1.50	5.00	3.70	—	1.00	3.40
Trapezoidal	1.00	2.00	—	0.2	0.40	0.70
	1.50	3.60	—	0.3	0.80	2.50
	1.50	5.00	—	0.3	0.80	3.64
	2.00	6.00	—	0.3	1.00	4.55

Vent channels	*W* (mm)	*R* (mm)	*r* (mm)	*DG* (mm)	Area (mm²)
Round	2.00	1.50	—	0.40	0.51
	3.60	2.50	—	0.80	2.07
	5.00	3.70	—	1.00	3.40
	6.00	5.00	—	1.00	4.14
	8.00	6.10	—	1.50	8.22
	10.00	9.00	—	1.50	10.56
	10.00	7.30	—	2.00	13.38
Trapezoidal	2.00	—	0.2	0.60	1.05
	3.60	—	0.3	1.00	3.13
	6.00	—	0.3	1.00	4.55
	8.00	—	0.3	1.50	10.73
	10.00	—	0.3	2.00	17.60

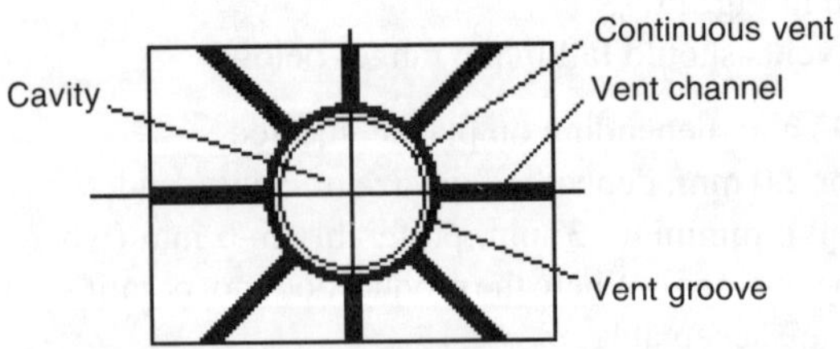

Figure 11.8 Suggested dimensions for a continuous vent, groove, and eight channels are shown (relative sizes are exaggerated).

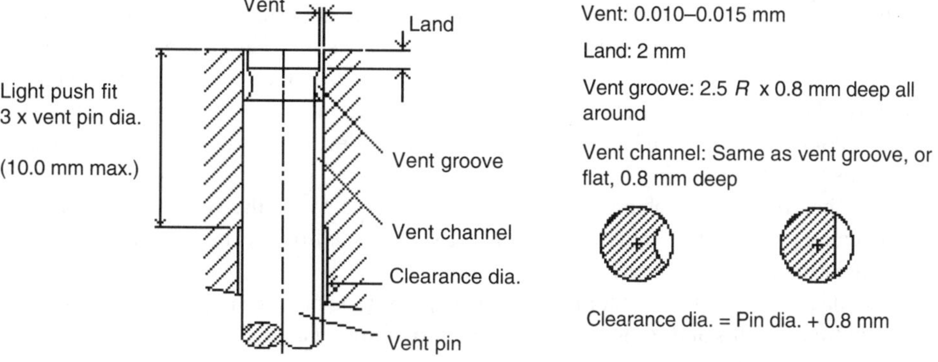

Figure 11.9 Typical cross section and dimensions for vent pin and land.

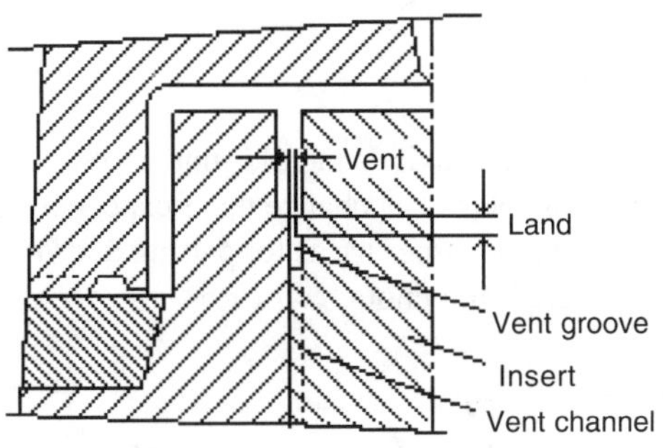

Figure 11.10 Insert vent for a rib.

11.4 Vent Pins and Insert Venting

11.4.1 Vent Pins

The standard design for vent pins has a light push fit (clearance from 0.000 to 0.005 mm on diameter) in the portion of the bore below the vent groove to ensure that the vent is concentric (evenly spaced) with the wall of the bore. Figure 11.9 shows the dimensions for the vent and land.

11.4.2 Insert Venting

This type of venting is absolutely necessary to ensure the filling of any deep (plastic) rib created in the insert (Fig. 11.10). With ribs produced from the solid, whenever the depth of the rib is more than 1.5 times its width, sufficient number and size of ejector pins or vent pins at the bottom of the rib must be provided to prevent trapping of air.

With regard to dimensions, if feasible the vent should be over the whole width of the insert to permit maximum air flow. The vent groove and vent channel dimensions should be as large as possible and consistent with the size of the insert. They are usually machined in the insert but could also be in the surrounding part (e.g., the cavity or core). The vent channel can be one or more drilled holes or can be created by a large chamfer in the insert as long as the cross section of the channel is large enough for the expected flow of air. In all cases, the vent channel must be connected to the outside air or to a pressure air line. For actual vent dimensions, see above discussion for parting line vents.

The shown vent dimensions (Fig. 11.10) are suitable where the vent is in line with the approaching plastic flow. Vents at right angles to the plastic flow can be larger than the above indicated dimensions without danger of flashing.

For cleaning of vents in vent pins and inserts, it is good practice to provide a tapped hole in the bottom of pins and inserts to facilitate removal for cleaning.

11.5 Miscellaneous Venting

11.5.1 Runner Venting

This practice facilitates filling of the cavities in cold runner systems to give some of the air pushed ahead by the inrushing plastic a chance to escape before it reaches the gates (Fig. 11.11). The vents should be the same size as for parting line vents, but they may also be narrower (1–2 mm wide) if there is not enough space.

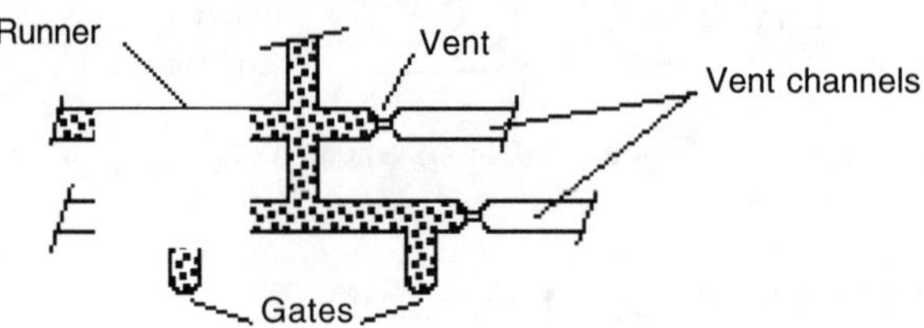

Figure 11.11 Typical runner vents in a cold runner system.

11.5.2 Venting the Bottom of a Cavity

The purpose of venting in the bottom of a cavity is to vent some of the air ahead of the inrushing plastic stream, similar to the venting of runners. Also, when opening the mold, the vent breaks the vacuum in the cavity which may otherwise keep (suck) the product into the cavity and prevent proper ejection. Often, such vents are connected to a compressed air line to ensure release of the product.

In Fig. 11.12 left, the vent is in line with the plastic flow; in Fig. 11.12 right, it is at right angles and, therefore, could be larger. The hole must be large enough not to restrict rapid venting.

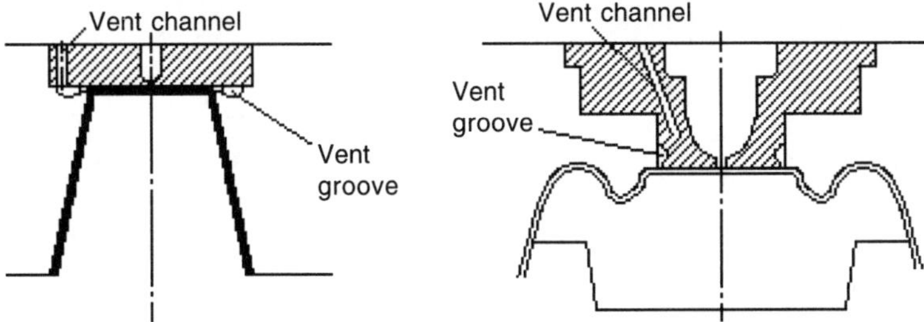

Figure 11.12 Venting in the bottom of a cavity in line with the plastic flow (left) and at right angles to the plastic flow (right).

12 Ejection

12.1 General Comments and Definitions

After the product has cooled in the cavity and the mold has opened, the product must be ejected from the mold. Not all molds, however, have mechanisms for ejection. Where a mold is used only to produce a few prototypes of a product, the product can often be removed manually to save the expense of an ejection mechanism. Also, removal of some large or odd-shaped products from the molding area is often handled by an operator or by a robot equipped with suction cups, and an ejection mechanism may not be needed at all, even in a production mold. Air blast may be directed at the core or the open end of the product to facilitate removal.

Some definitions and terminology necessary for the designer:

Automatic ejection: Most molds have ejection mechanisms to separate the product from the molding surface, usually from the core. The products are ejected and clear the molding area without need for manual labor. As a rule, all molds should have automatic ejection.

Semi-automatic operation: An operator must open the safety gate after the ejection mechanism has operated to remove the product(s) and re-close the gate before starting the next cycle. In cases where an operator must be there anyway (e.g., for subsequent work on the product), this usually does not add to the cost of the product.

However, different operators, and/or fatigue by the same operator, may vary the cycle time and affect the quality (different residence time of the plastic) and the size (different shrinkage) of the product.

There are two types of automatic ejection: free fall (occasionally called "random") and controlled removal. In free fall, the products fall freely into a box or onto a conveyor belt. For controlled removal, the products are removed from the molding area with the assistance of a robot or take-off mechanism. The advantages of automatic ejection are:

1. Consistent cycle time and, therefore, uniformity of products,
2. absence of operators near machine improves safety of plant operation, and
3. possibility of cycling faster than with operator. There is a limit to how fast an operator can handle a semi-automatic operation, without excessive fatigue. With very short cycles, it may be impossible.

12.2 Basic Requirements for Any Ejection Method

The one basic rule in ejection is: *As the mold opens, the products must stay on the side from which they will be ejected.* Unless the shape of the product (ribs, little draft, etc.) is such that it will stay naturally on that side, features such as undercuts may have to be used to ensure that the product stays where required for ejection.

Injection molding machines have usually an ejection actuator only on the clamp side of the machine. Most molds, therefore, have the ejection mechanism on the clamp (moving platen) side. Exceptions occur:

1. Where ejection must be from the same side as injection, for cosmetic reasons (e.g., inside gating of cups, underside of plates, trays, etc.), and
2. in stack molds, where one half of the mold has the cores on the injection side of the machine.

12.2.1 Ejection Features

A mold with ejector pins is shown in Fig. 12.1. If the mold must be inserted between the machine platens by sliding it into the machine along the platens, knockout pads are used to keep the machine ejectors level with the moving platen surface for ease of mold installation.

A mold with a stripper plate that is actuated from the ejector plate (Fig. 12.2) may be necessary if the ejector plate cannot be directly (Fig. 12.3) reached (e.g., because of unsuitable locations of the machine ejectors).

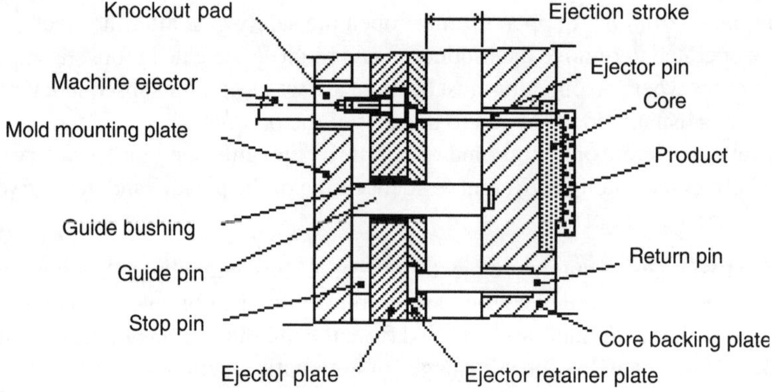

Figure 12.1 Typical mold with ejector pins.

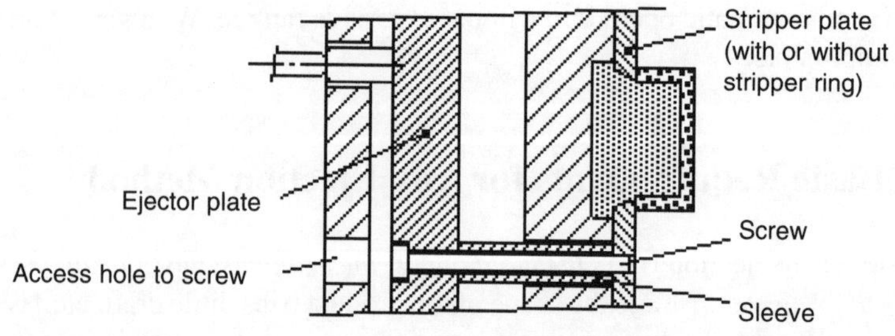

Figure 12.2 A mold with stripper plate actuated from ejector plate.

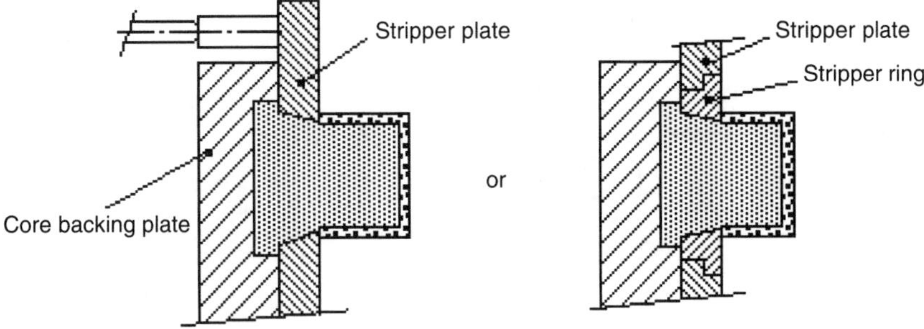

Figure 12.3 A mold with stripper plate and direct actuation.

12.2.2 Length of Machine Ejector

If the mold can be inserted between the machine platens by moving it toward the moving platen, there is no need for knockout pads, and the machine ejectors can reach all the way to contact the ejector or stripper plate (Fig. 12.4). This is less expensive than adding the knockout pads but may cause difficulties in inserting the mold between the platens.

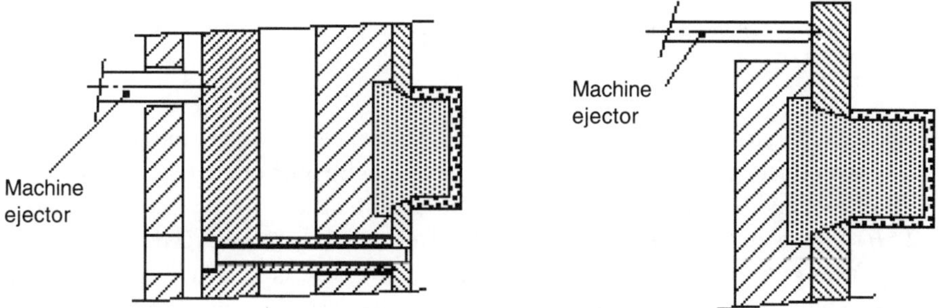

Figure 12.4 Knockout pads are not required where the ejectors can reach the ejector or stripper plate.

12.3 General Ejection Guidelines

These guidelines apply to any method of ejection. Note that they apply equally to single level and stack molds.

12.3.1 Stroke, Clearance, and Product Height

There must always be enough clearance between the cores and the cavities to permit the products to fall freely,without hanging up. With take-offs, the clearance must be also sufficient to permit the products on the take-off plate to clear the molding area, without interference with cavity and core.

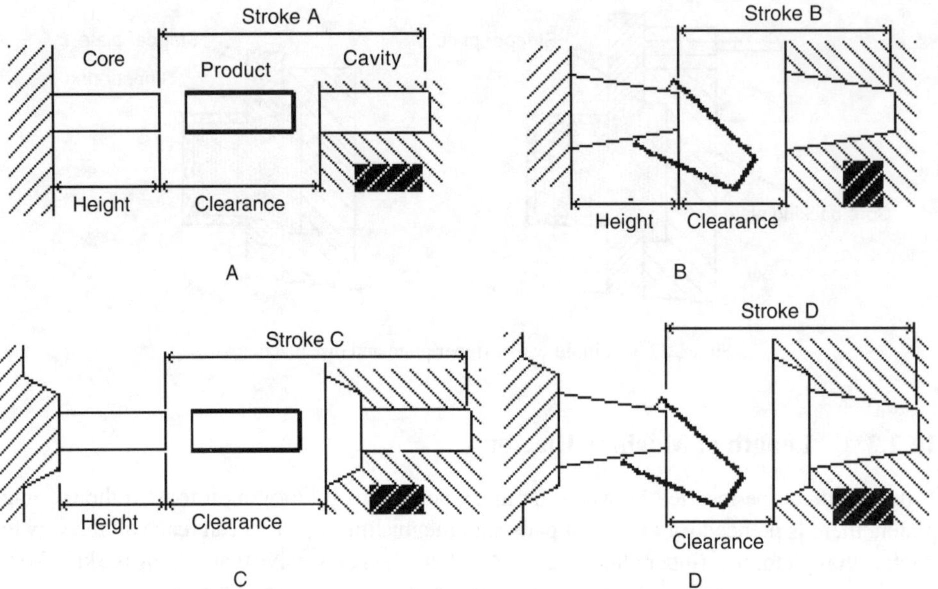

Figure 12.5 Mold stroke must allow clearance for the product to fall freely: A. a product with no draft requires clearance of its entire height, B. a product with much draft requires far less clearance, C. stroke must be increased by the height of the alignment taper, D. stroke must include alignment taper but higher draft of the product allows shorter clearance.

In general, the stroke (S) should be about $S = 2.5H$ for deep products, and somewhat more for shallow products (H = product height). A container with little or no draft must be ejected over all of its height. (Fig. 12.5A). Molds with take-offs must always use the full stroke ($S = 2.5H$) or greater. However, for molds without take-offs, there are exceptions to this rule.

Containers with considerable draft will fall free off the core much sooner and will, therefore, require less clearance (Fig. 12.5B). Because less clearance is required for the product to fall safely, less stroke is required, and the mold will run faster. The use of less than the suggested stroke of $S = 2.5H$ must be carefully considered.

The method of alignment can also affect the required stroke. In comparison with Fig. 12.5A and B, the examples in Fig. 12.5C and D show a method of cavity and core alignment which is sometimes selected. Because the face of the cavity is recessed by the height of the alignment taper, the stroke must be increased by this amount to prevent the ejected product from catching in the tapered recess in the cavity. Note that the clearances are the same in Fig. 12.5A) and C, and in 12.5B and D, respectively.

Excessive stroke usually presents no problems other than extending the cycle time. The unnecessary stroke slows down the cycle time and, therefore, reduces the productivity.

It is conceivable to have extremely small strokes (minimum stroke) for very flat products, especially when using methods where the products are removed from the molding area without the need for large opening strokes.

However, for servicing of the mold, usually during start-up, it is advantageous to keep the stroke large enough, at least 150 mm (6 in.), to allow the mold technician to have access for viewing (with a mirror) and cleaning such areas as the gates. The greater the molding area, the more clearance should be available. Occasionally, the machine can be set up and operated so that the minimum stroke required for ejection is used to operate the mold at maximum speed, but additional stroke can be selected during set-up and servicing.

If the mold uses features such as the patented (Husky) guide rails, the stroke can be as low as the above stipulated $S = 2.5H$, (Fig. 12.6A) or just enough to permit the guide rails (or take off) to enter for product removal, because the products will not tumble after ejection. If the products are ejected free fall, they must not be permitted to jam as they fall between cores and cavities, and the clearance must be greater than the maximum three-dimensional diagonal size of the product (Fig. 12.6B).

In single-level molds, the available opening stroke is usually ample, but with two-level (stack) molds, the total stroke, which is double that of a single-level stroke, must be within the maximum available stroke of the molding machine. However, special conditions will occur. An exception to the above stroke requirements happens in molds where the cavities split in a vertical plane (i.e., where the splits move horizontally).

In Fig. 12.7 (top view), the clearance between core tip and surface of cavity is less than the height of the product, but because the cavity is (partially) split, there is enough clearance between the slides to permit the product to clear the mold in vertical free fall. The clearance between the slides must be sufficiently large for the products to pass without getting scratched on the sharp edges of the slides.

Figure 12.6 Minimum stroke must allow A. just enough clearance for take-off removal, since products cannot tumble, or B. greater clearance than the 3-D diagonal of free falling objects to prevent jamming.

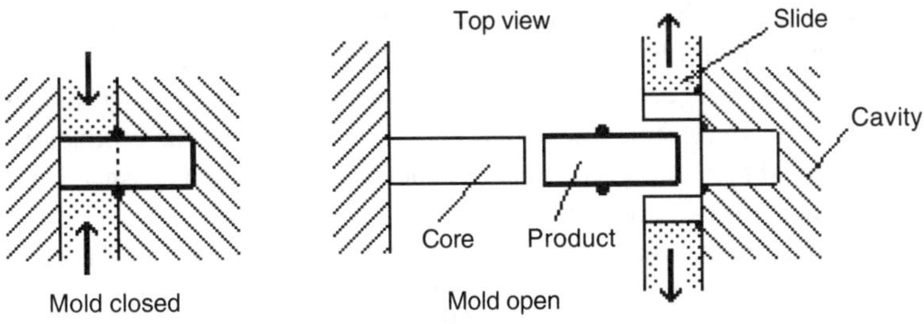

Figure 12.7 Mold cavities split in a vertical plane, allowing a stroke clearance of less than the product height.

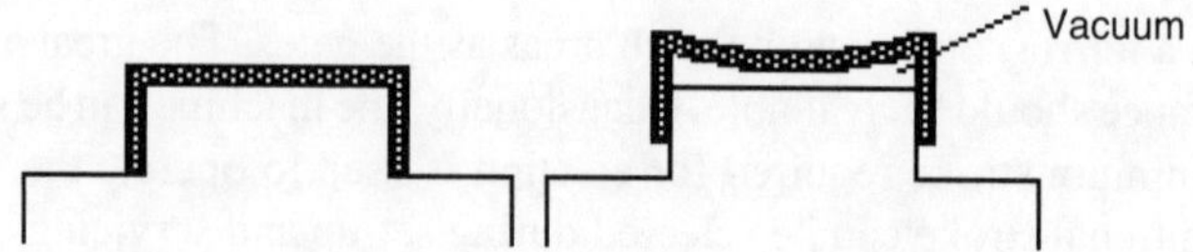

Figure 12.8 A cup-shaped product may be deformed by formation of a vacuum as the product is pushed off the core.

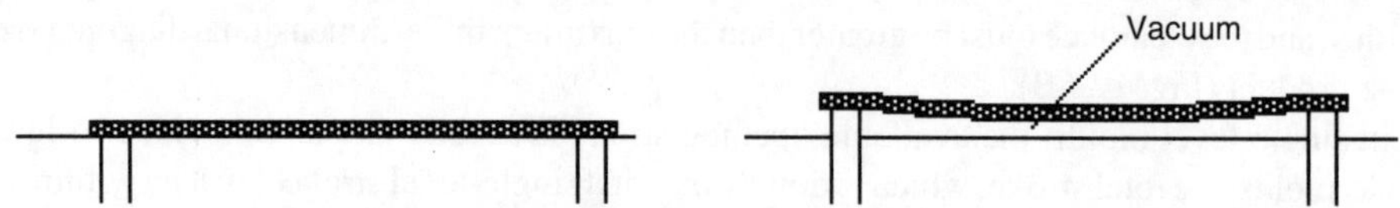

Figure 12.9 A flat product is ejected only near its edge, resulting in deformation.

12.3.2 Venting

Venting is very important with any product where a vacuum may be created during ejection between core and plastic. This applies to both cup-shaped and flat products.

Figure 12.8 shows that when a cup-shaped product is pushed off the core, the vacuum created will deform or crack the product during ejection. This is more severe with complete absence of side draft, but even with draft angles of several degrees, combined with fast ejection, air to break the vacuum may not enter fast enough at the rim and may cause an *implosion*.

If the molding material is flexible (e.g., PE), it may return to its molded shape; if the material is brittle (e.g., PS), the bottom of the product may break. Similar conditions may occur with flat products when ejected only near the edge (Fig. 12.9). (An ejector near the center would act as vent.)

12.3.3 Where to Eject Relative to the Product

For deep products (usually cup-shaped), the general rule is to eject at the points where the product is stiffest. In the left illustration in Fig. 12.10, it would be possible to eject the product even before it is fully cooled down and thereby reduce the cycle time.

The right illustration in Fig. 12.10 is incorrect, but it is surprising how many molds are built this way. There is no problem with venting, but because the plastic is relatively thin at the points where the ejectors contact the product, they often pierce it rather than eject it. To make the mold run, the cycle time must be increased to ensure that the product is cold (and strong enough) to prevent piercing. Also, the product will toe-in as it is pushed from the core, which makes the ejection more difficult.

For a product with internal, deep ribs, the same rule applies as above. The ejection force must be applied at the lowest point (Fig. 12.11), where the product is stiffest. This is usually done with ejector pins. If the rib is circular, it is best to use a sleeve ejector, of a standard size if possible, or a specially made sleeve. There is usually no need for ejectors at shallow ribs where the height of the rib is less than 1.5 times its thickness, but venting must be considered to prevent implosion.

For pin ejection of shallow products, with or without a rim, the following rules apply. Distribute the ejectors to ensure that the product is lifted without cocking. If there are ribs or bosses, as a rule, ejectors must be placed at their lowest point. Ejector pins and sleeves are "natural" vents (i.e., because of their operating clearances (fit), air can pass through). It is important that the land of any ejectors and sleeves be short enough not to create too much resistance for the air to pass through.

As a rule, the land should not be longer than 2*D* (*D* is the diameter of the pin or sleeve) (Fig. 12.12). Since the hole in the core is usually jig ground to provide accurate clearance, the reduction of the length of hole also saves time in manufacture.

Ejector pins and sleeves leave round marks called *witness lines*, on the product. Since ejectors are generally inside or on the underside of the product, such marks are not considered objectionable. In some instances, it may be necessary to discuss the location of the ejectors on flat surfaces with the customer to avoid the embarrassment of having ejectors in the wrong visual location.

Ejection of bosses depends on the shape and solidity of the boss. For solid bosses (round or any other shape), if the boss is relatively shallow (i.e., the depth is equal to or less than the diameter or width) and there is good draft angle (at least 5° per side), there is usually no need

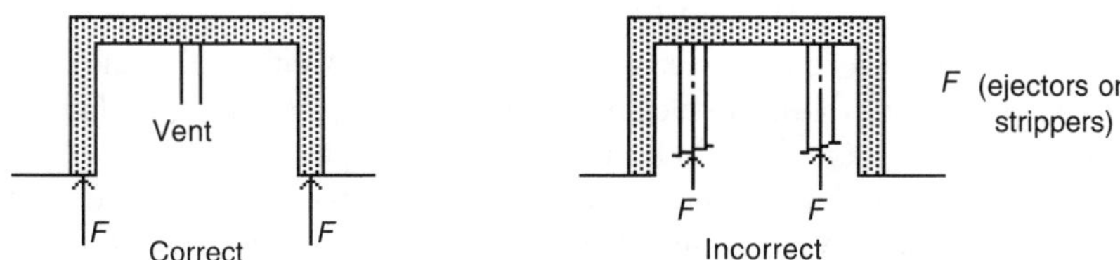

Figure 12.10 Ejection points are shown at the correct (left) or stiffest point of the product and at incorrect (right) points where the product is thinner.

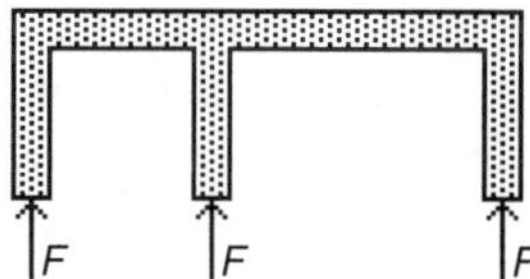

Figure 12.11 Ejector pins are applied to the lowest point of the ribs.

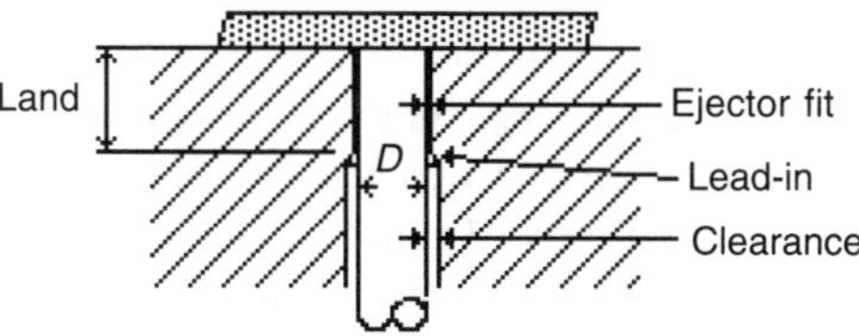

Figure 12.12 The land should be no longer than 2*D*.

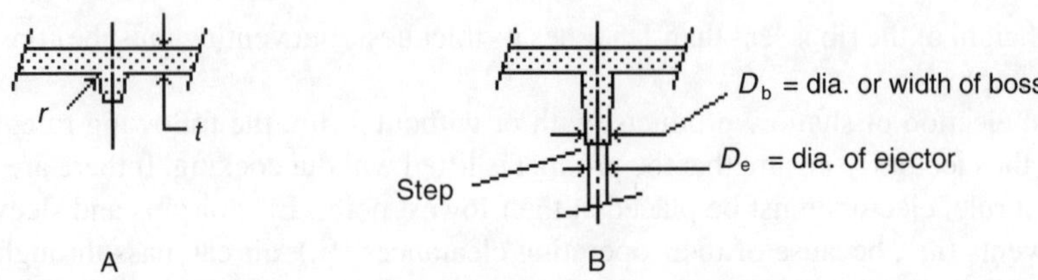

Figure 12.13 Ejectors under bosses: A. a shallow boss with good draft or small depth to diameter ratio requires no ejector, B. a deep boss requires an ejector not only for ejection but also for venting.

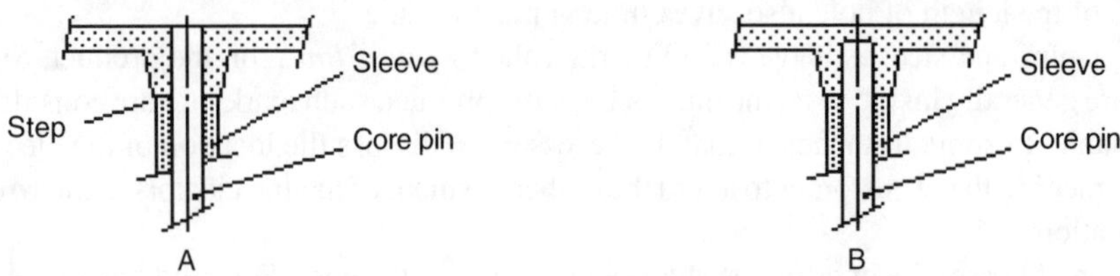

Figure 12.14 Tubular bosses: A. hole passes through the boss, and B. blind hole in boss.

for an ejector under the boss (Fig. 12.13A). If the ratio of depth to diameter is larger, or whenever the draft angle is small, an ejector will be required, not only to eject the product but also to ensure good filling of the product, by providing venting at a low spot (Fig. 12.13B).

The diameter of the ejector should be a standard ejector pin size and should always be somewhat smaller than the bottom diameter (or width) of the boss to ensure that the pin will not scratch the wall when advancing during ejection. The difference per side (step) should be at least 0.005 in. (0.12 mm). This step simplifies manufacture; also, it permits the use of oversize ejectors if required for repairs later on.

A problem with solid bosses is that the plastic forms a rather large volume at the base of the boss, which will create sink marks in the product. A good product designer will avoid solid bosses with a diameter or width larger than the wall thickness *t* of the product and design a tubular boss instead, to remove the bulk of material, which not only reduces the danger of sinks but also improves the cooling in this area.

A radius *r* at the point where the boss joins the wall, while advantageous for strength of the boss, increases the already larger-than-wanted volume at this point. If the radius specified on the product drawing is too large, permission should be requested to reduce it.

Sink marks can often be avoided by changing the molding parameters, but usually at the cost of slower cycle speed.

There are two kinds of tubular bosses: one with the hole passing through it and the other with a blind hole (Fig. 12.14). The advantage of the hole going right through is that there are no sinks marks on the surface of the product opposite the boss. A blind hole almost to the end of the boss greatly reduces or completely eliminates the likelihood of sink marks and helps provide cooling, thereby reducing the cycle time when compared to a solid boss.

In general, even with the smallest depth, regardless of the draft angles, the boss should be positively ejected. The use of a sleeve ejector is expensive and should be avoided (Fig. 12.15A).

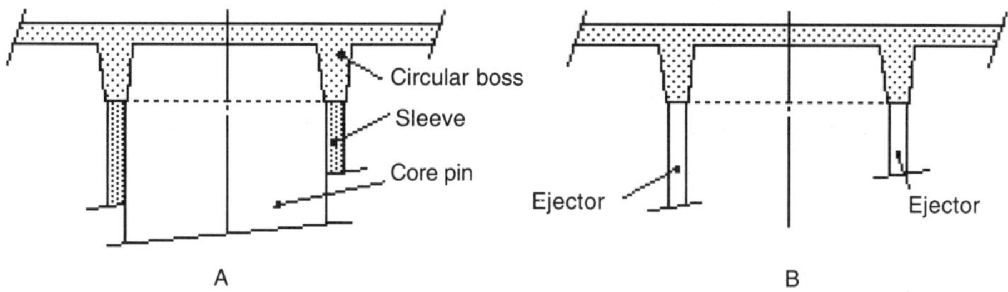

Figure 12.15 Positive ejection of bosses: A. using sleeve ejectors and B. using evenly spaced ejector pins.

If possible, place a few ejector pins evenly spaced at the bottom of a large boss, instead (Fig. 12.15B). However, very small ejector pins also should be avoided, as they require much maintenance.

If possible, standard, commercially available sleeve ejectors should be specified. The manufacture of small sizes requires specialized equipment. Larger sizes can be made in-house, but they are expensive to make and often quite fragile, requiring expensive upkeep of the mold.

12.4 Ejector Pins and Sleeves

Pins and sleeves can be used for any shape of product. The advantage of ejectors is that they are relatively inexpensive and easy to use. They provide good natural, self-cleaning venting in areas where the air would otherwise be trapped and would require vent pins.

The disadvantage of ejector pins and/or sleeves is that the area where the product is pushed is relatively small and the product must be well-cooled and stiff enough so that the forces exerted by the pins or sleeves will not deform or pierce the plastic during ejection. Also, ejector pins and sleeves leave witness marks that may be objectionable. Their location must be carefully considered and, if necessary, discussed with the customer.

Alternatives to ejector pins and sleeves may include strippers or air ejection. Strippers (see Section 12.5) may be plates, rings or bars, and are a common alternative to and preferable to ejector pins and sleeves because the surface where the product is pushed during ejection is relatively large. Also, the vestige left from the stripper is usually less noticeable. However, they should only be used where the product at the parting line has a contour that can be generated with machining.

Air ejection (Section 12.6) is preferable to both ejector pins and strippers because of simplicity of mold construction without the moving masses of ejector plate or stripper plate. This results in longer mold life when compared to the stripper or ejector pin methods. Air ejection is limited to certain cup-shaped products—usually, shallow or deep containers. There will be a witness line on the inside of the product, either around the side walls or at the bottom of the container.

12.4.1 Clearance (Fit) and Length of Land

Clearances for ejector pins and sleeves should be as specified by the pin manufacturer. In general, clearances correspond to the vent sizes shown in Chapter 11, Venting.

There are several reasons for limiting the clearances in diameter and the length of the land of ejector pins:

Clearance of pin in bore: Too tight a fit may bind the ejector, resulting in breakage, or may contribute to rapid wear of pin and/or bore. Too loose a fit may create flashing, especially when injecting with high pressure and when the plastic is very hot and has a low viscosity.

Length of land: As suggested above, the land should not be longer than 2*D* (the diameter of the pin for small sizes (up to 6 mm), and not longer than 1.5*D* for larger sizes.

If the land is too long, it is costlier to produce (more jig grinding or reaming time), and it offers more resistance to the escaping air and may result in poor venting. If the land is too short, there may not be enough sealing to keep plastic from flashing, especially when injecting with high pressure, and when the plastic is very hot and has low viscosity; also, the bore may wear out more rapidly.

12.4.2 Size, Finish, and Shape

Size of pins: The designer should always use standard (listed) sizes. Special sizes are very expensive to make or to procure from suppliers.

Finish of pins: To ensure long wear life, pins are made from H13 or similar steel, hardened and nitrided, with a surface hardness in the order of 70 Rc. They are finished to a very high polish and then covered with a solid lubricant.

Finish of bore: Bores must be smooth and free of grinding marks that could act as a file on the pins. Suggested finish is 0.4 μ or better (16 micro-inches). Bores are usually produced by jig grinding. This avoids the *bell-mouth*, which causes a very unsightly flash at the location of the ejector pin or sleeve (Fig. 12.16). Bell-mouthing is a funnel- or bell-shaped ending of the bore at the molding surface, usually caused by drilling when the drill enters the core from the product side.

If jig grinding is not available, the ejector holes must be reamed. To avoid bell-mouthing, the experienced mold maker will drill from the rear of the core or drill undersize before finishing by reaming from the underside of the core. Because it is usually easier to drill from the product side of the ejector bore, many designers who provide drawings to

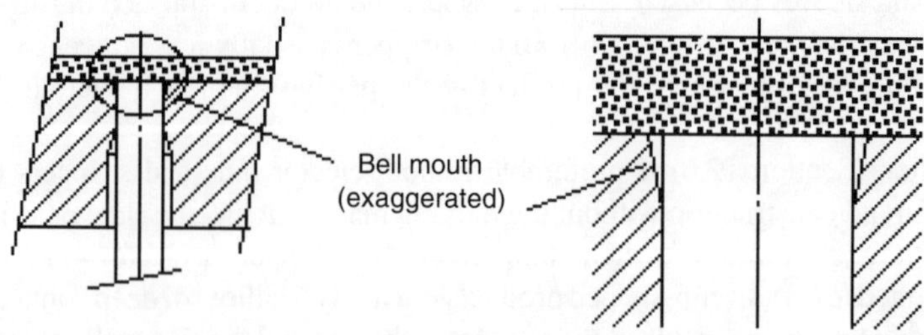

Figure 12.16 Bell-mouth occurs at the end of a bore at the mold surface, causing flash.

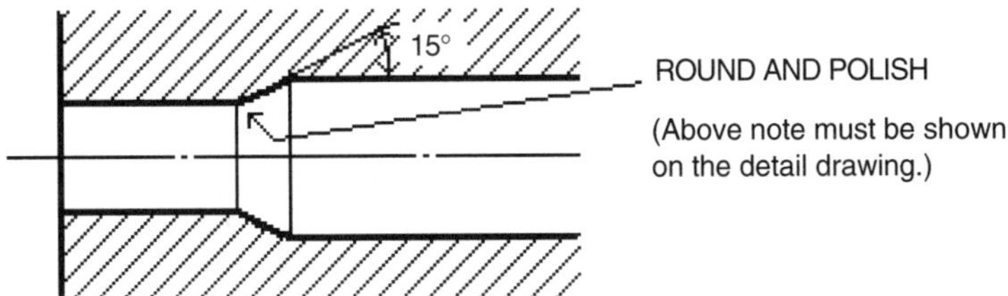

Figure 12.17 Ejector pin (or sleeve) holes in the core are funnel-shaped to guide pin entry during assembly.

shops where the bores are not jig ground put notations on these holes such as: "DRILL (OR REAM) FROM THE REAR".

Shape of ejector pin (or sleeve) hole in core: (This applies also to holes for return pins, discussed later). The holes must be funnel-shaped from the rear to facilitate entry of the pins (sleeves) during assembly (Fig. 12.17).

Selection of ejector pin sizes: As a general rule: Make the ejector pins as large as possible. (Note that bore diameters above 6 mm are cheaper to produce than smaller bores.) Sizes 3 mm (1/8 in.) diameter and smaller should be avoided, particularly if the pins are longer than 50 times the diameter. Smaller sizes than 3 mm (2.5, 2.0 and 1.5 mm) are stepped down from a 3-mm pin, which extends to a distance *E* of 75 mm.

The reason to avoid such long, slender pins is the danger of collapsing the pins under the ejection force; also, frequent replacement of these pins is required, as they wear more rapidly. Wear problems are particularly serious with abrasive (e.g., glass-filled) plastics.

12.4.3 Special Ejector Shapes

If very small ejectors are required in locations such as at the bottom of ribs, etc., it is better to provide (if possible) a flat ejector pin instead of a very small, round ejector pin.

Example: If the width at the bottom of the rib is 1.5 mm, it is better to have a flat ejector, 1.5×5.0 mm. The area facing the plastic in the first case is only 1.77 mm^2; in the other, 7.5 mm^2.

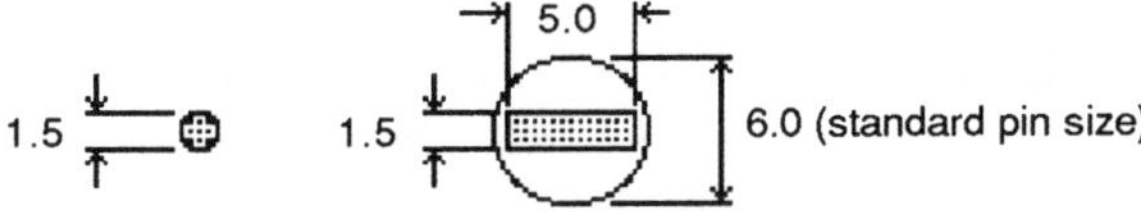

The much larger surface is less likely to penetrate the plastic during ejection, particularly if the plastic is still warm enough and has little resistance. This may translate to the possibility of saving a few seconds of cycle time. The pin is also stronger and will last longer. If possible, special pins should be made from a standard size ejector pin or core pin.

Mold hardware suppliers have listings of standard blade ejectors. Designers should always try to find a standard size before specifying a special part to be made in-house; such a pin would not only be very expensive but also not as good as a standard commercial part, which has the same nitrided finish and polish as round ejector pins.

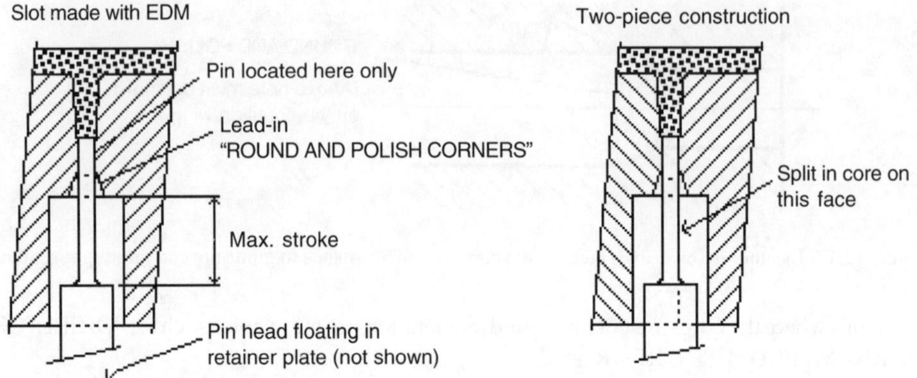

Figure 12.18 Specially shaped pin slots may be created with EDM (left) into the solid core or two-piece construction (right) of the core.

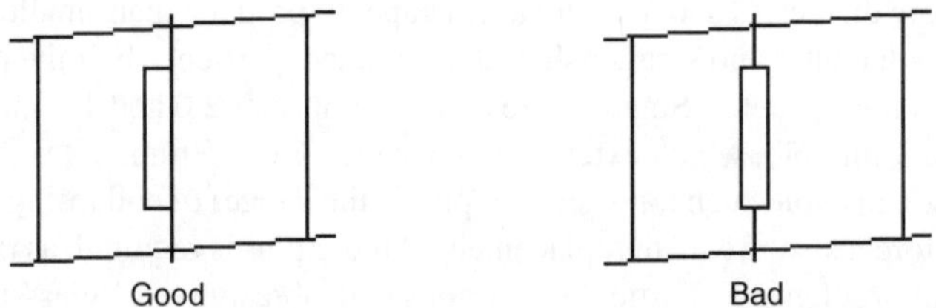

Figure 12.19 Two-piece construction of core: Good design (left) has machined groove on one side only, bad design (right) has two matching grooves, requiring unnecessarily tight tolerances.

When designing the slot for the special shape pin, the designer has the options to EDM into the solid core or composite core, or to use a two-piece construction. For EDM, the clearances for the pin and the length of land should be selected similar to the dimensions for a round pin of a size corresponding to the smaller dimension of the slot. This will not only ensure enough sealing and venting length but also facilitate manufacture and avoid excessive length of the special shape. The rear of the core under the slot is bored out for clearance of the shank of the pin from which the special ejector was made (Fig. 12.18 left).

With a two-piece construction, a relatively shallow U-shaped groove is machined into one portion of the core. The matching portion is flat. All four sides can be ground if necessary. The counterbore for the shank of the pin is similar to the solid core.

Do not try to match two U-shaped grooves opposite each other (Fig. 12.19). There is no need for it, and to ensure perfect matching, the tolerances would have to be unnecessarily tight.

Shapes other than rectangular are, of course, possible, but difficulties are usually connected with the making of the slot. With advances in EDM technology, even odd-shaped slots can now be made relatively easily. Even so, such pins and pin holes are expensive and should be avoided if possible. The tolerances for both pin and slot must be very closely specified to arrive at the desired clearances. For these clearances, use sizes similar to comparable ejector pin sizes.

12.4.4 Number and Location of Ejector Pins

There are no set rules as to the number of ejector pins per cavity. The more pins there are, the better the ejection and flatness of the ejected product. However, each unnecessary pin is an unnecessary cost. There should be as few ejector pins as possible, but good, reliable ejection must be the prime objective.

For example, in a 24-cavity mold, two unnecessary pins per cavity add up to 48 unnecessary pins per mold. This means not only the unnecessary cost of the pins but also the cost to produce bores in the core, in the core backing plate, and in the ejector retainer plate. It may also affect (complicate) the layout of the cooling lines in the core and core backing plate, which may be an additional unnecessary cost.

Pins must be located where they will provide necessary venting of areas which do not vent to the parting line:

- Pins should be located at the lowest points of the product, such as rim, deep ribs, or bosses.
- Pins may be required at or near the corners of the product.
- Pins should be located symmetrically and evenly spaced about the product, if possible.
- Pins should be located at intersections of rib and rib or wall and rib (Fig. 12.20).

By locating pins at intersections, the ejector size can be increased, as shown for the two cases in Fig. 12.20. By adding a small radius or chamfer in the corners, the pin size can be made still larger, which may be necessary to get to the next standard size pin.

Example: In Fig. 12.20, with 1.5-mm ribs, the diagonal dimension at the rib intersection would be

$$1.5\text{ mm} \times \sqrt{2} = 1.5\text{ mm} \times 1.4142 = 2.12\text{ mm},$$

so, a 2-mm pin would be okay without radii in the corners. At the wall,

$$1.5\text{ mm} \times 1.25 = 1.875\text{ mm},$$

which is not enough for a 2-mm pin, but radii of 0.15 mm in the corners would make it possible to use a 2-mm pin.

CAUTION: Too large a radius increases the bulk of the intersection and may slow down the molding cycle and/or create sink marks.

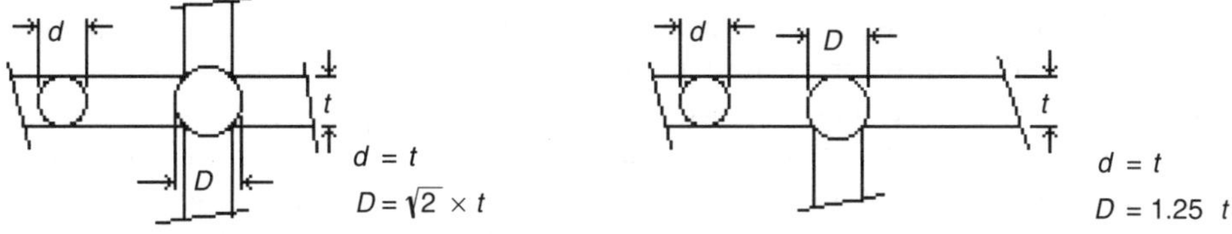

Figure 12.20 Pins located at intersections of rib and rib (left) and wall and rib (right) allow ejector sizes to be increased.

12.5 Ejector and Ejector Retainer Plates

There are two forces which tend to deflect the ejector plate: ejection forces and injection forces.

12.5.1 Ejection Force

The forces acting on the ejector plate are difficult to estimate. They depend mainly on the following elements:

Finish of the core: In general, the rougher the core, the more force is needed to push the product off the core. (Exception: PP and PE require some roughness for good ejection.)

Draft angle: The smaller the draft angle, the more force is required to remove the product from the core.

Undercuts required in the product: Some heavy undercuts (e.g., as on the inside of bobbins for textile spinning machines) require almost as much ejection force as clamping force. In this case, additional, high power hydraulic actuators must be added to the mold.

Undercuts not required in the product itself may have to be added to the core to ensure that the product will stay on the core for proper ejection.

Undercuts may be required when the product tends to remain in the cavity because of the following factors:

1. Vacuum under the bottom of the product in the cavity. This is frequently the case with containers.
2. Stickiness of the plastic. Some plastics adhere more strongly to the steel than others. If the "holding" surfaces in the cavity are equal to or larger than those in the core, the product may stay in the cavity instead of on the core .
3. Shrinkage forces in the cavity (e.g., when there are pins in the cavity). As the plastic shrinks, it "grabs" the pins and remains in the cavity. As a rule, the mold should be laid out so that the product will stay naturally on the core, but this is not always possible.
4. Symmetrical (or nearly symmetrical) products about the parting line (e.g., records, dinner plates, etc.).

All the above forces which resist the ejection must be considered. As it is virtually impossible to estimate these forces, we will assume that the ejector plate is subjected to the maximum force available from the machine.

Most injection molding machine specifications list the maximum available *hydraulic ejector force*. It is usually in the order of 6–10 tonnes (metric tons) for small and 10–16 tonnes for larger machines. Once this force is known, the required strength of the plate can easily be calculated. This ejection force data is available from the molding machine specifications.

In the case of fixed (bumper) ejection force, as the moving platen opens and carries the mold toward the open position, the ejector plate in the mold comes into contact with the bumpers that are solidly mounted on the clamp body. As the clamp continues to open, the ejector plate is moved forward to eject the products. In this case, the ejection force at the moment of impact depends not only on the force of the opening cylinder of the clamp but also on the speed of the moving masses at the moment of impact, which can be considerably higher than the hydraulic forces. This method is rarely used today.

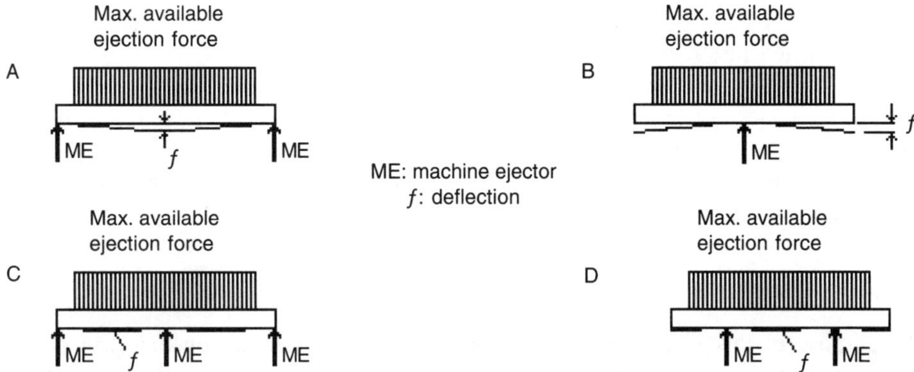

Figure 12.21 Examples of varying locations of machine ejectors: A. Outside machine ejectors only;deflection in center is large. B. Center machine ejector only; deflection is less, but still large. C. Outside and center ejectors; little deflection. D. Outside ejectors moved toward center. Little deflection. Deflection is lowest in C and D.

12.5.2 Injection Forces

The ejector pins are loaded at their face with the injection pressure during and at the end of the injection cycle. The greater the area subjected to the injection pressure, the greater the force transmitted through the ejector pins to the ejector plate.

By placing *stop pins* under or near heavily loaded pins, the effect of plate deflection due to these forces can be eliminated. Occasionally, the ejector plate of a mold rests directly on the machine platen (no mold mounting plate). In this case, the designer must make sure that the stop pins will be located so that they will not overlap (fully or partially) any ejector holes, tapped holes, or slots in the machine platen and, thereby, become useless.

The ejector plate is more often a plate rather than a simple beam, with the ejector pins well distributed over the entire surface. This would make calculations quite involved. By reducing the plate to a simple beam, and to be able to use simple formulae to calculate deflection, the result is more severe than when using a plate formula. But the difference is in the favor of the designer, resulting in a somewhat more solid plate, or less deflection.

The plate deflection f must be kept to a minimum. As a rule, $f < 0.1$ mm is acceptable.

12.5.3 Influence of Location of Ejectors Relative to Machine Ejectors

In the examples shown (Fig. 12.21), the number of ejector forces (which are more or less evenly distributed over the mold surface) are simulated by an evenly distributed load on a beam, with a total force equal to the available ejection force of the machine.

Design of ejector plates must take relative locations into account and must keep deflection as small as possible. Where possible use machine ejector locations evenly spaced under the load of ejector pins.

Make the ejector plate sufficiently thick, especially if the span between machine ejectors is large, to keep deflection to a minimum. Design the ejector as a beam. Do not skimp on plate thickness. If in doubt, calculate the deflection. Suggested maximum deflection $f = 0.1$ mm.

Remember, deflection f depends on the geometry of the plate and the modulus of elasticity E, not on the tensile strength of the steel.

It is preferable to use four or more machine ejectors, and not fewer than three. The use of more than four ejectors is more difficult in mold setup, since all ejectors must "hit" at the same time. However, occasionally it may be necessary if the span between four ejector locations is too large and the resulting plate deflection would be unacceptable. In small molds, one center ejector is often acceptable.

12.5.4 Ejector Retainer Plate

The purpose of this plate is to hold the ejector pins (or sleeves) on the ejector plate. Occasionally, the ejectors are threaded into the ejector plate and a retainer plate is not required. This is a bad practice because of possible misalignment with the bores in the core. Sometimes, several small retainer plates rather than one large plate is used on one ejector plate if, for example, the distance between ejectors is large.

The heads of the ejector and return pins, or sleeves, should float with lateral play in the retainer plate so that the pins or sleeves find their proper alignment with the bores in the core (Fig. 12.22). The forces on the plate are small, usually just the friction of the pins in their bores. The thickness of the retainer plate at the head (t) need never be more than 3 mm. Axial clearance should be very little, maybe 0.1 mm. Radial clearances HC and SC should be at least 0.5 mm.

12.5.5 Preventing Ejector Pins from Turning

Ejector pins must be prevented from turning when:

- The front of the pin in the cavity is shaped to match the shape of the product,
- it carries engraving, which must stay in a determined relation to the product, and
- a large pin is used partly for ejection and partly as a return pin.

A preferred method is to key the head of the pin. A groove in the retainer plate as illustrated in Fig. 12.23 can be used.

The location of the head at Fig. 12.23A is not as good as in Fig. 12.23B, but the milling cutter in view B is smaller and thereby slower. As an alternative, an inserted key may be used (Fig. 12.24). The disadvantage is that an additional mold part is required. The advantage is that

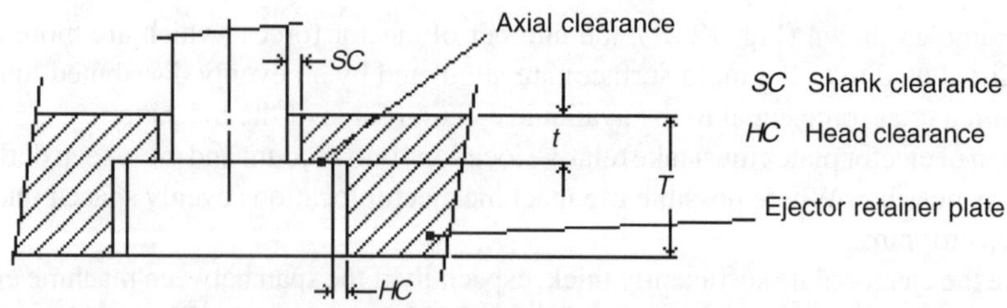

Figure 12.22 Cross section of ejector retainer plate shows clearances for pins and heads.

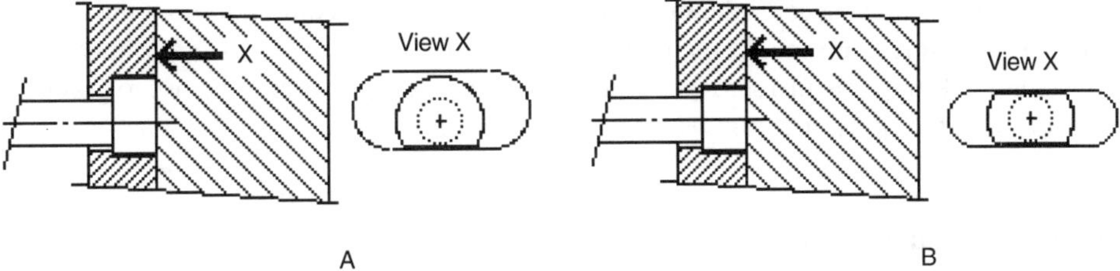

Figure 12.23 Grooves in the retainer plates in views A and B prevent the pin from turning.

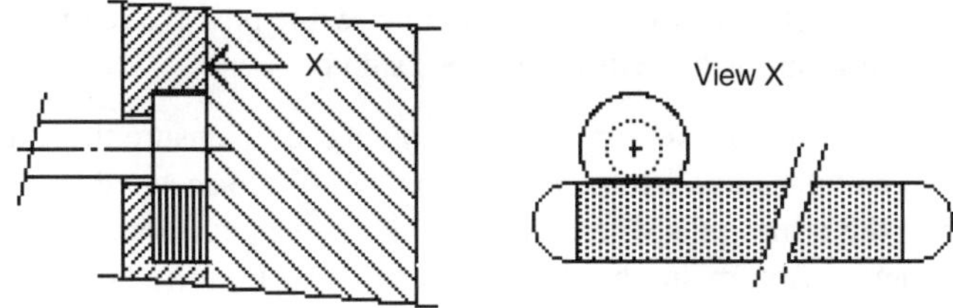

Figure 12.24 An inserted key prevents the pin from turning.

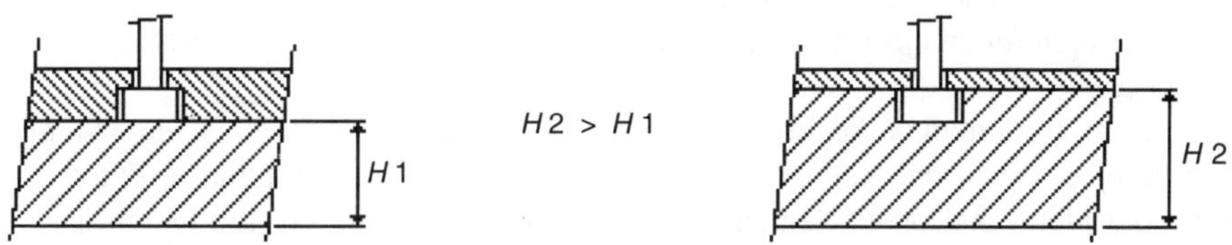

Figure 12.25 The left view shows the original ejector plate and retainer plate thicknesses; the right view shows the increased thickness *H* of the ejector plate as the retainer plate is reduced.

a larger key can be used to permit a large groove and faster milling. One key can be used with any number of pins, as long as they are more or less in line.

The thickness of the ejector retainer plate does not add anything to the strength of the ejector plate against deflection. In critical cases, it could be necessary to place the bores for the ejector and return pin heads into the ejector plate, and reduce the retainer plate to 4 mm (enough for the screw threads) but at the same time increase the stiffness of the ejector plate considerably because of the thickness *H* gained. As the stiffness increases with the third power of the thickness, the ejector plate at the right is much stiffer than the one on the left (Fig. 12.25). However, this method is only rarely used.

Example: If the ejector plate thickness $H = 25$ mm, and the head is 6 mm, the stiffness is proportional to 25^3 (to the third power) or 15,625. If the plate thickness is increased to $H = 31$ mm, the stiffness is proportional to 31^3 or 29,791—almost twice as stiff against deflection.

The designer must specify the number and size of screws required to hold the retainer plate. Since the forces are small, small screws are sufficient, such as 6 mm (¼ in.), but for ease of machining, 8-mm (5⁄16-in.) screws are preferred. By spacing the screws close to groups of ejectors and return pins, few screws are usually necessary.

12.5.6 Return Pins

Return pins ensure that all ejector pins are returned to their back position by the time the mold closes to prevent the pins from contacting (and possibly damaging) the opposite cavity surface. Return pins must always be provided, even if the ejector plate is positively returned by the molding machine or by any other method, in case the positive return fails.

Several design suggestions for return pins are listed below:

- There should always be four return pins, evenly spaced, to ensure that the ejector plate cannot cock. In circular molds, three pins at about 120° are acceptable.
- Do not position the return pins where they could hit a vent channel.
- Minimum diameter of return pin should be 12 mm (½ in.). Preferred sizes are 16 or 19 mm (⅝ or ¾ in.). The larger the pins, the less damage will be done on the opposing plate surface where they hit.
- Always use standard (commercial) pin sizes.

For proper clearance for a return pin in a mold plate, the passage through the plate can be much larger than for an ejector pin, to prevent wear and pick-up, since there is no molding surface nearby.

In general, the length of the return pin should be less than the theoretical length required to push the ejector plate back all the way. This accounts for the possible build-up of tolerances in the mold plates and other elements. A rule of thumb is to make the length less than theoretically required by 0.25 mm ± 0.05 mm (.010 in. ± .002 in.).

The most important feature of the return pins is to protect the cavity wall opposite an ejector pin from damage. Since the return pins are slightly "short" (see above), the ejector pins may stick out somewhat after being returned. This is of no consequence because the injection pressure acting on the face of the pins will seat them properly.

Usually, the return pin meets the cavity at the parting line. If, however, the return pin in its forward position would interfere with a take-off or with the free fall of the products out of the mold, the pin must be shortened and another pin placed opposite so that they meet below the parting line, in the core plate, when the mold is closed (Fig. 12.26).

A combination ejector and return pin may be used. Occasionally, it is practical to use large ejectors at the rim of a product with an odd-shaped circumference that makes it impractical to provide a stripper plate. This has the advantage that a large segment of the pin acts as an ejector, and the remaining, much larger surface, acts as a return pin (Fig. 12.27). In this case, the length of the pin must be so that it stops exactly at the P/L.

Note that in this case (Fig. 12.27), the pin must be keyed against turning. Where the pin hits the cavity over time, it will slowly hob in, and if allowed turn, it may cause flashing. With modular molds, a stop block can be mounted next to the cavity (Fig. 12.28).

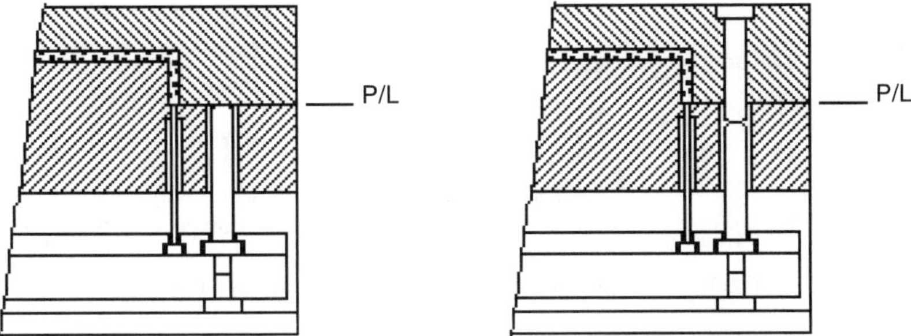

Figure 12.26 The return pin usually meets the cavity at the P/L (left), but if the pin interferes with a take-off or free fall, an opposite pin must be designed to meet the return pin below the P/L (right).

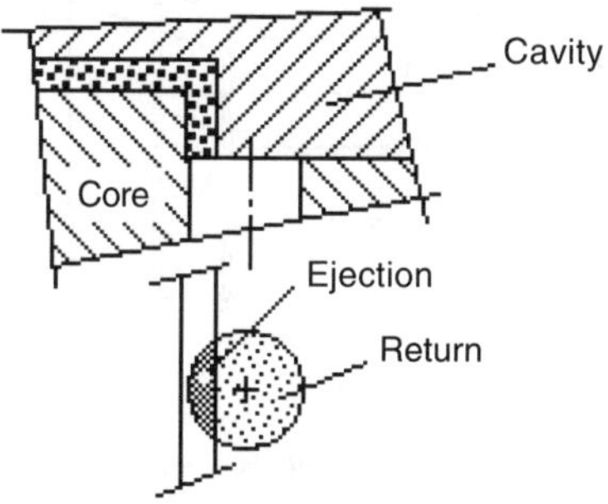

Figure 12.27 Combination ejector and return pin.

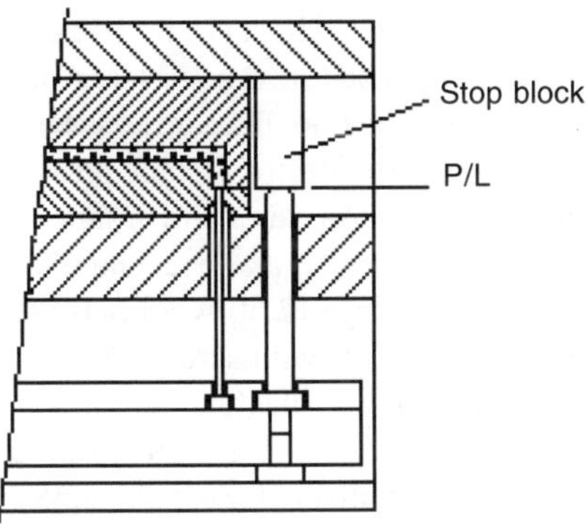

Figure 12.28 A stop block mounted next to the cavity .

12.5.7 Ejector Box

The ejector box is that portion of the mold assembly which surrounds, supports, and guides the ejector plate. Traditionally, the ejector box consisted of a mold mounting plate, two parallels, and, if the span between the parallels is great, additional support pillars (Fig. 12.29).

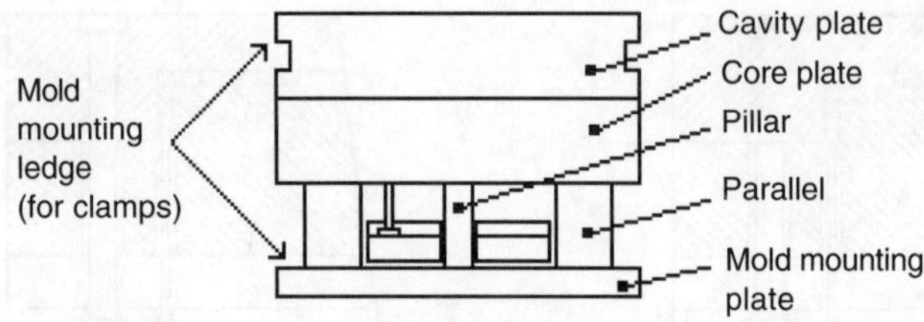

Figure 12.29 End view of mold with traditional ejector box.

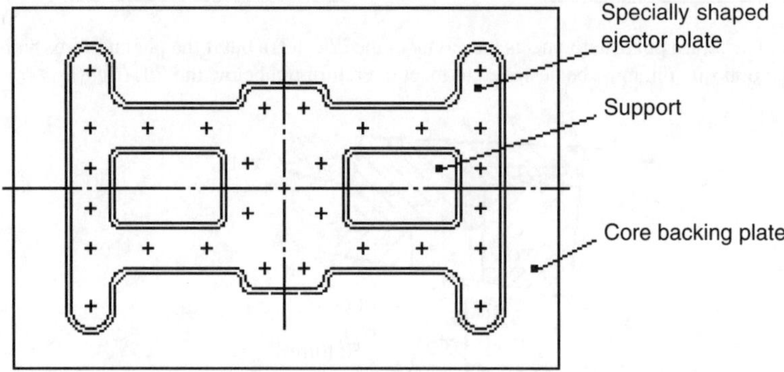

Figure 12.30 A specially shaped ejector plate allows ejector pins to be placed in odd locations.

While there are still many molds built this way, often using standard commercial mold sets, there is no need to follow this practice, which is based on using certain standard sizes of rectangular plates and round pillars, all held together by a large number of screws. This often requires that the designer must select layouts that are not the best for an efficient mold. Occasionally, the ejector plate must provide for ejector pins in odd locations. This fact, combined with the effort required to provide a maximum of support for the core plate, suggests the use of a specially shaped (Fig. 12.30), rather than a rectangular, standard size ejector plate.

The shaped ejector plate shown in Fig. 12.30 would also be possible with individual shaped supports and parallels held together with screws. Some mold makers find that it is cheaper, and results in a more solid mold, to machine all supports and the parallels into the core backing plate. Since there are only few or no assembly screws required to enter the core backing plate, there is also more space available for the planning of cooling and air channels, which is an important considerations for an efficient mold layout.

Tapped holes for mold mounting screws or slots for mounting ledges can be also machined into the core backing plate so that there may be no need to provide a separate mold mounting plate. The additional advantages are reduced mold height (shut height) and cost savings.

An important note regarding the machining of the ejector plate and ejector box is that, for milling, the larger the cutter diameter, the faster the milling operation. Depending on the depth of machining, the deeper the pocket, the larger should be the cutter.

Suggested inside radii for ejector plate and for pockets are based on the preferred standard size cutters: 1½ in., 2 in., 2½ in. diameter. (¾ in., 1 in., 1¼ in. radius, respectively). Smaller

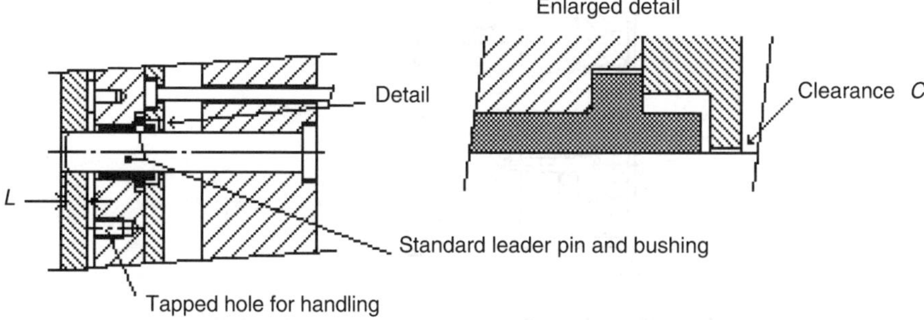

Figure 12.31 Typical method of guiding the ejector plate.

radii (smaller cutters) are, of course, possible but much slower and, therefore, more expensive to use.

The outside corners of the ejector plate can be rounded or chamfered to fit the pocket in the core backing plate. Minimum clearance around the ejector plate should be about 3 mm.

12.5.8 Guiding Ejector Plates

All moving plates must be guided in at least two, and preferably in four locations, evenly spaced. In older mold designs, support pillars were used for guiding the ejector plate. Today, usually, separate guide pins are used. Wherever possible, standard leader pins (LP) and LP bushings should be used to guide the plate. Clearance *C* (see detail in Fig. 12.31) should be about 0.1 mm (or .004 in).

At assembly, the retainer plate is inserted first, then the pins are inserted through this plate and into the core plate. Then, the ejector plate is lined up with the guide pins, pushed into position, and all the screws are installed to hold the retainer plate to the ejector plate. It is good practice to align the ejector retainer and ejector plate with two dowels at opposite corners.

When the ejector box is made from one piece, there is usually no access from the sides to help during assembly in the alignment of the ejector plate. For this reason, the guide pins must be long enough (see dimension *L* in Fig. 12.31 schematic sketch) to project beyond the bottom face of the ejector box so that the alignment bushings can enter the guide pins for aligning before the assembler loses hold of the plate. Also, because the plate is inside the box, there must be at least one tapped hole on the underside of the ejector plate to make it easier to pull it out of the ejector box.

In molds without a mold mounting plate, the ejector plate must be suspended from the core backing plate by stripper bolts (Fig. 12.32), and the guide pin will be cantilevered, supported in the core backing plate. For the use of stripper bolts to be acceptable, there must be clearance under the head of the bolts. During operation, the ejector plate rests on the stop pins. The stripper bolts never see any load except when preventing the ejector plate from falling out during handling. Two bolts are usually enough, for small plates; more should be provided for larger plates.

The stripper plate also can be suspended from the edge (for assembly), but a cutout must be provided in the ejector box for access to the hole and to clear the eye bolt (Fig. 12.33).

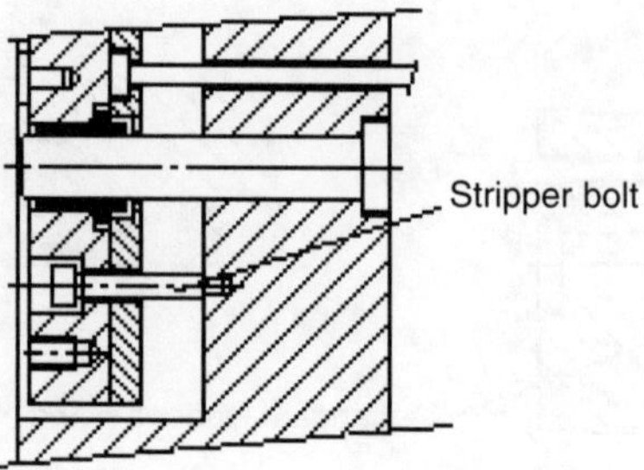

Figure 12.32 Stripper bolts hold the ejector plate in place when there is no mold mounting plate.

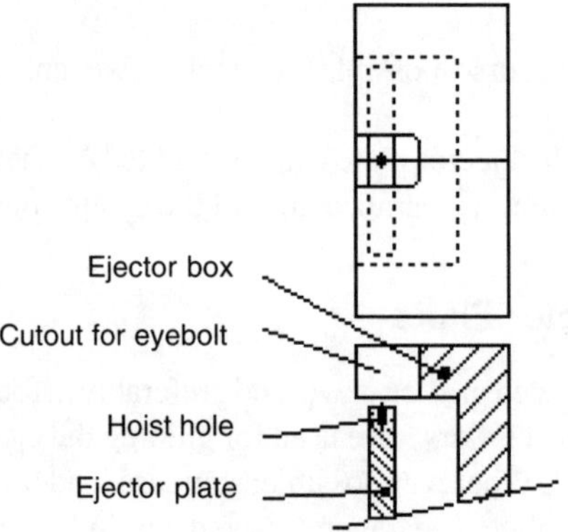

Figure 12.33 A cutout for the eyebolt in the ejector box.

12.5.9 Returning the Ejector Plate

Theoretically, with the use of return pins, there is no need to provide additional features to return the ejector plate. As the mold closes, the return pins would drive the ejector plate back into its "rest" position, touching the stop pins. However, this is only practical with very slow closing speeds of the clamp. It would be acceptable only if the ejector stroke is small, well within that portion of the final mold closing stroke where the mold is moving very slowly.

Normally, the ejector pins move so far out that the return pins would strike the cavity side of the mold while the clamp is still moving rapidly. The resulting impact would not only be noisy but also damaging to the return pins and the surface where they hit, particularly if the masses of the ejector system are large. For this reason, the ejector plate must be returned by an independent method before the return pins strike the opposing mold half. (The occasionally used term "safety pin" instead of "return pin" implies that return pins are essentially a safety feature, in case the regular return method fails.)

The following methods are used to return the ejector plate:

- Tying the ejector plate to the molding machine (M/C),
- using linkages attached to the mold,

- using return springs, or
- attaching air cylinders to the ejector plate.

12.5.9.1 Tie the Ejector Plate to M/C Ejector

Called "push-pull", this method is often used with hydraulic machine ejectors. Care must be taken in setup that the ejector plate is retracted and seats on the stop pins before the mold is fully closed. The disadvantage of this method is the additional connection of mold and machine, which slows down the mold installation. With fixed (bumper) ejection, this method is not possible.

12.5.9.2 Use Linkages Attached to the Mold

Both ejection and return motions of the ejector plate are linked to the mold stroke. To prevent damage to the linkage, this method requires a very exact and repetitive opening stroke. Care must be taken to ensure that the clamp does not open beyond what the links are designed for; machine clamp stroke limiters may have to be provided. If linkages are used to drive the ejector plate only forward, springs or air must be used to return the plate.

12.5.9.3 Return Springs

There are both internal and external return springs. When using internal springs, stripper bolts must be used to prevent the springs from driving the ejector plate too far during disassembly of the mold.

Placing springs between core plate and ejector plate is simple (Fig. 12.34) but has disadvantages, particularly with longer ejection strokes. To be effective in the returned position, that is, to have enough force to hold the ejector plate against the stop pins and to allow for spring tolerances, the open spring must be preloaded (*L*1) to about 10% of its maximum stroke to solid. A spring must not be compressed by more than 25% of this stroke for long life (*L*2). Therefore, the effective stroke must not be more than about 15% of the maximum stroke to solid. This means that for an ejection travel of 30 mm, a standard medium-force die spring should be about 200 mm long. Such a length is usually difficult to fit in the available space.

A good rule is to follow spring manufacturers' guidelines shown with standard die spring charts. Unfortunately, mold designers often ignore this simple consideration, with the result that the springs are over-stressed and break because of fatigue after a relatively short time of

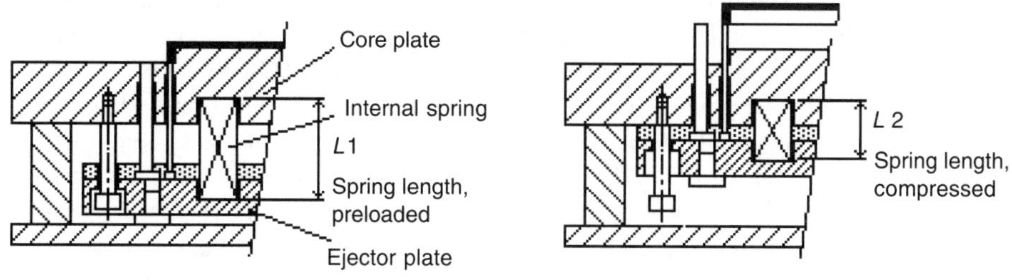

Figure 12.34 Internal return spring between the ejector plate and core plate.

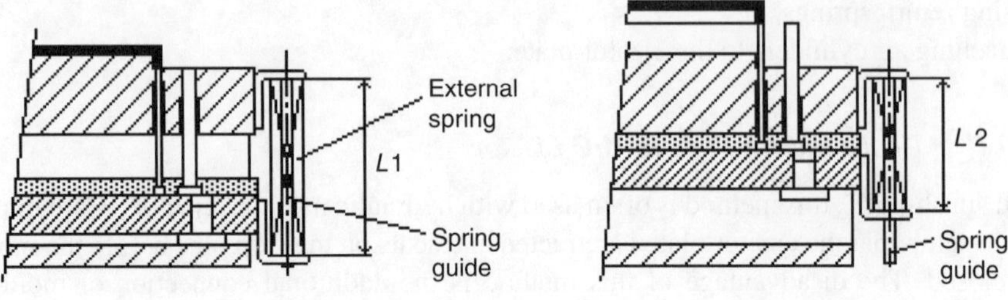

Figure 12.35 External springs mounted near the corners of the ejector plate.

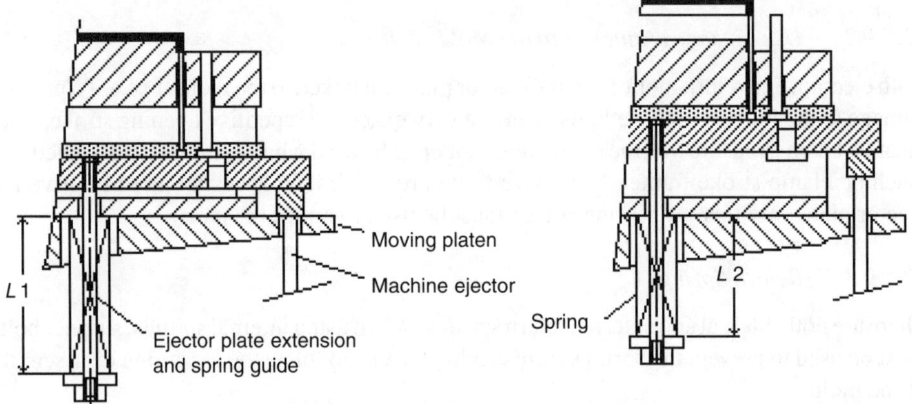

Figure 12.36 A return spring used with a central machine ejector must have a strong ejector plate extension and guide.

operation. Because the springs are hidden between the plates, broken springs are not easily noticed, and the job of returning the plates is then left to the return pins.

NOTE: The unguided (unsupported) length of any spring should never be more than its diameter. If the distance is larger, the spring must be supported (internally) with a rod.

When using springs with a long ejection stroke, they should be placed outside of the mold where there is enough space and where broken springs can be easily noticed. Four springs should be used, located near the corners of the ejector plate. Note that Fig. 12.35 is a schematic illustration. The actual construction may differ to suit the mold layout.

To prevent possible injury in case of spring failure, every external spring must be shielded, for example, with telescoping sleeves. To prevent injury, caution labels should be attached to the mold indicating "SPRING OPERATED RETURN OF EJECTOR PLATE."

For small molds, one single, centrally located spring may be used, provided that the moving platen has a suitably large central opening. A disadvantage of this system is that the projecting spring assembly is easily damaged during storage, or when hoisting the mold into the machine. Note that Fig. 12.36, too, is only a schematic illustration. The actual construction may differ to suit the mold layout. If the central machine ejector is used at the same time as a center return spring, the ejector plate extension (and spring guide) must be heavy enough not to buckle under the impact of the machine ejector.

The advantage of springs is that they are inexpensive and are simple to use, provided the basic guidelines are followed. In general, springs are quite appropriate for short strokes. For example, with a required stroke of 9 mm, a spring with a preload of 6 mm could be 60 mm long, for a maximum compression of 25%.

12.5.9.4 Air Cylinders

The main advantage of a force generated by an air cylinder is that such force is constant over the whole length of the stroke, unlike the springs, which are weaker in the expanded and stronger in the compressed state of the spring. Two typical examples of attaching air cylinders to the ejector plates are illustrated and described below.

Example 1: Four cylinders are located near the corners of the ejector plate to provide balanced force on the plate.

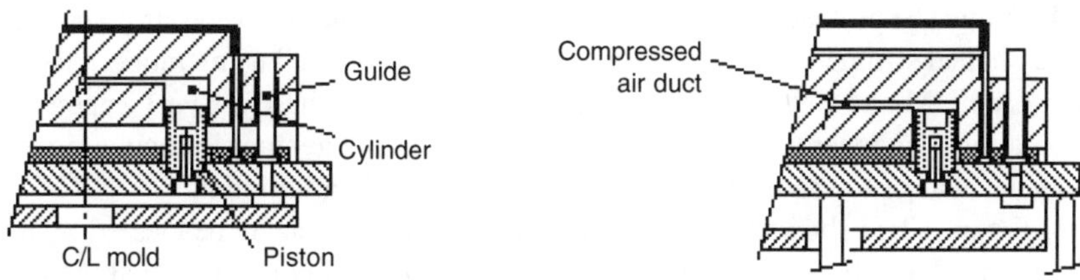

The example is a typical arrangement for four air actuators to return the ejector pate. In the left illustration, the piston holds the plate against the stop pins. In the right, the machine ejectors push the plate forward. The illustration shows both outside and center ejectors. Normally, only one or the other would be used.

Example 2: One air cylinder is centrally located (small molds only). In this case, the piston is fixed to the core plate, and the cylinder is fastened to and moves with the ejector plate.

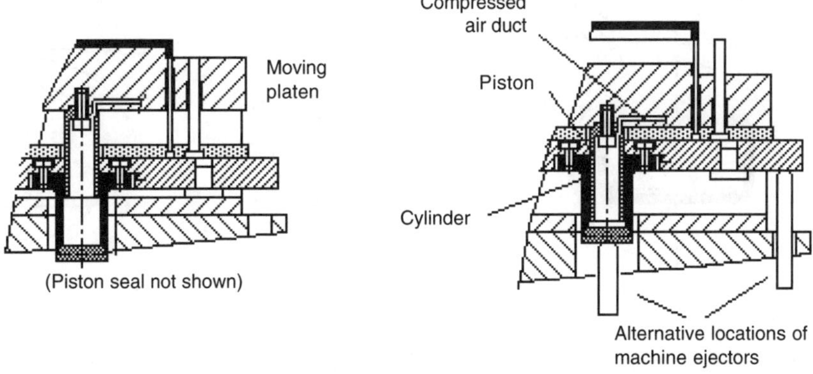

In either of the preceding examples, the compressed air may be connected permanently (without controls) to the air cylinders. The force of the air cylinders is considerably less than the machine ejector force and will, therefore, not significantly reduce the available ejection force. In either case, a "Caution" name plate must be attached to the mold to warn that air pressure must be released before doing work on the mold.

12.6 Strippers

Strippers (plates, rings or bars) are preferable as compared to ejector pins. The surface of the product where it is pushed during ejection is relatively large, and the ejection force is evenly distributed. Also, the vestige left from the stripper is usually less noticeable.

However, strippers can (and should) only be used where the product at the parting line has a shape which can be easily generated when machining. The stripper must seat on a taper, which must be preloaded to prevent flashing into the space between stripper and core. Today, odd-shaped strippers with tapered shut-off can be produced using modern technology, such as wire EDM and CNC milling and grinding; however, these are very expensive machining operations. In general, a stripper should have a simple geometric shape (circular or linear), which is easier to produce with conventional milling and grinding.

Figure 12.37 shows how the same noncircular product can be ejected by three different methods. The different complexities and difficulties involved in making these strippers should be obvious. For more discussion of preload on tapers, see Section 22.8.2.

Occasionally, a portion of the product may be molded on the stripper itself, but care must be taken to ensure that the product will not hang up on the stripper during ejection. This can be a problem if the product has little mass and will not separate from the stripper by its own weight (Fig. 12.38).

In Fig. 12.38, the large container will fall easily from the stripper plate. The small product may hang up on the tab, and some method, such as air blast, or even mechanical take-offs, must

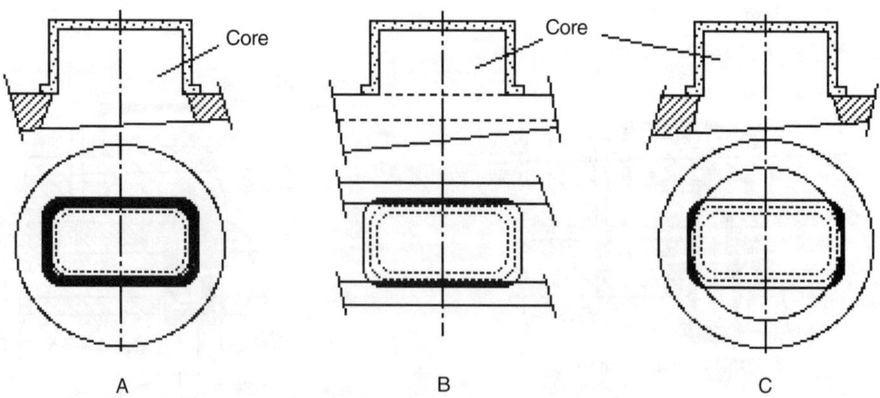

Figure 12.37 Three examples of strippers used to eject a product: A. a shaped stripper plate, B. stripper bars, C. round stripper ring, stripping at the corners only.

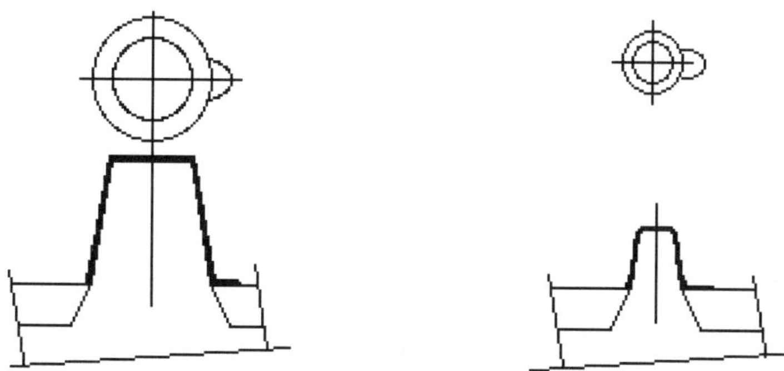

Figure 12.38 A large container (left) has more mass than the small (right) container, which may become hung up on the stripper plate.

be provided to ensure that the product will reliably separate from the stripper, every time. The problem may be worse if the projection (tab, etc.) does not seat flat on the stripper but is recessed in it.

12.6.1 General Rules for Strippers

There are several rules regarding stripper design in the mold. Many considerations and guidelines accompany each "rule" listed below:

1. The stripper must always clear the core.
2. All stripper must seat on a taper (tapered shut-off is recommended).
3. Cylindrical shut-off is not recommended.

As the stripper moves along the core, there must be a minimum clearance between stripper and core of 0.25 mm (.010 in.). (Less may be acceptable for repairs or in exceptional cases.) This clearance prevents scratching and damage to the core and to the edge of the stripper, in case of slight misalignment or excessive play in the guides of the stripper. This is particularly important with no or little side draft of the core, as shown in Fig. 12.39.

Commercially available ejector sleeves are the only exception to the rule that all stripper must seat on a taper. This is permissible because the sleeves are quite inexpensive and easily replaced. Also, their superior quality of surface finish will not easily wear or damage the bore in the core. Manufactured sleeves used in composite cores (i.e., for two-stage ejection of overcaps, etc.) must have a taper seat.

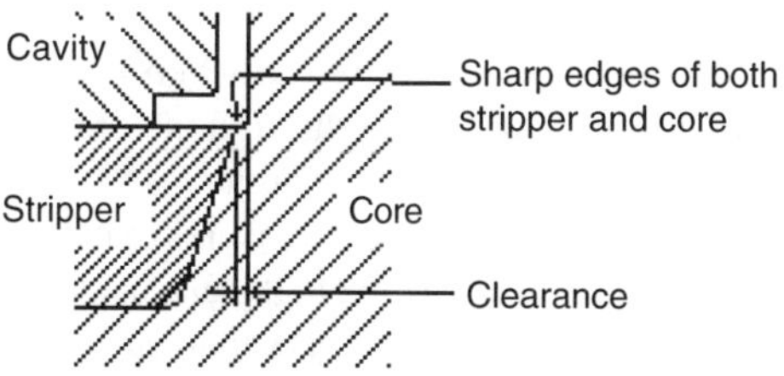

Figure 12.39 The stripper must clear the core.

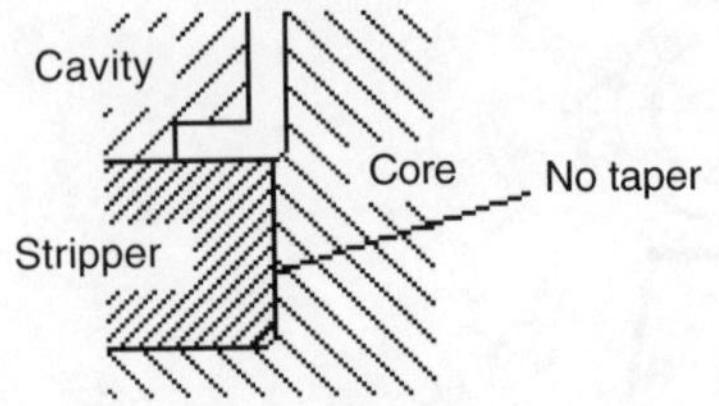

Figure 12.40 Cylindrical shut-off is not recommended.

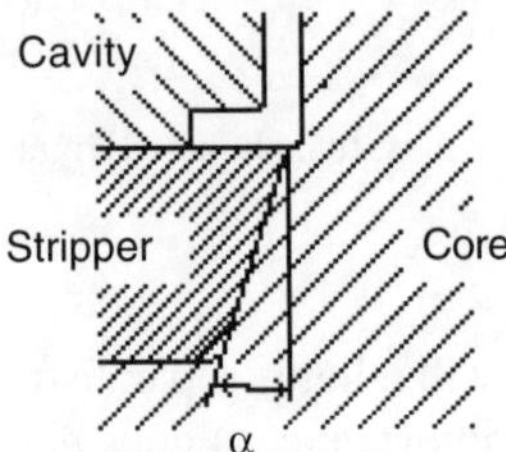

Figure 12.41 Tapered shut-off of the stripper.

There are several good reasons for not specifying cylindrical shut-off:

1. Cylindrical shut-off will wear as the stripper slides on the core (Fig. 12.40). Wear means flashing. But even a new, good sliding fit may be too large to prevent plastic from entering (flashing) between the two mating surfaces. Because the stripper and the core can move relative to each other, the gap on one side may end up at zero, and the opposing side will then have twice the intended radial clearance, and flash.
2. The cylindrical shut-off leaves the seat on the core if the product requires a long ejection stroke. When returning, the sharp edge on the core will be easily damaged.
3. Once the stripper is worn, it cannot be reseated but must be replaced.

The only advantage of vertical shut-off is that it is cheap. Cylindrical shut-off may be acceptable in prototype molds when used only to make a few samples.

The tapered shut-off method is standard practice on all good molds. The angle a is usually from 5 to 15° (Fig. 12.41). Smaller tapers may have to be specified if, for example, there is not enough thickness in the stripper to accommodate a standard taper without the steel becoming too thin at the end with the wider taper dimension. This can be the case with tapers in manufactured sleeves, as shown in Fig. 12.42, where there is not enough thickness in the sleeve to permit a large inside taper (see arrow).

Any stripper seat must always have some preload, even if it is only minimal. Without preload, the stripper will be flashing. For further discussion of preload and how to calculate it, see Section 22.10 and, also, Chapter 18, Metal Fatigue.

Advantages of tapered shut-off are that the taper ensures that the critical edges of stripper and core will not collide as the stripper moves toward its seat on the core and prevents damage to these edges. Also, if the sharp edge of the stripper is damaged, it is relatively easy to recreate a sharp edge by grinding over the surface of the taper or the parting line. However, this may require recutting of the core to avoid a step in the product (Fig. 12.43).

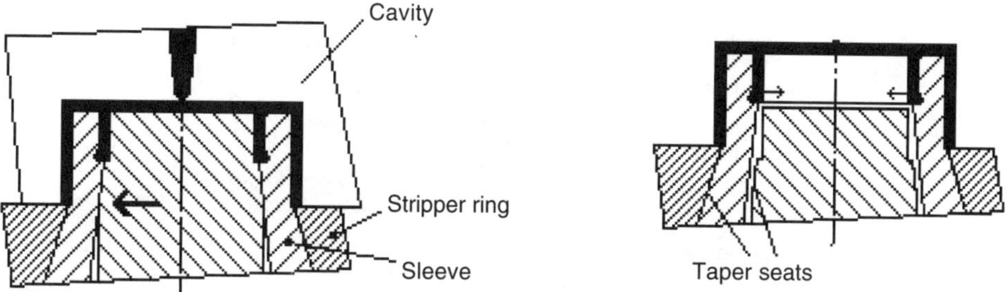

Figure 12.42 A manufactured sleeve (left) may not allow room for a standard taper, so a smaller taper (right) may have to be specified.

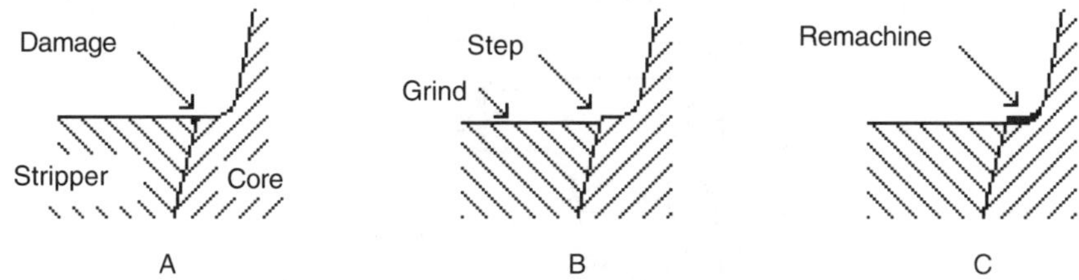

Figure 12.43 Damage to stripper and core can be corrected: A. damaged stripper, B. reground stripper results in step between stripper and core, and C. the core is remachined to correct the step.

If the taper is loose, it is also relatively easy to reseat it by grinding the underside of the stripper. The effect is similar to grinding the top surface in the previous example. It will create a step that may have to be removed by remachining the core.

Note that in both cases, the product length will change. If this is unacceptable, the core will have to be shortened by the amount the stripper was lowered. The decision must then be made as to whether it is not less expensive to replace the stripper than to remachine, polish, and possibly to rechrome the core. (If the core was chrome-plated, the plating must be stripped before remachining.)

Sometimes, not only the stripper but also the core (its sharp edge or the taper) is damaged and may require rework. It may then be necessary to make a new stripper.

12.6.2 Guiding the Stripper

The stripper must be guided in the axial direction of the mold. This ensures that the taper seat is moving properly into engagement without hitting the edge of the core. This guiding can be done by various methods.

One method is to use the leader pins of the mold to guide the stripper. This applies whether the stripper plate is one piece or carries stripper rings or stripper bars (Fig. 12.44).

The disadvantage of this method is that the alignment of the leader pins may “fight” the alignment of the stripper taper. To avoid this, the guide bushings on the stripper plate are made looser than standard, to provide an approximate location for the plate as it returns.

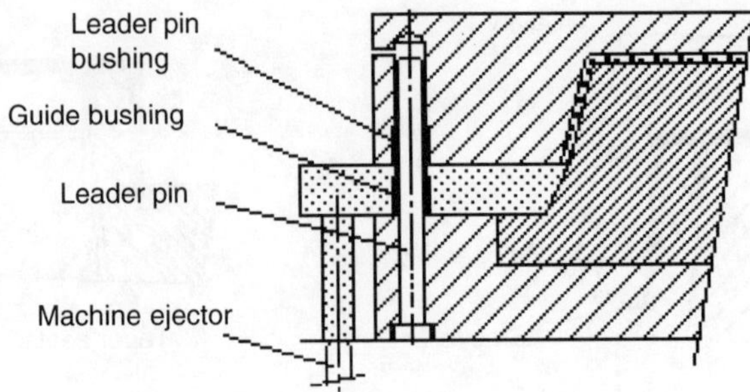

Figure 12.44 Mold leader pin used to guide the stripper.

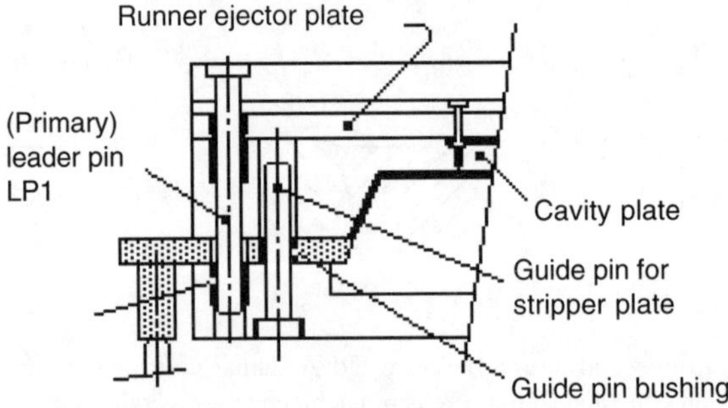

Figure 12.45 Separate guide pins are used to guide the stripper plate in three-plate molds.

This "looseness" or extra clearance in the leader pin bushings must not exceed the total of the positional tolerances of leader pins and cores. Floating core mounting will overcome this problem of aligning, but it could result in uneven wall thickness of the product. This method should only be used if side wall thickness tolerances are sufficiently large.

Because of this added looseness in the bushings, the stripper plate "hangs" on the guide pins, and the top of the alignment taper of the stripper will always be hit first as the mold closes and will, therefore, wear sooner. To reduce this problem, in hot runner or three-plate molds, use of the direct acting stripper plate should be avoided and floating stripper rings used instead.

Another method is to use separate guide pins for the stripper plate. This is usually the case in three-plate molds, or with stripper plates or stripper rings driven from an ejector plate inside an ejector box. The mold leader pins LP1 aligns the mold (Fig. 12.45) but also guides the third plate (cavity plate) and the runner ejector plate. The guide pins and guide bushings align the stripper plate and protect the core(s).

The problem is similar to when the stripper plate is guided by the mold leader pins. Floating stripper rings are required; otherwise, the taper fit will fight the alignment in the leader pin bushings.

Note that in both examples of stripper guides shown above, there must be stroke limiters to prevent the stripper plate from falling off the leader pins when the plate is pushed out too far (Fig. 12.46). This may happen accidentally when cleaning the mold behind the stripper plate or during handling (hoisting) of the open mold. The ejector stroke must be less than the stroke limiter stroke so that the stripper bolt will never see (be stressed by) the ejector force.

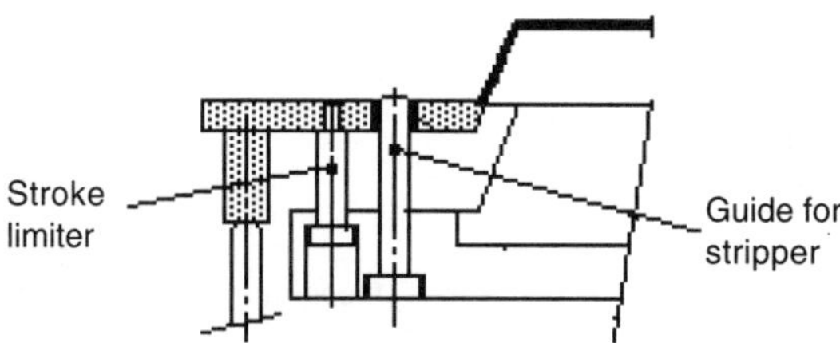

Figure 12.46 A stroke limiter prevents the stripper from falling off the guiding pins.

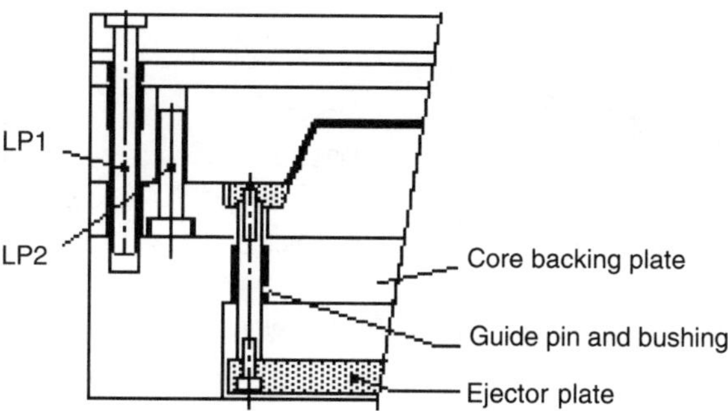

Figure 12.47 Stripper is guided from the ejector box in three-plate molds.

Yet another method of guiding the stripper is to actuate it from the ejector box (Fig. 12.47). In this case, mold leader pins LP1 have the same functions as in the above example. Good practice is to provide pins LP2 to protect the core(s); they do not require a bushing in the cavity plate. The stripper is guided by guide pins and bushings in the core backing plate. The stroke is limited in the ejector box, and additional limiters are not required.

When guiding in the ejector box, there is no need for stop buttons. If too high, they would prevent the stripper from seating properly; if too short, they would not touch (stop) anywhere.

The guide must be made from hardened steel, and the guide bushings, which are difficult to lubricate, should be permanently greased ball bushings, or plastic bushings. Except when floating stripper rings are used, the sliding fit must be large enough so that the taper in the stripper will not fight the alignment in the guide bushing.

Stripper plates without stripper rings are used today only when (e.g., in a large, single-cavity mold) the cost of supplying a hardened stripper plate is little different or could even be lower than the cost of a softer plate with a hardened, inserted stripper ring.

12.6.3 Stripper Rings

12.6.3.1 Fixed Stripper Rings

Fixed stripper rings are hardened inserts in an otherwise softer plate. Their advantage is that they will wear less than a softer plate and are easier to replace in case of damage. The disadvantage, as with the solid stripper plate, is that the alignment of the tapers can fight the alignment of the leader or guide pins.

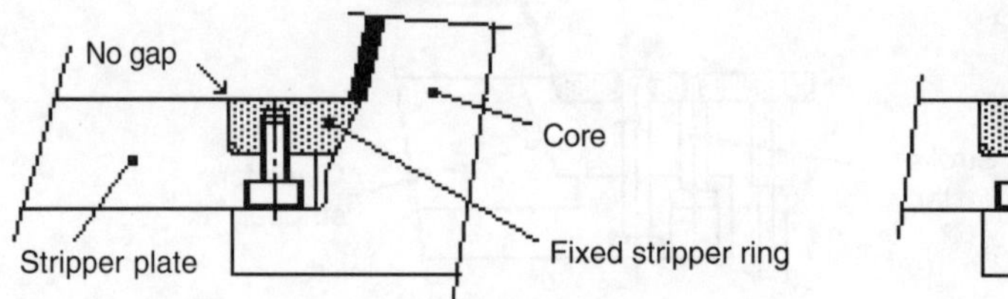

Figure 12.48 Fixed ring designs.

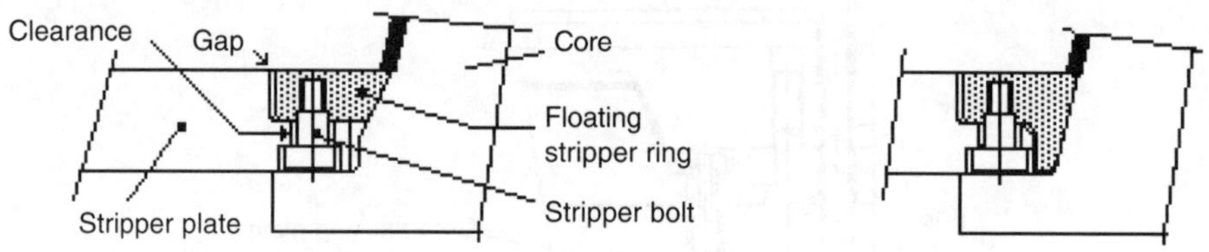

Figure 12.49 Stripper bolt mountings.

Fixed stripper rings are rarely used, except with large, single-cavity molds (e.g., pails, etc.) or in combination with floating cores. They are usually used in two-plate molds, where the runner must cross the gap between the plate and the ring. If the ring were floating, the gap would fill with plastic and void the floating feature. Also, the runner could hang up there and keep the mold from being automatic (Fig. 10.64 in Section 10.5.2).

In Fig. 12.48, the stripper ring rests in both designs on the stripper plate. The example on the left is less expensive, but the taper is very short; the right design with a longer taper is preferred.

The screws do not see the stripping force; they only hold the ring on its seat. Two, three, or more screws hold the ring from the bottom. There should be no holes in the parting plane where plastic from flashing (often during start up) could catch.

There are exceptions to this. Some molders like to be able to remove the stripper from the front for cleaning. However, there is the danger of screws coming lose during operation, which could cause severe damage to cavity and stripper. Screws should not be spaced more than 100 mm (4 in.) apart.

12.6.3.2 Floating Stripper Rings

Floating stripper rings will align with the taper on the core, provided they are free to move. The masses of the rings to be realigned are much smaller than the mass of the stripper plate, and the forces to realign are small compared to the force required to align the whole stripper plate. The disadvantage is that they cannot be used for two-plate molds because of the gap. Several examples of floating stripper rings are discussed and illustrated below.

For a stripper bolt mounting, the gap should be as much as the screw clearance shown in Fig. 12.49, in the order of 0.25 mm (0.010 in.). This method is simple and low cost but not recommended for high production molds because the stripper bolts cannot be properly

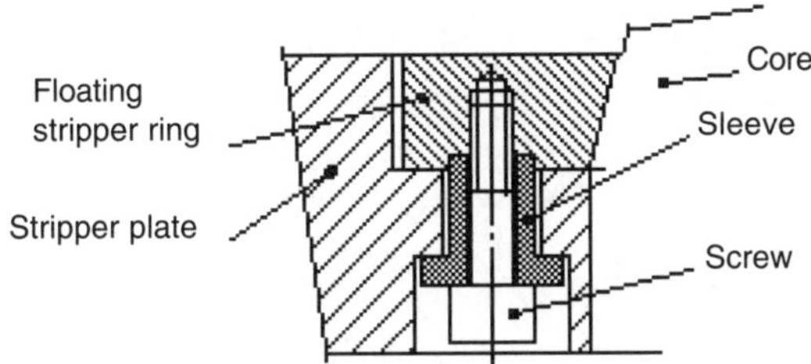

Figure 12.50 A sleeve and screw arrangement for a floating stripper ring.

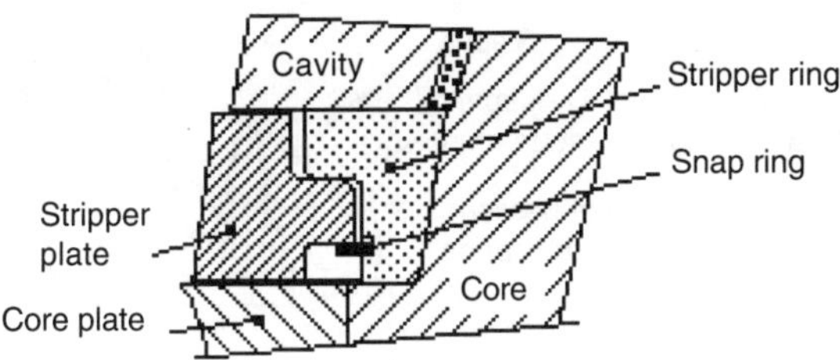

Figure 12.51 A snap ring holds the stripper ring by fitting within an external groove.

tightened, although in prototype or low production molds they could be secured with Loctite™ on the threads.

The length of the cylindrical section of the stripper bolt is such that the ring can float by the amount of clearance between bolt head and stripper plate, but without moving axially. For proper alignment, the stripper bolt should enter the stripper ring by at least 2 mm for accurate location because the thread portion of a stripper bolt is not always concentric with the cylindrical section.

A "sleeve and screw" arrangement (Fig. 12.50) is better than a stripper bolt because the screws can be properly tightened. This method may not be possible if space is limited.

Another method is a snap ring or similar method of holding the stripper ring by a groove, similar to holding a shaft laterally in a bearing. The snap ring in Fig. 12.51 is a standard, external type. The bore in the stripper plate must be large enough to accommodate the expanded snap ring for assembly.

For snap rings, follow manufacturers' design specifications for dimensions, clearances, and tolerances. Note the "important dimensions" in Fig. 12.52 for proper floating. If necessary, the snap ring may be ground flat to a close tolerance.

This method is very good, particularly with smaller sizes, up to about 150 mm (6 in.). Larger ring sizes are not as readily available, and the rings can be quite expensive.

Note that the stripper ring seats on the cavity and on both the flat and tapered areas of the core after overcoming the preload of the taper. The stripper *plate* is clear from both cavity and core and does not help in supporting the clamp force. If the seating areas of the ring(s) are too small to support this force safely, additional support areas must be added.

Disadvantages of the above methods are that the stripper ring is free to move sideways when affected by gravity and must align every time after it has moved. To overcome this

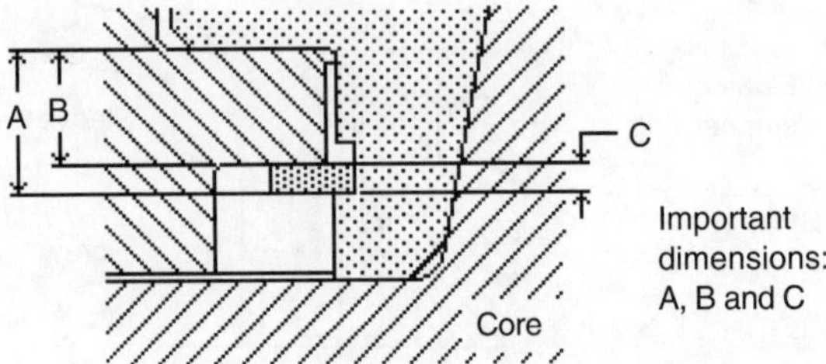

Figure 12.52 Schematic shows dimensional relationships for snap ring size.

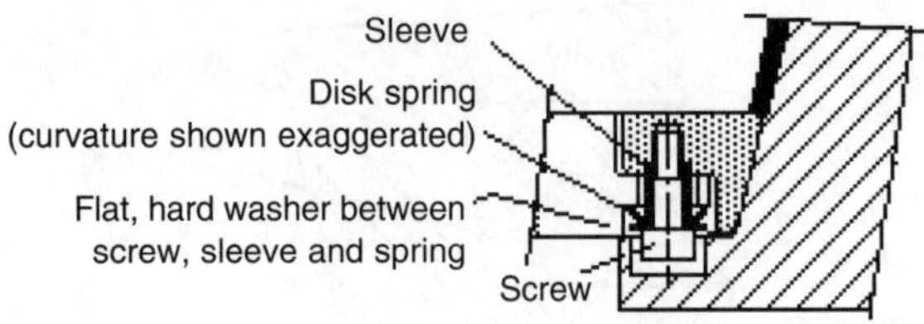

Figure 12.53 A disk spring and a washer between the screw and spring provide a slightly different floating core mounting.

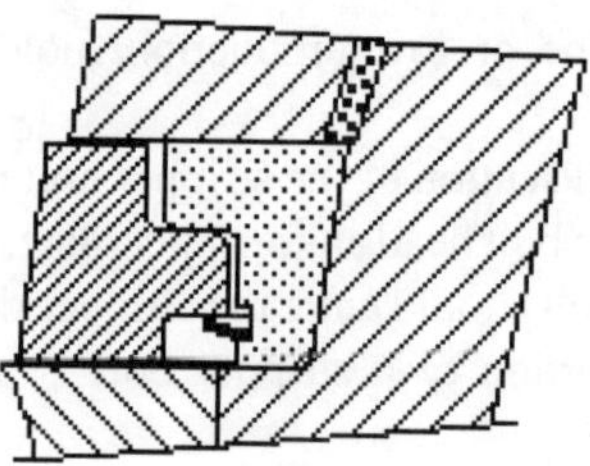

Figure 12.54 Dished snap ring.

problem, a spring force can be introduced which acts as "brake" and prevents the stripper ring from sliding too easily out of alignment.

A method similar to the floating core mounting is shown in Fig. 12.53. The problem could with this method could be lack of space.

A variation on the snap ring is a dished snap ring. The spring force created by the dish shape pulls the stripper ring onto its seat on the stripper plate (Fig. 12.54). This is much simpler than the previous design, but this type of snap ring, especially in large sizes, may be difficult to get; it is also expensive.

In the past, only circular stripper rings with tapered seats on the core could be produced. With modern technology, more odd shapes of core seat and stripper ring configuration are now possible to produce.

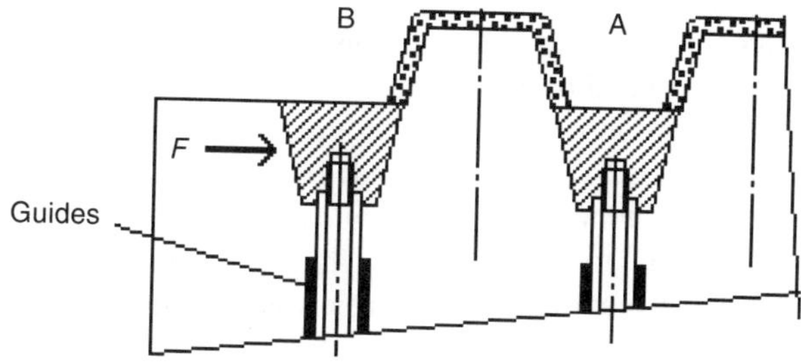

Figure 12.55 Stripper bars.

12.6.4 Stripper Bars

Stripper bars can be considered as stripper rings with an infinite inside radius ($R_i = \infty$). All rules for stripper rings apply, including guiding, limiting of stroke, taper seat. However, while the ring "hugs" the core naturally, a bar needs an additional force *F* to make it seat on the core. This force must be supplied by an outside taper that holds and seals the stripper bar on the core (Fig. 12.55). The guides must not fight the tapers, but floating as for stripper rings is not possible. There must be enough clearance in the guides.

In Fig. 12.55, at A, the stripper bar lifts two products, and the two opposing tapers ensure proper seating on the core. The forces between the two cavities are balanced. At B, the "outside" stripper bar lifts only one product, and the outside taper provides the necessary balancing force *F*. In this schematic illustration, the taper is cut into the core block, but it could also be a separate wedge, held in the core plate.

Stripper bar ejection is sometimes used where a number of elongated products must be stripped, rather than ejected by pins. The products can be arranged side by side, with the stripper bars between the cores, as shown in Fig. 12.55, or with only two outside bars lifting a series of products from their narrow ends. This method is also useful for products that must be stripped off the core, but for which a shaped stripper ring would be extremely difficult to produce or to maintain.

12.6.5 Stripping from the Injection Side

There are cases where it is necessary to gate from the inside (or underside) of the product, usually for cosmetic reasons (appearance) of the product, but also for some functional reasons. The outside of some overcaps must be smooth to the touch or will be decorated, so for *cosmetic* reasons they will not tolerate any gate mark on the top surface and require inside gating and ejection from the same side.

When molding dinnerware (plates, trays), for *functional* reasons, no gate vestige or ejector pin marks in the top of the plate are permitted. They must be injected from the underside; if they have any severe ribs, or undercuts to hold the product on the injection side, they must be ejected from under the ribs or near the undercuts.

The rules, whether using ejector pins or strippers, are similar to ejection from the core side. However, added depth to the injection side is necessary to accommodate the ejection mechanism. This increases the length of drops of hot runner or three-plate molds, or the length of sprues in two-plate molds. With the additional complications in the construction to activate ejection from the injection side, this will considerably increase the mold cost.

When specifying such products, the product designer (or customer) is sometimes unaware of these complications and added costs as compared to conventional ejection from the moving mold half. An alert estimator or mold designer should point this out to the customer and may be able to suggest some changes to the product (e.g., hiding of gate vestige in ornaments, etc.) to be able to build a more conventional mold with ejectors from the moving side. This could result not only in a lower mold cost but also in lower production costs (better cycles, less upkeep, longer mold life).

Air ejection from the injection side will be covered under air ejection.

12.7 Air Ejection

Air ejection can be used for practically any cup-shaped product. When compressed air enters the space between the core and the product, it does not immediately exhaust into the open, as it would with a flat shape, but has enough time to create pressure under the product to push it free from the core.

12.7.1 Problems with Air Ejection

A container with little draft and heavy undercuts that are holding it on the core may build up a heavy pressure inside the product (Fig. 12.56). The product may burst if it is not strong enough.

The product may suddenly let go from the undercuts and fly with force against the cavity side of the mold. The product may lodge itself inside the cavity and stop the machine at the next cycle, and/or the product may get damaged (scratched) as it hits the cavity side. Reducing the air pressure may solve one problem, but the product may not get free from the undercuts and may fail to eject. In such cases, mechanical ejection must be used to strip the product off the core far enough so that even low air pressure can eject reliably the rest of the way.

Containers with little or no draft, but without undercuts, are very successfully ejected with air as illustrated in Fig. 12.56, and many molds are in operation using this method.

The Bernoulli effect must be taken into account with air ejection. This happens within a certain range of angles of side walls, and if the air enters the space under the product at the

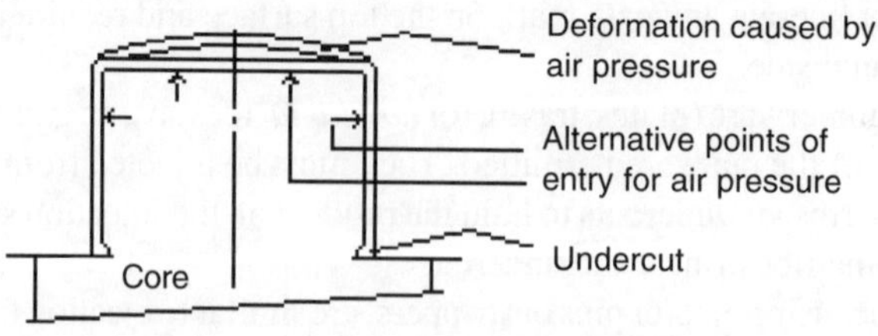

Figure 12.56 Heavy air pressure inside the product.

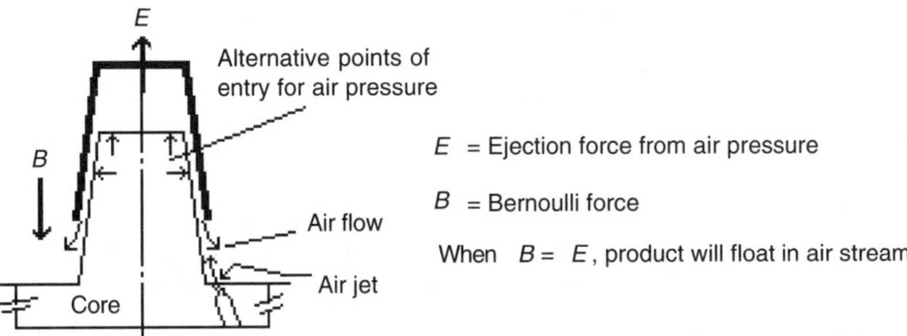

Figure 12.57 Bernoulli effect in products with angled side walls.

wrong spot. As the air flows out of the space between the product and the core, in the direction of the open end of the product, it can create a low pressure inside the product which counteracts the force exerted by the air pressure on the product (Fig. 12.57).

The result of this effect is that the product will eject some distance but then stop and float in mid-air without falling off the core. Proper location of blow slots can prevent this from happening, and changes in the air pressure and the resulting air flow may help. Blowing air through small jets (small, carefully sized holes) in the direction of the gap between core and product (as shown in Fig. 12.57) provides good ejection. NOTE: The timing of the blow-off sequence is very important.

In a typical setup, the duration of core air is 0.2 seconds. At the end of this 0.2 seconds, the blow-off jet is actuated for 0.3 seconds. These short blasts are sufficient for good ejection.

When there are openings in the bottom of a container, the designer must be guided by the following principles. Openings (holes) are acceptable as long as the amount of air entering the space between the core and the product is greater than the amount of air lost through the openings, so that some pressure can build up to eject the product. If there are undercuts, more force will be required, and mechanical assistance for ejection may be required.

Here, again, the location of the air slots or air entry is important. If the air enters right at the openings, no pressure can build up. With holes in the bottom of the container, there is no Bernoulli effect present.

12.7.2 Basic Requirements of Air Ejection

Air supply must be sufficiently large. The amount of air permitted to escape in each cavity must be proportioned so that in multicavity molds, even with some or most of the products ejected, there is still enough pressure in the channels to be able to eject the remaining products.

Dimensional uniformity of air slots or jets is imperative to avoid flashing into them, but at the same time to ensure sufficient air flow through the slots or jets.

Air must be clean (filtered) without dust, oil or water. Air pressure must be reliably controlled to provide continuous, trouble-free ejection.

Caution: In some medical and food applications, the ordinary (not sterile) compressed shop air must not come into contact with the product from the inside or the outside. In these cases, the designer must make sure that the selected method of ejection will function without air or air assist.

The advantages of air ejection without mechanical assist are:

- Simpler core half construction.
- fewer masses to be moved (power savings),
- less shut height,
- less mold weight,
- absence of impact noise and wear of ejector mechanism,
- no metal wear, which is always present with taper seats of stripper rings, and
- less down time required for maintenance, particularly on parting line.

Disadvantages of air ejection are:

- Much compressed air required, which can be expensive. Also, existing compressed air supply may not have enough capacity.
- balancing of air outlets, particularly in multicavity molds, requires intricate channelling in cores and/or core plate (Since this is in addition to the cooling channels, it makes the mold layout more complicated.), and
- controls such as additional (air) circuits, hoses, and valves are required.

However, the advantages usually greatly outweigh the disadvantages, and more and more molds are built with air ejection, many of them without mechanical assist.

12.7.3 Poppet Ejection

This is the oldest method of air ejection. Today, it is rarely used alone but often is used in conjunction with air slots or mechanical ejection.

During injection, the poppet is tight on the taper seat in the tip of the core and prevents plastic from entering into the air system. After the mold starts opening, the poppet moves forward by the amount *SE* (Fig. 12.58).

This forward motion of the poppet can come from the air pressure on the underside of the poppet, and/or from the pressure of the coolant (see below). In both these cases, the stroke *SE* will be small, in the order of 0.25 mm.

The poppet can also be actuated by the ejection mechanism of the mold (combined mechanical and air ejection). Some designs use a double acting air cylinder at the base of, and acting on, the valve stem. The area is larger than the poppet itself to give more force. This may

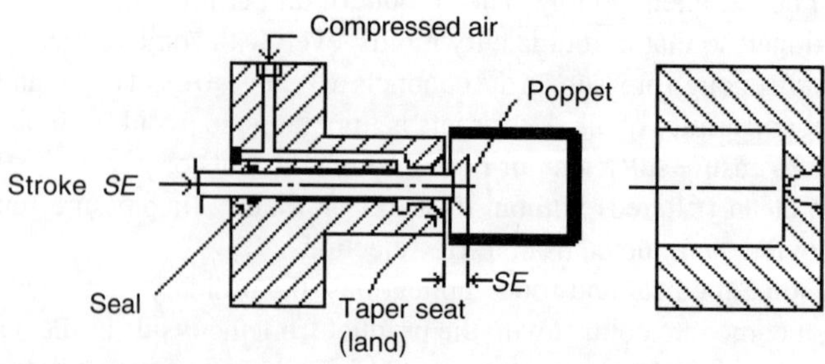

Figure 12.58 Poppet ejection.

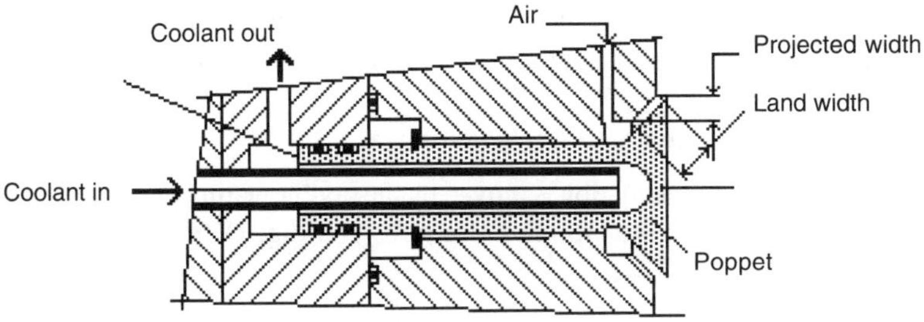

Figure 12.59 Poppet ejection with coolant provided.

eliminate the need for a separate mechanical actuation. Air seals around poppet stem prevent the air from escaping.

The main disadvantage of this design is that the poppet in not cooled where it is most needed (i.e., opposite the gate), resulting in slower molding cycles. A better, simple method is shown in Fig. 12.59.

The poppet in Fig. 12.59 is well cooled. The cooling water pressure acts on face A and keeps the poppet open even without air pressure. The plastic will close the poppet as the injection starts. The main advantage of this design is its compactness. Air and coolant are well separated in a small space. The poppet is held in place by a snap ring.

12.7.3.1 Rules for All Poppet Designs

The poppet should be located opposite the gate so that the inrushing plastic will close the poppet and prevent plastic from entering the air channels.

If the poppet is not opposite the gate, springs can be used to close the poppet to make sure that plastic cannot enter under the poppet. These springs must be properly designed with enough force to hold the poppet closed and with enough length to make sure that the spring will not fatigue.

The stroke *SE*, if the poppet is located opposite the gate, must be less than the 50% of the thickness of the product at this spot. There must be enough space for any cold slug in the gate to eject freely into the space between gate and the top of the poppet. The stroke *SE* is usually selected as 0.25 mm, unless returned by positive action.

Commercially available poppets—complete assemblies of poppet and spring in one unit (Fig. 12.60)—can be pressed into the core at the desired location. For removal, they must be drilled out.

The size of seat (land) of the poppet is similar to any other seat that comes under injection or clamping pressure. The load should be limited to 5% of the yield of the softer of the two materials contacting at the seat. For calculations, use the projected area, not the actual width (see "land width" in Fig. 12.59).

The designer must ensure that plastic will never enter the air channels. This could cause a major down time in production and a costly mold repair.

For positive return of poppets, if a larger poppet stroke *SE* is required, the poppet must be returned to its seat before the cavity approaches the poppet to ensure that no damage is caused to the poppet or to the surface of the cavity. The forward and return stroke of the poppet can be done either mechanically, by connecting the poppet to the mold ejector system, or by independent double-acting air (or hydraulic) actuators acting directly on the poppet.

12.7.3.2 Poppet with Double-Acting Actuator

A poppet with double-acting actuation is moved by the double-acting piston, which is held by a snap ring. The air sequence is as follows (see arrows in Fig. 12.61):

1. Piston forward,
2. blow-off air, and
3. piston return.

It is possible that steps 1 and 2 may act simultaneously.

Sealing elements must be properly selected for such designs. In the author's experience, T-seals are better than O-rings for sealing in axial motion, but Teflon™ - type seals, such as

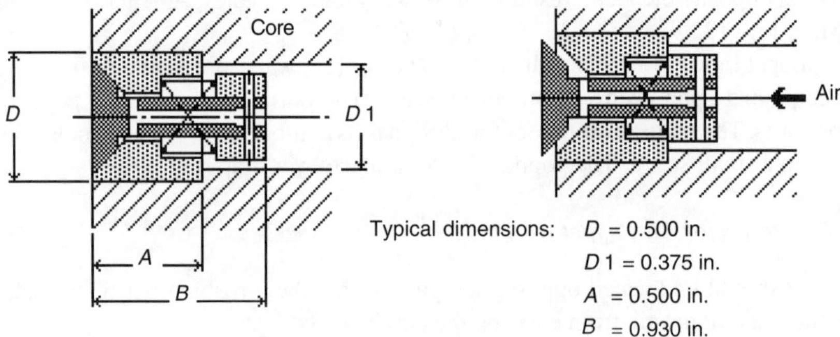

Figure 12.60 Typical commercial poppet.

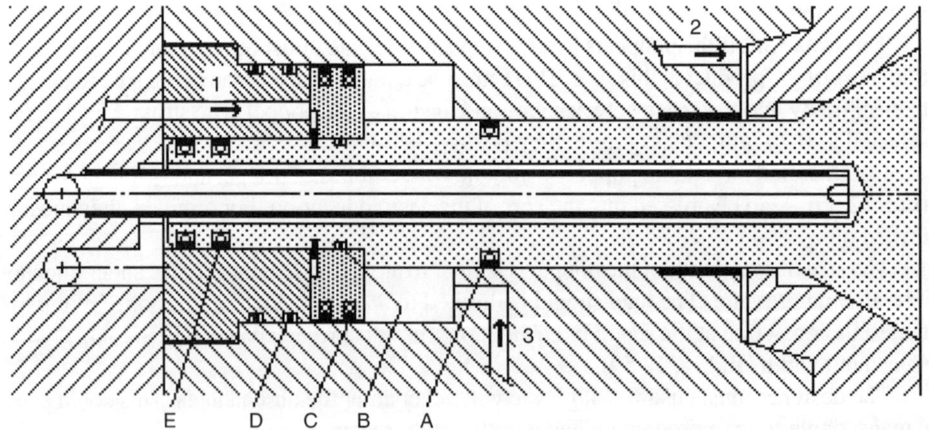

Figure 12.61 Double-acting poppet actuation. The air sequence is numbered: 1. piston moves forward, 2. blow air on, and 3. piston returns. Letters A through E indicate seals: A. double delta II ring, B. O-ring, C. Glyd ring, D. O-ring, and E. double delta II ring.

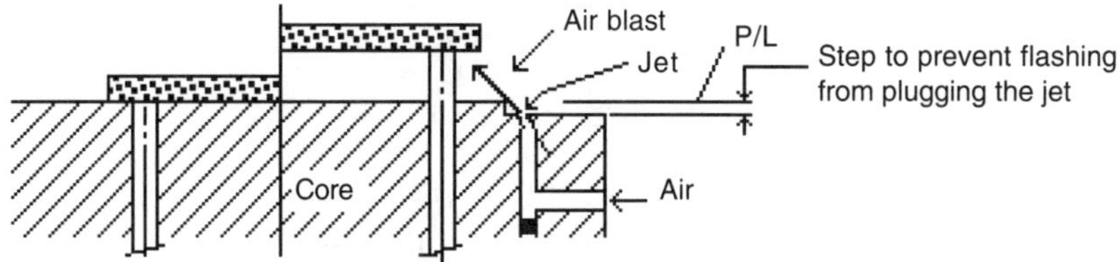

Figure 12.62 Blow jet prevents plastic from entering an orifice.

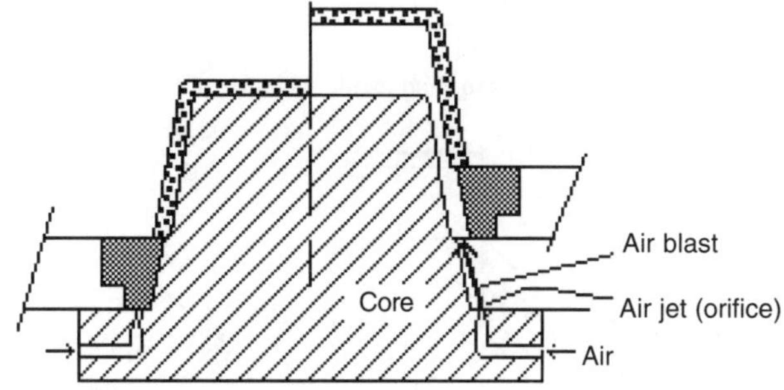

Figure 12.63 Blow jet directed at underside of product prevents it from sticking to stripper.

Glyd™ rings and Double Delta II™ rings, are still better. Seals for the design in Fig. 12.61 are recommended as follows:

A. Double delta II ring,
B. O-ring,
C. Glyd ring,
D. O-ring, and
E. Double delta II ring.

12.7.4 Blow Jets

The *intensity* or force of the air blast coming from a blow jet (orifice) decreases considerably with the *distance* from the orifice. The orifice must be far enough from the molding surface so that plastic cannot inadvertently enter and plug the orifice or, even worse, the air channel (Fig. 12.62). The orifice should be uncovered only after the cavity moves away, and/or after the stripper moves forward a certain distance.

Blow-off (directed at the underside of the product) is used to separate the product from the ejector pins or stripper, where it may stick (Fig. 12.63). Blow-off also permits limiting of the ejection stroke of the stripper to less than would be normally required to keep the shut height of the mold small in case the stripper is actuated from the ejector box. The stripper travels only a short distance, and the final ejection is achieved by blowing air against the underside of the product. The disadvantage is that the distance from the orifice increases as the stripper advances.

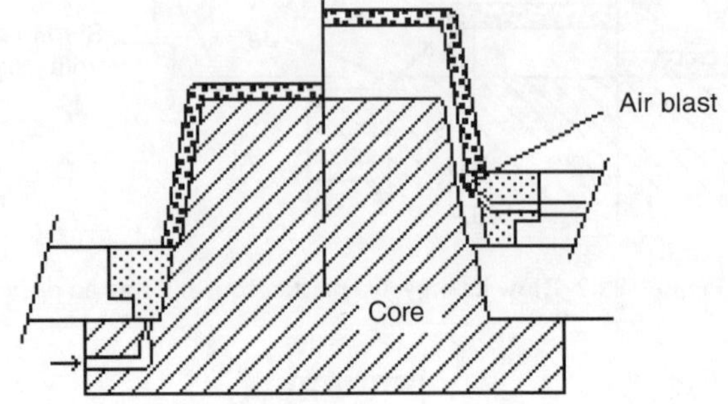

Figure 12.64 Blow jet is located inside the taper of the stripper ring.

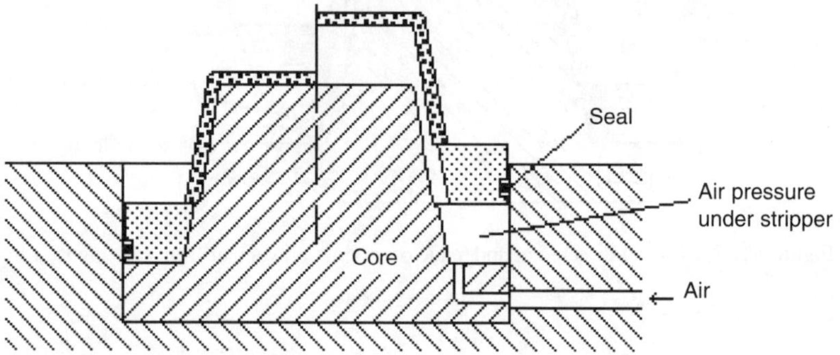

Figure 12.65 Air pressure builds up below stripper ring under seal.

In the following example (Fig. 12.64), the jet is inside the taper of the stripper ring, and the distance to the parting line does not change. This is only practical if the stripper is carried by a stripper plate where the necessary channelling and sealing (not shown) for the air can be located.

A similar effect can be achieved by pressurizing the underside of the product. The space below the stripper ring is sealed to ensure that air pressure will build up (Fig. 12.65).

12.7.4.1 Blow-Down

Blow-downs are air blasts or air streams which blow at the product from the outside, after the ejector (stripper or pins) have come forward, to remove the products if they stick to the ejectors. Previously, they were frequently installed after the mold had been tested. However, their need usually can be anticipated and become part of the mold design

There are two methods of blow-down commonly used: individual open jets or blow-down curtains. Individual jets directed against the tip of the product (sometimes adjustably mounted) are used to get the best leverage for knocking the products off the stripper. Previous designs used open jets, usually by squeezing the ends of flexible copper tubes. This method is simple but has some distinct disadvantages:

- Safety regulations do not permit pressurized air to escape freely if above 30 psi,
- the noise of escaping compressed air is excessive and often above legal limits,

- air through the jets is difficult to control,
- high operating cost.

Example: The amount of air escaping through a ¼-in. copper tubing at 80 psi is 33 ft^3/min. (cfm). At an approximate cost of $0.25 per 1,000 cf, one such continuously open nozzle costs:

$$\frac{33 \text{ cfm} \times 60 \text{ min./hr} \times 24 \text{ hr/day} \times 5 \text{ days/wk} \times 50 \text{ wks/yr}}{1{,}000 \text{ cf}} \times \$0.25 =$$

$$\$2{,}970/\text{year}.$$

The jet is usually smaller than the hole of a ¼-in. tubing, but it is still large enough to waste a large amount of energy (and money).

A better system is the Exair™ nozzle, which is described in Fig. 12.66. (The information given here is a condensation of Exair's data.) These nozzles utilize the "coanda" effect (wall attachment of a high velocity fluid).

Compressed air is ejected through a thin ring nozzle. As this ring of air travels along the outer wall of the nozzle, surrounding air is entrained into the stream. This entrained air stream combines with the air ejected from the center hole to produce a high volume, high velocity blast of air, at minimum consumption.

Exair claims that this nozzle uses as little as 20% of the air as an equivalent open nozzle. Also, the noise level is below 90 dBA, and the air pressure at the exit is below the permissible 30 psi. For more data on the Exair nozzles, see the manufacturer's data sheets (Exair Corporation, 1250 Century Circle North, Cincinnati, OH 45246).

Blow-down curtains are the second method of blow jet. Rather than using individual jets, a continuous sheet of air blows down over the surface of the mold. As the mold opens, the products arrive into a moving air stream and are blown down.

Air curtains are sometimes used even if the products eject easily to prevent them from hitting the cavity side of the mold. They also assist products sliding down a chute. There are two methods used:

1. High-volume, low pressure air curtain: A low-power air blower is mounted above the mold and blows a steady stream of air down over the mold. (Blowers are more economical than pressure blasts but may not be effective in some cases.) The advantages of this method are that it is practically noiseless; also, the power consumption is negligible. Since the blower runs continuously, timing devices are not required.

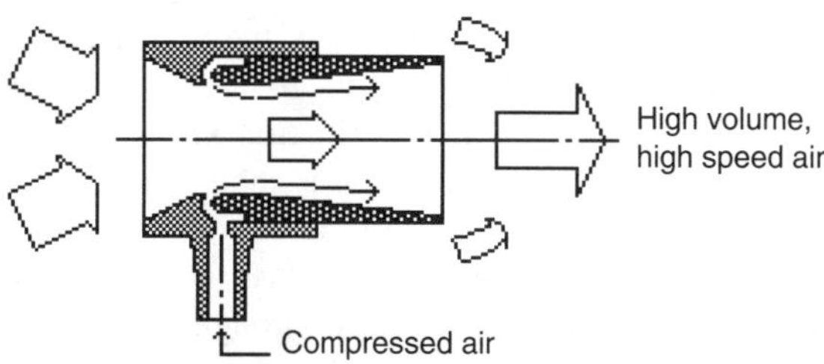

Figure 12.66 Cross section of an Exair™ nozzle.

The disadvantages are that the air pressure (and air velocity) is low and may be insufficient to remove some products. Also, the blowers are rather bulky and awkward to install. Today, this method is rarely used.

2. Exair Knife™: The principle of this method is similar to the Exair jets. It is applied to a continuous, very narrow gap in the order of 0.02 in. (0.5 mm) along a plenum chamber of standard lengths from 6 to 30 in. The air escaping the gap entrains the surrounding air and thereby creates a high-volume, high-velocity, relatively quiet air curtain. For more details, see data by Exair Corporation, 1250 Century Circle North, Cincinnati, OH 45246.

The general concern with all blow down methods is that the products are thrown down with a force greater than that created by gravity alone and are more likely to be damaged (scratched) than by free fall alone. As with air ejection, the use of air directed at certain food and medical products may not be acceptable because of the risk of contamination from the dirty shop air. This must be established before designing an ejection system for a mold. Filters, etc., may then be required, or another reliable ejection system must be selected which does not use air.

12.7.5 Blow Slots

Open slots stay open during injection even though they are in the cavity space and subject to injection pressure. If they are too large, plastic can enter into the slot and into the air channels. *Closed slots* stay closed during injection and open after the mold starts opening. There is no risk of plastic entering during injection, and they can open wider than the open slots to let more air pass through. The disadvantage of closed slots is that they require moving parts, which are subject to wear.

12.7.5.1 Open Blow Slots

There are three basic types of open blow slots:

1. Cylindrical,
2. peripheral near the core tip, and
3. peripheral near the core base.

A cylindrical open blow slot is shown in Fig. 12.67. Advantages are the simplicity of the design and the rigidity of the core.The main disadvantage is that the blow slot has a relatively small circumference and is located near the gate, where the plastic is hottest and least viscous. The plastic is more likely to enter the blow slot. The narrow gap and the small circumference reduce the total cross sectional area of the blow slot and the amount of air that can pass through it. This method is generally not recommended by itself, but it can be used to assist ejection by breaking the vacuum inside the product, in combination with stripping.

A peripheral blow slot near the tip of the core is shown in Fig. 12.68. Advantages of a peripheral blow slot is that the circumference of the blow slot is longer, and the plastic has already cooled down somewhat. With less danger to flash into the slot, it permits a wider gap. This helps to increase the amount of air that can pass through the gap. The disadvantage is that the core tip near the corners is less well cooled. Use of beryllium-copper for the core tip is suggested.

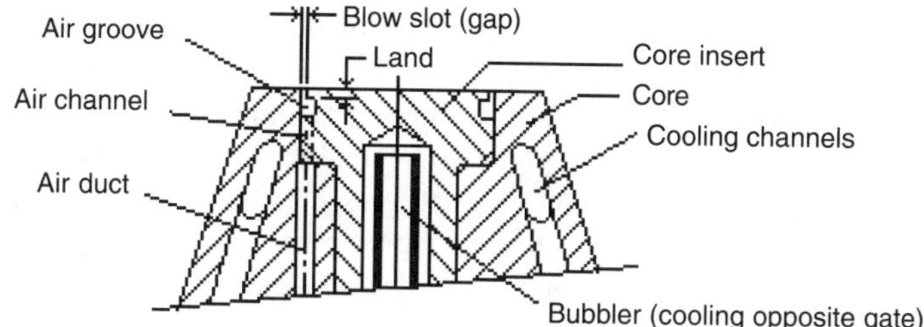

Figure 12.67 Cylindrical blow slot.

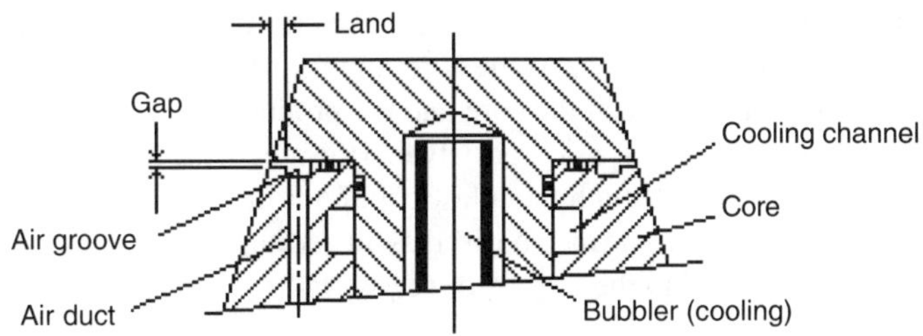

Figure 12.68 Peripheral blow slot near the tip of the core.

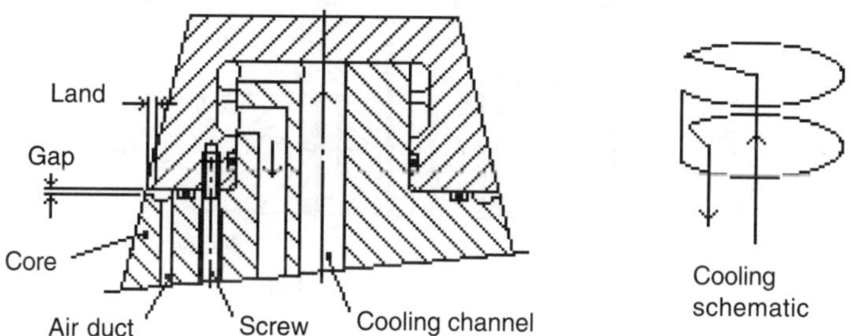

Figure 12.69 Peripheral blow slot nearer the core base.

The third method, a peripheral blow slot nearer the base of the core, is shown in Fig. 12.69. The advantage of this method is that the conditions for the blow slot are better than when near the tip; the tip of the core can be well cooled. The cap could be made from beryllium-copper for still better cooling. The disadvantage is that the slot should be in the upper third of the core height. If the slot is too near the open end of the product, the air will escape too soon and fail to eject the product. Suggested dimensions for blow slots are listed below:

Gap (cylindrical)	0.01 mm (0.0004 in.)
Gap (peripheral, near core tip)	0.015–0.020 mm (0.0006–0.0008 in.)
Gap (peripheral, near core base)	0.025 mm (0.001 in.)
Land	1.0 mm (0.040 in.)

The air groove behind the land should be shaped similar to vent grooves.

There are many variations of the above three methods, but the principle is always the same. The size of the gap (open slot) also depends on the type of plastic and the molding conditions (temperatures and pressures). With high operating temperatures and pressures, and with a plastic with very low viscosity, the gap must be smaller than with a plastic with high viscosity and/or with low operating temperatures and pressures. Because this is not always easy to control, the use of a closed blow slot is often preferred, even though the system is somewhat more complicated (and costly) and more subject to wear.

12.7.5.2 Closed Blow Slots

Figure 12.70 shows a typical schematic of a core with a closed blow slot. It is opened after the mold opens by the blow air, which acts on a large surface in the "large groove" inside the cap. The air pressure creates sufficient force to lift the cap to the end of its small stroke. Stroke is usually 0.125 mm (0.005 in.) As soon as the core ring lifts, the blow slot is open, and the air enters freely into the space between the core and the product.

In Fig. 12.71, some clearance is required between the moving cap and the fixed core insert. This clearance must be small enough that plastic will not flash into it, but large enough not to bind the ring. This clearance can also be used as an additional (open) blow slot, as shown, with an air groove connected to the large groove.

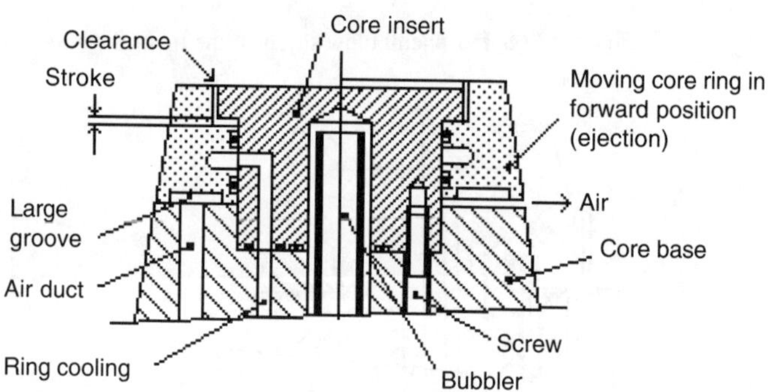

Figure 12.70 Closed blow slot.

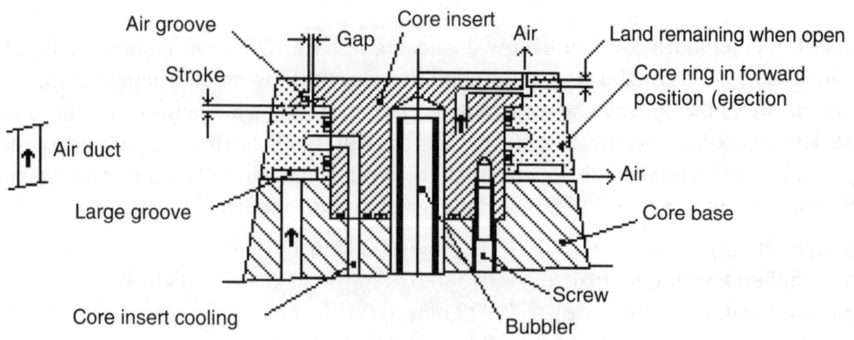

Figure 12.71 Closed blow slot with additional open blow slot.

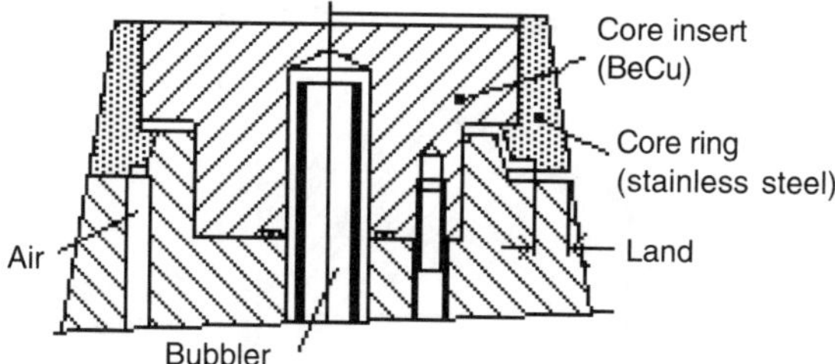

Figure 12.72 Closed blow slot with a thin, light core ring that requires no internal cooling.

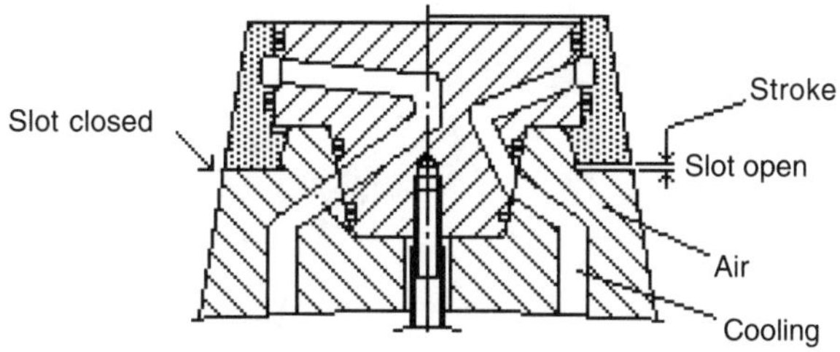

Figure 12.73 Closed slot design with cooled core cap.

This method could be necessary if vacuum in the space on top of the core causes a problem. The cap is returned to the original position by the pressure of the plastic during injection. The land must be wide enough to prevent plastic from entering the slot before the cap is seated.

The size of land must be determined by the designer. In closed slot designs, after opening, the land creates little resistance to the air flow as compared to the open slot design. However, the size of land is important, because it supports the injection pressure acting on the moving ring. As a rule, the size of land should be calculated so that the compressive load is less than 5% of the yield strength of the softer of the two parts, the ring or the core. There are many variations to this design.

In Fig. 12.72, the moving ring is very thin and light and has no internal cooling, which makes it easier to produce than a cooled version. The size of the blow slot is shown exaggerated. The ring is returned by the plastic pressure. Figure 12.73 is an example of the closed slot design with a relatively small moving ring, with the core cap well cooled. The core assembly is more complicated than the one in Fig. 12.72.

It is also possible to have a combination of open slot ejection and poppet ejection. The poppet design could be as shown earlier, operated with coolant holding the valve open (Fig. 12.74).

There is very little stroke of the poppet (0.25 mm), and the peripheral blow slot is always open at 0.025 mm (0.001 in.). Air groove sizes are similar to vent grooves.

Because the core cap (from the O-ring outward) overhangs the core base, it is suggested to make the blow slot interrupted (Fig. 12.75), so that at least a portion of the core cap can seat on the core base, beyond the air supply groove. This prevents flexing of the over cap and failure at the slot.

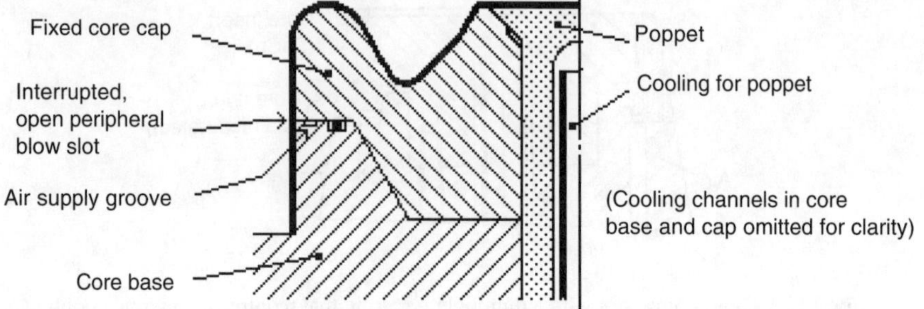

Figure 12.74 Combination open slot ejection and poppet ejection.

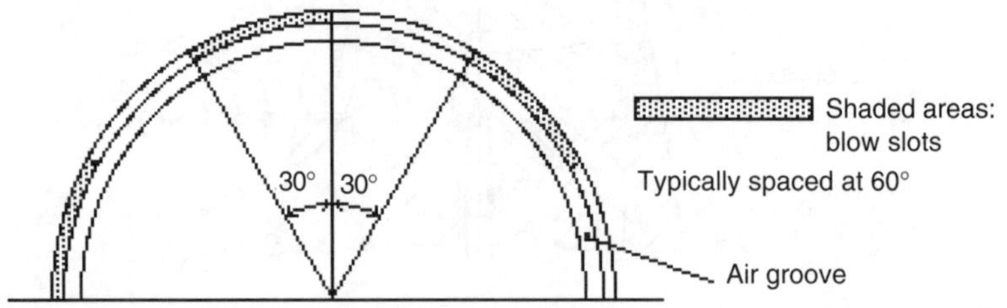

Figure 12.75 Plan view of interrupted blow slot: core cap seats on core base beyond the air groove.

12.7.6 Air Ejection from the Injection Side

This type of ejection is simpler than when compared to stripper or ejector pin ejection from the injection side and does not require much, if any, added shut height to accommodate the air channels. All basic rules of ejecting with air from the core side apply.

The additional air channels in the hot side, in addition to the runners and cooling channels, result in a more complicated cavity and gate pad layout. For example, air pressure may be directed on the product in the cavity side with a moving cavity. Figure 12.76 shows schematically how air pressure can be used to assist in freeing a severe outside undercut (rib) at the bottom of the product.

As the mold opens, the cavity follows the core and moves forward sufficiently to permit the outside rib to squeeze between the cavity walls, while holding the product on the core by the air pressure on the injection side. The product will then be ejected (by air poppet) from the core, and the outside rib will return to its molded shape.

The advantage of this design is a much simpler mold, which would otherwise require a split cavity to free the rib. The forward motion should not depend on the air pressure behind the cavities, but actuators must be added, which are not shown in Fig. 12.76.

12.7.7 Combined Mechanical and Air Ejection

The following discussion may repeat some statements made earlier, when discussing different methods of ejection. The basic rule for any automatic ejection requires that the products will

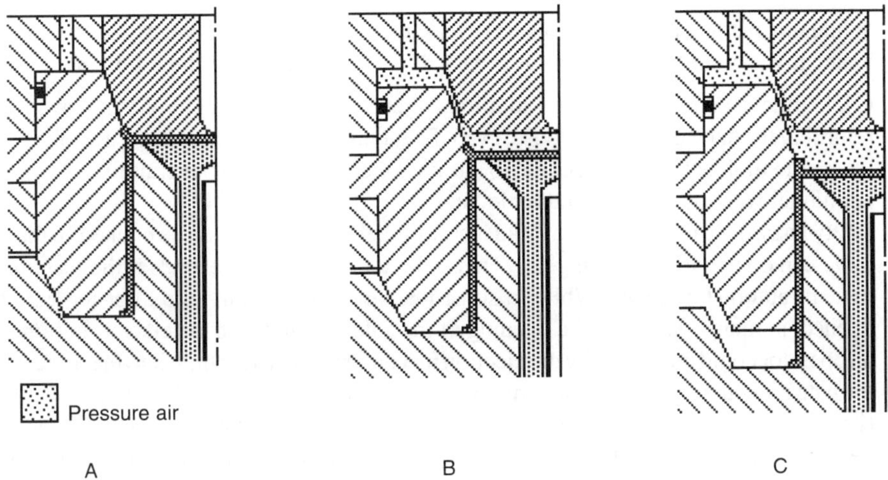

Figure 12.76 Air ejection from inside the cavity: A. mold closed, B. mold opens to permit rib to clear while air holds product on core, C. product follows core, with rib squeezed by cavity walls.

always (100% of the time) be reliably ejected, that is, they must clear the molding area before the mold recloses. If a product hangs up only once in 10,000 cycles, (a reliability of 99.99%), it is too often and not acceptable because it will, sooner or later, cause a stoppage of the machine. The designer must recognize the importance of this fact and ensure that the selected ejection method will work at all times. (See also Section 2.5, Ejecting the Product.)

Mechanical ejectors (ejector pins or strippers) often fulfil this requirement, particularly if the mass of the product is greater than about 10 g. With lighter products, and even with heavier ones where the shape of the product is such that it may catch somewhere, or where the plastic may stick to the ejector or stripper, some additional ejection method must be provided to ensure 100% reliability of ejection. Air ejection or air assist, such as air jets, air curtains, or air pressure under the product, are preferred solutions.

Air ejection may not work 100% in cases where undercuts on the core may prevent the product from coming off the core. In these cases, mechanical ejection (e.g., a poppet with sufficiently long stroke, or ejector pins or strippers), even for a short stroke only, will be required to assist the air ejection.

If neither of these combinations (air-mechanical or mechanical-air) will ensure proper ejection, two-stage ejection must be considered.

12.8 Multiple Ejection Stroke

Today, all molding machines are equipped with a feature that allows the machine ejector to stroke more than once after the mold has opened. Some molders use it to "shake" the product loose from the stripper if the product hangs up occasionally. Machines usually permit up to ten strokes.

While this method can sometimes be successful in running a poorly designed mold fully automatically (without an operator), it is basically a bad practice for the following reasons:

1. Cycle time is penalized with the additional ejector travel times of stroking forward and returning to shake the product loose, even if it has already ejected on the first stroke (Note that some molders set the machine to stroke not only twice, but several times.),
2. additional wear of machine and mold, and
3. safety hazard. Usually, after the mold has closed, the safety gate can be opened so that the mold can open with the open gate. (This is done to save a few seconds if the machine is in semi-automatic mode.) While in some machines all motions stop after the gate has opened, in many machines, the opening of the gate may not stop the ejector motion, and an operator could be injured by the ejection mechanism when reaching into the mold while the gate is open. When designing a mold for such machines with semi-automatic operation (i.e., if inserts must be placed into the mold or the product must be removed by hand), the designer must ensure that the machine and the ejectors cannot move with the open gate.

All products must eject 100% at the first ejection stroke.

12.9 Special Ejection Methods

Wherever possible, the designer should use one of the simple ejection methods (i.e., ejectors or strippers or air) in their simplest applications as discussed earlier. However, there are occasions where these methods are not suitable, or not reliable enough, and special ejection methods may be necessary.

There are no limits to the ingenuity of the designer, as long as the basic requirement—100% reliability of the molding operation—can be achieved at the lowest mold cost and with minimum cost of upkeep. The following examples show schematically some special designs.

12.9.1 Neck Ring Ejection

This method is required when no witness line is permissible at the face of the radius, indicated with X in Fig. 12.77. Standard methods of ejection would be sleeve or stripper ring ejection, which are not acceptable in this case.

Immediately after the mold has opened, the neck rings (which are connected to the ejector mechanism) advance (SE). Cams in the mold first separate the neck ring slightly while the core is still well within the product to ensure that the product will not stick to the neck ring. As the neck rings advance, they strip the product from the core (in Fig. 12.77, a preform for bottles). Nearing the end of stroke SE, the cams separate the neck rings fully in a radial motion (SR) to free the threads and the ring. The product can then fall freely or be removed by robot.

12.9.2 Tulip Ejection

In the following example, the requirement was to build a *small* multicavity mold for flash bulb cubes. The problem was that the necessary four-way split would normally require slides on four sides of the product and, therefore, use much space for conventional slide designs and

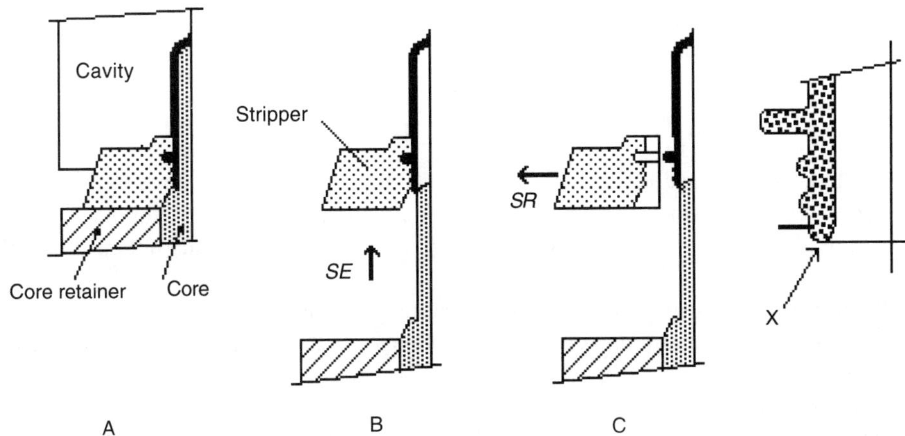

Figure 12.77 Neck ring ejection: A. closed mold, B. neck rings advance, C. neck rings separate fully.

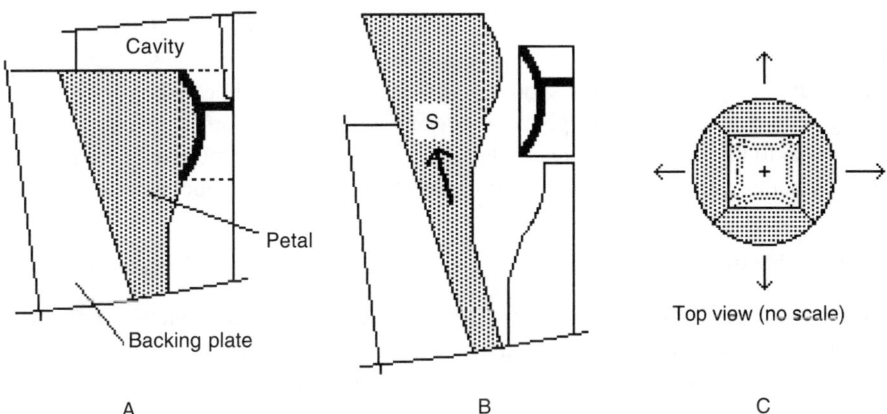

Figure 12.78 Tulip ejection of flash bulb: A. "petals" form product shape within backing plate, B. mold opening slides petals forward at angle, freeing product, and C. top view of closed mold.

their actuating mechanisms. A very large mold would be required to accommodate the planned number of cavities, far to large for the small machines required because of the very small shot size.

With the design in Fig. 12.78, there is little space required for opening up. Four "petals" are built so that they form the circumferential shape of the product while enclosed by the backing plate (Fig. 12.78A). After the mold has opened, the petals are pushed forward out of the backing plate, while sliding at an angle to the center of the core (Fig. 12.78B). This frees the undercut areas, and the product, which has already been pulled off the core by the motion of the petals, can now fall freely.

The mold making requires high precision to prevent flashing. Also, the petals cannot be cooled. However, because of intimate contact of the petals with the backing plate, the cooling was acceptable. This mold ran at about a 4-second cycle using crystal polystyrene. Theoretically, this design is good for any number of petals.

12.9.3 Core Pull Ejection

This method is occasionally used where the product must be split in a direction where the projected area is large when compared to the other direction, and where ejector pin marks are not permissible or, because of the nature of the plastic, the product would not eject reliably with pins.

In the example shown in Fig. 12.79, the deciding reason to select this design was that the plastic was soft and rubbery, and ordinary ejector pins into the side of the product would not eject reliably.

After the mold opens, the machine ejector drives the ejector plate forward to lift the products out of the core. Then, the air cylinder strokes to the top, and as the products contact the stripper they are pulled off the side cores. The mold is very simple, and there are no ejector marks on the product.

A similar design can be used for very long, small-diameter spools, which would require high, split cavities. The splits would have to be very solid to resist deflection created by injection pressure.

With the core pull design, the spools would lie in the parting line. When compared with the large side core forces in the split cavity design, the side core forces are relatively small, proportional to the size of the flange of the spool, or even only to the size of the hole in the spool. Also, there are no ejector marks in the working area of the spool.

12.9.4 Swing Arm Ejection

This method is applicable for products such as coffee mugs with a handle, or any similar product which requires that the cavity must be split. The split cavities are actuated by links. As the mold opens, the links swing around their pivots and move the splits away from the

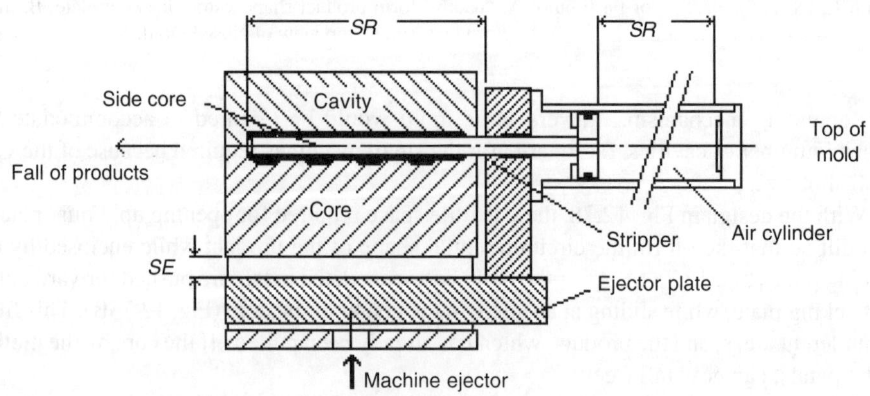

Figure 12.79 Core pull ejection.

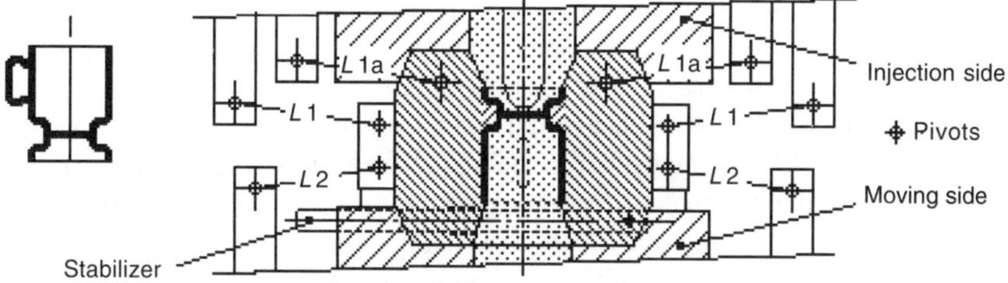

Figure 12.80 Swing arm ejection schematic (right) for product with handle (left).

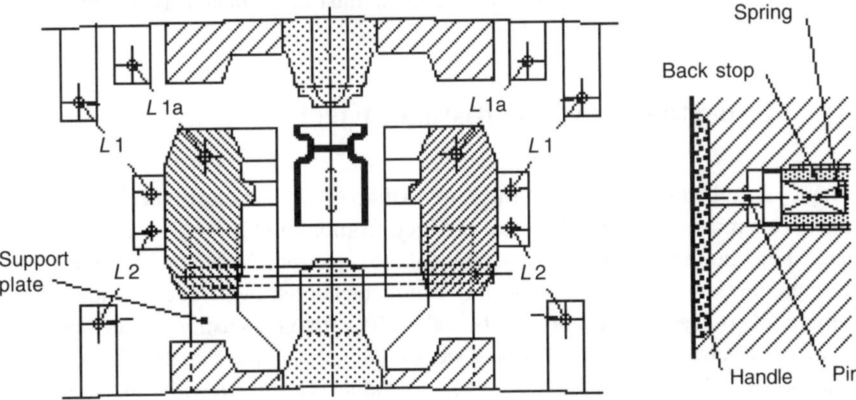

Figure 12.81 In swing arm ejection, a plate supports the heavy splits (left), while spring-loaded pins ensure that the handle does not follow the splits (right).

injection and moving sides, while at the same time separating. One additional link, L1a (same length as L1), is mounted on top of the splits to act together with the other links L1 as a parallelogram to pull the splits out in a straight motion (Fig. 12.80).

Two pairs of links mounted at each side of the splits have lengths L1 and L2. These lengths will be selected so as to provide the proper motion of the splits to pull the product off the core as the splits move away from the injection and moving sides. Lengths L1 and L2 can be found by computer simulation, but also by making a cardboard mockup of the essential elements (injection and moving sides, links, and the two splits) using pins at the articulation points.

It is important to watch that the corners of the wedges do not interfere as the splits swing to close. Spring-loaded pins may have to be used to ensure that the handle will not follow either one of the splits (Fig. 12.81). Any spring pin must be vented and easily accessible for maintenance (cleaning) to ensure proper operation. From experience, it has been found that it is better to place the mug handle at the top, rather than at the bottom, of each cavity.

To ensure parallel motion, a stabilizer (a leader pin and bushing) is mounted in one split, while sliding in a bushing in the other split. One pin only is required, and it must be positioned at the top of the splits to prevent the products from hitting it during ejection.

The heavy splits must be supported by additional support plates (shown in Fig. 12.81). These plates may have to be cut out so that they will not be in the way of the falling products.

Because of the sliding action of splits on top of these supports, they must be made from adequate materials, preferably self-lubricating materials (Teflon, Nylatron, etc.) covering the steel, or steel with a bronze surface or bronze ways, provided it can be lubricated.

The final alignment of the mold is assured when the splits come to seat on a cylindrical or tapered sections of the core plugs, before being locked by the wedges. The wedges must be preloaded enough that the injection pressure will not separate the splits. Also, the splits must be thick enough for minimum deflection during injection. Two leader pins (not shown in Fig. 12.80 or 12.81) near the top of the mold align the injection and moving side.

The advantage of this design is simpler construction as compared to cam-actuated side cores, which require additional ejection mechanisms. This method has been very successful with two- and four-cavity mug molds. The cavities must be vertically in one line.

12.10 Two-Stage (and Multistage) Ejection

Two-stage (or multistage) ejection is the combination of two (or more) mechanical ejection systems in one mold, acting on the same product. Combining blow-off or blow-down with mechanical ejection is not two-stage (or multistage) ejection. Also, combining air ejection with a single motion of strippers or ejectors is not considered two-stage ejection.

In this context, the use of robots or other take-offs with molds using single-stage ejection is also not considered two-stage ejection. Occasionally, due to the shape of the product two-stage ejection may be necessary, even where the product is removed with robots, because the mold must also be able to run automatically when operating without them.

Two-stage ejection can be combinations of these motions:

1. Ejectors–ejectors,
2. ejectors–strippers,
3. strippers–ejectors,
4. unscrewing–stripper,
5. unscrewing–ejector, or
6. side core–ejector.

The timing is usually such that both systems start together, and then one of them stops while the other continues, or that one starts after the other has started or has already completed its stroke.

The actuation of any of these systems can be by the following:

1. Machine ejector (usually hydraulic),
2. fixed (bumper) ejector (rarely used today),
3. cams,
4. air operators (cylinders), internal or external, or
5. hydraulic operators where large forces are required.

When using all-mechanical ejection, the force available is the machine ejector force. When using air actuators, the required forces must be calculated to provide sufficient force for 100% reliability of ejection.

Hydraulic actuators are used when it is not practical or possible to place large enough air actuators within the available space in or on the mold. They are also required where it is important to provide smooth operation of the actuator. There is no jerky motion (often associated with air cylinders); also, hydraulics can have good, accurate speed control. The disadvantage with any hydraulics is higher mold cost and the risk of contamination of the products with oil.

Control of two-stage actuation depends on the methods used. If both stages move mechanically (e.g., with methods using lost motion or dwelling on cams), no additional controls are required to operate the two-stage ejection. This is usually preferred but not always possible. If a mechanical method is not practical or possible to use for both motions, air or hydraulic actuators will be necessary with their additional piping and controls.

12.10.1 Why and When to Use Two-Stage Ejection

Two-stage ejection is always required when the product will not eject freely by being pushed off the core using standard (single-stage) ejection, regardless of whether the ejection is mechanical or by air. This may be the case for products with undercuts, as in the following typical example (Fig. 12.82).

The rim on the outside of the inner cylinder in Fig. 12.82 would hold tight and break or deform permanently if the product were ejected normally, as for example with stripper motion only. With two-stage ejection, the sleeve and stripper move first together (arrows 1 and 2), lifting the product and bringing the rim above the top of the core. At this point, the motion of the sleeve (1) stops, but the stripper (2) continues to push the product. This causes the inner cylinder (and rim) to be somewhat squeezed (indicated with arrows), which allows the rim to slide out of the groove where it was formed. It permits automatic ejection of the product without damaging it.

Even without undercuts, if the wall thickness is very small when compared with the depth of engagement between core and sleeve, and there is a risk of hanging up, the above or a similar method must be used. Also, two-stage ejection may be used whenever there is a possibility that stickiness of the plastic (or flash) would cause the product to adhere to the stripper or ejectors. Two-stage ejection ensures that a product which otherwise could be ejected with single-stage ejection will be positively pushed off by the stripper or ejectors.

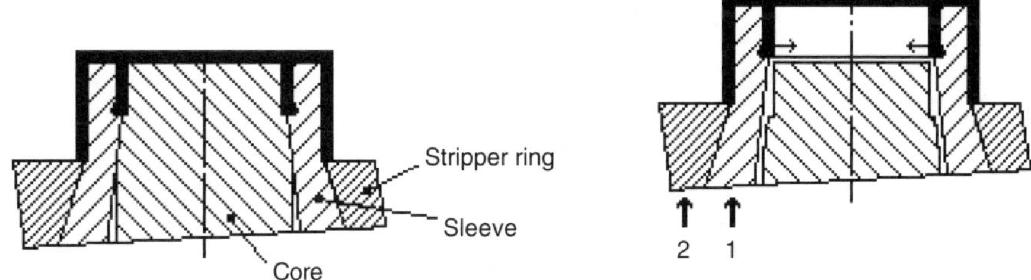

Figure 12.82 A product with undercuts is ejected in two stages: 1. sleeve and stripper lift product rim off core, and 2. stripper continues to push product out of groove, allowing automatic ejection.

12.10.2 Typical Two-Stage Ejection Actuations

Two machine ejectors may be used for the stripper and ejector. The stripper plate in Fig. 12.83 is actuated by one set of machine ejectors, and the ejector plate by another set; for example, by a fixed (bumper) and by a hydraulic ejector actuation. This is about the simplest arrangement possible but is rarely used today.

When using two machine ejectors similar to the method described above, but for two ejector plates, the ejectors act independently of each other (Fig. 12.84). Both ejector plates need return pins. The system is very simple, but it is rarely used today.

The following examples have two-stage ejection but are driven by one machine ejector system only. Shown are central ejectors, but a set of four or more outside ejectors could also be used. Shoulder bolts are shown for simplicity of illustration only. They must always be the screw-sleeve arrangements. Standard shoulder (or stripper) bolts must never be used in these applications.

Two-stage ejection combining springs and lost motion is shown in Fig. 12.85. When the machine actuator moves forward, it moves plate B; by the force of the springs, plate A is also moved forward. When A arrives at the core backing plate C, it stops, but B keeps moving

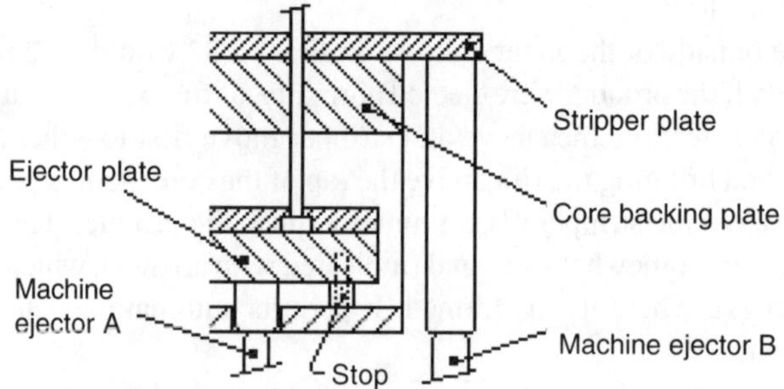

Figure 12.83 Two-stage ejection by two machine ejectors for stripper and ejector.

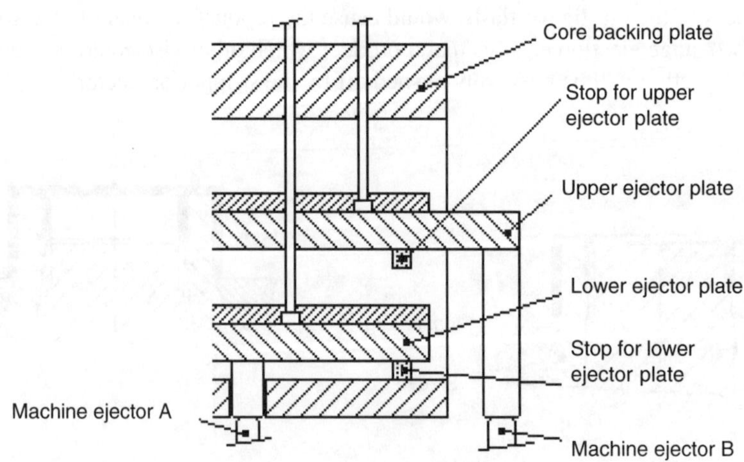

Figure 12.84 Two-stage ejection using machine ejectors for two ejector plates.

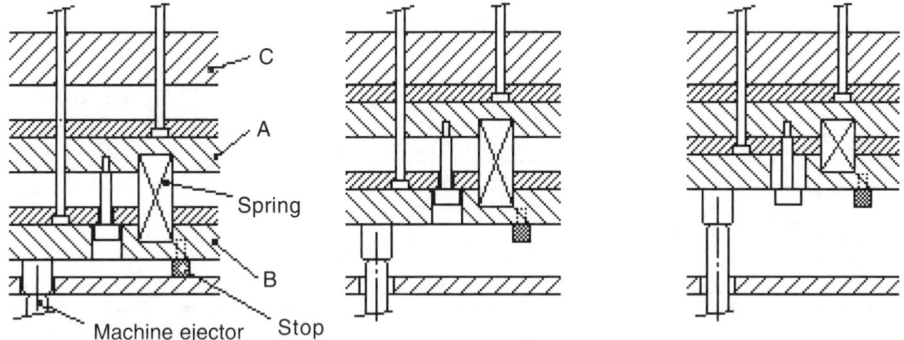

Figure 12.85 Two-stage ejection using springs and lost motion: A. ejection plate, B. ejection plate, C. core backing plate.

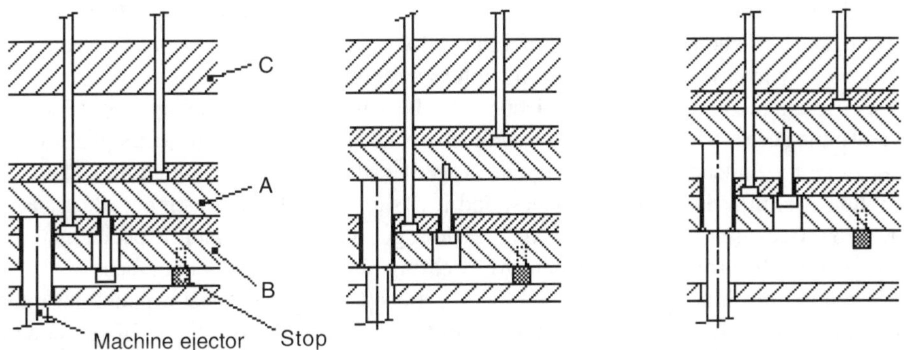

Figure 12.86 Two-stage ejection using lost motion and pulling: A. ejection plate, B. ejection plate, and C. core backing plate. (Note the different positions and movement of the three moving parts, as compared to Fig. 12.87).

forward while compressing the springs. This method is simple, but the springs must be strong enough to move plate A and to overcome the ejection forces. At least four springs must be used and evenly distributed. If the springs are too strong, they may use up all the force of the machine ejector to move plate B after A has stopped.

(Not shown in Fig. 12.85 are the screws holding the retainer plates to the ejector plates, or the return pins. Only plate B requires return pins. Plate A is pulled back by the shoulder bolts joining plates A and B. These bolts are also providing the necessary preload for the springs.)

Even though simple and frequently used, this method is not recommended because of the inherent danger when using a number of preloaded springs. Also, because of the springs, the travel is limited.

When ejection instead uses lost motion and pulling, as in Fig. 12.86, there is no need for springs. The timing is different from the spring actuation method. The machine ejector drives first plate A, while plate B rests. When the stripper bolt heads engage, plate B is pulled along with plate A until A (and B) stop.

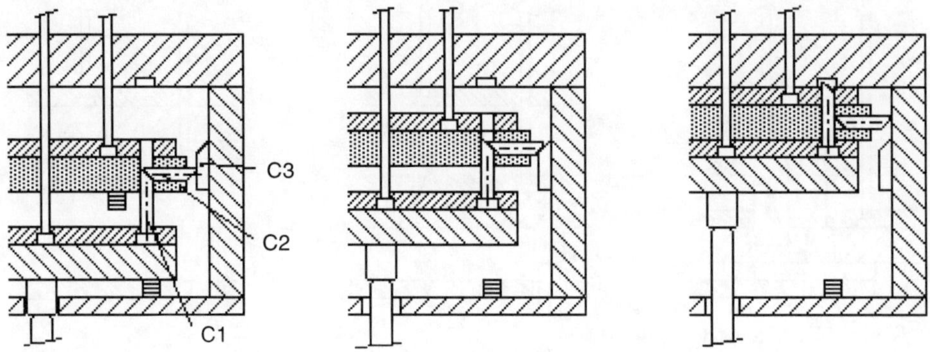

Figure 12.87 Two-stage ejection by interlocking cams that cause the ejector plates to stroke in sequence.

In all examples shown here, the ejector and/or stripper plates must be positively returned by springs or by air actuators, which is not shown for simplicity of illustration. The return pins provide safety in case of failure of such a return system.

A different method of two-stage ejection involves the use of interlocking cams. There are several mold accessories commercially available which interlock the two ejector plates so that they stroke in sequence. (First, both plates together, then the lower one alone.) The basic operation is similar in all these devices, and it is similar to the arrangement illustrated in Fig. 12.87. Because of the high loads on the cams, good lubrication and/or special materials selection must be provided to prevent early wearing out of the critical cam faces.

When the machine ejector moves forward, it drives the lower ejector plate, but it also drives the upper ejector plate through the cams C1, C2, and C3. As long as C2 rests on C3, it stops C1 and the upper plate is pushed forward. As soon as C2 comes to the slope of C3, it can move sideways and let C1 pass, so that the lower ejector plate can now approach the upper ejector plate. There must be at least four such cam systems per ejector plate to ensure even advance.

Such standard hardware units are available from vendors of moldmaker supplies, but they are usually much smaller than the design shown in Fig. 12.87. If there are problems with the available space in smaller molds, such standard units may be the answer.

Note that mold makers' supply houses also list a line of standard ejector return assemblies and accelerated ejectors, which are quite practical for use in some molds where the designer is faced with space problems in a layout, and/or where a designer prefers to use easily purchased items rather than to design one's own system. For extremely high production molds, which must run under severe operating conditions for long periods, the designer must make sure that any purchased units are sufficiently rugged to perform under these conditions without breaking down frequently.

Two-stage ejection using lost motion may also operate through the use of levers. Figure 12.88 shows schematically a typical arrangement. There is an infinite number of possibilities using links and levers, but the principle is usually the same. For simplicity of illustration, ejector retainer plates and other construction details are not shown.

After the mold has opened (at the P/L), the lever, pulled by the link, will contact the roller on ejector plate A and start lifting plate A and B together. As the mold continues to open, the

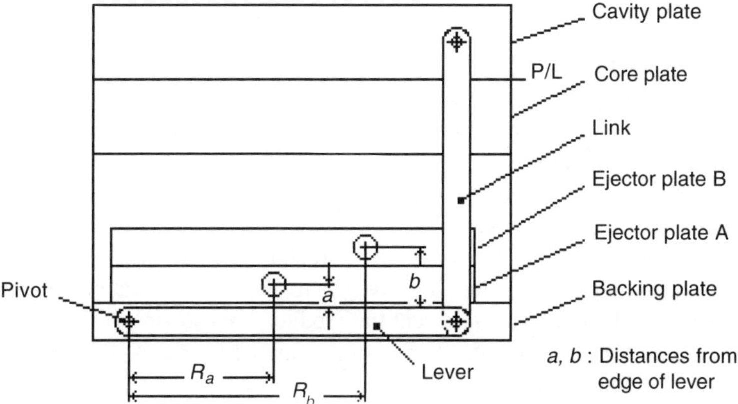

Figure 12.88 Two-stage ejection operating on lost motion with levers.

roller in plate B is contacted, and because of the larger distance from the pivot, plate B will move away from the (still advancing) plate A.

Note that by spacing the plates A and B, and by supporting plate B on its own stop in the rear position, the motions could be as follows: Plate A moves first, then plates A and B together, then plate B away from plate A. The shape of lever and link, and the position of the rollers a, b, R_a, and R_b, depend on the mold requirements and must be developed by the designer.

Rules to observe when working with links and levers:

1. There must be at least one, but preferably two, pairs of links and levers per side, acting on opposite sides of the plate (usually front and back), to ensure that the plates lift parallel to the backing plate.
2. The plates must be returned positively. Since the strokes are usually fairly short, spring return with springs within the plates is practical as long as the springs are properly designed and strong enough for the job. With linkages, there is usually not enough space to place return springs outside the mold. The alternative would be to use (internal) air actuators.
3. The rollers (usually cam followers) must be strong enough for the anticipated forces. If there are two rollers on each plate, each roller could see one-half of the mold opening force. While this is not likely to be the case, the rollers are often under-engineered (too weak).
4. Lubrication of rollers and articulations is very important:

   - Rollers: Use standard cam followers with permanently lubricated needle bearings.
   - Pivots: Use permanently lubricated needle bearings. If using bronze or steel bushings, they require lubrication, either through drilled bores in the links and levers, or with separate tubings.

   Lubrication of the faces where the rollers touch the levers is difficult to achieve. The rollers are hardened steel. The levers should also be hardened, at least at the contact faces, to reduce wear which in the long run could cause imbalance (locking) of the actuation of the ejector plates.

12.10.3 Moving Cavity

This method is not really a primary ejection method, although it could be combined with the final ejection when the product is ejected from the cavity side. In that case, it would constitute the first stage of a two-stage ejection from that side.

A "moving cavity" is essentially a limited motion of the cavity with respect to the gate pad. It is often used where the product presents a "false bottom" as illustrated in Fig. 12.89, especially if the ratio of wall thickness to depth at this point is greater than 1:1. This motion frees the product and prevents it from tearing or breaking. It is also used to free an undercut. Another, important feature of the moving cavity is to allow air to enter behind the cavity, to break the vacuum, which may also prevent the product from staying with the core.

Figure 12.89 shows the essential elements of a moving cavity. As the mold opens (right half of picture), the cavity plate (or module) follows the clamp motion for the distance *S*, which must be sufficient to release that portion of the product which was locked between cavity and gate pad. If there is little or no draft, as shown here, or if an undercut in the cavity requires the wall of the recessed portion to squeeze to come out of the cavity, the stroke *S* should be slightly greater than *F*. In many molds, the distance *S* need not be as large as the depth of the recess *F*, but a small motion of 2 or 3 mm will be sufficient.

The actuation of the moving cavity can be achieved by various means:

- Mechanically, by tying it in with the opening of the mold using latches outside the mold; but they may interfere with cooling hoses and electrical wiring, and with take-off mechanisms. Springs could also be used to create the motion of stroke *S*, but this may interfere with some mold protection methods. (Neither method is recommended.)
- Hydraulically, by providing (commercial) hydraulic actuators at the edge of the mold. The advantages are good control of the motion, especially if the stroke is large; also, it provides

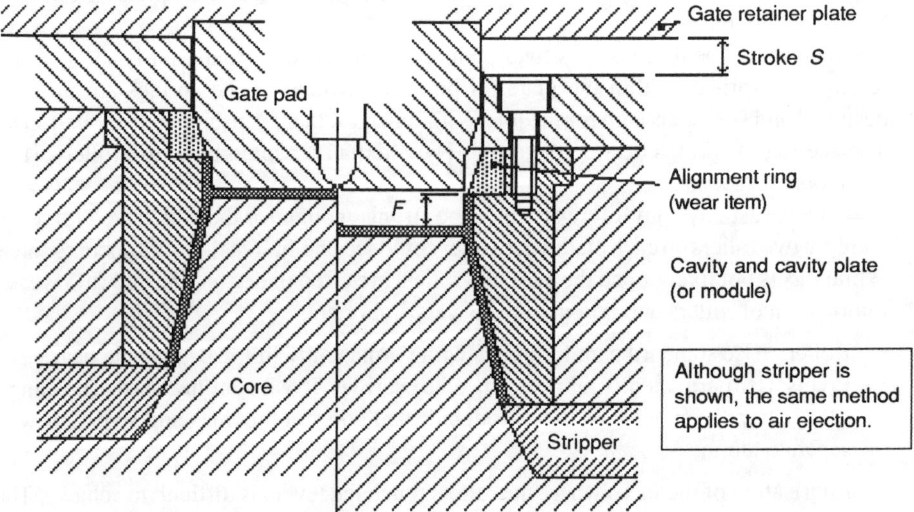

Figure 12.89 Schematic of a moving cavity in two halves: Left. mold closed, Right. mold opens and cavity plate moves.

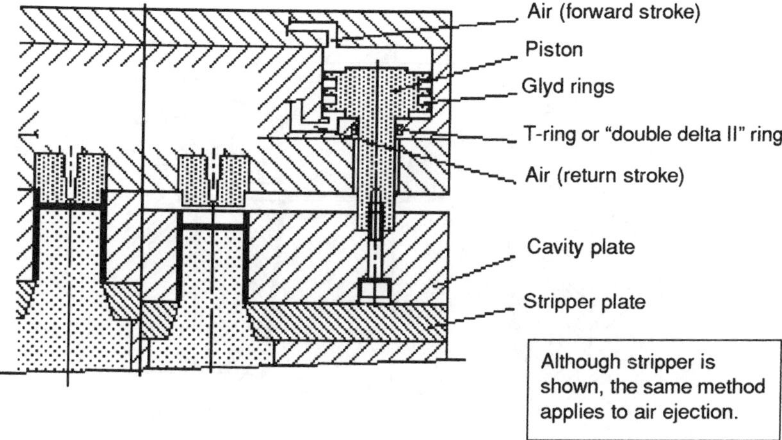

Figure 12.90 Actuation of a moving cavity using air actuators in the mold plates: Left. mold closed, Right. mold has opened and is at end of actuator stroke.

large forces with small actuators. The disadvantage is the difficulty to place the hoses, etc., so that they will not interfere, and the cost of flow and pressure controllers.

- Pneumatically, by designing air actuators into the mold. Two typical examples are outlined and illustrated below.
  1. Design four or more air actuators into the mold plates (Fig. 12.90).
     Figure 12.90 is a schematic sketch only. The runner system, cooling, etc., are omitted for clarity of illustration. At the left, the mold is closed; at the right, it is opening and has just reached the end of stroke of the actuator. The cylinders are bores in the manifold or gate retainer plate (usually at the four corners of the plates), and the piston rods are mounted in the cavity plate. In this design, the piston rods also act as stroke limiters to prevent the cavity plate from moving too far.
  2. Use all, or at least four gate pads as "pistons". The cylinders are the bores for the gate pads in the cavity plate (Fig. 12.91). In this system, air drives the cavity plate forward as the mold opens, but the closing mold force will bring the cavity plate back. This is usually acceptable if the stroke *S* is small, in the order of 6 mm or less. Separate stroke limiters are required.

     The air groove is important to provide a sufficiently large area for the air pressure to separate the mold sections. Air connections to the operators in examples 1 or 2 can be in the mold plates, as shown.

While the use of gate pads as pistons is simpler than adding operators in the corners, as in example 1, it is not recommended because if that one cavity does not fill, the air will escape at that cavity and the plate will not move, which makes it then difficult to remove the molded products from the other cavities.

General rules for the design of actuators for example 1 or 2 must be adhered to. Stroke *S* should be kept as small as possible but must be sufficient to ensure proper ejection.

Use suitable seals. Seals between moving shafts and bores should be of the Teflon type, such as Glyd rings or double delta II rings.

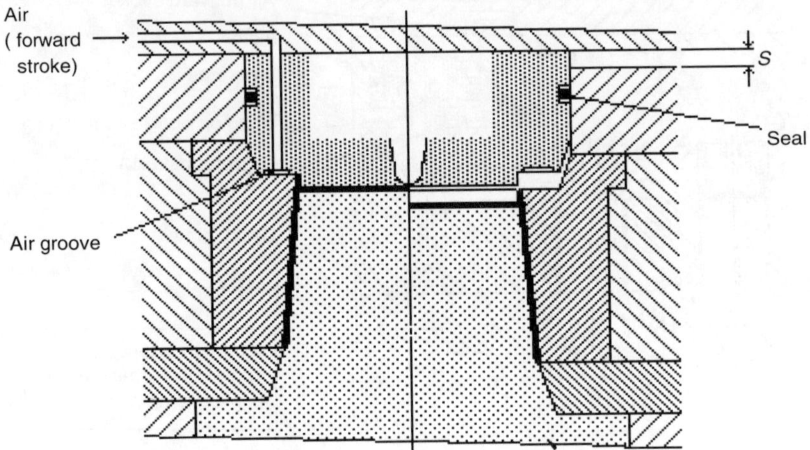

Figure 12.91 Actuation of a moving cavity using gate pads as pistons: Left. mold closed, Right. air drives cavity forward.

Use O-rings only for static sealing of air (wherever air channels in one plate connect to channels in another plate). As a rule, O-rings should not be used as dynamic seals. If air passes from one mold part to another, and the contacting surfaces are ground and touching each other, O-rings may be omitted. However, if there are coolant passages nearby, O-rings are recommended even though the coolant passages have their own, always mandatory, O-rings.

Wherever there is the possibility of two plates "breathing" (i.e., separating when clamping pressure is relieved), O-rings must be used. If O-rings are used, they must be placed as close as possible to the area subject to air pressure, to reduce the pneumatic forces generated by these areas.

The area of actuators must be sufficient to overcome the anticipated resistance against opening. This is difficult to forecast, and the forces are often underestimated and require costly rework after the mold is tested. It is usually very difficult to increase the size of the actuators or to add more actuators once the mold is completed.

Usually, four actuators with 75-mm (3-in.) diameters are sufficient for smaller molds, but possibly six or eight actuators may be required, with diameters up to 100 mm (4 in.). The number and size of actuators depends only on the force requirement. Only too often, too small or too few actuators have been designed because of the limited available space in the mold size that was originally planned. To avoid this common error, larger plates than planned may have to be selected, or hydraulic actuators must be used instead of air actuators.

The size of air lines (diameters and lengths) must be balanced so that all actuators are subjected to the same air pressure, at the same time, to prevent cocking or jamming of the plates. Typical layouts and channel sizes are shown in Fig. 12.92. If proper balancing is not possible, the main air supply lines in the mold must be much larger than the branch lines which connect the main lines to the actuators.

Diameters of 11.11 and 14.3 mm correspond to tapped holes for ¼ in. and ⅜ in. NPT. The diameter of the main lines need not be larger than the inside diameter of the air hoses

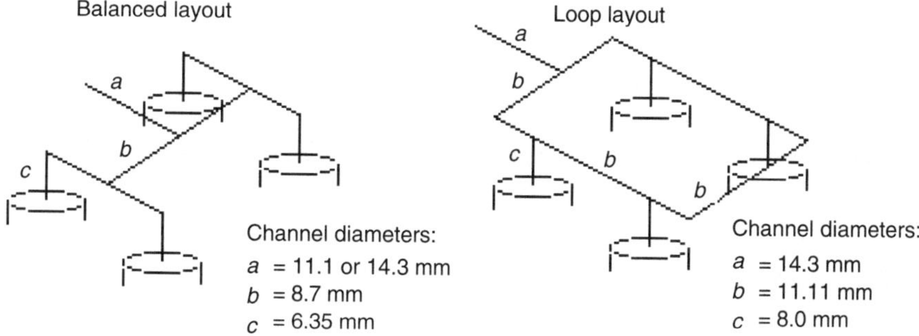

Figure 12.92 Typical layout of air lines and channel dimensions.

connecting the mold to the machine air supply. Corresponding hose lines are ⅜ and ½ in. nominal size. If air accumulators are used for faster action, use the next larger nominal sizes of air hoses (½ and ¾ in., respectively). The air accumulators and the controls should be located close to the mold to reduce the actuation time.

12.11 Molding Surface Finish

There is a relation between molding surface finish and ejection. The designer must be aware that the finer the finish, the more expensive will be the mold.

Even though many of the earlier methods of hand polishing are now replaced with much faster, and often better, mechanical methods or by fine EDM finishes, a finer surface finish still requires more time than a rougher finish. Unless the product requires a specific surface finish that is shown on the product drawing, the designer should question any finish that could impair easy release from the mold, and consult with the customer.

General guidelines regarding surface finish are listed below:

- Specify finishes that will permit easy ejection, such as grinding or draw stoning, in the direction of the withdrawal of the core, core pins, or side cores.
- Where it is expected that the product will not naturally stay with the core when the mold opens, a rougher finish should be specified, and deliberate grinding across the direction of withdrawal may be required to keep the product on the desired side. In extreme cases, undercuts that are not specified on the product may have to be added. Size, number, and location of such undercuts must be discussed with the customer.
- Use as rough an EDM finish as possible to permit faster EDM speeds and less machining time.
- Certain plastics (e.g., LDPE, PP) require a rough finish for easier release. Typically, the cores of LDPE over caps are vapor honed (sand blasted) to provide such rough surfaces. Often, after a time in operation, the mold surface may be roughed up by hand, using fine sand paper, to provide the necessary rough condition.
- Cavities may have to be chrome-plated and/or treated with Poly-Ond™ or Dicronite™ to ensure that the products stay with the cores.

12.12 Sequence of Ejection

(For further information, see also Chapter 7, Sequences of Operations.) The time period from the moment the mold starts opening until the mold is closed again is "lost time". The shorter we can make this time, the shorter will be the cycle time. This time depends on four elements:

1. Clamp speed,
2. opening stroke,
3. method of ejection, and
4. timing of ejection.

Speed of the machine motions (dry cycle) is a characteristic of the molding machine and usually can be adjusted within certain limits. A short dry cycle is particularly important where the molding cycle (time from mold closed to end of cooling cycle, Fig. 12.93) is short; the lost time is then a large percentage of the cycle.

The opening stroke required to permit proper ejection (Fig. 12.94) is usually a function of the product design, but also depends on the number of cavities (i.e., the distance the products from the uppermost cavities must fall to clear the lower edge of the mold). The space required for take-off plates entering the open mold to receive the products will also affect the mold opening stroke. The designer will usually specify a minimum stroke required for the mold, and this is also the optimum stroke when setting up the machine.

The method of ejection will also affect the cycle time. Some products can be ejected rapidly in the shortest possible time, while others may have to be ejected slowly, to prevent breakage or deformation of the products. This depends on the product design, and is outside the control of the mold designer.

As for the timing of ejection, as a rule for fast molding, the sooner ejection starts, the better. It could start as soon as the mold has opened sufficiently to ensure that the product will stay on the core and not be pushed into the cavity. In very fast operations, it is permissible or even necessary to time the ejection so that the mold is already closing while the products are still

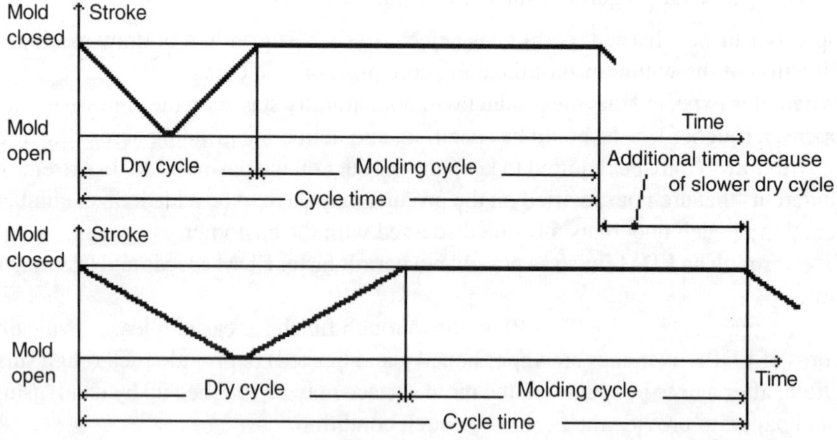

Figure 12.93 Timing diagram shows the effect of clamp speed on mold cycle time.

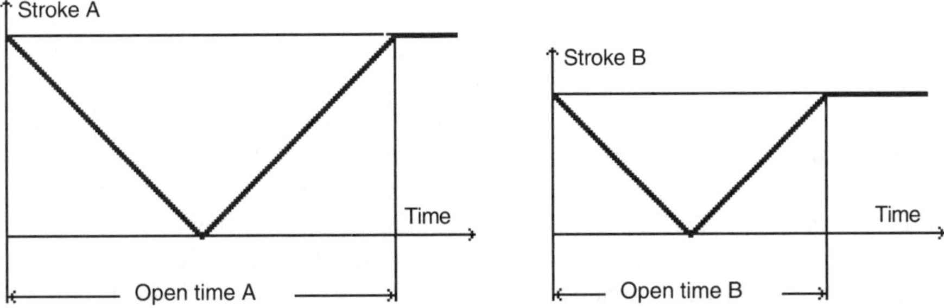

Figure 12.94 Timing diagram compares different opening strokes and their effect on mold open time.

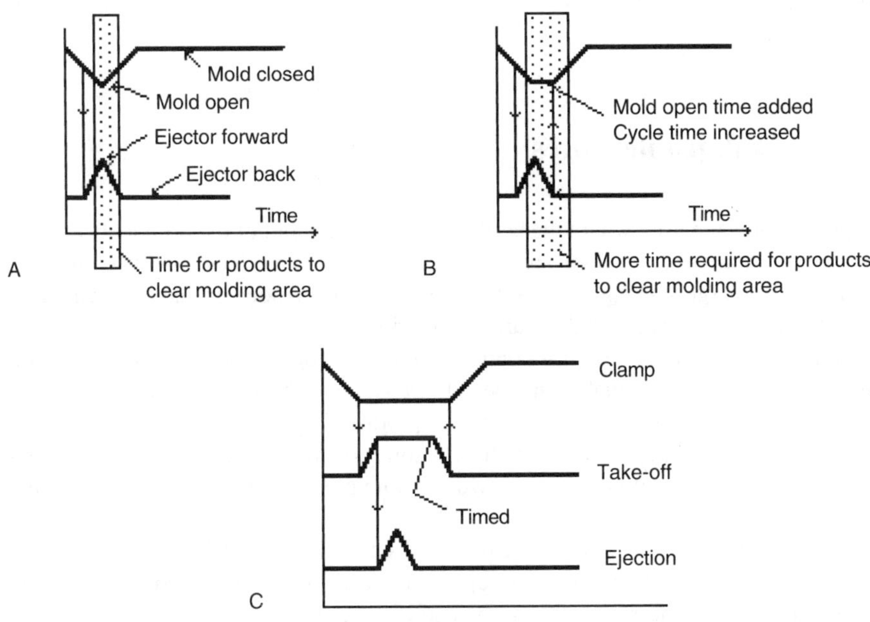

Figure 12.95 Timing diagrams show relationships between different mold actions: A. ideal cycle time in which ejection is complete and products clear mold without mold open time, B. some mold open time required for products to clear mold, and C. ejection starts only when mold is fully open and requires longer mold open time, for example, when using robots or take-offs.

falling down, as long as the molding area is clear of any product or take-off as it nears the mold closed position (Fig. 12.95A and B).

Figure 12.95A represents the ideal, where the ejection is completed and the products have cleared the molding area by the time the mold arrives in the fully open position. The mold can reclose immediately without stopping. This method is called “on-the-fly” cycling.

Figure 12.95B shows the usual situation, where the ejection is started before the mold arrives in the open position, but there is more time required for the products to clear the molding area. Some mold open time must be added before reclosing can start.

With mechanically synchronized (interlocked) take-off systems, the “in” and “out” motions take place during the opening and closing motions of the clamp, and timing as in Fig. 12.95A or B is possible. In some cases, however, the mold must first open fully before ejection can start (Fig. 12.95C), and all the ejection takes place during the mold open time. This is the least desirable condition, because it requires the longest lost time period. For example, with pneumatic, hydraulic, or servo-motor driven take-offs (TO), the risk of malfunction (“catching” the take-off plate between the mold halves) is great. It is, therefore, necessary to have the TO wait until the mold is open and stopped, and the mold must not start closing until the TO is in the “out” position.

It is clear that the most efficient method of molding is to operate “on the fly”, as shown in Fig. 12.95A. Other considerations may also enter the picture, such as ejection protection, where products must all be ejected before the clamp can reclose. All this must be considered by the designer. A timing diagram will clearly show how the machine will be set up for fastest and safest operation.

12.13 Collapsible Cores

Collapsible cores are used where the core cannot be pulled out of the product because there are projections into the core which are too large or too stiff to be ejected by stripping. Typical applications are large-size rigid screw caps, such as for threaded coffee jars, which do not require complete threads going 360° around the rim.

The design can also be used for containers that have deep inside projections or for any product having a large core that cannot be withdrawn through a narrower neck opening. This latter type of product could otherwise only be produced by blow molding, where the final (outside and inside) shape is provided without a solid core but by the inside blow-air pressure. The other alternative would be to mold two separate pieces, which must then be welded or glued together.

Figure 12.96 shows a mold for a threaded cap. The thread had to be modified by interrupting it in two spots because the cap would otherwise be damaged as the center core (CC) withdraws. The interrupted thread can end with a radius, as shown, or it can run out into the center core. This is easier to produce but will create sharp corners where it meets the center core. The thread usually covers only about 65–80% of the circumference, which is usually acceptable.

Fig. 12.96 shows only the basic principle of collapsible cores, without going into more details such as cooling, the actuation and guiding of the cores, and the stripping of the product. The design of such molds can be one of the more difficult tasks, even for an experienced mold designer.

In operation, as soon as the mold starts opening, the center core is withdrawn relative to the side cores, which permits them to move toward the center of the core and thereby release the undercut thread. The third motion is the stripping of the product.

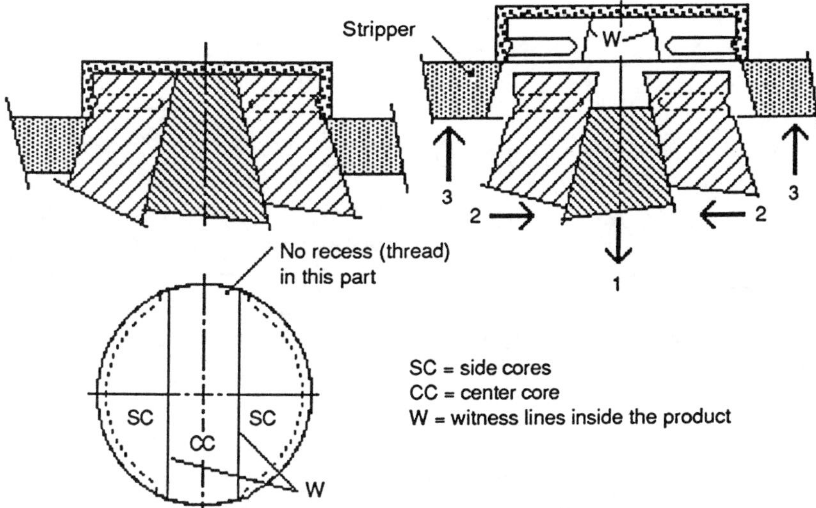

Figure 12.96 Mold for a threaded cap: 1. withdrawal of center core, 2. withdrawal of side cores, and 3. stripping.

In the illustrated example (Fig. 12.96), the core consists of only three moving parts; other cases may require five, or even more, parts to withdraw an undercut from the complete circumference. This is then much more complicated, since all these components must slide against each other with minimum wear and lock up completely without gaps where plastic could enter during injection. They must be well guided and solidly supported to withstand the injection pressure. All corners must be sharp to minimize the appearance of the witness lines where the core components meet. For best molding performance, there should be cooling channels inside each core section.

Collapsible core molds may be simpler than unscrewing molds, and the final shape of the product is much better defined than with blow molding. The disadvantages are:

- Complexity of the core assembly, which consists of at least three members;
- difficulties of actuation and locking of the components;
- wear of the sliding surfaces, with the inherent danger of flashing, resulting in high maintenance costs;
- difficulty of providing cooling to these moving core sections; and
- high mold cost with slow cycles.

A product intended for molding with collapsible cores may still require some design modifications to suit the planned sectionalizing of the core; this will require close cooperation with the product designer.

12.13.1 Angled Moving Core

This is not really a collapsible core, but the principle is the same. It is used often in molds where there are one or several severe undercuts on the inside which cannot be stripped with conventional methods.

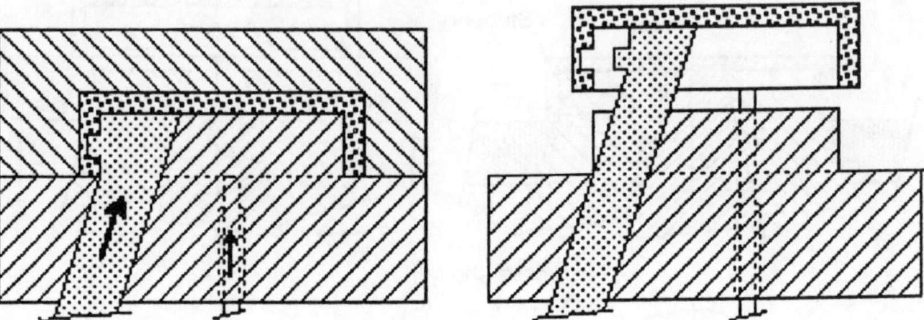

Figure 12.97 Schematic of angled moving core ejection: mold closed (left) and core moves forward at an angle while ejection is straight forward (right).

During ejection, the moving core moves forward and sideways, while the ejector pins go straight forward (Fig. 12.97). The stroke of the moving core needs to be just enough to get out of the undercut (not as far as shown here), and the ejectors may have to continue to move forward to eject the product without it hanging up on the moving core (two-stage ejection).

12.14 Unscrewing Molds

Usually, unscrewing refers to removing *inside-threaded* products from the core. There are, however, occasions where unscrewing is required to remove an *outside-threaded* product from the cavity when the accuracy of the threaded portion is so demanding that the splitting of the cavity is not acceptable. A typical example is car speedometer gears. There are two basic methods of unscrewing: conventional unscrewing methods, which rotate the core around inside the a stationary product, and the Husky™ unscrews the products from stationary cores.

12.14.1 Conventional Unscrewing Methods

The product is kept from rotating while the core rotates to unscrew from the product. There are several methods used to rotate the core and to eject the product. Figure 12.98 shows a system using a rack and pinion for the rotation of the core.

As the core rotates, the sleeve retaining plate and the stripper plate move forward by a light spring pressure under the sleeve backing plate. This motion can also be produced by mechanically synchronizing the forward travel with the travel of the unscrewing motion. In some systems, the core is held in a lead nut below the drive pinion, so that the core is positively retracted, rather than depending on a spring or other synchronized forward motion of the stripper assembly.

The product is kept from rotating by teeth (or other projections) in the end of the sleeve. At the end of the unscrewing, the stripper plate moves forward to eject the product from the teeth, which occasionally may still be keeping the product from falling free.

The rack is usually driven by a hydraulic cylinder outside of the mold. The earliest unscrewing systems, and some experimental molds, use gear drives with a hand crank for the

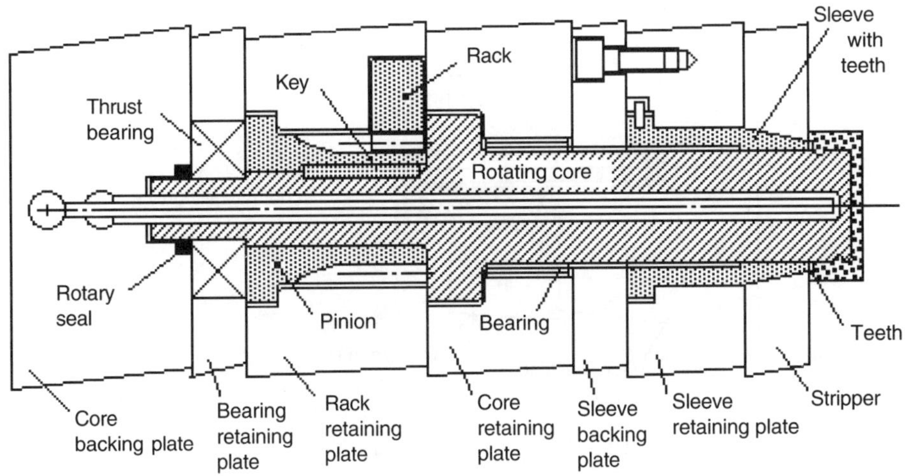

Figure 12.98 A rack and pinion system for rotating the core.

unscrewing motion. Some systems use sun gears in a circular layout of cavities, or a complex system of gears driven from a hydraulic motor. Other molds use the opening and closing motion of the mold to rotate the gears by using a steep lead-angle screw with a matching nut. Either the screw or the nut are connected to the gear that drives the cores.

The common problems with all these systems are:

1. Combining the mechanical drive system with a mold which is subject to temperature variations;
2. the bearings need lubrication, and the sealing gaps must be tight enough so that the plastic will not flash during injection, but wide enough to prevent galling of mold materials that cannot be properly lubricated because of the potential contamination of the products with lubricants; and
3. the coolant must be contained with rotary seals, which are, from past experience, a common source of leaks, even after relatively short periods of use.

Despite these problems, there are many good unscrewing molds in operation, and some mold makers are specializing in this field, using their own standards of core assemblies, drives, etc., so that these molds can be produced on a production basis very competitively.

The great advantage of these molds is that they can fit into any standard machine large enough to accommodate the mold.

12.14.2 Husky's Original Unscrewing Mold

This system is only applicable to the Husky™ unscrewing machine (Fig. 12.99). For further information, see Section 12.14.3. The system differs from the conventional methods by having the unscrewing process separated from the molding process. The core is stationary, and there are only two mold plates, compared to the many plates in a conventional unscrewing mold.

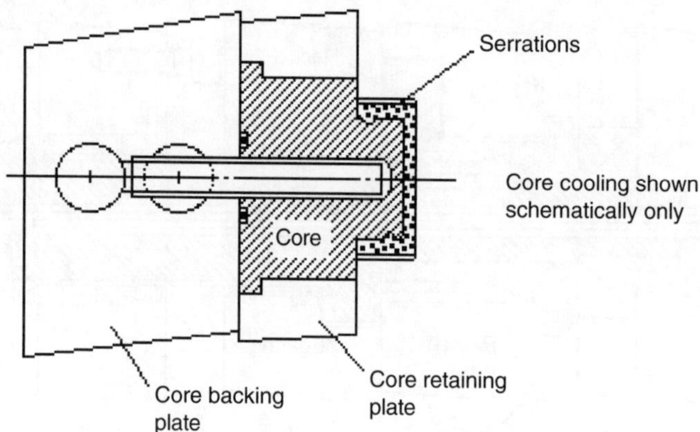

Figure 12.99 Husky unscrewing machine mold.

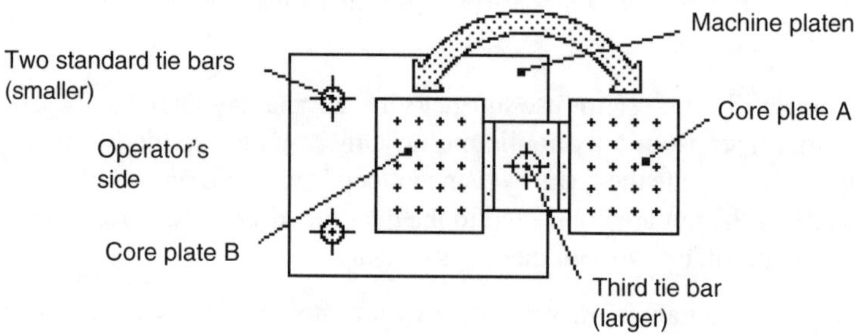

Figure 12.100 View into unscrewing machine with three-tie bar system.

These molds require a special, three-tie bar unscrewing machine (Fig. 12.100). Core plate A is in the unscrewing position, outside the molding area, while core plate B is in the molding position. Core plates A and B are identical and are positioned alternately into the molding position by rotating about the third tie bar.

While the cavities in one plate are injected and partly cooled, the plate in the unscrewing position faces an unscrewing head, with a pattern of unscrewing chucks matching the cavity pattern of the mold. After the products on plate B are cooled enough to be removed from the cavities, the machine opens, and the swing mechanism rotates the core plates by 180°. The products, which are still on the core, continue to be cooled from the inside where they have good contact with the core and, therefore, gain some additional cooling time.

As the mold closes again on the "inside" plate, which is in the molding position, the same mold closing motion brings forward an "unscrewing head" to the outside plate. Spring-loaded chucks slide over the serrations (or other projections) to engage the product to transmit the necessary torque to unscrew the products from the cores.

The unscrewing starts from a timer, which allows maximum cooling time to ensure the product is stiff enough to withstand the unscrewing force (torque). The unscrewing motion away from the cores is synchronized with the pitch of the thread being unscrewed. This function is part of the machine.

The chucks then move a short distance back from the core plate, and in so doing, the products are ejected by ejector rods inside the chucks. Note that the unscrewing chucks are matched to the product and belong to the mold. It is conceivable that the same unscrewing chucks can be used for several molds, provided that the critical dimensions are the same.

Advantages of the Husky systems are:

- The molds are simple, rugged, without moving parts, and do not require rotary seals. They are relatively inexpensive.
- Because of the absence of any unscrewing mechanism within the mold, it is easy to place many more cavities into a molding area than would be possible with conventional unscrewing molds.
- Molding cycles are much faster because of the excellent cooling of the cores. Also, the cooling time is extended, since the product stays on the core after mold opening during the additional time of swinging and until the unscrewing starts. This additional time, approximately 3–5 seconds, permits a reduction of the mold closed (cooling) time.
- Because of the larger number of cavities and faster cycles, the productivity of this system is considerably higher than that of conventional unscrewing systems and is mainly used where the quantities required are in the millions; for example, for tooth past caps, one machine can easily produce 42,000 pieces per hour, or 226 million, or more, per year.

Disadvantages of the Husky system are:

- It requires a special molding machine, equipped with three tie bars and a swing mechanism, in addition to an unscrewing mechanism (head) that matches the number of cavities and the cavity layout of the mold. This means that a different head is required for each different layout (pattern) of stacks.
- It requires for each mold one cavity plate and two identical core plates, but this is still much less expensive than a conventional unscrewing mold.
- There are restrictions to the shape of products that can be molded with this system. Generally, only products that can be grabbed positively (typically, with serrations) can be unscrewed easily with this method.

  Exception: There is a feature available on the machine that ensures that the unscrewing chucks will always stop at the same angular position, so that a non-round circumference or one or more specific points (holes, bosses, etc.) in the product can be used for unscrewing (e.g., speedometer gears, oval caps, etc.).
- Certain shapes, such as cylindrical caps, with smooth circumference cannot be unscrewed by this Husky method.

12.14.3 Husky's New Unscrewing Mold System

This is a rather new development (patented), which combines the advantages of the conventional unscrewing method in Section 12.14.1 and that of the Husky swing mold method, described in Section 12.14.2. Without going into detail, the system works as follows:

- As with the conventional mold, the product must have some teeth at the open end so that the caps can be turned and unscrewed from the core.

- The cores are fixed (not rotating), as in the swing mold system, and can therefore be well cooled, and without the risk of leaking connections—two of the greater disadvantages of conventional unscrewing methods with rotating cores.
- Teeth to grab the closures are on top of stripper rings; these rings are rotated after the mold has opened. While the stripper rings rotate to unscrew the products, the stripper plate on which the rings are mounted is moving forward, in synchronization with the pitch of the screw thread. The products fall free from the mold after they lose their engagement with the thread on the cores.

The major advantages of this system are:

- The mold will fit into a regular injection molding machine. (No special "unscrewing machine" is required.)
- The cores are stationary and well cooled, without the problems of sealing the coolant.
- Virtually every shape of threaded closure can be molded with this system, which was not possible with the system described in Section 12.14.2.
- The productivity is better than the conventional unscrewing mold and competitive with the original Husky unscrewing system.

The major disadvantage is:

- Because of the in-mold mechanism for rotating the stripper ring, the cavity spacings (and the mold layout) are not as small as is possible with the original Husky system.

13 Mold Cooling

13.1 Introduction

This chapter will provide the mold designer with guidelines for mold cooling. There is an infinite variety of applications a mold designer will face, and there are no simple rules. There are many molds which, for all intents and purposes, produce identical products; all appear to be good, sensible designs, but the differences in output can be astounding. The designer must from the beginning be aware of all parameters which affect an economical (not necessarily cheap) mold cooling design.

Typical parameters to be considered (not necessarily in this order or excluding others, not enumerated) are:

- Materials to be molded,
- mold material,
- machines where the mold is used,
- productivity expected,
- projected cost of molded product,
- shape of product,
- mold cost.
- maintenance cost, and
- personnel available to run and to maintain the mold.

In view of the above, simple, all-encompassing rules lose their meaning. It is more helpful to the mold designer to understand the basic principles of heat and heat exchange that take place in a mold, and the effects of heat inside a mold. Designers can then judge for themselves the value of a planned approach to cooling of a specific mold.

It must not be interpreted that the designer has unlimited freedom to proceed as he or she wishes to achieve the ultimate cooling design. Whatever the designer does must still be within the practical limits of the capabilities of the shop to *produce economically* the desired cooling system. It should be based on existing shop standards; it also must be practical for the end user.

For example, at a plastics show, a six-cavity mold was demonstrated with 52 hoses connected to the various cavities and plates! Certainly, it appeared to be a well-cooled mold, but the job of just connecting all the hoses, and the resulting inaccessibility of the mold during operation, showed clearly that the designer did not envision the results of what he was doing.

A very important consideration is the expected total productivity of the mold: Will the product last only for a short period (or season), or will the product be required for many years to come? Is the total output of the mold restricted to a few samples or to short runs, or will the product stay "as-is" for long periods?

There may have been other molds, or precedents, used to make the same or a similar product. But this does not mean that the cooling design used in these molds was appropriate.

It may have been right, but it could have been "overdone" (money wasted), or the mold may not have been well enough cooled and caused unnecessarily slow molding cycles.

For example, a mold runs at a 6-second cycle; however, with better cooling, even at a higher initial cost, it could run at 5 seconds. This amounts to a 17% improvement in output, and in a long running job, the money savings would be impressive. It therefore pays to consider all the facts before repeating the design.

There are a number of highly theoretical approaches used to determine the optimal conditions for cooling and the sometimes overlooked but extremely important strength of materials. Such mathematical exercises are usually done at some universities which do special research in plastics technology. Often, such studies have little practical significance for the mold designer, but they do help to highlight some of the areas of which a designer should be aware.

Also, many parameters used in such calculations, while theoretically significant, can not be used in practice because of some difficult-to-define influences that affect their values, such as batch-to-batch variations of the plastic. Also, the atmospheric differences in the plant environment, such as humidity and ambient temperatures in various locations (even in the same room), must be considered.

The designer must never lose track of the objective, which is to design a mold that will produce a desired product, to run in an indicated machine (or machines), in the most economical way. There is no point in wasting time and money designing and building the best cooled and highest quality mold that can last for many years (and produce the lowest cost products if used over a long period) when it is known beforehand that the total requirements (over the life of such product) can be produced in such an "ideal" mold in just a few hours or days.

13.1.1 Objectives of Mold Design

The main objectives of mold design are:

- To produce the desired product shape and accuracy,
- to provide the lowest cost product (including the mold cost and the machine hour cost attributed to the product),
- to produce only good moldings,
- to be economical to maintain,
- to run at the most economical cycle time, and
- to be economical to build.

Good (adequate) mold cooling is required to satisfy the last two objectives.

13.1.2 Available Assistance

For help in designing cooling systems, designers may rely on personal experience, computer programs, written references, or courses of study. Often, designers have their own preferences, sometimes based on past experience, but not necessarily in touch with the practices of the latest methods of mold making. Or, they have no experience at all in this field. What should they do? Where can they get help?

There are several computer programs available, but they will tell us only if the selected cooling is satisfactory for a certain application. There is nothing wrong with this approach as a checking method for a selected design.

Written references, such as textbooks deal only very generally with this problem. If they go into more details, they are usually highly scientific, requiring a great deal of mathematics, and cover only small, specific areas of the cooling design. Written references are helpful only after a design has been chosen.

Trade magazines have, from time to time, good articles on mold cooling for certain applications, and an experienced designer may want to collect such articles for future reference. However, there is a major problem with this type of information: It usually deals with experiences dating back several years before the article was written. New experiences and technologies may have made the content of such an article outdated or obsolete.

There are some universities or colleges that offer courses in this field, but much of what they teach may be discouraging to a practical mind. It is either too complicated, too mathematical, or not up to date in the area of practical (mold) application. However, they do teach the basics to an understanding of the cooling.

The intent of this chapter is to provide mold designers with the basic understanding of what is required in mold cooling and to combine, with a minimum of theory or mathematics, all important points one has to consider before laying out the cooling for a mold. Past experience indicates that a cooling layout, based on the principles shown here, has always been proven "well designed" when such a layout was checked by computer analysis.

13.2 Why Cool an Injection Mold?

Molds shape the hot, injected plastic into the desired product. To be able to be *ejected*, the product must withstand the forces during ejection without being deformed. For that, the temperature of the plastic at ejection must be lower than during injection, but not necessarily cold. In short: *The mold is a heat exchanger.*

The designer should be aware that a substantial portion, sometimes as much as 80% of the total molding cycle, could be required for cooling. Therefore, for high production molds, it is imperative that this "lost" time is reduced to a minimum. However, there are also a number of other considerations a designer must consider when laying out the mold cooling.

Required injection and mold temperatures for some materials are listed in Table 13.1, which shows that the temperature ranges even within a certain group of materials are wide. It may require the experience of a molder or a technical representative of the materials supplier to suggest the most suitable temperatures for a particular material and for a particular molded product. The reason these few examples are shown here is to point out why the cooling design of molds for different materials may be different.

An experienced molder (or mold designer) may know that the physical qualities and/or the appearance of the molded product often depend on the rate of cooling. For example, a product may become brittle or lack the required gloss when cooled too fast, or at too low a temperature, or it may show unwanted crystallization when cooled too slowly or not cold enough.

There are also other considerations, such as humidity in the air, which may cause condensation on the mold and even on the surfaces of cavity or core if the mold stays open

Table 13.1 Required Injection and Mold Temperatures (°C) for Various Molding Materials

Material	Injection temperature (°C)	Mold temperature (°C)
Polyethylene (PE)	170–320	0–70
Polystyrene (PS)	200–250	0–60
Polyamide (Nylon)	240–320	40–120
Acrylonitrile-styrene	230–260	50–80
Polycarbonate (PC)	280–310	85–140
Polyacetal (Delrin)	180–230	70–130
Polypropylene (PP)	180–280	0–80
Acrylonitrile-butadiene-styrene (ABS)	180–240	50–120

long enough while the cold cooling medium is circulated through the mold. This is important where the molding room is not air conditioned or humidity-controlled.

13.3 What Affects the Cooling of a Mold?

Many factors affect the cooling of a mold:

- Temperature increase (ΔT) of the coolant, from “in” to “out”;
- flow of coolant, from “in” to “out”;
- chemical composition of coolant;
- thermal conductivity of mold parts;
- temperature drop of the plastic from injection to ejection;
- runner system (size, layout);
- type of runner system (hot or cold);
- cooling channels (layout, size) in the components in contact with the molded product;
- layout and sizing of cooling channels in the mold plates; and
- size (diameter) and number of supply and return lines (hoses).

In the following explanations there are some statements that are based on theoretical (physical and mathematical) fundamentals and should be accepted as stated. If deeper insight is desired, refer to textbooks.

13.3.1 Temperature of the Cooling Medium

The cooling medium is usually water, preferably conditioned so as to minimize contamination of the cooling system with lime deposits or other substances present in the water which can corrode the cooling channels. These contaminations also create deposit on the walls of the cooling channels which reduces the heat transfer and can also block the passages.

Where water temperatures are low or the supply lines are exposed to freezing temperatures, an antifreeze is added. However, in most cases, the cooling water is kept at or above 5 °C

(40 °F) and antifreeze is not required. The addition of antifreeze changes somewhat the capability of the water to remove heat, but this can usually be ignored.

In rare instances, the cooling water comes directly from wells or from the municipal water supply, at the temperature of these sources. This is usually very expensive.

In most molding installations the coolant to the molds is supplied either from a water reservoir and a cooling tower, or from a water chiller system; this can be either a central system, or individual chillers supplying one or a small group of molding machines. Ideally, a whole plant would run on one cooling water temperature, supplied from one source. However, in many cases this is not practical because a plant may be molding products from plastics requiring different cooling conditions. The designer should know the cooling system used in the location where the mold is going to be installed.

Note that with some materials, such as ABS or PC, the cooling water is actually heated to the required temperature, usually by using portable heating units. In some cases, especially with higher temperatures, oil or a nonflammable oil substitute is used instead of water. We will ignore such specialized cooling, and base our considerations on a plant with a central supply of about 5 °C water.

13.3.2 Temperature of the Plastic—Heat Balance

To heat the plastic from the cold granules to the temperature at which it becomes a viscous fluid which can be injected, a certain amount of energy must be added to it. This takes place in the extruder of the molding machine. The energy is supplied mainly by the work of the screw, which transforms mechanical energy into heat, and with some heat added from the heaters surrounding the extruder barrel.

The heat supplied to the plastic during plasticizing must be removed inside the mold. This is achieved by the mold cooling during the cycle time.

Note that heat is extracted from the mold during the total cycle time t, not only during the cooling time C (Fig. 13.1). This can be also explained as follows: When the mold is mounted in the machine, and connected to the services, it is at room temperature. As the cooling supply is turned on, and before injection starts, the mold will gradually reach the temperature of the coolant (colder or warmer than room) and then stay at this temperature until hot plastic is injected.

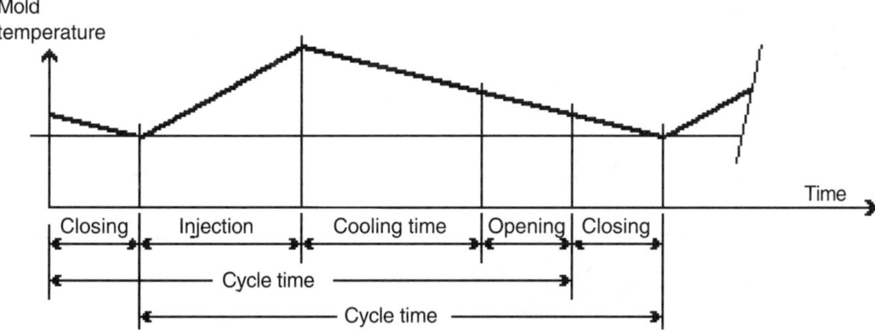

Figure 13.1 Timing diagram showing mold temperature changes during the mold cycle time.

During the injection cycle, the mold temperature will rise while coolant continues to remove heat from the mold. As the injection is completed, no more heat will be added to the mold, but the coolant continues to remove heat from the mold until the next injection takes place. It is important that, once the mold is on cycle, the amount of heat entering the mold is the same as the heat removed from the mold.

Note that heat is not only removed by the coolant but also, in part, by the products, since they usually leave the mold while they are still hot and, therefore, carry a certain amount of heat, which will be radiated or conducted into the plant environment. Also, a certain small amount of heat is radiated directly from the surface of the mold into the plant.

If the cycle time is too short to remove the heat, the temperature of the mold will gradually rise and will eventually be too hot, creating unusable products. Cooling time must be added (Fig. 13.2).

If the cycle time is too long (i.e., the heat is removed before ejection starts), the mold wastes time and operates inefficiently. Cooling time can be reduced (Fig. 13.3).

To shorten the cycle time, various options are available and are illustrated in Figs. 13.4-13.7, which are described in the following as schematic examples of four cases of molding cycles. Note that the coolant temperature is the same in all four examples. The following variables apply:

t Total cycle time,
A mold closing time,
B injection time (heat input),
C cooling time, and
D mold opening time and product ejection.

A depends on the machine, or the closing speed setting. *B* depends on plastic material, machine, and mold. *C* depends on mold (cooling layout, mold materials selection), and on the plastic molded. *D* depends on the machine speed setting and the method of ejection from the mold. In the following figures, it is assumed that ejection is completed during mold opening.

In Fig. 13.4, *A* and *D* are shown relatively long (slow machine). *B* is selected (arbitrarily) at a certain length shown. *C* is also selected at a certain length, shown.

Figure 13.5 shows the same mold, but a faster machine is used. *A* and *D* are shorter. Cooling time *C* must be extended to allow enough time for heat removal during cycle time *t*, which will remain unchanged. There is no gain in cycle time by selecting a faster machine.

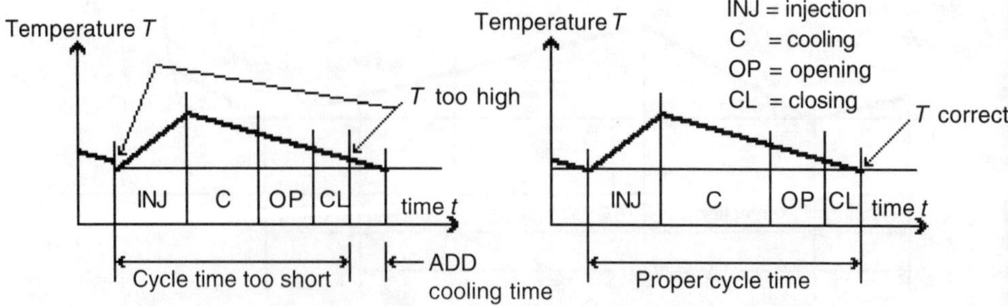

Figure 13.2 Timing diagrams show that when proper cooling has not occurred by the end of the cycle time (left), cooling time must be added (right).

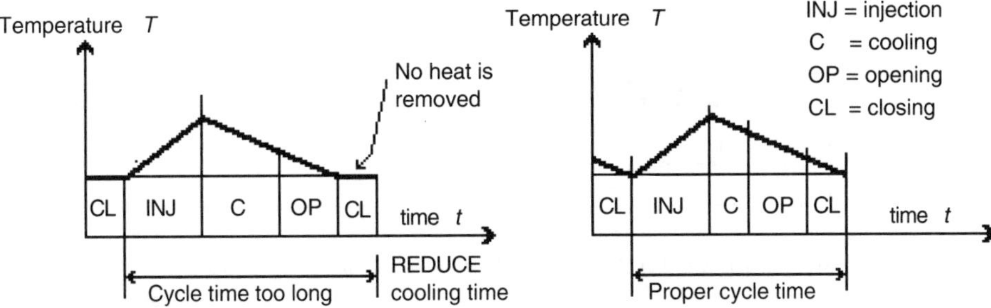

Figure 13.3 Timing diagrams show that when cooling is already completed (left), cooling time maybe reduced (right).

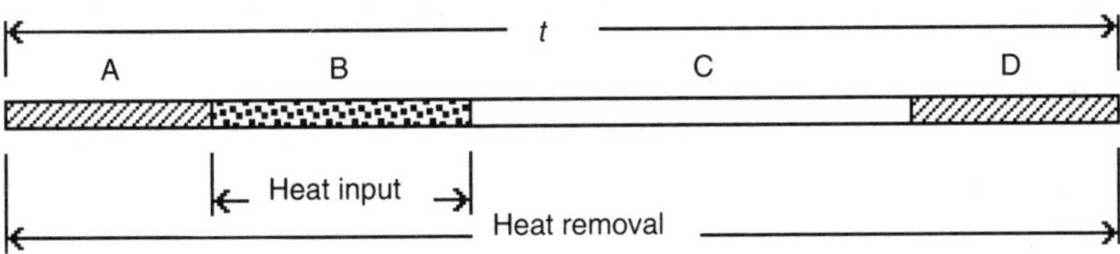

Figure 13.4 Typical molding cycle for a slower machine.

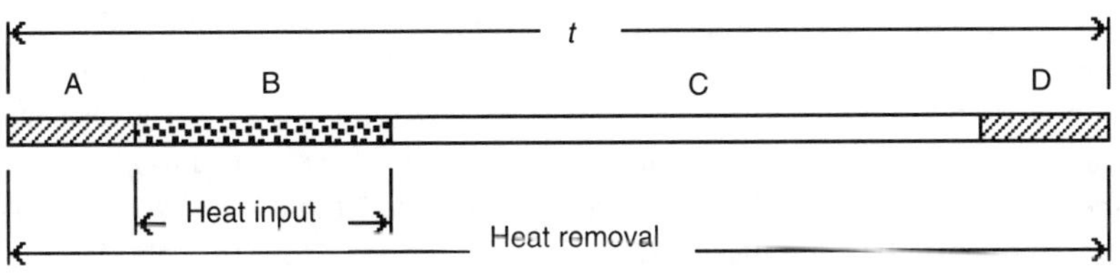

Figure 13.5 Molding cycle for same mold using a faster machine.

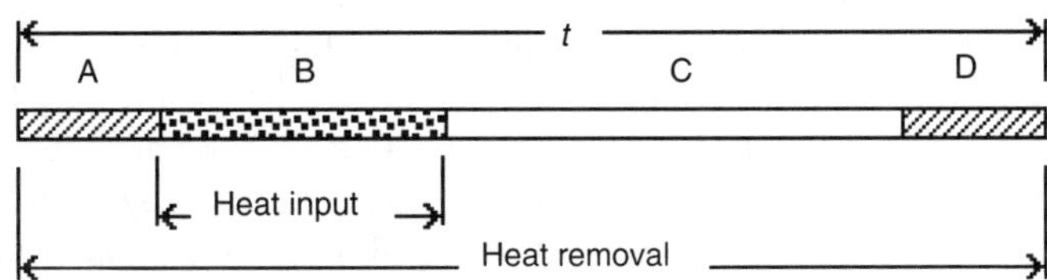

Figure 13.6 Molding cycle for same mold but with more efficient cooling.

Figure 13.6 shows the same mold again, but with an improved cooling layout. C is shorter than in Fig. 13.5, and as result of the greater cooling efficiency, *t* will also be reduced.

Figure 13.7 again shows the same mold, but it is now running on a machine with a better injection unit, which allows injecting at lower temperatures. The heat input and, therefore, the cooling requirement is greatly reduced, so *C* and *t* are reduced.

General comments on the above figures must include closing and opening times, injection time, and cooling time. Closing and opening times should be at a minimum. However, the

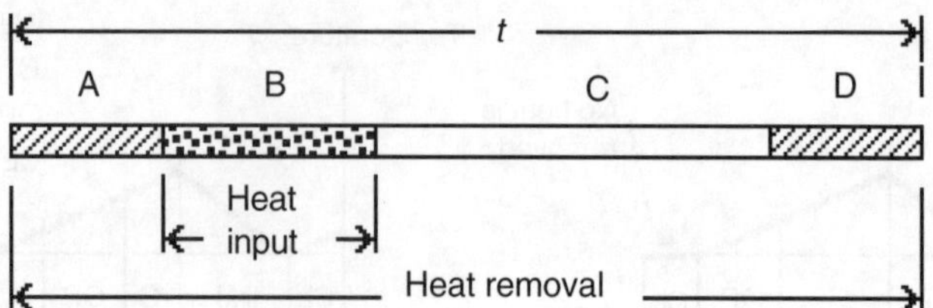

Figure 13.7 Mold cycle for the same mold with a better injection unit.

opening time is limited by the time required for the products (and runners, if any) to eject and to clear the molding area.

Injection time is affected by shot size, viscosity of plastic, the injection unit used, and the runner system. Shot size will depend on the plastic mass. The greater the mass of plastic, the longer it takes to inject.

Viscosity and other characteristics differ between plastics and by batch. Selection of a different grade and/or supplier of the same material can permit a reduction of the injection temperature.

The injection unit should be selected that best suits each mold. Higher injection pressures and speeds available help to reduce the injection time. Properly engineered extruder screw designs permit plasticizing more efficiently and faster.

The most efficient and economical runner system should also be selected. Cold runners (two-plate or three-plate design) need more time because more material enters the mold (products plus runners) and, therefore, more heat must be removed. Also, the resistance to the plastic flow in the cold runners reduces the pressure to push the plastic through the gate when filling the cavities. Whenever product configuration and molding materials permit, hot runners should be used, economics permitting.

From the above, it becomes clear that as the mold closing and opening times and the injection times are reduced, the efficiency of the cooling system becomes more important because comparatively less cycle time is available for cooling in relation to the injection time. This is critical in the molding of thin-walled articles, where high-speed machines are used and the injection time is very short. Even though such products are usually very light, the total throughput (mass) per unit of time is very high; such molds require maximum cooling.

In some applications, very long cooling cycle times can be offset by shifting a portion of the cooling time so as to run simultaneously with the following cycle. In machines with *swing plates* or with *turrets*, the molded product is cooled to where it can be removed from the cavity, but it is not yet cold enough to be ejected from the core. The core plate with the product still on it is removed to a different position, and an alternate plate with identical cores is positioned in the mold so that the next molding cycle can start. The cooling of the products on the core plate outside of the mold continues during the cycle until the products can be ejected. A typical example is the Husky method for screw cap molding.

With *multistation* machines, sets of cavities and cores, still clamped together, move away from the injection station while a new set of cavities and cores enters this station. The mold opens at the end of the cooling cycle, and the products are ejected. This requires a number of mold sets, sometimes as many as 30 or more, arranged in a circle.

Such methods of extending the cooling time are used quite frequently for elastomers in the plastic shoe industry, where the cooling times required are often several minutes rather than seconds, but because of the complexity of the molding machine, they are used mainly for such specialty fields, and will not be further covered here.

There are other methods of after-cooling, where the products are ejected while still very hot (but manageable). Some operate by removing the molded products with a take-off and placing them on a cooling conveyor, or the take-off plate itself may provide additional cooling for the hot products (e.g., PET preforms). Other methods remove the products manually (or with a take-off) and place them on cooling (and/or shrinking) fixtures. Still other methods involve dropping the products into a cooling water container (trough), which is quite efficient but requires drying of the wet products. It is sometimes done with thick-walled Nylon™ products, which besides cooling must absorb water to gain strength.

13.3.3 Basic Principles of Heat, Temperature, and Energy

13.3.3.1 Heat

Heat is *energy in transit* from a hot to a cold body. Temperature is a *property* of a body which determines the direction of the heat flow from (or into) it when placed into contact with another body.

Heat is measured in calories (cal) or in British thermal units (BTU). One cal is the amount of heat required to raise the temperature of one gram of water by one degree Celsius at atmospheric pressure. Temperature is measured in degrees, preferably in degrees Celsius (°C). Thus, the amount of heat, measured in Calories, must not be confused with temperature, which is measured in degrees (°C). In other words, you raise the temperature from a lower value T_1 to a higher value T_2 by adding heat (a number of calories).

Temperature difference is expressed as ΔT (delta T) in °C. The same holds true for cooling, which removes a number of calories from the plastic (by transferring it to the mold and the coolant), to reduce the temperature of the plastic. This amount of heat removed from the plastic raises the temperature of the coolant and the air around the mold and machine.

Energy is the capacity to do work. Heat is one form of energy. Other familiar forms are measured in horsepower-hours, kilowatt-hours, etc. They all are used interchangeably, using the appropriate conversion factors, such as:

$$1 \text{ kw-hr} = 1.34 \text{ (US) Hp-hr} = 860 \text{ Kcal} = 3415 \text{ BTU} = \ldots, \text{etc.}$$

Conversions can be found in technical handbooks.

The amount of heat (q) flowing from one location to another is "proportional to" (gets larger with) 1) ΔT, 2) the thermal conductivity, and 3) the cross sectional area (A) through which it flows, and is "inversely proportional to" (gets smaller with) the distance of the flow (L). This is expressed in the following formula (f means "proportional to"):

$$q = f \frac{A \, \Delta T}{L}. \qquad (13.1)$$

In a mold, the heat from the plastic flows into the mold, and from the mold into the coolant and, to a small amount, into the air and into the machine platens.

13.3.3.2 Temperature Differences

The mold designer is concerned about three different ΔTs: that of the plastic from injection to ejection, of the cooling water from the time it enters the mold to the time it leaves, and that between the plastic and the cooling medium as cooling occurs.

The ΔT of the plastic from injection to ejection depends on the plastic (see 13. 1), the shape of the product, and on molding conditions. For example, small cross sections or passages may require that the plastic be heated to a higher temperature to reduce its viscosity and to allow the material to flow easier.

The ΔT of the cooling water from "in" to "out" occurs as the coolant flows through the mold and picks up heat, and its temperature rises. For many "general purpose" molds, this ΔT should be kept to not more than 5–6 °C. A greater ΔT can result in uneven mold cooling and longer molding cycles.

In high-production molds, such as molds for preforms for bottles, the ΔT should be held below 3 °C, better yet, between 1–2 °C. This must be critically considered, because to achieve such small ΔTs, huge quantities of coolant must pass evenly through the mold to be effective. The energy to maintain such flows and the required pumps and piping may not be worth the achieved benefits. Specialized applications, particularly if they involve huge production quantities such as bottle preforms must, therefore, be carefully analyzed and should be done with computer simulation.

Finally, the ΔT between the plastic and the cooling medium must be considered The greater the ΔT between the plastic and the coolant, the more energy will flow from one to the other in a certain amount of time. So, if the difference is large, the plastic will cool faster than if the difference is small. As the plastic cools down, the ΔT gets smaller and smaller, but we will ignore this for practical purposes.

13.3.3.3 Heat Conductivity

Heat conductivity is a measure of the rate at which a material conducts the heat from hot to cold. This rate is different for various materials, and the actual figures can be looked up in reference books. The mold designer is concerned about heat conductivity of plastics and of mold materials.

Plastics are generally poor conductors, and there can be significant differences between various plastics. However, in most cases we can ignore these differences.

It is important for the designer to realize that the layer of plastic touching the cavity walls has a very low conductivity compared with the conductivity of the mold material; it forms an insulating layer between the mass of the plastic and the cold mold steel. This influences the rate of cooling to a much greater degree than the temperature of the mold or the material of the cavities. This is of particular significance when designing molds for products with heavy walls, or when there are isolated areas in the product which are thick and cannot be cooled properly.

The heat conductivity of mold materials is also important. For practical reasons (cost, strength, wear resistance, ease of machining), molds are usually built from steel. Here, again, there are some differences in the conductivity of the various steel alloys that are used for the molds, but, except in some special applications, the differences are not significant.

Typically, thin-walled products are a special application. The plastic is so thin that the insulating effect of the plastic as it cools down (see above) becomes insignificant; this makes it important to efficiently "carry" the heat away from the cavity walls. This may mean introducing mold materials with significantly higher conductivity than steel, such as beryllium-copper alloys. Alternately, the number and/or size of cooling channels may be increased above that which is normally considered adequate, or they could be brought closer to the molding surfaces than would normally be done. Any one of these solutions adds to the cost of the mold but is usually well repaid by increased productivity.

13.3.3.4 Heat Content

To raise the temperature of any material, a certain amount of heat (energy) must be introduced into it. The amount of energy is different for different materials; to get the plastic to its molding temperature, a certain amount of heat energy must be introduced into the plastic. This heat content flows with the plastic into the mold cavity and must then be removed so that the plastic returns to its rigid state for ejection.

This heat removal takes place through the walls of the cavities, mostly into the cooling medium—some as radiation into the air, and some conducted to the molding machine itself. For practical purposes, we will ignore the latter two, and concern ourselves with the coolant only.

Designers should be well aware of the above facts. For example, a prototype mold that runs only rarely and only to produce a few samples at a time may not need any cooling, since the mass of plastic (and its heat content) injected is insignificant when compared to the amount of heat that would be required to heat up the mass of the mold.

At this point, we should understand that the amount of heat required to melt the plastic is directly proportional to the mass of the plastic to be melted. The amount of heat to be removed from the molded plastic by cooling is identical to the amount that was introduced into the plastic. This assumes that the temperatures of the raw material and of the ejected products are the same. But, most of the time, the products are ejected fairly hot and, therefore, still contain some of the heat added during plasticizing. This heat will now be radiated from the products into the surrounding plant air.

The designer should also realize that amorphous (not crystalline) plastics, such as PS, vinyl, ABS, etc., increase their temperature (and start melting) proportionally to the heat added, and cool down proportionally to the heat given up to the coolant (Fig. 13.8). Crystalline materials, such as PE, nylon, PET, etc., require a certain amount of heat at a certain

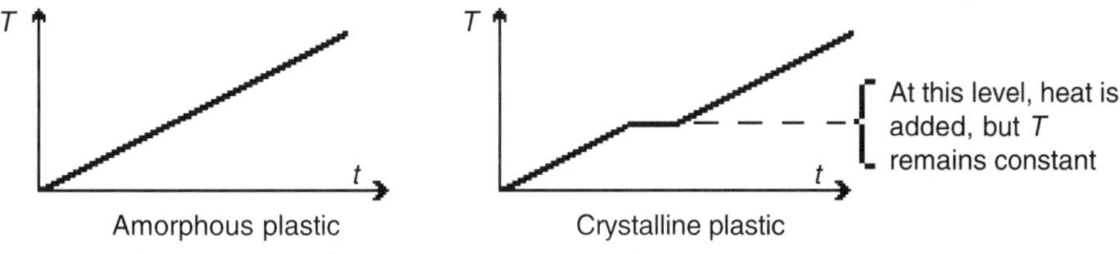

Figure 13.8 Increasing temperature is plotted against time for amorphous plastic (left) and crystalline plastic (right).

temperature level just to melt the crystals without increasing the temperature of the plastic. Only after the crystals have melted will the now amorphous plastic continue to increase its temperature in proportion to the heat input (Fig. 13.8).

The cooling process is similar. As the plastic passes through the crystallization range, heat is given up while the temperature remains the same. Only after crystallization will the plastic cool down further. For practical purposes, this means that a crystalline material needs more heat (energy) to raise a certain mass to the desired melt temperature than an amorphous material and, therefore, also more cooling (energy) to remove its heat content.

13.3.3.5 Coolant Flow

Various factors affect the flow of the coolant through the mold. The amount of heat removed from the mold is proportional to the ΔT between the temperatures of the plastic and the coolant. Unless the heated coolant is removed and replaced by fresh coolant, there is no cooling of the mold. Therefore, we are most concerned about the *flow* of the coolant and what affects it.

Generally, these factors affect the flow of the coolant:

1. Pressure differential (Δp) of coolant between supply and return lines,
2. cross section of passages,
3. length of passages and their layout,
4. viscosity of the coolant,
5. condition of the coolant channels, and
6. Reynolds number.

13.3.3.5.1 Pressure of Coolant Supply and Return Lines Basically, this is a difference of pressures, expressed as Δp (delta p). The pressure in the return lines is usually only slightly above the atmospheric pressure—just enough to ensure the flow back to the source (reservoir). Usually, the coolant supply at the inlet of the mold is in the order of 5–6 kg/cm^2 (70–85 psi). Higher pressures would provide higher flow (which is desirable), but for practical reasons the industry has adopted this range.

13.3.3.5.2 Cross Section of Channels It is quite easy to understand that a larger cross section lets more fluid flow than a smaller one. The flow rate is roughly proportional to the square of the diameters, that is, a 10 mm Ø channel passes roughly four times as much fluid as a 5 mm Ø channel ($10^2 = 4 \times 5^2$).

It is very important to consider this when laying out cooling channels. Designers and molders are mostly concerned about the quantity of coolant in the vicinity of the cavities and cores to remove the heat in that area. There must be enough coolant where it is needed and that it is evenly distributed. "Starving" the supply to some areas results in hot spots or uneven cooling in the mold.

13.3.3.5.3 Cooling Lines The effect of length on the flow is linear, that is, the resistance against flow is proportional to the length of the channel. However, except in rare cases, the influence of length can be ignored because, except in very large molds, the total length of channels in a mold is not very great. Changes in direction of flow such as those created where bores of coolant lines intersect add to the flow resistance and reduce the coolant flow. The "equivalent length" means that, for example, a 90° bend in the flow (most common) is the equivalent of a certain length of straight flow path; this equivalent length is dependent on the diameter of the channel and the volume of flow. It is good to remember that fewer changes of direction of flow are better for the flow, especially with small cross sections of the flow path.

13.3.3.5.4 Coolant Viscosity The effect of the viscosity of the coolant, although theoretically affecting the flow, especially with high percentage of antifreeze or other additives, can for practical purposes be ignored.

13.3.3.5.5 Condition of Channels Dirt and contaminations in the cooling channels can seriously affect the cooling efficiency. It can be easily understood that such deposits have a doubly negative effect on the performance of the cooling system. Firstly, the deposit reduces the cross section of the passage and thereby reduces the flow of coolant. Secondly, the deposit usually has a much lower heat conductivity than the mold material and reduces the amount of heat passing from the plastic to the coolant. This is of particular importance where small (round or flat) cooling channels are used. They should then be so designed that the channels can be cleaned from time to time.

13.3.3.5.6 Reynolds Number (Re) Without going into theory, this (dimensionless) number is a measure of the effect (on the cooling) of the velocity of the fluid in a channel, of the diameter, and of the specific density and the specific viscosity of the coolant. It has been found that the heat transfer from the walls of the channels to the coolant is greatly improved when the Reynolds number is above 3,200; below 3,200, we speak of laminar flow, and above 4,000, of turbulent flow condition. Desirable and suggested by various computer programs is Re > 4,000, better yet, Re > 10,000.

In the schematic illustration (Fig. 13.9), it can be seen that in the laminar flow only the coolant next to the wall removes heat, while the remainder of the flow just passes through the channel. With turbulent flow, the entire flow of coolant participates in the cooling created by eddy currents.

The Reynolds number is indirectly affected by the surface condition in the flow channel. The rougher the surface, the higher the number, which is desirable for good cooling. However, this roughness is also conducive to corrosion and retention of sediment. Generally, a bore should be left as drilled and preferably protected against corrosion by plating, such as by the electroless nickel process (ENP), or by selection of stainless steel for the mold parts.

Cooling channels in cores and cavities, as well as inserts, must have turbulent flow. In large diameter passages, the flow may become laminar, but these larger passages are used mainly to distribute the coolant to the actual cooling channels, which are then much smaller and provide the desired turbulent flow.

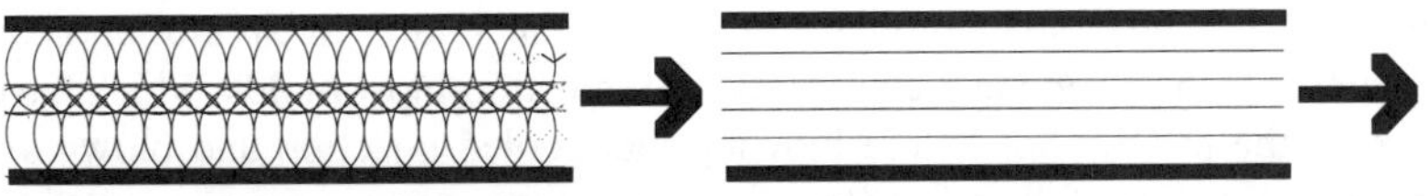

Figure 13.9 Schematic of cooling during turbulent flow (left) and laminar flow (right).

13.3.4 Runner Systems

Hot and cold runner systems produce different effects on mold cooling. In a *cold runner* system, the plastic on its way from the machine nozzle (or from the hot sprue bushing) to the cavities must be cooled in a manner similar to the molded product itself. The cold runner is then ejected with the molded products. The plates in which the runners are located must, therefore, be cooled to avoid unduly long cycle times. However, over-cooling of these plates must also be avoided to assure that the plastic is allowed to fill the mold cavities before the runner freezes. A proper balance must be struck between the size of the runners, which affects the plastic flow to the cavities, and the amount of cooling around the runners, which depends essentially on the mass of the runners.

Generally, the runners should be as small as practical for proper mold filling and to allow proper ejection of the runners. This, in turn, will reduce the requirement for intricate cooling of the plates that surround the cold runners.

In a *hot runner* system, the plastic is kept hot (melted) from the machine nozzle to the cavities. The manifold, which contains the flow passages for the plastic, must be kept at a temperature at which the plastic does not lose or gain any heat on its way to the cavities. Because the manifold and associated hot runner hardware are supported by the surrounding mold plates, a certain amount of heat is lost to these plates and increases their temperature. This makes it necessary to cool these plates. The better the design of the hot runner system, the smaller the losses, but they are never so small that they can be ignored.

13.3.5 Temperature of the Mold Shoe

The designer must consider proper cooling of all plates that make up the mold. The purpose of these plates is to support and hold the cavities and cores; to house the runner system (cold or hot) and the ejection mechanism, including side core mechanisms, etc.; and to provide proper alignment.

As has been shown, most of the heat in the mold is removed by the cooling of the cavities and cores, but a certain amount of heat travels to the plates that support the cavities, cores and, especially those that surround the hot runner components. Some of this heat travels into the machine platens; from there, heat is radiated into the plant.

Both mold halves should be kept at essentially the same temperature. Differences in the temperature between them can seriously affect the alignment of leader pins and bushings or taper locks and result in rapid wear of these components or the mold itself. (For more detail, see the example in Section 14.4.)

In many cases the feeder lines for the coolant to the cavities and cores are drilled in the mold plates and provide some natural cooling for these plates. However, if the cavities are fed directly (by hoses) and the leader pins and bushings are mounted on uncooled plates, the designer must make sure that there will be no excessive temperature differences between the two mold half plates.

Sometimes, it is unavoidable to have different temperatures in cavities and cores. One typical example is the design and construction of injection-blow molds, which have heated cores to keep the parison hot before blowing, but cooled injection and blow cavities. In addition to good cooling of the plates (for alignment), the support and mounting of the heated components must be well insulated to reduce the heat losses to the cold plates.

13.4 Laying Out Cooling Channels

13.4.1 Distance Between Drilled (Bored) Passages and Other Passages or Walls

There are no fixed rules possible regarding the distance between bored passages; just use common sense. A few practical recommendations will help.

13.4.1.1 Length of Bore

The longer a drilled hole, the more the drill may wander off. It is recommended that holes be drilled at least 6 mm away from other openings in the case of long drills, but for short holes ($L \leq 6D$), 4–5 mm may be sufficient. Using gun drilling reduces the risk of wandering drill bits considerably.

Because of strength considerations, closer distances are not advisable, especially in hardened steels. If the cross section "X" (Fig. 13.10A–C) is too small, the stresses induced during molding can be very high. Corrosion inside the bore may cause stress raisers and cracks in the steel (see also Chapter 18, Metal Fatigue). In such thin spots, heat cracks may develop during quenching in the heat treatment of the mold steel.

13.4.1.2 Drill Size

Use standard, available drill sizes.

13.4.1.3 Fitting Size

Use standard fitting sizes at the end of a bore. The designer may consider (national) pipe threads (NPT) and fittings, or O-ring threads and fittings.

13.4.1.3.1 Pipe Threads and Fittings For straight adaptors, the minimum distance from obstacles is governed by the size of the wrench required to tighten the adaptor and the swivel hose fitting. These two wrenches are not necessarily the same size. For elbow adaptors, consider the swing of the elbow when screwing it into the mold part. For plugs, there is no such problem.

We must consider that the outside diameter of the thread is larger than the bore. The combination of notching caused by the cutting of the threads and the wedging action of the

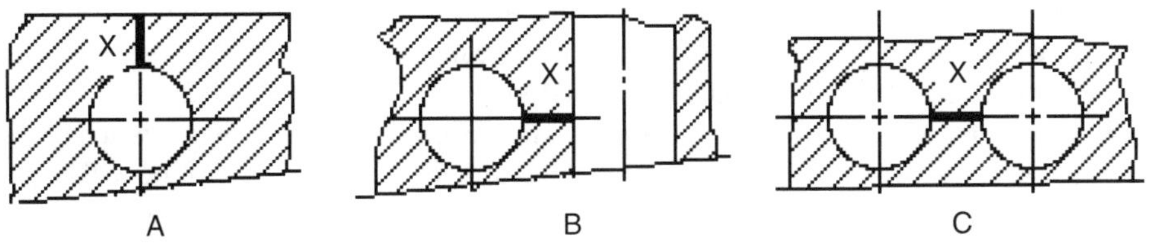

Figure 13.10 Cross sections show the distance X between A. drilled hole and wall, B. drilled hole and passage, and C. two drilled holes.

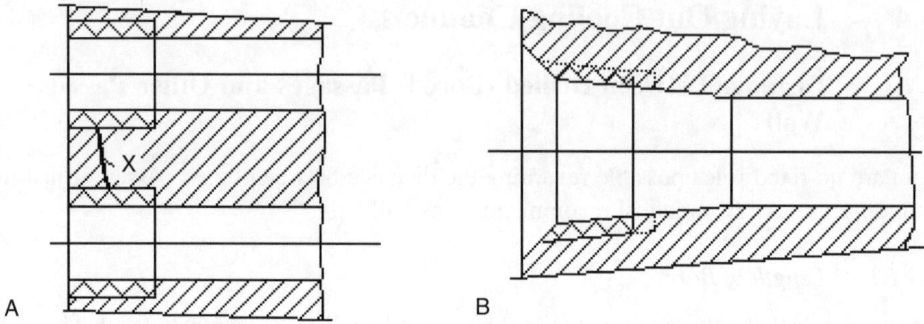

Figure 13.11 Pipe threads: A. distance X between threads is critical, and B. large threads should be taper reamed.

plug when tightening exert tremendous forces that tend to split the steel if not enough material is left around the bore (Fig. 13.11A).

To improve the quality of the seal, the thread portion of the bore should be taper reamed before tapping, especially with larger sizes. Standard taper is approximately 3.5° per side (Fig. 13.11B).

13.4.1.3.2 O-Ring Fittings O-ring fittings are sometimes used in place of the tapered pipe fittings. They are more expensive, and the cost of drilling and thread cutting is also slightly higher because of the greater precision required. However, there are at least two advantages to using them:

1. Since the thread is not tapered, there is less wedging action when tightening, so the holes can be closer to the walls or other bores.
2. There is less risk of leaking, especially under vibration, for example, when attaching hoses to moving slides, etc.

This type of fitting is quite common in (oil) hydraulic manifolds but seldom required for cooling connections and, in general, is not recommended.

13.4.1.4 Distances of Cooling Channels

Distances of cooling channels from the molding surface and from each other must be considered (Fig. 13.12). The recommendations are based on theory, to produce optimal, almost even surface temperatures. However, there are many times when the distances shown here are not adhered to. But it should be understood that as the cooling channels are moved closer to the surface, the cooling effect on the surface is greater near the channels and smaller in the space between them. If the channels are moved closer together, the cooling will be more even, but the added cooling channels will increase the cost of the mold and require more (and possibly unnecessary) flow of the coolant.

Recommended distances are that $B \approx 2.5\text{–}3.5D$, and $A \approx 0.8\text{–}1.5B$.

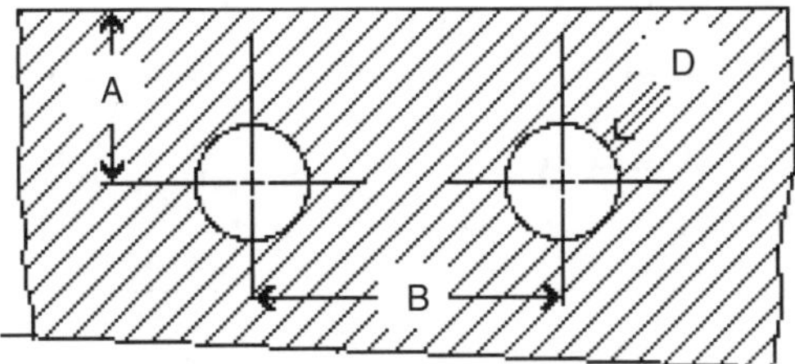

Figure 13.12 Cross section of cooling channels shows distances A. between cooling channel and molding surface and B. between cooling channels.

13.4.1.5 Strength of Mold Material

The mold cavity (or plate) must withstand the forces created by the plastic during injection. These forces are very high and can be in the order of 1,400 kg/cm^2 and even higher in thin-wall molding. Also, these forces are cyclical (i.e., they change from zero to full to zero at every molding cycle).

Under cyclical conditions, the safe, permissible stresses in the steel (for long mold life) should be held below 10% of the yield stress of the mold material. When following the recommendations (for steel) shown in Fig. 13.13, it will not be necessary to calculate the stresses. For other materials and other proportions, stress calculations should be done to avoid unpleasant surprises such as caved-in or cracked surfaces.

For Fig. 13.13A, the recommended dimensions are $S \geq L$ and $A \geq 0.66L$. Note that this applies to a square channel. A rounded channel formation as shown in Fig. 13.13B is considerably stronger, and S can be as small as $0.6L$.

13.4.1.6 Efficiency of Cooling Channels

Figure 13.14 B shows the same channel cross section and the same distance from the molded product in both A and B. However, the active cooling surface in Fig. 13.14A (shown in heavier lines) is about three times the active surface shown in Fig. 13.14B. The reasons for this is that 1) there is about three times as much cooling channel surface near the molding surface and 2) only one side of the channel is in direct contact with the steel of the mold cavity, and the

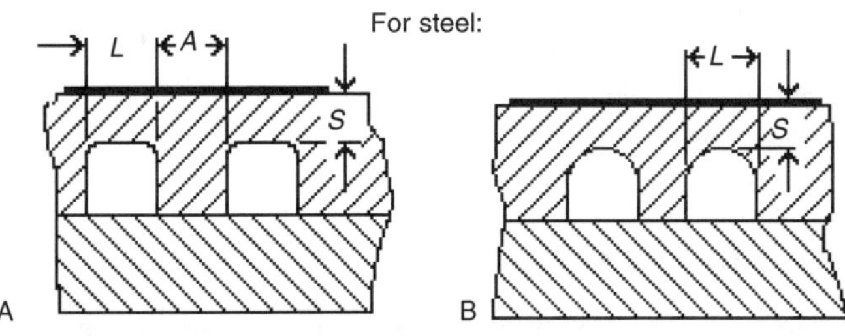

Figure 13.13 Recommended dimensions for cooling channels in steel molds. A. square channel cross section, B. round channel cross section.

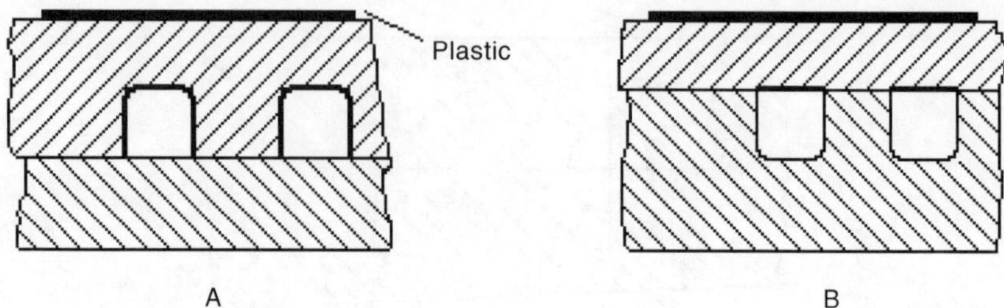

Figure 13.14 Two different cross sections for the same channels at the same distance from the molded plastic: A. three times as much cooling channel surface area in the mold cavity, B. only one side of the channel in contact with the mold cavity.

heat must travel through the area where the steel of the mold cavity touches the underlying mold plate. The conductivity at the contact area is at best (with clean metal, under load) only 50% of what would be if there was no separation. In general terms, it means that the cooling lines should be, wherever possible, in the mold parts that are in contact with the plastic, rather than in adjoining plates.

13.4.2 Layout and Sizing of Cooling in the Mold Plates

The first practical consideration for the cooling of the mold plates is whether the plate really needs separate cooling. Where does the heat originate? Is there already sufficient cooling because the supply of water to the cavities and cores flows through the plate?

If the cavities and cores are cooled only by contact with the adjoining plate, such a plate (usually a mounting plate) must be well cooled and in very good contact between cavity or core. The plate contact must be created by using a sufficient number of mounting screws or by tight fit when set into the plate, or a combination of both.

Other considerations for mold plate cooling are:

- Simplicity of layout,
- symmetry, and
- number of circuits required to ensure even cooling.

Usually, plate cooling represents a compromise between:

1. Size and available space on the machine,
2. the number and size of screws used in the plates,
3. the position and size of leader pins and bushing,
4. optimum cooling, and
5. cost.

13.4.2.1 Size and Available Machine Space

The size of the cooling layout is often limited because of the size of the machine platen and the location of the tie bars. This is important when determining the points where the cooling lines are to enter and leave the mold.

13.4.2.2 Number and Size of Screws

The number and size of screws required to hold plates together and to mount cavities and cores depends on such forces as the opening force of the mold and any warp or deflection caused by heat expansion. If cavities and cores are pulled apart as the mold opens, especially if the mold is overpacked, the pulling forces can be considerable. This applies particularly to molds for deep draw products, such as containers with a very small draft angle.

Warping as a result of heat expansion usually occurs with large plates and large, flat cavities and cores. A sufficient number and adequate location of screws must prevent plates (or cavities) from separating (because of plate deflection or warping) and creating leaks in the coolant circuits.

There should be no compromise in this area. The mold must function properly, without plates being pulled apart and without leaking. Screw strength can be figured out by estimating the separation forces.

13.4.2.3 Position and Size of Leader Pins and Bushings

While the selection of leader pins and bushings has nothing to do with the mold cooling, their location can influence the cooling layout.

13.4.2.4 Optimum Cooling

Ideal cooling is impossible because of the possible restrictions outlined in the above basic rules, and possibly other restrictions. This is the area where the ingenious designer can come up with the cooling layout (circuit) that provides the optimal conditions.

13.4.2.5 Cost

Obviously, cost is affected by the complexity of the circuit. As pointed out in the introduction, all parameters must be considered to decide on what quality of cooling will be provided and at what cost.

It is up to the designer to make the proper compromises while considering the above listed points.

13.4.3 Series and Parallel Cooling

The coolant, as does any fluid, always flows in the path of the least resistance. Figure 13.15 shows a typical series arrangement. The flow depends only on the pressure differential Δp between the "in" and "out" ports and on the resistance within the channels. The longer the channels, the more resistance; the greater the channel diameter, the less resistance.

As the coolant flows through the channels, it picks up heat from the surrounding metal and thereby increases in temperature. Therefore, the cooling efficiency is greater near the inlet port (greater ΔT from hot plastic to cold metal) than near the outlet, where the coolant is warmer. This is the reason why it is recommended to keep the maximum temperature difference between "in" and "out" to less than 5–6 °C.

For a parallel arrangement of cooling channels, it is important that the cross section of the "in" and "out" channels (D) is sufficiently larger than the sum of the cross sections (d) of the

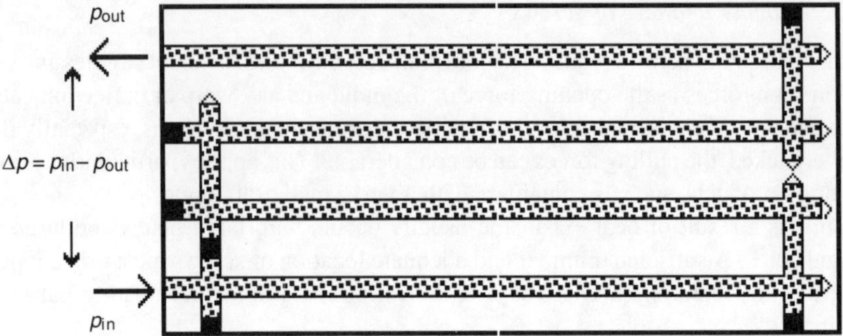

Figure 13.15 Typical series arrangement of cooling channels.

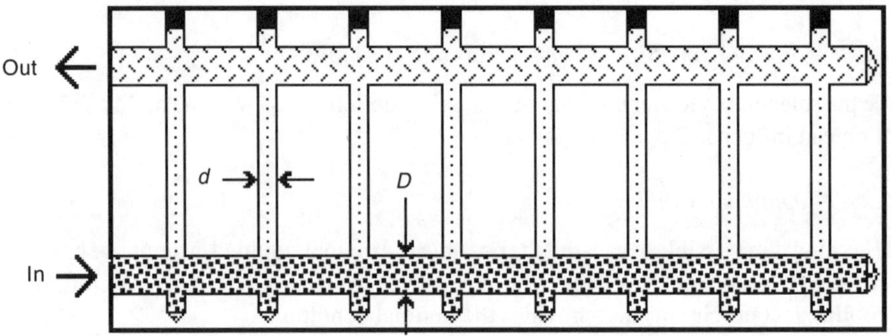

Figure 13.16 Typical parallel arrangement of cooling channels.

parallel branches. If d was the same as D, the coolant would take the shortest route (first branch) from "in" to "out" and provide no cooling for the rest of the plate.

Remember that the cooling efficiency depends not only on the ΔT but also on the amount of coolant flowing through each channel and on the area touched by the coolant. The amount of coolant depends on the cross section of the channel. In case of round channels, it is proportional to d^2.

The area touched is proportional to d. In other words, the cooled area increases only with the diameter, while the cross section (and the flow) increases with the square of the diameter. For example, by doubling the diameter of a channel, the flow increases four times (approximately), but the surface only doubles in size. This is another reason (see Reynolds number) why more small lines provide more efficient cooling than fewer, larger lines for the same amount of coolant flowing through the system. The selection of the channel (bore) size also depends on the requirement to use standard drill sizes where possible.

Figure 13.17 shows a schematic layout of a combined parallel and series cooling layout and a fairly simple method to calculate the relative sizes of the branch channels. In this example, the number of the branches N_1, N_2, N_3 is 1, 2, and 6, respectively. Their diameter is D_1, D_2 and D_3. To arrive at the approximate diameter of the two smaller lines, D_2 and D_3, use the following equations:

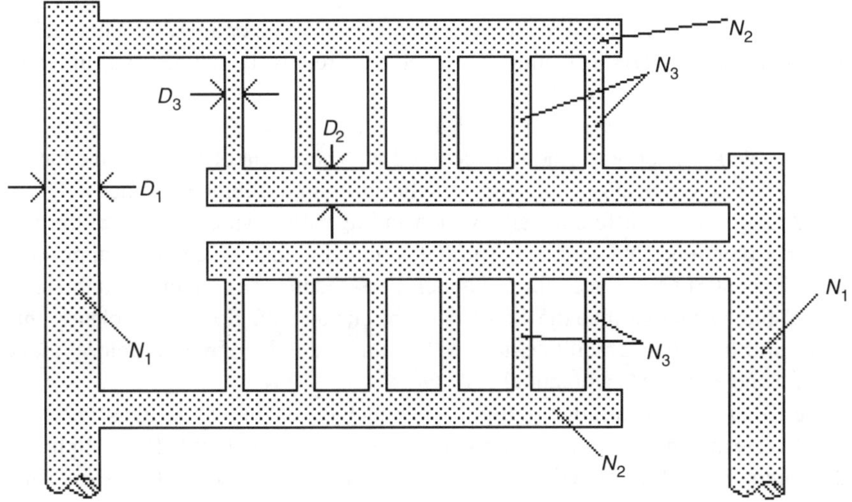

Figure 13.17 Schematic layout of combined parallel and series cooling channels.

$$D_2 = \sqrt{\frac{D_1^2}{N_2}} \text{ and } D_3 = \sqrt{\frac{D_1^2}{N_2 N_3}}, \tag{13.2}$$

where D_1 = input line diameter, with 12 mm < D < 50 mm. D_2 and D_3 should be, preferably, larger than 6 mm for ease of manufacture.

Parallel and series cooling are used in combination in many molds. For example, the plates may be all series cooled, and the cavities and cores may be series and/or parallel cooled. The decision usually depends on the following considerations:

- Quality of cooling required,
- space available,
- strength of mold material,
- number of cooling circuits available at the machine, and
- number of cooling circuits desired for simplicity of installation.

13.4.3.1 Baffles in Plates

Cooling channels in plates are usually produced by drilling or boring; therefore, they have a circular cross section. The entry spot of the bore must then either accept a connector or be plugged.

To achieve a flow path for the coolant within the plate, the bores must intersect. This usually occurs in the same plane of drilling, but in some cases it is necessary to offset the center lines of the bores. It is then important to make sure that the passage from one bore to the other does not restrict the flow and that burrs from drilling can be removed from such offset intersections.

Figure 13.18 shows a typical (partial) end of a plate with plugs at the end of bores and baffles to direct the flow. The arrangement of baffles shown is the most commonly used method. The baffles are fastened to the rod with pins or screws, or are soldered to it. The rod ensures that the baffles stay in position. This method is relatively inexpensive, and standard, commercial parts are available.

A disadvantage of baffles and rods is that their fit in the bore must be loose so that they can be assembled. The bore is gradually covered with sediment or rust, and the baffles are difficult to remove for cleaning after the mold has been in use for some time.

Figure 13.19 shows a different method for installing baffles which eliminates the problem of removal as mentioned above. The baffle is positioned so that its centerline approximately intersects the centerline of the cooling channel. By selecting the diameter and the length of the baffle somewhat larger than the diameter of the channel, the flow is completely shut off. A threaded hole in the baffle facilitates removal. A plug in the side of the plate closes the access to the baffle and eliminates a potential leak.

Figure 13.20 shows a variation of Fig. 13.19 which is used in critical areas where there is not enough plate thickness for pipe thread and plug. The diameter of the baffle is slightly larger than the bore it must block, and the end should be at least 1 mm beyond the bore. The bore for the plug should be smooth, and the entry chamfered to protect the O-ring on assembly.

In both Figs. 13.19 and 13.20, the fit of the baffle in its bore should be a sliding fit (h6H7) to facilitate removal of the plug. Both of these methods are somewhat more expensive to produce than the rod and baffle arrangement but are often used where the baffle and rod arrangement is not wanted or possible. (See also Section 22.13 on Drilling in Molds.)

Figure 13.21 illustrates the commonly used practice of drilling from both sides of the plate to avoid inserting baffles; however, this method should only be used if there are other holes to be drilled from both sides to save the extra machine set-up time, which may cost more than a baffle.

Figure 13.22 shows a commonly used drilling arrangement where a cooling line passes through an opening, such as a bore for a leader pin or bushing. It is not absolutely necessary

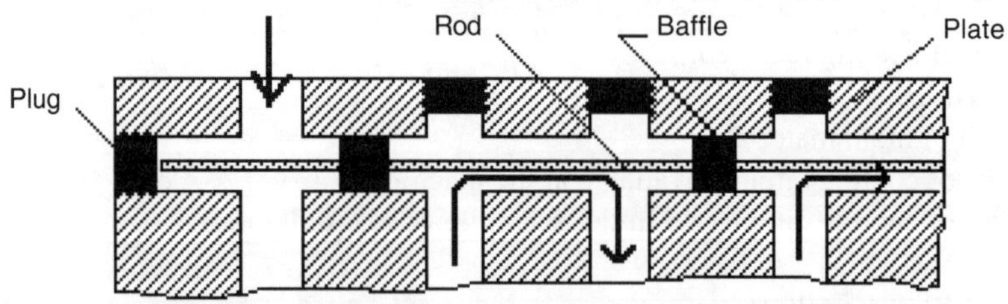

Figure 13.18 Typical arrangement of baffles in a plate.

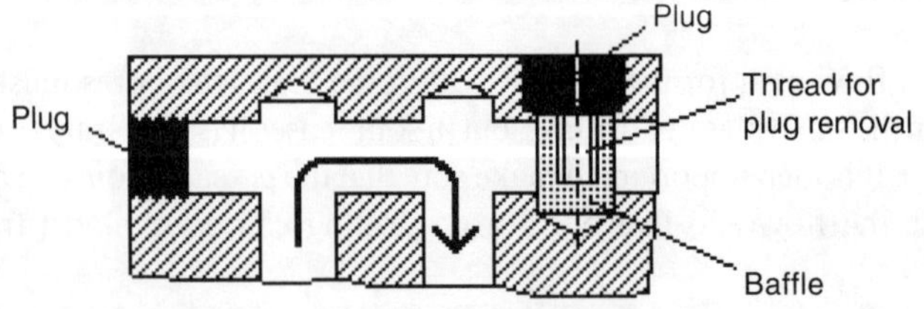

Figure 13.19 Baffle centerline intersects channel centerline and shuts off flow.

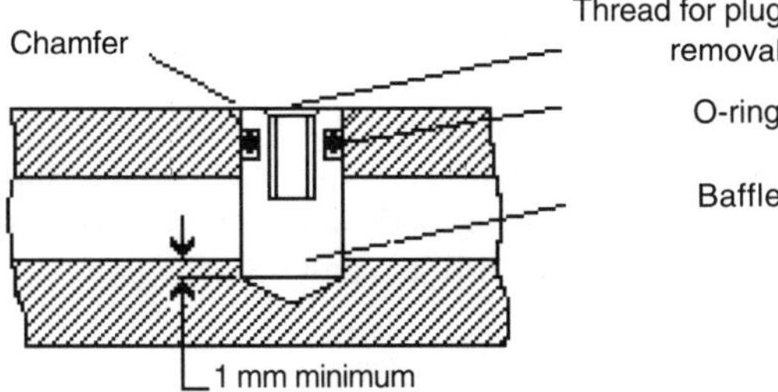

Figure 13.20 Variation of baffle uses O-ring fitting and chamfered entry for installation where threaded plugs will not fit.

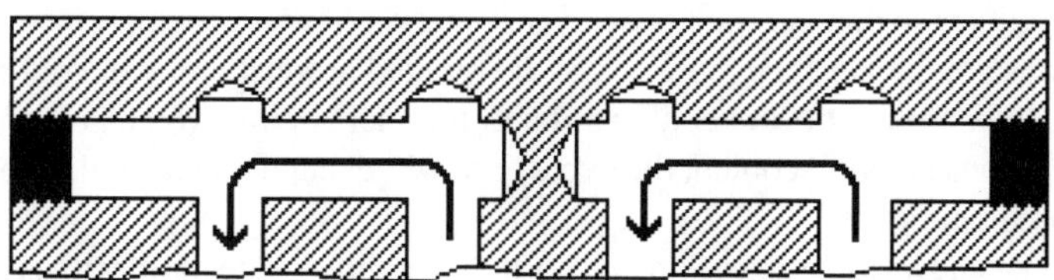

Figure 13.21 Channels are drilled from both sides of plate to eliminate need for baffles.

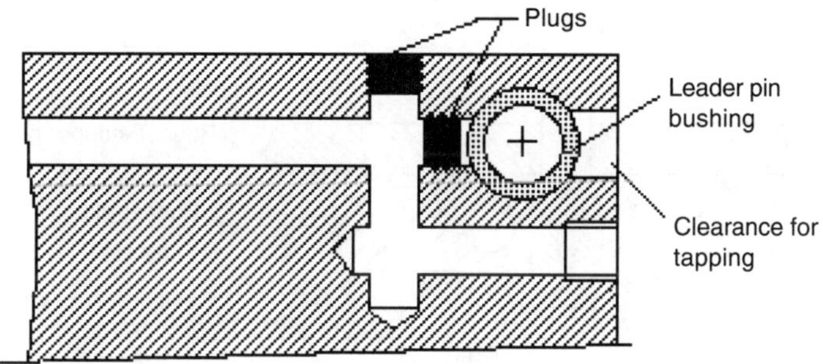

Figure 13.22 Cooling line passes through a leader pin bushing.

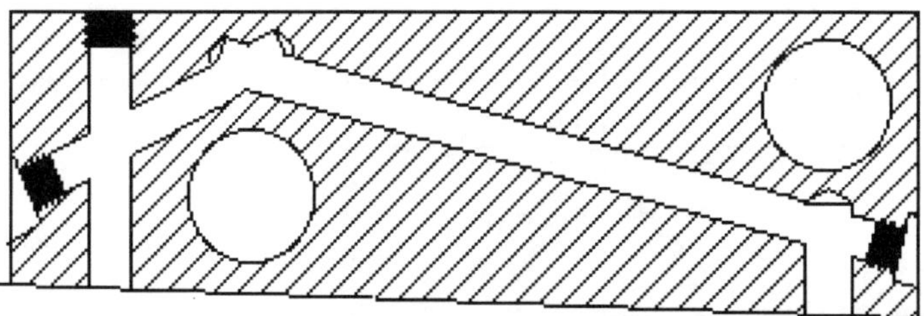

Figure 13.23 Cooling channels directed around obstacles in the plate.

that the centerline of the cooling bore and the centerline of the opening (in the illustrated case, the bore for the leader pin bushing) intersect. The illustration is self-explanatory, but one should remember to apply all the rules regarding flow of coolant and strength of material.

Figure 13.23 shows a method sometimes used to go around obstacles (leader pins, etc.). As this method requires additional set-ups for drilling, counterboring, and tapping at an angle, it is more expensive than square (90°) drilling, and should only be used if no other solution is possible.

13.4.4 Runner and Cavity Plate in Three-Plate Molds

The cooling of this plate is a special case, as the plate contains both the runners (with drops to the cavities) and the cavities. It is important to cool the runners and the drops as quickly as the molded products so that they do not delay the ejection. This is of importance since, in most cases, the drops are thicker than the walls of the molded product and will cool slower. Also, there is some bulk around the dovetailed point of the sucker pins, which hold the runner and drops. If this area is not well cooled, the cycle time will depend on the cooling of the runners rather than on the product.

Figure 13.24 shows a cross section of a portion of a three-plate mold with separate runner and cavity retainer plates. As shown, the runner plate must be cooled, especially around the drops.

Figure 13.25 shows a similar arrangement to Fig. 13.24, except that the cavity and runner plates are in one piece. The designer should try to bring cooling channels as close as practical

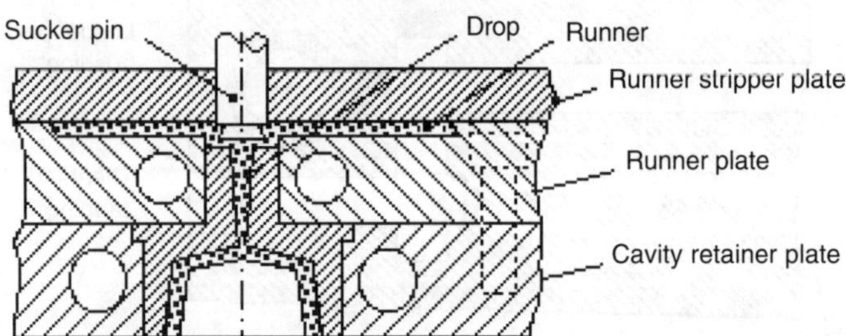

Figure 13.24 Cross section of a three-plate mold and location of cooling lines in both runner and cavity retainer plates.

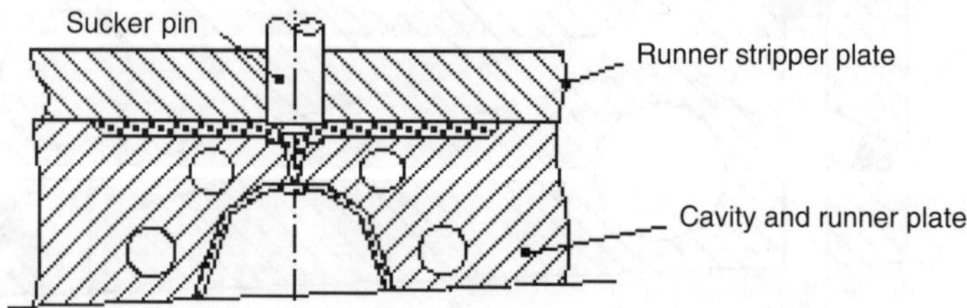

Figure 13.25 Cross section of a three-plate mold show cooling lines around cavity and runners in one plate.

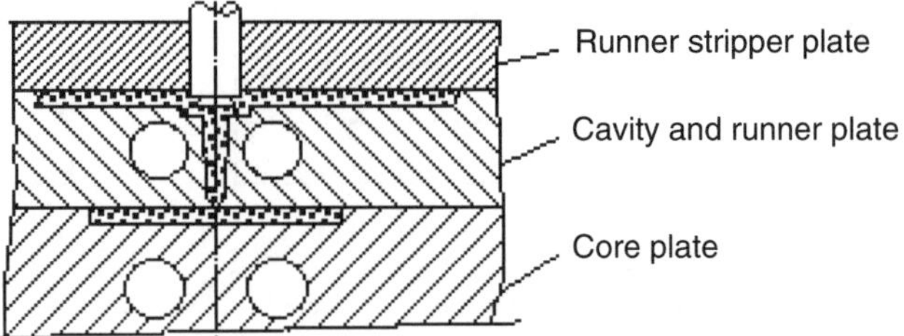

Figure 13.26 Cross section of a mold for a flat product showing cooling lines in the core plate and cavity and runner plate.

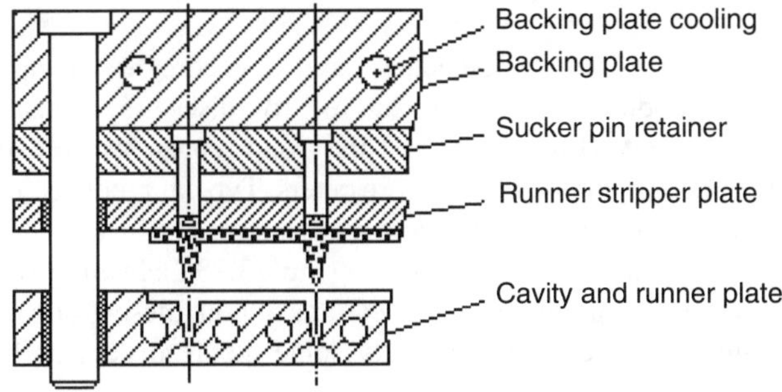

Figure 13.27 Cross section of mold showing cooling in the backing plate as well as the cavity and runner plate.

toward the drops to assure good cooling of this area. This design is used in relatively cheap molds, for short runs, and where changes or repairs are not anticipated. However, it is also used when the product is of a shape that would not allow the use of inserts.

Figure 13.26 shows molding of a flat product. It is important to ensure that both sides of the product are equally well cooled. If cavity and core inserts are used, make sure that both inserts and the plates are well cooled because of unbalanced or unequal heat expansion.

Heat expansion of all plates must be considered. The leader pins are often anchored in an uncooled backing plate. It is, therefore, important that there is enough cooling provided in the cavity and runner plate so that this plate will not expand and cause binding of the leader pins during opening and closing of the mold. If there is not enough cooling in these plates, it will then be necessary to provide cooling to the backing plate as well, as shown in Fig. 13.27

The sucker pin retainer plate and the runner stripper plate are kept as thin as possible (because of mold shut height, weight, and cost) and are usually not cooled. The backing plate should always be cooled if it surrounds a *hot runner* manifold.

13.4.5 Layout and Sizing of Cooling Channels in Mold Components Contacting Molded Products

There is an infinite number of mold configurations possible, depending on product shape, size, plastic, expected productivity, and degree of automation. Considered here are the cooling features common to all molds.

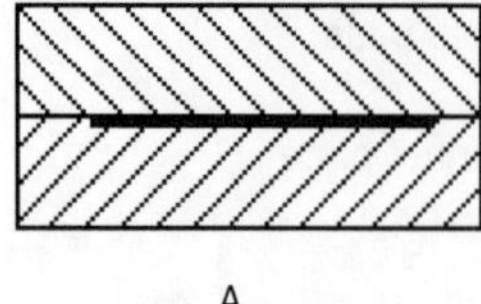

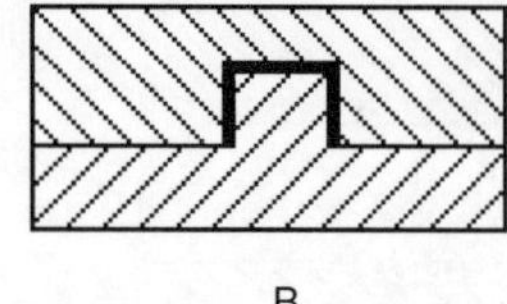

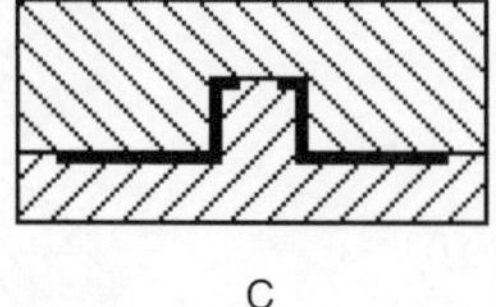

Figure 13.28 Typically shaped products: A. flat , B. cup-shaped, and C. combination flat and cup-shaped.

It is possible to divide molds into three basic groups, according to the shape of the molded product:

1. *Flat products* (Fig. 13.28A): Both sides (cavity and core) are of approximately the same shape. Typical products in this group are records, discs, trays, lids, test pieces, etc.
2. *Cup-shaped products* (Fig. 13.28B): This group represents a majority of molds. The cavity is a depression into which the core enters. Typical products in this group are cups, pails, closures, preforms, cassettes, and many more.
3. *Combination flat and cup-shaped products* (Fig. 13.28C): In this group are some products which, for cooling purposes, combine features of both these groups. Typical products in this group are often "technical" products, some spools, disks with hubs, etc.

The first two groups will be considered in more detail to highlight the features that deal with proper cooling. The lessons learned from these groups do apply to the third group, as well. Note that all basic rules apply equally to the cooling of cavities, cores, gate pads, etc.

13.4.5.1 Compressibility and Shrinkage of Plastic

The understanding of these two properties is necessary before going into more details. In Chapter 8, Shrinkage, both properties are explained in detail and will not be redefined here.

As the hot plastic enters the cavity, it is hot (expanded) and under more or less high pressure (compressed). As the injection is cut off and the plastic cools down, the plastic shrinks to return to its smaller, cold volume, but at the same time increases in size as the pressure on the plastic is decreasing.

Usually, the effect of shrinkage is the greater of these two influences, with the result that the plastic, as it cools down, loses its close contact with the cavity walls but maintains its contact with the core as it "shrinks on the core." This condition has a great influence on the cooling of the molded product.

Referring to Fig. 13.28A above, it can be seen that a flat product is free to shrink in all directions. If the thickness is relatively small, shrinkage and compressibility have little effect on the contact with the flat surfaces on the cavity and core, and the cooling efficiency will be the same on both sides of the flat product. The length (or diameter) of the product will reduce as the product shrinks.

When designing cooling for a flat product, it is important that both sides of the product are equally well cooled. This can be seen in Fig. 13.29, which shows one side of a flat product with less cooling than the other side. To achieve short cycle times, the product is usually ejected as soon as it is rigid enough for ejection. At this moment, the side in contact with the poorer

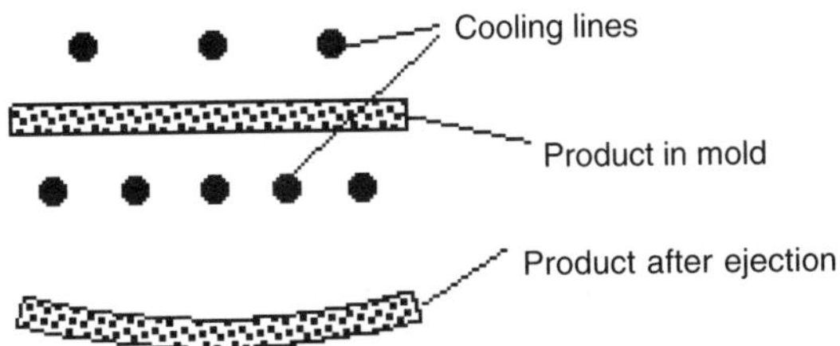

Figure 13.29 Uneven cooling for flat product (top) results in warped product (bottom).

cooling is still hotter than the other side. As the product cools down outside the mold, the still warmer side continues to contract, and warps the product.

Note that Fig. 13.29 is greatly exaggerated, but it clearly indicates what can happen with flat products if they are unevenly cooled. The same applies to the third case (Fig. 13.28C), combination of flat and cup-shaped products.

There are cases when such differential contraction is designed into the mold, such as in the case of snap-on lids. The surface must be flat, but the rim, being heavier, cools down slower and creates a reverse taper which is desired for the product. The amount of "toeing-in" can then be controlled with the amount of cooling and the cooling time; the longer the cooling time, the less the toe-in.

Now, referring to Fig. 28B and applying the same rules for shrinkage, we see that the product will contract as it cools and shrinks onto the core while losing its contact with the cavity wall. This improves the cooling efficiency of the core while reducing it for the cavity. The result is that the cooling of the cavity is not as effective as the cooling of the core.

With thin-walled products, the difference in efficiency between cavity and core is not as noticeable because the cooling time is very short. Shrinking onto the core ensures that the product will stay on the core for ejection purposes. Both cavity and core should be equally well cooled.

For heavy-walled products, the condition is different. The product, as it cools, loses contact with the cavity wall so that the cavity cooling becomes less efficient. The molding cycle will be much more dependent on the quality of the core cooling. This is unfortunate, because it is usually much easier to find room for cooling in the cavity, which is larger than the product, than in the core. The smaller the core is, the more difficult the cooling becomes.

In its smallest form, the core could be just a small pin. It is then up to the designer to find a way to provide maximum cooling for such a small core while still making it strong enough to withstand the cyclical loads during the molding cycle, which could cause cracking, bending, or caving in of the core. As in all design problems, the final answer will be a compromise between cooling efficiency and selection of material and shape of the mold component.

These statements apply not only to cores but also to core inserts and to (movable) side cores. The cycle time of the mold depends on the cooling of the whole product; therefore, any poorly cooled portion of the product will be the controlling factor for the productivity of the mold.

As there is an infinite number of possibilities for the cooling of cavities and cores, we will not go into specific mold designs. There are books that show some typical designs, but the designer should be cautioned not to follow such examples blindly. What is good for one product design and one material may not be so good for a similar product and a different material. Before copying any design, it is always worthwhile to investigate, if possible, and to determine any potential problems with regard to the cooling shown.

13.4.5.2 Cavity, Core, Insert, Gate Pad and Stripper Cooling

This cooling applies to all components in touch with the molded product. The following describes some specific factors and arrangements applying to any product, either flat, cup-shaped, or combinations of both, with illustrations to highlight these features.

13.4.5.2.1 Series, Series-Parallel, and Parallel Cooling Layouts

Series Cooling Figure 13.30A shows schematically a series cooling arrangement. Provided there is enough flow to keep the ΔT between "in" and "out" to about 5 °C, there is little argument against this layout, which is simple and inexpensive to produce.

Series–Parallel Cooling Fig. 13.30B shows a similar mold with a series-parallel cooling layout. The feeder and return channels must be larger than the branch channels to provide equal flow through each of the branches. This method requires more drilling, but the cooling efficiency is better than with series cooling. It is often used for multicavity molds.

With very large numbers of cavities, if possible, the feeder channels should be split to provide more even supply to the branches, such as that shown in Fig. 13.31A. If the feeder

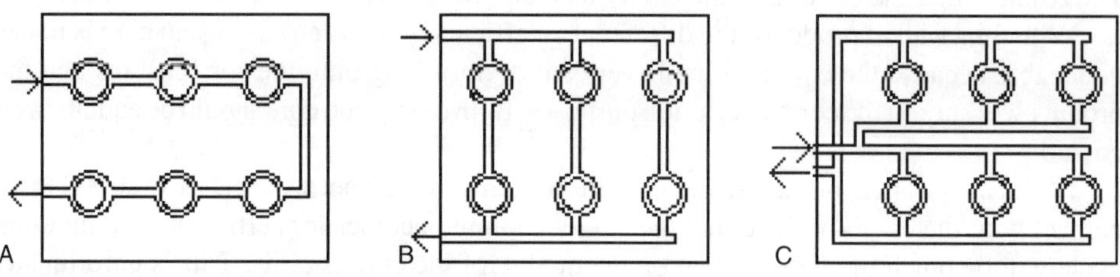

Figure 13.30 Cooling layouts for components in touch with the molded product: A. series cooling, B. series–parallel cooling, and C. parallel cooling layout.

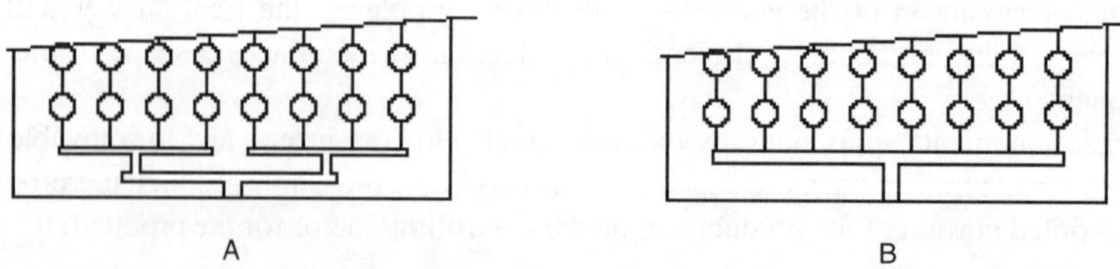

Figure 13.31 Feeder channels for cooling layouts in multicavity molds: A. split feeder channels provide more even coolant supply, B. larger channels may not need to be split.

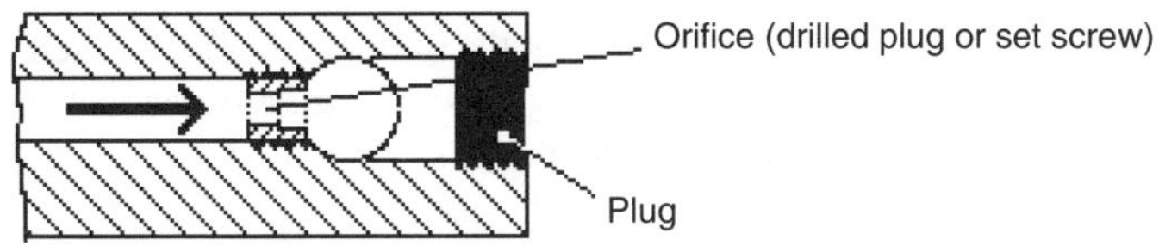

Figure 13.32 Simple fixed restrictor in a flow channel.

channels can be made sufficiently large, a cooling arrangement as per Fig. 13.31B is acceptable.

It should be understood that the cooling arrangement of the cavity and the core side do not need to be the same. It is often quite easy to provide parallel cooling for the cores, while it is more difficult to do so for the cavities. A very common approach is to use series or series–parallel cooling for the cavities and parallel or series–parallel cooling for the cores. This approach is quite acceptable since in most molds (except for thin-walled products and flat disks) the cooling of the core is more important than the cooling of the cavity.

Cavities and cores could also be cooled indirectly only by way of a cooled backing plate. This method is less expensive, but such a mold will run much slower than one with individually cooled cavities and cores. However, there are many cases, usually for technical products that require many inserts, when it would not be practical or possible to provide cooling in the cavity or core blocks, which often contain not only many inserts but also the mounting screws, ejector pins, and the drops for the gate. A common cooling arrangement for such blocks is to have cooling channels around the circumference of each block, but most cooling for the center portion and the inserts is obtained from contact with the cooled backing plate.

Parallel Cooling Figure 13.30C shows a typical parallel cooling layout. All cavities are equally well cooled, provided the branch channels are sized so that the flow to each cavity (or core) is the same. This is the most efficient but most costly arrangement.

In the area of high-productivity, high-speed molding, such as thin, disposable products, the use of parallel cooling must always be the first choice. Because of the thin wall thickness of the product, the heat must be removed equally fast from both cavity and core to permit fast cycles. The poorest cooled cavity will always control the length of the molding cycle; parallel cooling supplying each cavity and core with coolant of the same temperature at the inlet is the most effective cooling system.

Restrictors in the Flow Channels Restrictors are sometimes used to achieve equal flow through a number of parallel branch channels. Generally, the use of such restrictors can be avoided by properly sizing the flow channels in the first place. However, this is not always possible, and restrictors may be the only way to correct a bad flow situation. Figure 13.32 shows a simple and effective design for a fixed restrictor.

Note that a fixed restrictor is preferred to an adjustable one not only because it is simpler but mainly because it keeps unauthorized personnel from changing the adjustment.

The size of the restriction should be developed during the test run, to optimize the mold cooling, and should not be changed thereafter.

Note that the restrictor is always inserted at the outlet end of the branch of the flow path to ensure a back pressure to the coolant flow. This, in turn, ensures that the coolant in the channels stays in full contact with the channel walls.

Sealing the Passages from Plate to Plate This is commonly done by inserting O-rings at the sealing points. This applies to all locations in a mold. All O-ring applications for cooling circuits are considered static seals. For O-ring groove dimensions and finishes, use the established standards supplied by the O-ring manufacturers. O-rings may provide flat seals or circumferential seals.

Flat Seals O-rings are located between flat surfaces. They are compressed as the plates are joined. Figure 13.33A shows the preferred arrangement. Some material (at least 1 mm) is left standing between the O-ring groove and the coolant passage.

Figure 13.33B shows an acceptable method for cases in which there is not enough space available for material to be left between the O-ring groove and the coolant passage. The bottom of the groove is tapered to prevent the O-ring from slipping into the coolant passage. This method is frequently used in hydraulic circuits. A special form tool to machine the correct shape of the O-ring groove is commercially available.

Figure 13.33C is an example of a poor method that must not be used. The O-ring may be sucked into the passage and disappear with the coolant.

Figure 13.34A shows an arrangement where two or more passages are sealed by one common O-ring. This is quite acceptable because any leaking from one passage to the other can usually be ignored. A disadvantage is the corrosion that takes place in the "wet" area within the area bordered by the O-ring. If this should be a service problem, the plates could be nickel-plated, or rust-free materials should be selected.

Screws within a "wet" area must be sealed with O-rings to prevent corrosion of the threads (Fig. 13.34B). In this case, an O-ring groove similar to that shown in Fig. 34C is acceptable.

Circumferential Seals O-rings are located between cylindrical surfaces. They are compressed as the insert is assembled.

The main problem is the possible damage to the O-ring during assembly. Any sharp edge, for example, at the entry of a drilled passage into the bore for the cavity insert, which lies in the path of the O-ring during assembly, must be well rounded or chamfered.

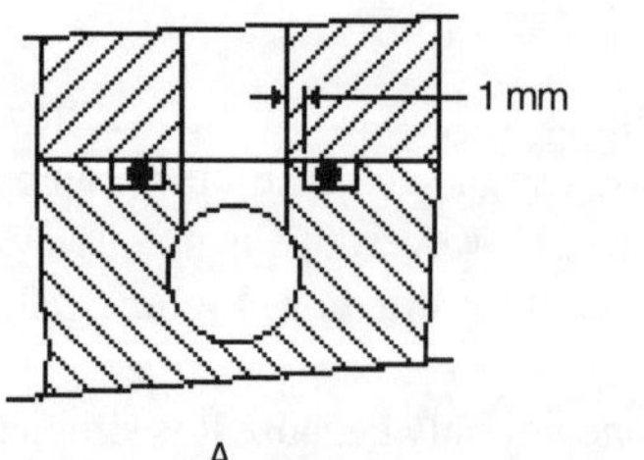

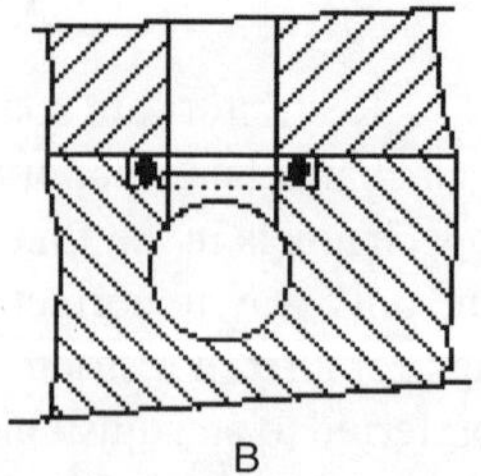

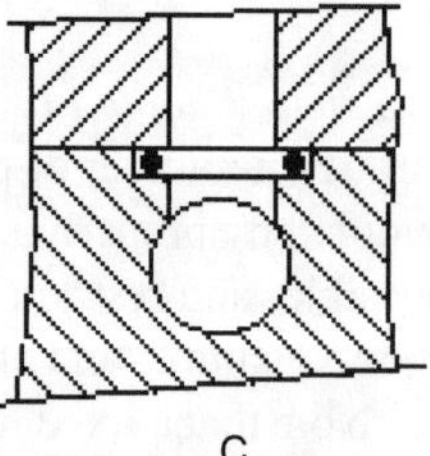

Figure 13.33 Three examples of flat O-ring seals: A. preferred arrangement, B. tapered O-ring groove bottom to prevent seal from slipping, and C. poor arrangement wherein seal may slip into channel.

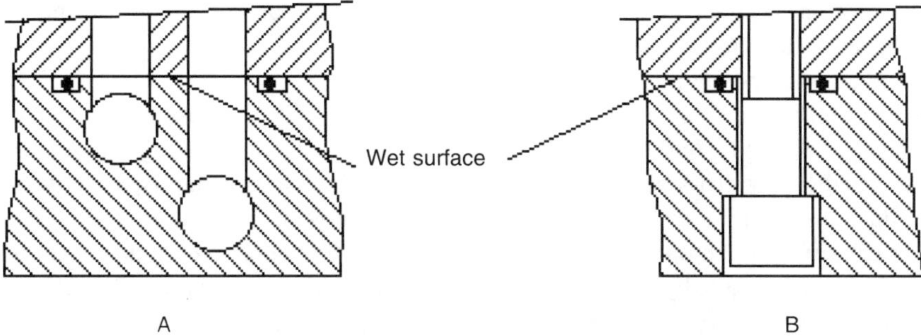

Figure 13.34 Wet surfaces in two seal arrangements: A. one common seal around two coolant passages—there is a wet surface within the seal, and B. screws within a wet area are sealed against corrosion of threads.

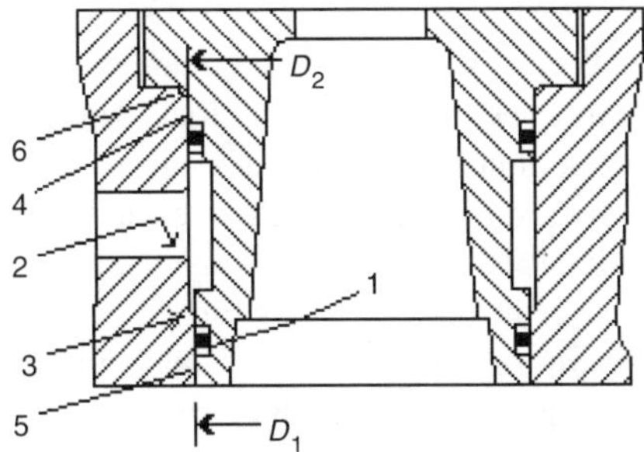

Figure 13.35 Cross section shows circumferential seals for the cavity insert bore.

Figure 13.35 shows the correct method for using two diameters to fit the insert into the plate. D_2 is large enough for the uncompressed O-ring in groove (1) to pass almost to the end of its insertion travel without contacting the surrounding wall of the bore or the entry of the supply channel (2). A chamfer (3) ensures smooth transition from D_2 to D_1. There must be a radius under the flange and a chamfer in the bore (6). The accurate location areas of the insert within the bore—(4) and (5)—are thereby kept dry and free of corrosion. (See also Section 22.12, O-Ring Seal Installation.)

13.4.5.2.2 Cavities and Cores for Flat Products Flat products are considered all those whose cavities and cores are virtually identical (e.g., records, flat plates, test pieces, etc.) or where a large portion of the projected area consists of a flat surface that is expected to remain flat after the product has cooled. Any cooling arrangement must satisfy the requirement that the heat will be equally well removed from both sides of the molded product. Difficulties usually fall under the following considerations:

- The gate area is a source of additional heat input, whether in a three-plate mold or in a hot runner mold.
- The ejector mechanism, whether stripper plate or ejector pins, requires space that competes with cooling lines.

- The ejector mechanism, whether stripper plate or ejector pins, requires space that competes with cooling lines.
- Mounting screws require sealing to prevent leaks.
- Inserts and/or core pins often make it difficult to provide a symmetrical, equal, and balanced cooling layout.
- Localized heavy cross sections of the molded product should be discussed with the product designer to make him or her aware of how such features will affect the productivity of the mold. Some simple product modifications to eliminate such heavy sections, for example by coring out or ribbing, may help to reduce the wall sections and improve the productivity of the planned mold.

Coolant supply to flat cavities (or cores) can be achieved in three basic ways or in a combination of them (Fig. 13.36). Figure 13.36A shows the cooling channels in the cavity block piped directly from the outside, usually from the edge. This is only practical with a small number of cavities. Figure 13.36B. shows cooling of the cavity block through bores (with seals) from the backing plate. This is a common arrangement.

Figure 13.36C shows an inserted (round) cavity block. This is rather rare for a completely flat product as shown but very common for small flat or cup-shaped products. The cavity block is cooled from a circular groove around the circumference of the block. The coolant enters and exits through the cavity retainer plate. O-rings seal against leaks.

The flow pattern is similar to the ones shown in Fig. 13.30A–C, with the coolant splitting into two symmetrical paths around the cavity block. It is important that the flow volume divides evenly around each side. Sometimes, the flow pattern resembles the Greek letter omega (Ω), that is, the "in" and "out" channels are close together and only one stream of coolant flows around approximately 80% of the circumference. A baffle in the circular cooling groove directs the flow around the insert.

Coolant Circuit Layouts in Flat Cavities When mounting flat cavity blocks, it is important to provide sufficient mounting screws to hold the blocks solidly to the backing plate. These screws should be located so that they will not seriously affect the layout and the symmetry of the cooling channels. The designer must compromise between the number of screws used and their size and location, and the location of the channels. If open channels are used (see Fig. 13.37A), leaks of the coolant through the screw holes must be prevented by the use of O-rings or plugs, as shown in Fig. 37B and C. Plugs require less space, but the screws are "wet" and will corrode over time, making removal more difficult.

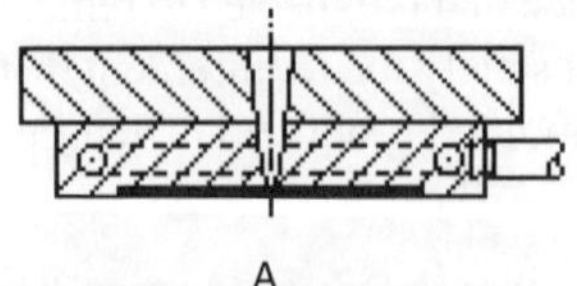

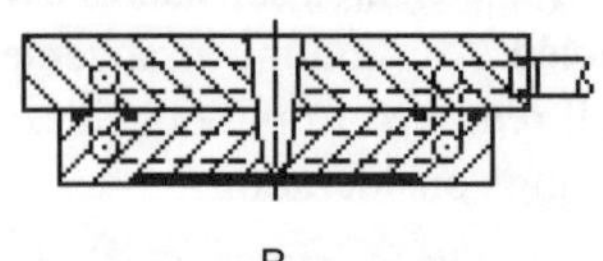

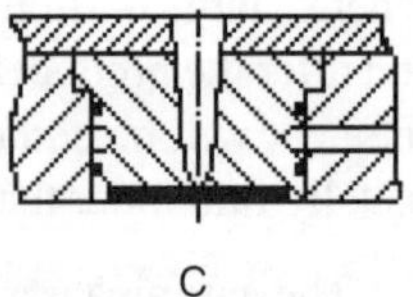

Figure 13.36 Coolant supply to flat cavities: A. cooling channels in cavity block piped directly from outside, B. cooling channels to cavity block through backing plate, and C. circular cooling groove around cavity through cavity retainer plate.

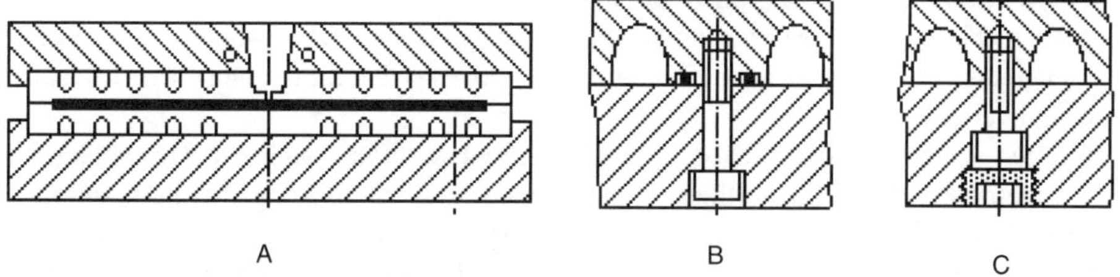

Figure 13.37 Coolant circuits in flat cavities: A. symmetrical open channels, B. O-ring seal prevents coolant leaks through screw hole, and C. plug prevents coolant leak through screw hole.

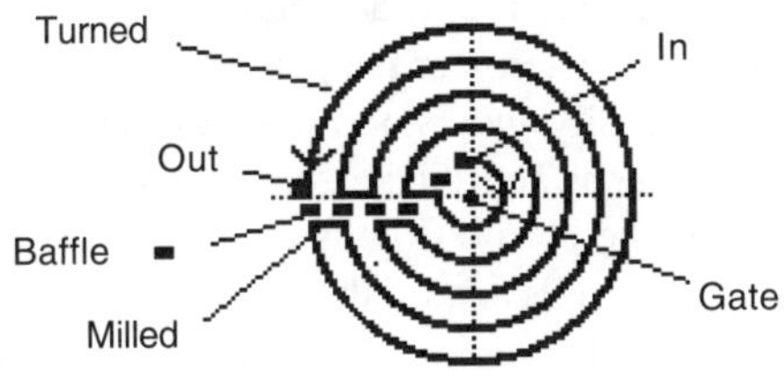

Figure 13.38 Common concentric cooling circuit in mold cavity.

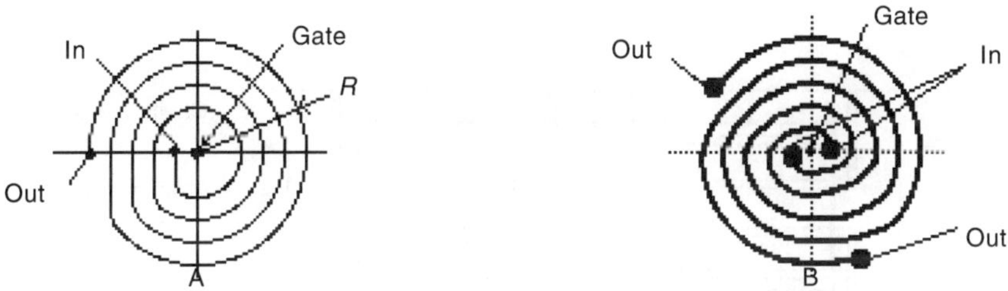

Figure 13.39 Spiral cooling circuits: A. single spiral, B. double spiral.

The following shows several commonly used cooling circuits for mold cavities (or cores). The decision of which to choose rests usually with the designer, and it depends on the available mold making equipment and the expected productivity of the mold. As a rule, the better the cooling, the more complicated and expensive it is to fabricate the mold, even with the most sophisticated equipment.

Figure 13.38 shows a frequently used arrangement of concentric cooling circuits. The circular grooves are easily produced on a lathe or milling machine. The cross-over channels are milled. If the rings are turned, it is necessary to insert baffles (shown schematically) to direct the flow. With milled grooves, such baffles can be avoided by milling only 330–340° (approximately) of the circumference. However, milling is slower than turning. This cooling is quite acceptable but not as good as in spiral arrangements (Fig. 13.39).

Figure 13.39A shows a typical single spiral. This can be produced on a duplicating (milling) machine or on CNC equipment where the spiral path can be numerically specified. A disadvantage of both circular and single spiral grooves is that the coolant enters offset from the gate location.

Figure 13.39B shows a typical double spiral layout. In this arrangement, the coolant inlets are symmetrically located about the gate and provide better cooling than the single spiral. The number of spirals and their distance from each other depends on the cavity size, on strength of material, and on the number, size, and location of screws. The use of milled slots also permits the designer to create detours around screw locations, but in critical molding applications this may affect the evenness of cooling desired.

Figure 13.40 shows a simple cooling layout for a large flat product. This design is often used; it is simple and low cost, but there is a disadvantage.

In all molds, the cooling should enter near the gate location and spread out toward the outside of the product. This can be easily understood since the plastic is hottest where it enters and will give up more of its heat content, and quicker, because the coolant at this spot is coolest (ΔT max.). At the same time, as the plastic flow is cooled by the mold, it becomes more difficult to push between the cavity walls toward the edges of the product, and it is therefore advantageous for easier mold filling to provide less cooling near the end of the flow path of the plastic.

The foregoing applies to any shape of mold cavity, whether flat or not.

Figure 13.41 shows two commonly used layouts, with the cooling entering closer to the gate, but the cost is higher. (Drilling, baffles, and plugs are omitted in these two schematics.)

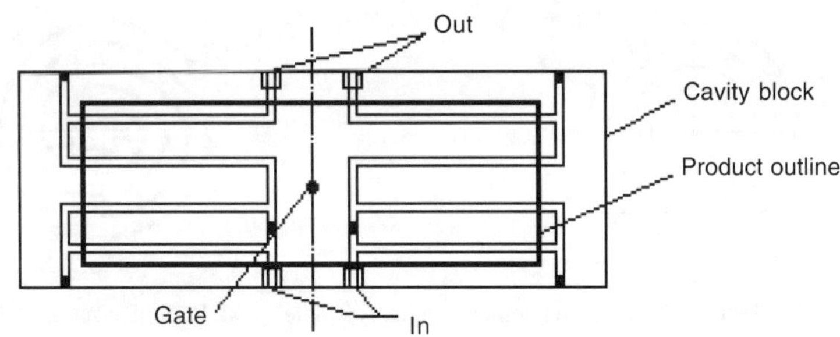

Figure 13.40 Simple cavity block cooling layout for a large flat product.

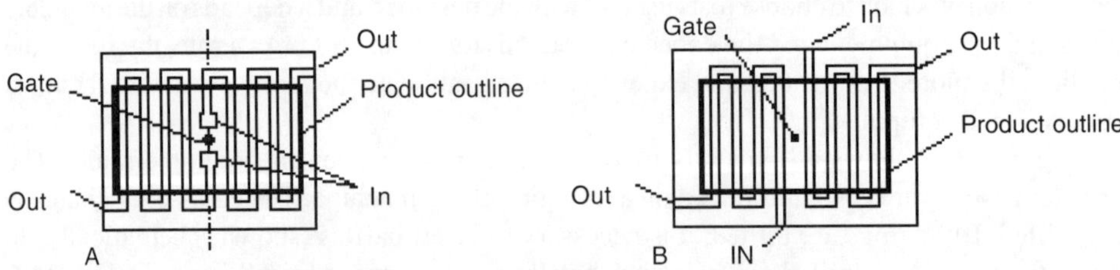

Figure 13.41 Common cooling layouts: A. coolant enters very close to gate, B. coolant enters near the gate from different sides of the mold wall.

13.4.5.2.3 Cavity Cooling for Cup-Shaped Products Two methods are commonly used:

1. Cavity inserts in a plate (cavity retainer plate) or
2. "modular cavities" mounted on a backing plate.

For cooling, the main difference is in the method of supplying coolant to the cavity blocks or inserts. For relatively shallow products (cassettes, etc.), there is no difference from the methods shown in Fig. 13.36. The following examples will deal with deep products such as closures, cups, containers in general, vials, and preforms for bottles.

Cavity Inserts They are used for closures, vials, preforms, etc., which are often but not necessarily round products, and usually but not necessarily with large numbers of cavities.

Figure 13.42 shows typical sections through the three most common arrangements. The retainer plate provides cooling for the cavities. [NOTE: the backing plates for the three-plate and hot runner system must be cooled; the backing plate for the cold runner (tunnel gated) cavities needs no cooling. The same applies to cold runner edge gating, which is not illustrated.]

The arrangement of cooling channels in the retaining plate only is inexpensive but not very effective. This can easily be seen in Fig. 13.43, which shows a section of a plate taken through the cooling channels.

Such an arrangement provides fairly good cooling for the plate itself, but the heat from the cavity must pass a heat barrier where the cavity is inserted in the plate. Also, the distance from the molded product to the cooling channels is unequal around the circumference.

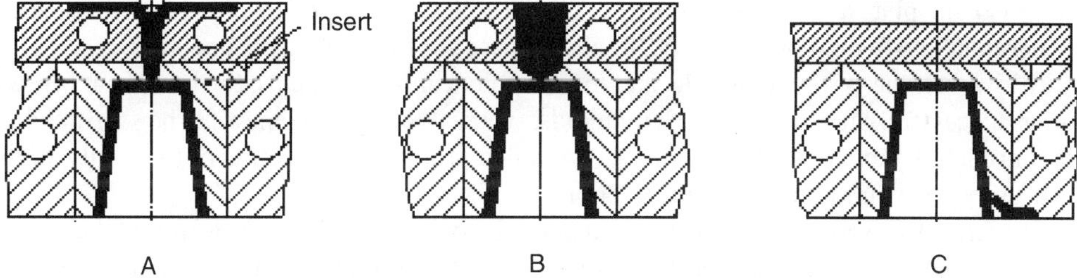

Figure 13.42 Cross sections of three common cavity insert cooling layouts: A. three-plate mold, B. hot runner mold, and C. cold runner mold (tunnel gated).

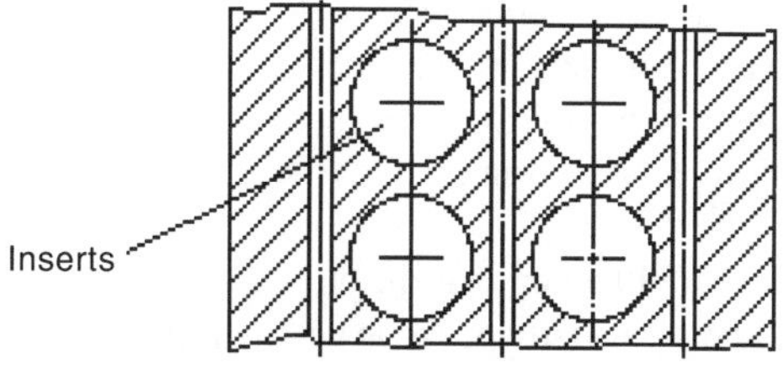

Figure 13.43 Top cross section through cooling channels in retaining plate.

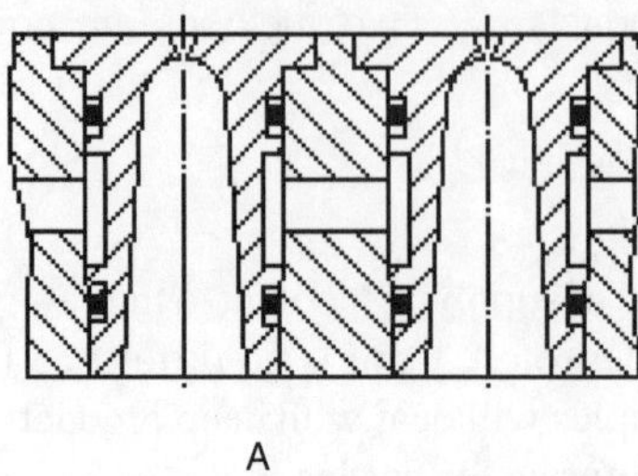

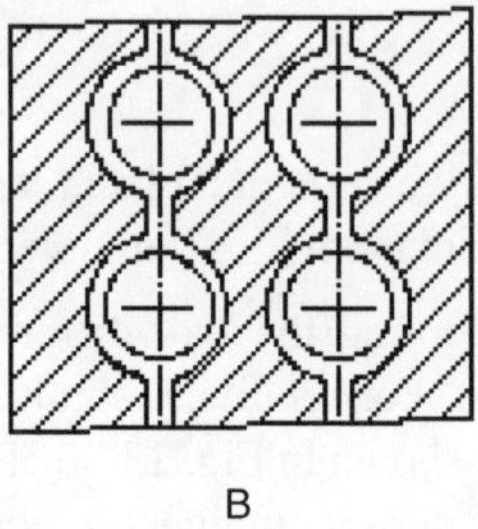

Figure 13.44 Circular grooves around cooling channels in cavity insert: A. cross section , B. top view.

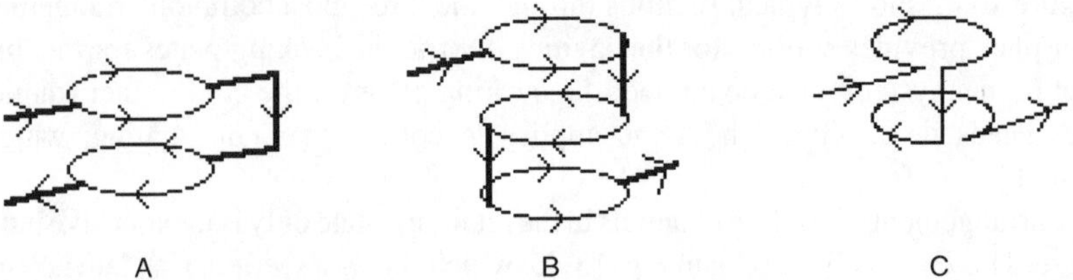

Figure 13.45 Arrows show path of coolant through cooling grooves around the cavity insert: A. two ring grooves connected by riser, B. three ring grooves connected by risers, and C. one coolant path for two incomplete rings flows around the insert.

Also, the plate area is poorly utilized. Cavities (at least in one direction) must be spaced wider than they would otherwise be, to allow space for the cooling channels.

However, the system is simple, and there are no leaking problems. Such a layout is used quite successfully for low productivity molds, where cooling efficiency can be sacrificed in favor of a low mold cost. Also, with heavy-walled products, the cooling of the cavity walls is usually not as important as the core cooling.

Figure 13.44 shows an often used and better layout. A circular groove around the cavity insert divides the flow of the coolant into two branches. It is important that the sum of the cross sections of the branches be equal or less than the cross section of the drilled lines to prevent one half being less cooled than the other. There may be more than one cooling groove around the cavity, and the feeder and return channels may be at different levels, as shown schematically in Fig. 13.45. Milled slots and baffles in the grooves and in the drilled channels of the plate are required to direct the flow as desired.

Cavities for vials and preforms can have a greater number of cooling grooves than those shown in Fig. 13.45. Sometimes, spiral grooves are used instead of ring grooves. This eliminates the need for milling connecting grooves (risers) from ring to ring, and the insertion of baffles.

Preferably, the coolant should enter near the closed end of the cavity (near the gate) and flow toward the open end. This is usually possible with parallel cooling, where the supply line is at the "In" level and the return line at the "Out" level (Fig. 13.46A). With series cooling (two or more cavities in one branch), for practical reasons, the entry to the cavities is alternating;

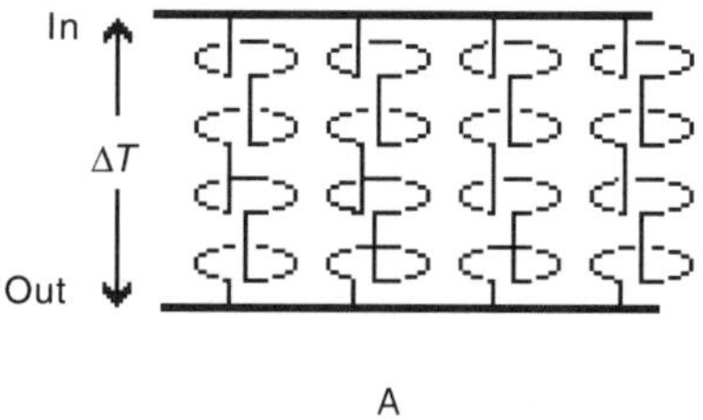

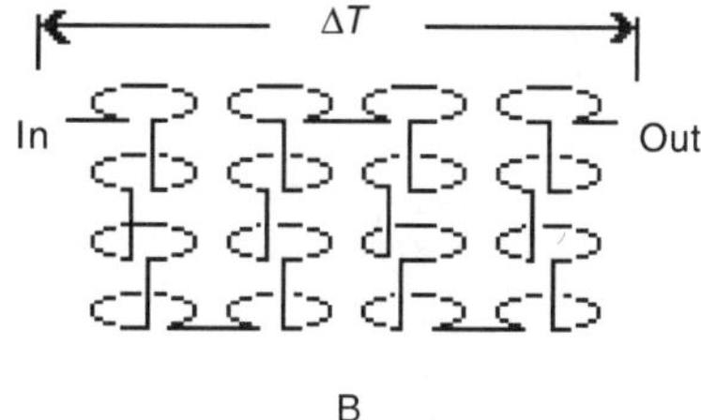

Figure 13.46 Coolant enters at In level and exits at Out level: A. parallel cooling enters closed end of cavity, and B. series cooling that alternates entry from top and bottom of cavities.

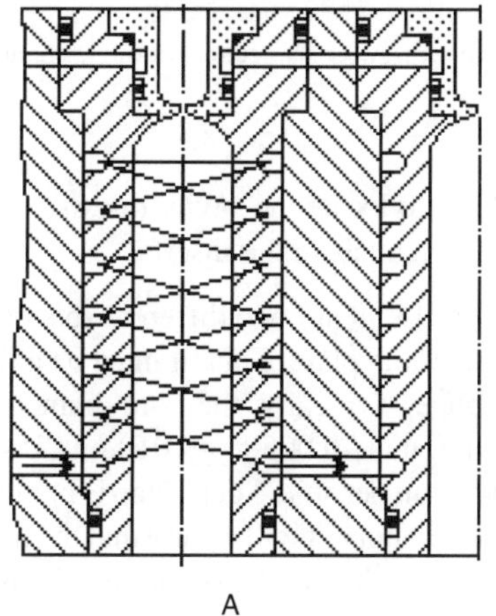

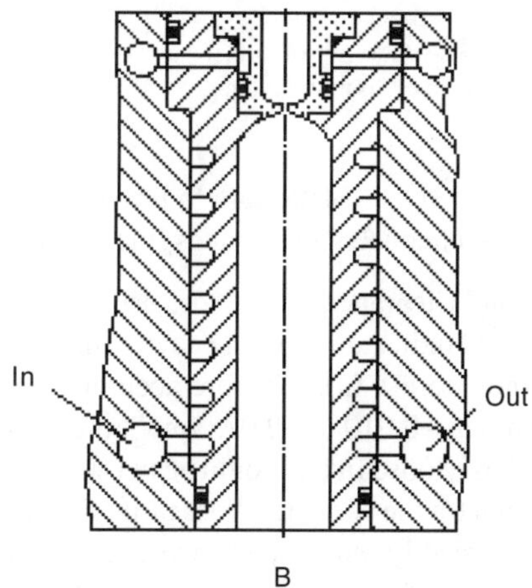

Figure 13.47 Double spiral cooling: A. series cooling and B. parallel cooling.

that is, the coolant goes in one cavity from the top, and in the next from the bottom, and so on (Fig. 13.46B).

If the ΔT is held to not more than 5 °C, both the arrangements shown in Fig. 13.46 are good. The cross section of the grooves in parallel cooling (Fig. 46A) must be smaller than that for series cooling (Fig. 46B) so that the total flow from "In" to "Out" is the same in both arrangements and that all the cavities get the same amount of coolant.

Figure 13.47 shows examples of double spiral cooling. Both the "In" and "Out" are at the same level. Figure 13.47A is series cooling, Fig. 47B parallel cooling. As explained previously, the groove cross sections must be smaller for parallel cooling.

Occasionally, cavity inserts are cooled with drilled channels. This could be the case if the cavity is offset within the cavity insert, to provide optimal cooling as may be required for thin-walled products—in this example, a disposable cup with handle (Fig. 13.48A). Figure 13.48B shows a schematic layout of the cooling channels.

There are other arrangements possible and often used. It is up to the designer to select the most suitable one, while always keeping in mind the basic requirements for proper, sufficient, and equal flow for all cavities at a reasonable mold cost.

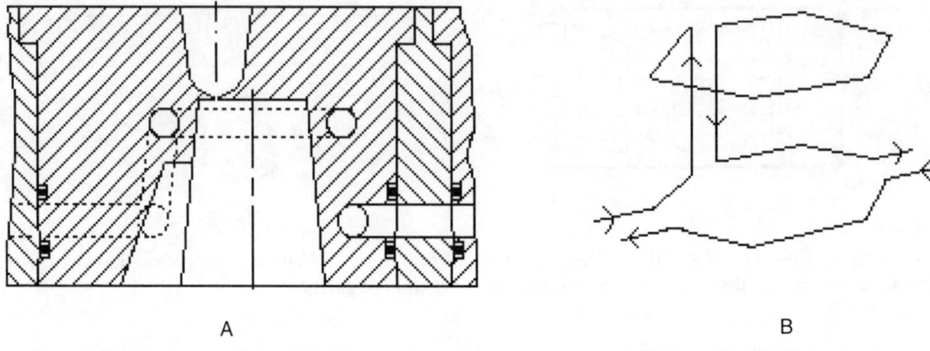

Figure 13.48 Cooling of offset cavity insert with drilled channels: A. cross section through channels, B. schematic of cooling layout.

Orientation Make sure to provide dowel pins or keys to locate the cavity inserts so that cooling channels in the plates line up properly with the channels in the inserts.

Fits and Clearances Cross sections illustrated in Fig. 13.49 are two different sets of fit. A and C indicate where the cavity insert is located in the mold plate. This fit must be a tight fit (not a press fit) to ensure accurate location in the plate. Typically, the bore diameter is "nominal +0.010 to +0.015 mm"; the cavity O.D. is "nominal –0.005 to –0.010 mm".

B is a loose fit and could be between 0.015 and 0.045 mm per side. Such a gap is not large enough for the coolant to pass through in any significant amount but will ensure that the insert can be readily removed even after a long time of operation of the mold. For the same reason, there is no O-ring required to separate the cooling circuits of cavity and gate pad.

Orifices (Restrictors) The restrictor in Fig. 13.49 is a possible solution to ensure equal flow through all the parallel branches; however, this is not a good solution for a distribution problem. The main objection is the fact that the flow in the channels surrounding the cavity is slowed down by the insertion of the orifice, and this lowers the Reynolds number and the efficiency of the coolant. It is, therefore, better to calculate and specify the grooves correctly for equal flow in all cavities and cores. (See also Section 13.4.5.2.1.)

Modular Cavities Modular cavities are usually round, square, or rectangular blocks mounted on a backing plate. Each block contains one cavity but may occasionally contain a cluster of two or more cavities.

Coolant Supply for Modular Cavities

1. *Coolant supply with hose lines.* This method is used with a smaller number of cavity blocks, usually one to four blocks, rarely more. A possible advantage is the cost saving by avoiding drilling in the backing plate, which can then be thinner—a saving in weight and shut height of the mold. Also, it permits individual flow control for each cavity, although experience has shown that this should not be necessary.

   The obvious disadvantage is the large number of hoses cluttering the mold (inaccessibility) and the time required to make the connections at setup. A noted exception is when

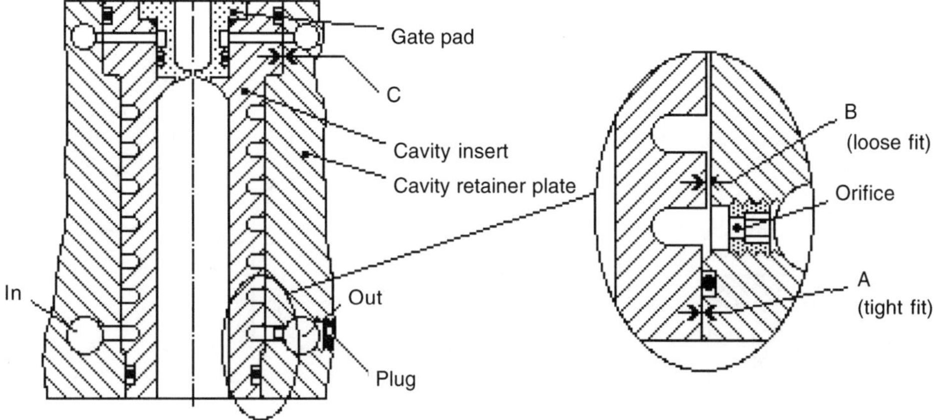

Figure 13.49 Two sets of fit: Arrow A. tight fit of cavity insert, B. loose fit between cooling channels, C. tight fit of cavity insert near gate.

each block is permanently piped to manifolds mounted at the edge of the mold. There is a limit to the number of cavities that can be so connected by the number of rows of blocks. Where there are more than two rows, the center rows cannot be reached easily or at all from the outside with piping.

2. *Coolant supply through the backing plate.* Connections are with drilled passages, sealed with O-rings. The backing plate is cross drilled for all supply and branch connections. The flow to the cavities can be arranged either for series or parallel cooling. There are two common methods of cooling the modular blocks:

   - Cavities are inserted. The modular block is similar to a retainer plate for cavity inserts. This arrangement is used in cases when maximum cooling efficiency is required, as in the case of some thin-walled containers.
   - Cross drilling. The modular block is drilled in a pattern which should surround the cavity as symmetrically as possible. Two basically different methods a and b are both used. See the following examples.

     a. Channels are parallel to the parting plane (Fig. 13.50). In square blocks, there is only drilling in a square pattern. In round blocks, the drilling can be in a square or hexagonal pattern. For cooling, the hexagonal pattern (Fig. 13.50B) is somewhat better than a square pattern (Fig. 13.50A).

        When several layers of cooling channels are used, the connections (risers) from one layer to the other are with vertical bores, with baffles where required, and plugs at the start of the drilled channels. These connections can create additional uneven cooling, but in many cases it is only important to keep the block at an average, even temperature.

     b. Channels are vertical to the parting plane. The channels are approximately at right angles to the parting plane. Which of the two methods to use depends on the cooling efficiency required. The first method is less expensive; the second method gives better cooling, especially for deeper cavities such as containers, but is not

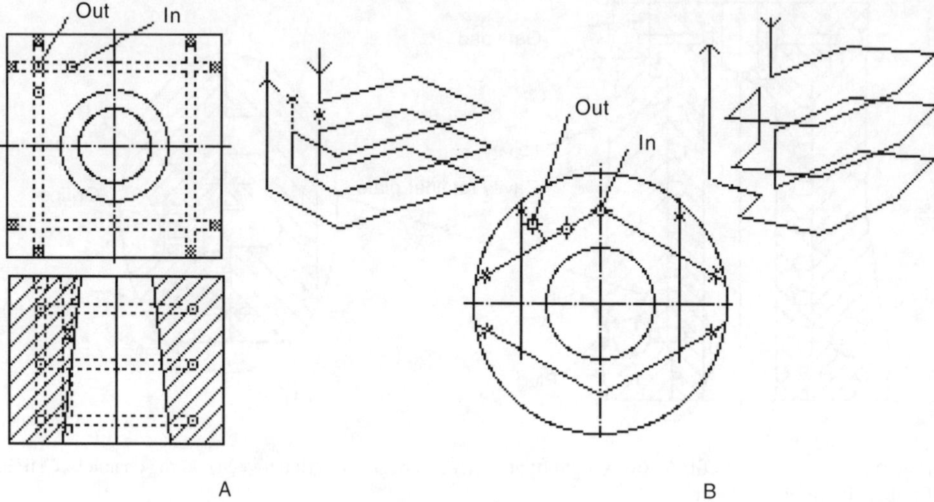

Figure 13.50 Cooling channels parallel to parting plane in modular cavities: A. square drilling pattern around a round block, B. preferred hexagonal drilling around round block.

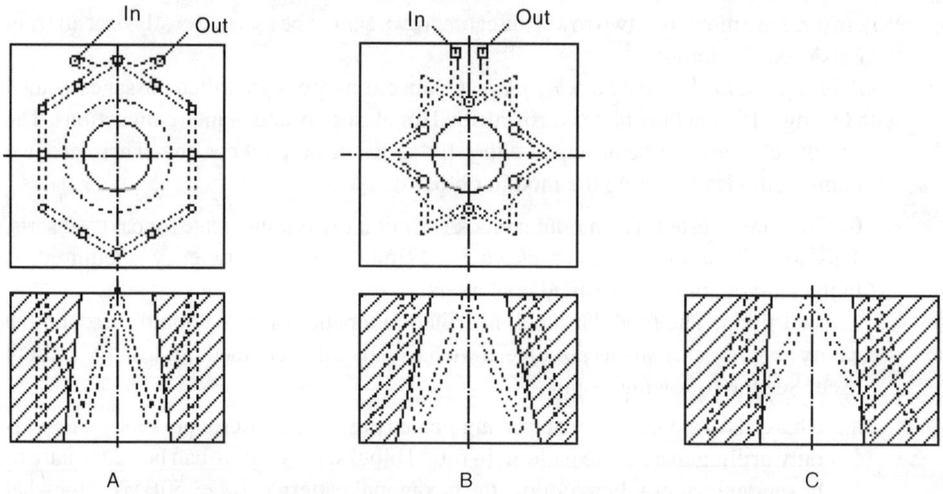

Figure 13.51 Cooling channels vertical to the parting plane in modular cavities: A. channels at right angles to the parting plane, B. channels parallel to cavity wall at angles to parting plane, and C. channels diverge from cavity wall.

always practical to achieve, or may not even be necessary. These channels are usually used for larger, deep cavities, and the differences are only of significance if the best cooling is needed, as in the case of thin-walled containers.

In Fig. 13.51A, the channels are at right angles to the parting plane and are, therefore, a further distance from the plastic near the closed end of the product and closer near the rim of the product. The cooling efficiency is not as good as in Fig. 13.51B or C, but often good enough. This method is the simplest of the three. There are no compound angles.

In Fig. 13.51B, the cooling channels run parallel to the cavity wall (tapered container, pail, etc.) and are at an angle to the parting plane. This arrangement provides better, closer spacing of the coolant over the whole length of the cavity; it is commonly used.

In Fig. 13.51C, channels diverge from the cavity wall. This (patented) method is rarely used but could be considered in a case where good cooling is required near the closed (gate) end of the product but less cooling near the open end, to facilitate the plastic flow toward the rim.

NOTE: For simplicity, the above sketches show only six pairs of "zig-zag" bores in each block. The number can be and usually is much larger and depends on the size of the block (and the quality of cooling required).

13.4.5.2.4 Core Cooling

Flat Products Core cooling for flat products is essentially the same as the cavity cooling. However, an added difficulty arises if ejector pins must pass between the cooling lines. If the cooling channels are milled (see Fig. 13.37), the ejector pins must be sealed against the coolant with O-rings, similar to screws (see Figs. 13.34 or 13.37B).

It is desirable and often mandatory to provide additional "spot" cooling opposite the gate, especially if the cooling circuits (drilled or open channels) cannot be routed close enough to the gate area. Such spot cooling is usually done with bubblers. Fig. 13.52 shows a typical arrangement of cooling for a flat part in a modular core with drilled cooling and a bubbler.

The cooling arrangement is preferably parallel, that is, each core is supplied from a common cooling supply channel and empties into a common return channel. The bubbler in Fig. 13.52A is generally the same as that later described in Fig. 13.53, with a tube (fountain) directing the flow toward the gate area and returning it around the outside.

As the designer is only concerned with cooling a small area opposite the gate, a cheaper arrangement is shown in Fig. 13.52B. This arrangement also permits the placing of several bubblers in one cooling channel in series. The baffle or divider (a flat strip) is locked in place

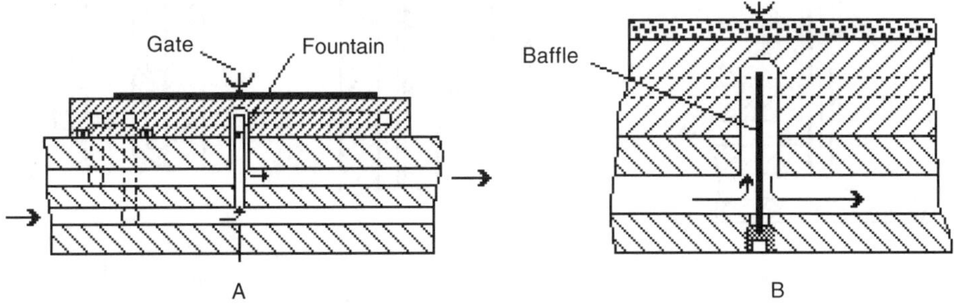

Figure 13.52 Typical drilled cooling for flat modular core with bubbler: A. spot cooling near gate using a bubbler, B. baffle mounted in a plug directs flow near gate.

with a plug, and it is important to ensure that the baffle is properly oriented in the bubbler to provide the desired flow. In this method, the flow passages to the right and the left of the baffle must be the same as the drilled passages of the plate (series cooling).

Cup-Shaped Products Cup-shaped products can be divided into three groups, according to size: small products (vials, closures, preforms, etc.), medium-sized products (cups, small containers, etc.), and large products (pails, boxes, etc.).

Cores for Small Products Because of the size of these products (usually, but not necessarily round), the choice of cooling is very restricted. In most applications, it is possible to cool such cores with bubblers, using the conventional cooling medium, water. However, there are cases where the size will not permit use of water, and other, later described methods are used.

Figure 13.53A and B show a typical bubbler arrangement in a small, elongated core. There are several things to observe:

1. Stability of tube inside the bore. Especially with long cores, it is important to provide standoffs, or to flare the tip of the tube outward so that the tube is held centrally in the bore.
2. Figure 13.53B: The inside area A of the fountain (supply) must be at least equal to, but better if larger than, the annular area F (return) between the O.D. of the tube and the I.D. of the core. If this is neglected, the coolant will not flow symmetrically around the tube and will create unequal cooling of the product. This is especially important if the coolant is close to the outside wall of the core and therefore close to the plastic. Figure 13.53C shows a solution where the coolant is forced to return along a helical path which is so dimensioned that the area A of the tube is equal or larger than the area F of the helix.

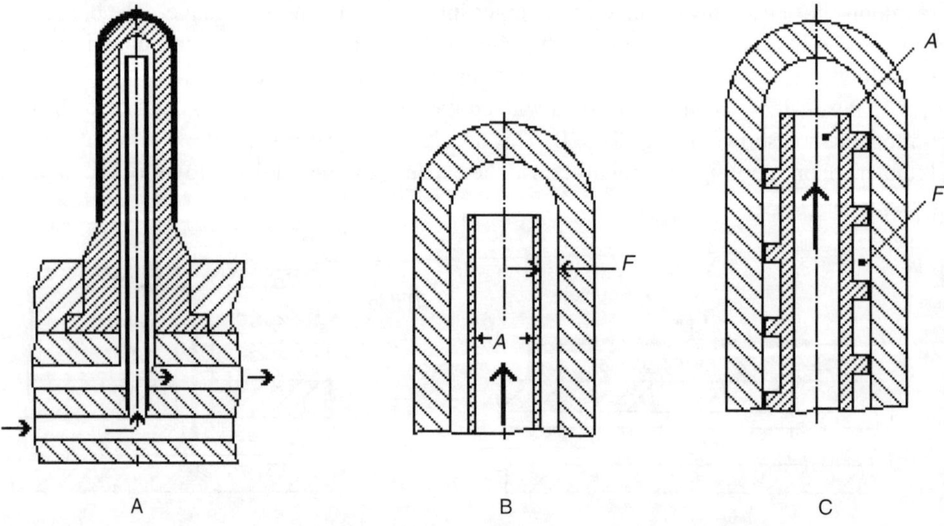

Figure 13.53 Typical bubbler arrangements: A. entire cross section through bubbler, B. inside area *A* must equal or exceed annular area *F* for even cooling, and C. coolant return is forced along a helical path, dimensioned so that $A \geq F$ for even cooling.

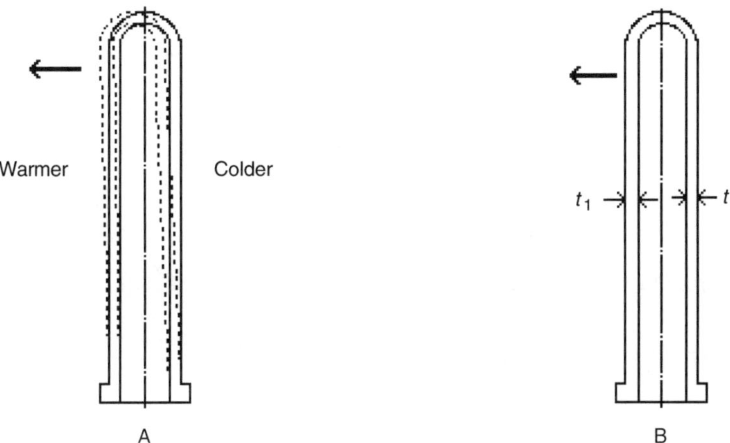

Figure 13.54 Cooling in products with a long core: A. uneven cooling causes distortion, B. uneven wall thickness $t_1 > t_2$ causes similar distortion.

Effect of Uneven Cooling in Long Core on the Product Effects of unequal cooling can be seen schematically in Fig. 13.54. When the product is ejected before it is completely cooled down, the warmer side will continue to shrink after ejection while the colder side will not shrink or will shrink less. This results in a distorted part, or "banana".

Note that the effect is similar to a product which is molded with uneven wall thickness (Fig. 13.54B), caused by poor workmanship or by core shift because of uneven filling of the part. Because of its greater mass, the thicker section t_1 retains more heat than the thinner section t_2 and will continue to shrink after the product is ejected; it will distort the product in the direction of the arrow.

Dividing Baffles in Bubblers This type of bubbler (Fig. 13.52B) is satisfactory for the cooling of the tip area of a core opposite the gate (as shown in Fig. 13.52A), but as a rule should not be used for cooling of long cores because of the inherent cooling differences around the circumference of the core as schematically shown in Fig. 13.55A. However, there is the need for such bubblers where other methods of cooling are impossible or economically not viable.

Figure 13.55A shows in schematic a cross section of the flow pattern through a core with a divider. The main flow (oval shape) moves faster and provides better cooling to that portion of the product which is nearest to it. The partial, slower flow in the corners near the baffle will have less cooling effect on the product. Note that for many applications such uneven cooling may be of no significance at all, and this arrangement would be perfectly acceptable.

Figure 13.56 shows an example of a small, elongated, cup-shaped product. Figure 13.57 shows three typical methods for cooling such a product. The first, Fig. 13.57A, shows an arrangement with two bubblers.

Figure 13.57B illustrates a commonly used method with two drilled channels converging toward the gate area. The advantage of this method is that the gate area is better cooled than with two bubblers, provided the enclosed angle between the two drilled holes is not too acute. As shown, the main flow of the coolant will find the path of least resistance and flow around the lower portion of the intersection of the two bores, leaving the coolant stagnant near the

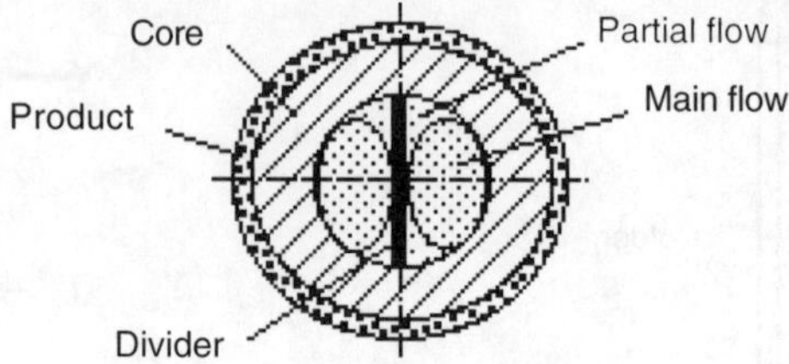

Figure 13.55 Cross section of coolant flow pattern through a divided core.

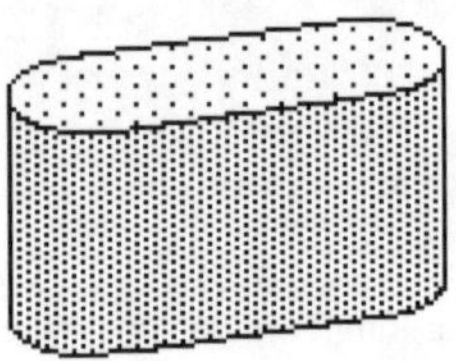

Figure 13.56 Small, elongated cup-shaped product.

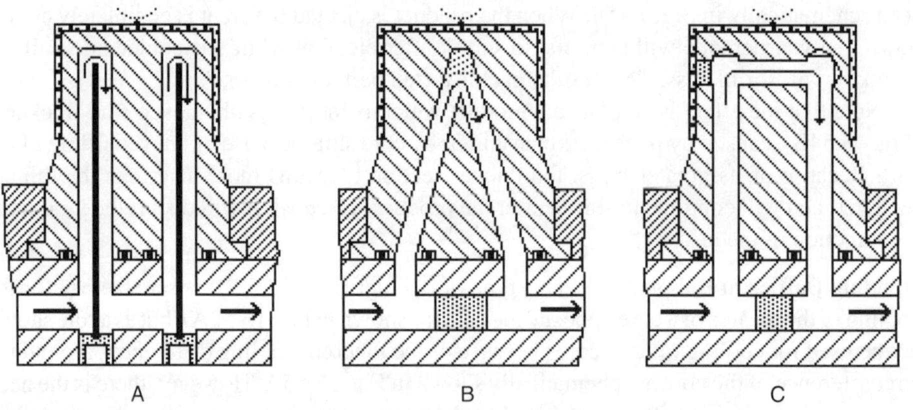

Figure 13.57 Three typical methods for cooling an elongated cup: A. two bubblers, B. two drilled channels converging toward the gate, and C. cooling channels with plugs on molding surface.

gate. Also, the total channel surface exposed to the coolant is smaller than in the other two examples shown.

Figure 13.57C shows another good method of providing cooling channels in such a core, but it requires blocking of a channel (usually by welding or soldering a plug) on the molding surface, which may leave a witness mark on the surface. This may not be acceptable in some products.

Placement of O-rings: In all three cases, individual O-rings are shown at each passage, which is good practice to reduce the area exposed to corrosion. However, in small cores there may not be enough room for two O-rings and one, surrounding both passages, will have to be used.

Cores for Very Small Products The difference between "small" and "very small" products is hard to define, but in the context of cooling, we will consider a product as being very small if it is not practical to provide cooling by conventional coolants and using cross drilled passages or bubblers.

Limits are usually set by the strength of the core and the available sizes of cooling tubes inside the bubbler.

Drilling out the core will:

1. Reduce its stiffness against deflection, but by only a relatively small amount. (Remember: stiffness is proportional to the fourth power of the outside diameter, or the difference of the fourth power of the O.D. and I.D. of the core), and
2. reduce the resistance against crushing under the injection pressure.

Stainless steel tubing used for hypodermic needles can be used as small tubing for the bubbler. However, there is a limit to how far the size of the tubing can be reduced before other problems become critical, such as contaminations in the coolant, which may require frequent cleaning of such tubing.

Practical alternatives may involve the use of other coolants such as compressed air, carbon dioxide, or nitrogen. The cooling is affected not so much by the flow of such gases, since the mass of the gas (per unit of time) to remove the heat is small, but by the deep cooling effect of a compressed gas when it exits from a nozzle into a space at much lower, in this case, atmospheric pressure.

It has even been suggested that all or parts of the mold be transformed into an "evaporator" similar to the one used in refrigerators. For example, liquid, cold refrigerant under pressure could be ducted into the mold and passed through nozzles inside the cores where it would expand rapidly, thereby cooling the mold. The now gaseous refrigerant is piped back to a compressor, where it is again liquefied and, after passing through a cooler, pumped back into the mold. This is a completely closed circuit; the main problem would be the connecting, sealing, and charging with gas every time the mold is changed. This would create serious environmental considerations.

Another, proprietary cooling system (Logivac™) is mainly applicable to very long slender cores, such as for pen barrels, where the end of the core is touching or entering the cavity to create a tubular product. The coolant flows through the core, from the core into the cavity, and does not return within the core. The flow is under vacuum so that as the mold is opened no coolant will leak out. There are some applications where this system can be used to advantage.

Another, occasionally used method is to direct a jet of cold air from outside of the mold against the core. This is more practical in the case of turret molding, for example, where a set (or sets) of cores remains a considerable time outside the molding area in a waiting mode.

Other methods of cooling of very small cores are similar to the cooling of inserts, and will be discussed later.

The above described methods of cooling small cores leave the designer a number of alternatives to evaluate and to consider which method to select. Sometimes, a combination of some methods could be the proper answer for a problem, for example, combining cross drilled channels and bubblers.

Cores for Medium-Sized Products Medium-sized products are considered those to have cores that are too large to be cooled by a simple bubbler as for small products. In this group are typically cups, containers, etc., whether round, oval, rectangular, or odd-shaped, and the depth of draw is generally from one-half to double the diameter of the open end.

General rules:

1. The area opposite the gate must always be properly cooled. Wherever possible, the cooling supply for the core cooling should be directed first against the gate area before being channelled to cool the rest of the core. If for some reason this is not possible, a separate bubbler should be directed at the gate area.
2. Provide adequate cooling for the rest of the core. There are a number of methods to provide cooling for the core, and their selection depends on the cooling efficiency required. Generally, the best cooling is needed for high-speed molding of thin-walled products where the time available to cool the core is short and the total mass of plastic per unit of time is high. In these cases, the cost of "exceptional" cooling is justified.

Thick products or products with bosses, which preclude fast cooling, will usually not need an elaborate cooling layout. As long as there is enough cooling capacity in the circuit to remove the heat of the plastic during the projected cycle time, the mold will work well without wasting money on an excessively complicated cooling arrangement.

Venting Through the Core (See also Chapter 11, Venting.) Before continuing, an important feature in the mold, the venting of the inside of a cup-shaped product during ejection, must be covered. This is absolutely necessary in high-speed molding, where the molded product must be ejected as soon as it is rigid enough to withstand the forces exerted during ejection.

What has Venting to do with Mold Cooling? When laying out the cooling channels inside the core, allowance must be made to allow space for vent pins and/or vent slots and for the connecting air lines between the drilled coolant lines or open channels. Space must also be allowed for sealing with O-rings against leaks into the vent lines.

The following illustrations show schematically some examples of core cooling for medium-sized products. It should be understood that there are many ways of executing these designs, and that the designer should always refer back to examples of good designs on file or in some reference books.

One-piece cores are generally used where a witness line on the core is not acceptable. The internal plug should be made of stainless steel and provided with tapped holes to facilitate extraction. Figure 13.58A is a simple design, similar to the bubbler method. The groove in the plug can be a helix or consist of a number of circular grooves connected with milled slots and provided with baffles to create the desired flow pattern. The method is cheap but not very efficient for cooling (see also Fig. 13.14).

The cross section of groove C must be equal to or smaller than B. The relation between B and A depends on the number of cores supplied in parallel from A. If core cooling is in series, A = B = C. The fit of the plug in the core is a light sliding fit. Clearance F (per side) = 0.010–0.015 mm.

Figure 13.58B is a reversal of Fig. 13.58A that provides improved cooling efficiency but at a higher cost. Cross section of channels and fit of plug in core is the same as for Fig. 13.58A.

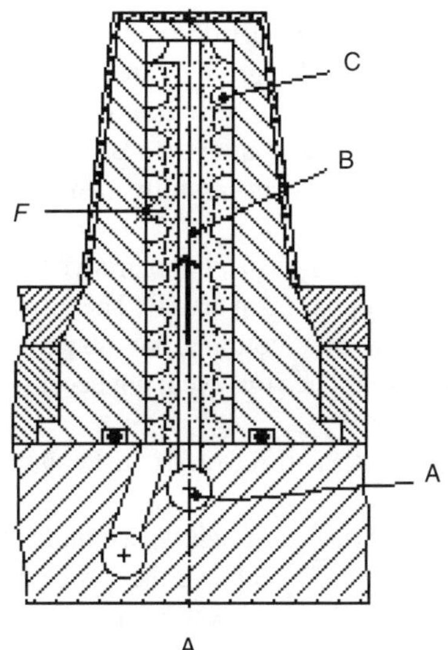

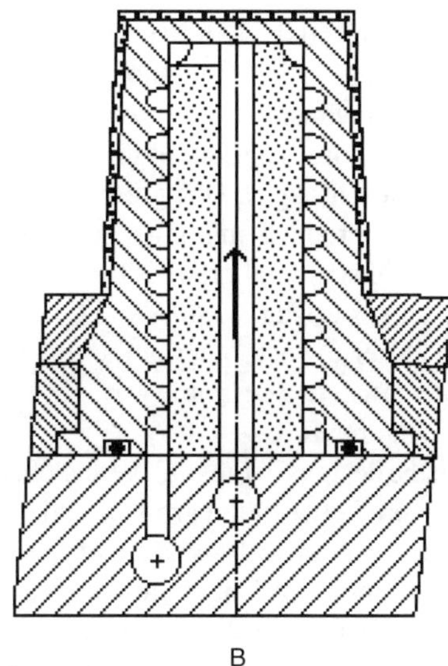

Figure 13.58 One-piece cores: A. internal plug with helical grooves connected by milled slots and baffles to direct flow, and B. reversal of A with most cooling surface area in the core.

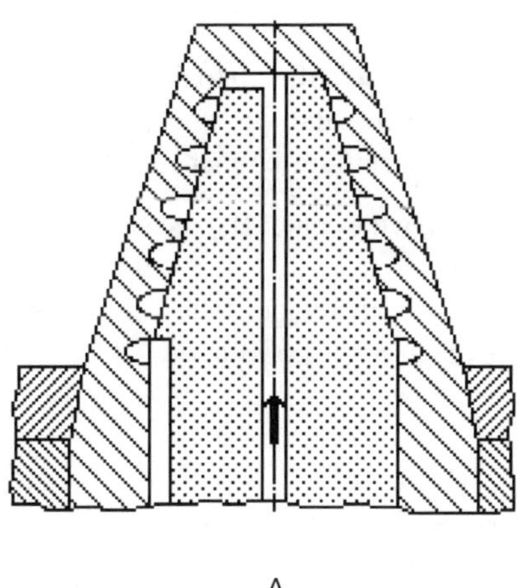

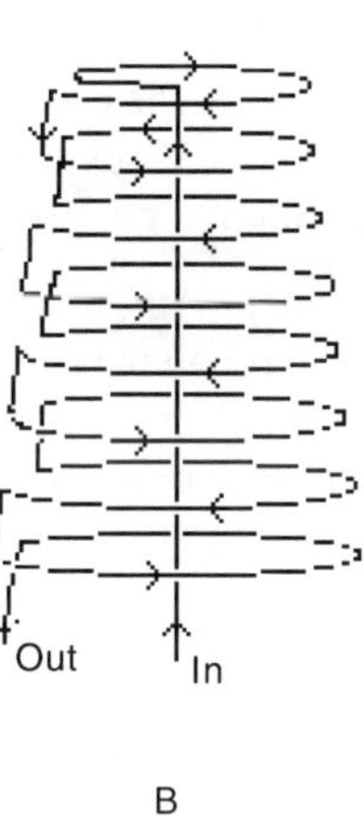

Figure 13.59 Variation of one-piece core: A. cross section shows the distance of cooling from molded part is kept constant, and B. schematic shows flow path through grooves as directed by baffles.

Figure 13.59 shows a variation of this design. While in the previous two examples, the cooling channels were farther away from the molded part near the open end than at the closed end of the mold, the distance is kept constant in this example. The same comments regarding flow cross sections and fits apply. Figure 13.59B shows schematically how the circular grooves are connected and baffled to obtain even flow.

Figure 13.60 shows a different design, with grooves running parallel to the wall of the core. This is a very efficient method but very expensive to produce. The grooves are made with EDM into the inside of the core, and as close to the wall and to each other as permissible for strength of core. The sum of the cross sections C must be equal or smaller than B so that the flow passes through all the grooves. This design has been used very successfully in molds for disposable products in high-speed molding.

Four or six radial grooves D connect the supply with a circular distribution groove G and the channels C (Fig. 13.60A and C). The top of the core is supported at area R. The grooves C are closer to each other at the top and more distant toward the bottom, thus varying the cooling effect in a desirable way. However, there is only little space for a vent pin because of the small area R.

In the designs in Figs. 13.59 and 60, the tapered plug must fit snugly to assist in the support of the core between the grooves and over the area R.

Two-piece core designs are of special significance in the design of high-speed molds for thin-wall molding. There are many different designs, but all try to achieve the same goals: good cooling of the area opposite the gate and good cooling along the wall of the core, at a reasonable cost. A split line along the side or the bottom of the core must be acceptable.

Figure 13.61A shows an inexpensive design that has been used successfully but is not recommended. The beryllium-copper tip insert is held only by light press fit. The pressure of the coolant is not enough to dislodge it, and the injection pressure holds it in place.

Figure 13.61B and C show an efficient cooling design that is quite common but which has two disadvantages:

1. The threaded tube used to supply the coolant and to fasten the tip to the core is subject to tremendous stresses and requires frequent replacement, and
2. the distribution grooves D make the center of the beryllium-copper insert weak.

Note that in both of the above examples, the flow through the center B must be at least as large as the flow through the sum of the channels (bores) C.

Figure 13.62 shows a layout that avoids the two disadvantages listed above. A central, solid bolt instead of a tube holds the tip, which is (as in Fig 13.60B and C) usually (but not

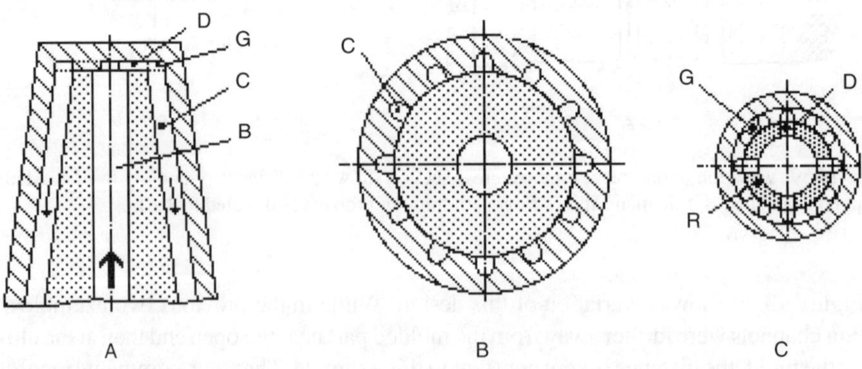

Figure 13.60 One-piece core: A. cross section shows cooling channels close to core wall, B. Top view shows spacing of channels around core, C. position of radial and circular grooves for supply distribution.

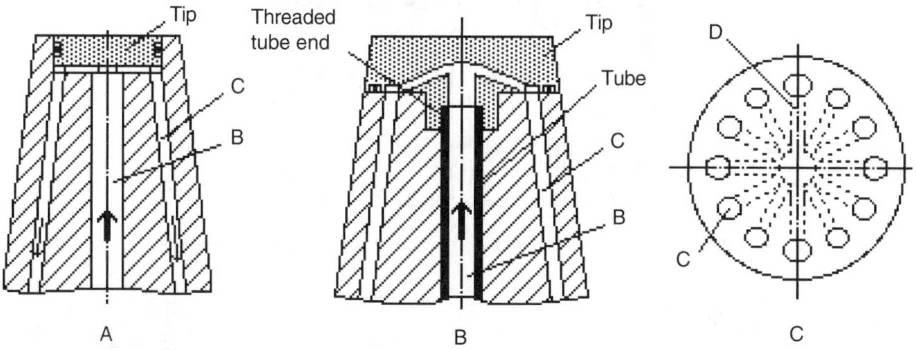

Figure 13.61 Two-piece cores: A. tip insert held by light press fit, B. tip insert held by threaded tube that supplies coolant, and C. cross section through channels.

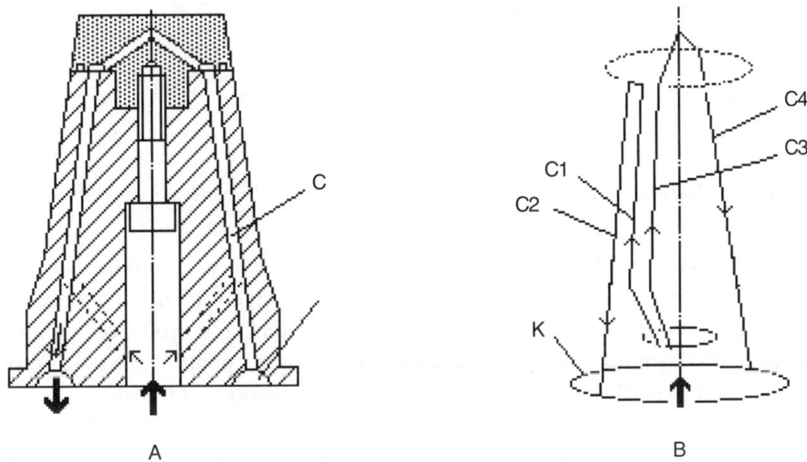

Figure 13.62 Two piece core layout: A. central bolt holds the tip insert, and B. flow is divided into two groups of cooling circuits.

necessarily) made of a beryllium-copper alloy. The flow through the channels (bores) C is divided into two groups (Fig. 13.62B): The first group consists of four pairs of separate circuits C1 and C2 (only one shown), while the second group consists of only two circuits C3 and C4 (only one shown). This way, the tip has only four bores, which cross opposite the gate, while the sidewall is still cooled by twelve channels.

The foregoing examples were introduced to show that the designer has much freedom in designing core cooling and that there may always be better solutions. In all cases, the designer must consider venting the mold, and the number and the distance of the cooling channels from the core wall and their distance from each other. There must be enough space for air (vent) channels and for the necessary placement of O-rings to prevent the coolant from leaking into such channels, as well as into the outside of the core.

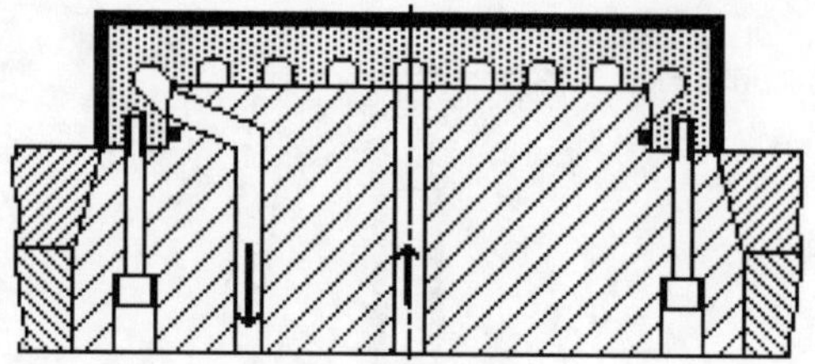

Figure 13.63 Typical over cap cross section showing spiral grooves for two-piece core for flat products.

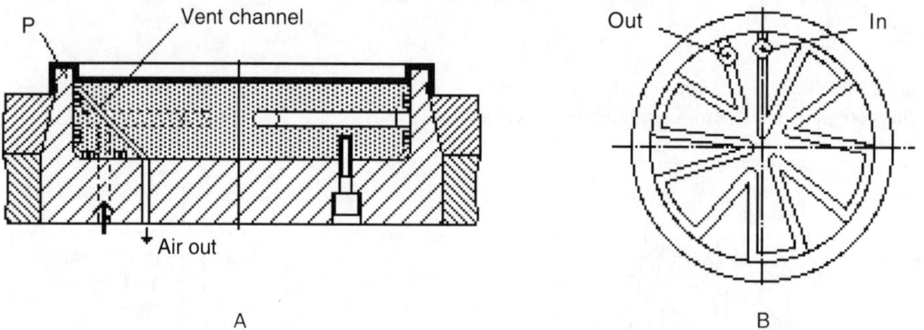

Figure 13.64 Variation of over cap design: A. cross section of core , B. pattern of cross drilling in insert.

Another arrangement of two-piece cores is used for relatively flat products, such as lids of containers, which require very efficient cooling. The products are usually made from PE or PP, which are crystalline materials with a high heat content. The products are thin-walled and are expected to be molded at high speeds; therefore, large amounts of plastic are processed per hour. The materials shrink much and hug the core for good heat transfer.

Figure 13.63 shows a typical design of an over cap for the core. It is usually made from beryllium- copper alloy B-25 for good heat conductivity. In larger products, the over cap is sometimes pressure cast, but there could be a problem with porosity of the casting. If available, it is preferable to machine the cap from solid, drawn rods or forged bars. The channels are in spirals similar to that for flat products shown earlier.

Figure 13.64 shows a variation of the over cap design. The projection P may be so thin that it would be risky to make it out of beryllium-copper. The insert may be cross drilled in a pattern such as shown in Fig. 13.64B. Note how some of the zig-zag lines are converging close to the gate area. Also, whenever beryllium-copper over caps are used, the stripper ring seat should be on the steel base of the core, not on the beryllium-copper.

Cores for Large Products There is no clear demarcation between medium and large products. All methods used with medium-sized products can be applied to the cores for large products. If there is a notable difference, it is the speed of molding, which is often much slower due to the heavier wall thickness of the larger products. Although the molded products are heavier, the overall amount of plastic per hour is often less. However, this does not imply that cooling is not important. Even though molds for large products run slower than those for smaller products, poor cooling will certainly make the condition worse.

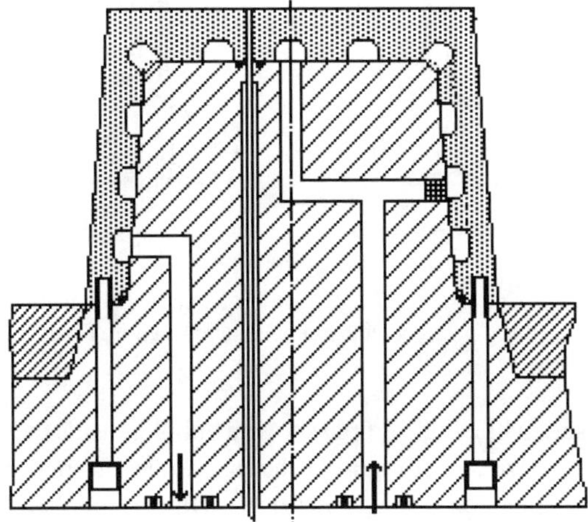

Figure 13.65 Typical over cap with spiral cooling grooves and moving vent pin.

Figure 13.65 shows a typical over cap for a large product. Coolant flows spirally (in concentric grooves with connecting slots and baffles) from the center toward the rim of the product. Venting is provided by the vent pin.

Many large cores, especially if they are not round, are cross drilled for cooling and must have a bubbler opposite the gate. Cross drilling is quite satisfactory if the entrances of the bores are properly welded shut or blocked using pressed-in plugs, if the product permits witness lines at the welds.

13.4.5.2.5 Insert Cooling For cooling purposes, we can define an insert as a small core. It could be located in the cavity or in the core, and in either case, it could be fixed or movable. The reasons for choosing inserts rather than machining the shape out of the solid are that it is easier to replace a broken pin and to change the insert shape or size after the mold is finished.

Usually, the small mass of the insert will cause its temperature to rise rapidly, unless provision is made to remove the heat. Where size permits, cross drilling or bubblers should be used, similar to the cooling of small cores, with conventional coolants or with gases, as described earlier.

In some applications, the use of a "heat pipe" may be considered. Usually, however, if there is enough space for a heat pipe, a bubbler can be installed, which is much more efficient.

Beryllium-copper alloys are sometimes used to help remove the heat from the insert, but lower strength and higher wear of these materials are serious considerations. While the heat conductivity of beryllium-copper is about four times that of steel, its modulus of elasticity, E, is only half that of steel, and its shock resistance is much lower than for tool steel.

Fixed Inserts Figure 13.66 shows some typical examples of insert cooling. In Fig. 13.66A, the insert is cooled only by conduction to a cooled plate. The method in Fig. 13.66B is somewhat better; it cools the pin directly, with the coolant flowing around the base of the insert. Figure 13.66C shows a bubbler inside the base of the insert, reaching as close as

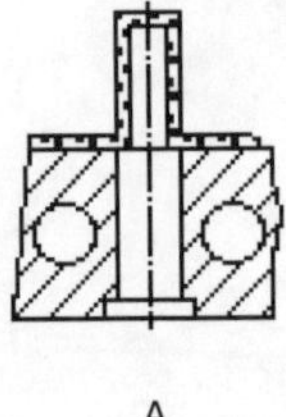

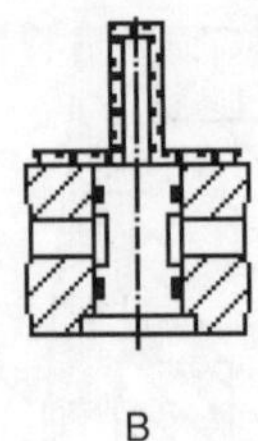

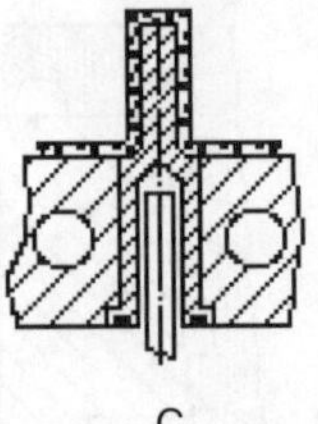

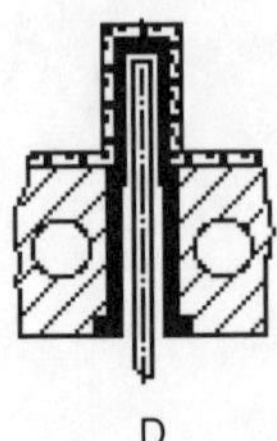

Figure 13.66 Typical examples of insert cooling: A. insert cooled by conduction to a cooled plate, B. pin cooled directly at the base, C. insert base cooled by bubbler, and D. entire insert cooled by enclosed bubbler.

possible toward the insert, while Fig. 13.66D shows a bubbler inside the insert. This is the most efficient method, but the size and strength of the insert may not permit it.

If internal cooling is planned through the inserts (Fig. 13.66C and D), it is important to make sure that there is enough flow available through the small passages in the inserts. If necessary, additional cooling circuits must be provided.

The penalty for poor cooling of the insert is a slower molding cycle to allow time for the insert to cool down after every shot. Otherwise, the plastic will not have cooled enough around the insert, even though the rest of the product may be ready for ejection. Premature ejection will deform the shape of the opening made by the insert; typically, some melted plastic will string and/or obstruct the poorly molded opening.

Movable Inserts, Including Side Cores Movable inserts can be in line with the axis of the mold, as in two-stage ejection. They have to be withdrawn from the plastic during the mold opening motion; they also may be at (usually, but not necessarily) right angles to the axis of the mold, as side cores either in the cavity or in the core. The same rules apply for their cooling as for fixed inserts.

Composite Inserts These are inserts within inserts, or where several inserts are packed side by side. There are no fixed rules regarding their cooling. The cooling depends entirely on the product design and the practicality of special cooling for the inserts. If the mold becomes too complicated because of extraordinary cooling layouts, it may be too expensive to build and to maintain. It may then be better to sacrifice some cooling efficiency at the expense of a slower cycle.

Examples shown in Fig. 13.67 show cooling of some frequently appearing shapes of inserts. The designer must realize that it is impossible to give examples for all the possible insert or mold shapes.

Figure 13.67A shows a cylindrical insert with a cylindrical cooling channel, created by welding in two levels (W) a cylindrical sleeve around the insert. Coolant supply is through drilled channels from the bottom. Care must be taken to create a proper flow by the use of grooves and/or baffles to avoid stagnant pockets of coolant. Figure 13.67B shows a similar arrangement, but the cooling channels are drilled zig-zag from the bottom to avoid welding. Figure 13.67C shows a typical flat blade cooled by cross drilling and welding a plug. If welding is not permissible, zig-zag drilling from the bottom (not shown) can be used.

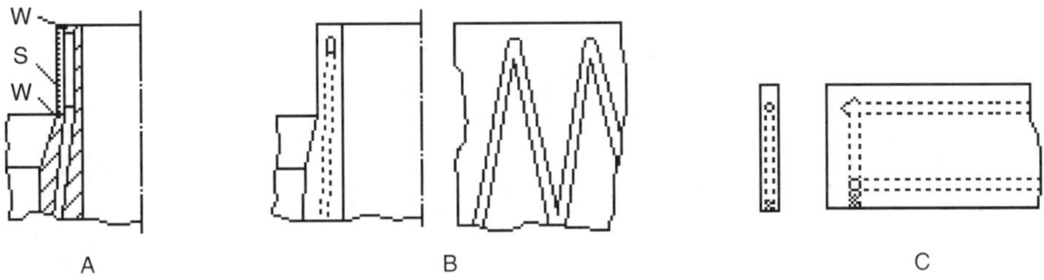

Figure 13.67 Typical cooling for different shapes of inserts: A. cylindrical insert with cylindrical channel, B. cylindrical insert with zig-zag channels, and C. flat insert cooled by cross drilling and welded plug.

13.4.5.2.6 Gate Pad Cooling The gate pad is an insert in the cavity, surrounding the gate area. This area is specially mentioned for the reasons outlined below. In hot runner molds, the plastic is kept hot up to the gate. With open gates (not valve gated), the gate must freeze as the product cools down. If the area is not properly cooled, the gate will not freeze (close) and the hot plastic will drool out of the gate while the mold is opened. In thin-walled products, such drool may be enough to produce bad products or to stop the machine.

In valve-gated molds, the tip of the valve pin (where it shuts off the gate) must be cooled so that the portion of the product opposite the gate will have time to solidify. Cooling of the valve pin is provided by contact with the cooled gate pad. In three-plate molds, the heat content of the heavy drop must be removed quickly so that the runners and drops are cold enough for ejection when the product is ready.

Why have Cavity (Gate Pad) Inserts? For some products, it may not be permissible to have a witness line where the gate pad is located in the cavity. In such cases, the cavity must provide sufficient cooling in the gate area, usually by cross drilling or with circular grooves. However, there are several reasons why a gate pad is desirable:

- *Wear of the gate.* It is usually more economical to replace the smaller gate pad than the whole cavity, although it is sometimes feasible to repair a damaged gate by welding and remachining.
- *Material of the gate pad* can be selected different than the material of the cavity. For example, to improve cooling, the gate pad is made from BeCu, while the cavity may be made from hardened mold steel. In some special cases, the gate pad may be made of a good wearing tool steel, while the cavity is made of another, softer material, such as cast BeCu, prehardened steel.
- *Venting.* Cavity venting in high-speed molds is just as important as venting of the cores. A circular vent slot around the gate pad insert is easy to produce and to maintain.
- *Differential cooling is possible.* Instead of having separate gate pad cooling circuits, the cooling for the cavities and gate pads can be apportioned properly, either by design of the size of the passages or by permanent adjustments with restrictors, during the initial mold tests. This can eliminate the need for separate controls for gate pad and the cavity cooling circuits. Start-up problems caused by overly cold gates can be avoided by shutting off all cooling for a short time.

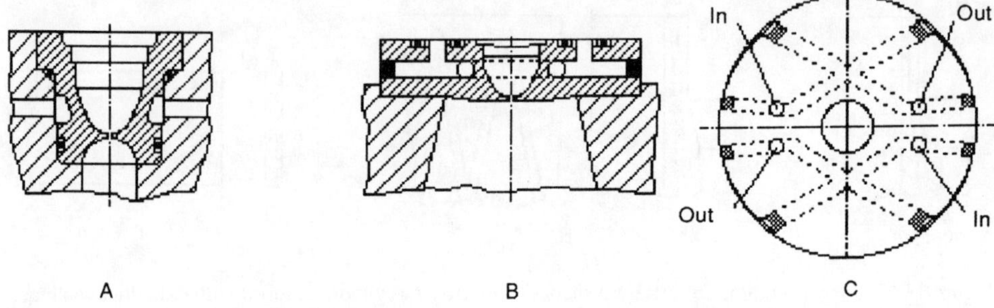

Figure 13.68 Examples of gate pad cooling: A. gate pad with cooling groove, B. gate pad with cross drilled channels, C. pattern of cross drilled channels in B.

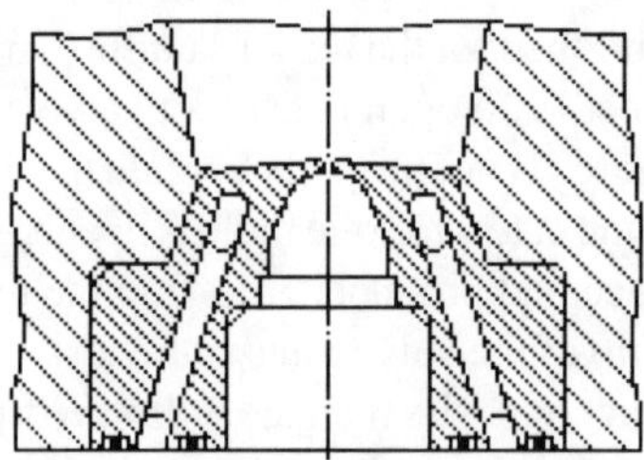

Figure 13.69 Cross drilling pattern for gate pad cooling that brings cooling close to the gate.

Figure 13.68A shows a gate pad with cooling groove. Figure 13.68B and C show a gate pad with cross drilled cooling channels.

Figure 13.69 shows another drilling pattern, which brings cooling close to the gate. The pattern is similar to a cavity cooling, as shown in Fig. 13.51B and C. There are many variations of these designs, and it is up to the designer to come up with the most suitable arrangement.

13.5 Size and Number of Supply Lines (Hoses)

Up to now we have considered only the cooling of the various mold elements. We shall now look at the whole mold. A single-level mold consists of a stationary mold half with the cavities (injection side) and a moving mold half with the cores (ejection side). A stack mold has an additional, floating mold portion, which carries two levels of cavities between the stationary and the moving mold sections, which each carry one level of cores. Special molds, such as used on turret-, shuttle-, or swing-plate machines, use two or more injection or ejection mold portions. In all these cases, cooling "In" and "Out" lines must be connected to each mold portion. These cooling lines must be sufficiently large to supply each mold portion with the required amount of coolant.

We know from experience or can calculate how much coolant per hour is required (see Section 13.6). The amount of coolant per hour is directly proportional to the amount of plastic processed in the mold per hour.

Nowhere is it suggested or implied that we must have separate controls for each cooling circuit within the mold, as it has been the practice for so many years. It is just a question of rearranging the thinking about the method of connections. Instead of having manifolds on or near the machine connecting with numerous hoses to the mold (resulting in ungainly and impractical hose "jungles"), manifolds can be part of the mold portion they are to supply. It is important that the hoses to these mold manifolds are adequately sized for the required total flow of all the branch circuits.

Such manifolds can either be integrated (machined) right into the mold plates to supply the various channels leading to the cavities or cores, or they can be auxiliary manifolds (usually made from aluminum) bolted to the sides of the mold plates and connected to the drilled passages in the plates using O-ring seals or using fixed, external piping.

The advantages of this method are:

- The number of hose connections that have to be made at every mold change is reduced to the minimum: one "in" and one "out" per mold portion. This could reduce the total number of hoses for the example in the introduction from 52 to 4. While there are few molds with so many hose connections, molds with 12 to 16 hoses are quite common.

  A frequently used argument for the large number of hoses is that the operator can, with taps, individually adjust the coolant flow to the various mold elements. With knowledge of the heat exchanges taking place inside the mold, one can, as a rule, calculate the cooling channel sizes and the distribution ratio between the various cooling branches. In rare cases, adjustments can be made with fixed restrictors at the time of the initial mold tests. After that, there should be only two modes of operation: cooling is either "on" or "off".
- Mold operators cannot alter the cooling flow pattern. A mold should be installed, connected, started, and run. Mold cooling is too important to be left to adjustment by people who do not understand the cooling needs. Should there be some problems with the cooling in operation, such as plugged channels, the mold must be repaired.
- Automatic mold installation. This is an important consideration. Even with fewer hoses, the time for connecting hoses to the mold takes still a considerable portion of the set-up time. With only two lines per mold portion, connections can be made much faster, especially when using "quick connectors." After the mold is moved into position, the installation is much easier, and it can even be automated.
- Accessibility. This is of special importance where automation in product handling requires the placing of conveyors below the molding area, or where take-offs move in from above or from behind the mold. There are also air lines and electrical connections to place which can restrict accessibility.
- The actual cost and upkeep of a number of small hoses and joints is larger than for only two to four larger ones. Of course, there is additional cost for the manifold, but overall, it is better to keep the number of hoses to a minimum.

13.6 Coolant Requirements for a Mold

Frequently asked questions are:

- How much coolant (liters, gallons) does a mold require?

- What is the coolant temperature required?
- Will I need a chiller?
- How large a chiller will I require for this mold?
- How large do the cooling lines to the mold have to be?

13.6.1 Introduction

A mold is a heat exchanger. Energy (in form of heat and electric energy) put into the plastic on its way from the machine hopper to the machine nozzle is used to bring the plastic from room temperature (T_r) to the temperature required for injection (T_i). This energy (in the form of heat) must be removed from the plastic while it is cooling in the mold so that the finished products become stiff enough and can be ejected.

If we assume that there are no heat losses due to radiation and conduction away from the machine, the actual amount of energy used for plasticizing can be measured and calculated. Important considerations for such calculations are:

1. The actual wattage (kW) of the heaters used at the extruder (plus the shooting pot in a two-stage injection system), and
2. the actual amount of energy (in kW) used in the extruder to plasticize the resin and used during the time the heaters and the drive motor are "on" during each cycle, or more accurately, during an hour of operation, once the machine is "on cycle" (i.e., running and producing without further adjustments).

Such a calculation is quite easily taken from an electrically driven extruder by reading the current to the electric motor. This is more difficult for a hydraulic-motor screw drive because of the losses between the hydraulic pump motor in the pump and in the hydraulic screw drive motor. It would be wrong to use the total connected load of heaters plus motor in a calculation without allowing for the time they are "off". Most of the energy is supplied to the plastic by the screw drive motor, and only a smaller amount by the heaters surrounding the extruder barrel. This ratio depends to a great extent on the design (configuration of flights) of the extruder screw. The better the screw design, the less heat will be required from the barrel heaters.

Note that the heater capacity in the hot runner system (manifold and nozzle heaters) must not be added in these calculations. In well-designed hot runner systems, the heaters are used only to maintain the melt temperature on its way from the machine nozzle to the gates. (The reason for the often large ratings of manifold and nozzle heaters is to be able to start up the mold faster, from cold. Once the mold is running, the heaters use less than 10% of their rated capacity, and in fast running molds, they can even be turned "off" altogether.)

The above measurements and calculations can only establish the power consumption and heat load for a cooling system after the mold has been built.

13.6.2 Calculation of Heat Input

The method explained here is very accurate but depends on several factors which have to be assumed, mainly from experience, with similar products in similar molds and plastics.

1. Estimated cycle time depends on:
   - Shape of product,
   - wall thickness,
   - method of ejection,
   - machine characteristics (clamp speed, injection speed and pressure), and
   - type of plastic (melt flow index, heat content, heat conductivity).
2. Estimated efficiency of cooling in mold depends on:
   - Design (cooling layout),
   - mold materials selection,
   - estimated injection temperature (T_i),
   - estimated temperature of product at ejection (T_e),
   - coolant temperature and pressure, and
   - estimated ambient temperature of mold plant.

Example: A certain mass of plastic will be heated from T_r to T_i (ΔT_h). Subsequently, in the mold, it will be cooled from T_i to T_e (ΔT_c). The mass of plastic (M_s) (shot size) per cycle equals the mass of the product (M_p) multiplied with the number of cavities (n):

$$M_s = M_p \times n. \tag{13.2}$$

If we estimate the cycle time (t_c) in seconds, the total mass per hour (M_h) equals the shots size (M_s) times the number of shots per hour:

$$M_h = M_s \left(\frac{3{,}600 \text{ sec/hour}}{t_c} \right). \tag{13.3}$$

13.6.2.1 *Specific Heat*

The specific heat (C_p) can be expressed in two ways:

1. It is the ratio of the heat required to raise the temperature of a certain mass of the given material by one degree Celsius (°C) to that required to raise the temperature of a similar mass of water one °C. (The specific heat of water is 1.) Most common plastics have a specific heat (ratio) between 0.25 and 0.55.
2. It can also be expressed in definite units, either in BTU/lb-°F or in cal/g-°C.

Values for specific heat for materials of interest are shown in Section 22.11.

Since the specific heat is not constant at all temperatures, it is commonly assumed that it is determined by raising the temperature from 14.5 to 15.5 °C. One (1) calorie (cal) is the energy required to raise the temperature of one gram (g) of water from 14.5 to 15.5 °C (1 kcal = 1,000 cal). One cal is also 4.1868 Joule (J). For other conversions of energy, see Section 22.13.

Example: A product made from PS ($C_p = 0.34$) weighs 35 g. It is to be molded in a 16-cavity mold at a 6-second cycle.

Shot size: M_s = 35 g/cavity × 16 cavities= 560 g

Shots/hour: $\frac{3{,}600 \text{ sec/hour}}{6 \text{ sec/shot}}$ = 600 shots/hour

Mass molded/hour: 560 g × 600 shots/hour = 336,000 g/hour (or 336 kg/hour)

Calories/hour: 336,000 g/hour × 0.34 cal/g-°C = 114,240 cal/hour for 1 °C rise

Heat energy/hour: 114,240 cal/hour × 4.1868 J/cal = 478,300 J/hour This is the heat (energy) required to raise the tem perature of this mass by 1°C.

Since the plastic must be melted so that it can be injected, its temperature must be raised from room temperature, T_r = 22 °C, to an estimated injection temperature, T_i = 280 °C. The temperature difference is $\Delta T = T_i - T_r$. In this example, ΔT = 280 °C – 22 °C = 258 °C. Multiplying the total mass per hour with ΔT yields the total energy input required for the plastic during one hour:

478,300 J/hour-°C = 123,401,400 J/hour, or 123 MJ/hr. (MegaJoule/hour).

One MJ = (1/3.6) kWh (kilowatt-hour); therefore, the total energy required is

$$\frac{123 \text{ MJ/hour}}{3.6 \text{ MJ/kWh}} = 34.17 \text{ kW}.$$

13.6.3 Cooling Requirements

The heat added to the plastic must be removed from it during the molding process so that the molded (relatively stiff) products may be removed from the mold. The same equations shown above can be used for the calculations.

In most cases, it is not necessary to cool the plastic down to T_r, but only to a suitable (estimated) T_e. The T_e can be much higher than T_r, as long as the plastic is stiff enough to resist the forces of ejection and will not lose its shape after being ejected while still warm. Therefore, not all of the heat added to the plastic will be removed within the mold by the cooling water. Most of the remaining heat will be dissipated into the surrounding (plant or storage) air space. Some of that heat will be radiated by the mold and the machine into the plant air space.

The hot product will continue to shrink after ejection until it is fully cooled down. This will affect the dimensions of the product; the T_e may be critical in cases where accurate product dimensions are important, as in the case of fit of lids and containers. The effect of shrinkage after ejection is of particular concern for materials with large shrinkage factors (PE, PP, etc.).

Example: To continue the example above, the 336 kg of hot plastic, now in the form of products, will cool from a T_i of 280 °C to a safe ejection temperature T_e of 50°C:

$$\Delta T = T_i - T_e, \text{ or } 280\ °\text{C} - 50\ °\text{C} = 230\ °\text{C}.$$

As before, calculate the heat removed in the mold:

$$478{,}300 \text{ J/hour-°C} \times 230\ °\text{C} = 110{,}000{,}000 \text{ J/hour},$$

or 110 MJ/hour.

The amount of heat which must be removed from the coolant, either in the cooling tower, in the chiller, or in the combination of both, is:

$$\frac{110 \text{ MJ/hour}}{3.6 \text{ MJ/kWh}} = 30.55 \text{ kW}.$$

In molding and refrigeration technology, the amount of heat is often indicated in "tons" of refrigeration. This is defined as the amount of heat required (in BTU) to melt one ton (2,000 lb.) of ice at 0 °C into water of 0 °C. The conversion factors are:

1 ton refrigeration = 12,000 BTU/hour, or

1 ton refrigeration = 3.516 kW (1 kW = 3,413 BTU/hour)

In our example, we would need 30.55 kW ÷ 3.516 kw/ton refrigeration = 8.7 tons refrigeration.)

13.6.4 Temperature of Cooling Water

The heat to be removed from the plastic within the mold must be carried away by the coolant flowing through the mold and back to the cooling system, such as a chiller and/or cooling tower. The temperature difference, ΔT, of the coolant between "in" and "out" ports of the mold must be assumed, and is based on past experience. In many cases, a ΔT_c of 5 °C is a good choice, but occasionally a ΔT of 1–2 °C may be necessary. In some molds, a ΔT of up to 10 °C is quite acceptable. See also Section 13.6.7, Efficiency of Cooling.

It is not very important whether the coolant is very or only moderately cold. The flow of heat from the hot plastic to the coolant depends largely on the temperature difference between the injected plastic and the coolant, on the design of the cooling channels, and the material selection for the mold components. Assuming (as in the example) T_i = 280 °C:

With coolant temperature T_c = 5 °C, the ΔT_c = 280 °C – 5 °C = 275 °C;

With coolant temperature T_c = 10 °C, the ΔT_c = 280 °C – 10 °C = 270 °C.

This represents a difference of only about 2% at the high end of the cooling. When ejecting at an estimated 50 °C, at the low end of the cooling (i.e., when the plastic has almost completely reached its final (in mold) temperature), the differences are:

With coolant temperature T_c = 5 °C, the ΔT_c = 50 °C – 5 °C = 45 °C

With coolant temperature T_c = 10 °C, the ΔT_c = 50 °C – 10 °C= 40 °C.

This represents a difference of only about 10% at the low end of the cooling.

In fact, using higher T_cs will only slightly increase the time required to cool the plastic, and in many cases, the added cost (equipment and power) of using very cold coolant may not be justified economically.

13.6.5 Dew Point

The dew point is the temperature at which the water vapor in the air (expressed in relative humidity) will condensate. The higher the relative humidity (in %) of the plant air, the higher is the dew point. If the coolant temperature is below the dew point, water condensation will form on the coolant lines and on the mold itself, and cause dripping and corrosion.

Coolant below the dew point can also cause molding trouble: Assume that the air is very humid and the coolant is below the dew point. After the products are ejected, during the short time the cores and cavities are exposed before the mold is closed again, enough water droplets can condensate on the cold molding surfaces to cause severe blemishes on the finished products, resulting in rejects. This actually happened with an unscrewing mold where one core plate, after unloading the screw caps, remained another 3 seconds in the open before swivelling and arriving in the molding position. On some days, the molds worked well; on other days, the products were unusable. Investigation showed that these quality changes coincided with the weather. On dry days, the products were fine; on humid days, there was trouble. After increasing the cooling water temperature by only a few degrees, to bring it above the dew point, everything worked well, regardless of the humidity in the air. The only sacrifice was that the cycle had to be increased by less than one second.

13.6.6 Quantity of Cooling Water Required

Continue the calculations using the same formulae as before, but inverted, and use the example started above. It is known that 110,000,000 J/hour (or 110 MJ/hour) heat must be removed from the mold every hour. If the coolant is assumed to enter the mold at 5 °C, and to rise in the mold to 10 °C, $\Delta T_c = 5$ °C. How many liters of water (or kg, which is for water about the same as liters) are required per hour for this example?

$$\frac{110{,}000{,}000 \text{ J/hour-°C}}{5 \text{ °C}} = 22{,}000{,}000 \text{ J/hour} = \text{Energy to heat X liters of water.}$$

Converting this into calories:

$$X = \frac{22{,}000{,}000 \text{ J/hour}}{4.1868 \text{ J/cal}} = 5{,}254{,}609 \text{ cal/hour.}$$

Since the specific heat of water (C_pw) equals 1:

$$\frac{5{,}254{,}609 \text{ cal/hour}}{1 \text{ cal/g}} = 5{,}254{,}609 \text{ g/hour} = 5{,}254 \text{ L/hour of water.}$$

Since the flow of cooling water is usually measured per minute:

$$\frac{5{,}254 \text{ L/hour}}{60 \text{ min./hour}} = 88 \text{ L/min. or } 23.3 \text{ U.S. gal/min.}$$

This is only a theoretical figure, which must be adjusted to a realistic amount, as explained in Section 13.6.7, Efficiency of Cooling. Before discussing efficiency, however, we will consider some comments regarding the above theoretical amount of coolant (water):

- If the selected ΔT_c of the cooling water between "In" and "Out" of the mold was just 1.5 °C instead of 5°C, the amount of coolant per hour would be greater by the ratio 5:1.5, or

$$23.3 \text{ U.S. gal/min.} \left(\frac{5\ °C}{1.5\ °C}\right) = 77.6 \text{ U.S. gal/min.}$$

- If the selected ΔT_c were 10 °C, the required amount of coolant would be smaller by the ratio of 5:10, or

$$23.3 \text{ U.S. gal/min.} \left(\frac{5\ °C}{10\ °C}\right) = 11.65 \text{ U.S. gal/min.}$$

- Similar considerations apply to the ejection temperature, T_e. If T_e were higher, say 60 °C instead of 50 °C, less cooling water would be required. Conversely, if T_e were lower, say 40 °C, more cooling water would be required.
- Plastics with higher specific heat factors, such as PP and PE (C_p = 0.50–0.55) require about twice the amount of coolant as plastics such as PVC (C_p = 0.25). See Table 22.3 in Section 22.11 for values of various plastics and other materials, or check with plastic supplier.
- If the cycle were not 6 seconds as estimated (example above) but 9 seconds, the number of shots/hour would be only 400 instead of 600, and only two-thirds of the heat input would need to be removed.
- Conversely, if the mold proved to run faster, say at a 4-second instead of 6-second cycle, there would be 900 shots/hour, and the plasticized material would have a mass of 504 kg, or 50% more than estimated; therefore, there would be a need for much more coolant per hour.

13.6.7 Efficiency of Cooling

The theoretical amount of heat to be removed from the mold is actually the true amount of heat which must be removed from the coolant, regardless of the quantity of coolant flowing through the cooling system of the mold. But it would indicate the correct quantity of coolant only if the transfer of the heat from plastic to coolant will take place in the stipulated cycle time.

In a mold with a poor cooling layout, heat will travel only slowly from the hot plastic to the coolant, and will require much more time than anticipated and used in the calculation for the heat exchange outlined above. In a mold with a good cooling layout, the time for the heat exchange could be much shorter. It could be as anticipated, or even shorter, thereby permitting a faster cycle time.

However, even in a mold with a good cooling layout, it is unrealistic to expect that the theoretical amount of coolant, as it flows through the channels, will pick up all the heat generated in the anticipated cycle time. Therefore, the amount of coolant required depends very much on the cooling efficiency of a mold (E_{cm}) to remove heat from the mold.

E_{cm} depends on various characteristics of the mold construction, on the chemistry of the cooling medium selected, on the conductivity of the coolant, and on the speed and the flow behavior (turbulent or laminar) of the coolant within the mold. Note that the flow (speed) of the coolant depends primarily on the pressure drop between the pressure of the incoming coolant and the pressure of the return line back to the source of the cold coolant.

To summarize, the cooling efficiency E_{cm} is affected by:

- The mold material (heat transfer);

- the proximity of the cooling channels to the molding surfaces;
- the sizing and balancing of the cooling channels, to ensure that all channels remove the heat efficiently and evenly from the areas they are supposed to cool;
- the surface condition (rust, dirt) of the cooling channels;
- the type of flow in the channels (turbulent or laminar), which in turn depends on the size of channels and the pressure difference between "in" and "out" of the coolant; and
- the chemistry (specific heat, viscosity, etc.) of the coolant.

It is not possible to assign definitive values to the cooling efficiency of a mold, E_{cm}. We only know that a definite (calculable) amount of heat enters the mold with the hot plastic, and that another definite (calculable), usually smaller, amount of heat must be removed by the coolant. In practice, the flow of the coolant through the mold will be larger than the calculated, theoretical amount. An actual E_{cm} could be determined by testing various completed molds by calculating the theoretical amount of coolant for a certain material and cycle time, and then measuring the required amount of flow through the mold once it is on cycle at the stipulated cycle time.

$$E_{cm}\ (\%) = \frac{Q_c}{Q_a} \times 100, \tag{13.4}$$

where Q_c = calculated quantity of coolant and Q_a = actual quantity of coolant, or

$$Q_a\ (\%) = \frac{Q_c}{E_{cm}} \times 100. \tag{13.5}$$

Using the example from earlier on, the calculated amount of coolant was Q_c = 88 L/min. Assuming an E_{cm} of 50%:

$$\frac{88\text{ L/min.}}{0.50} = 176\text{ L/min., or }46.6\text{ gal/min.}$$

Note that the effect of increasing the volume of the coolant will reduce the temperature increase of the coolant between "in" and "out", but as stated earlier, the amount of heat transferred to the coolant remains the same as calculated.

13.6.8 Cooling Water Supply

The questions regarding the cooling water supply are usually about whether a chiller is needed for a mold, and what size chiller should be used. These questions will be answered to give the designer some basics on this subject, but the serious student should refer to special literature dealing with cooling.

If the water supply to the plant were sufficiently large, at sufficiently low temperature and at low cost, there would be no need for chillers. It is conceivable that a molding plant may be erected close to a large, cold lake that has a fairly constant water temperature, and the plant may be equipped with large pumps to provide the necessary flow, but this scenario is highly unlikely. An alternative, the use of "city" water, would fill the requirements of temperature and flow, but usually at a very high cost. However, this is occasionally done by a small molder or for an experimental setup. It is not practical for industry, in general.

13.6.8.1 Cooling Towers

As the hot water cascades in small streams through the cooling tower, it is exposed to the (cooler) outside air, which is blown over the surface of the water. This causes some of the water to evaporate; the evaporation heat cools the remaining water. (This process is similar to cooling a spoonful of hot soup by blowing over it.) Most, if not all, mold plants use cooling towers for cooling and recirculating the water required for the cooling of the hydraulic oil in molding machines. In many plants where the total load from mold cooling is not high, or where some variations in temperature of the water to the molds can be tolerated, molds too can be cooled directly with tower water. The amount of water evaporating in the towers must be (automatically) replaced from the plant water supply (city water or well). There usually is a large water storage tank that is kept at a certain level, or the water is stored in a lagoon near the towers, which also will cool the water by evaporation.

While it is quite easy to pump large volumes of water to flow through one or a series of towers, the cooling process depends also very much on the outside temperature and humidity, so water from the cooling tower(s) cannot supply a constant water temperature as required for many molds. These temperature differences are usually acceptable for machine cooling and when supplying water-cooled chillers, which adjust automatically to compensate for such variations.

Cooling towers are relatively inexpensive. They require electric power for the blowers (usually about ½–2 hp motors, one or more per tower) and for the circulating pumps to keep the water lines in the plant at the required pressure, usually at 40–50 psi (3–4 kg/cm^2). The more water required, the larger must be the pump(s). In many mold plants, there are several pumps in parallel, and usually a standby pump in case of peak load requirements or breakdown of one of the other pumps, so that the plant will not be interrupted in case of a pump failure. Pump motors are usually 30–100 hp, each.

13.6.8.2 Chillers

All chillers have:

1. A cooling system, which exchanges heat coming from the mold(s) to the "tower" water or to the plant air;
2. a circulating pump, which provides the necessary flow (pressure) of chilled water to the mold, and
3. a control station to regulate flow and temperatures with the necessary sensors, switches, and valves.

There are two basic differences in chillers: some are air-cooled chillers and others are water-cooled chillers, usually using tower water. In addition, the chillers can be grouped in two classes:

1. Dedicated chillers (i.e., one chiller (located beside the machine) for one mold) or
2. central chillers (i.e., one or several chillers in one location supply a group of machines or the whole molding plant).

Both types (dedicated or central) can be either air- or water-cooled. Central chillers are usually water-cooled. With air-cooled chillers in summer, the air is ducted outside; in winter, the hot air from the chillers is ducted into the plant heating system. Dedicated chillers can be either air- or water-cooled, but usually, chillers with larger than a ten-ton capacity are water-cooled to avoid the addition of too much heat to the plant air.

There are many reasons for and against using either dedicated or central chillers. The advantages of dedicated chillers are:

- Each mold can have to its own cooling water temperature. This may or may not be really necessary, and this need should be carefully considered.
- Some chillers are equipped with heaters and can supply coolant at elevated temperatures, as may be required (e.g., for ABS, PC, etc.).
- Need for costly (heat-insulated) piping for chilled water (and return) is reduced.
- In case of chiller breakdown, only one molding machine is affected.

Disadvantages of dedicated chillers are:

- Additional electric power is needed at molding machines.
- They occupy plant space at the machines, which adds to the congestion in the plant and reduces accessibility to the machines.

The advantages of central chillers are:

- With proper installation, the risk of complete breakdown is negligible. There is always at least one chiller in the "standby" mode, in case of breakdown of another chiller. Most plants have two or more chillers in parallel, plus the standby unit.
- Central location of chillers, pumps for the towers, and air compressors remote from the molding machines makes for an easily accessible plant. By proper planning, additional units can be easily added without interruption of the plant.
- Per ton of cooling output, stationary (central) chillers are less expensive than movable, dedicated chillers.

The disadvantages of central are:

- Plant space for a separate room is required.
- Cost of piping is high. In addition to the two lines for tower water that are required to supply the machines, two more well-insulated lines are required to keep heat losses to a minimum as the chilled water is ducted through the hot plant environment.
- All molds will run on the same coolant temperature. Any adjustment to the mold temperature can be done only by regulating the flow of the coolant.

Many plants with central chillers use additional dedicated chillers for some molds to ensure that all molds are operating at optimal coolant supply.

13.6.8.3 Sizing of Chillers Required

In the preceding example, a cooling capacity of 30.55 kW (or 8.7 ton refrigeration) is required. Chillers come in certain standard ratings: 3, 5, 10, 15, 20, . . . , etc., ton capacity, according to the various manufacturers. In addition, dedicated chillers are equipped with circulating pumps with electric power ratings that can be specified when ordering (e.g., 1, 2, 3, 5, 10, . . . , etc., hp or gal/min. flow at desired pressures). Each chiller has its own data curves indicating relationships between pump volume, pressures, and chilling capacities. This field is outside of the scope of this book, and it is suggested to consult specialists in this area when deciding which size chiller and pumps to order.

NOTE: Regarding the sizing of chillers, it is usually better to have some reserve capacity rather than to work a chiller to its rated capacity. There are additional sources of heat which must be removed from the mold and the cooling water, as follows:

1. The heat input from the hot runner system. It was stated earlier that the heaters in the hot runner will add only very little to the cooling requirements. (As the mold is still cold, the heaters are "on" fully, until the plastic in the channels is melted. But during this time, the machine is not operating yet, and no heat will be removed from plastic coming from the injection unit.) However, even if a small percentage of the rated heater capacity remains "on" during operation of the mold, it is still only a minor amount. For example, if the heaters in the 16 cavities of the example are 250 W each, and the total of the manifold heaters are 6,000 W, the total [(250 W/cavity × 16 cavities) + 6,000 W = 10,000 W, or 10 kW. A reasonable estimate is that these heaters are "on" only 10% of the time, so there is a total of 1 kW coming from these heaters, small compared to the 30.55 kW from the plastic.
2. The circulating pumps also add heat to the water as it is pumped through the mold, but this is also relatively small.

By adding a safety factor of 25% to the figure derived from the earlier calculations, the result will be about right. In this example, 8.7 ton refrigeration x 1.25 = 10.87 ton refrigeration. Actually, a 10-ton requirement would be about right; otherwise, a chiller with 15 tons would have to be selected. With central chilling, the 10.87 tons will be added to the already existing capacity of the plant. If there is no sufficient reserve, another unit may have to be added to cover this requirement and to have, again, some reserve for future additions.

It is also important for the plant management to total up the requirements for all the different molds operating at the same time to see if the cooling water requirement is within the chilling capacity of the central system.

13.6.9 Size of Cooling Lines

This applies not only to the cooling lines that distribute the water to, and collect it from, the machines but also to the lines that connect the main piping with the hoses to and from the mold. Too much detail is beyond the scope of this book, but the designer must understand that the lines—pipes, hoses, fittings and any shut-off valves—must all be large enough to permit the required coolant volume to reach the various mold cooling circuits in the amounts planned and arrived at when designing the mold.

In many mold plants, depending on the number of molding machines, the supply lines from tower to the machines are 5–8 in. pipe size and larger, and the branch lines to each molding machine (depending on the machine size) are in the order of 1–2 in. or larger. Similar sizes will be required for supplying the chilled water lines. The return lines are usually one or two sizes larger than the supply lines to ensure that there will be little back pressure in these lines, which could affect the flow through the mold or the machine cooling circuits.

Recommended Reading

1. Bernhardt, E. C. (1987) Applications of Computer Aided Engineering for Injection Molds, SPE/Hanser Publishers, Munich.
2. Tseng (1984). Computer Cooling Analysis to Select Bubbler or Baffle. ANTEC.
3. Singh (1984). Computer Software for Plastic Cooling Analysis—A new approach. ANTEC.
4. Menges, G.; Mohren, P. (1992). How to Make Injection Molds. 2nd Ed. Hanser Publishers, Munich.
5. Rubin, I.I. (1972). Injection Molding: Theory and Practice, J. Wiley and Sons, New York.
6. Oberg, E.; Jones, F.; et al. (1994). Machinery Handbook, Industrial Press Inc., New York.
7. Modern Plastics Encyclopedia. McGraw-Hill, New York.
8. Mold Cooling Analysis Program (MCAP), General Electric Company.

14 Heat Expansion

14.1 Variables Affecting Heat Expansion

Every material expands as its temperature rises. The amount of this expansion in one direction is called "linear" expansion; the amount of this expansion is given in either mm per 100 mm length per °C or in inches per inch length per °F. Table 14.1 lists values *k* for some materials used in molds in the temperature range from room temperature to 290 °C, approximately. Values *k* for other materials can be found in reference books.

Forces generated by heat expansion must always be considered when temperature differences in a mold are unavoidable (e.g., between hot runner manifold, nozzles and surrounding plates). Dislocation due to heat expansion must also be considered in relatively cold plates, where the temperature differences may be small but the dimensions are large; this can result in considerable increase in size of the warmer plate in relation to the colder one, and lead to mismatch between mold parts.

Mathematical relationships between strength of materials and heat expansion are such that a body will expand (linearly) according to the formula:

$$e = L \times k \times \Delta T, \tag{14.1}$$

where e = expansion, L = length, and ΔT = temperature increase.

A body will elongate (or compress) by the amount f when subjected to an external force F, according to the formula:

$$f = \frac{F \times L}{A \times E}, \tag{14.2}$$

where L = length of body, A = cross sectional area, and E = modulus of elasticity.

Table 14.2 lists approximate values for modulus of elasticity, E, for several mold materials. Values of E for other materials can be found in reference books; if more accurate values are required, check with manufacturers spec sheets.

Table 14.1 Linear Heat Expansion k for Mold Materials from T_r to 290 °C

Material	Linear heat expansion *k* cm/cm/°C	in./in./°F
H13	0.00001224	0.0000068
P20, 4140	0.00001139	0.00000633
420 (SS), A2	0.00001152	0.0000064
Cast iron	0.00001179	0.00000655
Aluminum	0.0000224	0.00001244
Brass	0.000018	0.00001
Copper	0.0000162	0.000009
BeCu 25 & 3	0.0000176	0.0000098

Table 14.2 Approximate Values of Modulus of Elasticity *E* for Various Mold Materials

Material	Metric (kg/cm^2)	Imperial (psi)
Carbon steels	2.0×10^6	30×10^6
Stainless steels	$1.9–2.0 \times 10^6$	$28–29 \times 10^6$
Cast Iron (wide range)	$0.8–1.4 \times 10^6$	$12–20 \times 10^6$
Malleable iron	1.7×10^6	25×10^6
Cast steel	2.0×10^6	30×10^6
Aluminum	0.7×10^6	10×10^6
Copper, BeCu	1.2×10^6	17×10^6

The stress s to which the body is subjected can be expressed as

$$s = \frac{F}{A}, \text{ or } F = A \times s. \tag{14.3}$$

When substituting F in equation 14.2 with *F* from equation 14.3,

$$f = \frac{A \times s \times L}{A \times E} = \frac{s \times L}{E}. \tag{14.4}$$

Example: A block is placed between two plates that are held together with screws (Fig. 14.1).

Assume that the height (*L*) of the block is exactly the same as the space where it is placed between the plates. When the block is heated while the plates are maintained at the original temperature, the heat expansion due to the ΔT will generate a force *F* in the direction of the arrows. This force is taken up by the screws. If the strength of the screws is greater than the force *F*, the distance *L* will be maintained while the heated block is compressed by the amount of the heat expansion; if not, the plates will separate. (This assumes that the plates are strong enough that they will not stretch or bend.)

The heated block will expand by an amount *e*, which can be calculated equation 14.1. But because the cold plates surrounding this block will not yield to this force, the force will compress the block by an amount *f* (see equation 14.2, which is the same as *e*.

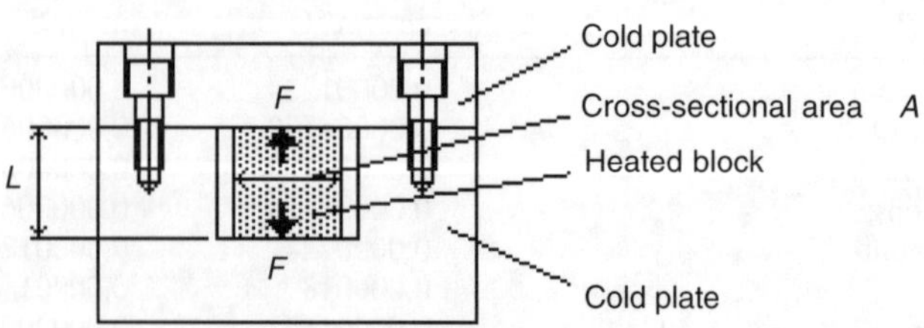

Figure 14.1 Cross section shows a heated block mounted between two cold plates.

Table 14.3 Units of Measure in Metric and Imperial Systems

Unit	Metric	Imperial
Temperature T	°C	°F
Coefficient of heat expansion k	cm/cm/°C	in./in./°F
Modulus of elasticity E	kg/cm^2	psi
Stress s	kg/cm^2	psi
Area A	cm^2	$in.^2$
Length L	cm	in.
Force F	kg	lb.

Conversion: 1 kg/cm^2 = 14.7 psi.

By equating equations 14.1 and 14.4,

$$L \times k \times \Delta T = \frac{L \times s}{E}, \quad \text{or } s = k \times \Delta T \times E . \tag{14.5}$$

Equation 14.5 shows that neither the length L nor the area A have any effect on the stress seen by the block. In other words, increasing the area A or changing L will be meaningless if the permissible stress s_p is exceeded. Only the material selection, which affects E, the permissible stress s_p, and the temperature difference ΔT are important.

NOTE: Before doing any calculation, the designer must select the measurement system that will be used, metric or imperial. Units are compared in Table 14.3. Note that many reference books show the metric values in kg/cm^2 rather than in kg/mm^2, but all technical drawings are in mm (1 cm = 10 mm).

Example: E = 2,000,000 kg/cm^2, k = 0 0.0000114 cm/cm/°C, ΔT = 150 °C. Calculate stress s:

$$s = 2{,}000{,}000\ kg/cm^2 \times 0.0000114\ cm/cm/°C \times 150\ °C = 3{,}420\ kg/cm^2$$

or 50,274 psi. Depending on the material selection, this may or may not be an acceptable stress.

14.2 Factors in Acceptable Stress

14.2.1 Material Selection

It is sometimes possible to select a hardened steel with a high yield point. However, the designer must consider that by operating at elevated temperatures, many mold steels show a sharp reduction in tensile (and yield) strength above a certain temperature. This can be readily seen in the data supplied by the steel manufacturers. If the material is stressed beyond its yield point, it will deform permanently.

The designer must also consider that, often, the stressed part is also subject to a cyclical rather than a steady load, which induces fatigue in the steel and must be considered by applying a generous safety factor to the theoretical yield point (see Chapter 18, Metal Fatigue).

A safety factor of (at least) five (5) should be used; if this is not practical, any lower safety factor must be carefully considered as to how it will affect the life of the mold. Substantial redesign may be required to provide the proper strength for the mold components.

14.2.2 Creation of a "Cold Clearance"

A cold clearance is created when the length of the heated part is less than the length between the supporting cold plates while the entire system is cold. As the system heats up, the heated part grows until it touches the plates and starts compressing. Because some of the increase in length is unopposed, only a portion of the expansion force will be stressing the part when it reaches its final (operating) temperature.

In many hot runner systems, this force is required to provide sealing so that the plastic cannot escape through the contact areas where the plastic passes through under pressure. As long as a sufficient number of screws are tightened to a preload sufficiently strong to contain the expansion, the parts will not separate. During injection, while the mold is clamped up, the clamping force acts in the same direction as the screws to hold the plates together, but once the clamping force is removed, the screws must hold the plates together against the heat expansion forces. (Refer also to Chapter 19, Screws in Molds.)

An understanding of the above relationships is most important when designing hot runner systems or other assemblies where mold parts have different temperatures. Many problems in molds are caused by poorly understood effects of heat expansion.

Calculation of heat expansion and the need for cold clearance must be done every time parts of different temperatures are present in one assembly. The biggest problem is usually to estimate what the temperatures are going to be.

Example: A mold part (e.g., a solid or tubular pillar of uniform cross section over its length) is locked between two plates. The cold distance L_c between the plates is 100 mm, the same as the length of the part L_p, and the expected temperature of the part T is 250 °C. The material selected has a tensile strength (at 500 °C) of 17,000 kg/cm^2 and a yield strength of 14,000 kg/cm^2. With a safety factor of 3, the permissible stress is limited to 4,667 kg/cm^2.

Using equation 14.1, the heat expansion e for a part of 100 mm length is

$$e = 0.0000114 \text{ cm/cm/°C} \times 100 \text{ mm} \times 250 \text{ °C} = 0.285 \text{ mm}.$$

If $L_p = L_c$, the stresses in the part would be

$$s = 0.0000114 \text{ cm/cm/°C} \times 250 \text{ °C} \times 2{,}000{,}000 \text{ kg/cm}^2 = 5{,}700 \text{ kg/cm}^2.$$

This is greater than the selected permissible stress (4,667 kg/cm^2). Therefore, the length of the part must be reduced to provide a cold clearance. If the length L_p is reduced by 60% of the calculated expansion of 0.285 mm to L_p1, the part would be

$$L_p1 = 100 \text{ mm} - (0.285 \text{ mm} \times 0.6) = 99.829 \text{ mm long}.$$

This should result in 40% of the stress calculated above, or 5,700 kg/cm^2 × 0.4 = 2,280 kg/cm^2, which would be acceptable.

The stress can also be calculated by equating equations 14.1 and 14.4, that is, e (the portion of heat expansion still active for compression of the part) = 0.285 mm × 0.4 = 0.114 mm, which is the equivalent of f, the amount by which the part is compressed.

$$e = f = \frac{L_p 1 \times s}{E}. \tag{14.6}$$

Now, solving for s:

$$s = \frac{f \times E}{L_p 1}, \text{ or} \tag{14.7}$$

$$\frac{0.114 \text{ mm} \times 2{,}000{,}000 \text{ kg/cm}^2}{99.829 \text{ mm}} = 2{,}280 \text{ kg/cm}^2.$$

Note that the dimension for the space between the plates (100 mm) and the length of the part $L_p 1$ (99.829 mm) are subject to manufacturing tolerances. The designer must calculate the worst case (i.e., when the cold clearance is at a minimum) to ensure that the stress is still acceptable, and also the opposite worst case (i.e., when the cold clearance is at the maximum) to ensure that there is still enough force to maintain a seal, such as may be required for a hot runner component.

14.3 Heat Gradient

If a mold part is between a hot (heated) and a cold (cooled) plate, the area where the part touches the hot surface will have the same temperature as the hot part, and where it touches the cold surface it will have the same temperature as the cold part. For simplicity of calculations, it is acceptable to ignore radiation losses and other factors and to assume that this part has a temperature T_{part}, which is the arithmetic mean of the temperatures of the hot and cold plates (T_{hot}, T_{cold}):

$$T_{part} = \tfrac{1}{2}(T_{hot} + T_{cold}). \tag{14.8}$$

An example is shown in Fig. 14.2. Equation 14.8 is a very simple approximation. However, with parts of rather complicated shapes and cross sections, such as hot runner

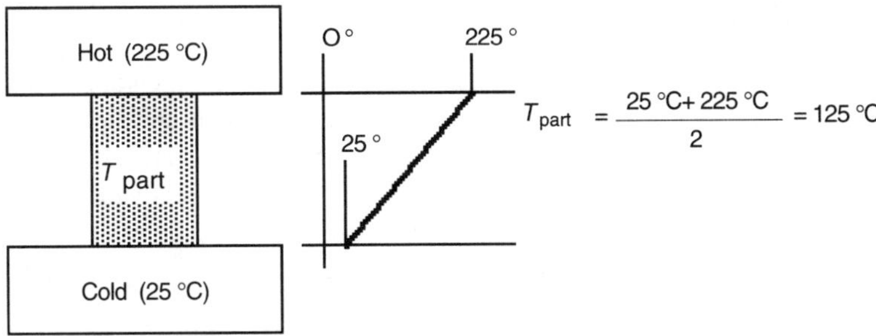

Figure 14.2 Mold part between a hot and cold plate, with diagram showing heat gradient.

housings, which have a number of different thicknesses, a more sophisticated approach may be required to determine the cold clearance.

14.4 Plates

Temperature differences between plates making up a mold assembly can be another source of problems. Ideally, all mold plates should have the same temperature to ensure that all alignment elements will be in the location where they are supposed to be.

A typical example will explain this statement.

Example: In a cavity plate with two leader pin bushings and a matching core plate with two leader pins, the leader pin centers are 400 mm apart. If, after the mold has reached operating temperature, the cavity plate is 20 °C warmer than the core plate, the bushings will be farther apart than the leader pins, by (using equation 14.1):

$$0.000011394 \text{ cm/cm/°C} \times 400 \text{ mm} \times 20 \text{ °C} = 0.091 \text{ mm}.$$

Considering that the regular clearance between bushings and leader pins is only about 0.025 mm (on diameter), the leader pins in our example will not enter freely into their bushings; the closing clamp will force them in and gradually wear the bushings out.

To be able to use these pins for proper alignment, the temperature difference between the plates should not be more than 5 °C. If the plates are larger, the effect of heat expansion will be proportionally larger.

To solve the above problem, consider the following choices:

1. All plates within a mold which support any alignment elements (leader pins, taper pins, dowels, etc.) are kept at the same temperature. This could be impossible to achieve.
2. Do not depend on widely spaced leader pins for alignment. Instead,
   a. Use a set of four interlocks, as described below in Fig. 14.3, or
   b. Align each cavity (or core) independently and mount them so that they float on the respective backing plates.

14.4.1 Interlocks

The interlocks in Fig. 14.3 are in fact "rectangular" leader pins, which are closely (slide-) fitting in one axis, and free to move in the the other axis. Both the male and the female interlock parts are held with close tolerances and tight fit in their respective plates. If the temperature varies between the plates, the hotter plate will increase in size (move outward from the center) without causing the interlocks to bind, as round leader pins would do under similar conditions. The central X and Y axes of the mold will remain in place, regardless of the temperature difference.

Typically, $M = F = 25$ mm (nominal), and $C = 1$ mm or more. The total clearance between F and M would be 0.045 mm (max) and 0.020 mm (min).

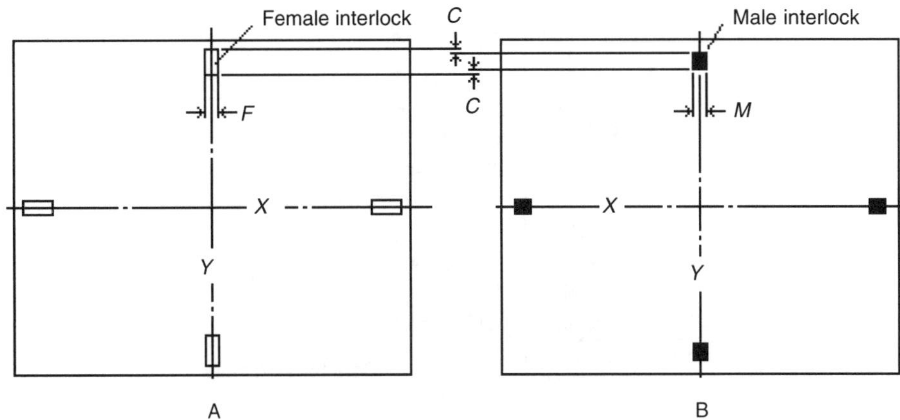

Figure 14.3 Typical interlocks mounted on cavity and core plates, respectively: A. female interlock and B. male interlock.

14.4.2 Individual Alignment of Cavity Stacks

The usual methods of alignment of individual stacks (modules) are:

- taper fits of cavity, core, and/or stripper ring;

Also, usually for small, technical cavity stacks:

- dowels or small leader pins, or
- wedges, or a combination of wedges and dowels or leader pins.

Alignment in the above cases depends on the accuracy of matching pairs of elements.

In taper locks or wedges, any temperature rise in the outer taper will reduce and possibly eliminate the intended preload and thereby lose the alignment feature of the taper; any temperature rise of the inner taper will increase the preload and may damage the mold or keep it from closing.

When using dowels or small leader pins, the distance between pin and bushing may change if one part is warmer than the other. In most molds these distances are small, but in all cases, the cooling channel layouts must be carefully considered to provide both cavity and core with adequate cooling to keep the temperature differences to a minimum, particularly if one side is in proximity to a hot runner.

In each of the above cases, the cores (rarely the cavities) are mounted so that they can be pushed by the closing mold into alignment by the use of floating mounting. This mounting is designed so that the core is held on its backing plate; thus, it can move sideways for a small distance but is restrained from lifting off the plate as the mold opens (Fig. 14.4).

In Fig. 14.4, the amount of float C = 0.1 mm (0.004 in.) per side. *LT* is locational transition fit. Usually, the cooling water supply enters the core from the backing plate; when using this method, if the core were to lift off the plate, water could leak out. The force holding the core must be sufficient that the core will not lift off the backing plate (because of the mold break force) as the mold opens. The limitation against floating too far sideways, usually not more

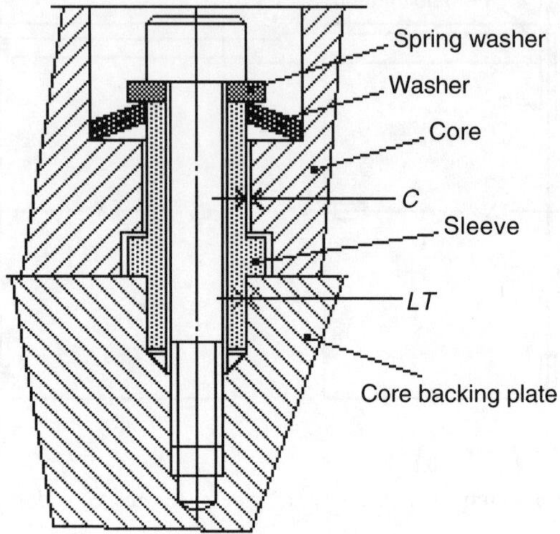

Figure 14.4 Floating mounting of core in backing plate.

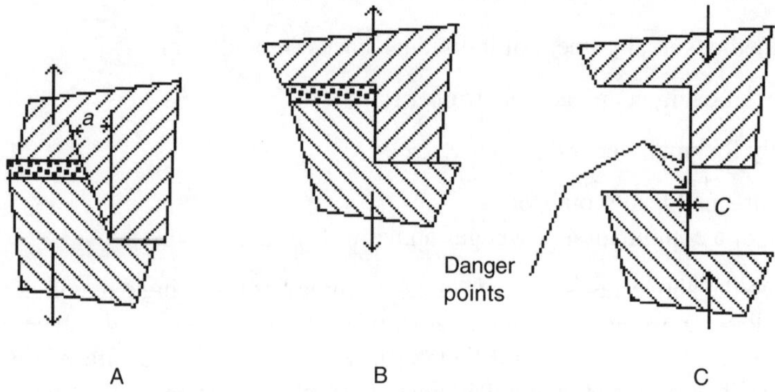

Figure 14.5 Shut-off in molds: A. tapered shut-off and B. vertical shut-off, and C. danger points in vertical shut-off are indicated by center arrows.

than 0.1 mm (0.004 in.) per side, is important to ensure that the alignment member can bring the core back into position without damaging the core or cavity.

14.4.2.1 Vertical Shut-Off

Vertical shut-off may be used in molds using floating mounting, and also where accuracy of match between core and cavity is important (even without a vertical shut-off). With a tapered shut-off, any alignment element is suitable; if a taper is used for alignment, the taper angle must be less than the shut-off angle *a* so that it acts before the "danger points" arrive at the same level (Fig. 14.5). If a vertical shut-off is unavoidable, it is imperative that the alignment elements act before the danger points come into contact. With leader pins, the leader pin clearance must be less than the clearance *C* of the shut-off between core and cavity with the closed mold, but in general, vertical shut-off should be avoided.

14.5 Hot Runner Manifolds

The hot runner technology today is such that, normally, the mold maker would not design a hot runner system but rather would buy all the components from an established manufacturer specializing in hot runner systems. (Practically the same thing is true in that nobody today would try to design their own hardware such as screws, valves, etc , but would buy them as standard parts available from the hard ware supplier.)

Hot runner suppliers also provide sophisticated engineering help to make sure that the planned mold will be equipped with the most suitable hot runner system. Since the design of hot runners is a very specialized field of the plastics engineering, a mold maker not familiar with and not equipped for the engineering and manufacture of the parts required in such systems should stay away from "home-made copies" and work together with the specialists in this field. The mold maker should concentrate on the other aspects of the mold, which are in his line of business.

We will here consider only one area which applies not only to hot runners but also to other components that may be subject to heat expansion, and where heat expansion must be calculated before designing a part. In many hot runner systems, the hot runner nozzles are centered in the cooled manifold plate, and the air pistons (for valve gating) in the cooled backing plate. The hot runner manifold is located in a pocket between these two plates and is held so that its center is in line with the machine nozzle.

When heated, the manifold will expand from this central point outward, toward the nozzle locations. To achieve acceptable alignment of the centerlines of the holes where the plastic exits from the manifold with the centerlines of the nozzle and the air actuators, the distances in the manifold (when cold) must be less than when at the planned operating temperature.

It is important to know before the designing stage:

1. Mold temperature (cooling) and
2. plastic temperature.

14.5.1 Mold Temperature

Some products or some molds will operate with chilled water at about 5 °C; with others, the mold may have to operate at elevated temperatures, maybe at 50 °C or even more. Since the interest is in the ΔT between cooled plates (and cavities) and the heated hot runner manifold, such a difference is important.

14.5.2 Plastic Temperature

This is probably the most difficult of the assumptions to be made by the designer. This temperature may be different even for "identical" materials from different suppliers.The molder may want to use different plastics in the same mold. The design characteristics of the hot runner system may affect the injection temperature. The production rate may affect this temperature. The product size may require a different temperature because of shrinkage factors.

These possible conditions and resulting temperature differences may result in a mismatch of the above cited centerlines; it may cause hang-ups of plastic and slow degradation, which will eventually contaminate the products. With some materials (PE, PP), this may not be of much concern, but with heat-sensitive materials, such as PET and other engineering materials, this can be serious, and it is important that the manifold designer has as much data as possible before designing the manifold.

14.6 Injection Blow Molds

Injection blow molding (IBM) consists of making preforms in one stage of the mold and transferring them (on the core) to a second stage of the mold, where they are blown into the final shape. In this system, the cavities and the cores are heated where the plastic remains hot for blowing, but the portion of the core producing the rim of the preform is chilled to produce a cold neck before transferring the preform to the blow station. The injection side is relatively hot, and the core side is supplied both with hot fluid and with cold water.

Heating media may be hot water for temperatures below 100 °C or hot oil for higher temperatures. This is mentioned here only to point out that while in ordinary molds, the endeavor is to keep the plates at the same temperature, usually using the water supply to the cavities as cooling medium for the plates, in IBM it is unavoidable to have different coolant temperatures in the plates of the stationary and moving mold halves.

The main problem for the designer is then to keep plates and parts of different temperatures well insulated from each other and to calculate the various expansions so that in the running mold, all alignment points are in their correct location.

15 Stack Molds

15.1 Theory of Stack Molds

An understanding of the stack mold technology will greatly facilitate the design of stack molds. In theory, the clamp force F equals the sum of the reaction forces R in the tie bars; this creates in any mold the necessary locking force. If there is more than one set of cavity and core plates (Fig. 15.1 shows two), the sum of the forces p in the left core block equals F, the sum of the forces p in the right core block equals the sum of the reaction forces R; the forces p within the cavity block to the left and to the right balance each other. The effective molding area is doubled, and the rated machine clamp force is available for each level (stack) of cavities and cores.

There could be any number of stacks in a molding machine. In some rubber molds, there are as many as ten stacks.)

15.2 Definitions and Nomenclature

For the purpose of injection molding, a stack mold is usually a mold assembly consisting of only two single-face molds, mounted back to back. This book will discuss only hot runner stack molds, although this technology originally started with cold runner molds; however, they were not economical for high-speed operation.

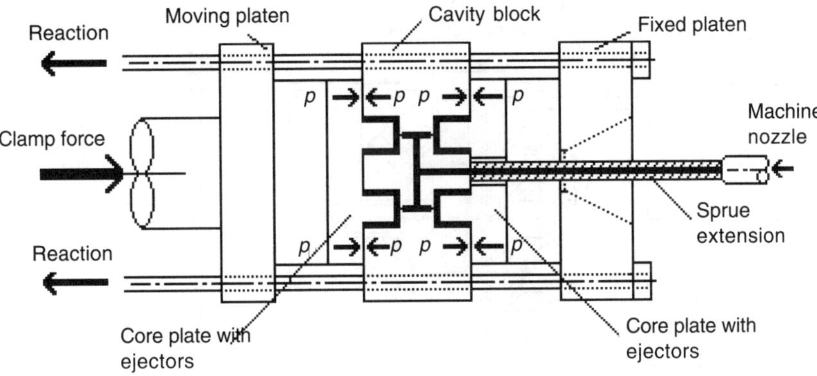

Figure 15.1 Schematic drawing of a typical stack mold layout showing clamp, reaction, and locking forces.

15.3 Arrangement

One of the two core sections is mounted on the moving platen (as in a single-face mold), the other is mounted on the stationary platen (Fig. 15.2). Note that "core sections" mean the portions of the mold opposite the section where the gates (cavities) are located.

The cavities (with the gates) are mounted back to back on the center section or "floating" section of the mold, sandwiched between the two core sections. The floating section is usually supported on the lower tie bars.

Plastic enters the mold from the machine nozzle through an extended sprue bushing, the *sprue bar*, which is solidly mounted in the hot runner manifold. The sprue bar passes through that core section, which is mounted on the stationary platen. The heated sprue bar is in contact with the machine nozzle only when the mold is closed.

The sprue bar is guided in the area of the mold locating ring. During operation, the sprue bar must never pull out of the guide. This could be caused by excessive clamp opening stroke.

15.4 Stroke

The mold opening and closing stroke of the center section is exactly one-half that of the moving platen. Synchronization of the travel is achieved by a rack and pinion arrangement, as shown in Fig. 15.3, or by some balancing lever mechanism shown in Section 15.9.7.4.

As the mold opens, the lower rack moves to the left at the speed of the clamp. The upper rack is held without moving on the stationary platen. This causes the gear to turn and makes the center section move to the left at half the speed of the clamp motion.

There must be two sets of racks, one in front and one in the rear of the mold. The position of the racks in the rear are inverted to that of the front so that the left rack in the rear is at the top, and the right rack at the stationary side is at the bottom.

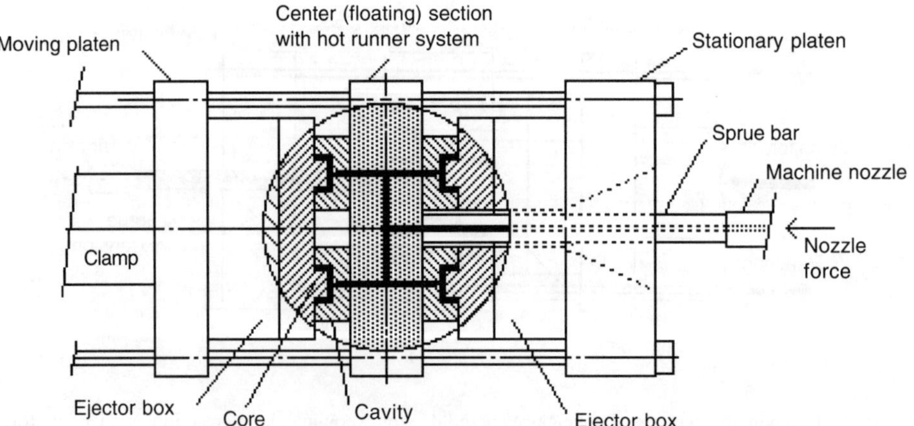

Figure 15.2 Cross section of typical stack mold for injection molding.

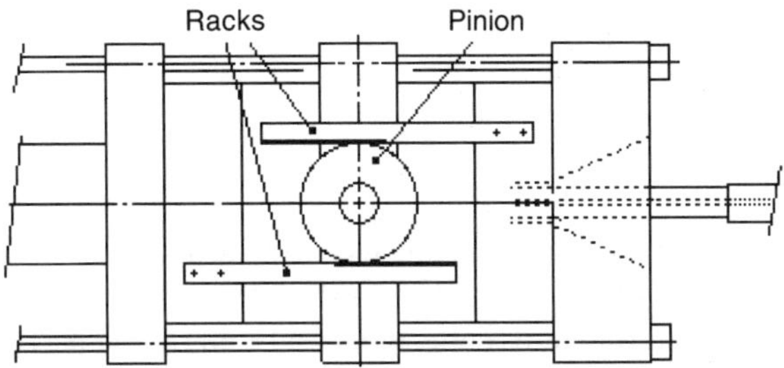

Figure 15.3 Rack and pinion system used to control platen movement during mold opening and closing.

There are variations to this arrangement; for example, to improve accessibility to the molding area, a set of idler gears are introduced between the pinion and the racks to keep the racks near the tie bars, out of the way of operators. In another variation, in combination with universal mold supports, the racks are mounted on the machine rather than on the mold.

15.5 Support

Adequate support of the center section (with the cavities and the hot runner), which can easily have a mass of 2,000 kg (4,400 lb.) or more, is very important to ensure proper alignment with the core sections. Any deflection of the supports may seriously affect the life of the mold alignment features (leader pins and bushings and/or taper locks). It may cause damage to the cavities and cores, particularly with thin-walled containers, or even with heavier walled products with deep draw. An example will best illustrate the need for a stiff (rigid) molding machine.

For the purpose of illustration only, assume that the four tie bars of a hypothetical machine are supported on the base at the clamp housing and at the stationary platen. The distance between the supports is in one case $L = 72$ in., in the other case $L = 96$ in. (Fig. 15.4) f_1 is the deflection caused by the weight of the tie bars, and f_2 is the deflection caused by the weight of the moving platen plus the moving mold half at about half way between the support. This may approximately correspond to the position of the moving platen before the mold halves meet.

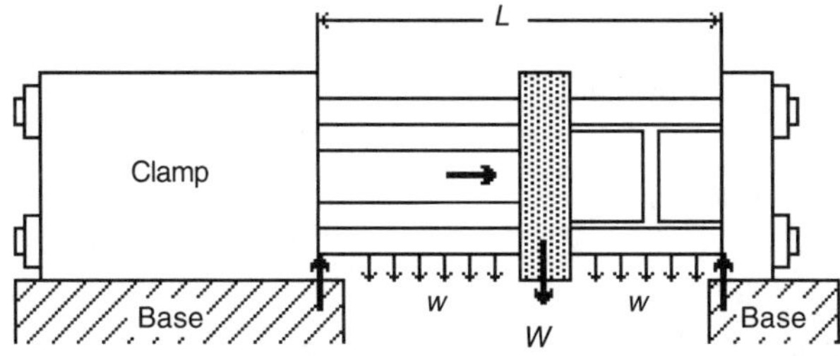

Figure 15.4 Tie bars supported on the base at the clamp housing and at the stationary platen.

Deflection caused by tie bar weight is

$$f_1 = 0.208 \times \frac{w \times L^4}{E \times d^2} \text{ inch,} \tag{15.1}$$

where w is the weight of the material per cubic inch (for steel, ≈ 0.3 lb/in³), E is the modulus of elasticity (for steel, ≈30,000,000 psi), and d is the tie bar diameter. The value 0.208 is a formula constant from the formula for deflection of a beam (round bar).

Deflection caused by the weight of the moving platen and mold half is:

$$f_2 = \frac{W \times L^3}{48 \times E \times I} = \frac{W \times L^3}{2.35 \times E \times d^4} \text{ inch.} \tag{15.2}$$

where W is the weight of the mold and I is the moment of inertia for a round bar. Table 15.1 shows calculated values for 3-, 4-, and 5-inch diameter tie bars for the hypothetical cases of $L = 72$ in. and $L = 96$ in. For simplicity, it is assumed that the moving platen plus mold half weighs 2,000 lb. Each tie bar is therefore loaded with only 500 lb. The total tie bar deflection is the sum of the values f_1 and f_2.

As can be seen from the values in Table 15.1, the amount of deflection and the resulting misalignment when core half meets cavity half can be substantial. It is important for the designer to be aware of these calculations and to know the approximate weights involved. The best mold design cannot work if the machine deflects excessively. Larger diameter tie bars deflect less, but there may be still too much deflection for good alignment. What usually happens is that the leader pins or the taper locks must lift the incoming mold half into alignment. This causes wear, and with long cylindrical products or products with little draft, there may also be damage caused to the molding surfaces when the core scrapes over the cavity wall. With thin-walled products the alignment may be too late to prevent collision between core and cavity.

Some molding machines support the moving platen directly on the machine base to minimize this deflection by sliding it on top of supporting ways (Fig. 15.5). Note that both the upper and lower tie bars are supported indirectly by the moving platen.

Table 15.1 Calculated Values for Tie Bar Diameters Given $L = 72$ in. and $L = 96$ in.

	Tie bar diameter d (in.)		
Tie bar deflection	3	4	5
f_1 (from their own weight)			
L = 72 in.	0.006	0.0035	0.002
L = 96 in.	0.019	0.011	0.007
$f2$ (from 500 lb. load)			
L = 72 in.	0.033	0.010	0.004
L = 96 in.	0.077	0.024	0.010
$f1 + f2$ (total)			
L = 72 in.	0.039	0.0135	0.006
L = 96 in.	0.096	0.036	0.017

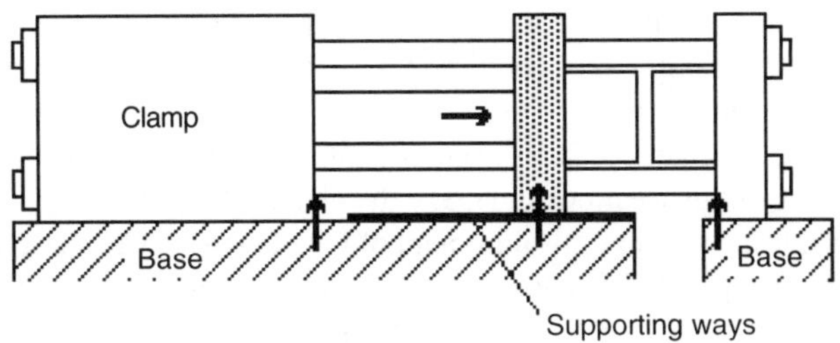

Figure 15.5 Moving platen is supported directly on the machine base and slides on supporting ways.

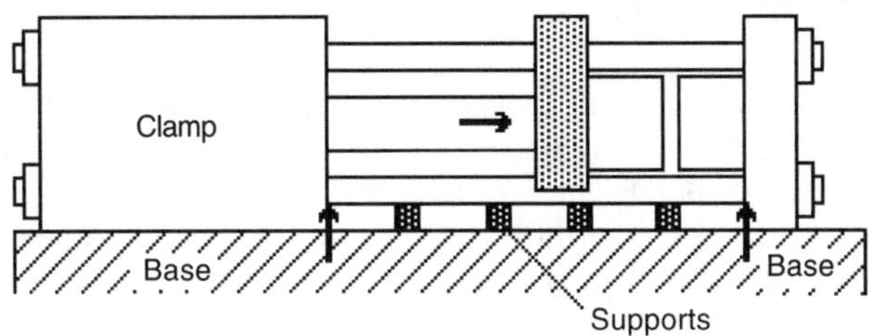

Figure 15.6 Supports for lower tie bars enable moving platen to slide on them as supporting ways.

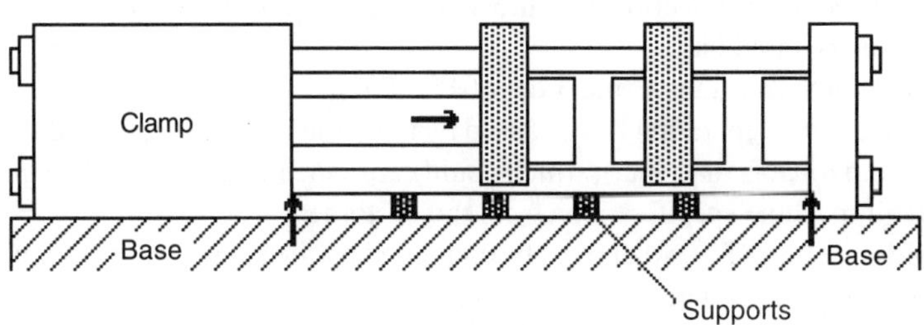

Figure 15.7 Supported lower tie bars act as slide supports for the moving platen and center mold section.

In some machines, the lower tie bars themselves are supported by a number of supports. The moving platen slides on top of the lower tie bars, rather than passing through the platen. In fact, in this case, the lower tie bars are supporting ways. The upper tie bars pass through the platen (Fig. 15.6).

By this method of support, the alignment and engagement of moving and stationary mold halves is considerably improved as compared to a nonsupported moving platen. This ensures that the mating mold parts will meet properly and will not wear prematurely.

Stack molds are even more susceptible to misalignment caused by tie bar deflection because the distance between the stationary and moving platens is much greater than for a comparable single level mold for the same product. By supporting the lower tie bars (Fig. 15.7), there is virtually no deflection of tie bars. The moving platen and the center section mold supports slide on the tie bars.

The tie bars of a molding machine should only guide the platens and provide (by stretching) the required clamp force. They should not be subjected to deflection as a result of supporting the platens or the molds.

Molds producing products of high accuracy should never depend on the tie bars for alignment. Tie bar bushings wear out and make the platens "sink down" with respect to their theoretical center lines, and cause misalignment of the platens. When designing stack molds for machines with nonsupported tie bars, it is important to provide the lower tie bars with some solid supports to avoid excessive deflection of the tie bars.

15.6 Clamping Force

The force required to prevent flashing for each level of the stack mold is the same as the force for a comparable single-face mold. Since both levels are back to back, the (compressive) forces within the center section balance each other.

In single-face molds, the nozzle force is taken up by the sprue bushing, which is solidly mounted in the stationary mold half and fastened to the stationary platen. Therefore, the nozzle force does not affect the clamping force.

In stack molds, the total clamping force is about the same as that for a single level mold with the same projected molding area; however, the nozzle force acts on the sprue bar, which is mounted on the center section. This force is then transmitted to the moving platen and adds to the force created by the injection pressure. It tends to separate the center section of the mold from the stationary mold half.

The nozzle force is usually in the order of 15–20 tons. This additional force may become significant if the mold separating force caused by injection is close to the rated clamp tonnage (when the clamp force is marginal), which could cause the mold to flash. The mold designer must obtain all machine specifications from the customer.

The mold separating force that must be contained by the clamp force, to prevent flashing at the parting line, can be calculated as the total of the larger projected total cavity area in one of the two levels, multiplied by the estimated injection pressure in the cavity. The nozzle force must then be added.

The projected molding areas may not be equal. If the larger projected area is on the clamp side, and the force generated by the area difference is greater than the nozzle force, the nozzle force can be ignored. This is important in family molds, where the products molded in one face are different from the products in the other face (e.g., in Petri dish molds where one face molds the base and the other face molds the [larger] cover).

The injection unit force is usually up to 10 tonnes for small machines and up to 20 tonnes for larger ones. As a rule of thumb, it is safe to assume that the clamping force required for a stack mold is about 10% greater than the clamping force for a single-face mold for the same projected area.

If there is no larger machine (with more tonnage) available, this can be overcome with either of two methods:

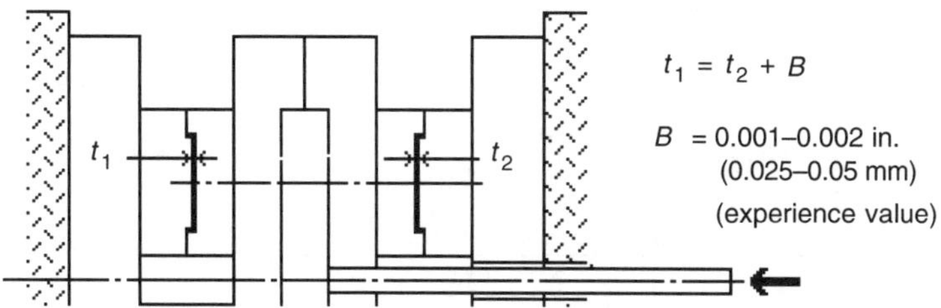

Figure 15.8 Differential cavity space at the bottom of the product.

1. Differential hot runner nozzle tip hole diameters. The tips near the clamp side have a larger hole (e.g.,0.060 in.), and the holes near the injection side are smaller (e.g., 0.045 in.). This usually can be done easily after the mold is finished.
2. Differential cavity space at bottom of product. Slightly increase the bottom space t of the cavities in the face near the clamp side to ensure that the products on both levels have the same bottom thickness (Fig. 15.8). Even though the cavity spaces are different, the products will have the same bottom thickness on both levels.

The possible need for either differential cavity space or differential hot runner tips, or none of the above, must be carefully considered before being incorporated into a design, since they require different mold components, affecting interchangeability.

15.7 Asymmetrical Layouts

Most stack molds are laid out symmetrically (i.e., the center lines of the cavities in both levels are in line). However, with certain valve gated molds, because of the valve actuation mechanism, it may be necessary to offset the location of center lines (Fig. 15.9)

While the forces are not in line, the total of forces is still the same. The effect will be a slight off-center loading of the platens and a bending moment $M = F \times A$ within the hot runner manifold. As the offset is usually small (in the order of less than 50 mm), the bending moment can be ignored. However, good support between the plates through the hot runner manifold is important to minimize the effect of this moment.

As an alternative, it is possible to avoid off-center loading of the platens by arranging the cavities symmetrically to the centerline of the machine (Fig. 15.10). The platens are loaded on center, but the cavity and mold layout may become more difficult. This method is suggested if the offset A is large.

15.8 Other Important Machine and Mold Features

15.8.1 Shot Volume

The shot volume must be twice that required for the cavities in a single-face mold for the same product, with the same number of cavities per face.

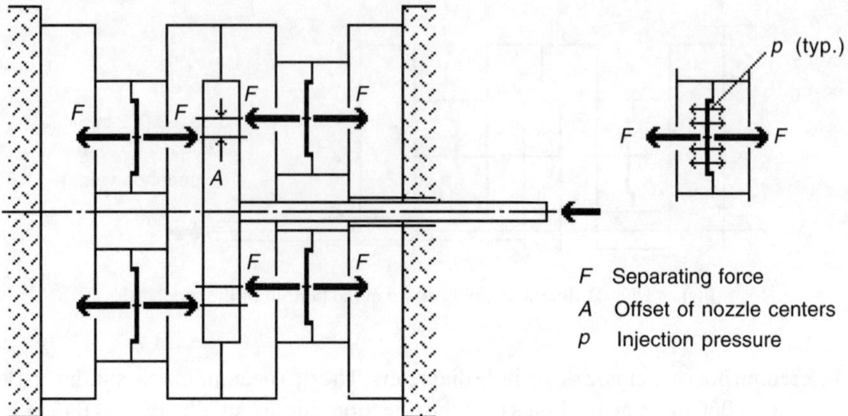

Figure 15.9 Offset centerlines in a stack mold with valved gates.

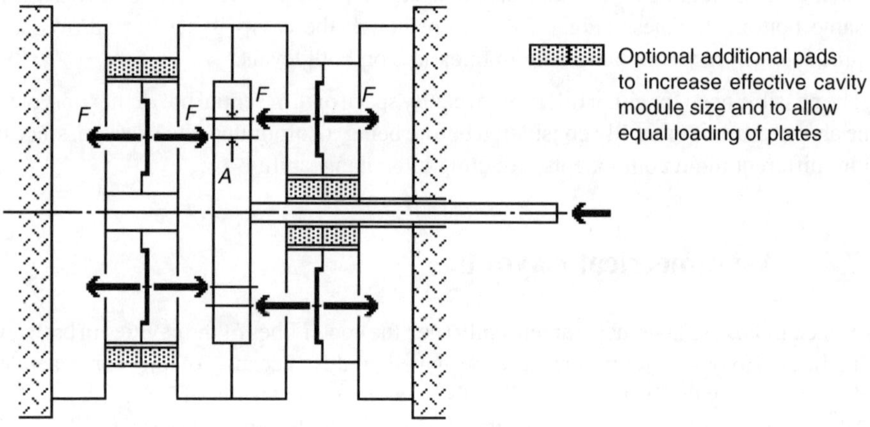

Figure 15.10 Offset cavities arranged symmetrically around the machine centerline.

15.8.2 Recovery Rate

The plasticizing unit must be of sufficient capacity to recover fast enough to be ready for the next shot, even though twice the volume is required of a comparable single-face mold. Otherwise, the molding cycle must be slowed down to allow time for recovery, and the full benefit of running twice the number of cavities in the same machine will be lost.

15.8.3 Injection Rate

Injection rate is an important consideration. To fill the same shot volume (to fill all cavities with optimum speed and pressure) in the same time as a single-face mold, requires twice the injection speed at the machine nozzle. Machines need a sufficiently large oil capacity (pump and accumulators) in their injection circuit to achieve this.

If the injection rate is too small, it may not be possible to successfully mold certain thin-walled products, because the plastic will freeze before filling the cavities, or it may require molding with higher stock and cooling temperatures, which will slow down the molding cycle. Increasing injection pressure may help to mold a thin-walled product, but such an increase will require higher clamp forces to keep the mold from flashing, and this may require placing the mold in a larger machine.

15.8.4 Repeatability of Shot Size and Product Quality

To ensure these two requirements, the required recovery rate should be not more than 90% of the rated capacity of the extruder for the material molded. (Note that this applies to all molds, not only stack molds).

15.8.5 Sprue Break

Sprue break, or "jump back", of the molding machine pulls the sprue away from its seat on the sprue bar after injection before the mold opens. This motion is required to protect the sprue bar from colliding with the machine nozzle when the mold closes, and to allow decompression. After the mold is clamped up for the next cycle, the injection unit moves forward by the amount of the sprue break to contact the sprue bar and to seal the plastic against leaking at the nozzle seat.

15.8.6 Decompression of the Hot Plastic Volume

The large channels in the hot runner manifold plus the long sprue bar are filled with plastic, which is compressed during injection. (Plastics are compressible fluids, and under the usual injection pressures may be compressed by 1–2 percent of their volume.) At the moment of sprue break, when the machine nozzle pulls away from the sprue bar, the pressure on the plastic within the hot runner system is relieved, and the plastic will expand by the amount it was compressed. This excess volume will come out of the hot runner system in two ways:

1. If the mold has open gates, some of this plastic may drool out into the cavity while the mold is open. This affects the product quality. The rest of the relieved plastic would drool out at the nozzle seat and waste plastic.
2. If the mold has valve gates, the gates are closed while the mold is open, and the compressed plastic would drool only at the nozzle seat.

To avoid drooling, molds can be equipped with "anti-drool bushings" at the end of the sprue bar. As the injection unit retracts (sprue break), the bushing is forced back to follow the machine nozzle (for a limited distance) by the pressure of the plastic within the hot runner system. As the bushing moves in the sprue bar, additional volume is created in the sprue bar which can contain much or all of the expanding plastic. However, there may still be a small amount of plastic that will escape through the plastic passage opening in the bushing and drool.

"Suck-back" is a machine function that assists in decompressing the system. The extruder screw (or the injection piston in a two-stage injection system) is pulled back before the sprue bar and the nozzle separate, before the machine nozzle is pulled back. This method acts similar to the anti-drool bushing. It removes the pressure from the plastic in the machine nozzle and in the extruder barrel and increases the volume available for the plastic to expand back into the injection unit.

15.8.7 Injection with Open Nozzles

The mold must be closed until the screw has recovered. The screw is now pulled back (decompressed), and only then can the mold open. This is quite satisfactory when molding with a long cooling time after injection hold time is ended. The extruder must be stopped while the mold is open.

15.8.8 Injection with Shut-Off Nozzles

In this case, the injection unit can be pulled back (sprue break) much sooner. After injection hold times out, the shut-off nozzle will be kept open for a additional short time to allow decompression of the plastic. After this short time, the nozzle can be shut. The sprue break moves the injection unit back and recovery can commence. The extruder can operate while the mold is open and until the mold is closed again, ready for the next shot. This method greatly extends the time available for the screw to recover, and is indispensable for molding with short cycles.

Not all machines are equipped for all of these sequences. If not, they may need to be modified to suit, or they will lose productivity or quality of products.

15.8.9 Day Light, Shut Height, and Clamp Stroke

Figure 15.11 illustrates the relationships between the thicknesses of mold parts and required clearances in the mold, and the clamp stroke and shut height. The variables are defined adjacent to the schematic; *D* is the day light of the mold.

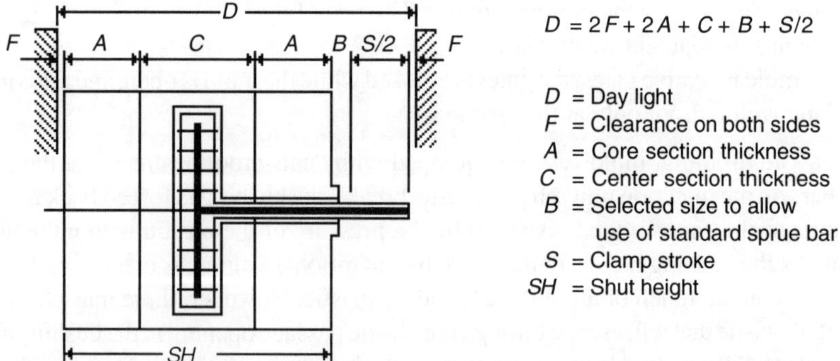

Figure 15.11 Schematic of mold closed position of platens shows sum of thicknesses and clearances that make up day light.

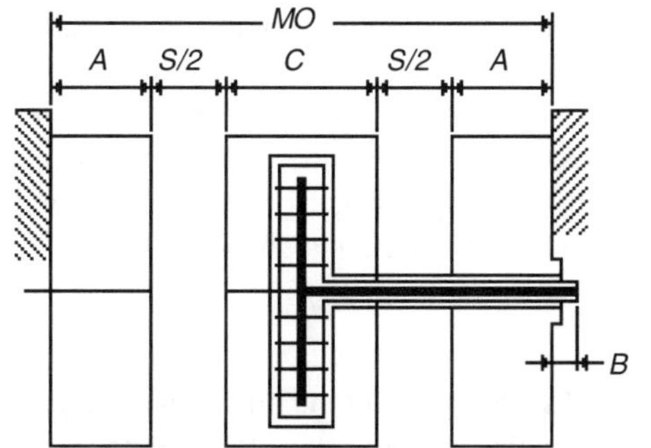

$MO = 2A + C + S$

To allow easy installation of mold, *MO* must be greater than *D*

$MO - D = S/2 - (B + 2F) > 0$

or $(B + 2F) < S/2$

F should be at least 10 mm

B maximum size allowable for selecting standard sprue bar

Figure 15.12 Schematic shows mold open position of the platens with MO > D (see Fig. 15.11).

15.8.9.1 Day Light

Day light is the maximum spacing between the moving and stationary platen. It is often large enough to permit the assembly of the whole mold, but occasionally, the mold is too large and must be mounted in sections.

Day light of the machine must be slightly larger (by about 20–25 mm) than MO (see Fig. 15.12). This will ensure that under normal cycling of the mold, the mold open position will always be reached *before* the maximum daylight position of the clamp. Dimension B is selected by the mold designer to permit use of standard size sprue bars, where such standards exist.

15.8.9.2 Stroke Limiters

The molding machine should be equipped with mechanical stroke limiters to prevent inadvertent, excessive mold opening.

15.8.9.3 Shut Height

Shut height is the closed (shut) mold height *SH*, excluding the height ($B + S/2$) shown in Fig. 15.11 ($SH = 2A + C$).

15.8.9.4 Clamp Stroke

The total stroke of the mold equals a minimum of twice the stroke required to permit ejection of the products for a single-face mold for that product. Excessive stroke may damage the mold. Insufficient stroke may not permit proper ejection or access for controlled product removal with robots, etc.

15.8.10 Injection Carriage Travel

With this (preferred) design of stack molds, the point where the machine nozzle contacts the sprue bar is much farther back from the face of the stationary platen than with single-level molds. This requires that the carriage must be able to move back much more than may have been provided for by the molding machine manufacturer.

It is important to ensure that there is enough travel and support available for the extruder. If necessary, additional rails must be provided to allow the injection unit to move back sufficiently far. In some cases, it may even be necessary to add extra supports under the extruder, or to lengthen the machine frame to ensure that the extruder will not topple off the machine when moved to its extreme rear position during servicing. The location and the plumbing of hydraulic actuator(s), etc., may also have to be changed to suit.

15.8.10.1 Injection Carriage Forward Stop

It is important that the machine has a mechanical stop to prevent the injection carriage from moving too far in the direction of the mold. In the absence of such a stop, the nozzle could collide catastrophically with the sprue bar as the mold closes.

15.8.11 Purge Guard

In most machines, the guarded area is only close to the rear of the stationary platen. Because of the extended length of the sprue bar, the new point of purging may be only marginally within, or even outside, the existing purge guard. The purge guard must then be extended to cover the danger area properly (i.e., to prevent plastic from spraying outside the guard during purging).

15.8.12 Ejection

There is little difference between ejection for a stack mold and a single-face mold, except that the stack mold has two sets of cores, and, therefore, two sets of ejection mechanisms. With mechanical ejection, this requires some actuation method of the ejectors in the core section on the stationary platen, which is usually not equipped with machine ejectors. Machine ejectors on the moving platen side can be used, and integrated air or hydraulic actuators are required on the stationary platen side section. Connections to the control circuits of the machine must be shown on the assembly drawing. In some molds, the motion of the ejector plates can be combined (with mechanical levers and cams) with the actuation of the center section (see Section 15.9.7.4).

15.8.13 Center Section Supports

There are four different methods for center section supports used:

1. The mold plates themselves are supported by the tie bars. This requires pulling of tie bars and is generally not recommended, although many molds have been built this way.
2. Vertical mold supports. The center section is supported by two separate vertical beams held accurately between the lower and upper tie bars, one on the operator's side and one in the rear. These supports are part of the mold assembly but are set into the machine before the mold is lowered into position. The mold is then fastened to these supports. It is used mainly when the mold can pass between the upper tie bars.
3. Horizontal mold support. The center section is supported by one horizontal beam resting on the two lower tie bars. On the top of the mold, two sliding shoes align with the upper

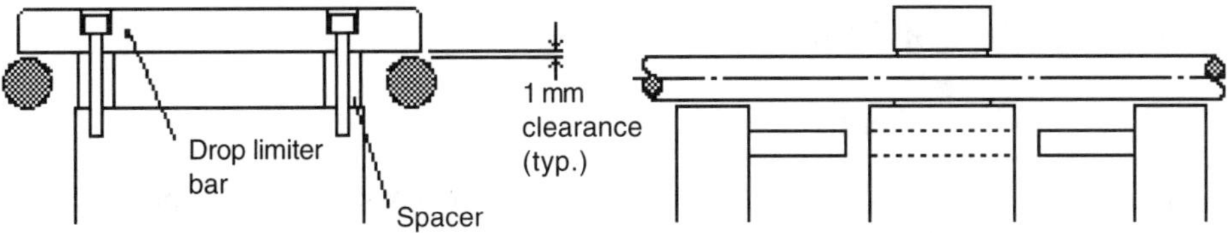

Figure 15.13 Drop limiters mounted on top of the plates prevent center section from "falling" between the tie bars.

two tie bars. This support is part of the mold assembly but is placed into the machine *before* the mold is slid into position. The mold is then fastened to this support. It is used mainly when the mold cannot pass between the upper tie bars but must be installed from the side of the machine.

4. Universal mold support. This is a patented Husky™ feature. Functionally, it is similar to the vertical mold support, except that it is not part of the mold assembly but is a machine accessory, complete with the actuation system (racks and pinions or levers) for moving the center section. It can be used for other stack molds designed to be used with this support, thus saving on mold cost.

15.8.14 Drop Limiters

Each center section should be equipped with a pair of "wings" mounted on top of the plates. These wings do not touch the tie bars but are close to them (within 1 mm or 0.040 in.). Their purpose is to prevent an inadvertent fall of the center section between the tie bars, in case of operator failure to adhere to the assembly instruction or if, for example, the hoist supporting the center section is lowered too far before all mounting screws are in place and tightened. A typical method is shown in Fig. 15.13.

Blocks and their mounting screws must be strong enough to carry the mass of the entire mold, not just the center section. This may be required during mold setup, when the whole mass of the mold may come to rest on these blocks while the hoist cable goes slack.

15.8.15 Mold Break Force

The mold break force is the force required to separate cavity and core after the products have cooled and the mold is ready to open. Unlike the clamp force (compressive), which is the same for a single-face and an equivalent projected area stack mold, the (tensile) mold break forces are additive; that is, a stack mold requires twice the mold break force of an equivalent single-face mold. This may be important with molds for products with deep draw and little draft angle, or for products that have undercuts both in cavity and core, thus requiring more than the usual mold break force. This force is not the same as, and must not be confused with, the ejection force required to eject the product from the core or out of the cavity. Some molding machines may have insufficient mold opening force, and additional actuators (hydraulic cylinders) may be required at the mold or on the machine to assist in opening the clamp.

15.9 Rules for Stack Mold Design

15.9.1 Cavity Layout

The problem with the sprue bar method is that the bar cannot pass through the product on the injection side face, unless the product has a horseshoe shape (open toilet seat, etc.), with the open end such that the product can fall freely (Fig. 15.14 left), or opening on that side that permits a robot to withdraw the product without touching the sprue bar (Fig. 15.14 right).

For molds unloaded by robots, mold testing without the robot engaged could result in the product hanging up on, or at least touching, the hot sprue bar when clearing the mold. In exceptional cases, this may be acceptable for mold testing, but it should be discussed before starting design. As a rule, it is not acceptable.

Generally, for symmetry, there will be at least two cavities on the sprue bar side in layouts as schematically illustrated in Fig. 15.15 for two to twelve cavities. As a rule, the horizontal layout for six cavities is not recommended.

Figure 15.14 Horseshoe-shaped product: Opening on top allows free fall (left), opening on side allows horizontal take-off (right).

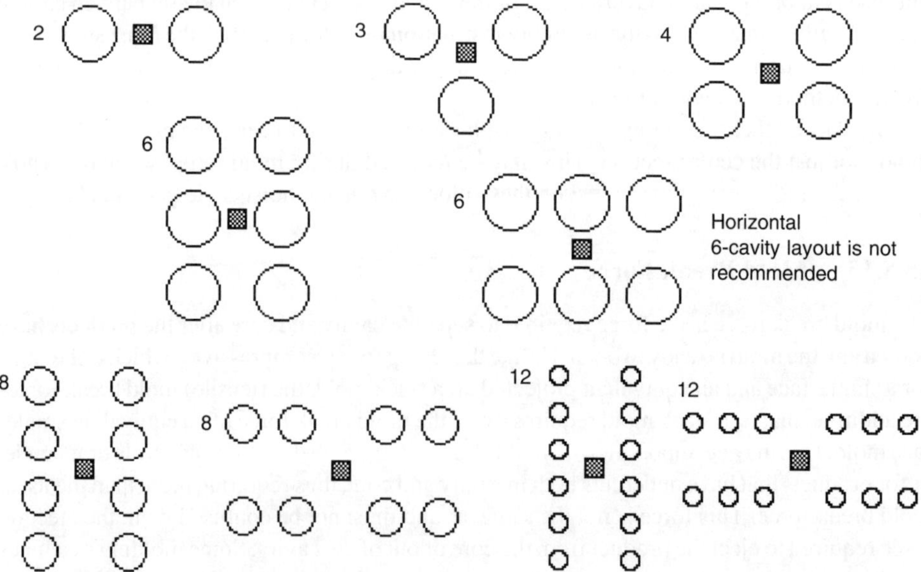

Figure 15.15 Schematic layouts showing cavity arrangements around sprue bars for two to twelve cavities.

Note that the horizontal layout for larger numbers (from eight up) is smaller in a vertical layout, but the mold will run faster with a horizontal layout because the time for the topmost products to fall clear from the mold is shorter.

For very large products permitting only one single cavity on the injection side, the sprue bar may be offset in the mold, or the whole injection unit must be offset so that the sprue bar enters outside the cavity. (It is always acceptable to have only one cavity in the center of the clamp side of the mold.) These rare, special cases will not be further described.

15.9.2 Length of Sprue Bar

The sprue bar must not be too long, to ensure that with a closed mold, the machine nozzle does not project too far toward the injection unit. In extreme cases, the injection unit cannot move far enough away from the mold, and may require modifications to the machine to enable the mold to be installed and run. This must be established at the time the mold is designed.

The sprue bar must not be too short, to ensure that with the mold open, the seat of the anti-drool bushing is still in the tapered section of the stationary platen so that any drooling from the sprue will run down, away from the nozzle seat. The tapered section in the stationary platen is different in various machines; in some machines, there is no taper at all. During operation, the sprue bar must never leave the sprue bar guide.

15.9.3 Heating of Sprue Bar

The sprue bar requires very little heat for maintaining the temperature of the plastic during operation of the mold, but requires high heat for start-up of a cold mold to melt the plastic within the sprue bar. The sprue bars should have two heaters, one on each side of the plastic channel. This is necessary because heating of one side only results in bending of the sprue bar because of heat expansion of the steel.

15.9.4 Protection from Accidental Contact with Sprue Bar

A portion of the sprue bar (on the injection side) is exposed when the mold is open. To protect operators from contacting the very hot sprue bar, it must be shielded. A heat shield also prevents products falling from the upper cavities in the injection side from touching the hot sprue bar.

There are specific regulations covering these heat shields:

- Europe (EN 201-1985): Temperature of heat shield must be less than 80 °C.
- USA (ANSI B.151-1, 1984): Heat shield must prevent injury.

15.9.5 Guiding the Sprue Bar

Because of the considerable length of the sprue bar, which is solidly mounted only on the manifold, it must be guided close to the point where it leaves the mold (Fig. 15.16).

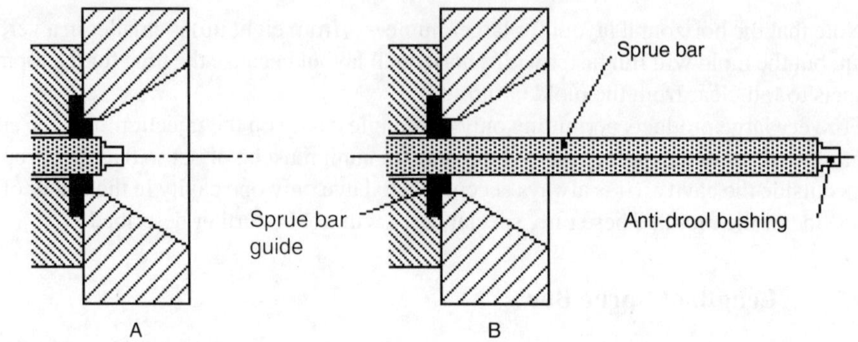

Figure 15.16 Sprue bar guide: A. mold open and B. mold closed position.

15.9.6 Clamp Stroke

The required clamp stroke is established similar to a single-face mold, except that the total clamp stroke is twice the amount required for one face. The stroke in both levels must be exactly the same, even if it is not required for ejection purposes of one of the products. For example, if a box in one face is 40 mm high and the lid for this box, in the other face, is only 20 mm high, then the stroke for each face must suit the larger one required for the box. It would be possible to have different strokes (e.g., by using a rack and pinion arrangement with different gear ratios, or by using one of the other methods described in Section 15.9.7, Actuation of Ejection Mechanisms, using different ratios of levers.)

15.9.7 Actuation of Ejection Mechanisms

This applies to all ejection methods except for molds with air ejection only. Where possible, air ejection should be used to avoid the need for any mechanical ejection method.

Unlike compression molding machines, which come equipped with ejection actuation from both sides, injection molding machines have ejectors only on the moving platen side.

Therefore, stack molds require for the cores mounted on the stationary mold half some sort of ejection mechanism on the stationary platen to activate a stripper plate or an ejector plate. There are several possible methods for this actuation.

15.9.7.1 Chains or Pull Rods

Chains or pull rods may be fastened to the moving platen and connected to the ejector or stripper plate on the stationary platen with lost motion so that this plate is pulled only after the mold has opened sufficiently to permit ejection from the stationary side. This is a cheap but very crude method and is seldom used. It makes the mold even more inaccessible than it already is, considering the other mechanisms attached to the mold.

15.9.7.2 Ejection System Added to Stationary Platen

Ejection systems can be added to a stationary platen by mounting commercial actuating cylinders (air or oil) on two opposing edges of the stationary platen and connecting them to the stripper or ejector plate, or by placing (usually four specially designed) air cylinders inside the platen in a suitable location so that the piston rods act similarly to the machine ejectors on the moving platen. This would be of advantage if the machine is used regularly for different stack molds which do not have actuators integral with the mold.

Cylinders operated with air are much weaker than those operated with oil and must be much larger. Also, it is difficult, if not impossible, to achieve a smooth, even motion with air. Oil cylinders are generally preferred.

15.9.7.3 Actuating Cylinders in Mold Plate

This is a common method used as an alternative to linkages. The cylinders are integral with the mold backing plate on the stationary mold half.

15.9.7.4 Ejection Linked with the Mold Movement

There are numerous methods of ejection linkages used. All are based on the utilization of the machine motion to actuate the ejectors.

Figures 15.17–20 show schematically a few linkage methods. There is a wide range of different executions of the linkages. The drive links are sometimes provided with lost motion elements (links with slots and/or springs) to delay the ejection as required.

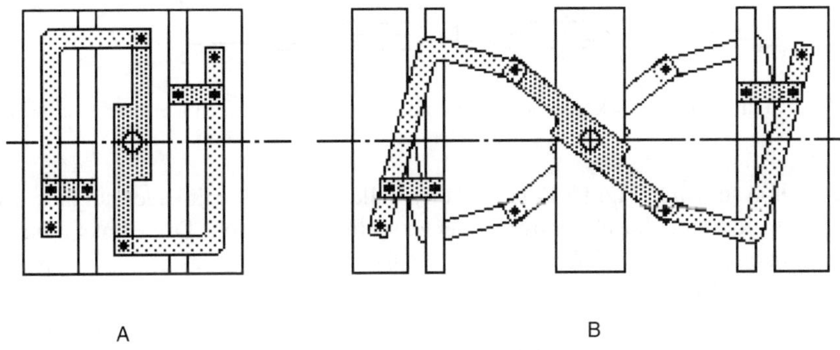

Figure 15.17 Ejection linkages for a small mold also move center section: A. mold closed, and B. mold open.

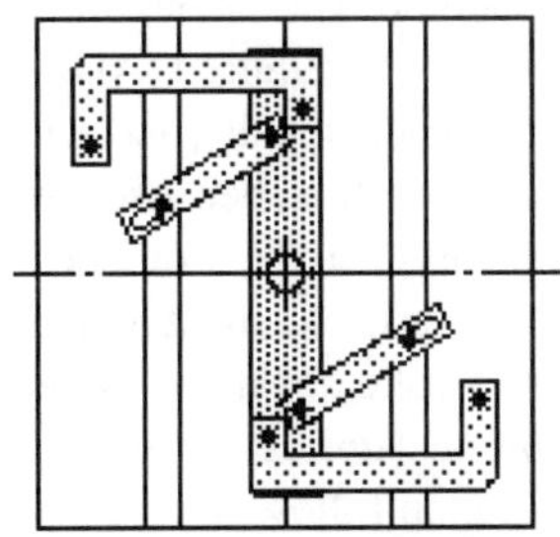

Figure 15.18 Ejection linkages with drive links connect swing arm to core plates and to the ejector or stripper plates.

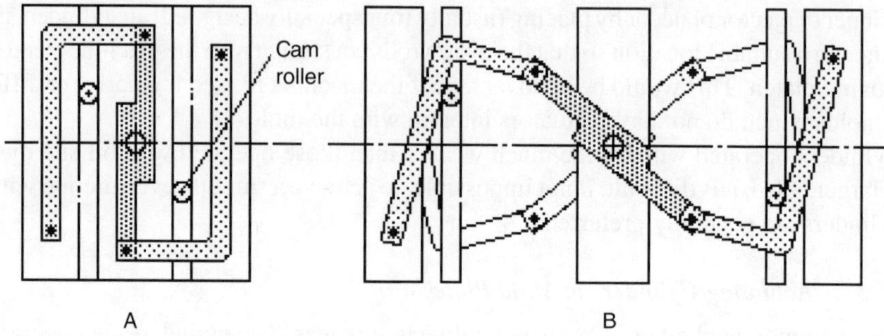

Figure 15.19 Combination of links and cam followers (rollers): A. mold closed and B. mold open.

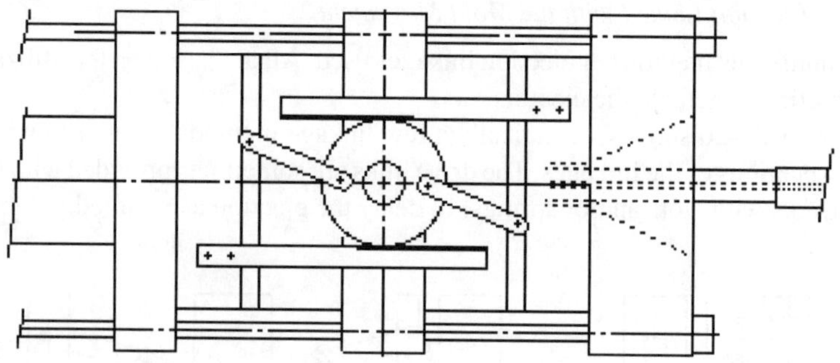

Figure 15.20 Combination of rack and pinion with links connecting pinion with ejector or stripper plate.

The method shown in Fig. 15.17 is used on smaller molds. The linkages not only actuate the ejection on both sides but also move the center section of the mold. Drive links, connecting the swing arm on the center section with the core plates, and with the ejector or stripper plates are shown in Fig. 15.18.

A combination of links and cam followers (roller) are shown in Fig. 15.19. Also, a combination of rack and pinion with links connecting the pinion with the ejector or stripper plate is shown in Fig. 15.20.

The advantage of these systems is that, usually, there is enough force for ejection. Even if the actual opening force of the machine is not enough, the levers in the linkages can be laid out so that they can provide great force if required. Also, this system does not require any timing devices, since the opening is mechanically tied with the machine stroke.

Disadvantages of linkage methods are that, because the ejection is tied with the motion of the machine, the linkages cannot be used for ejection once the mold has stopped its motion. Also, the length of the machine opening stroke must be well controlled to prevent damage to the linkage as a result of excessive stroke or failure to eject because of insufficient stroke. Other problems are the need for proper lubrication at wear points, such as the pivots of the various levers and sliding points of cam followers.

15.9.7.5 Two-Stage Ejection

Two-stage ejection can be provided by using some of the above systems, in combination, similar to a single-face mold. Air cylinders can be designed into the moving mold half, similar to the cylinders on the stationary half, to provide two-stage actuation or to provide ejection after the mold has stopped opening. However, for most efficient mold operation, the ejection should be finished by the time the mold arrives in the mold open position. The mold open time should be 0 (zero).

15.10 Water, Air, and Power Supply to the Center Section

It is very important that the center section is at the same temperature as the rest of the mold to prevent uneven expansion of the plates, which may result in different expansion of these plates and cause excessive wear of alignment features such as leader pins and bushings and taper locks.

The center section, which contains the hot runner system, is liable to heat up more than the other plates, and force the tie bars outward, causing the mold motion to jam. The problem is to provide good locations for water hoses and for power connections without hindering the ejection of the products.

In free fall ejection, the products fall freely, usually onto a chute or a conveyor. There should not be any hoses or electric wires at the bottom of the mold. The same applies to guide rails, which also have the products coming out at the bottom, into some form of conveying device that could interfere with hoses and wires.

With robots, the products are picked up by a take-off plate that usually moves in from the non-operator's side of the machine to receive the products, and moves back out again with the products. In this case, the points of interference with hoses and electric wires are in the rear of the machine; occasionally, even the top of the mold may have to be free because of some mechanisms located there.

There is no general rule on where to put these connections, but it is suggested that the designer position electrical connections for the hot runner heaters at the top of the mold. Water connections for the center section cannot enter from the sides, because of the mold carriers and plate moving mechanisms, and must, therefore, also be put on top of the mold.

This can create problems, because water leaks during setup or operation could then enter electrical connections. The use of watertight electrical fittings and connection boxes is, therefore, recommended. The designer must always ensure that in case of water leaks, usually during mold installation, water should not enter electrical connections, as this could cause damage to the electrical and electronic equipment, and in case of high voltage (110 V or more), that there is the added risk of personal injury.

15.10.1 Mold Services

General safety rules apply for routing, supporting, and protecting services for air, water, and electrical supply lines. However, the conditions are somewhat more complicated because of two facts:

1. All electrical power and thermocouple cables for the hot runner system, as well any air or hydraulic lines to operate a valve gate system, lead to the center section. Also, any air lines required for venting the cavities are usually on the center section.
2. Because the operating mechanism for center section motion, and often also the ejection, are on both front and rear of the mold, all supplies enter from the top of the mold.

To prevent these lines from colliding with the mold and other moving parts, they should be suspended from specially designed cable hangers.

Part I Mold Engineering

Section 3 Specific Subjects for the Mold Designer

16 Mold Materials Specifications

16.1 Material Comparisons

The following two charts (Tables 16.1 and 16.2) are intended to give the designer an overview of some common mold material specifications. The data are approximate and may vary slightly from one manufacturer to another.

Table 16.2 shows the average of some of the properties of the materials listed in Table 16.1 which interest the mold designer. Item numbers 1–14 refer to the material types in Table 16.1.

Generally, the raw material cost (steel, Be-Cu, etc.) amounts to about 10–15% of the total mold cost. "Savings" that can be achieved by selecting an inferior grade where a better grade is available usually cost more in the long run, particularly if the mold must have long life.

But this should not be taken as a "licence to squander" by selecting a superior grade where one is not required. It is, therefore, very important that the designer understands the basic requirements for the mold and selects the materials accordingly.

Steel costs in the following chapters are relative. The cost for a piece of steel depends on many "extras" added by the supplier to the base price. There are extra charges for piece weight,

Table 16.1 Material Specifications for Common Mold Materials

Item	recommended	Type designation	AISI material no.	Steel code DIN	Hardness (RC)
1	Prehardened	4140	1.7225	42CrMo4	30–35
2		P20	1.2330	40CrMnMo7	30–35
3	Stainless Steel Prehardened	420SS	1.2083	X42Cr13	30–35
4	Carburizing Steels	P5			59–61
5		P6	1.2735		58–60
6	Oil hardening	O1	1.2510	106WCr6	58–62
7	Air hardening	H13	1.2344	X40CrMoV5 1	49–51
8		A2	1.2363	X100CrMoV5 1	56–60
9		D2	1.2379	X155CrVMo12 1	56–58
10	Stainless Steel (SS)	420SS	1.2083	X42Cr13	50–52
11	Maraging	250			50–52
12	Maraging SS	455M			46–48
13	High-speed	M2	1.3343	S-6-5-2	60–62
14	Beryllium-Copper	BeCu			28–32[a]

[a] It is customary to indicate hardness of machinery steels and bronzes in the Brinell scale. Table 16.1, however, uses equivalent Rockwell "C" values to give a better comparison with the hardness of tool steels.

Table 16.2 Properties of Mold Materials Rated (P = poor, F = Fair, G = good, VG = very good, and E = excellent)

Property	Item # from Table 16.1													
	1	2	3	4	5	6	7	8	9	10	11	12	13	14
Wear resistance	F	F	F	VG	VG	VG	G	E	E	G	G	G	E	F
Toughness	VG	E	E	G	VG	F	VG	F	F	G	E	E	E	P
Compressive strength	F	F	F	G	G	E	VG	E	VG	G	G	G	E	F
Hot hardness	F	F	F	G	G	G	VG	VG	VG	VG	VG	VG	E	F
Corrosion resistance	P	F	G	F	F	P	F	F	F	VG	F	E	F	G
Thermal conductivity	G	G	F	F	F	G	F	F	F	F	F	F	F	E
Hobbability	P	P	P	E	VG	G	G	F	F	F	F	F	F	E
Machinability	G	G	F	EE	E	VG	E	VG	F	VG	F	F	F	E
Polishability	G	VG	E	VG	VG	VG	VG	VG	G	E	VG	VG	G	E
Nitriding ability	F	G	VG	VG	VG	F	E	VG	E	VG	E	N/A	E	N/A
Weldability	F	F	F	E	VG	F	G	F	P	G	G	G	F	VG

size, cutting, treatment, etc. The larger the piece ordered, the lower the cost per weight unit. Larger, annual quantities required can be negotiated with the supplier to bring the steel cost closer to the base price.

16.2 Guidelines for Selection of Mold Materials

In mold making, it is not always possible to say which material is the most suitable for any specific part. Much in the materials selection is based on past experience of the mold maker and on availability, either from in-house stock or from suppliers.

Materials must be selected to prevent galling and seizing; materials of equal molecular structure must not be used side by side if they move in relation to each other. However, there may be exceptions if, for example, at least one of the parts is plated or nitrided so that one surface structure is different from the other (e.g., commercially sold ejector pins, made from H13 and nitrided, are perfectly suitable to slide in an H13 bore).

Table 16.3 lists some preferred materials selection for mold parts, with recommended hardness, heat treatment (HT), and other treatment, if applicable. Comments give some of the reasons for the suggested selection. Table 16.4 provides a comparison of the actual material specifications.

There is continuous research by suppliers in the field of mold materials. New materials appear frequently with claims of improved performance. If the claims are interesting for mold performance and life, or for easier machining and or lower steel cost, they should be investigated. Feedback from such new materials may take a few years to develop, and should be considered before new materials should be added to any list of standard mold materials.

Table 16.3 Materials Selection Guide

Mold part	Material and properties	Comments and suggestions
Mold stack parts		
Cavities, cores, inserts, pins, etc.	AISI H13 Vacuum melted 49–51 Rc (2× draw)	Hot-work die steel, through hard Very tough, low movement in HT Good for delicate sections Good polishability
	AISI 420PQ Vacuum degassed Stainless steel 49–51 Rc (StavaxESR)	High corrosion and wear resistance Good polishability Use for corrosive materials, such as PVC, PC, Acrylic, etc.
	Dievac™ (AISI H13) 49–51 Rc (2× draw)	For best polish (optical quality) Vials, Petri dishes, etc.
	High-speed steel (CPMRex T15) 61–63 Rc Triple draw	For small, long cores (e.g., test tubes) Stiff, not as brittle and lower cost than tungsten-carbide
Cavity inserts Core inserts Core caps	BeCu B25 36–41 Rc	Best heat transfer Fast cycling molds (and others) for rapid cooling of specific areas.
	AISI H13 Vacuum melted 49–51 Rc (2× draw)	See Cavities, Cores above.
Core plugs	SS Type 420 Pre-hd 260–320 BHN	High corrosion resistance Use where corrosive action of water, etc., may destroy the seating, and prevent easy removal of the plug.
Gate inserts	BeCu B25 36–41 Rc	Best heat transfer
	AISI H13 Vacuum melted 49–51 Rc (2× draw)	If gate wear is a problem Requires better cooling layout
Lock rings	AISI A2 56–59 Rc	High strength, best wear resistance Large rings
	AISI H13 49–51 Rc (2× draw)	Smaller rings
Neck rings	AISI H13 49–51 Rc (2× draw)	General use

(continued)

Table 16.3 Materials Selection Guide *(continued)*

Mold part	Material and properties	Comments and suggestions
Neck rings	AISI 420PQ Vacuum degassed Stainless steel 49–51 Rc (StavaxESR™)	Corrosion resistant
Slides with molding surfaces	AISI A2 52–55 Rc	Cold work tool steel Good wear resistance, tough, stable Little movement in HT
	AISI H13 49–51 Rc (2× draw)	Good alternative to above, lower cost
Sliding inserts		Refer to 16.2, second paragraph, regarding dissimilar materials.
Stripper rings	AISI A2 56–59 Rc	Cold work tool steel Good wear resistance,tough, stable Little movement in HT
	AISI S7 56–58 Rc	Tougher material, use for fitted rings Shock resistant
Thin-walled sleeves	AISI D-2 58–60 Rc	Delicate parts, subject to heavy wear Good stability
Mold plates		
General use	AISI 4140 Pre-hd 260–320 BHN	Good strength Low cost for general purpose
	AISI P20 Pre-hd 260-320 BHN	Same strength Slightly higher cost than 4140 Less risk of distortion if much material is removed, as in hogging of large pockets
	AISI 420F-Mod Pre-hd 270–340 BHN Ramax S™ or Thyroplast 2316™	Good strength Slightly higher cost Good corrosion resistance Machining same as P20 May be HT to Rc 40–42
Mold hardware		
Blow-off pin, including air piston, at base of core	AMPCO 18™ bronze	Good wear without lubrication
Cams	AMPCO 18™ bronze	If strength calculations permit its use
	AISI 4140 Pre-hd 260–320 BHN	Requires wear plates for higher forces
	AISI M4 62–64 Rc	High cost

(continued)

Table 16.3 Materials Selection Guide *(continued)*

Mold part	Material and properties	Comments and suggestions
Cooling manifolds	Alum 6061-T6	Light weight, easy machining
Ejector rods (M/C) Air pistons Pillars Gibs Slide tracks	AISI 4140 Pre-hd 260–320 BHN	Tough Fair machinability Low cost Only for nonsteel slides With steel slides, use wear strips
	AMPCO 18™ bronze	One-piece gibs, high cost
Horn pins to move slides	Industry standard	Same as leader pins
Leader pins	Industry standard	Good abrasion resistance with tough core
Leader pin bushings	Industry standard	Same as leader pins
Lift bars	CRS	no HT
Locating rings	AISI 1015 or 1020	Low cost, any steel will do
Slides for severe service	AISI 4140 Pre-hardened 38–40 Rc	Requires wear pads of AMPCO 18™ bronze applications or Nylon 6 for light loads
Spacers (low strength)	Alum 6061-T6	Light weight, easy machining
Wedges and wear plates	AISI M4 62–64 Rc	Very high abrasion resistance Stable in HT Low toughness Poor machinability

Table 16.4 Specification Comparison for Mold Materials

DIN		AISI/SAE	
Material no:	Classification	Composition %	Classification
1.1401	C15	C 0.15, Si 0.25, Mn 0.55, P≤0.045, S≤0.045	1015
1.1730	C45W3	C 0.45, Si 0.30, Mn 0.70	1042/1045
1.2080	X210Cr12	C 2.1, Si 0.3, Mn 0.3, Cr 12.0	≈ D3 (30403)
1.2083	X42Cr13	C 0.42, Si 0.4, Mn 0.3, Cr 13.0	ER 420
1.2162	21MnCr5	C 0.21, Si 0.25, Mn 1.25, Cr 1.2	5120
1.2210	115CrV3	C 1.20, Si 0.25, Mn 0.30, Cr 0.60, V 0.10	50100
1.2311	40CrMnMo7	C 0.40, Si 0.30, Mn 1.50, Cr1.90, Mo 0.20	P20
1.2312	40CrMnMoS86	C 0.40, Si 0.40, Mn 1.50, Cr 1.90, Mo 0,20, S 0.06	≈ P20 (mold shoes)

(continued)

Table 16.4 Specification Comparison for Mold Materials *(continued)*

DIN		AISI/SAE	
Material no:	Classification	Composition %	Classification
1.2343	X38CrMoV51	C 0.38, Si 1.0, Mn 0.40, Cr 5.30, Mo 1,20, V 0.40	H11
1.2344	X40CrMoV51	C 0.40, Si 1.0 Mn 0.40, Cr 5.10, Mo 1.20, V 1.0	H13
1.2436	X210CrW12	C 2.10, Si 0.30, Mn 0.30, Cr 12.0, W 0.70	D3
1.2516	120WV4	C 1,20, Si 0.23, Mn 0.28, Cr 0.20, V 0.10, W 1.0	O7
1.2601	X165CrMoV12	C 1.65, Si 0.33, Mn 0.3, Cr 11.5, Mo 0.6, V 0.3, W 0.5	D2
1.2764	X19NiCrMo4	C 0.19, Si 0.25, Mn 0.40, Cr 1.25, Mo 0.20, Ni 4.0	P6
1.2767	X45NiCrMo4	C 0.45, si 0.25, Mn 0.40, Cr 1.35, Mo 0.25, Ni 4.0	4340 H
1.2826	60MnSiCr4	C 0.60, Si 0.90, Mn 1.0, Cr 3.0	S4
1.2842	90MnCrV8	C 0.90, Si 0.25, Mn2.0, Cr 0.35, V 0.10	O2
1.4541	X10CrNiTi189	C ≤0.08, Si ≤1.0, Mn ≤2.0, Cr 18.0, Ni 10.5	321

16.3 Heat Treating

Heat treating (HT) is a science in itself and should be left to specialists. One of the reasons that the designer should stay with a minimum selection of tool steels is to avoid the many heat treat specifications, which are usually different for each steel; this way both the designer and the heat treater become familiar with the steel and what to expect. Most, but not all, mold makers use outside suppliers for heat treat.

Sometimes two steels, even with an identical AISI number, may require significantly different HT (as shown in each supplier's specification sheets) to achieve the optimum properties expected from the steel. The designer need only indicate the required hardness of the part; as a rule, the designer need not indicate how this is to be achieved.

There may, however, be some other specifications which should be shown on the drawing, especially when the same hardness could be achieved with different methods of HT. For example, the wear characteristics or toughness of the steel may be improved by specifying double or triple drawing, as may be suggested by the steel manufacturer, for specific applications. Single draw(ing) is the most generally used and the cheapest method in HT. ("Drawing" means that the steel, after quenching, is heated to a specified temperature and then slowly cooled at a precisely controlled rate of cooling.)

16.3.1 Stress Relieving

This too should be specified by the designer who should be familiar with the risk of deformation of usually larger and/or "odd-shaped" pieces of steel after rough machining. The designer must also indicate on the drawing any other action which is not normally expected by the heat treater.

16.3.2 Carburizing

Steel can be hardened only if it contains at least 0.35% carbon. Mild steels, and specialty mold steels with less than 0.15% C but which are annealed, must not be used in molds, since they are too soft. The carburizing process adds carbon by penetrating the surface of the (machined) part; again, the designer need not indicate how this is to be done but will specify the carburizing depth required. Usually, the depth is from 0.5 to 1.5 mm, occasionally even as much as 2 mm.

16.3.2.1 Depth of Penetration

The effect of carburization diminishes with the depth of the carbon penetration; therefore, the hardness obtained after heat treat will also diminish with the depth of penetration.

Problems with carburization may be:

1. Large parts tend to grow and/or warp in HT, depending on their shape; if large portions of steel have been "hogged out", as may be the case in large cavities, stress-relieving before final machining of the soft steel is absolutely required to reduce the chance of warping in HT. Some heat treaters put warped parts under hydraulic presses to squeeze them back "approximately" into shape, or to straighten flat pieces. This requires considerable operator skill; also, it may set up new stresses which are then released during grinding, or even during operation of the mold. It may be possible to prevent this with another stress relieving after straightening, but this method is risky and not desirable.

   A production print must show how much steel must be added to critical dimensions before HT so that there is enough steel left to arrive at the desired size after HT and grinding. This requires considerable experience and some guessing; this is more art than science.
2. Grinding after HT removes some of the hardest skin. For larger pieces, more will have to be removed to bring the part to its required dimensions; in doing this, the hardest, most desirable layer will be more or less removed. This suggests that smaller pieces with shapes that are not expected to warp much may require less carburization (less time, less cost) than larger pieces.
3. Low carbon alloy steels such as P5 or P6 are relatively soft, suitable for hobbing, and are easily machined, but they have an ultimate strength of only 30–40 Rc. Because the hardened surface extends only as far as the carburization has penetrated, the steel below the "skin" remains relatively soft. Therefore, even with a skin of up to 61 Rc, the part will collapse much easier under heavy loads than a through-hardening steel treated to about 50 Rc.

16.3.2.2 Partial Carburizing for Thin Sections and Threads

A mold part may need to be hard in some areas of its surface but could, for the performance of the part, be soft in others areas; for example, additional machining (after HT) could be required to work through a hard surface. Before carburizing, areas can be covered (masked) wherever the carbon enrichment of the surface should be omitted. Subsequent HT will then have no hardening effect on the masked areas. Masked areas must be shown on the mold part drawing.

In thin cross sections of the steel, the carburizing penetrates from both sides that are exposed to the carbon-rich atmosphere. If the steel is very thin (less than 2 mm), and the penetration is 1 mm or more, there will be no soft core to support the hard skin; after HT, this area becomes through hard and brittle, thus losing one of the major attractions of using case-hardening steels.

This applies also to threads in mold parts. Their cross section is usually so small that there would be no soft core left between the flanks of the screw; the threads would become too brittle and could easily fail in service. Therefore, threads in mold parts made from carburizing steels should be masked.

16.3.2.3 Why Use Carburizing Steels at All?

The main reason for using carburizing steels is that these steels are a carry-over from times when there was no selection of suitable mold steels as there is today. Also, the machining is easy, the steel is relatively cheap, readily available in many sizes, it has excellent polishability and is very suitable for applications where good wear resistance is required, as with leader pins and bushings, or with cams, where a hard skin is required for good wearing, with on a softer, ductile core, which is more resistant against shock loads.

For the above reasons, the list of recommended mold materials omits carburizing steels, even though many (often large) mold cavities and cores are made from these steels. An important usage of carburizing steels used to be for hobbing (see Section 32.1, Hobbing).

In general mold making experience, the savings realized by lower steel cost and better machinability of carburizing steels can be easily lost as a result of the higher cost of HT (because of the additional steps required in HT) and the additional risks from warping after HT and collapsing under heavy clamping and injection forces.

16.3.3 Nitriding

In this method of HT, the hardened work piece is placed at an elevated temperature (≈500–550 °C) into a nitrogen-rich atmosphere. The nitrogen penetrates the surface (pores) of the steel and forms very hard (up to 70 Rc) nitrides. The added hardness and the change in crystallinity makes it possible that two similar steels, one nitrided and the other not, can slide against each other without seizing.

Note that the work piece must be hard. If the base steel is soft, the nitrided surface will collapse under load, similar to a case-hardened steel.

Some alloy steels are easy to nitride; others cannot be nitrided at all. Nitriding steels, such as AISI H13, top the list of steels that can be nitrided easily. It is important that the characteristics of the steel are such that it will not anneal at the nitriding temperature.

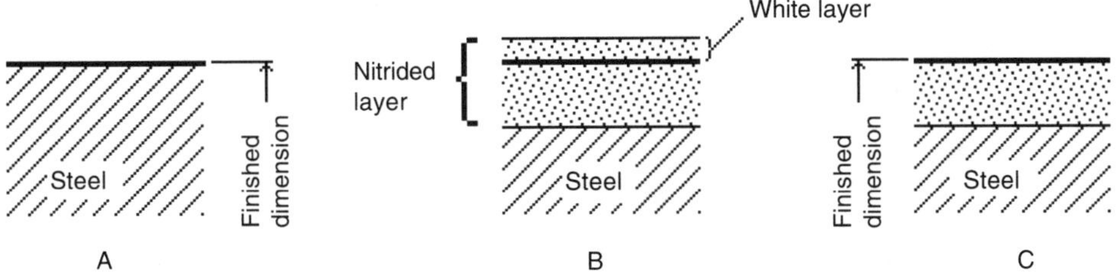

Figure 16.1 Stages of nitriding: A. finished dimension before nitriding, B. white layer on surface after nitriding, and C. finished dimension after nitriding and grinding.

The depth of penetration of nitrogen can be controlled by the length of time the steel is kept in the furnace. The designer should only specify the depth required but should know that a depth of 0.3–0.5 mm requires about 30 hours, and 0.7 mm requires 100 hours. Obviously, there is more cost in the longer use of the furnace. A depth of 0.3–0.5 mm, or even less, is usually sufficient for mold parts.

As the nitrogen penetrates, the surface "swells" slightly, forming a relatively soft layer above the original surface, thus changing the dimensions of the work piece. (An opening becomes smaller, an outside dimension increases.) This swelling, which can be easily seen as a white layer, is about 0.02 to 0.03 mm thick and relatively soft. It must be ground off or honed to revert the piece to the original (finished) dimension (Fig. 16.1).

Removal of the softer white layer not only restores the original size but exposes the hardest layer of the nitrided skin just below the white layer. The hardness of the nitrided skin also diminishes gradually as it approaches the base steel; or, the more that is ground off after nitriding, the less is the hardening effect of nitriding. As with carburizing, thin sections and threads must be masked to prevent nitriding from occurring on both sides and creating an extremely brittle cross section.

16.3.4 Flame Hardening, Induction Hardening

In flame hardening, a gas flame is directed at that surface of a steel part which needs hardening. As soon as the steel (containing at least 0.35% C) reaches a certain temperature (checked usually by the color of the steel under the flame), this area is chilled with a jet of water and hardens. The remainder of the steel is left relatively cold and soft. This method is rather crude and is not suitable for mold parts. The flame can again be directed at this quenched area to draw the steel down to the required hardness.

With induction hardening, the work piece—usually a cylindrical or flat piece of steel—moves along the center of an electric coil. An alternating current passing through this coil sets up, or "induces", very high currents in the work piece which heat the surface of the steel. The temperature and the depth of the hardened skin can be closely controlled by varying the current and the gap between the coil and the work piece. As the work piece moves, it passes immediately after heating through a ring from which water is sprayed uniformly at the hot surface, thereby quenching it and providing the desired surface hardness. Another pass through a coil may be used to draw the full-hard steel surface.

The method is fairly accurate, inexpensive, and does much that has been done before with carburizing for straight, symmetrical parts such as leader pins. The designer must specify the depth of hardening required. In most cases, a depth of 0.8–1.0 mm is sufficient. Induction hardening is not used for parts with molding surfaces.

16.4 Mold Finishing

16.4.1 Molding Surface Finishes

All mold part drawings must show the required finish specifications and any additional finishing requests such as sand blasting, vapor honing, buffing, etc. Table 16.5 lists many industry finishes and their related manufacturing methods and typical applications.

Table 16.5 Typical Manufacturing Methods and Applications of Various Mold Part Finishes

Finish required			Manufacturing method	Specify	Typical applications
Microns	Micro inches	SPI # equivalent			
0.025	1		Lapping 8000 diamond	0.025	Petri dish Optical quality
0.05	2	1	900 stone 8000 diamond polish	0.05 and buff	Test tube
0.08	3		900 stone 3000 diamond polish	0.08 and buff	Crystal tumbler
0.1	4	2	600 stone 3000 diamond polish	0.1 and buff	Opaque, shiny surfaces
0.1–0.2	4–8	3	900 draw stone	0.1–0.2	Opaque surfaces
		5	900 draw stone and vapor hone	0.1–0.2 and vapor hone	Matte surface
0.10–0.15	4–6		900 draw stone and vapor hone buff	0.10–0.15 and vapor hone and buff	Semi-opaque finish
0.2	8		600 draw stone 3000 diamond buff	0.2 and buff	General purpose
0.2–0.3	8–12	4	400-600 stone	0.2–0.3	Technical products, unspecified finish
		6	400-600 stone and sand blast	0.2–0.3 and sand blast	Texture finish

16.4.2 Molding Surface Finishing Symbols

Even where only one or just a few areas require finishing specifications, symbols and notes are used to indicate the finish, rather than placing written specifications directly on the affected surface. It also may be necessary to show limits for a finished area, as shown in Fig. 16.2.

If the part has several areas that require finishing specs, each surface to be finished will require a symbol pointing at it. The shape of symbols is left to the designer, and the meaning of each symbol must be explained in notes. Many typical symbols are shown in Fig. 16.3. There is no (standard) meaning attached to these symbols. Each must be explained every time it is used on a drawing. Other shapes may also be used, as long as it becomes clear to which surface they apply.

16.4.3 Special Textures

Textures such as basket weave, leather grain, etc., are usually produced by suppliers specializing in texturing, which is a chemical etching process involving the removal of

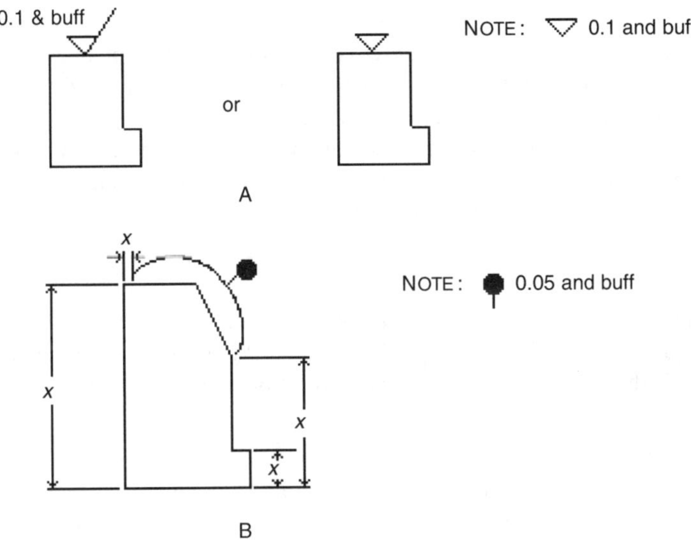

Figure 16.2 Finishing symbols on a drawing: A. symbol at surface area and types of notations, and B. limitations accompanying finishing notation.

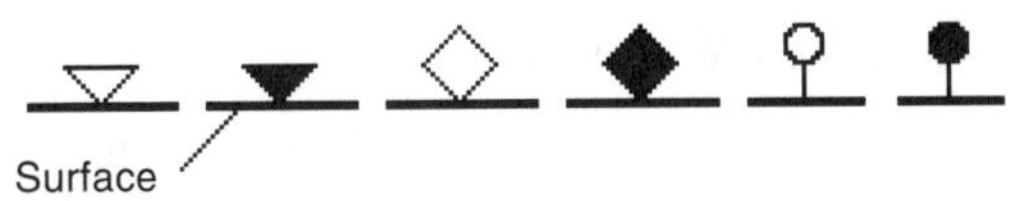

Figure 16.3 Typical finishing symbols used on mold drawings.

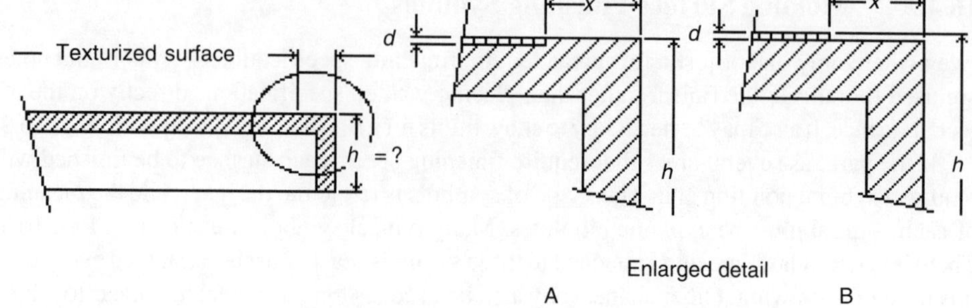

Figure 16.4 Part drawing does not indicate height *d*, and two results are possible: A. texture height *d* may be included in height *h* or B. texture height *d* may be added to height *h*.

material from the surface on which the texture is applied. The texture itself is defined by referring to sample chips, identified by names and/or numbers. Unless it is obvious, the designer must specify the area limits of the texture and show the depth of the texture from the surface to which it is applied.

If not properly indicated on the product drawing, Fig. 16.4 shows clearly what can happen. In Fig. 16.4A, the height h of the product includes the texture; in B, the texture is added to the height h. Since the depth (height) of textures is usually 0.05–0.10 mm, or even more, an error in specifying the depth of the etching may affect the product height and appearance significantly, and may lead to scrapping of the mold part.

Note that the etching process starts from the surface of the mold part and sinks into the steel. This depth d must be specified. It can be shown with dimension lines, as in Fig. 16.4, or with a notc ncxt to the texture specification.

16.4.3.1 EDM Texturing

For texturing, or any other pattern that can be machined into the electrode with EDM, the electrode requires the finish as specified for the molding surface. During the EDM process, the electrode is then sunk into the (often already hardened) mold steel.

16.4.3.2 EDM Finish

This finish is produced by a smooth electrode approaching a work piece with a matching smooth surface. The finish is produced uniformly over the whole area touched by the electrode. The grain of the finish is controlled by the intensity of the current used. To specify and check the finish, comparison chips issued by VDI (a German engineering organization) are used (Fig. 16.5). This is a subjective method of specifying and inspecting, but it is satisfactory for most applications.

16.4.4 Sand Blasting and Vapor Honing

This method of finishing is used after machining and heat treating to impart a dull finish to the molding surface. On vertical or near-vertical molding areas, such roughness may prevent easy ejection.

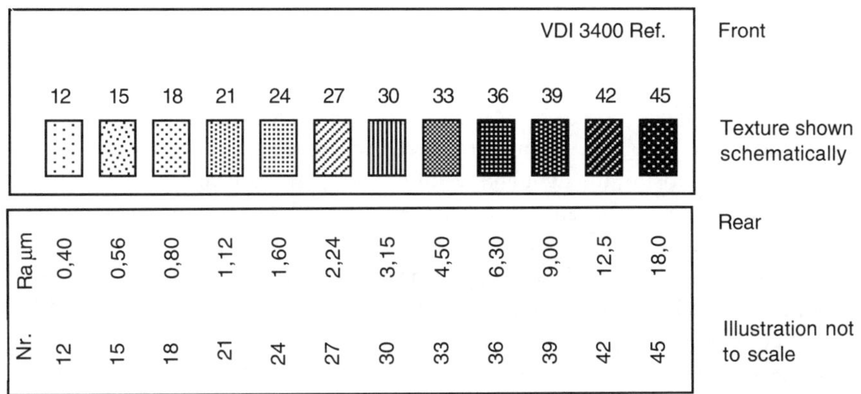

Figure 16.5 Comparison chip for EDM finish.

With products made from low-density polyethylene (LDPE) and a few other plastics (e.g., polyurethane (PU), etc.), ejection is actually improved by some roughness of the molding surface. If a shiny surface is required, buffing after blasting is required. Such roughness wears off with use, and the molds will require occasional roughing of the surface to keep operating without ejection problems.

16.4.4.1 Sand Blasting

Four grades of aluminum oxide (grit) for sand blasting are commonly used; they are #80 (coarsest), #120, #180, and #240 (finest). Other grades are also available. (Sand blasting is regularly used to remove oxides after heat treat.)

16.4.4.2 Vapor Honing

The term "vapor" is actually a misnomer, since no vapor is involved. Very fine "glass shot" is blown with a hand-held nozzle against the molding surface, using compressed air of ≈80–100 psi pressure. The process is visually controlled. When the whole surface is matte (dull), the blasting is finished. Areas to be blasted or honed must be clearly defined (dimensioned) on the part drawing.

16.4.5 Polishing and Buffing

Polishing is essentially the removal of metal by the action of free abrasive grains carried to the work piece by an agent or tool. The difference between these methods is only the degree of removal.

Polishing used to be an "art". Polishers spent many hours polishing every mold component by hand. Today, much is done mechanically, and only parts which, because of their shape, cannot be finished on the mechanical polishing equipment or where the finish is exceptionally fine are done by hand.

Note that in general, polishing and honing machines produce more regular (finer, flatter or cylindrical) surfaces, while hand finishing, particularly with unskilled operators, tends to increase waviness or even impart waves to a previously straight, but rough, surface.

16.4.5.1 Why Polish?

Several reasons for polishing are listed below:

1. Polishing may achieve special optical properties in a product. When light beams pass through the product wall (Petri dishes, vials, compact discs, mirrors, etc.), any major unevenness or waviness will distort the light and result in poor performance of the product.
2. Polishing may enhance the appearance of the product. This is the most common requirement in plastic molding. It should be noted, however, that the appearance often is required for the visible outside only. Often, no polish is required on the (not normally) seen inside (core) of the product, and the finish there need not be better than that required for easy ejection of the product from the mold.

   Exception: For clear plastic products where the inside finish will affect the appearance of the outside, the inside must have the same finish as the outside of the product.
3. Polishing may facilitate ejection. This applies generally to the finish of the core, including side cores. In cases where the product has projections (ribs, etc.) produced by deep recesses in the cavity and/or where there is little draft angle specified, the cavity must also be free of any roughness that may prevent easy ejection. (See Draw polish, below.)
4. Polishing may prevent stress risers in a product. Any sharp indent or corner in hardened mold steel, usually created during grinding, will increase the risk of breakage caused by stress concentration. Such corners must be polished to remove the sharp "gouges". See Chapter 18, Metal Fatigue, for further information.

16.4.5.2 Cost of Polishing

Even with increasing mechanization of the polishing process, there is still a considerable time required to "finish" a mold. The graph in Fig. 16.6 shows schematically the relation between quality of finish and time required to do it.

It can be seen from Fig. 16.6 that the best (ideal) finish takes infinitely long. The polishing time depends on the skill of the polisher, the quality of finish, the preceding machining operation, and on the hardness and quality of the material to be polished.

Often, the "as machined" surface may be perfectly suitable for appearance of the product and for ejection. It is the responsibility of the designer to specify the minimum finish consistent with operation and product requirements for each mold part. A good (fine) grinding finish or the surface as created by EDM is often all that is required. Too fine a finish may be

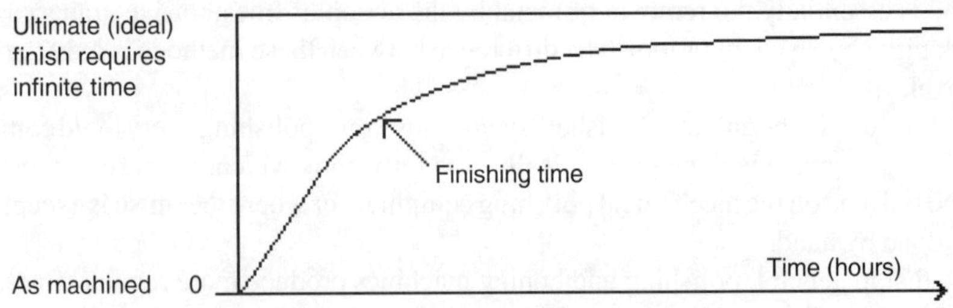

Figure 16.6 Plot of quality of finish against time required to product the finish.

Table 16.6 Polishing Grit Sizes

Grit no.	End-use operation	Micron range	Mesh approximation	Color guide
1	Super finish	0–2	14,000	White
3	Very high finish	2–4	8,000	Yellow
6	Mirror finish	4–8	3,000	Orange
9	High polish	8–12	1,800	Green
15	Fine finish	12–22	1,200	Blue

a waste of money and even be counterproductive; also, overpolishing may actually destroy a previously flat surface by entering softer areas of the surface and creating waves, called an "orange peel effect".

16.4.5.3 Draw Polish

Ridges (grooves) from turning, milling, and grinding in a direction about at right angles to the ejection may create serious problems in ejection, and may even retain and break pieces of plastic in deep grooves. This must be overcome by polishing in the direction of ejection any mold part which could create these problems. Ordinary polishing may smooth down the ridges but still maintain waviness, which affects ejection.

Draw polish affects mold performance. Drawings for mold surfaces with minimum draft must specify draw polish, in addition to the finish specified for the product.

16.4.5.4 Diamond Grit

Polishing is the removal of metal by the action of free abrasive grains carried to the work piece by an agent or tool. The free abrasive agents are usually *diamond grit* (crystals), which is graded by mesh size. Table 16.6 lists grit sizes commonly used for polishing.

Buffing has two stages. The first is "cutting down", the refining of a surface by removal of surface imperfections or scratch lines left behind from polishing to make a relatively smooth surface smoother. The second stage is "coloring", a further refinement of the cut-down surface to bring out maximum luster. The penetration of diamond grit with different tools is shown schematically in Fig. 16.7. Figure 16.8 shows schematically the working of various grits with felt backing.

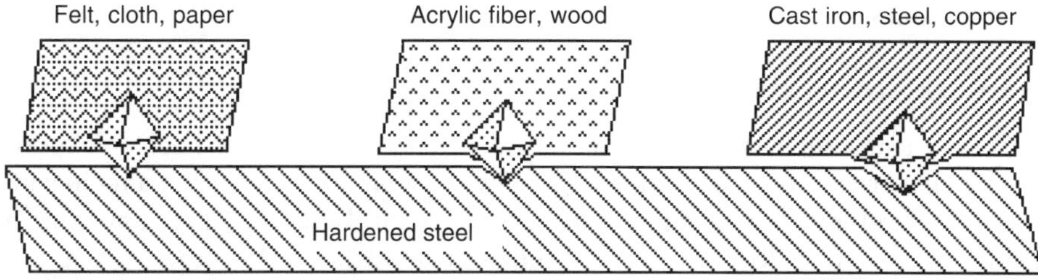

Figure 16.7 Surface penetration in hardened steel by diamond grit in soft (left), medium (center), and hard (right) "tools."

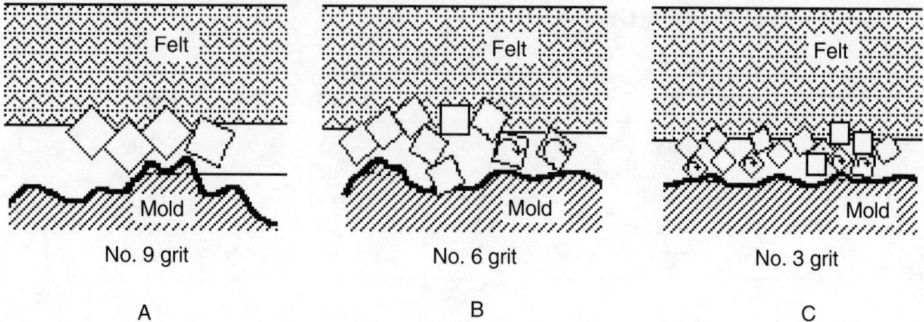

Figure 16.8 Different grits with felt backing are used for different stages of polishing and buffing: A. polishing, B. cutting down, and C. coloring.

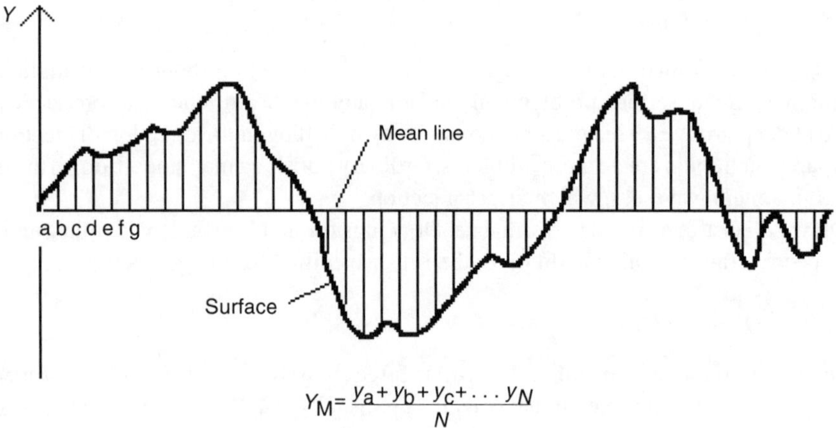

Figure 16.9 Graph of the measured surface of a mold part.

The amount of material removed from the surface depends on the size of grit embedded in the "tool." As the crystals cut, they rotate and loosen from their bonding to the backing, and fall away with the metal dust. Grits embedded in a harder matrix, such as wood, copper, iron, or steel, remove more of the mold surface before being shed but are more likely to cause surface scratches.

16.4.5.5 Measuring Roughness

Roughness is a measurement of the maximum deviation y from the mean line. It is expressed in μ (microns) or micro-inches. Figure 16.9 shows a graph of the surface of a mold part. A conversion chart for microns and micro-inches is shown in Table 16.7. The conversion rule is that the micron value divided by 0.0254 equals micro-inches RMS equivalent.

Three methods are generally used to measure roughness:

1. Optical interference techniques,

Table 16.7 Conversion Chart for Microns and Micro-Inches

Microns	Micro-inches	Microns	Micro-inches
0.025	1	0.3	12
0.05	2	0.4	16
0.08	3	0.8	32
0.1	4	1.6	63
0.2	8	3.2	125

2. Pieco-electric stylus, or
3. visual inspection.

Methods 1 and 2 are objective; however, the visual method is in practically all cases acceptable, and much faster.

For visual inspection, there can be a collection of typical finishes produced on mold steels and labeled with the finish designations. If required, the inspector compares the produced finish on a work piece with the appropriate sample finish.

16.4.5.6 Polishing Terms and Sequences

Automatic stoning: This method is used wherever possible. The (round) mold part is rotated while a polishing stone is moved in an axial direction over the tapered or cylindrical working surface. Because the work piece rotates, the actual stoning takes place in a direction at a slight angle to the direction of the draw. In many cases, this is acceptable; if not, a final hand stoning in the direction of the draw is required. Bottoms of cavities and tops of cores are also finished mechanically, using a "spinning machine". The work piece is fastened to a rotating spindle, and the operator uses small sanding (grinding) discs.

Hand stoning: This is usually done with an oscillating hand tool (Diprofile™) in areas that are not accessible for mechanical polishing. Stroke and speed of oscillation can be varied by operator as required.

Vapor honing: Sharp edges may be affected by the blasting and may need to be stoned to remove any burr.

Brushing: To remove lines from stoning, a small hand-held rotating brush is used, made of horsehair, brass, or steel, with different bristle lengths, and using diamond paste #9.

Buffing: A hand-held tool with a rotating felt wheel is used to provide the required fine finish, using diamond paste #15, #9. #6 and #3. The work piece may rotate or may be stationary.

Draw polish: A hand-held felt stick, using the same type of diamond paste as for buffing, is moved over the molding area in the direction of the draw with the work piece stationary.

The cost of "automatic machine finishing" is much less than that of hand finishing; the mold designer must be critical of demanding polish specifications made by the product designer. The best way is always to ask: *"Does the specified finish really make sense, considering the product?* Much time can be wasted by specifying excessive finishes.

16.4.6 Nonmolding Surface Finishing Specifications

In general, the surface finish of areas of mold parts which do *not* contact plastic are not specified except by the use of a general notation:

Unless otherwise shown, all
surfaces to be finished 1.6

A better finish than the 1.6 μ (63 micro inches) shown above would be 0.8 μ (32 micro inches), and 0.4 μ (16 micro inches) (See Table 16.6). For example:

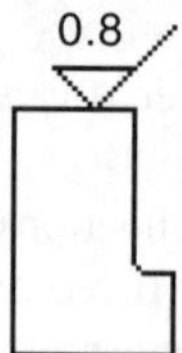

Standard mold parts may require a better finish than 1.6 μ (64 micro inches) or even a molding surface finish. This better finish may also be required to reduce stresses in stress concentration points of any mold part (see also Chapter 18, Metal Fatigue). Such better finishes must be noted directly on the part drawing in the critical location, similar to molding surface finishes.

16.4.9.1 Polishing of Sharp Corners

While this is not really part of this section (mold finishing), the designer is reminded of the fact that any sharp corner may start a crack when under stress and/or through corrosion, particularly in hardened tool steels. Even if a small radius is specified, turning or milling grooves or scratches produced from grinding a corner are just as bad as a sharp corner. Grooves or scratch marks must be removed by polishing. The higher the expected specific loads or the risk from corrosion, the better the polish must be to ensure lowest risk of cracks. See Chapter 18, Metal Fatigue.

17 Mold Plates

17.1 Purpose and Material Selection

A mold plate is any plate required to build a complete mold. The assembly of mold plates (without the cavities and cores) is often called the “mold shoe”.

In the past, and even today in low cost molds, mild steel plates molds were used. As a rule, these plates are made from hot or cold rolled AISI 1015 or, even better, from free machining steels such as AISI 1130, etc. This is the cheapest type of steel (boiler plate, etc.) and is readily available. These plates are often too soft for the requirements of high production molds

As a rule, mold plates have no molding surfaces cut into them. However, in very large molds, or in molds for low production, cavities and occasionally cores are cut directly into the mold plates. In these cases, the quality of steel may be important. It must have good polishability and be free of impurities (dirt) which might leave marks on the product.

In all plates used for molds, the main requirements are:

- Good machinability (less machining time),
- adequate hardness for the requirements of the mold,
- homogeneity of steel (e.g., no hard spots detrimental to machining operations),
- good corrosion resistance,
- easy availability,
- cleanliness (free from enclosures or dirt) if used for cores or cavities, and
- low cost.

An improvement over the mild steels is the use of a “machinery” steel such as AISI 4140, supplied prehardened to about 260–320 BHN (equivalent to about 130,000–170,000 psi tensile strength). This steel meets many of the above requirements. However, it often contains impurities, and rusts easily.

A better steel is a machinery steel that was developed specially for molds, P20. It is essentially a clean equivalent of AISI 4140 which can be used also for cavities and cores. P20 meets most of the requirements, but it too rusts easily. To prevent corrosion it must be plated, preferably with the electroless nickel process (ENP) to ensure that the inside channels are also protected. Cost for plating adds to the cost of the steel. Some molders just paint the shoe from the outside with oil paint, but this does not prevent corrosion in the cooling channels.

Stainless steels (SS) such as AISI 420 F Mod., supplied prehardened to about 270–340 BHN, do not require protection against corrosion and meet all the other requirements as well. Although the cost of SS is usually considerably higher than that of P20, by selecting certain standard plate sizes and by committing to buy large quantities, it is possible to reduce greatly the steel price (when compared to the equivalent P20). Also, there are savings because no additional plating is required. To take full advantage of low steel prices, the designer must specify sizes which can be cut from standard mill sizes.

17.2 Deflection of Mold Plates

It is suggested that the designer become familiar with the basics of strengths of materials, from any source, such as a machinery handbook or any engineering textbook dealing with this subject. Some important statements regarding deflection of plates are listed below:

1. Any plate will deflect under load. The mold designer is mainly concerned that the deflection is so small that it will not affect the alignment of the mold.
2. If a plate deflects more than desirable, it is preferable to make the plate thicker, rather than adding another plate. This can be readily understood by looking at the formulae for deflection. The resistance to deflection is proportional to the *third* power of the plate thickness.

Example: Compare two setups: One plate of 100 mm, or two plates of 50 mm stacked together. With the 100-mm thick plate:

$$100^3 = 1{,}000{,}000.$$

With two plates each 50 mm thick:

$$2 \times 50^3 = 2 \times 125{,}000 = 250{,}000.$$

Two 50-mm plates are only one-fourth as stiff as one 100-mm plate, even though the total thickness is the same. Under the same load, the two thinner plates will deflect four (4) times as much as the one, thicker plate.

3. Only the modulus of elasticity (E) affects deflection. The hardness of the steel has no influence on deflection. If a plate is deflected under load below its yield strength, it will return to its original shape after the load has been removed. Because the yield strength of a harder steel is higher than that of a softer steel, the same load will create a permanent deflection in the softer steel, while the harder steel, under the same load, will return to its original shape after the load has been removed.

   All steels have approximately the same modulus of elasticity E, within ±5%, with the exception of tungsten carbides, which may have twice the stiffness of tool steels. Exact figures are given with every material specification sheet. E for steel is ≈ 2,000,000 kg/cm^2 (30,000,000 psi).
4. To prevent failure resulting from cyclical loads, the maximum allowable stress in a plate must not exceed the fatigue strength of the plate material. Data are always shown at room temperature. At elevated temperatures (200 °C or more), the strength is reduced; these data are found in specification sheets for each material.
5. It is very difficult to calculate the exact deflection of a plate supported in several locations. But for practical purposes, this is usually not required. To ensure that the mold is "stiff" enough, assumptions can be made using the "worst case scenario"; that is, a) assume single-point loading (at the center of the cavity) rather than spreading the load over an area, and b) assume that the plate is a simple beam, freely supported at two ends only. With these two assumptions, the result is much more severe than arrived with other, more accurate methods, and provides additional safety against excessive deflection.

6. It is better (more supporting areas can be left standing) to cut out a plate in the shape required (e.g., for the ejector plate or hot runner manifold), resulting in a much stiffer mold, rather than using round or square cut-outs and adding pillars to support excessive spreads. While it may require removal of considerable amounts of material, it is in the long run less expensive, and the mold is easier to assemble than if it had many loose parts (pillars, etc.).

   This does not mean that support pillars must never be used. The decision whether to "hog out" a plate or to provide a simple frame (with parallels and/or large cut-outs) and support pillars rests ultimately with the designer, preferably with input from production engineering. Sometimes, support pillars are used for guiding the motion of the ejector plate; in this case, the pillars must be hardened and ground similar to leader pins. Make sure that all support pillars, when seated directly on the machine platen, are placed over solid steel of the platen and not over holes or cut-outs. (Check the platen layout of the machine !!)

17.3 Forces Affecting Mold Plates

The plates are "loaded" by the clamp force, which must be greater than the separation force created by the injected plastic trying to open the mold. The separation force is equal to the total of projected cavity area(s) multiplied by the injection pressure of the plastic inside the cavity.

The pressure inside the cavity is often considerably less than the injection pressure at the machine nozzle, because of the pressure drop in runners and gates. This pressure is difficult to estimate and varies considerably for various types of molds and plastics. It is practically impossible to give exact figures; the pressure inside the cavity depends on many factors, such as flow resistance within the runner system, plastic viscosity, melt temperature, cooling temperature, injection speed, gate size, etc. If in doubt, the estimated cavity pressure must be established by talking with people familiar with the products.

(Additional discussion of this topic is included in Chapter 18, Metal Fatigue, in Section 18.1).

17.4 Simplified Calculation of Plate Deflection and Stress

17.4.1 Deflection

Note in Fig. 17.1 and the related equations that the formula for I (moment of inertia) is based on a "unit width" of 1, which extends "into the paper". The load W must also be reduced to the unit corresponding to a width of 1 (not the total load); that is, the total load is divided by the width of the plate.

When calculating, make sure to use the appropriate measurements, either in the inch or the metric system. Deflection f of any mold plate should never exceed 0.05 mm (0.002 in.).

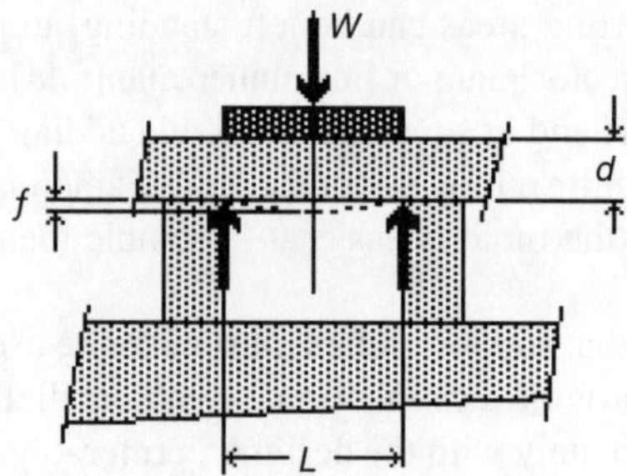

Figure 17.1 Deflection *f* in a mold plate.

Deflection *f* in a mold plate is illustrated in Fig. 17.1 and calculated in units of mm or in.:

$$f = \frac{W\,L^3}{48\,E\,I}, \tag{17.1}$$

where *E* is the modulus of elasticity and *I* is the moment of inertia in mm^4 (or $in.^4$), for which the formula is:

$$I = \frac{1 \times d^3}{3}. \tag{17.2}$$

17.4.2 Stress

The stress calculations are shown below and illustrated in Fig. 17.2. The maximum stress S_{max} under load *W* is:

$$S_{max} = -\frac{W\,L}{4\,Z}, \tag{17.3}$$

where *Z* is the section modulus, which must be based on the same "unit width" of 1, as explained for the moment of inertia *I*:

$$Z = \frac{1 \times d^2}{3} \; (\text{mm}^3 \text{ or in.}^3). \tag{17.4}$$

S_{max} must be equal to or less than the allowable fatigue stress for the steel.

17.4.3 Design Check

With the above two very conservative formulae, the designer can easily check whether the plates are thick enough for a specific mold. Minimum deflection results in long mold life and accuracy of products, and is particularly important for thin-walled products.

(CAUTION: Make sure to use proper units. If *E* and *S* are given in kg/cm^2, use cm (not mm) for *d* and *L* (1 cm = 10 mm). In the inch system, use psi and inches, as in the example that follows.)

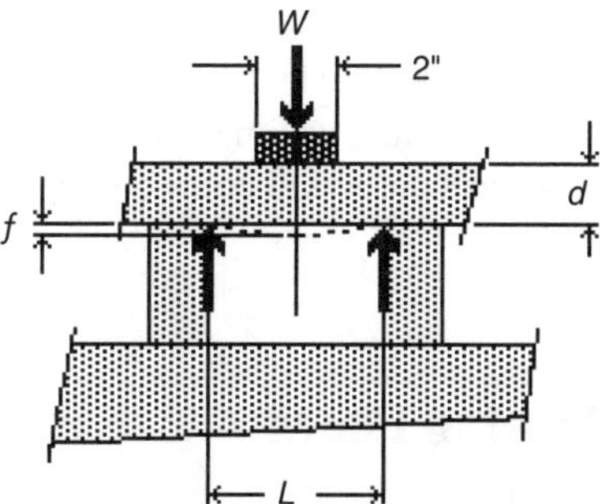

Figure 17.2 A typical design check for a mold, showing calculations for deflection and stress.

Example: Assuming the cavity area is 2 in. by 4 in. (into the paper), and the injection pressure is assumed 10,000 psi:

$$2 \text{ in.} \times 4 \text{ in.} \times 10{,}000 \text{ psi} = 80{,}000 \text{ lb.}$$

Dividing this by 4, the designer arrives at a unit load W of 20,000 lb. Substituting in equations 17.1–17.4 for deflection and maximum stress shows that, for this example, both the deflection and the stress are well within acceptable limits, given that the allowable stress, $S_{allowable}$, is 30,000 psi. From eq 17.4,

$$Z = \frac{1 \times 2^2}{3} = 1.33 \text{ in.}^3,$$

so from eq 17.3, $S_{max} - \dfrac{20{,}000 \text{ lb.} \times 4 \text{ in.}}{4 \times 1.33 \text{ in.}^3} = 15{,}038$ psi.

Since 15,038 psi < 30,000 psi, the S_{max} is acceptable. The deflection of the plate must be less than 0.002 in. From eq 17.2:

$$I = \frac{1 \times (2 \text{ in.})^2}{3} = 2.67 \text{ in.}^4,$$

so from eq 17.1, $f = \dfrac{20{,}000 \text{ lb.} \times (4 \text{ in.})^3}{48 \times 30{,}000{,}000 \text{ psi} \times 2.67 \text{ in.}^4} = 0.000333$ in.,

which is less than 0.002 in. and, therefore, acceptable.

In the example shown, the designer could make a decision whether to accept the selected design on considerations such as expected life of the mold, measured in molding cycles. If the figures for maximum deflection or stress were higher than permissible, the designer would have to make some decisions. First, the designer could argue, if the difference between the acceptable limit and the calculated value is little, that there is enough reserve in the strength because the assumptions are in the designer's favor. The designer could also argue that if the mold is planned only for a limited number of products and will not cycle even close to the fatigue limit (much fewer than 2,000,000 cycles), the steel will not fatigue in this time. But if the mold must be built for a long productive time, the only proper way is to redesign (i.e., to make the plate stronger and to recalculate until both conditions of allowable deflection and stress are filled).

17.5 Guiding of Moving Plates

17.5.1 Guide Supports

All plates which move during operation of a mold (opening, closing, sliding) must be properly guided. It is not acceptable to use essentially weak mold components (e.g., ejector pins, return pins, piston rods or moving inserts) to guide any plate.

Independent plate guiding must ensure accurate positioning for these weak components. Whenever one or a number of such components are located on a plate, they must be mounted "floating" so that each such component can assume its own proper location with respect to the matching mold part in which it slides. If there is misalignment in the absence of float, there is danger of severe side forces, resulting in excessive wear on these mold components and possible jamming of the motion, and breakage.

This "float" is acceptable because, usually, such parts have very low mass, and the effect of gravity on these parts to pull them out of alignment is small. However, if a sliding mold component is of substantial mass, it must be guided; it is usually fastened to a plate, which is then guided by pins, etc.

Guiding of moving plates is usually achieved by one of the following methods:

1. Mold leader pins,
2. auxiliary leader pins,
3. supporting pillars, used as guide pins, or
4. gibs.

Guiding of plates by their actuators (air pistons, cams, etc.) is not acceptable. Such plates must be supported by independent guides so that there is no side load on the actuators (as a result of misalignment or gravity) which could cause premature wear of seals or of the cylinder walls. Some moving plates and components that require guiding are:

1. Ejector plates,
2. stripper plates,
3. moving cavity plates,
4. sliding plates that carry mold inserts,
5. robot take-off plates,
6. side cores and slides,
7. moving cores and core sleeves, and
8. collapsible cores.

In general, a plate should be supported by three or four pins and bushings, although only two may be acceptable for smaller plates or components.

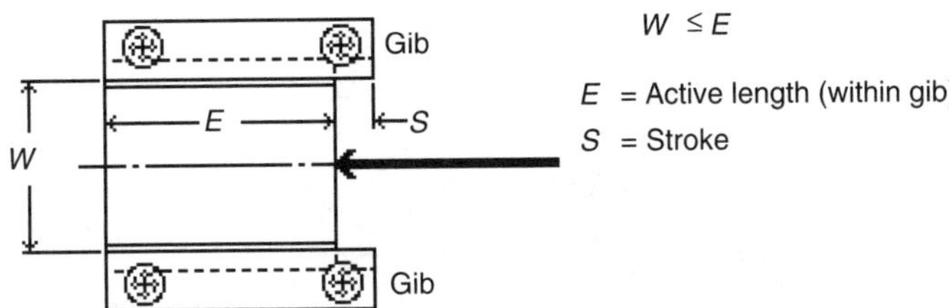

Figure 17.3 One actuator, centrally located, guides the moving plate.

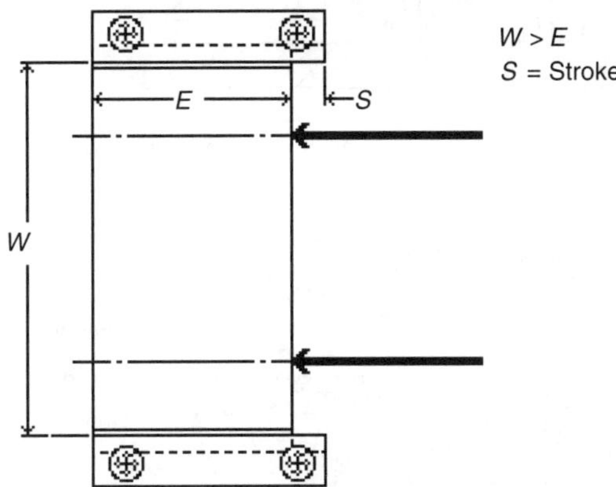

Figure 17.4 Two identical actuators placed near each gib provide smooth sliding of the (moving) plate.

17.5.2 Guiding Plates

If gibs are used to guide slides, their active length should be at least as large as their distance from each other, particularly if there is only one actuator (piston, cam, etc.) to move the sliding plate (Fig. 17.3). If the distance between the gibs is larger than the active length, two identical actuators must be used, one of each near one of the gibs, to avoid cocking and jamming (Fig. 17.4). Figure 17.5 shows guiding for an ejector plate. Usually, there should be four such guides, but in exceptional cases, three or even two are acceptable.

Note that the ejector pins in Fig. 17.5 are floating (i.e., free to move sideways) both in the ejector retainer plate and in the core backing plate. The float is usually about 0.5 mm (0.020 in.) radially; therefore, a misalignment of about 0.013 mm would not jam. Under the heads, there should be a small axial clearance of about 0.025–0.050 mm (0.001–0.002 in.).

It is important to align the *retainer* plate with the core backing plate, to facilitate mold assembly, and with the ejector plate, to ensure that the ejector pins, etc., maintain their concentricity within the holes of the ejector retainer plate and maintain their float. Alignment with the core backing plate can be done, as shown in Fig. 17.5, with short bushings with a radial clearance of about 0.13 mm (0.005 in.) in the guide pillar, so as not to "fight" the basic alignment provided for the ejector plate. Alignment with the ejector plate also is achieved. Other methods of ensuring proper float follow (The decision between methods is usually affected by the available space.):

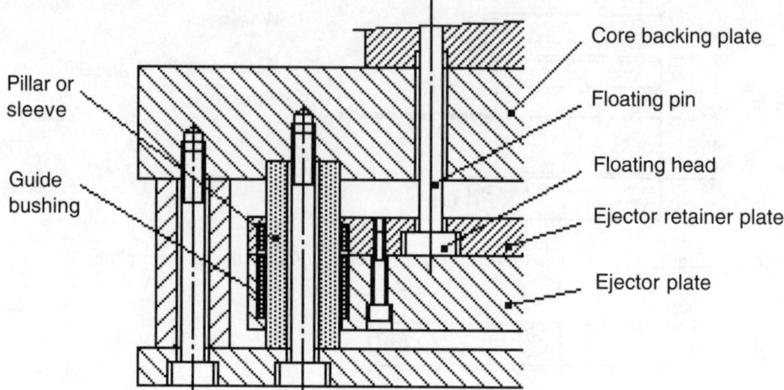

Figure 17.5 Cross section showing floating ejector pins, and also guiding of the ejector plate.

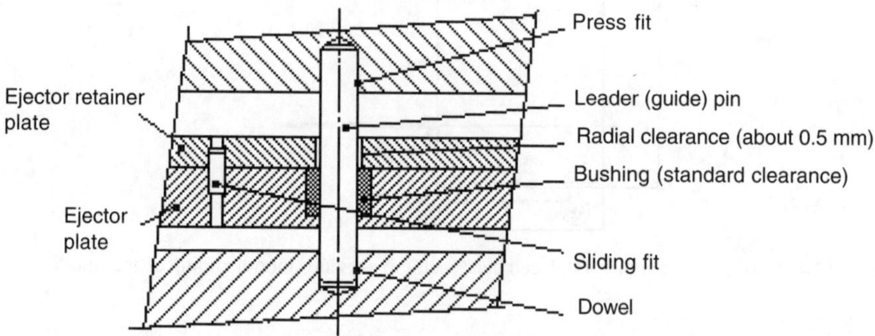

Figure 17.6 Dowels align the retainer plate and ejector plate.

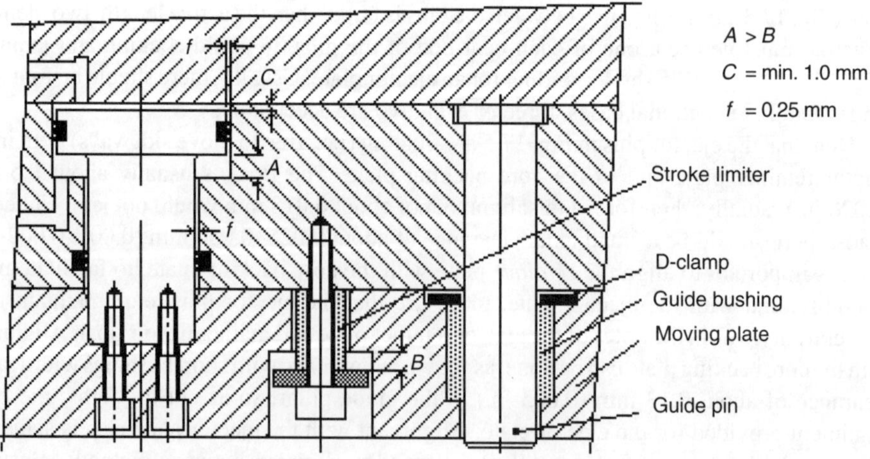

Figure 17.7 Typical moving plate guided by pin and bushing, with actuator and stroke limiter.

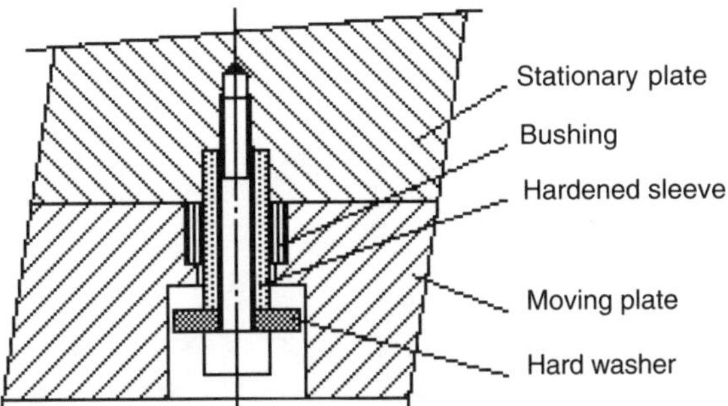

Figure 17.8 Stroke limiter acts as a leader pin in a small mold or for short strokes.

1. Two (2) dowels may be loose in the core backing plate and tight in the retainer plate. During assembly, these dowels enter the core backing plate while the ejector pins are inserted. The ejector plate is then screwed tight to the retainer plate (not illustrated).
2. The bushings in the ejector plate can reach into the retainer plate, and the outside of the bushing is loose with a clearance of 0.13 mm (0.005 in.) in that plate.
3. Two (2) dowels, loose in the retainer plate and tight in the ejector plate, ensure that the retainer plate will be in the correct location (See Fig. 17.6).

Figure 17.6 illustrates a good and simple design for guiding plates. Note that any bushing used for guiding a moving plate must be retained on both ends by recesses, adjoining plate(s), D-clamps, etc. *Press fit of a bushing without securing it is not acceptable.*

The next illustration (Fig. 17.7) shows a typical moving plate with actuator (air cylinder), stroke limiter, and guide pin and bushing. There should be at least three, but preferably four, actuators, at the corners (as far apart as possible) to minimize the possible cocking of the plate.

Note that there must be always a minimum clearance (*C*) of at least 1.0 mm between the top of the piston and the end of the cylinder. Also, the piston must *never* be used to limit the stroke of the plate. As shown, separate stroke limiters (minimum three, preferred four) must make sure that the motion stops on these limiters and not on the pistons: A must always be larger than B (Fig. 17.7).

The clearance *f* between air piston and piston rod from surrounding plate must be 0.25 mm (0.010 in.), regardless of the recommendations by the seal manufacturer, to prevent rubbing and galling between piston, rod, and plate. (This added clearance has been proven by experience in molds to be without effect on the seal.) Note that the diameters of the seal grooves in the plate (for the rod seal and in the piston) and the corner radii must be in accordance with seal specifications to avoid premature failure of the seals.

Guide pins and stroke limiters should be placed as near the actuators as possible. There must be at least three, but preferably four guides to ensure that there is no side load on the piston rods and seals.

For smaller jobs, and with short travel, if there is no space for separate leader pins, the stroke limiters can be designed so that they perform the job of leader pins, as shown in Fig. 17.8. The sleeve acts as leader pin, with the bushing in the moving plate. The length of the sleeve entering the stationary plate must be at least as large as the sleeve diameter and is assembled with light interference fit of the same magnitude as a leader pin of the same size.

18 Metal Fatigue*

18.1 Mechanical Properties of Materials

Before discussing metal fatigue and how it can affect mold construction and mold life, we will review some basic principles of material strengths. The relationships and terms apply to practically all materials.

In molds, the designer is mostly concerned with steel, occasionally with beryllium-copper and aluminum, and also with some bronzes and certain plastics that provide special properties, such as wear resistance. The most important relationship is shown in the stress-strain curve (Fig. 18.1). Factors involved in this relationship are defined below:

Stress: Force per unit area (in psi or in kg/cm^2).

Unit strain: Change in length of a specimen, when subjected to a force, divided by the original length (commonly simply called "strain").

Proportional limit: The point in the stress-strain curve where it begins to deviate from the straight-line relationship between stress and strain.

Elastic limit: The maximum stress to which an object can be subjected and still return to its original shape when the force is released. For steel, the proportional and elastic limits are considered to be the same.

Yield point: The point on the stress-strain curve where there is a sudden increase in strain without a corresponding increase in stress. For steels, the yield point is about the same as the elastic limit. Not all materials have a yield point.

Yield strength: S_y (in psi or kg/cm^2), the maximum stress that can be applied without permanent deformation. This is the value of the stress at the elastic limit of any material which has an elastic limit.

All mold calculations should be based on the values of yield strength, S_y.

Ultimate strength: S_u (in psi or kg/cm^2), the maximum stress value on the stress-strain curve. S_u is called the "ultimate tensile strength" (UTS).

Modulus of elasticity: E (in psi or kg/cm^2), the ratio of stress to strain, below the proportional limit.

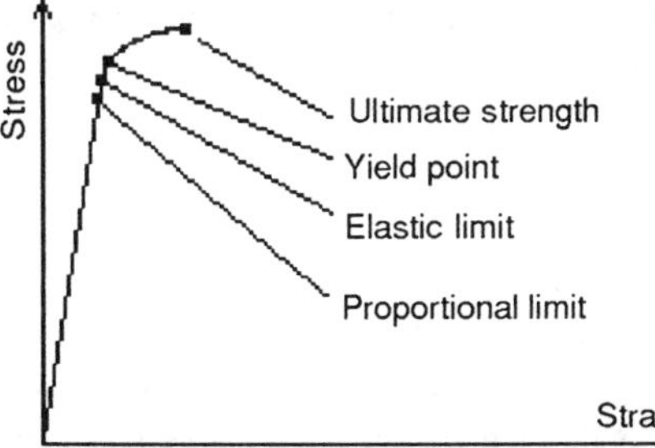

Figure 18.1 Stress-strain curve for steel

* Portions of this chapter and a number of illustrations are based on the book *Engineer to Win,* copyright Carroll Smith, Motorbooks International.

For most steels, the value of E is about the same, whether the steel is soft or hard, with the notable exception of tungsten-carbide steels, which have a much larger value. Values for E are given on material data sheets.

The important difference between soft (mild) and hardened steels is that the tensile strength (and the yield point value) of hardened steels is much higher. In other words, a hard steel can be stressed much higher, and it will still return to its original shape after the force acting on the steel is relieved; a mild steel subjected to a similar force will be stretched out of shape and will not return to its original shape. Note that for hard steels, the values for ultimate and yield strengths are about the same (i.e, the hard steel will usually break just after reaching its yield point).

Other definitions of significance for mold designers are those regarding compressive strength and shear properties:

Compressive strength: From compression tests, compressive yield stress (S_{cy}) and compressive ultimate stress (S_{cu}) can be determined. For hard materials, these values are usually the same as the values for tensile stresses, (S_y) and (S_u). Ductile materials, when loaded, will swell or buckle without breaking.

Shear properties: Shear tests determine the shear yield strength (S_{sy}), shear ultimate strength (S_{su}), and the modulus of rigidity (G), which is the ratio of stress to strain in shear, below the proportional limit. For most steels, G is about $0.4E$. Actual values of G are shown on material data sheets.

18.2 Metal Fatigue

Many mold failures are the result of metal fatigue; they could have been prevented if the designer had applied some of today's well understood practices to avoid metal fatigue.

Metal fatigue is well understood by metallurgists and taken into consideration by all designers of equipment such as airplanes, cars, etc., where metal failure may have severe consequences. Fatigue applies to any product made from metals and, particularly, from steel.

The following will explain metal fatigue, using a minimum of theory. In some cases, the designer will have to accept the statements made and can study the subject in more detail in a multitude of books and papers on the subject.

The following uses an approach similar to that in *Engineer to Win* [1] (written for designers of racing cars) to explain this complex subject in simple language and with easy-to-understand graphs, as it applies to mold design.

18.2.1 What is Fatigue?

When a structure is loaded steadily, and the components are properly sized, they will last practically forever. However, when subjected to repeated, even apparently insignificant, changes of load, a component may fail early.

"Rule: Under repeated ('cyclical', as opposed to continuous or 'static') stress, the capacity of a metal to withstand stress gradually diminishes and (in most cases) cannot be restored" [1].

Metals subjected to fluctuating loads break after a finite number of load cycles, even though the stress is well below the ultimate strength of the metal. This is called "fatigue failure". However, tests have shown that a steel part which under a certain load has not broken after about 2 million cycles, will not fail at all under this load.

How does this relate to molds? As was shown in Chapter 8, Mold Forces, mold parts are loaded in:

- Compression,
- tension,
- bending,
- shear, and
- torsion.

This chapter will not discuss combinations of stresses, such as tension and shear, or tension and compression, which are very complicated. Note that any screw is loaded in tension and torsion (shear) during torquing. However, as soon as the torquing is stopped, the torsional force ends, and the screw is solely loaded in tension. This explains why a screw should be well lubricated to reduce the torsional stress.

Rule: For fatigue failure to occur, the total elapsed time, or the time interval between loads, is immaterial. Only the number of cycles is significant.

For example, a "slow" mold, running at a 15-second cycle, will complete one million cycles in 173 days (of 24 hours), but a "fast" mold, at a 3-second cycle, will produce the same number of cycles in only 34 days. The effect of fatigue on mold components is the same in both these examples.

18.3 Four Basic Types of Loads

18.3.1 Axial Load

Axial loads are illustrated in Fig. 18.2. By convention, compressive stress is termed negative (–), and tensile stress is termed positive (+). Note that the cross sectional distribution of stress is uniform.

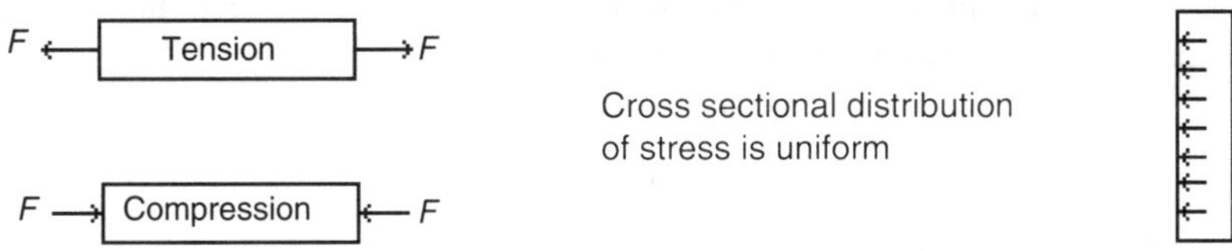

Figure 18.2 Axial load on a mold part: tensile stress (top) and compressive stress (bottom), with uniform cross sectional distribution of stress (right).

18.3.2 Bending Load

Examples of bending loads are shown in Fig. 18.3. Note that the cross sectional distribution of stress varies from zero at a neutral axis to a maximum at the surface.

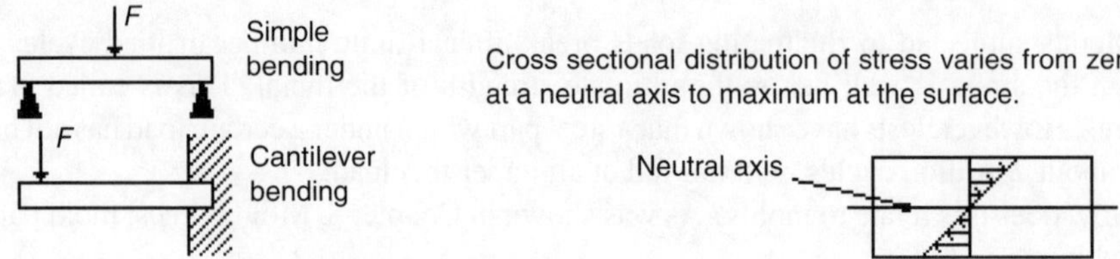

Figure 18.3 Bending loads: simple (top) and cantilever (bottom), and cross section of stress distribution (right).

18.3.3 Shear Load

Figure 18.4 shows shear stress on a bolt fastener. Cross sectional distribution of stress in the bolt is uniform. Sections A are in shear.

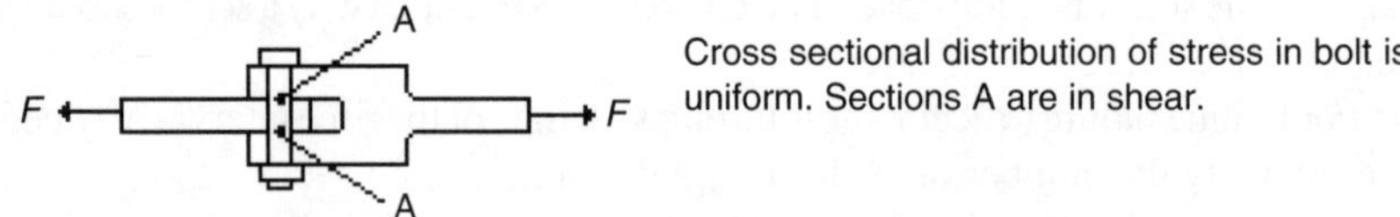

Figure 18.4 Shear stress on a bolt.

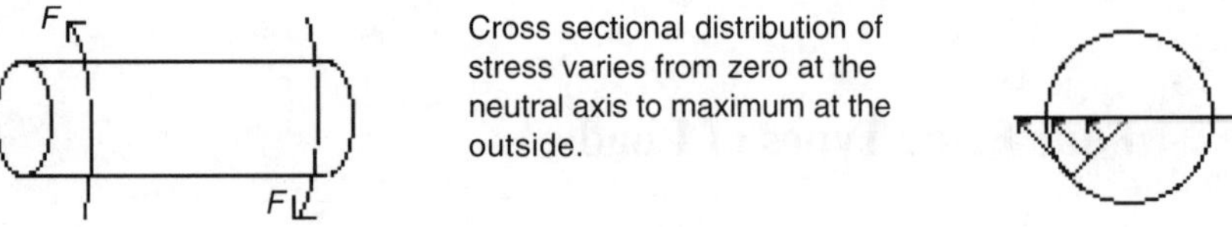

Figure 18.5 Torsional load is shown schematically (left) and in cross section (right).

18.3.4 Torsional Load

Torsional stress on a turning part is illustrated in Fig. 18.5. The cross sectional distribution of stress, in this case, varies from zero at the neutral axis to maximum stress at the outside of the part.

18.4 Load Fluctuation, Size and Direction

It is important to understand that the grain structure of the metal (steel) is affected differently if the load varies from 0 to 100%, for example, or from 50 to 100%, or from full tension to full compression (+100% to −100%), etc. This difference is shown in Fig. 18.6 with six different examples. The stress ratio, R, is equal to the minimum stress divided by the maximum stress. The greater (more positive) the ratio, the better (as shown in the graph (Fig. 18.7). *Fluctuations from 0 to 100% occur regularly in every mold.*

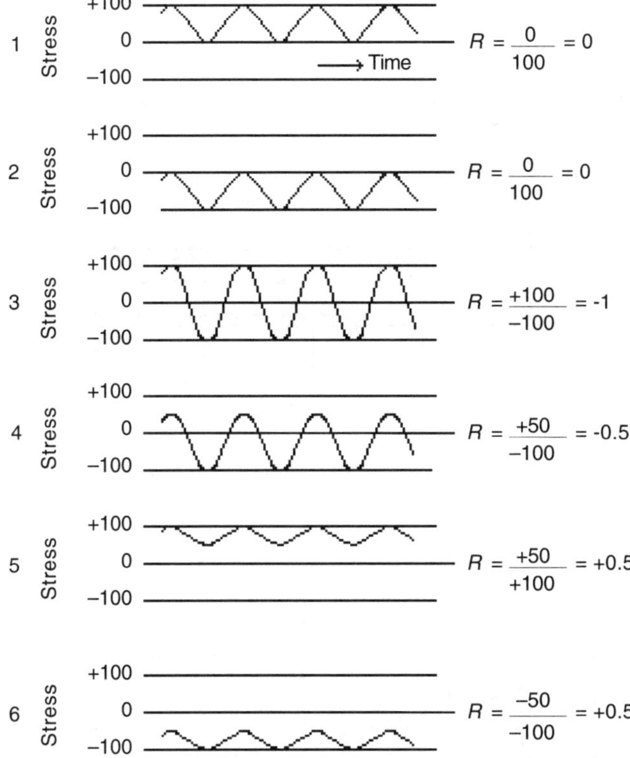

Figure 18.6 Six examples of load fluctuation yield different stress ratios, *R*.

Fig. 18.7 shows (for one specific steel, treated in one specific way) the relation between stress (*S*) and the number of cycles (*N*) to failure (*S–N* diagram) for various levels of stress ratios, *R*. The best condition is when the stress does not vary at all (steady load) or when the stress level fluctuates little. The worst condition is when the stress changes from positive (+100) to negative (–100), which is "stress reversal" ($R = -1$). This condition can, for example, occur in robot arms, as a result of forces generated by accelerating and decelerating masses, such as the take-off plates.

18.5 Fatigue Life Curves

In the example in Fig. 18.7, at about two million cycles or more, the curves for stress ratios of $R = +0.5$ and $R = -1$ shows permissible stresses of 82,000 psi and 62,000 psi, respectively (a rather dramatic drop in the fatigue strength).

Figure 18.8 explains the effect of fatigue. An AISI 4130 steel bar, with a cross section of 1 square inch is tested. The ultimate strength of this bar is 90,000 psi, so the bar will fail suddenly when a load of 90,000 lb or more is imposed even just once (point A). Assuming the

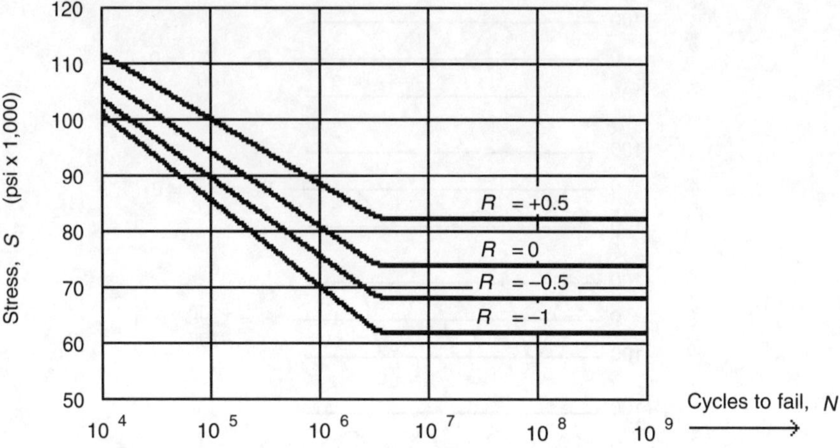

Figure 18.7 Fatigue life curves showing fatigue life at varying levels of stress for identical test specimens at different stress ratios, *R*.

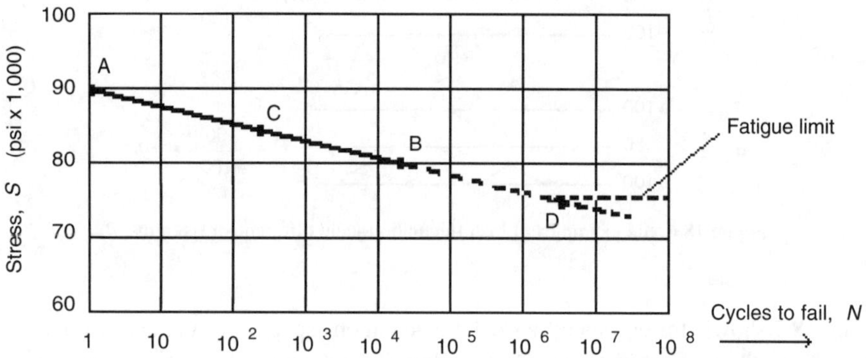

Figure 18.8 Fatigue life curve showing effect of fatigue on a steel bar, for $R = 0$.

bar is loaded cyclically, with the load varying between zero and 100% (stress ratio $R = 0/100 = 0$), and the load cycling at a stress level of 80,000 psi, the bar will fail soon after 10^4 cycles (point B).

By interpolating between A and B shows that this bar, when stressed to 85,000 psi, will fail after only about 200–300 cycles (point C). (Note: The scale of cycles-to-fail is logarithmic to show large numbers of cycles in the available space on paper.) By extrapolating (dotted line), one can also see that at a stress of 75,000 psi, the bar appears to last about 4,500,000 cycles before failing. In fact, once the part has been loaded for more than 2 million cycles, it will not fail at all, as long as the stress (point D) is less than 75,000 psi (horizontal, dotted line).

The fractured bar will exhibit the typical smooth and progressive "beach marks" of a fatigue failure, along with some rough, crystalline surface where the failure finally occurred (Fig. 18.9). If the test had been stopped before 10^4 cycles occurred, and the bar inspected, the fatigue crack where the failure would eventually occur could have been seen on the surface.

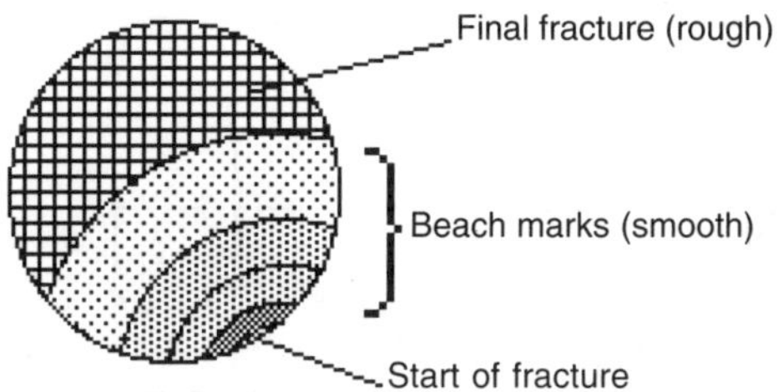

Figure 18.9 Fractured bar carries beach marks showing where the fracture started and the direction in which it progressed to the rough section where the failure occurred.

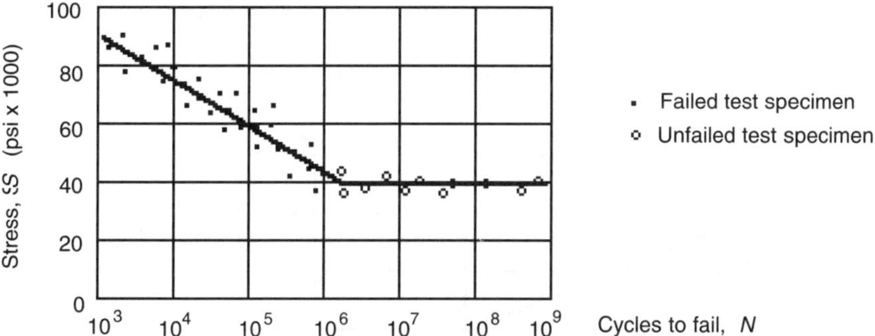

Figure 18.10 Fatigue life curve drawn through scatter plots of various test samples.

In molds, a water leak may indicate that a crack has started from a water channel and has reached the surface of the part.

All data designers get from suppliers are based on laboratory tests, under well-controlled conditions. But even so, the results of identical fatigue tests are different, as shown by the "scatter" on the graph in Fig. 18.10. Usually, data "curves" are arrived at by averaging a number of test results from a fairly small sample of test specimens, which means that there can be better, but also worse, results in performance of the material.

From experience, we know that the fatigue life of "real" parts is often less than that derived from standard test specimens. For example, a low alloy chromium-nickel steel quenched and tempered to an ultimate tensile strength (UTS) of 120,000 psi was tested in reverse torsion ($R = -1$). The standard size test specimen (1 in^2) had a fatigue limit of 36,000 psi. However, a larger bar of 1.2-in diameter (1.13 in^2) of the same steel exhibited an actual fatigue limit of 31,000 psi; a still larger bar, with a 1.75-in diameter (2.41 in^2) had a fatigue limit of only 26,000 psi.

It is of the utmost importance that the designer recognizes the fact that laboratory tests and "real life" performance are different. This does not mean that the laboratory tests are useless. On the contrary, they are an excellent way to compare various materials, but the results shown in manufacturers' data sheets must be used with discretion.

Steel fatigue curves always look similar to the ones shown so far; that is, they show a straight line decrease until they reach a point after which the strength will not reduce (horizontal line), regardless of the number of cycles. Fatigue curves for nonferrous metals are

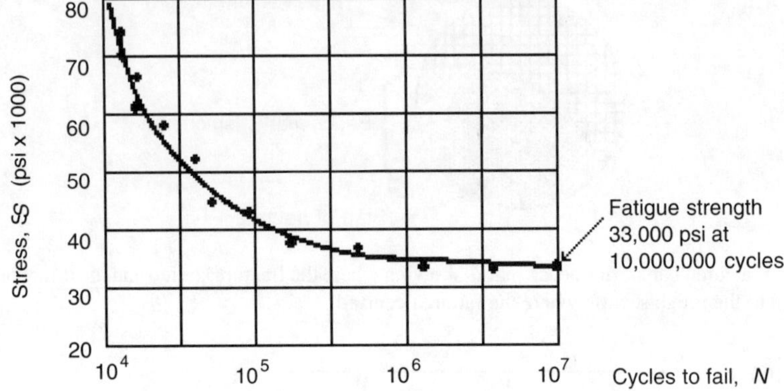

Figure 18.11 Fatigue life curve for a nonferrous material.

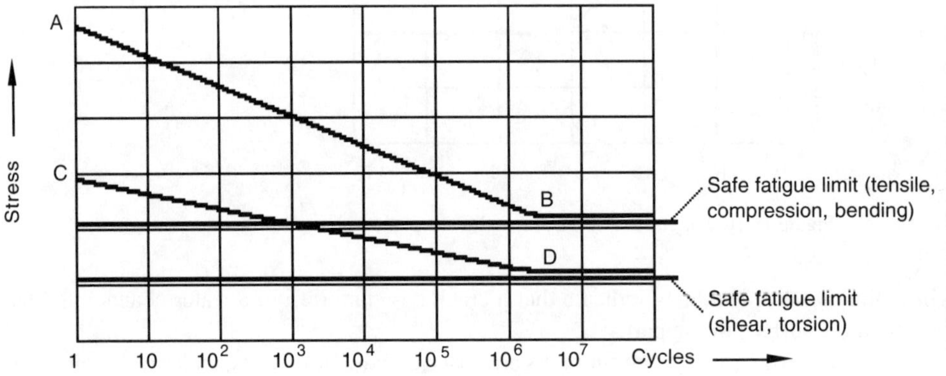

Figure 18.12 Typical *S–N* curves for steel.

more like a hyperbola; that is, they never reach a point where the curve is completely horizontal. In other words, the fatigue stress keeps diminishing as the number of cycles increases (see Fig. 18.11).

In fast-cycling molds, one million cycles can be reached very soon. If a steel part has safely reached about 2,000,000 cycles without failure, it should last “forever”.

To summarize the foregoing “theoretical” introduction, every material has certain strength characteristics which can be found in handbooks, but it is better to get them from materials suppliers’ data sheets. (If fatigue data are not specified, the supplier can be asked to supply them.)

Figure 18.12 shows schematically, without values given, typical *S–N* curves for steel. At point A, at one cycle, the fatigue stress is equal to the UTS (ultimate tensile stress). The curve is a straight line (on a logarithmic scale) and reaches a minimum at about 2 million cycles at point B. From here on, the steel will not fail, regardless of the number of cycles. Slightly less than this stress is the safe fatigue stress (or fatigue limit) in tension, compression, or bending.

The curve starting at point C applies to shear stress (and torsion). The ultimate shear stress is in general much less than the UTS. This stress also follows a straight line to point D; from there, the stress will remain constant, regardless of the number of cycles. Slightly less than this stress is the safe fatigue stress (or fatigue limit) in shear or torsion.

The large difference between UTS and safe stress is the reason why designers use a safety factor (SF), depending on the characteristics of the material and on the type of loading. An SF of 2–6, or even greater, based on yield strength, is often used when calculating the strength of a part to ensure it will not fail in service. But the problem with large safety factors is that the parts will be very heavy, which is often undesirable or impractical.

18.6 Avoiding Fatigue Failure in Molds

All of the above examples were based on simple, round test pieces. In practice, mold parts have odd shapes with corners, notches, drilled holes, threads, engravings, and other design features. Also, the part surface is affected by machining and grinding marks, heat treat stresses, scratches, etc. All this affects the fatigue life of the part.

18.6.1 Material Selection

Harder materials usually have a (relatively) lower fatigue strength than softer materials (see Fig. 18.16). On the other hand, hard surfaces are often necessary for other requirements in the mold (e.g., wear, abrasion resistance, resistance against hobbing-in, etc.). To decrease the stress level below the fatigue strength, the designer must therefore increase the steel cross section.

18.6.2 Finishing

Finishes as created by EDM and also electroplating adversely affect the surface of a part and thereby its fatigue life. Since these processes cannot be avoided, the designer must strive to reduce the stress level in the part, usually by increasing its thickness.

18.6.3 Product Shape

Usually, the designer cannot change the shape of the product, which is specified by the customer. But this does not mean that changes in the product design should not be suggested if the designer believes that a feature such as a sharp corner or a rough surface finish will seriously affect the mold life. Customers will usually understand such a problem and agree to any proposal which enhances the mold life.

It is most important for the designer to establish how each mold part is stressed (size, direction, frequency of load) so that the effect of the load on the part to be designed can be foreseen. Many mold components, such as plates, do not need special consideration, because they are sufficiently large and thus only lightly stressed. Also, they are usually not hardened and, therefore, better in fatigue. However, if a plate is criss-crossed with drilled channels and has other pockets, it must be checked for fatigue life to prevent it from caving in. All components such as cavities, cores, inserts, and all small parts which are loaded cyclically, especially if they are through-hardened (hot runner parts, stripper rings, locating rings, inserts, etc.), must be checked carefully for fatigue life.

18.6.4 Stress Concentrations

All mold parts are affected by stress concentrations, or "stress risers". Such concentrations are responsible for most mold part fractures. An important job for the designer is to recognize stress risers and to make sure they are avoided or reduced in magnitude.

Even if the designer has selected a safe stress level (below the fatigue limit), the part is still seriously affected by the influence of stress concentration. Remember, the tests are made with smooth, polished specimens, and the data reflect only the results of these tests. Mold parts always have corners, notches, holes, etc., and are not necessarily polished where it counts for better mold life.

Stress risers can originate in a number of areas: they may be within the steel itself or be caused by bad design. First, review a few typical examples (Fig. 18.13).

A test piece is stressed to a value *s*, as shown in the left and center illustration of Fig. 18.13. When a hole is added, the stress *s* near the hole is shown schematically (right) to be about three times larger than the evenly distributed stress that was present before the hole was added, as shown schematically in the center illustration.

The actual value (3*s*) is not as important as the fact that the stresses at the edge of any hole are much larger and will, therefore, initiate failure. In this example, the fatigue limit of the piece would be much lower (worse) than without the hole.

In mold design, it is practically impossible to avoid holes in stressed parts (e.g., waterlines, mounting holes, etc.). But designers can do something to improve the (bad) condition by controlling the finish of the hole. This can be clearly seen in the graph in Fig. 18.14, which shows *S–N* curves for a 4140 steel strip, drilled with a 0.25-in. hole. (The figures show actual test results.)

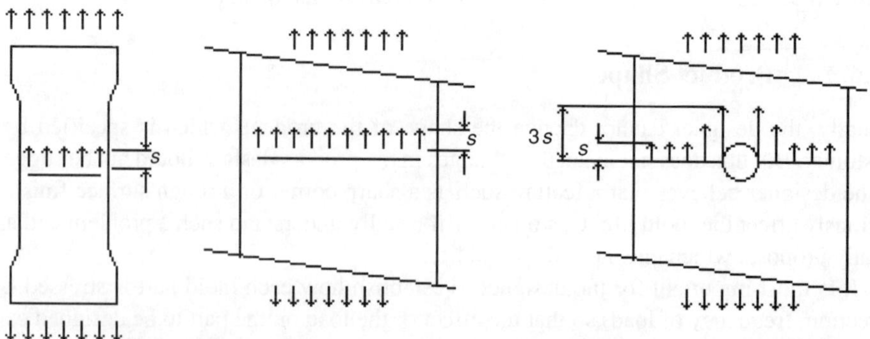

Figure 18.13 Typical examples of stress risers in a test piece stressed to a value *s* (left and center), and when a hole is added (right).

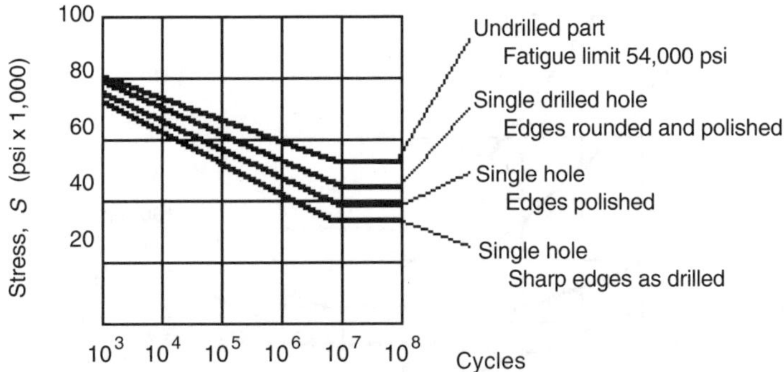

Figure 18.14 *S–N* curve for 4140 steel strip showing fatigue limits for the undrilled part (top), and with a drilled 0.25-in. hole with various finishes.

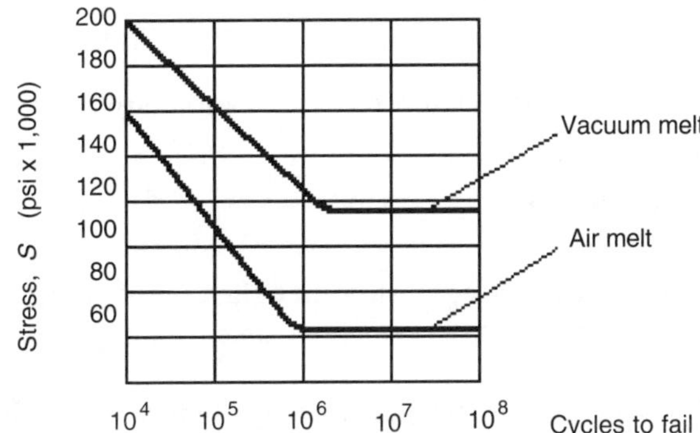

Figure 18.15 *S–N* curves for 4340 steels vacuum-melted (top) and air-melted (bottom).

But there are other, less easily seen problems, in the steel itself. There are impurities (inclusions) in some steels which can affect the fatigue life even more than a drilled hole. As an example, the graph in Fig. 18.15 shows actual *S–N* curves for similar (4340) steels, one air-melted, the other vacuum-melted. The differences are dramatic. In fatigue, the vacuum-melted steel is almost twice as strong as the air-melted steel.

18.7 Effect of Hardness on Fatigue Limit

In "mild" steels such as SAE1020, stress risers are less significant. In comparison, the fatigue limit of a hardened steel diminishes so much that little would be gained in fatigue life by its use. The graph in Fig. 18.16 is based on actual fatigue (laboratory) tests. The heat-treated steel is (in fatigue) only a little stronger than the cold drawn steel SAE1020. Mild steel is not used in molds, but Fig. 18.16 illustrates that to improve fatigue strength, increasing hardness is not much help.

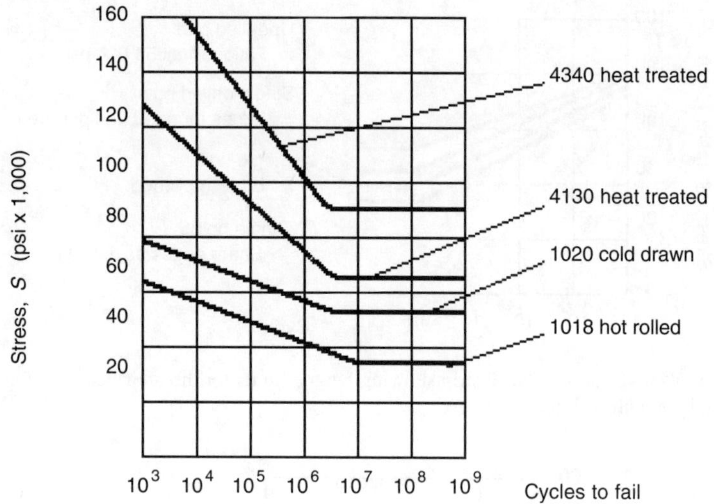

Figure 18.16 Comparison of *S–N* curves for several grades of steel.

18.8 Effect of Finish

Figure 18.17 shows the effect of surface finish on the fatigue life of a hardened round bar, with 230,000 psi UTS, when subjected to the worst possible case of loading any part, that is, reversed torsional stress.

As the graph shows, thc fatigue strength of a rough-turned test specimen is reduced to about 18% of UTS, and even a highly polished specimen has only 22% of UTS. In both cases, a safety factor of at least 5–6 would be correct. If there are additional stress risers, a higher safety factor will be required.

Figure 18.18 shows the effect of surface finish on the fatigue life of test specimens of alloy steels which have been treated to various strength and hardness levels. Note: In the range used in molds, hardness is roughly directly proportional to tensile strength at a ratio of 1 Rc point to 5,000 psi (e.g., 30 Rc equals about 150,000 psi, or 45 Rc equals about 225,000 psi, etc.).

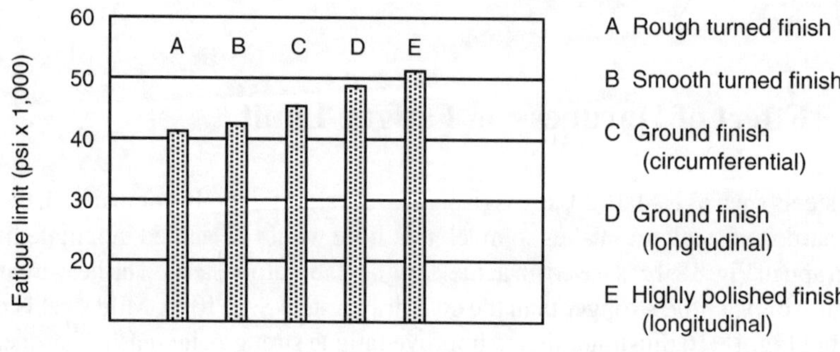

Figure 18.17 Effect of varying surface finishes on fatigue life of a hardened round bar.

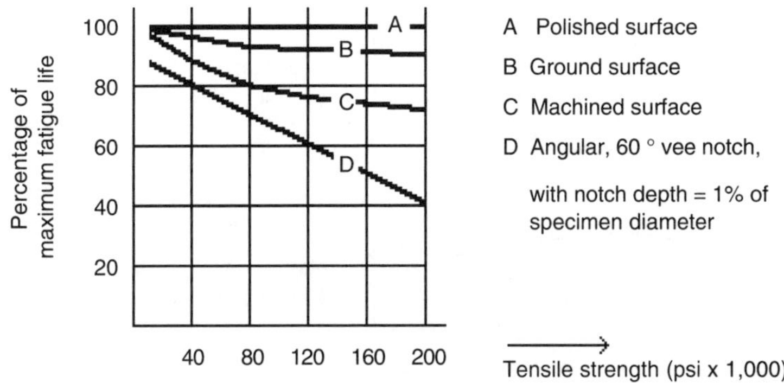

Figure 18.18 Effect of surface finish on alloy steels of varying strength and hardness.

Also, tool marks from grinding or stoning in the direction of stress have less damaging influence than those across the direction of stress.

18.9 Effect of Stress Risers on Fatigue Limit

18.9.1 Section Changes

Figure 18.19 shows the effect of changes in cross section on fatigue limit. The three examples shown apply to any part, round or not.

18.9.2 Stress Risers in Corners

Stress diagrams in Fig. 18.20 show the effects of stress risers on corners. The stress relationships shown apply to any change in section of a part. The values are approximate but highlight the importance of the need to provide proper radii in corners. The quality of surface finish has an additional effect on fatigue life. If the only concession available from the product designer is a small radius in a corner or section change, at least the surface in this risk area must be highly polished to improve the fatigue life. As for O-ring and snap ring grooves, a manufacturer's suggested finish and minimum size of radii in grooves must be followed.

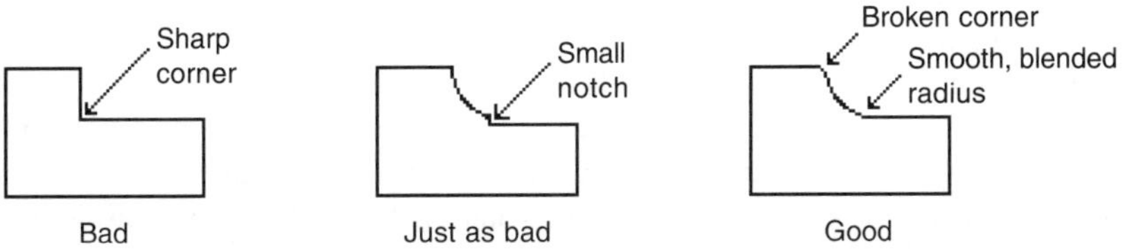

Figure 18.19 Effect of sectional change on fatigue limit.

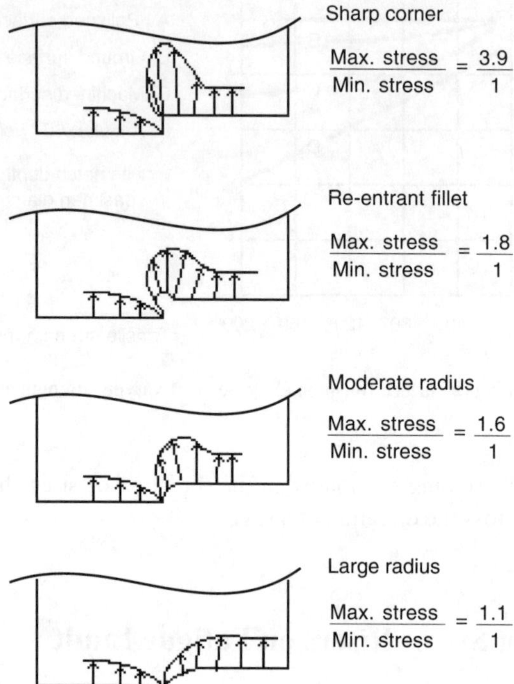

Figure 18.20 Stress diagrams for different corner stress risers.

18.9.3 Tool Marks

Holes and drilled bores must be clean, without tool marks. This applies particularly to water and melt channels.

Gun drills produce smoother walls than twist drills. Reaming after drilling is more expensive. Honing is very expensive and rarely used, but fluid-bed honing, typically inside melt channels, is regularly used, not only for better plastic flow but also to improve the fatigue life.

The corrosive effect of water on channel walls creates conditions similar to a bad finish. Corrosion will start at scratches (tool marks) and "eat" into the steel, creating a sharp notch. What is worse is that it is on the inside and cannot be seen until the part fails (water starts leaking).

Cavity walls are always highly stressed. At every cycle, the walls are stressed by the pressure of the injected plastic. Channels or screw holes in the wall can make the problem very complicated, because each drilled hole adds its own stress riser (Fig. 18.13) within the steel. If the channels are too close together, they act like perforations and have a severe influence on the fatigue life. In many cases, by "overdesigning", the designer can ensure proper performance of the part. But in critical areas, where space and weight considerations are important, the necessary calculations to arrive at the minimum safe size of the part must be made.

It is often difficult to arrive at acceptable stresses in a part because of the lack of data about the melt pressure within the cavity which causes these stresses. The designer must, therefore:

1. Use common sense when calculating wall thickness,
2. apply ample safety factors, and
3. avoid:
   a. excessive hardness,
   b. holes too close to each other,
   c. poor hole finishes, and
   d. sharp corners or notches.

It is seldom possible to avoid threads in mold parts, but the location of threaded holes can be selected so that the unavoidable stress risers (caused by thread cutting) are located in areas of low stresses. Screws must have clean threads and be free of scratches.

It is good practice to place, whenever possible, the threads in the softer material (e.g., into the plates at 35 Rc) rather than into the (usually much harder) stack part (cavity, core, etc.). Rolled threads in screws are better (fewer stress risers) than cut threads.

18.9.4 Splines and Key Ways

Splines and key ways are sometimes used in robots, or in drive mechanisms. The same considerations apply; avoid sharp corners, do not place them in highly stressed areas, and use good finish. Avoid creating a "perforating" effect created by locating internal grooves (or splines) too close to external grooves or to channels.

18.9.4.1 External Threads or Splines and Snap Ring Grooves

Examples of threads, splines, and snap ring grooves are shown in Figs. 18.21–23. Figures 18.24–26 show how fatigue life can be increased by removing material to change the part shape, depending on the load fluctuation.

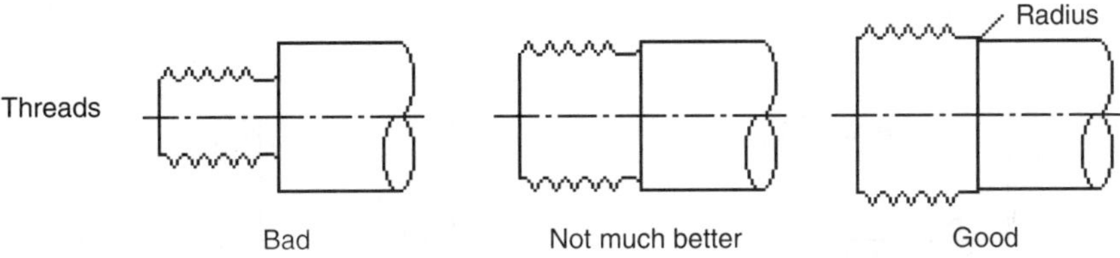

Figure 18.21 Effects of external threads on fatigue life.

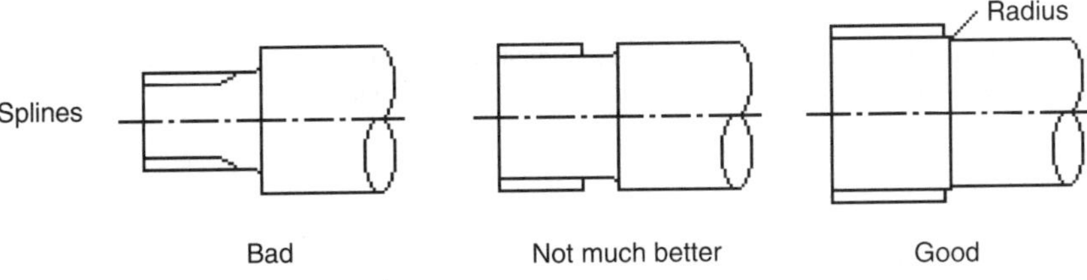

Figure 18.22 Effects of various splines on fatigue life.

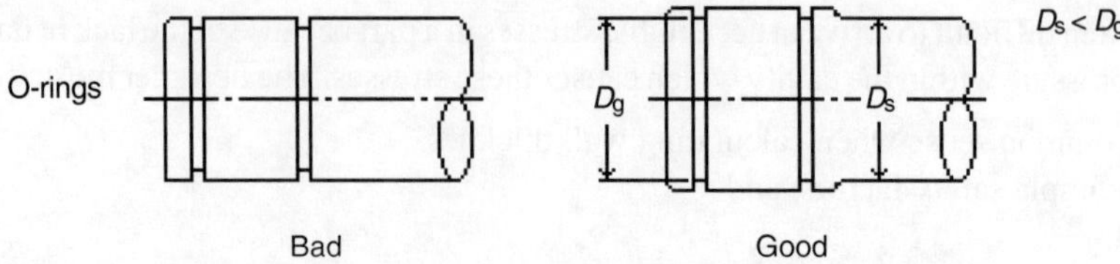

Figure 18.23 Effects of O-ring groove design on fatigue life.

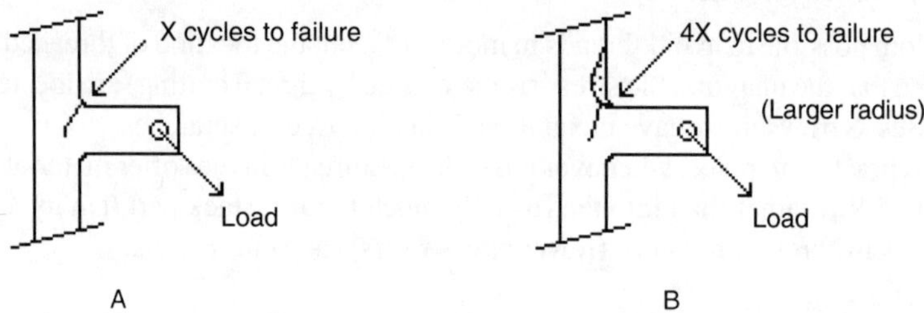

Figure 18.24 Load changes from Ø to 100%: A. original part shape, B. increase in fatigue limit by removing corner material to create larger radius.

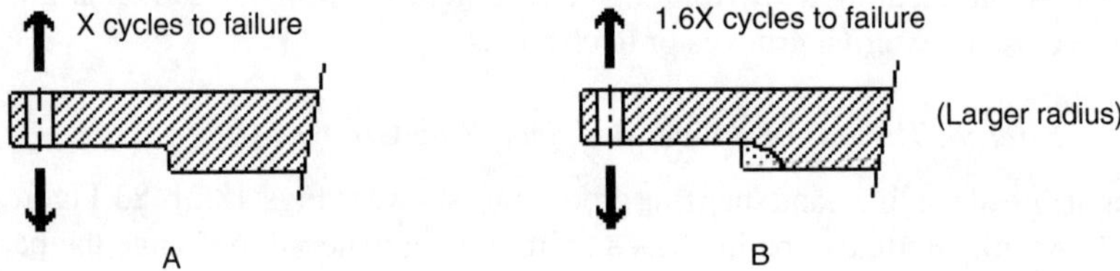

Figure 18.25 Load changes from –100% to +100%: A. original part shape, B. increase in fatigue limit by creating larger radius

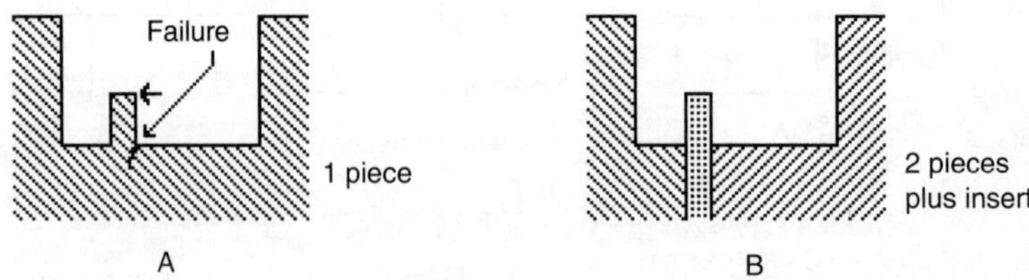

Figure 18.26 Projecting pin in a cavity block: A. pin made from solid block has too small a radius, and will fail, and B. a two-piece cavity with an insert for the pin results in no failure.

In a cavity or core block (body), if a projecting pin or blade made from the solid block cannot have a large enough radius where it joins the body, the body should be split in two pieces and an insert be used (Fig. 18.26). An alternative would be to make the body from one piece and machine (e.g., with EDM) a slot for the insert. With small round pins, or small blades which could be machined on top of round pins, a one-piece body with a bored hole and round insert could be used. Note that the bore (for the inserted pin) also acts as a stress riser in the one-piece body and will be worse if the bore for the pin is small. It is then preferable to make the bore much larger than the pin.

18.10 Fretting

Fretting refers to damage caused to the metal surface by a minute motion between two parts pressed against each other, such as by screwing the parts together or by shrinking one onto the other. One of the parts may be subjected to stresses which make the surface move ever so little but enough to cause friction between the crystals of the parts. The result is invisible surface cracks, and the transfer of metal in very small amounts from one part to the other.

Cracks are severe stress risers, and will make the stressed part fail if the stress level is too high. The only practical solution in such cases is to reduce the stress level by increasing the critical cross section. To use keys or dowels would be counterproductive, because of the stress risers associated with these elements. It is also important to use different alloys, with different grain structures and different hardness, to reduce the effect of fretting.

18.11 Welding and Fatigue

In general, welding is not used in molds, except occasionally for repairs. There are no data available for fatigue of welds in molds.

In general, welding should only be used as an expedient to keep production (molding) going until proper repairs can be made. "Proper repair" usually means building a new part and/or redesigning to provide an insert instead of the fragile, integral part that broke.

18.12 Other Areas Affecting Fatigue Life

18.12.1 Grain Structure

Figure 18.27 shows how direction of grain structure will affect the fatigue strength when testing a typical sample of a low-alloy steel bar with UTS of 120,000 psi.

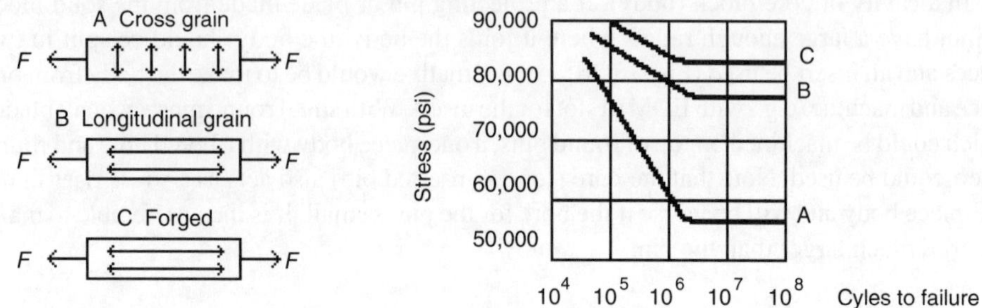

Figure 18.27 Effect of grain structure (left) on fatigue strength shown by S–N curves (right).

18.12.2 Location of Engraving

Every engraving "damages" the part surface and reduces its fatigue strength, just like a notch. The location of engraving must, therefore, be in an area where only minimum stress will be expected.

If the location cannot be changed, at least the finish of the engraving must not cause stress risers. Shallow depth, rounded inside corners, and even polishing will cause fewer stress risers. When using raised lettering (depressed in product), the corners where the engraving meets the surface of the part must be well-rounded and polished. A typical, sensitive engraved area in molds is at the gate pads.

18.12.3 Shot Peening

Shot peening is a technology used to increase fatigue strength of metals, particularly steel. Basically, it is a process similar to sand or shot blasting, where tiny hard steel balls are blasted at high speed against the part surface. The result is compression (preload) of a very thin top layer, which reduces the risk of surface cracks starting when under fatigue loading.

18.13 Examples of Fatigue

The following examples point out critical areas that affect the longevity of mold parts. The designer may say "but I have done this before and did not have any breakage." This may be true, but it is usually based on three facts:

1. There was no information available about the mold after it left the shop;
2. The mold was not old enough (not enough cycles) for the trouble to show; and
3. The pressures and forces were not high enough; therefore, the mold was suitable for the job.

This last comment needs some explanation.

When using high injection pressures at the machine, the pressure drop from the injection, through the machine nozzle, the (hot) runner system, and the gate is assumed to account for about 10,000 psi. As the plastic flows away from the gate, the pressure will drop even more as the plastic is pushed through the (narrow) space between cavity and core. During the filling of the cavity, there is much less pressure than after the cavity is filled (which would stress the steel).

The designer must now consider two possibilities:

1. *Injecting just the right volume (starve feeding)*: Since the volume injected is equal to the cavity space, there will be minimal pressure at the end of injection; therefore, the cavity will see little or no bursting pressure at all.
2. *Molding with "cushion"*: Pressure starts building up in the cavity after it is filled. Holding pressure will continue filling the cavity to compensate for shrinking.

With thin-walled products, the plastic, as it touches the cold cavity and core walls, freezes and restricts the passage for the plastic so that there is little bursting pressure left. The high initial pressure is used up rapidly to fill the mold. (This is the reason why high filling speed is so important.) Holding pressure is of no use once the plastic in the gate is frozen.

With heavy-walled products and/or with poorly cooled or with heated molds, there will be some restriction of the passages, but there will be enough "hot core" (viscous) plastic between the cooled layers to propagate the injection pressure to the cavity wall. The bursting pressure is only a little less than the cavity pressure after the gate.

This is a condition to watch for when considering the fatigue strength of a cavity. The fact that the high pressure may last only a short while, or that the holding pressure is lower than the injection pressure, is of no consequence. Fatigue failure is only based on the highest pressure (forces) the cavity regularly sees and on the number of cycles, not on the duration of these forces.

At the time of designing, it is difficult to calculate or to foresee which of the above conditions applies to a mold. The designer must use past experience and common sense. Computer simulations can give good forecast of pressures expected. Starve feeding is unusual and should not be used for assuming forces when designing a cavity.

18.13.1 Cavity Inserts

Cavities are stressed by the injection pressure p of the plastic at every cycle, from zero to full load. This is a typical loading with a stress ratio $R = 0$). The main problem is that we do not know exactly what p will be. At this time, it must be assumed, from past experience.

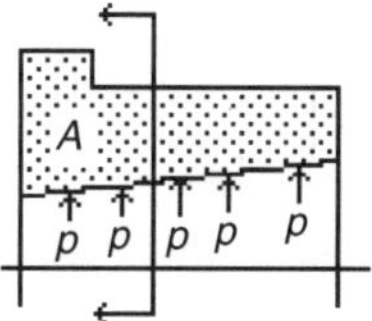

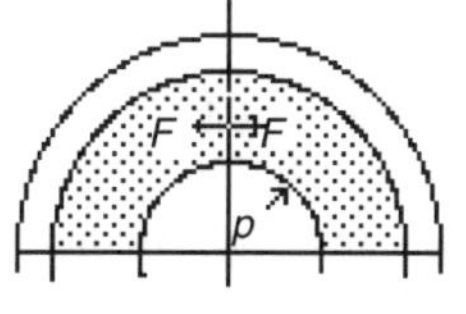

p = Injection pressure
F = Longitudinal cross section
A = Separating force caused by p
S = Stress in area A

$$F = K \times p \qquad S = \frac{F}{A}$$

Figure 18.28 Side (left) and top (right) cross sections of a cavity insert show relationships between pressure, separating force, and stress in area A.

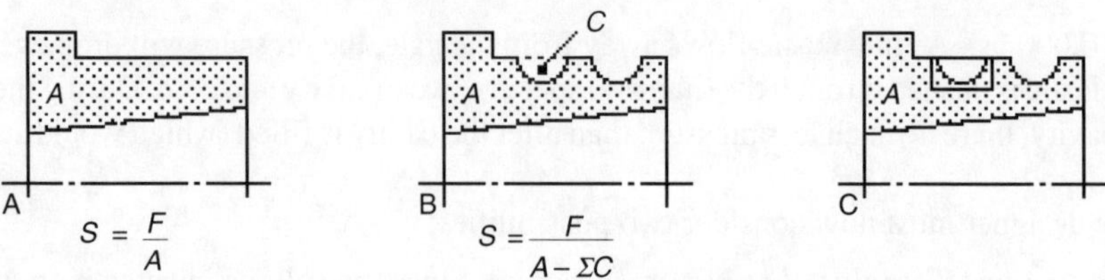

Figure 18.29 Three cavity wall cross sections and effects on fatigue strength: A. strongest, B. strong, C. weak.

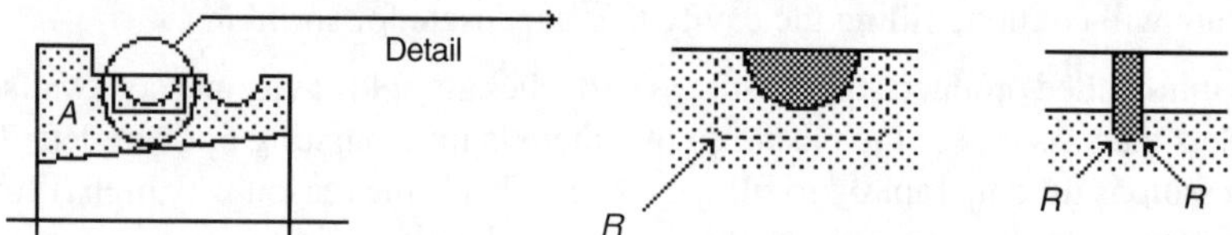

Figure 18.30 Cross section of cavity wall with baffle arranged in cooling channel.

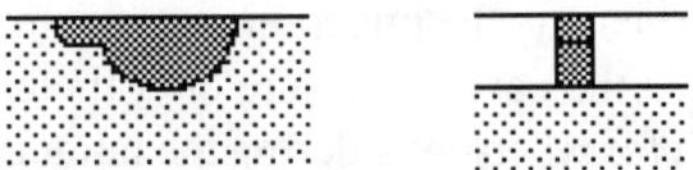

Figure 18.31 A shallow, rounded slot at the circumference of the insert holds a baffle adequately in place.

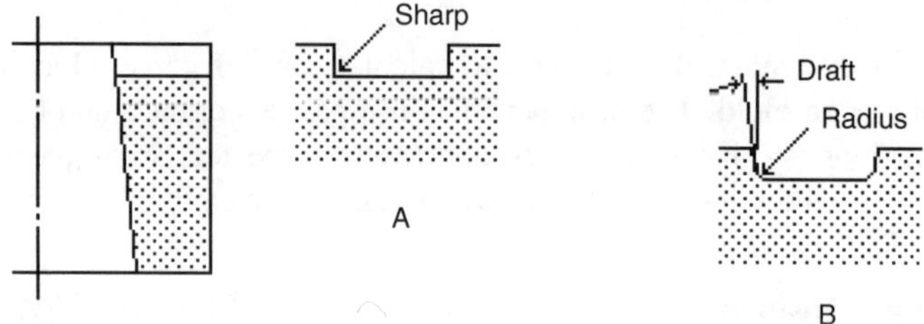

Figure 18.32 Notched cavity wall for receiving insert: A. sharp corners are large stress riser, and B. adding larger radius reduces stress.

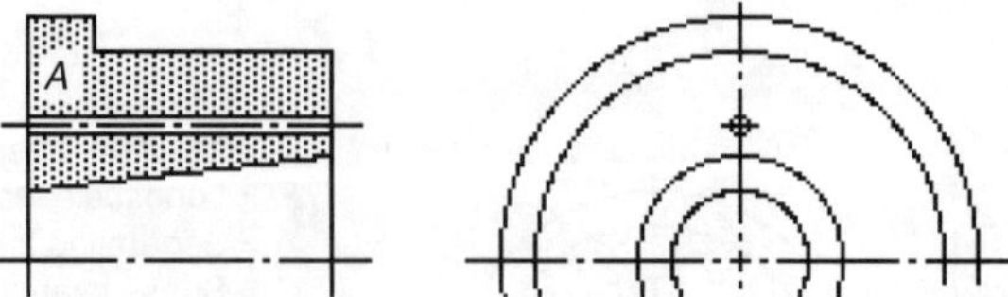

Figure 18.33 Side (left) and top (right) cross sections of cavity wall show location of drilled hole in area *A*.

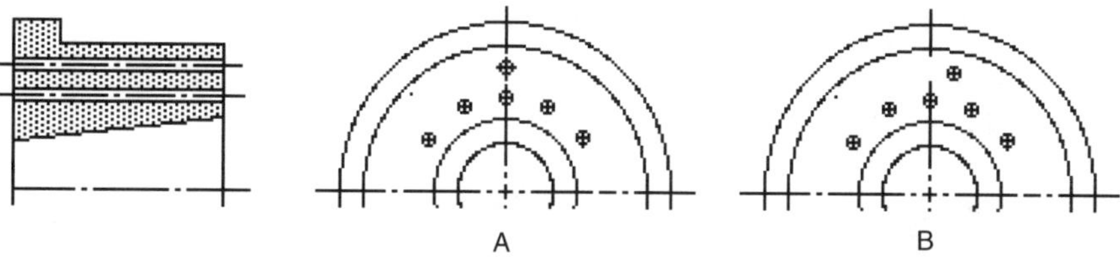

Figure 18.34 Two holes in-line with each other drilled in the cavity wall: A. bad, and B. better but not desired.

K in Fig. 18.28 represents a factor. The true relationship may be quite difficult to establish, but the fact is that the greater *p* is, the greater *F* (the force trying to expand the cavity insert) will be. *F* creates a stress *S* in the section A. The greater *F*, the greater is *S*; the smaller A, the greater is *S*.

The three examples in Fig. 18.29 compare the fatigue strength of similar container cavity walls. The cavity wall in Fig. 18.29A is strongest. The wall in Fig. 18.29B is only slightly less strong. Not much of the area A is lost, since the sum of groove cross sections C (ΣC) is usually much smaller than A. Provided there are no scratches in the grooves to create stress risers, *S* will be only slightly greater than in Fig. 18.29A. Because of the notching effect in Fig. 18.29C (for a baffle or any other reason, such as an insert), the wall is severely weakened (see Fig. 18.30).

Figure 18.30 shows a common baffle arrangement (exaggerated) in a cooling channel. Any slot cut into the side of the cavity insert (or any highly stressed area) creates a severe notch, and will reduce the fatigue life considerably. A radius *R* will help, but the area A is (in this example) also reduced quite a bit, which increases stress *S*. Additional stress may be induced by hammering the baffle (or any insert) into the slot.

Figure 18.31 is a much better design. A baffle does not need to enter into the wall of the channel, it can be held in position with a relatively shallow, well-rounded slot at the circumference of the insert. There is no need for perfect, watertight sealing-off of the channel by the baffle. A little seepage around the baffle is insignificant in relation to the main coolant flow.

Figure 18.32 shows a frequent feature, a notch in the cavity to receive an insert. The notch reduces the area A but also induces a tremendous stress riser in the sharp corners. Sharp corners (Fig. 18.32A) may be desirable for easier mold making but must be avoided if long mold life is important. Figure 18.32B shows a much better method with a radius as large as possible, and with a draft angle on the sides. Any insert must fit easily, without force needed to drive it in.

18.13.2 Drilled Holes in Cavities

Holes in cavities cannot be avoided, but the designer must realize that these holes act as stress risers. The smaller the hole, the higher are the local stresses. Also, the worse the finish of the hole wall, the higher are the stresses. The influence of these stress risers is greater than the effect of loss of area in section A of the cavity wall (Fig. 18.33), from the diameter of the hole.

The worst possible case is when two or even more holes are in line, which not only reduces the area A but puts more stress risers in the area of the highest stresses. In Fig. 18.34, case A is really bad and should be avoided; case B is somewhat better, but not ideal.

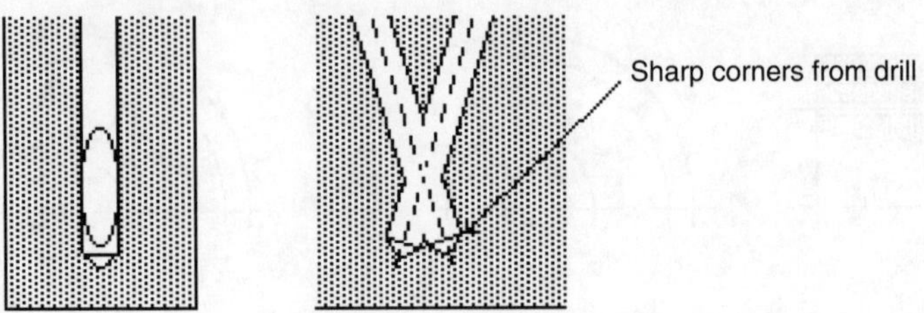

Figure 18.35 Sharp corners or tool marks should be smoothed.

Another little appreciated fact is the bad influence of the finish at the end of drilled holes, as shown in Fig. 18.35. If sharp corners or tool marks inside a steel block cannot be avoided, they should at least be smoothed out after machining.

References

1.Smith, C. (1984). *Engineer to Win*. Motorbooks International, Osceola, WI.

19 Screws in Molds

19.1 Threaded Fasteners

In industry, terms "screws" and "bolts" are used indiscriminately. Both screws and bolts are used to hold (clamp) together two or more parts of an assembly. It is preferable to call a screw a fastener, which is threaded into another part, while a bolt requires a nut to complete the assembly (Fig. 19.1).

Molds generally use only screws. The following considerations apply to both screws and bolts, but only the term "screw" will be used. If screws thread into soft materials (e.g., aluminum) and cannot be torqued sufficiently, (steel) nuts may be used rather than threading the soft material.

19.2 How Screws Work

The following section is not a highly scientific approach to the understanding of how screws work, but it should help the designer to select quality, size, and number of screws.

Essentially, every "cap" or "hex(agonal) head" screw consists of a head and a threaded shank. When the screw is turned into the receiving (female) thread in a part (or in the nut), there is at first no resistance except friction, until the joint touches and the head comes to rest on the part. Further turning (torquing) of the screw will begin to stretch the screw, as a result of the wedge action exerted by the threads. Wedge action can be simply explained when projecting the thread onto a flat surface (Fig. 19.2). As the screw is torqued, the wedge action pulls on the shank (Fig. 19.3).

Using the example and variables shown in Fig. 19.3, the following equations are derived for torque, *FW*, and *FR:*

$$\text{Torque} = F \times R = FW \times r. \tag{19.1}$$

FW is the tangential force (at radius *r*) exerted on the screw by the torque used to tighten the screw. Therefore, the smaller the screw diameter, the less torque is required to tighten a screw.

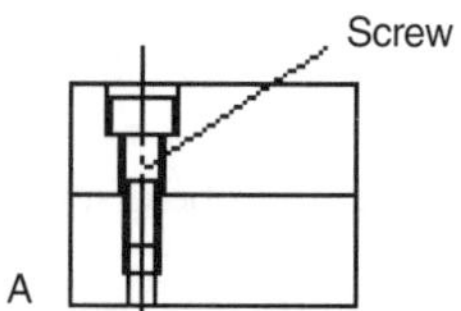

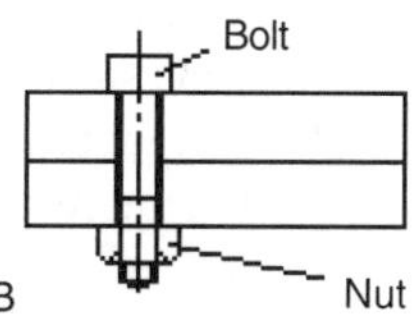

Figure 19.1 Threaded fasteners: A. screwed joint, and B. bolted joint.

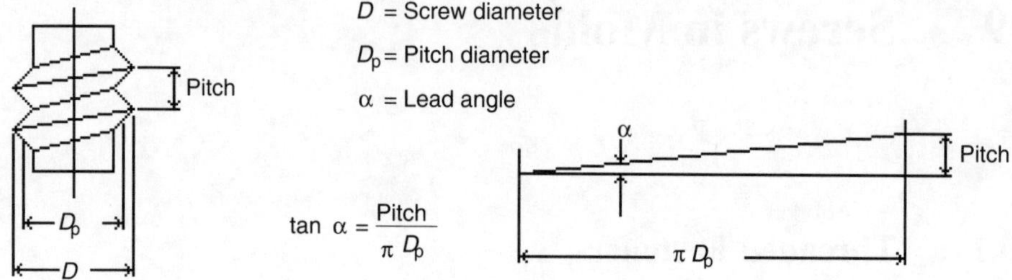

Figure 19.2 Screw diameter and pitch diameter calculations.

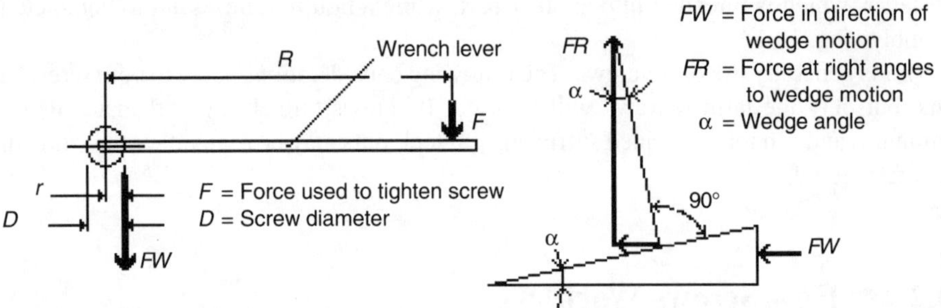

Figure 19.3 Torque and the tangential force (*FW*) exerted on screw by torque (left), and wedge angle and wedge motion (right).

Since $r = \frac{D}{2}$, substituting in equation 19.1 gives the equation for *FW*:

$$FW = \frac{2(F \times R)}{D}. \tag{19.2}$$

The smaller the wedge angle, the greater is the axial force *FR*, which acts in the screw. In other words, with a finer thread (on the same diameter), a greater force can be exerted on the joint.

From the relationships shown in Fig. 19.3 between the wedge angle and wedge motion, the equation for *FR* is:

$$FR = \frac{FW}{\tan \alpha} = \frac{2(F \times R)}{D \tan \alpha}. \tag{19.3}$$

All screws used for fastening have a lead angle or wedge angle smaller than the friction angle between steel and steel, even when lubricated, and the screw is inherently self-locking (i.e., it will under normal circumstances not unscrew after tightening). However, there are conditions which do require securing screws after tightening; they will be discussed later.

19.3 Types of Screw Threads by Method of Manufacture

19.3.1 Rolled Threads

Rolled threads are made using rollers to force (squeeze) the steel into the required thread form (Fig. 19.4). This produces stronger threads, with good flow of grain within the steel; the threads are without tool marks which could induce notching and premature failure. Before rolling, the diameter of the length of the screw to be threaded is approximately the same as the *pitch* diameter.

All "property class 12.9" screws up to 20 mm Ø and up to 150 mm long have rolled threads. All screws over 20 mm Ø and, also, all those over 150 mm long may be rolled, cut, or ground at the manufacturer's option.

19.3.2 Cut Threads

Cut threads are machined into blanks. There is always the danger of tool marks (scoring) at the root of the thread and at the run-out which may become the start for progressive failure of the screw. The *shank* diameter is approximately the same as the major thread diameter (OD) of the screw (Fig. 19.4).

19.4 Types of Thread Pitch: Fine or Coarse Threads

In the inch-based measurement system, the pitch is given by the number of threads per inch (TPI). For example, a screw thread specified as ½-13 means: ½-in. major screw diameter, with a pitch of (1 in. ÷13), or 0.077 in.

In the Metric system, the pitch is given in millimeters. For example, a screw specified M12-1.75 means 12 mm major diameter, with a pitch of 1.75 mm. With coarse screws (standard), the pitch need not be shown (e.g., M12 would be sufficient).

With screws of the same major diameter, the finer thread screw has a larger root diameter than the coarser thread and is, therefore, somewhat stronger (Fig. 19.5). Also, because the lead

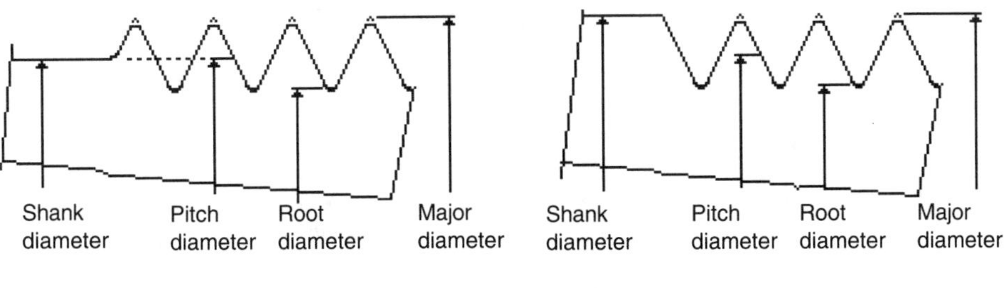

Figure 19.4 Schematics of diameters for A. rolled thread and B. cut thread screws.

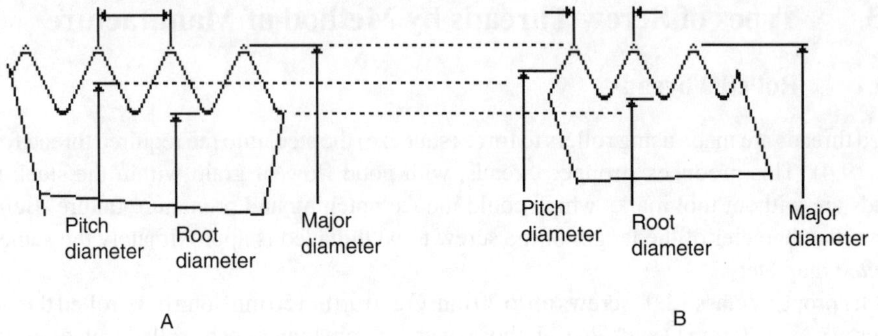

Figure 19.5 Relationship of diameters in A. coarse thread and B. fine thread screws.

angle is smaller, the force generated by torquing is greater with fine threads. Coarse threads are cheaper to produce and more resistant to damage and corrosion.

As a rule, the designer should always use standard, easily available screws. The use of specially made screws is not recommended. In molds, generally, only coarse thread screws are used.

19.5 Holding Action and Preload of Screws

The following schematic (Fig. 19.6) and simple formulae will help the designer understand the principle of fastening with screws.

The stretching of the screw f is expressed by the formula:

$$f = \frac{\Sigma F \times L}{n \times X \times E}, \tag{19.4}$$

where n is the number of screws and E is the modulus of elasticity (other variables are defined in Fig. 19.6). The stress in the screw S is calculated using the formula:

$$S = \frac{\Sigma F}{n \times X}, \tag{19.5}$$

and S must be less than the permissible stress, S_p ($S < S_p$).

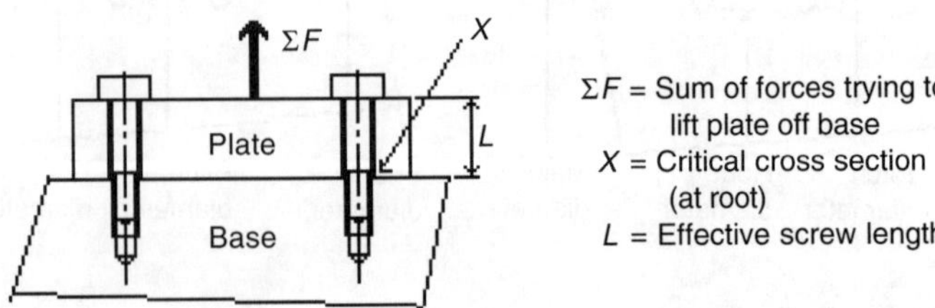

Figure 19.6 Schematic of screws fastening a plate to a base, with variables defined and equations shown.

In discussion of the foregoing schematic and formulae, many points may be made:

1. The sum of forces $\sum F$ may be an external (pulling) force or may be generated by internal (e.g., plastic) pressure between plates.
2. Assume the screws are just hand-tightened, where the heads touch the plate, but they are not torqued. As the external force $\sum F$ acts, the screws will be pulled and thereby stretched, as per step 1; the plate will lift off the base.
3. To prevent the plate from lifting, the screws are tightened before applying load $\sum F$. By use of a wrench, the screws will be torqued to a suggested value, which in fact stretches the screws within the effective length *L*. This stretching is equivalent to a preload on the plate.

   As long as the preload is greater than the load $\sum F$, the plate will not lift off the base. If the force $\sum F$ reaches the preload, the forces in both directions are balanced. When $\sum F$ becomes greater than the preload, the plate will lift off the base.
4. The amount of preload must be such that the stresses (*S*) in the screw are less than the permissible stress (*S*p). The permissible stress is usually selected to be about 75% of the yield stress of the screw material used. Recommended torque for each size screw is shown on screw charts and data files on screws (see also Section 19.10).
5. Elongation (*f*) is solely related to the modulus of elasticity (*E*) and to the sum of forces $\sum F$. Hardness or tensile strength do not affect elongation. (Soft screws will stretch as much as hard screws.) However, the yield strength of hardened screws is much higher than that of soft screws, so that hard screws can be preloaded much higher than softer ones.
6. The critical point (*X*) of a screw is the cross section of the first thread root, as indicated on Fig. 19.6. The whole screw force goes through there, and this is the section that will break when the screw is overloaded.
7. If screws are not tightened at all, there is no force to clamp the parts together. This may be required for special applications where two parts must be held together without being clamped. In this case, the screws must be secured from loosening, since there is no force to provide self-locking of the screw.
8. It is important that the designer has a good estimate for the amount of force ($\sum F$) that must be secured by screws. Often, there are more screws used than really necessary, which increases unnecessarily the cost of the mold.
9. If more than one screw is used ($n > 1$), it is important that all screws are torqued equally to share the load. If not, some screws may bear more than their share of the load and be stretched beyond their yield points; these screws will be permanently stretched and completely loses their holding force. The other screws will, therefore, have to take more than their share and may also fail.

   This fact also implies that screws should be located as symmetrically around the load as possible to prevent uneven loading of these screws. Forces acting on a plate are usually distributed evenly on the plate. By adding extra screws at the point of maximum deflection, the plate deflection can be greatly reduced (Fig. 19.7). Note that the plate thickness has a large influence on the selection of number of screws.

   The deflection of a plate is inversely proportional to the third power of its thickness *t*, and directly proportional to the distance *M* between supports or screws (Fig. 19.8).

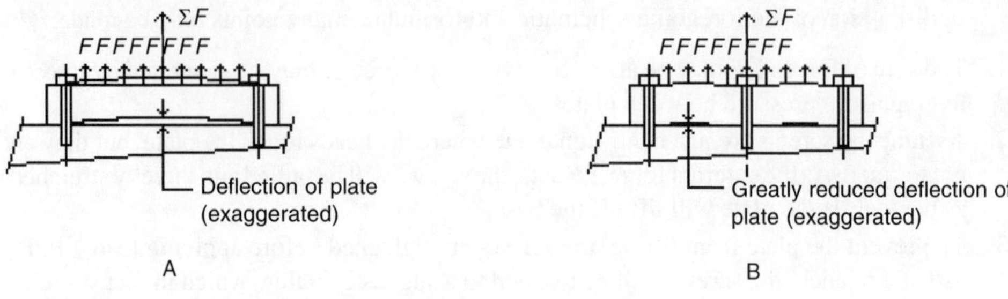

Figure 19.7 Forces acting evenly on the plate result in A. much deflection when not enough screws are present and B. little deflection if screws are placed at the point of maximum deflection.

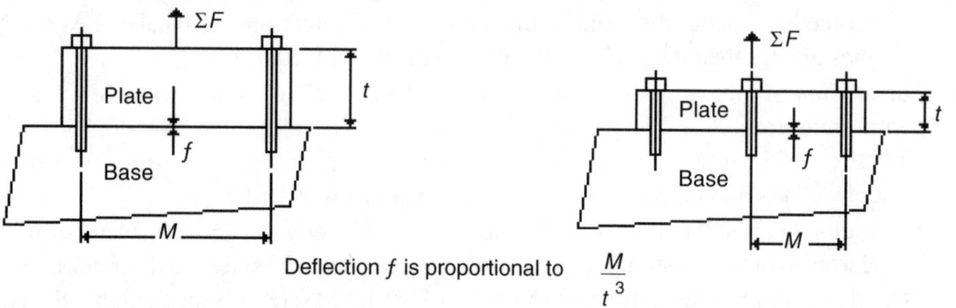

Figure 19.8 Example of thick (left) and thin (right) plates where deflection of thin plate is four times that of the thick plate.

Using the relationship illustrated in Fig. 19.8, in a thick plate, given that $M = 8$ and $t = 4$,

$$f \text{ is proportional to } \frac{8}{4^3}, \text{ or } \tfrac{1}{8}.$$

Similarly, in the thin plate, given that $M = 4$ and $t = 2$, f is proportional to ½. This shows that the thin plate will deflect four times the amount of the thick plate, even though there are more screws. This clearly indicates that with a thinner plate, many more screws will be required to hold the same force $\sum F$ than when using a thicker plate.

10. Note that the cost of a screw joint is not only the cost of the screws, which is usually not very high, but also the cost of drilling and tapping one plate, drilling (and counterboring) another plate, and, possibly, drilling additional plates held together by the same screws. Add to this the fact that the space for unnecessary screws could be used to simplify locating other mold parts or water lines, etc.; and note that this could all add up to considerable savings.

    A larger screw takes not much more space than a smaller one. For example, by selecting 12-mm instead of 10-mm screws, the number could be reduced from 6 to 4, from 12 to 8, or from 18 to 12 screws, which would constitute considerable savings.

    Also, larger drills and taps operate more efficiently than smaller ones and do not break as easily, particularly for thread sizes below 8 mm.

19.6 Influence of Temperature on Screws

Two models, A and B, are used to illustrate the influence of temperature on screws. In model A, a heated "body" is held between a cold plate and a base; the screws are at the same temperature as the plate (Fig. 19.9).

By preloading screws to achieve a tight joint, the screws will be stretched and created within each screw will be a stress S (equation 19.5), which must be smaller than S_p, the permissible stress. If the body is now heated, it will expand in all directions. The designer is concerned about the expansion in direction L_H, which will apply additional forces to the (cold) screws. Preloading one screw, to ensure a tight joint, will create a stress

$$S = \frac{\Sigma F}{X} < S_p. \tag{19.6}$$

As heat is applied, the body expands to create an additional force F_h as its temperature rises to T_f, while the surrounding plates (and screws) remain at their original temperature T_0. The force F_h is proportional to the expansion of the body f_h, which in turn is proportional to the coefficient of heat expansion K for the material of the body, the length L_H, and the temperature difference ΔT between T_f and T_0 ($\Delta T = T_f - T_0$).

$$f_h = K \times L_H \times \Delta T \tag{19.7}$$

f_h will add to f resulting from the preload and may stress the screw beyond its permissible stress S_p.

From equations 19.4 and 19.5 can be obtained the equation:

$$S = \frac{f \times E}{L}, \tag{19.8}$$

where f is stretching caused by torquing. When adding f_h (stretching caused by heat) to f, the total stretch is

$$f_t = f + f_h \tag{19.9}$$

and the total stress will be

$$S = \frac{(f + f_h) \times E}{L} < S_p. \tag{19.10}$$

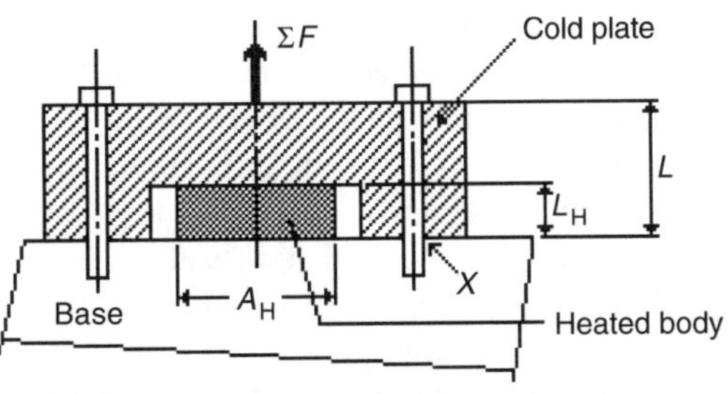

X = Critical cross section of screw (at root diameter of thread)
A_H = Cross section of heated body

Figure 19.9 Model A: a heated body held between a cold plate and a base with screws at plate temperature.

From equation 19.4, f is inversely proportional to $n \times X$. Heat expansion f_h is independent of the number or size of screws. The sum $(f + f_h)$ is, therefore, only partially dependent on the size and number of screws.

As long as $n \times X$. is much smaller than A_H, increasing the diameter of the screws or their number has no significant effect on the stress in the screws. If A_H becomes smaller, the $\sum F$ will not only stretch the screws but also act increasingly as a compressive force on the body, and when A_H becomes much smaller than $n \times X$, compression may crush the body.

To ensure that the screws (or the body) are not stressed beyond their permissible stress (yield), the designer has three alternatives:

1. The preferred method is to reduce the cold length L_H; in other words, create a cold clearance between the heated body and the plates. This clearance (cold gap) will disappear as the body heats up. Since the size of f_h is usually very small, it is important to check the tolerances so that under maximum conditions, the loads will not be exceeded, and under minimum conditions, the body is not loose between the plates.
2. Another method is to reduce the original preload so that even with the added stress caused by heat expansion, the permissible stress is not exceeded. If this route is chosen, the (exceptionally low) torque on these screws must be clearly indicated on the assembly drawing.
   NOTE: Torque figures are not normally shown on the drawings, except:
   - where they deviate from standard torques for the screws,
   - when requested by the customer, or
   - when deemed critical for any reason.
3. A third method is to increase the length L of the screws. Because of the lack of available space, this may not always be possible.

In model, B, the conditions are the same as those in the first model (A), but the whole system works at an elevated temperature. In this case, the designer must be concerned with:

1. The effect of higher temperatures on the yield strength of the screws, and
2. the effect caused by heating up at different speeds.

High quality, "12.9 screws" are tempered at a minimum of 180 °C (SI), or 345 °F, and the yield strength is virtually unaffected as long as they operate below this temperature. Above this temperature, the yield strength can diminish rapidly; therefore, either the rated load on these screws must be reduced proportionately or special, high-heat screws must be specified.

The use of special screws, for example for high-heat, should be used only in exceptional cases, and when they are physically (and in appearance) distinct from the regular screws within the mold, because of the possibility that the customer may mix the screws when servicing the mold. This should also be shown prominently on the assembly drawing.

The effect of heating up at different speeds will be clear with an example, shown in Fig. 19.10. The screw is heated by the surrounding hot body; however, the heat flow is restricted to pass through the small areas at a and b, while little (radiated) heat passes through the gap around the screw, over the length L. This creates a slow rise in temperature of the screw, while the heated body heats up and expands more rapidly.

If the screw was tightened with recommended torque, the additional stretching caused by the expansion of the body can raise the stress in the screw beyond the permissible limit, and

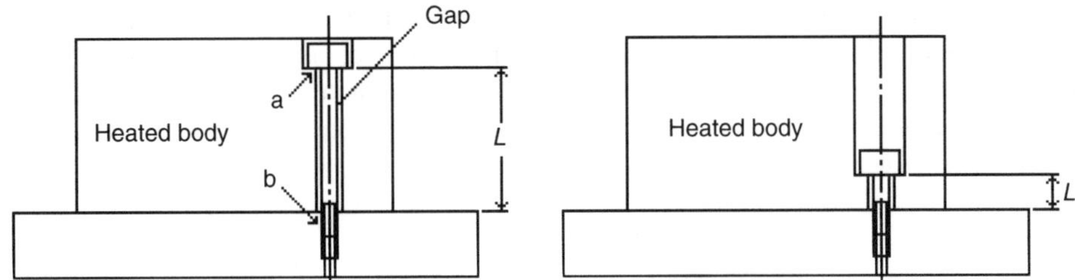

Figure 19.10 Stretching of a screw because of heat expansion of the body may cause a long screw (left) to fail; a shorter screw (right) heats faster with the heated body.

the screw will fail. To preload the screw to a less than recommended value will not help, since once the screw reaches the temperature of the body, there will be not enough preload to hold the heated body on the base securely.

In this case only, the recommended solution is to shorten the screw, so as to reduce the (not contacting) length L, which can then more rapidly follow the temperature rise in the body. This is particularly important if the heating-up time is short, as is sometimes the case in hot runner systems.

If the whole system heats up slowly, it is considered to be equivalent to a cold system, except that high-temperature screws may be required if the temperatures are at or above the annealing temperature of standard screws.

19.7 Effect of Cyclical Loads on Screws

Preloaded screws will stretch by f, but the plate will be compressed by the amount f_p because of the effect of this preload. However, because the cross section (A_p) of the plate (or in general the body which is held down) is usually much greater than that of the screws (A), or $f >> f_p$, (">>" means "much greater than"), the effect of compression by the screws can be ignored.

When the plate is subjected to a large clamping force F_c, while the mold is closed, F_c will compress the plate by the amount f_c. If this compression (f_c) becomes equal to or larger than the stretch caused by preload (f), the screws are temporarily unloaded. The screw head may even lose contact with the body held down. This means that during clamping, the screws could become "not loaded" and the threads not "self-locked", and the screws are, therefore, free to loosen, which may happen gradually. They may then gradually lose their preload completely. Also, the screws are now subjected to a cyclical load and can fail much sooner, since they are less tightened.

To prevent this full unloading and subsequent unscrewing, the designer must make sure that the screws are long enough to stretch when torqued (f) more than the expected compression (f_c); or, $f > f_c$. This is even more important when the loading is applied suddenly (e.g., by shock). If shocks are unavoidable and of a force which may tend to unload the screws, the screws must be secured from unscrewing by some reliable method. Short screws will loosen more readily than longer ones because they cannot be stretched enough. Heavy vibrations have a similar effect to that of frequent shocks and may cause the screws to loosen, particularly if the screws are short and it is not possible to stretch them sufficiently.

In molds, the plate sizes and cross sections A_p are usually in proportion to the clamp size and force, and the effect of plate compression can be ignored, but with the full clamp force acting on smaller units such as cavity modules, the compression f_p may be significant. Note that plate compression f_p must be calculated to include cored-out areas and steps, to arrive at a reasonable value.

19.8 Lubrication of Screws

Lubricated screws are self-locking. The coefficient of friction of steel on steel, well-oiled, is greater than the tangent of the lead angle of the screw fastener.

It is good practice for all screw threads to be oiled before assembly and, when used in hot applications, to be treated with lubricating antiseizing compounds containing aluminum, copper, or other powdered metals to create a barrier between the threads, because:

1. A screw should offer minimum resistance to inserting, to show clearly where the torque begins to be applied to stretch the screw. (Torque should not be used up just to overcome friction). It should be possible to assemble by hand any screw until its head comes to rest.
2. Lubrication will help prevent the threads from seizing, particularly after long periods of use.

Mold assemblers usually know the need for lubrication of screws. However, it is necessary to indicate on the assembly drawing where antiseizing compounds are required.

19.9 Steel Screw Standards

19.9.1 Screw Steel Property

There are standards for many types and qualities of screws, but the designer is essentially concerned with screws used in molds. Today, most molds use only metric screws to ISO standards of quality, or better. The screws are made from medium carbon steel with additives, quenched and tempered.

SI standard hardness is 39–44 Rc; U.S. standard hardness is 36–38 RC for larger screws and 38–40 RC for smaller sizes. Steel (property) class is expressed with a number which must be stamped in the screw head. Recommended are socket head cap screws (SHCS) property class 12.9. The numbers signify as follows:

The first number (12) equals 1/100 of the tensile strength, expressed in N/mm^2 (e.g., 12 means a minimum of 1200 N/mm^2, or $\approx$120 kg/mm^2, or an equivalent of 170,000 psi).

$$1\ \text{N} = \frac{1}{9.806\ \text{kg force}} = 0.102\ \text{kgf}.$$

The second number (9) indicates 10 times the ratio between tensile and yield strength. In the example class 12.9, it means that the yield strength is 90% of the tensile strength, or

$$170{,}000\ \text{psi} \times 0.9 = 153{,}000\ \text{psi}.$$

If the screws are made from low-carbon, martensitic steels, the symbol on the screw head is underlined, (e.g., class 12.9).

In wet areas, stainless steel screws should be used, which have a lower yield strength. Areas where the required screw strength is high must be kept dry by use of O-rings. The finish of SHCS is preferably black, and the heads of SHCS are preferably knurled.

Nonferrous screws (brass, nylon, etc.) are used only in exceptional cases. Brass is often used for electrical connections.

19.9.2 Thread Tolerances for 12.9 Screws

Standard tolerances for ISO threads are:

- External: major diameter, g6, and pitch, g5
- Internal: minor diameter and pitch, H6

19.9.3 Length of Thread Engagement

The strength of the screw joint is also determined by the length of the thread engagement with its mating part, or with a nut. The threads themselves, at the bottom of the thread (root diameter) are loaded by bending and shear. (For practical purposes, the effect of bending is neglected.) The area (A_s) resisting shear is a cylinder with a diameter D_r (root) and length L_t (the active thread length), or

$$A_s = \pi \times D_r \times L_t. \qquad (19.11)$$

The maximum strength of the screw joint is then

$$S_{max} = S_s \times A_s \times L_t. \qquad (19.12)$$

S_s (shear strength) is ≈50% of S_y, the yield strength of the screw (or the female thread) material, whichever is lower. Assuming $L_t = D_r$,

$$S_{max} = 0.5\ S_y \times D_r^2 \times \pi, \text{ or}$$

$$S_{max\ (shear)} = \frac{\pi}{2} \times S_y \times D_r^2.22 \qquad (19.13)$$

For the same screw, the maximum permissible tensile strength equals

$$S_{max\ (tensile)} = \frac{\pi}{4} \times S_y \times D_r^2. \qquad (19.14)$$

These equations show that with $L_t = D_r$, a screw is twice as strong in shear than in tension. For this reason, standard nuts have a thread length (nut thickness T) of only 0.8D, which gives them ample reserve, even if the nut material has a lower yield strength than the screw.

For safety, the length of the screw engagement in a mold part (with a hardness of at least 30 Rc) should be at least the same as the diameter (D) of the screw, but not more than 1.5D.

The length of the first thread (pitch) of both screw and female thread is lost because of chamfers in both parts to permit easier assembly. Excessive engagement is costly because of added, unnecessary machining time and higher screw cost.

In practice, to accommodate standard screw lengths, the designer will select a length of engagement between 1 and 1.5 of the screw diameter as the active length. Exception: If the female thread is cut in softer materials (aluminum, copper alloys, cast iron, etc.), with a yield strength substantially less than that of the screw, the engagement must be proportionately larger to avoid deforming or stripping of the thread when torquing the screw to its limit. An engagement of twice the screw diameter or even more may be required.

It should also be noted that most screw joints will corrode, especially if they are not opened for a long period of time, and become difficult to unscrew. In such cases, excessive length of engagement aggravates the situation and may cause the screw to break during unscrewing. It may then be difficult and costly to remove the broken piece of screw.

19.9.4 Additional Suggestions Regarding Screws and Screw Joints

19.9.4.1 Odd Screw Sizes

Use standard size screws wherever possible. It is costly and, therefore, bad practice to modify a screw to fit an application.

The following “Don’ts” describe three typical and frequently done modifications which should be avoided, and which should be used only if absolutely necessary.

1. Don’t cut a screw down from a standard length. Usually, by changing the depth of the counter bore for the head, a suitable standard length can be found.
2. Don’t machine the thread to increase the length of the threaded section. This is bad because cut threads are weaker than rolled threads. The threaded portion of a screw is always at least 2.5 times larger than its diameter and should always be suitable. If necessary, use a larger screw or redesign this area.
3. Don’t reduce the size (diameter and/or height) of the screw head. This can seriously weaken the screw joint. Reduce the screw size and increase the number of screws, or redesign the area requiring the smaller-than-standard head.

19.9.4.2 Depth of Counterbore

The depth of the counterbore (C/B) should be approximately 1 mm more than the height of the screw head so that the head will not protrude above the surface of the mold part. It can be deeper to allow use of a standard size.

19.9.4.3 Space for Wrenches

Most screws in mold assemblies are tightened with free access and present no problem. However, if a screw must be tightened after other parts are in place, and/or where an overhang covers some of the area above the screw head, the design must provide sufficient vertical and lateral clearance so that a “hex” (hexagonal) driver can be inserted and rotated by at least 90°. Note that ratchet driven hex drivers require more space than the slower to operate L-shaped (Allen) keys.

19.9.4.4 Other Screw Head Types

A screw derives its holding force from being stretched as it is torqued. The torque is transmitted to the screw through a hex key, a screw driver, or any of a variety of methods used to drive screws (Phillips, etc.).

The hex key sockets in screw heads for standard socket head cap screws (SHCS) are designed for the recommended torque required to tighten each size. Special, low-profile heads of the same screw size cannot be tightened as much, because of the reduced depth of the socket and the lower rating (class 10.9) of the steel. Similarly, the sockets in round, flat, oval, etc., heads are usually one size smaller than the sockets for the equivalent size SHCS and cannot, therefore, be tightened as much as a standard SHCS. Note that these screws are also rated at 10.9, 8.8, or less. These screws may be used in areas where loads are small, but they may then require some method to prevent loosening, particularly in areas with shock and/or vibration.

Hex head screws are not used in molds. The strength of a hex head screw is less than that of a SHCS, and more screws must be used to compensate for this deficiency. Also, because of the lack of length, these screws must be prevented from loosening, usually with lock washers. They also require more space for access by wrenches or spanners than a SHCS.

19.9.5 Shoulder Screws (Stripper Bolts) and Set Screws

The use of shoulder screws is discouraged except (occasionally) in applications where they may be used to limit a motion (e.g., of a plate) without contacting the plate at every cycle during normal mold operation. They are meant to limit the stroke only during servicing of the mold (e.g., to prevent a plate from falling during handling of the mold). The reason is that these screws cannot be tightened sufficiently to prevent them from coming loose, because there is no substantial length *L* to be stretched (see Fig. 19.11A) Actually, there is a slight length *L* because the screw threads always terminate about 2 mm below the shoulder. It is good practice to use Loctite™ to prevent a shoulder screw from loosening. If cyclical action of a shoulder screw is required, a combination of sleeve and long screw must be used to achieve a long *L* (Fig. 19.11B).

Set screws may be used when it is not practical to use a standard SHCS (e.g., when fastening a rod to a plate). Because there is no length *L* to be stretched (Fig. 19.12 A), it is good practice to counterbore one side (or both) so that the threads are free (not engaged) for a

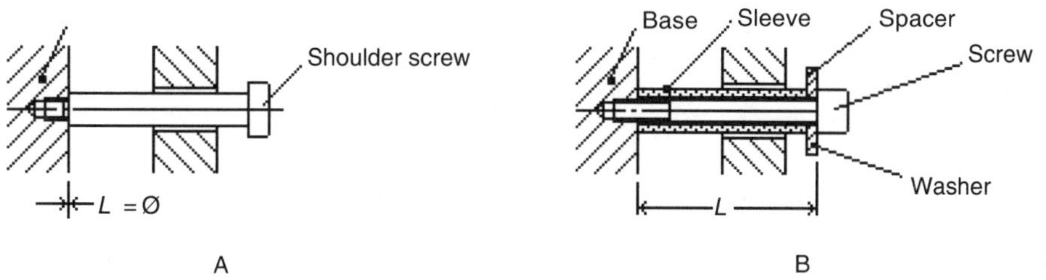

Figure 19.11 Shoulder screws: A. holding plate to base for mold handling only and B. a longer screw with a sleeve where cyclical action is required.

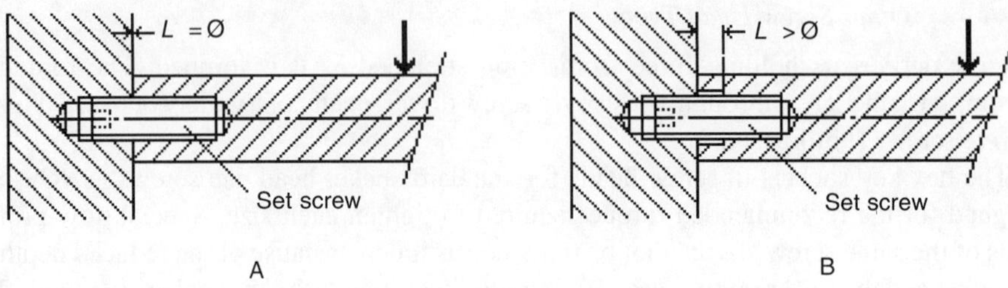

Figure 19.12 Set screws: A. $L = Ø$, and B. counterbore creates $L > Ø$.

distance of at least 3–4 mm, thus creating a length $L > Ø$ which can be stretched (Fig. 19.12B). This improves greatly the strength of the joint. It also prevents the screws from breaking in case of a blow hitting the rod sideways, as shown with the heavy arrow. This method for reducing the thread engagement can also be used with a standard SHCS if the length L appears to be insufficient to ensure a good joint.

19.10 Screw Data

Tables 19.1 and 2 show a proof load, a test of the actual screw, which is about 10% less than the yield strength of the bolt. This proof load is established by testing a machined test specimen [1].

Table 19.1 Specifications for Metric Socket Head Cap Screws (Coarse Thread) of Class 12.9 for Thread Sizes M3–M10

Specifications	Thread size					
	M3	M4	M5	M6	M8	M10
A. Design dimensions						
Thread (nominal)	M3	M4	M5	M6	M8	M10
Pitch (mm)	0.5	0.7	0.8	1	1.25	1.5
Thread length (mm)	18	20	22	24	28	32
Head height (mm)	3	4	5	6	8	10
Head dia. (mm)	5.5	7	8.5	10	13	16
Hex key size (mm)	2.5	3	4	5	6	8
B. Basic thread dimensions, tap drill sizes						
Major diameter (mm)	3	4	5	6	8	10
Min. pitch diameter (mm)	2.675	3.545	4.480	5.350	7.188	9.026
Min. minor diameter (mm)	2.458	3.242	4.134	4.917	6.647	8.376
Tap drill size (mm)	2.5	3.3	4.2	5.0	6.8	8.5
C. Tolerances (mm): Screw shank Ø and head Ø—h13, hex flats—D12						
Shank Ø (all + 0)		−0.018			−0.027	
Head Ø (all + 0)	−0.018		−0.022			−0.027
Hex flat minimum	+0.02			+0.03		+0.04
Hex flat maximum	+0.12		+0.15		+0.19	

(continued)

Table 19.1 Specifications for Metric Socket Head Cap Screws (Coarse Thread) of Class 12.9 for Thread Sizes M3–M10 *(continued)*

Specifications		Thread size					
		M3	M4	M5	M6	M8	M10
D. Bore diameters for screw heads and drill sizes for screw clearance							
Screw heads (mm)		6	8	10	11	15	18
Drill sizes:	Fine fit (mm)	3.2	4.3	5.3	6.4	8.4	10.5
	Medium fit (mm)	3.4	4.5	5.5	6.6	9	11
E. Tightening torque—Recommended maximum (≈ 50–60% of yield)							
N-m		2.1	4.6	9.5	16	39	77
kg-m		0.22	0.5	1	1.6	4	8
ft-lb		1.6	3.4	7.1	12	29	57
F. Proof load—maximum load without permanent deformation							
kN		4.88	8.52	13.8	19.5	35.5	56.3
lb		1,097	1,915	3,102	4,384	7,980	12,656

Table 19.2 Specifications for Metric Socket Head Cap Screws (Coarse Thread) of Class 12.9 for Thread Sizes M12–M42

<table>
<tr><td colspan="2">Specifications</td><td colspan="6">Thread size</td></tr>
<tr><td colspan="2"></td><td>M12</td><td>M16</td><td>M20</td><td>M24</td><td>M30</td><td>M42</td></tr>
<tr><td colspan="8">A. Design dimensions</td></tr>
<tr><td colspan="2">Thread (nominal)</td><td>M12</td><td>M16</td><td>M20</td><td>M24</td><td>M30</td><td>M42</td></tr>
<tr><td colspan="2">Pitch (mm)</td><td>1.75</td><td>2.0</td><td>2.5</td><td>3.0</td><td>3.5</td><td>4.5</td></tr>
<tr><td colspan="2">Thread length (mm)</td><td>38</td><td>44</td><td>52</td><td>60</td><td>72</td><td>96</td></tr>
<tr><td colspan="2">Head height (mm)</td><td>12</td><td>16</td><td>20</td><td>24</td><td>30</td><td>42</td></tr>
<tr><td colspan="2">Head diameter (mm)</td><td>18</td><td>24</td><td>30</td><td>36</td><td>45</td><td>63</td></tr>
<tr><td colspan="2">Hex key size (mm)</td><td>10</td><td>14</td><td>17</td><td>19</td><td>22</td><td>32</td></tr>
<tr><td colspan="8">B. Basic thread dimensions, tap drill sizes</td></tr>
<tr><td colspan="2">Major diameter (mm)</td><td>12</td><td>16</td><td>20</td><td>24</td><td>30</td><td>42</td></tr>
<tr><td colspan="2">Min. pitch diameter (mm)</td><td>10.863</td><td>14.701</td><td>18.376</td><td>22.051</td><td>27.727</td><td>39.077</td></tr>
<tr><td colspan="2">Min. minor diameter (mm)</td><td>10.106</td><td>13.835</td><td>17.294</td><td>20.752</td><td>26.211</td><td>37.129</td></tr>
<tr><td colspan="2">Tap drill size (mm)</td><td>10.2</td><td>14</td><td>17.5</td><td>21</td><td>26.5</td><td>37.5</td></tr>
<tr><td colspan="8">C. Tolerances (mm): Screw shank Ø and head Ø—h13, hex flat—D12</td></tr>
<tr><td colspan="2">Shank Ø (all + 0)</td><td colspan="2">-0.027</td><td colspan="3">-0.033</td><td>-0.039</td></tr>
<tr><td colspan="2">Head Ø (all + 0)</td><td>-.027</td><td colspan="2">-0.033</td><td colspan="2">-0.039</td><td>-0.046</td></tr>
<tr><td colspan="2">Hex flat minimum</td><td>+0.04</td><td colspan="2">+0.05</td><td colspan="2">+0.065</td><td>+0.08</td></tr>
<tr><td colspan="2">Maximum</td><td>+0.19</td><td colspan="2">+0.20</td><td colspan="2">+0.275</td><td>+0.33</td></tr>
<tr><td colspan="8">D. Bore diameters for screw heads, and drill sizes for screw clearance</td></tr>
<tr><td colspan="2">Screw heads (mm)</td><td>20</td><td>26</td><td>33</td><td>38</td><td>48</td><td>66</td></tr>
<tr><td>Drill sizes:</td><td>Fine fit (mm)</td><td>13</td><td>17</td><td>21</td><td>25</td><td>31</td><td>43</td></tr>
<tr><td></td><td>Medium fit (mm)</td><td>14</td><td>18</td><td>22</td><td>26</td><td>33</td><td>45</td></tr>
<tr><td colspan="8">E. Tightening torque_Recommended maximum (≈ 50–60% of yield)</td></tr>
<tr><td colspan="2">N-m</td><td>135</td><td>330</td><td>650</td><td>1100</td><td>2250</td><td>6270</td></tr>
<tr><td colspan="2">kg-m</td><td>14</td><td>34</td><td>66</td><td>114</td><td>228</td><td>627</td></tr>
<tr><td colspan="2">ft-lb</td><td>100</td><td>242</td><td>480</td><td>810</td><td>1660</td><td>4630</td></tr>
<tr><td colspan="8">F. Proof load—maximum load without permanent deformation</td></tr>
<tr><td colspan="2">kN</td><td>81.8</td><td>152</td><td>238</td><td>342</td><td>544</td><td></td></tr>
<tr><td colspan="2">lb</td><td>18,250</td><td>34,170</td><td>53,500</td><td>76,880</td><td>122,300</td><td></td></tr>
</table>

The designer will use a reasonable safety factor (sf), which should be at least three, and preferably 5, when establishing the size and number of screws. In case of pulsating loads (both mechanical and thermal), select an sf of at least 5. Table 19.3 shows notable differences between metric and inch screws which are believed to be of similar strength.

More information on inch screws, comparable to data on metric screws shown in Tables 19.1 and 2, is shown in Tables 19.4 and 5.

Table 19.3 Approximate Equivalent Screws: Metric and Inch (U.S. Unified) (All sizes shown in mm.)

Metric (nominal)	M3	M4	M5	M6	M8	M10
Diameter	3	4	5	6	8	10
Minimum pitch diameter	0.5	0.7	0.8	1.0	1.25	1.5
Min. minor diameter	2.458	3.242	4.134	4.917	6.647	8.376
Inch (nominal)	#5-40	#8-32	#10-24	¼-20	5/16-18	⅜-16
Diameter	3.175	4.166	4.826	6.350	7.938	9.525
Minimum pitch diameter	0.635	0.794	1.058	1.270	1.411	1.588
Minimum minor diameter	2.395	3.193	3.528	4.793	6.205	7.577
Comparison: Critical cross section at root (minor diameter) (mm^2)						
Metric	4.74	8.25	13.4	19.0	34.7	55.1
Inch	4.11	8.04	9.79	18.1	30.2	45.0
Metric > Inch	13%	3%	27%	5%	13%	18%
Metric (nominal)	M12	M16	M20	M24	M30	M42
Diameter	12	16	20	24	30	42
Minimum pitch diameter	1.75	2.00	2.50	3.00	3.50	4.50
Minimum minor diameter	10.106	13.835	17.294	20.752	26.211	37.129
Inch (nominal)	½-13	⅝-11	¾-10	1-81	1¼-71	1½-6
Diameter	12.7	15.875	19.05	25.40	31.75	38.10
Minimum pitch diameter	1.954	2.309	2.540	3.175	3.629	4.233
Minimum minor diameter	10.302	13.043	15.933	21.504	27.242	32.845
Comparison: Critical cross section at root (minor diameter) (mm^2)						
Metric	80.2	150.3	234.9	338.2	539.6	1,083
Inch	83.5	133.8	199.2	363.5	583.4	847
Metric > inch		11%	15%			21%
Inch > metric	4%			7%	8%	

Table 19.4 Specifications for "Inch" Socket Head Cap Screws (Coarse Thread) of Class 12.9 for Thread Sizes #5-40 to 3/8-16 (Allen™, Unbrako™, Holo-Krome™ Quality)

Specifications	Thread size					
	#5-40	#8-32	#10-24	1/4-20	5/16-18	3/8-16
A. Design dimensions						
Thread	#5-40	#8-32	#10-24	1/4-20	5/16-18	3/8-16
Pitch (mm)	0.635	0.794	1.058	1.270	1.411	1.588
Thread length (mm)	19	22	22	25	28	32
Head height (mm)	3.18	4.17	4.83	6.35	7.94	9.53
Head diameter (mm)	5.21	6.86	7.94	9.53	11.91	14.29
Hex key size (in.)	3/32	9/64	5/32	3/16	1/4	5/16
B. Basic thread dimensions, tap drill sizes						
Major diameter (mm)	3.175	4.166	4.826	6.350	7.938	9.925
Minimum pitch diameter (mm)	2.715	3.594	4.074	5.453	6.944	8.410
Minimum minor diameter (mm)	2.395	3.193	3.528	4.793	6.205	7.577
Tap drill size (mm)	2.58	3.45	3.80	5.11	6.53	7.94
C. Tolerances						
To ANSI Standard B18.3, and ASTM A574;						
Screws up to and including 1" diameter—Class 3A threads						
D. Bore diameters for screw heads and drill sizes for screw clearance						
Screw heads (mm)	6.35	8.0	9.5	11.1	13.5	15.9
Drill sizes: Fine fit (mm)	3.4	4.4	5.1	6.8	8.3	10.3
Medium fit (mm)	3.6	4.9	5.6	7.1	8.7	11
E. Tightening torque—Recommended to between ≈ 60 and 75% of yield)						
kg-m(at≈75%)	0.3	0.6	1.0	2.2	4.6	8.2
ft-lb(at≈75%)	2.0	4.5	6.5	16	33	59
ft-lb (at≈60%)	1.7	4	5	12	25	45
F. Average screw force—to yield (from manufacturer's chart)						
kg	500	890	1,115	2,020	3,330	4,930
lb	1,110	1,960	2,450	4,450	7,330	10,850

Table 19.5 Specifications for "Inch" Socket Head Cap Screws (Coarse Thread) of Class 12.9 for Thread Sizes 1/2-13 to 1 1/2-6 (Allen™, Unbrako™, Holo-Krome™ Quality)

Specifications	Thread size					
	½-13	⅝-11	¾-10	1-8	1¼-7	1½-6
A. Design dimensions						
Thread	½-13	⅝-11	¾-10	1-8	1¼-7	1½-6
Pitch (mm)	1.954	2.309	2.54	3.175	3.629	4.233
Thread length (mm)	38	44	51	64	79	95
Head height (mm)	12.70	15.88	19.05	25.40	31.75	38.10
Head diameter (mm)	19.05	23.81	28.58	38.10	47.63	57.15
Hex key size (in.)	⅜	½	⅝	¾	⅞	1
B. Basic thread dimensions, tap drill sizes						
Major diameter (mm)	12.7	15.875	19.05	25.4	31.75	38.1
Minimum pitch diameter (mm)	11.336	14.272	17.287	23.208	29.149	35.082
Minimum minor diameter (mm)	10.302	13.043	15.933	21.504	27.242	32.845
Tap drill size (mm)	10.72	13.89	16.67	22.22	28.18	34.00
C. Tolerances						
To ANSI Standard B18.3, and ASTM A574; Screws up to and including 1" diameter—Class 3A threads						
Larger screws —Class 2A threads						
D. Bore diameters for screw heads and drill sizes for screw clearance						
Screw heads (mm)	20.6	25.4	30.2	41.3	50.8	60
Drill sizes: Fine fit (mm)	13.5	16.7	19.8	26.2	33.3	41
Medium fit (mm)	14.5	17.5	21	28	36	43
E. Tightening torque—Recommended to between ≈ 60 and 75% of yield)						
kg-m	20	39	70	145	290	505
ft-lb (at≈75%)	144	284	500	1,044	2,088	3,630
ft-lb (at≈60%)	108	210	367	867	1,750	3,040
F. Average screw force—to yield (from manufacturer's chart)						
kg	9,020	14,390	21,250	33,000	53,000	77,000
lb	19,850	31,650	46,750	72,700	116,300	168,600

19.11 Screws and Studs for Adjustment

Screws should not be used for adjustment but solely for fastening. The mold designer must foresee (calculate) areas which may need adjustment, and if adjustment is unavoidable, they must provide for a substantial spacer (i.e., at least 2 mm thick) to be positioned so that the final location can be fixed solidly with screws. These spacers can then be machined to a definite size and become a part of the assembly, and do not need further adjustment.

The use of thin shim stock is not a good practice because it is easily lost or overlooked during maintenance. A substantial (thick) spacer rather than a shim will be obvious at reassembly of the mold.

Exception: If a spacer is not practical, screws or studs may be used for adjustment. Because such screws cannot be tightened (stretched) sufficiently to prevent loosening, they must be prevented from turning by one of the methods, shown later under Section 19.12, Securing Screws.

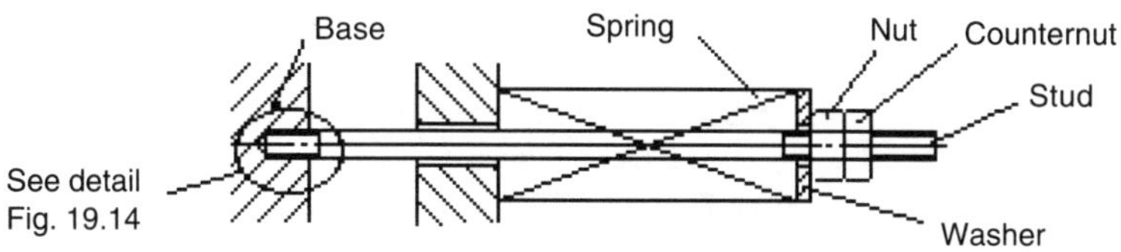

Figure 19.13 Stud used to retain a return spring.

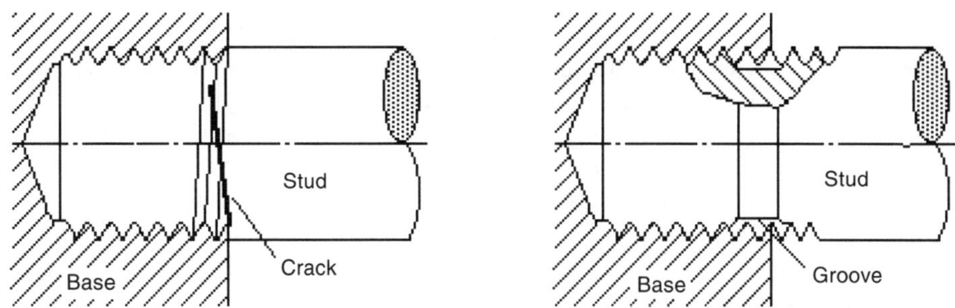

Figure 19.14 Enlarged detail of the stud thread at the base: Poor design (left) and proper design with longer thread on stud and groove at minor diameter (right).

Correct use of studs is shown in the above example of retaining a return spring (Fig. 19.13). This (inexpensive) method is frequently used but is subject to unauthorized adjustment (fiddling) and invites operating problems.

A properly selected (or designed) stud must have a special screw thread (stud thread) where it screws into the base. Stud threads have interference rather than clearance threads found on standard screws. Note the design of the detail (Fig. 19.14), where the thread enters the base. The length of the thread in the stud should be longer than the thread in the base, but grooved to the minor diameter where the surface of the base intersects with the stud to prevent stress cracking as a result of possible shear loads on the stud.

There is also a difference between studs used to mount into steel or into cast iron, or other softer materials. If such threads are not available, Loctite™ must be used to prevent the stud from unscrewing.

(In Figs. 19.13 and 14 above, the length of the engaged thread is shown shorter than it should be (i.e., ≈ 1.2–1.5 the thread diameter). It should be more if the base is made from softer material.)

Studs usually bottom in the threaded hole; they are installed with special "stud drivers" or by using two nuts locked against each other at the other end of the stud. As a rule, studs should not be removed once they have been installed, to prevent wearing down the thread and thereby reducing the amount of interference with the base.

Commercially available studs are property class 5.8; therefore, less than half the strength of a SHCS. Commercial threaded rod is still softer (class 4.6), and its use instead of studs is not recommended.

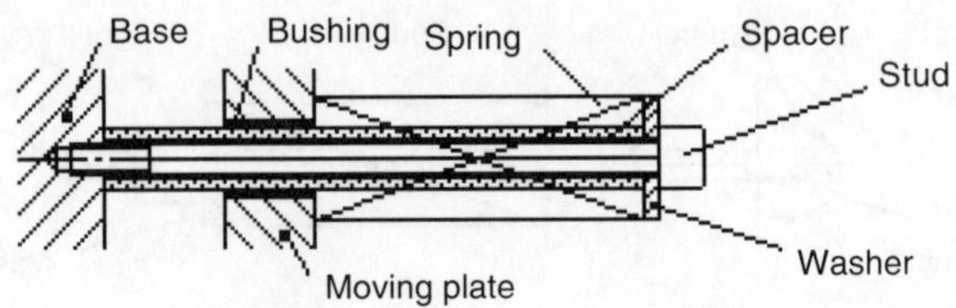

Figure 19.15 Cross section of spacer with spring and bushing through moving plate.

The best design (Fig. 19.15) uses a spacer (sleeve) of calculated length or, if necessary, a length that has been established after mold testing. With this method, a (standard) screw can be properly tightened, and unauthorized adjustments are made difficult.

The spacer (sleeve) should have a cross section of its wall of at least the same, but preferably twice the cross section of the screw. It should be made of steel, preferably hardened to Rc > 40, and ground to size on the OD where the spring rubs on the sleeve.

A good practice is to have the moving plate fitted with a standard size bushing. With light side loads, use either an oil-impregnated bronze bushing (e.g., Oilite™) or a Teflon-type, nonlubricated bushing (e.g., DU™ bushing). With heavy side loads, the bushing should be lubricated bronze or steel, to be periodically greased, either manually or from an automatic system. It is important that any oil dripping from the bushing will not contaminate the molded products; this may affect the selected location of the springs.

SAFETY NOTE: A spring as shown in Figs. 19.13 and 15 must always be enclosed, to prevent injury if the bolt should break and the broken fastener or the spring flies away.

19.12 Securing Screws

Properly selected and applied screws are self-locking and do not need prevention against loosening. Exception: For enclosed mechanisms where the loosening of a screw could result in catastrophic loss, it may be necessary to add a "100%" safe method of securing, such as wiring of screw heads or installing cotter pins. However, this is seldom necessary in molds.

In general, short screws (i.e., where the shank length is insufficient to be stretched to provide safety against unscrewing) should be prevented from loosening by some added means.

19.12.1 Lock Washers

Lock washers, including "star washers" may be thought of as spring washers. The principle of these "spring washers" is that they are compressed to a flat shape under the head of the screw. Teeth (one pair for an ordinary lock washer, many for star washers, dig (bite) on one side of the washer into the part to be fastened, on the other side into the screw head (or nut), such that they oppose the unscrewing of the screw (or nut).

To be effective, the hardness of the screw and the fastened part must be less than that of the washer. Since the hardness of lock washers is between 44 and 51 Rc, and of star washers between 44 and 47 Rc, they will not bite into the underside of a standard (property class 12.9)

SHCS, nor into a hard (49–51 Rc) mold part, although they would bite into a prehardened (≈ 30–35 Rc) plate. They are, therefore, not recommended for use with 12.9 SHCS, or on hardened steel parts.

19.12.2 Loctite™, Permaflex™, Etc.

These products are anaerobic liquids which are applied to the (clean, oil-free) threads at the moment of assembly, instead of oil. After application, the liquid hardens and bonds to the threads in the mold part and the screw. This bond is sheared off as the screw is removed, and the liquid must be reapplied for the next assembly.

This ensures a safe joint; however, it is important that the mold assembly drawing contains a clear note (warning) that Loctite must be applied to the designated screws when the mold is assembled and serviced. The following lists some Loctite products, with their characteristics:

Loctite #242 (blue)	Medium strength; minimum thread length $L = 1.5D$.
Loctite #271 (red)	High strength (2.5 times #242); use for studs.
Loctite #262 (red)	15% more strength than #271.
Loctite #290 (green)	Same strength as #242 but has "wicking" property (i.e., it will enter and lock an already tightened thread). This is a preferred product.
Porosity sealant #290	Similar to Loctite #290, as it enters (by wicking action) into porous metal.
PST pipe sealant w/Teflon	For pipe threads;does not require Teflon tape.
Gasket eliminator #515	Use instead of gaskets; fills voids up to 1.25 mm.
Nickel antiseize #771	Antiseizing compound for screws.
Retaining compound #601	To retain bearings in damaged or oversized bores; shear strength 3,000 psi, max. 300 °F.
Retaining compound #620	Same strength as #601, but up to 450 °F.
Retaining compound #680	4,000 psi, max. 300 °F.
In aluminum, all grades:	Strength is reduced to 600 psi.

CAUTION: Do not use Loctite or any similar products on air lines, where a spilled drop of such liquid could be carried by the air stream into narrow (blow) slots and plug them, requiring costly repair work.

19.12.3 Deformed Threads

Some screw manufacturers supply screws whose threads are slightly deformed (oval, tri-lobed, etc.). As the female thread is slightly deformed by driving in these screws, the screws are said to be prevented from loosening. These screws are *not* recommended for molds.

19.12.4 Wiring of Screw Heads

This method is frequently used in machine design, but rarely in molds. The screw heads are drilled in three directions at 120° to each other. After tightening, a wire is threaded through

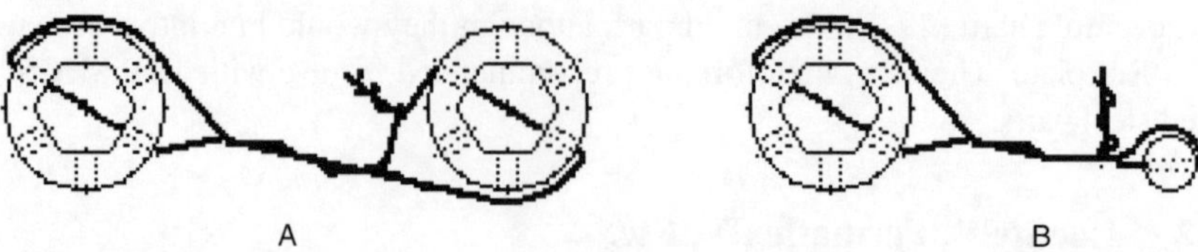

Figure 19.16 Wiring of screw heads: A. two screws wired to each other, and B. screw wired through an auxiliary hole.

the most suitable hole so that the head is prevented from unscrewing. The wire is then either threaded through one or more other screw heads, which also are to be prevented from turning, or it may be threaded through an auxiliary hole created solely for this purpose (Fig. 19.16). The wire between screw heads is twisted to hold the wire in position.

This method is expensive (time consuming) and should be used only where absolutely necessary. If used, it must be shown on the assembly drawing.

19.12.5 Screws Secured by Plastic Inserts (Nylok™, etc.)

Screws are secured by a small plug (cylindrical or rectangular) inserted into the thread section of the screw (Nylok™ screws, etc.). When the screw is turned into the matching part, the (slightly projecting) plastic is deformed by the female thread; this deformation creates an area of high friction, which prevents the screw from loosening. Such a screw will stay in any position where it was placed, even without being tightened, despite vibrations or shocks.

These screws are often quite useful, but to ensure that they function properly, they should be replaced after every removal, especially if they have been in place for a long time, because permanent set in the plastic may reduce the friction force. The assembly drawing must carry a note to this effect.

Typical application is in locations where short screws, such as round or button head screws, etc., are required but which cannot be secured sufficiently by torquing (e.g., when fastening sheet metal enclosures subject to heavy vibration or shock). An alternative would be to use Loctite™ or an equivalent.

19.13 Nuts

19.13.1 Securing Nuts

Nuts are rarely used in molds, but they may be necessary in some related equipment, such as take-offs.

19.13.2 Cotter Pins

Cotter pins may be used with castellated nuts and drilled bolts. This method is used with bolts or with studs. The method is secure, but the problem is that the hex nut has a limited number

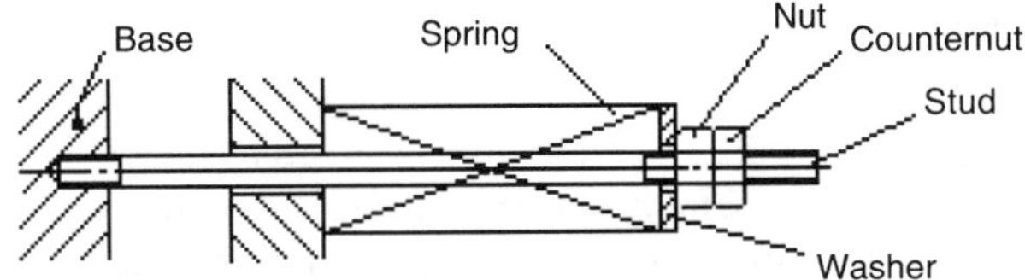

Figure 19.17 Jam nut used as a counternut to hold nut in position.

of slots, usually three, at 120°. Because there is only one matching hole in the bolt, the nut will be rarely tightened "just right" when the hole and one of the slots match. It is usually necessary to either not tighten the bolt enough, or to overtighten it to be able to insert the cotter pin.

19.13.3 Jam Nuts

Also called counter nuts, jam nuts are used with bolts or studs. One nut (the jam nut) is screwed on top of the nut which must be secured and which was tightened first. The principle is that the length of the bolt between the nuts is stretched. In fact, this length is very small, but by tightening the jam nut while holding the primary nut, the threads will be slightly deformed (within their elastic limit) and provide some holding power (Fig. 19.17). In general, this method is not recommended in molds.

If the design in Fig. 19.17 (using a stud and nuts) should be necessary in a mold (e.g., in case a very long spring would require a nonstandard (and not available) length of screw), the design should be modified to use a long sleeve between the washer and the base. A single nut can then be used to tighten the stud, and a counter nut will not be required; this is similar to the design shown in Fig. 19.15.

19.13.4 Nuts Secured by Plastic Inserts (Nylok™, etc.)

Nuts with plastic plugs or collars (made from Nylon, etc.), either inside the thread or at the end of it, are so designed that they are deformed by the male thread of the stud or bolt, as it passes through the nut. This deformation creates an area of high friction and will prevent the nut from unscrewing. Such a nut will stay in any position where it was placed, without being tightened, despite vibrations or shocks.

This is a good method, but the nuts should be replaced after every removal, especially if they have been in place for a long time, because permanent set in the plastic may reduce the friction force. The assembly drawing must carry a note to this effect.

19.13.5 Other Methods for Securing Nuts

19.13.5.1 Loctite™, Permaflex™, etc.

These products can be used the same as for screws.

19.13.5.2 Lock Washers

Nuts (property class 10.9 or less) are usually softer than SHCSs. If the part that is secured by the nut is as soft or softer than the nut, lock washers are acceptable.

Note that in a nut and bolt assembly, the washer is usually inserted under the nut. In rare cases, it may be required that another lock washer is inserted under the screw head, provided its hardness is at least 10 points less than the hardness of the lock washer. (The first lock washer prevents the nut from turning, but not the bolt).

19.13.5.3 Tab Washers

The tab washer is essentially a flat, soft, steel washer with a small tab projecting at right angles to the washer. To prevent the washer from turning, this tab enters the part which is to be fastened, or an opening (groove) in the stud. After the nut is tightened, a segment of the washer is pried away from the fastened part and hammered tight onto one of the hex flats of the nut. This bent-up section and the tab of the washer prevent the nut from turning.

This method is very good and safe. However, it requires that a new tab washer be installed every time the nut is removed, because the bend in the washer which secures the screw from turning is usually fractured when turned back to its flat position. The need for a new washer must be clearly stated with a note on the assembly drawing.

References

1. Internaltional Standards Organization. (1988). *ISO Standards Handbook for Fasteners, Screws and Threads*, 2nd ed. , International Standards Organization, Geneva.

20 Mold and Mold Parts Handling

20.1 Safety

Safety must be a primary consideration of the mold designer. How can a mold, or any heavy mold part, be handled safely? To avoid injury or permanent disability, such as a strained back, no person should repeatedly lift parts heavier than 20 kg (44 lb.) or any part which is awkward to handle, regardless of weight, without some mechanical assistance, such as a hoist.

Not only should nobody get hurt when handling a mold or mold part, but the mold maker has a legal responsibility for any accident that may arise because of an omission or poor design. Where it is not possible to provide a suitable safety feature, at least a clearly visible warning label or name plate must be provided to inform of the possible hazard and its consequences.

There are four situations during which part or all of the mold is handled:

1. Machining a mold part (plates, heavy cavities and cores),
2. assembling the mold (including servicing and maintenance),
3. mounting the mold in a machine, and
4. operating the mold.

Each situation may require special considerations. During the design phase, the designer must try to foresee how a mold (or mold part) will be handled (or mishandled) during all of these situations, and where the handler (machinist, assembler, or molding technician) may cause injury to himself or to others. The designer must consider in all cases how size, location, and number of tapped holes in general and for handling devices will affect the strength and the fatigue life of the part.

Note: In the illustrations throughout this section, the sizes of lifting holes are shown out of proportion in size compared to the plates, etc., illustrated.

20.2 Lifting Plates

20.2.1 Machining of a Mold Part

From the onset, a block for any heavy mold part (usually a plate, but also cavities and cores) must have one or more threaded holes to permit the machine operator to utilize eye bolts and a hoist for handling (Fig. 20.1). Lifting with eye bolts is preferred to lifting with magnets, particularly when lifting a finely ground plate, which could be scratched when using a magnet.

The location of these holes can only be determined at the completion of the mold design, after the location for all necessary cooling channels and mounting holes and the shape of the mold part are finalized.

It is usually best to provide the handling holes as close as possible above the center of gravity (CG) of the heavy part while avoiding an area which is already occupied by holes, channels, etc. (Fig. 20.2). It is important that these holes do not weaken the cross section of the mold part by reducing its wall thickness or by creating stress risers, which are unavoidable with tapped holes.

It is sometimes impossible to reconcile both above-mentioned conditions, and a compromise will be required. The compromise may be that the lifting hole is not exactly above the CG, and the part may not hang squarely when handled, or it may be that two or three holes are provided in such a way that a chain can be utilized to connect two or three eye bolts, with the hoist attached to the center point, to provide an approximate location above the CG.

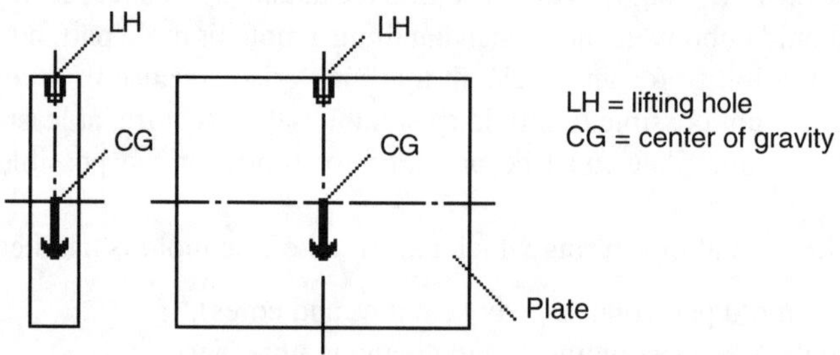

Figure 20.1 Lifting hole (LH) location above center of gravity (CG) in a plate for handling.

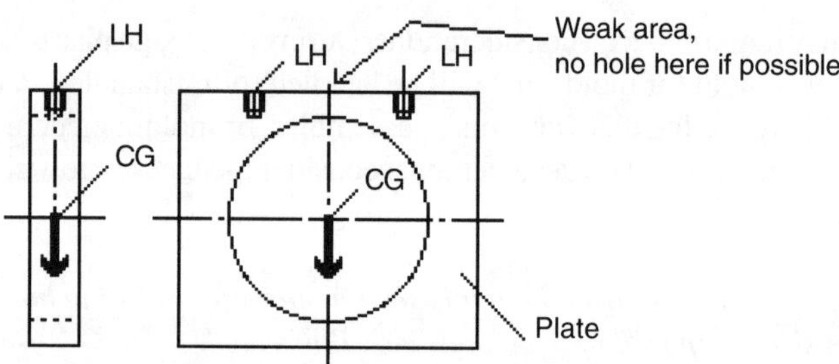

Figure 20.2 Lifting holes near center of gravity but away from weak area of thin wall.

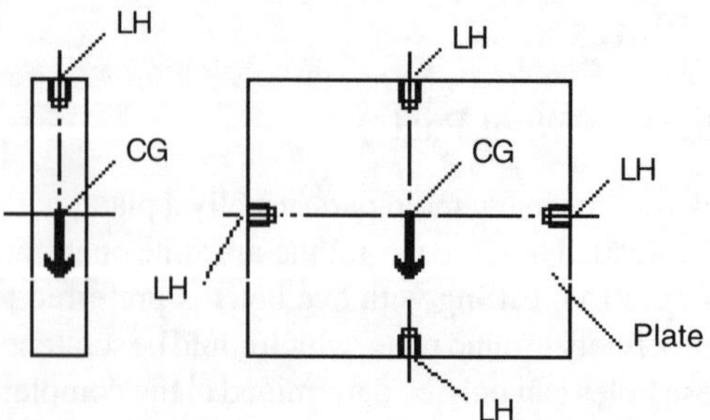

Figure 20.3 Lifting holes on all edges of plate evenly distributed around center of gravity.

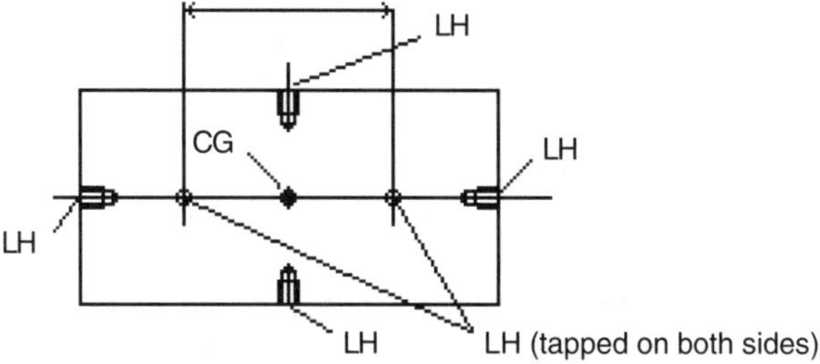

Figure 20.4 Lifting holes tapped in all edges and on both faces of the plate along the centerline.

Usually, plates will be handled and lifted from more than one side. There must be lifting holes on all edges to accommodate lifting of the plate (Fig. 20.3). The same rules for location of lifting holes apply as shown above. If doubt arises regarding lifting hole location, or when a situation arises where a part cannot be made with suitable lifting holes, persons who may be expected to lift the mold part must be consulted to plan an alternative method for its lifting and handling.

20.2.2 Lifting Holes on the Faces of a Plate

There are basically two reasons for placing lifting holes on the face of a plate:

1. The work piece is cut out in its center to such an extent that a lifting magnet cannot touch on a large enough area to hold the plate safely. The diameter of a magnet is usually between 250–300 mm (8–10 in.). This applies specifically to large rings or to large plates.
2. All heavy plates should have lifting holes in the faces (largest plane portion of the plate), unless there are already suitably tapped holes in the face required in assembly of the mold.

The location of such additional holes should be established in cooperation with the operators required to handle these parts, particularly for grinding of large and heavy plates. The holes should be placed symmetrically about the CG and are best located on the centerline of the longer axis of the part (Fig. 20.4). The exact location of the two holes for lifting can only be determined after the plate is completely designed, to ensure that these holes do not interfere with other features of the plate.

As a rule, all plates over 50 kg (110 lb) mass must have lifting holes. If possible, lifting holes should be located so that they are not covered by other components and can be used for lifting even after the plate is partly or completely assembled.

20.3 Eye Bolts

20.3.1 Lifting and Slinging Theory

When lifting a mass straight up, the force *F* goes through the center of gravity CG and through the axis of the eye bolt. The angle of attack is the angle between the direction of hoist pull and the surface on which the hoist pull is acting.

The eye bolt is loaded only in tension. Figure 20.5 illustrates the simplest case. By placing the hole over the CG, a relatively small eye bolt could be used.

Although many types and strengths of eye bolts, hoist rings, etc., are available on the market, it is best to use commonly available shoulder-type eye bolts. This allows the customers to work safely using ordinary shop equipment or tools.

Eye bolts with shoulders have better load carrying capacities for angle loads than eye bolts without shoulders. Also, the operator can see whether the eye bolt is fully screwed into the plate. Without the shoulder, the operator may stop too soon and create a hazard by not having enough threads engaged to carry the load.

Assume that the same plate illustrated in Fig. 20.5 is now lying flat and must be lifted. There are two extreme possibilities: the ring may be in a vertical plane or in a horizontal plane (Fig. 20.6).

The ring in Fig. 20.6A is in a vertical plane. A bending moment, $F \times l_1$, adds to the forces acting on the cross section of the bolt, A. There is also a certain amount of shear in this section. Because the eye bolt rests tightly on its shoulder, there is some additional strength. With an angle of attack of 0°, there is a more complex system of loading the eye bolt, more severe than that of the straight-up (tensile) load with an angle of attack of 90°.

The ring in Fig. 20.6B is in a horizontal plane. The bending moment is now $F \times l_2$, which is larger than l_1 in Fig. 20.6A. The hook of the hoist must attach at this point, or it would twist or loosen the eye bolt. Also, the additional bending moment about the shoulder will be larger. In this case, the operator must use a shim washer to stop the eye bolt in the correct (vertical) orientation to the lift direction, as in Fig. 20.6A.

Force calculations are illustrated using variables in Figs. 20.7 and 8. As soon as the hook starts to lift the plate, the actual force required is only about 50% of the force required to lift the total mass of the plate.

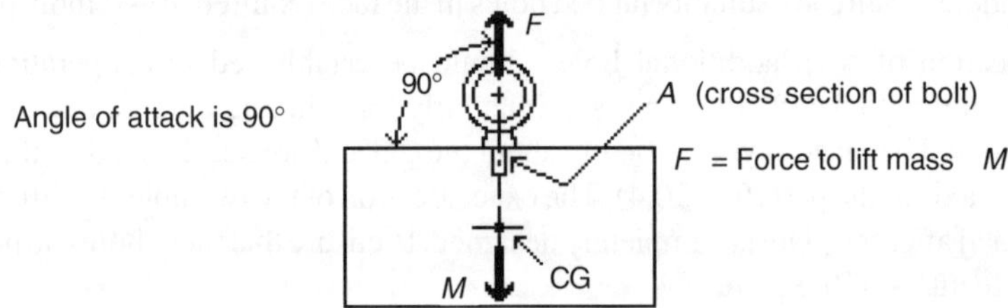

Figure 20.5 Cross section shows eye bolt placed directly over the CG, and arrows show direction of force away from the mass.

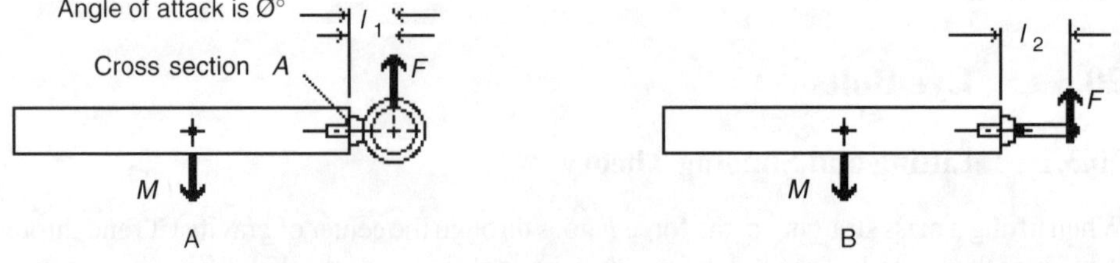

Figure 20.6 Eye bolts used to lift plate while lying flat may be A. in vertical plane or B. in horizontal plane.

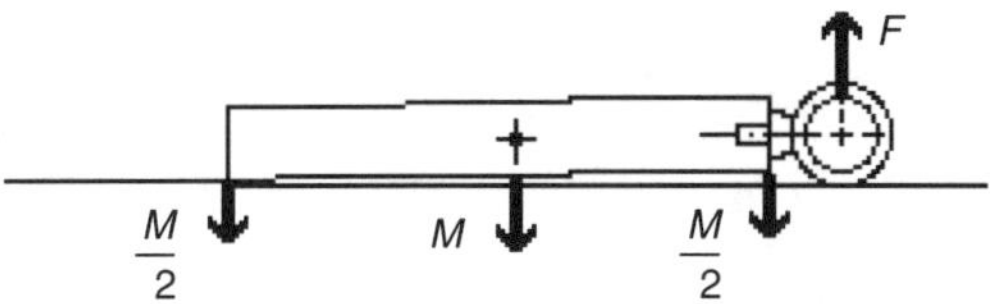

Figure 20.7 As soon as lift begins, the force required to lift the total mass is halved.

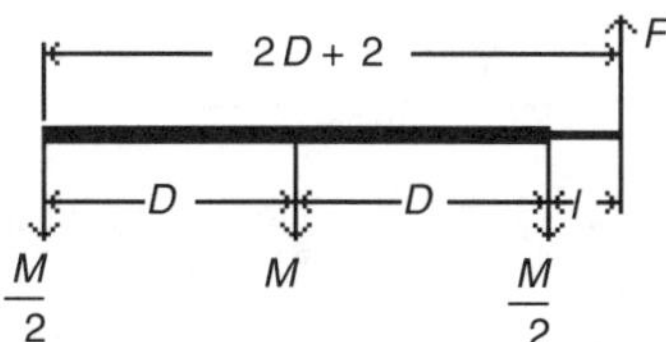

Figure 20.8 Forces and calculation for bending moment of bolt.

Actually, the force F is even less than $M/2$, because of the additional lever created by l_1 or l_2. The larger l is, the smaller the force F. F_{max} would be $M/2$. The force that will tend to bend the bolt at the surface where the eye bolt is attached is proportional to the force $M/2$. The longer l, the smaller is F, but the bolt inserted there must be strong enough not to bend when subjected to F_{max}.

The calculation for the bending moment about the left end of the plate in Fig. 20.8 is

$$F \times (2D + 2) = M \times D \text{, or } F = M \times \frac{D}{2D + 2}. \tag{20.1}$$

Down-rating the eye bolts in relation to the angle of attack becomes a factor in eye bolt selection. Use safe load values found shown in manufacturers' catalogs.

If a finely ground plate lies on top of a very smooth surface, the effect of vacuum under the plate may be considerable. For example, a steel plate 20 × 20 × 1 in. weighs about 120 lb. The area is 400 in^2, or, at 14.2 psi (atmospheric pressure), the force holding the plate on the surface as a result of air pressure (vacuum) is about 5,680 lb!

20.3.2 Slinging of Loads

The term "slinging" applies when a load is lifted with two eye bolts. A simple example is shown in Fig. 20.9. (NOTE: The eye bolt location shown is a worst-case scenario and should be avoided.)

Example: Two eye bolts are used to hoist a mass M of 2,000 kg. The resulting load forces for various sling angles are listed below:

Sling angle α	Load F_α (kg) in each chain
90°	1,000
75°	1,040
60°	1,150

(continued)

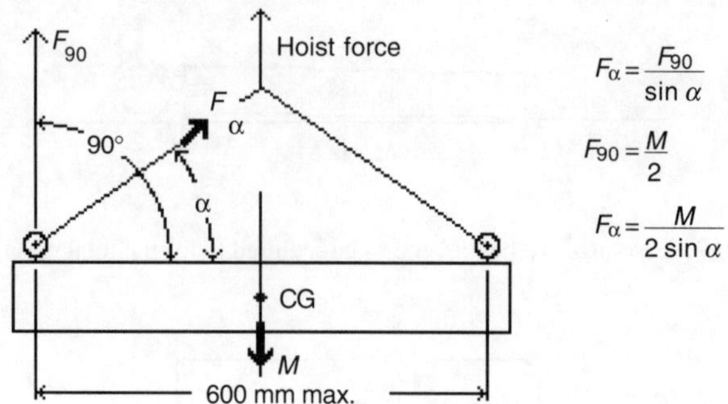

Figure 20.9 Two eye bolts used to hoist the plate via a chain.

(continued)

Sling angle α	Load F_α (kg) in each chain
45°	1,410
30°	2,000
15°	3,800
5°	11,480
0°	∞

From the above example, it is clear that the hoist operator also has a responsibility to sling the chain or rope to no less than 60° to keep the forces on the eye bolts within safe limits. A distance between lift holes of no more than 600 mm will satisfy most requirements.

Responsibilities of the designer are:

- If it is not possible to lift a plate from the center (above the CG), select the distance between eye bolt holes to be equal or less than 600 mm (the length of the sling) symmetrically about the CG (Fig. 20.9).
- Provide tapped hoist holes for eye bolts of a size with adequate strength.

Standard eye bolts are designed for a safety factor of about 5. It is relatively inexpensive to provide an even higher safety factor by selecting the next larger size eye bolt. This ensures that whatever type of eye bolt is selected, there is less risk of failure.

Lifting with eye bolts is considered safer than lifting with magnetic chucks. Even though electromagnetic chucks are usually protected with power-failure devices, failure of the magnets has occurred.

Another important designer responsibility is that the maximum crane hook height above the top tie bar of the molding machine must always be considered. The molder may have limited space above the machine, and there may not be enough space for lifting the mold out of the machine, particularly if the mold requires a lift bar. The mold may then require a method of hoisting different than the standard. The designer must always ensure that the lifting methods designed are suitable for the mold, and for the molder's plant and equipment.

20.3.3 Do's and Don'ts of Eye Bolt Usage

Attention to these rules is particularly important if the applied load approaches the rating of the eye bolt.

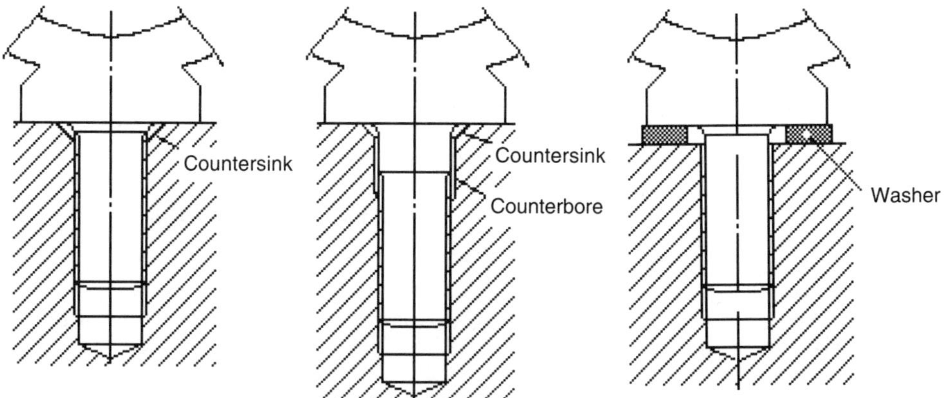

Figure 20.10 Seating the eye bolt shoulder against the load: A. hole is countersunk, B. hole is countersunk and counterbored, and C. a spacer washer is used.

1. Eye bolts must never be ground, machined, stamped, or marked with a sharp tool or instrument.
2. Standard shoulder eye bolts must never be undercut to seat the shoulder against the load. Instead, the receiving (threaded) hole should be countersunk and/or counterbored, or a spacer washer should be used (Fig. 20.10). Ensure that at least 90% of the threads are engaged in the threaded hole.
3. Ensure that the eye bolt is screwed down all the way and is properly seated. The shoulder should bear firmly against the mating plate surface or a spacer washer.
4. Loads should always be applied in the plane of the eye, not at an angle to this plane.
5. Never exceed the recommended capacity of the eye bolt.
6. Do not paint, plate, or galvanize an eye bolt. This could easily hide flaws (cracks) and/or affect fits.
7. Do not use M20 tapped holes for lifting purposes. They may appear to accept a 3/4-10 UNC eye bolt, but they will engage only for a small depth in the thread root. If left with this small engagement, the thread will be damaged, possibly stripped, and the plate can drop.
8. An inspection program for eye bolts should be established and implemented on an ongoing basis. It is recommended that an inspection program be instituted in every plant operation using eye bolts, hoist slings, and hoisting equipment. To ensure safety, all such equipment must be regularly checked for possible stress signs (cracks) and wear. Each item should be identified by number, type, etc., and a report on its state be made periodically. Questionable items must be discarded and destroyed, not repaired or welded.

20.3.4 Examples of Eye Bolt Use

Lifting holes must also be provided for handling of other items (not only for plates) if they are heavy, unwieldy, and in particular if they must be removed for servicing while the mold is in the machine. This applies to cavities, cores, and locking and stripper rings.

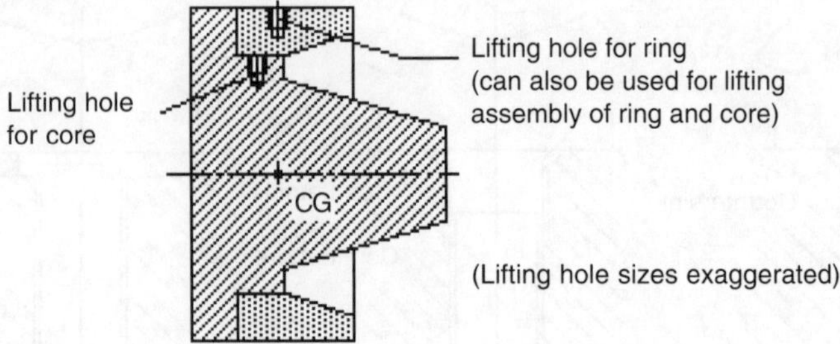

Figure 20.11 Lifting holes in core and in stripper ring (hole in ring can also act as assembly lifting ring for both ring and core).

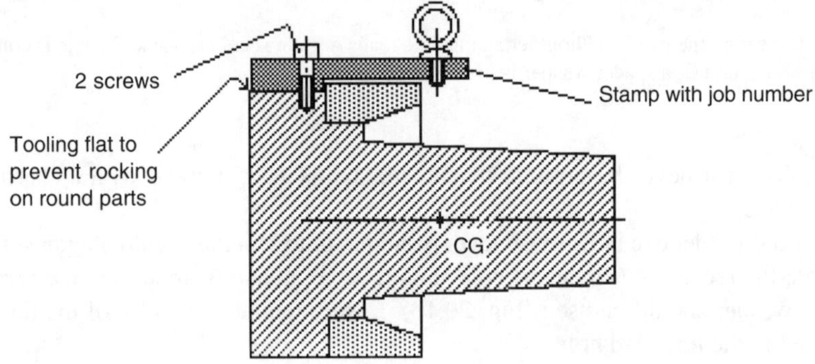

Figure 20.12 Lift bar used to handle the mold part.

Each of these elements must have at least one lifting hole above the CG. Also, an assembly should have a lifting hole, if possible. (Of course, such an "assembly lifting hole" could coincide with a part lifting hole [Fig. 20.11].) The designer must ensure that such lifting holes will not weaken the part excessively, particularly in the case of a ring-shaped part. The location of lifting holes in mold parts other than plates needs the same considerations as outlined for lifting holes of a plate, and is the responsibility of the mold designer.

If the CG is not through an area where a tapped hole can be installed, a lift bar may be required for handling of the part. A lift bar as shown in Fig. 20.12 must be shown on the assembly drawing, and must be identified with its own part number and the mold number. This is important, because it will be used only when servicing the mold, and the molder may store it remotely from the mold and would otherwise have difficulties identifying the proper lift bar when needed. There must be a name plate on the mold shoe to advise that the lift bar is to be used for safe handling of this assembly.

A work piece which had a lifting hole above the CG (in the center of its side) before machining may have a different CG once the plate is finished, after substantial amounts of material have been removed from one face of the plate. Lifting holes LH1, in Figs. 20.13 and 20.14 should be considered for handling the work piece.

The new CG is more important for the next stage of manufacturing and assembly, where "square" handling of the parts will facilitate work. It is, therefore, preferable to determine the

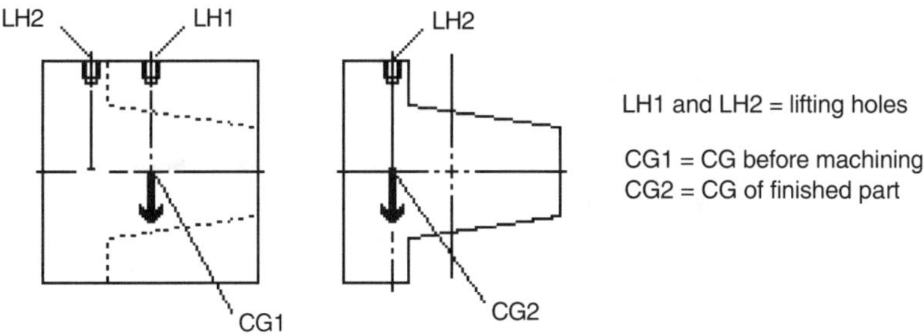

Figure 20.13 Lifting holes before (LH1) and after (LH2) machining in a heavy core.

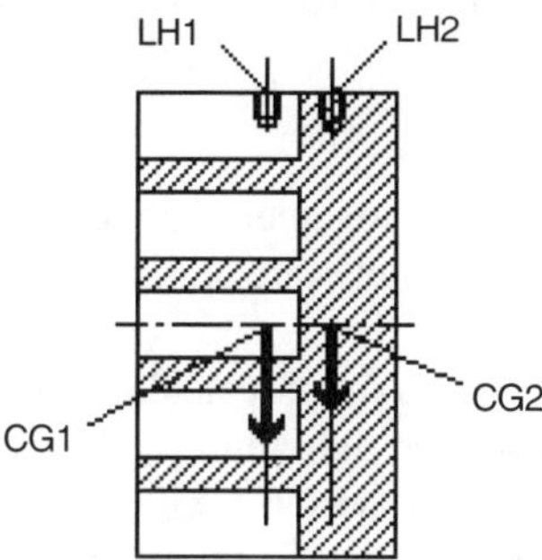

Figure 20.14 Lifting holes before (LH1) and after (LH2) machining of an ejector box.

location of the CG of the finished part rather than the blank, and locate the lifting hole(s) above it. Otherwise, the plate or part may hang at an angle during handling after machining the bulk of material.

Once several plates, each with its own CG, have been assembled, the resulting CG of the assembly will probably be at a new location. To achieve square handling of the assembly, a new lifting hole may be required. The exact location of the new CG can easily be calculated, once the average weight of each part has been estimated and the location of the CG for each component has been established.

To determine the CG of a part (plate, etc.), a part which is crisscrossed with holes and small pockets can be assumed to be solid. However, a part with large pockets and/or with irregular shape may require some amount of calculation to arrive at a better estimate of the location of its CG.

20.4 Lift Bars

If it is not possible to provide an additional lifting hole to hoist the whole mold, because of interference with another hole, etc., a lifting bar must be introduced with holes spaced to pick up at least two lifting holes of the parts making up the assembly. The lift bar has then a threaded hole for an eye bolt above the new CG.

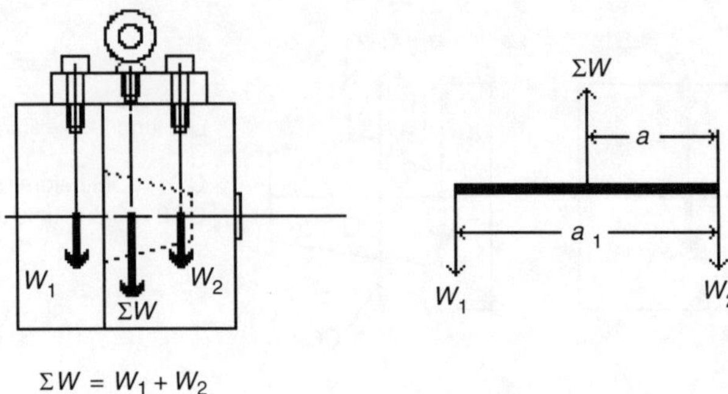

Figure 20.15 Proper lifting: A. shows mold and centers of gravity (CG) for weights W_1 and W_2 of the mold halves, and B. shows distance (*a*1) of CGs from each other and the distance (*a*) of the CG of the whole mold from CG for *W*2.

The lifting holes should be as close together as possible, to reduce the span on the lift bar which will bend under the load. If the holes are closer together, the bar can be lighter than if the holes are farther apart.

Figure 20.15 shows schematically a simple calculation for the proper lifting hole location. In this example, this location would be right in the area where two plates meet; therefore, a lift bar must be provided. Occasionally, a lift bar will span over the whole length of the mold (particularly in stack molds), and the holes farthest from the center are used to hold the complete mold together. The lift bar must be strong enough to support the whole mold.

The formula below permits the calculation of *a* (i.e., where to locate the hole for the eye bolt for the whole mold:

$$a = \frac{a_1 W_1}{\Sigma W} \tag{20.2}$$

The lift bar or eye bolt hole must always be located so that it is possible to handle a complete mold unit (injection side, ejector half, center section of a stack mold), without the possibility of parts (e.g., stripper plates) falling off during handling. It is sometimes necessary to have more than one lift bar for one mold: one for the whole mold, and one or more for the subassemblies. A lift bar must never be screwed to a moving part, such as a stripper plate.

The lift bar (usually cold rolled steel) must be strong enough that its yield strength is not exceeded; a safety factor of 5 (five) is adequate. The mold number and weight capacity must be shown on the lift bar. This is important, because it will be used only when servicing the mold, and the molder may store it remotely from the mold. All lift bars should be painted "safety yellow". *The lift bar(s) must be shown on the assembly drawing.*

To determine the CG of a complete mold, the designer must use reasonable approximations. Drillings and other hollows within the mold, plates, and other large parts can be assumed solid, to simplify the process of determining the CG. The following two examples illustrate such calculations.

In the "mold" shown schematically in Fig. 20.16, to get the total mass *M*, simply calculate the volume *V* and multiply it by the specific gravity of the mold material, usually steel. The position of the CG is at *L*/2. For the calculation of the lift bar, the total mass *M* is used.

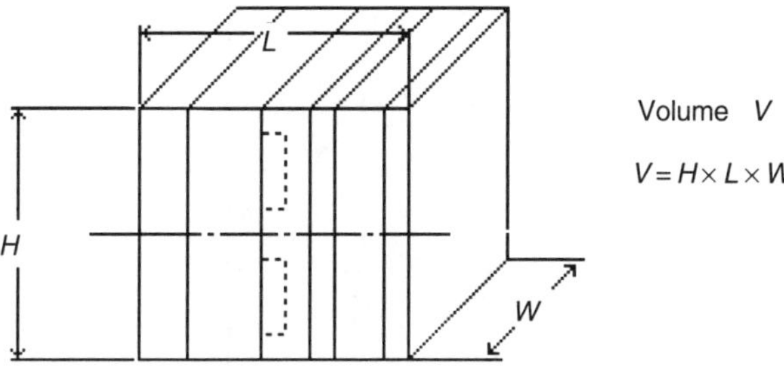

Figure 20.16 Mass M of mold is the product of the total volume and the specific gravity of the mold material.

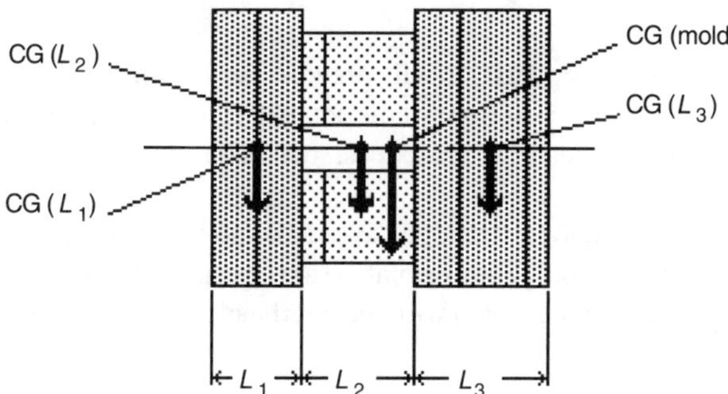

Figure 20.17 Volumes (masses) of each mold section are used to determine CG for entire mold.

In the example in Fig. 20.17, a "modular mold" is divided it into three sections. The volumes are calculated over the distances L_1, L_2 (the volume of the total of all modular parts), and L_3. The volume (mass) is then calculated for each section, and the location of the CG is determined as shown in Section 20.4. If a part "hangs" at an angle of not more than 5° from the vertical, this is sufficiently accurate for handling. (This statement should prevent the designer from wasting time on lengthy calculations which will result in little appreciable benefit for handling.)

20.5 Mounting of Mold in Machine

20.5.1 Larger Molds

This section applies to large molds or to any mold where each half is mounted separately. If a mold is mounted into the machine one half after the other, the same lifting holes that were used for the subassembly are appropriate.

However, there is an exception to the "5° rule" cited above. As the mold (or mold half) is pushed toward the stationary platen, the locating ring (or other alignment means provided) must enter the matching hole in the platen. If the angle is so large that the bottom or top edge

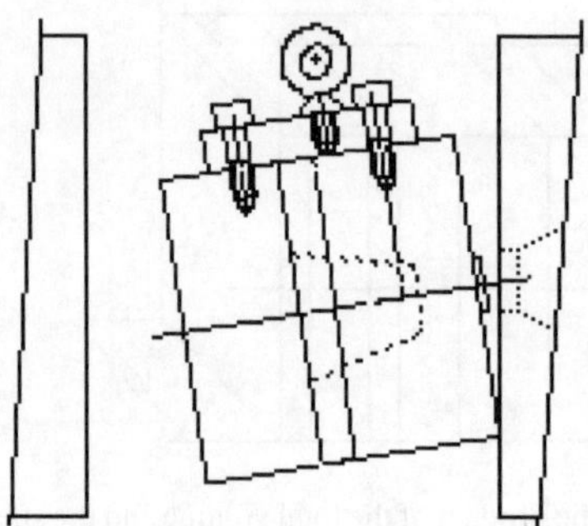

Figure 20.18 Hanging angle is too large and bottom of mold touches the platen first during mounting.

of the mold touches the platen first, it will prevent the locating ring from entering (See Fig. 20.18). In this case, a closer estimate of the actual CG is required, and appropriate lifting bars must be provided to prevent unsafe practices, such as forcing the mold by hand into proper attitude.

Note that the hanging angle (if it can't be avoided) shown in Fig. 20.18 is better than if it were in the other direction (top edge hitting platen first), because it enables the installer to push the mold from the top against the platen with the overhead crane, which is safer than pushing it by hand from the bottom.

20.5.2 Quick Mold Changers

In molds designed for quick mold changing by lowering the mold into the machine, it is important that the mold hangs virtually vertical to ensure that the mounting ledges of the mold enter readily under the mold clamps.

Smaller molds are usually lifted into the machine with both halves strapped (latched) together. As for half-mold installation, the lifting hole location must be such that the locating ring enters more or less squarely into the platen.

20.6 Latches

Any mold which depends solely on tapers for alignment, without the use of close-fitting leader pins, could open up at the bottom during lifting. This could cause damage to the tapers and to the surfaces of cavity and cores, particularly in molds for thin-walled products. This condition, as shown exaggerated on the left illustrations in Fig. 20.19, can be avoided by providing latches.

Even though most molds have leader pins which prevent the mold from opening, it is good practice to provide latches for shipping and handling. This applies also to molds where the lifting bar cannot be mounted directly on the plates but must be spaced with pillars, in which case the screws holding the bars could bend (right illustrations in Fig. 20.19).

It is good practice, and strongly recommended, to provide latches on all molds. There should also be a location *right on the mold* where these latches can be stored when not in use so that they are readily available the next time they are required.

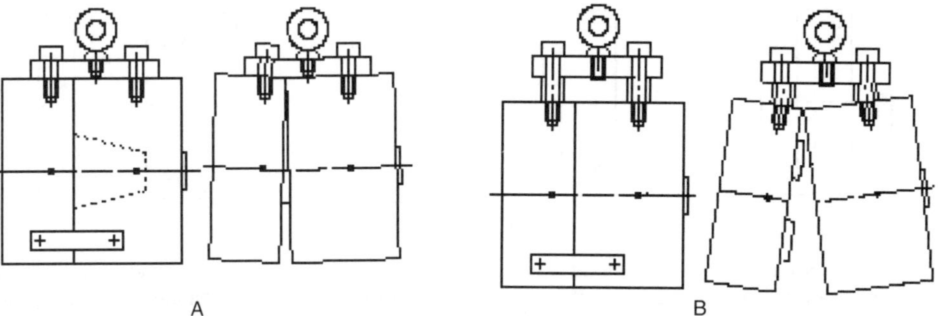

Figure 20.19 Latches prevent opening of molds: A. latch (left) prevents opening of mold with taper alignment (right), and B. latch (left) prevents opening of mold with lifting bar mounted on pillars with screws that could bend (right).

Latches must always be placed on the sides of the mold, accessible from the operator's side and from the rear of the mold when the safety gates are open. Never put latches on top or bottom faces of the mold, where they may expose the operator to unsafe situations.

20.6.1 Latches for Mold Servicing

In some molds, it is necessary to gain access to areas between plates while the mold is mounted in the machine. Typically, it may be required that access be provided to:

1. *Hot runner components* by separating the cavity plate from the manifold plate to gain access to nozzle tips, band heaters, thermocouples, etc., for inspection, cleaning, or replacement;
2. *Tapers at the base of cores* where the strippers seat for cleaning of vents, etc., by separating the stripper plate from the core plate; and
3. *Tapers at the base of cavities* in moving cavity molds, also for cleaning purposes, by separating the (moving) cavity plate from the stationary cavity backing plate.

As a safety precaution, if two (or groups of) plates as listed in the above examples must be separated while the mold is opened to remove the screws holding these plates together, latches must be attached to ensure that portions of the mold which become loose after removal of the screws cannot inadvertently separate from the plates to which they were held by these screws. These latches must be easily accessible while the mold is in the machine, and not hidden (e.g., behind the tie bars).

The "latching process" is as follows:

1. The latches are attached on that side of the mold where the screws which need to be removed are (i.e., the stationary side).
2. The mold is opened, and the screws which hold the plates (or groups of plates) are removed. The latches hold the assembly together, and there is no risk of inadvertent separation. In the case of moving plates, such as strippers and moving cavity plates, the screws to be removed may be stroke limiters, spring return mechanisms, and screws which hold the plates to their actuators.

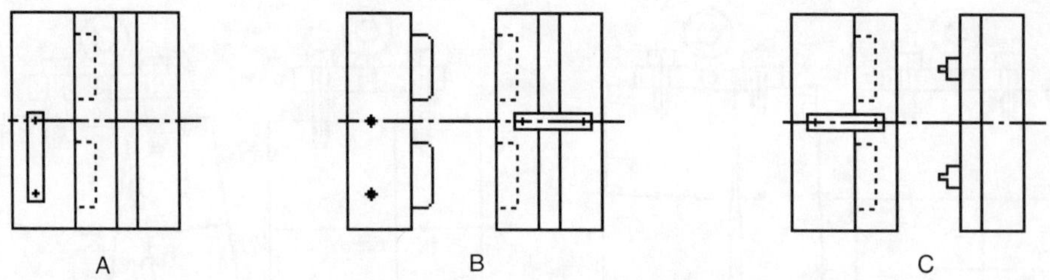

Figure 20.20 Sequence using latches to attain access to hot runner nozzles: A. mold closed and latches parked, B. mold open with latch holding cavity plate, providing access to screws, and C. after mold is closed again, latch is mounted to hold cavity plate to moving plate. When mold is open again, access to hot runner system is provided.

3. The mold is closed again. Now, the latches are unscrewed from the stationary portion of the mold and mounted to connect the portion of the mold which will be moving away as the mold is opened.
4. As the mold is fully opened, the separation of plates to give access to the desired working area is complete. The latches prevent inadvertent separation from the moving side.

   To reassemble the mold, the latching process is reversed:

1. Close mold.
2. Switch position of latches and fasten them.
3. Open mold.
4. Insert and fasten all screws.
5. Remove latches.

If possible, keep the latches fastened to the mold in suitable "parking places" so they will not get lost. If they remain on the mold, each latch must be held by two screws to ensure that they will not come loose during the cycling of the mold.

All latches must have the mold part number and mold number stamped on them so that they can be easily identified if kept in a separate location while not in use. They are also shown on the assembly drawing.

The sketches (Fig. 20.20) show graphically some of the sequences for removing a plate to get access to the hot runner nozzles. To ensure square pulling off, without cocking, the latches must always be used in pairs, diametrically opposite from the center of the mold. Two latches are usually sufficient; for heavy molds, four latches may be required. The size of latches must be adequate for the size of plates moved. For molds which fit tightly between the tie bars, the location of latches should be planned so that they need not be removed or do not require pulling of the tie bars when lowering the mold into the machine.

Note: Latches must never be parked by joining two plates which separate during operation of the mold (e.g., between cavity plate and manifold plate in a moving cavity system, or between stripper plate and core retaining plate). This may seem obvious, but it has been done!

Latches may have two holes if the distance between the two locations can be made the same. If the distance must be different, the latches must have three holes to suit the different distances. If the latches are also required to steady another plate, such as a stripper plate, an additional hole will be required.

21 Air and Oil-Hydraulic Actuators

21.1 Air Cylinders and Pistons

Air actuators are used when a motion within the mold is required which cannot be achieved either 1) by mechanical linking with the motion of the mold with cams or links, or 2) where spring action is required. Air actuators do not have the disadvantage of springs, which require preloading that is often undesirable or unsafe (e.g., springs could counteract the operation of mold protection built into the machine).

With actuators, usually the piston moves while the cylinder is stationary. But, occasionally, the piston is stationary while the cylinder moves. The forces in either case are the same. The force F generated by the cylinder (and piston) is equal to the area A which is pressurized, multiplied by the available air pressure p at the cylinder (Fig. 21.1).

21.1.1 Single-Acting Actuator

In the example shown in Fig. 21.1, the piston acts only in one direction. The return must be provided by some other means. The pressure p may be kept ON, and create a permanent force F, similar to a spring. It should be noted that the force of a spring increases as it is compressed, but with an "air spring", at a given air pressure, the force is constant over the whole stroke.

The air supply could also be switched ON and OFF if the force is required only at a given time period. If forces are required in both directions, a double-acting cylinder is used.

21.1.2 Double-Acting Actuator

When the air in a double-acting actuator actuates the piston from its rod end, the effective area is the difference between piston area A_1 and rod area A_2. The return force F_r (on the rod end)

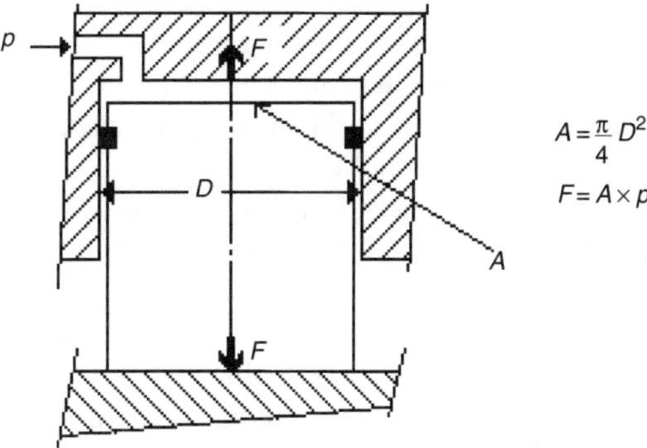

$$A = \frac{\pi}{4} D^2$$

$$F = A \times p$$

Figure 21.1 Cross section of cylinder and piston depicting variables for calculation of force.

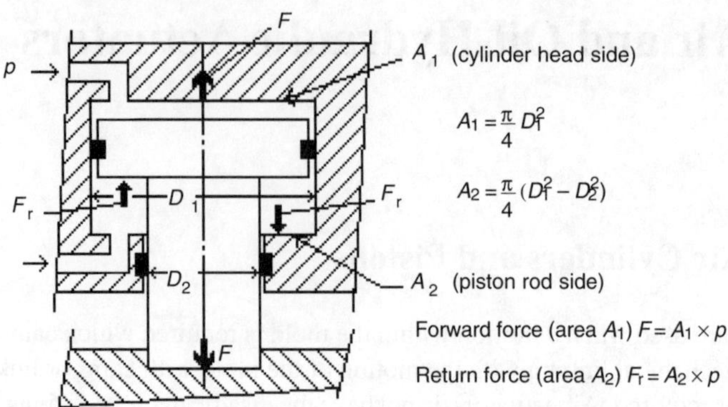

Figure 21.2 Double-acting actuator in cross section showing forward and return forces.

is, therefore, less than the force on the cylinder head end (Fig. 21.2). Sometimes, this difference is desired; at other times, it may be necessary to have both forces equal, in which case an additional "rod end" is introduced on the cylinder head end to provide equal areas on both sides of the piston.

21.1.3 Speed of Build-Up of Force *F*

The air pressure at the cylinder builds up gradually to the line pressure, depending on the pressure drop in the line leading to the mold and the valve controlling the influx and the exhaust of the cylinder, and the size of the air passages within the valve. Balancing of airlines is important, since all air cylinders acting simultaneously on a part (plate) must receive the same air pressure at the same time. In other words, the distance from the controlling valve to all cylinders, and the resistances (restrictions, bends, etc.) within the piping, must be as equal as possible (balanced).

If airlines are not balanced, the piston first pressurized will push ahead of the other pistons and could cause cocking and binding of the plate, which would be noticed as jerky motion. If such jerky motion is unavoidable and unacceptable, oil-hydraulic actuators must be selected instead.

If rapid actuation is required, the pressure drop must be reduced as much as possible by selecting shorter and larger diameter lines (pipes, hoses) from the pressure air supply to the control valve and from the valves to the mold, making sufficiently large air channels within the mold, and by providing larger valves. An additional method to accelerate the motion of air actuators is to speed up the air exhaust from the cylinder. This can be achieved with the introduction of quick exhaust valves mounted close to the cylinder. This way, air need not return and exhaust through the control valve.

The diagram in Fig. 21.3 shows a typical example of a 3-way arrangement. When the solenoid is energized, it shifts the valve spool so that the air in the cylinder can escape, and the force is removed from the piston so that an external force (mechanical, hydraulic, etc.) can push the piston back. When the solenoid is de-energized, a spring in the valve shifts the valve spool back to the position shown, the cylinder is pressurized, and the piston moves forward.

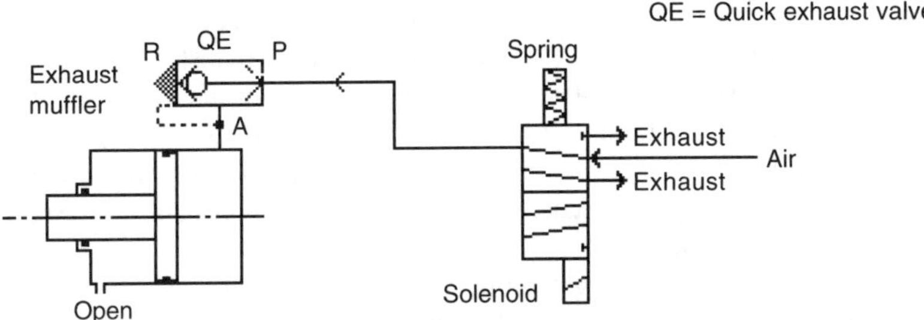

Figure 21.3 Schematic of a typical 3-way solenoid valve arrangement.

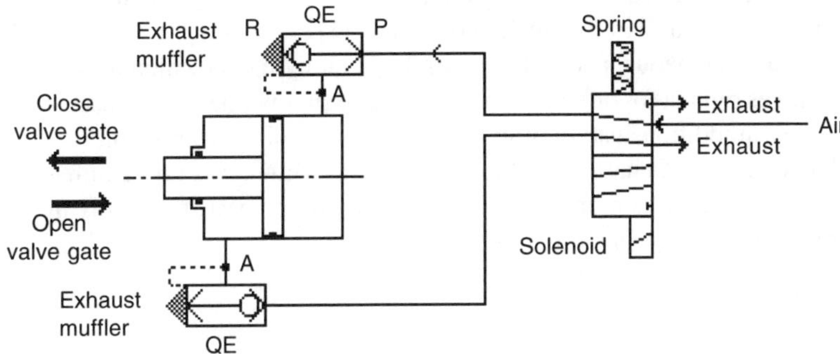

Figure 21.4 Schematic of a typical 4-way solenoid valve circuit (or "arrangement").

It should be well understood that Fig. 21.3 shows just one of many examples of possible arrangements of a 3-way valve.

For quick exhaust, a ball check (as shown in Fig. 21.3) or diaphragm check in the quick exhaust valve is closed while air pressure is ON. As the air pressure is shut OFF, the residual pressure in the cylinder lifts the check and lets the air escape without needing to return through the control valve.

Three-way valves are used often as shown in Fig. 21.3 when pressure is to be exerted on one side of a piston acting on a plate or other mold element, but when this pressure must be eliminated at every cycle, or from time to time. Three-way valves can also be used in lieu of two-way valves. In the ON position, air is supplied; in the OFF position, air is shut off (e.g., for air blow-off, for cleaning vents, for blow-down, etc.).The valve can be either automatically cycled or operator controlled by push button. All air valves (2-, 3-, or 4-way) are similar in construction, and are used for the purpose needed by closing certain ports and by connecting the supply, exhaust, and cylinder lines to suit the application.

A 4-way valve arrangement is used when both the opening and the closing force (e.g., for valve gates) is provided by air. As long as the valve is energized, the valve gates are open; when de-energized, the spring return of the valve directs air to the other side of the piston, and the valve gates are closed.

In Fig. 21.4, a 4-way valve is shown in the position where the spring has pushed the valve spool so that air is supplied to the head side of the cylinder. The air on the rod side is exhausting through the quick exhaust valve. When the solenoid is energized, the spool will shift, the rod end will receive pressure air, and the cylinder head side will exhaust.

In some applications, instead of spring actuation to return the valve spool, it is also possible to use a second solenoid, which would be energized for shifting while the first solenoid is de-energized. There are a large number of possibilities for valve applications; in the foregoing, only some very basic schematics are shown. For more data, the designer should consult with valve suppliers.

In the most commonly used single-acting (3-way) air system, only one line is required between the control valve and the actuator. Double-acting actuators require 4-way valves and two lines to the actuator, requiring more drilling for air channels in the mold plates (or parts). But in either case, there is only one line connecting the plant air system to the control valve. The "return" air is exhausted into the atmosphere.

It is often advantageous, particularly with large air cylinders which must act rapidly, to mount the valves right at (or on) the mold, using large supply lines (e.g., hoses) to the valves, and sometimes, in addition, installing an air pressure reservoir (accumulator) next to the intake of the valve(s), which is charged by the line pressure while air is not required. This added nearby air supply increases momentarily the capacity of the basic air supply line to provide the required amount of air in a minimum amount of time.

21.1.4 Seals

In general, only good quality seals for axial motion (T-seals, Glyd™ rings, etc.) should be used. In exceptional cases, O-rings may appear to be necessary because of space problems. If this is the case, a seal manufacturer may help in avoiding the use of a seal in the wrong application.

O-rings are best used in static applications (i.e., when permanently squeezed between two surfaces to prevent gases or fluids to pass). In dynamic application, where one surface moves relative to another, O-rings for sealing between these surfaces roll in their grooves and will be destroyed rapidly. This is particularly the case when the motion is more than about 1 mm and if the ring is not, or only poorly, lubricated. In small motions, where the ring can just "rock", O-rings can be tolerated, even when dry.

The clearances between cylinder wall and piston, and between piston rod bore and piston rod, can usually be slightly larger than those recommended by the seal suppliers. This is based on experience and will not affect the life of the seals. It often prevents scraping (and damage) of the cylinder bore and piston rod as a result of misalignment (e.g., caused by leader pin bushing wear after a long time of mold operation). However, the diameters where the seals seat, the width of the seal groove, and the shape and finish of the grooves must be to the manufacturer's specification.

21.1.5 Location and Number of Cylinders

Cylinders can be used either to move a plate forward, to return it after it was moved forward by some other means (e.g., machine ejector), or for both motions. In general, cylinders should be arranged symmetrically about the point(s) where force is required. If a plate must be moved, the best arrangement is four cylinders, spaced in a rectangular pattern so that the actuated plate will deflect a minimal amount while ensuring parallel motion (Fig. 21.5).

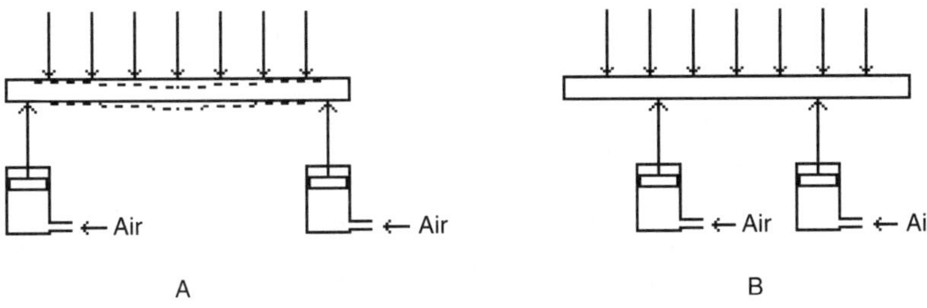

Figure 21.5 Spacing of air cylinders: A. widely spaced cylinders cause much deflection, and B. closely placed cylinders reduce deflection of plate.

There are no restrictions on the number of air cylinders used within a mold. The location and number of cylinders depend on the application and the forces required, and is left to the judgment of the designer.

21.1.6 Alignment

The piston and piston rod must never be used as alignment elements of a moving plate (see also Section 17.5, Guiding of Moving Plates). There must always be a system of guides such as leader pins or gibs for this purpose.

21.1.7 Finish of Walls and Piston Rod

To protect the seals from excessive wear, the finish of the sealing surfaces must be as specified by the seal manufacturer. This information must be shown on the detail drawing for the mold part.

21.1.8 Force Required

It is difficult to determine how much force will actually be required by the air cylinder and piston. The required force depends on many factors and on the application.

For example, if the force is required to strip molded products from cores (e.g., on the stationary half of a stack mold), it depends on:

- The draft angle of the side walls,
- the finish of the molding surface,
- the presence and severity of undercuts,
- the molding material injected (adhesion to the mold steel and shrinkage),
- the temperature of the product at ejection (when hot, it will stretch easier over undercuts or rough finish than when cold), and
- the number of products to be stripped.

If the force is required to move a mold insert out of a pocket in which it was sitting during injection, it depends on:

- The wedge angle of its fit with the matching part,

- the preload of such a wedge, and
- the finish of the surfaces.

As a rule, air cylinders should not be used to move parts out of a preloaded (wedge) position. The unlocking force is difficult to predict, and air cylinders may not have enough force (in the restricted space usually available) and may act jerky or not move at all.

The required force may also be influenced by the operating clearance of ejector pins, particularly if the number of pins is large. The designer will have to estimate the required force, or try to get an input from more experienced designers and mold technicians.

For high forces, large cylinder areas will be necessary. It is very important to get a good estimate of the forces required to move the plates or other mold elements before sizing and designing the cylinder. The size of the cylinders may affect the size of the plates (and thereby the size of the mold).

There may be limits to the available force. In practically all molding plants, air pressure is about 6–7 kg/cm^2 (80–100 psi). In exceptional cases, 8 kg/cm^2 (115 psi) may be made available. It is, therefore, important for the designer to ascertain what pressure is available at the plant for which the mold is to be designed.

21.1.9 Exhaust Noise

Air actuators can create a considerable noise problem in a plant. To reduce this noise, mufflers should be used at the exhaust port of any air valve.

21.1.10 Cost of Air Operation

This cost factor is often overlooked at the design stage of a mold. Compressed air is relatively expensive, not only regarding the energy cost but also the capacity of the installed services. If, for example, a mold requires four cylinders having a total volume of 0.1 cu. ft. and the mold cycles 10 shots per minute, this mold will require 0.1 cu. ft./shot × 10 shots/minute = 1 cu. ft. of pressure air per minute. In small plants, this may constitute a sizable portion of the available air capacity.

For this reason too, it is important to make a realistic estimate of the required actuator force and not to design the air cylinders unnecessarily large, particularly with fast-cycling molds. The amount of air required for cylinders may be in addition to other (often high) air requirements of the mold (e.g., for ejection and blow-down of the products with an air curtain or with air jets).

21.2 Oil-Hydraulic Actuators

There are three basic advantages for selecting oil-hydraulic actuators instead of air actuators:

1. More force is provided than is possible with air pressure;
2. smooth motion of the mold part, if this is essential; and
3. good rate control of the piston speed, which is difficult with air cylinders.

Whenever possible, commercially available cylinders should be used.

Hydraulic actuators should be used outside of the mold, preferably at the sides of the mold, to:

- Reduce the risk of oil leaks contaminating the mold (and the products), and
- minimize the risk of fire in case oil should contact hot parts of the mold, such as the hot runner parts, sprue bar, etc. Because this is not always possible, the need for well designed, "service free" cylinders is extremely important.

It should be noted that in molds for products subject to "food and drug" regulations, the type of oil may have to be approved for this purpose. Seals, etc., selected for these actuators must be compatible with any special oils.

The number and the arrangement of oil-actuators driving a plate are subject to the same considerations as described above for number and location of air cylinders. Because the hydraulic pressure available is usually considerably greater (up to 20 times) than the available air pressure, the hydraulic cylinders can be much smaller than equivalent-force air cylinders.

It is usually desirable to provide pressure reducing valves ahead of the control valves. However, it is recommended that cylinders be selected small enough, and the mold plates sufficiently strong, that even in case of failure or bad adjustment of the pressure reducing valve, the forces generated by the actuators will not cause damage.

Most injection molding machines have a system pressure of about 145 kg/cm^2 (2,150 psi). The oil volume required for the actuators is usually very small, and the pressure oil can be easily tapped off the molding machine's pressure system without affecting the machine. If the hydraulic supply is taken from the machine, a pressure regulator will be necessary.

Some molders use a separate hydraulic oil pump at about 500 psi pressure. In such cases, all components can be provided for the lower pressure, which may be substantially less expensive than using high-pressure actuators and the required high-pressure hoses, fittings, and valves.

From a design point of view, a disadvantage of hydraulic systems is that there are two lines (pressure and return) required between each control valve and its associated actuator(s), whether single- or double-acting (3-way or 4-way valve). Also, there is always a return to tank line required for the control and pressure reducing valves.

21.2.1 Hydraulic Oil Lines

With hydraulic actuators, the cross drilled holes must be plugged with suitable steel (pipe) plugs, not with the standard brass plugs used for water or air channels. Any piping between actuators and valves which are directly attached to the mold must be done with steel tubing and fittings of suitable pressure rating. Moving hoses must be also of appropriate quality and pressure rating, similar to the ones used for the machine.

All piping must be shown on the assembly drawing of the mold. Water hoses or fittings must never be used for hydraulic lines. Balancing of oil lines should be attempted, although, because of the small volumes, the high pressures, and the incompressibility of oil, it is not as important as with air lines. Noise is no problem with oil-hydraulic operators.

21.2.2 Stroke

Unlike air actuators, hydraulic actuators are usually designed to bottom out on the forward stroke, but this is not a requirement. Standard stroke lengths available from the supplier can usually be accommodated in a design.

21.2.3 Bleeding of Air

Every hydraulic cylinder in a mold must have, at its highest point, a "bleed hole", that is, a small hole which is normally screwed oil-tight with a ball and screw or with a special bleeder screw. The bleed hole permits air in the cylinder to vent out; it also permits the filling of the actuator with oil before the mold is put into service.

21.2.4 Interface with the Molding Machine

All actuators require limit or proximity switches to signal ends of stroke for interfacing with the electronic machine controls.

21.3 Mold Safety

When using air- or hydraulically actuated plates or other moving elements in a mold, it is the responsibility of the designer to ensure that they cannot endanger an operator when power is off and/or a safety gate is opened. The designer must refer to ANSI B151.1-1985 (North America) and EN201 (Europe) for regulations regarding the interlocking of power circuits in injection molding machines.

Air circuits are usually not interlocked with the machine's guarding system and can be operated regardless of the position of the safety gate, which is sometimes desirable. They may remain under pressure (charged) when a gate is open.

In all such cases, a clearly visible nameplate must be installed, right on the mold, to warn the operator of the possibility that a plate (or other part) can move, even though the front gate is open. Also, opening of most machines' rear gates shuts the motors off and empties the hydraulic accumulator. Any actuator which is lifted by hydraulic pressure can then slide down (by gravity) as the hydraulic supply goes to zero pressure. This could create pinch points, which must be properly guarded.

Wherever possible, any pinch point should be guarded with suitable (telescoping, or overlapping) shields, so that operators are protected from unexpected motions. Guarding and the name plate location must be shown on the assembly drawing for the mold.

22 Rules and Calculations for Designers

The following calculations are required for every mold. The assumptions are mainly based on experience; the designer is advised to look for precedents as recorded after tests of similar molds or to consult with more experienced designers who may have an input. If there is no precedent, the designer must search for similar applications and use common sense. In very difficult or risky cases, an experimental mold may be required to get the answers before committing to a design which may be risky, or even fail.

22.1 Projected Area (A_{pp}) of Product

The area of the product projected onto a plane usually is considered to be the area at right angles to the motion of the clamp. It is used to calculate the required clamp force to hold the mold closed, or the forces required to hold side cores or neck rings clamped together, during injection. All of the examples below are illustrated in Fig. 22.1.

Example 1: The projected area A_{pp} of a disk of Ø 200 mm is (in the direction of clamping)

$$\frac{(200\ \text{mm})^2\pi}{4} = 31{,}416\ \text{mm}^2.$$

Example 2: If the disk is 2 mm thick, the projected edge area is (at right angles to the direction of clamping, or in "perpendicular direction")

$$200\ \text{mm} \times 2\ \text{mm} = 400\ \text{mm}^2.$$

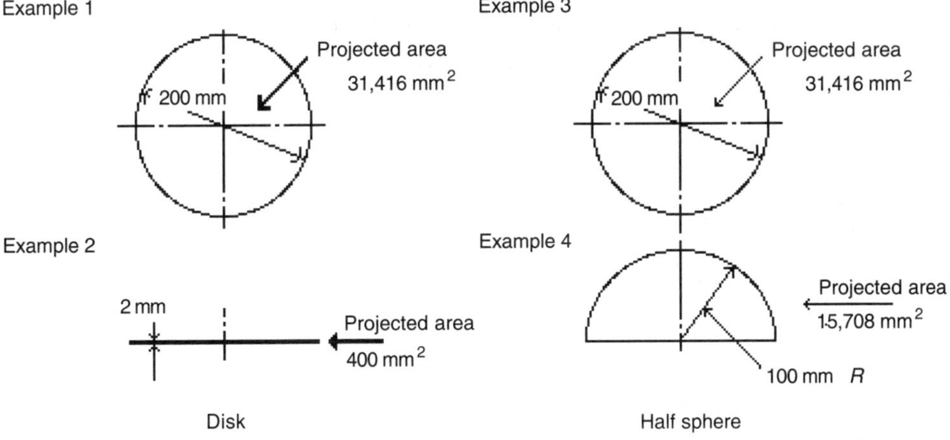

Figure 22.1 Sketches show projected areas of products described in examples 1-4.

Example 3: If a product has the shape of a half sphere with a 200 mm diameter, the projected area A_{pp} in the clamping direction is the same as that for a disk, or 31,416 mm^2.

Example 4: In the perpendicular direction, A_{pp} is one-half of this amount, or 15,708 mm^2. This may seem obvious, but is often missed.

Rule: The projected area is not affected by the configuration (shape) of the product within its outline, except that the area of any openings within the outline can be deducted from the overall area.

Rule: Theoretically, for these calculations the steel sizes should be used (i.e., the product sizes plus the shrinkage allowances), not the product sizes. However, since 1) the difference is not more than about 2% and there should be always ample reserve in the machine clamp to hold the mold closed during injection to prevent flashing, and 2) the plastic pressure inside the mold is only estimated and is certainly not closer than 2%, it is in general acceptable to use the projected area of the product.

22.2 Shut-Off Area of Steel (A_{ss})

The shut-off area of steel is the area where cavity and core halves touch as the mold is clamped shut. This is commonly called the parting line (P/L) area. The actual shut-off steel area A_{ss} is shown in Fig. 22.2.

22.3 Clamping Force F_c

To hold the mold closed (without flashing) during injection, clamping force F_c (as exerted by the molding machine) must be greater than the separation force F which is generated when the injection pressure p acts on the projected area App of the product. This pressure p is always smaller than the injection pressure as measured at the machine nozzle:

$$F_c > A_{pp} \times p. \quad (22.1)$$

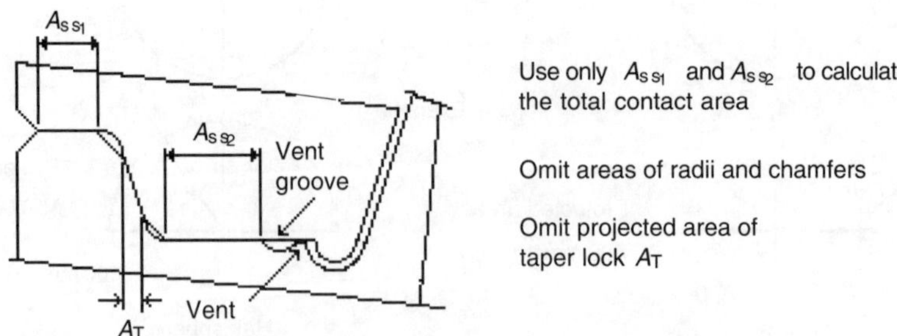

Figure 22.2 Schematic illustrates shut-off area A_{SS} of steel.

The suggested calculation is

$$F_c = K \times A_{pp} \times p. \qquad (22.2)$$

A_{pp} can be calculated from the product drawing (see Section 22.1). (Factor) K must be greater than 1, to prevent the mold from flashing. K may be (conservatively) assumed to be 1.2 before the mold is designed.

Pressure p must be assumed. It depends on many factors, such as injection pressure, speed, temperature and viscosity of plastic, etc. Pressure p may vary with the method of feeding. It is lower with starve feeding and higher when the product must be well "packed out" to reproduce the exact surface definition of the cavity (e.g., the "signals" in the grooves of records, etc.).

Actual pressure p is essentially based on experience with similar molds. To start the calculations, assume $p = 360$ kg/cm² (5,000 psi). Some molds may need more, some less.

The shutoff area of steel (A_{ss}) must be large enough to prevent collapse of the surfaces when the mold is clamped up. The harder the steel, the less the risk.

22.4 Fatigue

Since the clamp force F_c is often applied for millions of cycles, the fatigue strength of the steel must be used rather than its ultimate (compressive) strength S. While in general machine design the fatigue strength of steel S_f is assumed to be about 10–20% of its yield strength, it is better to use more conservative values, usually $S_f = 420$ kg/cm² (6,000 psi) for hard mold steel, and one half these values for plates, etc. If, after the first layout and calculations, the value S_f is too high, the shut-off area (parting plane) must be increased. If this is not practical, load-carrying pads of suitable shapes must be provided outside the molding area to share the load exerted by the clamping force with the parting plane. Note that the lesser of the values must be used if, for example, a hard pad sits on a softer plate.

$$A_{ss} \geq \frac{F_c}{S_f} \qquad (22.3)$$

Example: Given: A 4-cavity mold for a round PE container. OD = 10 cm

Assumed: Injection pressure (inside the cavity) = 360 kg/cm².
Mold material H13; $S_f = 420$ kg/cm²
Shrinkage = 2%
$K = 1.2$

Calculated (rounded figures):

$$\text{OD} = 10\text{ cm} + (10\text{ cm} \times 0.02) = 10.2\text{ cm}$$

$$A_{pp} = \frac{\pi}{4} \times (10.2\text{ cm})^2 = 81.7\text{ cm}^2 \text{ x } 4\text{ (cavities)} = 327\text{ cm}^2$$

$$F_c = 327\text{ cm}^2 \times 360\text{ kg/cm}^2 \times 1.2 \approx 141{,}000\text{ kg or} \approx 141\text{ t (metric tons).}$$

$$A_{ss} \geq \frac{141{,}000\text{ kg}}{420\text{ kg/cm}^2} \geq 336\text{ cm}^2 \text{ is required.}$$

The A_{ss} in the above example is based on $K = 1.2$, but now assume that only 300 cm² can be provided, which is less than the 336 cm² required. The designer must recheck K for 300:

$$420 \text{ kg/cm}^2 \times 300 \text{ cm}^2 = 126{,}000 \text{ kg, so}$$

$$K_{\text{actual}} = \frac{126{,}000 \text{ kg}}{117{,}700 \text{ kg}} = 1.07.$$

Since $1.07 > 1$, the design is acceptable. If the result were $K < 1$, the shut-off faces must be redesigned (increased) and/or pads must be added. Figure 22.2 shows the areas A_{ss1} and A_{ss2}, which may be used in this total. In general, both these faces can have any shape; they usually follow approximately the shape of the outline of the product.

This conservative approach to area loading ensures that vents will not be crushed after a short time of operation. From experience, and to simplify the calculations, the area of the taper seat is also omitted, partly because the taper wears after a time of operation and it is assumed that it carries less and less clamping load as it wears.

(For further discussion of fatigue, see also Chapter 18, Metal Fatigue.)

22.5 Compression of Stack

Steel is *compressible.* A measure of the compressibility is the modulus of elasticity E, expressed in kg/cm² or psi. All steels have approximately the same modulus E, regardless of hardness or composition. The only exception are tungsten-carbides, which may have an E which is two to three times that for other steels.

This compressibility affects the mold part dimensions after the full clamp tonnage is applied to the mold. It must be considered when dimensioning the bottom of thin-walled containers, or in other areas where very small differences may overstress a mold part.

For a simplified method of calculations, refer to Fig. 22.3. When the mold is clamped up, assume a "tube" defined by diameters D, d, and height h is compressed by the clamp force F_c. The size B (after clamp-up) will be smaller (by the amount f) than the original, unloaded size B_o. To compensate for the weakening effect of the cross drilling within the cavity, the calculated amount f should be increased by about 15%. (This is an experience factor.)

Stacking height in the product, and other distances which require accurate dimensions from the open end of the container, must also reflect the size reduction caused by compression f of the stack. If the outside is square, as with square or rectangular blocks in modular molds, or with any other shape, the whole projected area of the cavity block minus the projected area of the product (including any vents) will be used in this calculation. This adds to the area under compression, thus reducing f. This amount of compression f must be considered in the dimensioning of the mold parts, especially in products with very thin bottoms.

Units must never be mixed:

Metric:	$E = 2{,}100{,}000$ kg/cm²	F_c (kg)	d, h (cm)
American:	$E = 30{,}000{,}000$ psi	F_c (lb)	d, h (in)

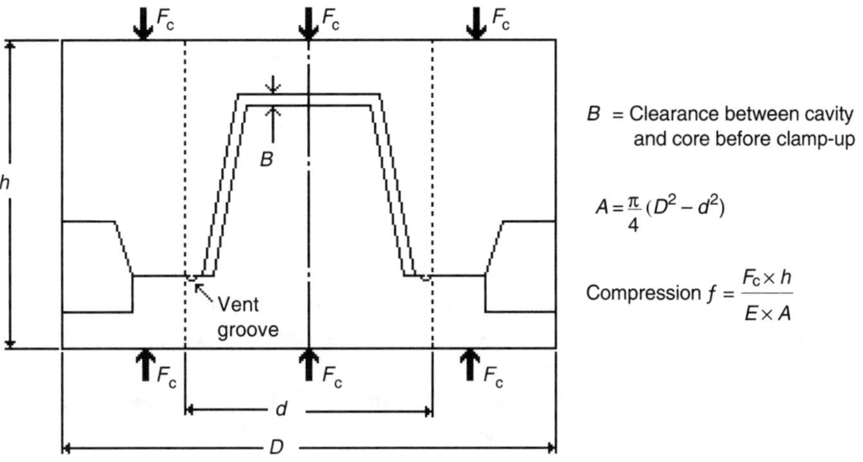

Figure 22.3 Compression of stack in container mold.

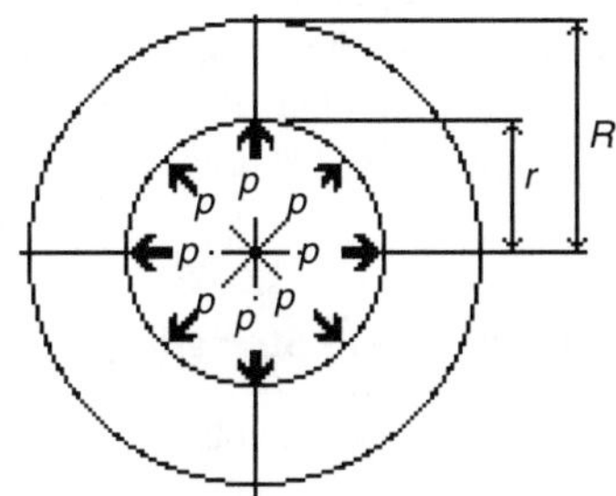

Figure 22.4 Hoop stress and expansion of thick-walled, tubular cavity.

22.6 Hoop Stresses and Cavity Expansion

Hoop stress and cavity expansion applies primarily to thick-walled, tubular cavities (Fig. 22.4). The formulae below are approximations but are suitable for mold design:

$$\text{Stress at the inside of hoop: } S = p\,\frac{R^2 + r^2}{R^2 - r^2} \text{ and} \tag{22.4}$$

$$\text{Stress at the outside of hoop: } S = 2p\,\frac{r^2}{R^2 - r^2}, \tag{22.5}$$

where S is the stress, p is pressure, R is the outer radius, and r is the inner radius of the tubular cavity. Both R and r will increase as a result of the pressure p:

$$\text{Increase of inner radius } r\text{: } f = p\,\frac{r}{E}\left(\frac{R^2 + r^2}{R^2 - r^2} - V\right) \tag{22.6}$$

$$\text{Increase of outer radius } R\text{: } f = p\,\frac{R}{E}\left(\frac{2r^2}{R^2 - r^2}\right) \tag{22.7}$$

where V = Poisson's ratio for steel (V = 0.3). Units are as follows:

Metric: E = 2,100,000 kg/cm² p, S (kg/cm²) R, r (cm)
American: E = 30,000,000 psi p, S (psi) R, r (inches)
Pressure p: Assume same as before (360 kg/cm²)

The expansion can be considerable, and S must be calculated to ensure that the walls are thick enough for the job. Note that stress S must be smaller than the fatigue stress S_f for the cavity material, for long life. Fatigue must be considered, since the stretching (loading) of the cavities occurs cyclically (i.e., at every molding cycle).

After the stack has been laid out, the designer must calculate the stresses and ensure that the steel is not only strong enough to reduce the stretching to a minimum but also to prevent the cavity from bursting (Table 22.1).

In the example shown in Fig. 22.5, the stresses are well below the 10% of the S_u of any mold steel selected. Also notice that the inside of the cavity will stretch by the amount f_r = 0.02 mm (0.0008 in.), which may be enough for the plastic to flash into the gap indicated with an arrow, or it may make the taper seat lose its fit with the matching mold part.

When the cavity is inserted into a mold plate, the expansion of the cavity "cylinder" will be reduced somewhat by the surrounding steel; however, the plate itself will expand in the

Table 22.1 Ultimate Tensile Strength S_u of Some Steels

Steel	Metric (kg/cm²)	U.S. (psi)
P6	11,600	165,000
AISI 4140	15.500	220,000
H13	17,600	250,000

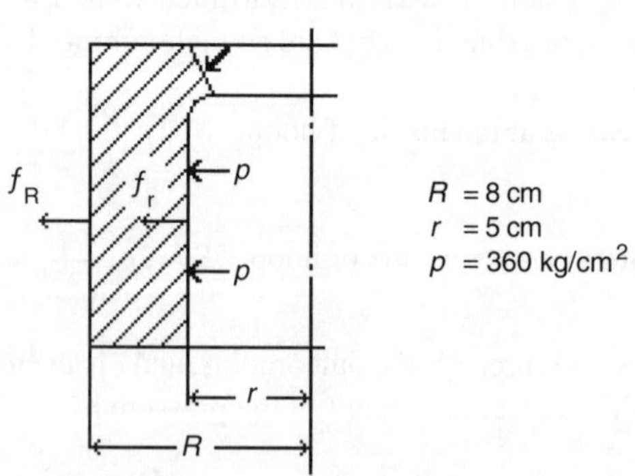

Figure 22.5 Hoop stress and expansion may allow flash at gap indicated by arrow.

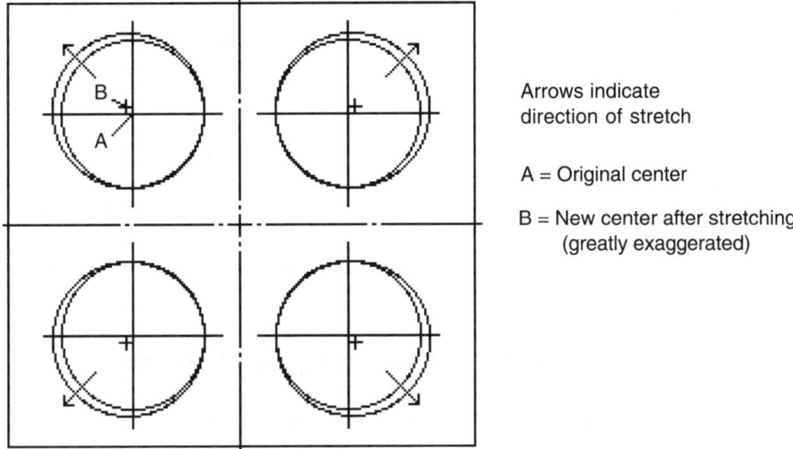

Figure 22.6 Expansion of four cavities inserted into a mold plate away from each other results in thinner product walls toward the center of the plate.

direction where the inserts are nearer to the edge of the plate. Under pressure, the centers of the inserts will move outward from the original centers of the bores in the plate. The sketch in Fig. 22.6 illustrates this movement. It is important that this movement is considered, especially with high injection pressures.

If the core is held solidly but the cavities move away from each other, the wall of the product will be thinner toward the center of the plate and thicker in the direction of the arrow. The use of floating core mounting overcomes this problem.

22.7 Springs

Springs fatigue (and break) over time, particularly if they are stressed excessively, and more so if they are exposed to high temperatures. For a reasonable spring life, the sum of stroke and preload should not exceed 20% of the maximum compression of the spring. If a greater operating range is required, the spring must be longer (Fig. 22.7, next page). Overloaded springs are a frequent cause of mold breakdowns. Follow spring manufacturers' recommendations.

22.8 Wedges

Figure 22.8 (next page) shows relationships for *FR* (the force at right angles to the wedge angle) and *FW*(the force in the direction of the wedge). The *FR* is calculated below:

$$FR = \frac{FW}{\tan \alpha}. \tag{22.8}$$

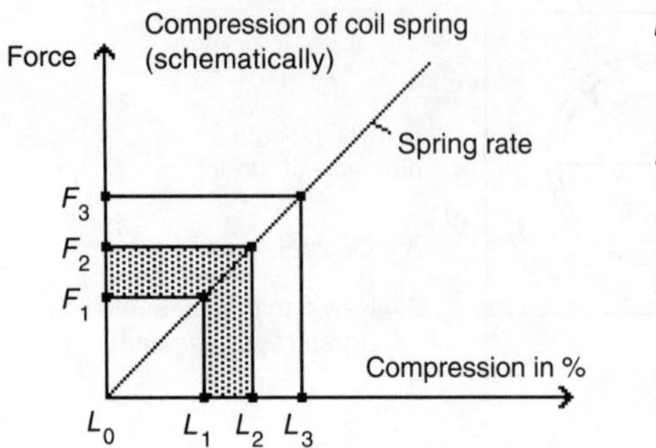

Figure 22.7 Graph showing spring operation range.

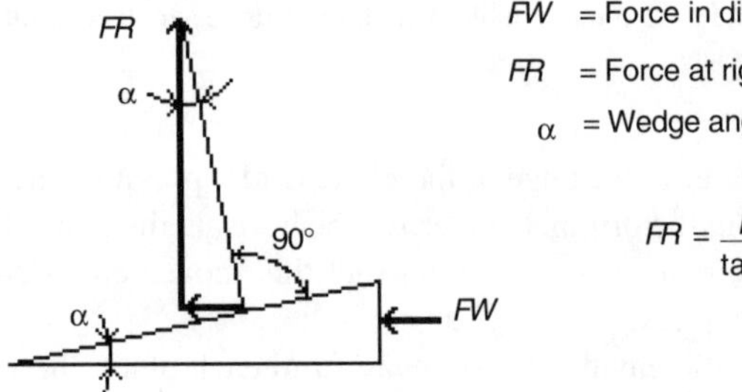

Figure 22.8 Wedge angle and forces *FW* and *FR*.

22.9 Miscellaneous Calculations

22.9.1 Dimensioning over Rollers or Balls

Calculations for dimensions at external and internal taper over rollers or balls are shown in Fig. 22.9.

22.9.2 Detent

Detents, or ball springs, are seated in an indent. The deeper the ball enters into the indent, the smaller is the angle φ and the greater the holding force created by the spring (Fig. 22.10). It is important that the angle φ is not so small that the ball locks the slide. The less the ball enters, the greater is φ and the less the holding power.

Detents are occasionally used to prevent slides from losing their position. In general, they are not recommended because of the risk that someone will inadvertently push the slide out of position, since the holding force is usually not very great and definitely not positive enough to ensure the proper position.

Detents should never be used to prevent vertical motion. For horizontal slides, they may be used, with caution.

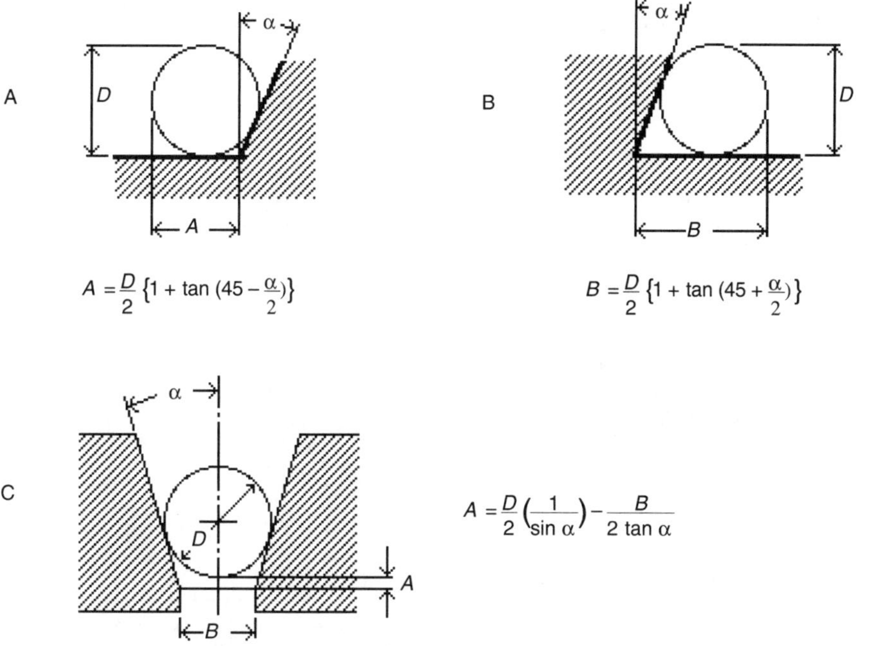

Figure 22.9 Calculating dimensions over rollers or balls: A. external taper, B. internal taper, and C. small internal taper.

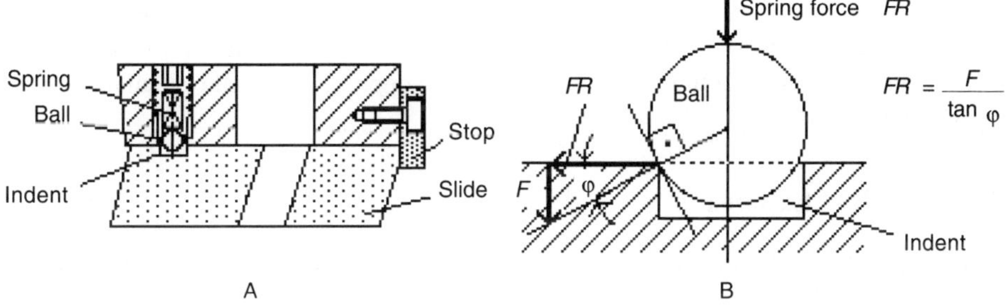

Figure 22.10 Holding force in a detent: A. cross section showing detent in slide, and B. detail of ball in indent and *FR* at angle φ.

22.10 Heat Conductivity

Thermal (or heat) conductivity measures the speed at which heat travels through a material. It is the amount of heat (expressed in calories or BTUs) which passes through a specified cross section (m^2 or ft^2), subjected to a temperature gradient in °C (or °F) in a specified unit of time (hour).

Each material has a coefficient of thermal conductivity. Shown in Table 22.2 are approximate values of some materials for comparison purposes only.

Table 22.2 Coefficient K of Thermal Conductivity for Various Materials

Material	Coefficient K of thermal conductivity (approx.)
Silver	360
Pure copper	350
Aluminum	270
Beryllium copper (depends on alloy)	100–250
Brass and bronze	50–100
Nickel	50
Steel (mild)	40–50
Steel (alloyed), hardened	10–30
Stainless steel (40% nickel)	10
Titanium alloys	2–4
Glass	1.0–1.5
Plastics (not burnt or degraded)	1.0–1.5
Air (not moving)[a]	
Insulating boards (asbestos-cement)	< 1
Insulating boards (Brandenburger)	< 1
Insulating boards (plastic)	< 1

[a] Air is an ideal insulator, but radiation through air will also create heat losses. Even so, air gaps should always be considered for insulation.

It must be well understood that the values for heat conductivity shown in Table 22.2 give the relative conductivities for various materials; for example, a beryllium-copper alloy is about 4–5 times as good a conductor as steel. Exact values for heat conductivity vary for each material (and alloy) and should be obtained from the supplier, if accurate calculations are to be performed.

22.10.1 Heat Expansion

For values of linear coefficient of heat expansion f for steel and other mold materials, see Chapter 14, Heat Expansion.

22.11 Specific Heat

Specific heat (of a substance) is the ratio of the heat required to raise the temperature of a given substance by one degree to that required to raise the temperature of the same amount of water one degree. Specific heat is *not* constant at all temperatures. It is generally assumed that it is determined by raising the temperature from 15 to 16 °C (62 to 63 °F).

The values for specific heat shown in Table 22.3 are typical for a resin class. For exact values for a specific resin, the materials supplier of the resin should be consulted.

Table 22.3 Specific Heat Values for Typical Materials

Substance	Specific heat (BTU/lb/°F)
Aluminum	0.214
Brass, copper	0.094
Glass	0.194
Graphite	0.201
Iron (up to about 300 °C)	≈ 0.115
Lead	0.031
Oil (machine)	0.400
Silver	0.056
Steel	0.116
Water	1.0
Wood	≈ 0.600
Plastics	
ABS	0.40
Acetal	0.35
Acrylic	0.35
Cellulose acetate	0.36
Cellulose proprionate	0.40
Ionomer	0.54
Nylon	0.40
Polycarbonate	0.30
Polyester (PET)	0.40
Polyethylene (LD or HD)	0.55
Polypropylene	0.50
Polystyrene	0.34
PVC (rigid)	0.25

[a] To convert the above values from BTU/lb/°F to J/kg/°C, multiply the value by 4,186.

22.12 Wattage per Length of Heater

22.12.1 Tubular Heaters

The wattage of a heater should be limited to a maximum of 10 W/cm^2 of heater surface area, or (65 W/in^2). This figure applies to good heat sink condition, that is, where the heater is cast into a heated body or pressed into steel of the hot runner manifold.

It should be less, only 5 W/cm^2 (33 W/in^2), when the tubular heaters are lying in a groove, with good contact on about 50% of its surface, and poor or no contact at the rest of the surface.

22.12.2 Cartridge Heaters

The use of cartridge heaters should be avoided if possible, mainly because of poor heat distribution within the heater and difficulties in achieving uniform heat transfer to the surrounding steel. There are also other mechanical and insulating problems with these heaters which make them more susceptible to damage. In general, they have a shorter life than tubular heaters but are less expensive and easy to obtain, as most dealers carry a wide variety in size and wattage of these heaters. Cartridge heaters usually have a surface wattage of about 5 W/cm^2 (33 W/in^2).

22.12.3 Heat Input (in W) per Mass of the Hot Runner Manifold

A simple, useful rule for designers of hot runner manifolds is that the heater capacity required for a hot runner system, provided it is well insulated from, and has little contact with, the surrounding cooled mold plates is about 2–3 W/ cm^3 (30–40 W/in^3) of total mass of manifold. Using a larger wattage than suggested above may overload the control system and is usually not required. Also, it could increase considerably the cost of the system by requiring larger controllers. There is also the danger of severe overheating in case of loss of control. The disadvantage of using a lower wattage than suggested is only longer start-up time from cold.

In a well-designed hot runner system, once the plastic in the manifold and in the nozzles is at the molding temperature, the only heat supplied by the hot runner heaters (manifold and nozzles) is to make up for the heat losses to the surrounding mold plates. The plastic is heated in the extruder to the required melt temperature, and its temperature is maintained as it flows through the hot runner system without requiring any additional heat input from the hot runner heaters.

22.13 Frequently Used Conversion Factors

Table 22.4 lists frequently used conversion factors for many units of measure. Other conversions not included on the list may be found in textbooks and engineering handbooks.

Note to designers: When typing formulas, leave one (1) space between numbers and units, except for symbols " (inches) and ^{2}, ^{3}, . . . (e.g., 1234 kg, not 1234kg; 1234^2, not 1234 ^{2}; etc.).

22.14 Preload of Taper Fits

22.14.1 Purpose of Taper in Molds

The most important function of tapers is to provide good alignment between matching mold parts or assemblies which move in relation to each other.

In molds, leader pins and bushings are the customary method for alignment. These leader pins and bushings are simple and low cost, but they have the disadvantage that there must be some clearance between the pins and their bushings to permit them to slide in and out without wearing. Inherently, this clearance is the cause of some (even if minor) misalignment. Even if there is very little clearance in the beginning, there soon will be enough materials worn away from the pins or bushings, especially with poor lubrication, to increase the clearance. Some minor misalignment may be insignificant for the required accuracy of the product, particularly in molds for heavy-walled products; the use of leader pins and bushings is then quite acceptable.

Table 22.4 Conversion Factors for Units of Measure

Multiply	by	to obtain:
Mass		
Short ton (2,000 lb.)	0.9072	t
Long ton (2,240 lb.)	1.0160	t
Metric tonne (t)	1,000.0	kg
Kilogram (kg)	2.2046	lb.
Pound (lb.)	0.4536	kg
or	16.0	oz.
Ounce (oz.)	28.3495	g
Gram (g)	0.03527	oz.
Length		
Meter (m)	39.3700	in.
or	3.2808	ft.
Millimeter (mm)	0.03937	in.
Inch (in.)	0.0254	m
or	25.4	mm
Volume		
Cubic centimeter (cm^3)	0.0610	$in.^3$
Cubic inch ($in.^3$)	16.3871	$cm.^3$
Cubic foot ($ft.^3$)	0.0283	m^3
or	28.3168	L
Ounce (Imperial, liquid)	28.4131	cm^3
Ounce (U.S., liquid)	29.5735	cm^3
Gallon (Imperial)	4.5461	L
Gallon (U.S.)	3.7854	L
Liter (L)[a]	1,000.0	cm^3
Density (mass density) [b]		
$lb./in.^3$	27.6799	g/cm^3
g/cm^3	0.03613	$lb./in.^3$
Force		
Newton (N)	0.1020	kgf
or	0.2248	lbf
Kilogram-force (kgf)	9.8066	N
Tonne-force (tf)	9.8066	kN
Pound-force (lbf)	4.4482	N
KiloNewton (kN)	101.972	kgf
Velocity		
Foot/second (ft./s)	0.3048	m/s
Acceleration/deceleration		
Gravity (g) (average)	32.1740	$ft./s^2$
or	9.8066	m/s^2
Foot/second square ($ft./s^2$)	0.3048	m/s^2
Torque (moment of force)		
Pound force (lbf)	1.3558	N.m
or	0.1383	kgf.m
Newton.meter (N.m)	0.7376	lbf.ft
or	0.1020	kgf.m
Kilogram.force.meter	7.2330	lbf.ft
or	9.8066	N.m

(continued)

Table 22.4 Conversion Factors for Units of Measure *(continued)*

Multiply	by	to obtain:
Pressure		
Poundforce/inch^2 (lbf/in.^{2}, psi)	0.006895	MPa
or	6.8948	kPa
or	0.0703	kgf/cm^2
Kilogramforce/cm^2 (kgf/cm^2)	14.2233	psi
Pascal (Pa)	1.0	N/m^2
KiloPascal (kPa)	0.1450	psi
or	1.0	kN/m^2
or	101.972	kgf/m^2
or	0.01020	kgf/cm^2
Column mercury inch	3.3741	kPa
Column water inch (at 20 °C)	0.2486	kPa
Atmospheric pressure (bar)	0.9869	kg/cm^2
or	100.0	kPa
Degrees (heat)		
Temperature difference (ΔT)		
Degree Celsius (°C)	1.0	°Kelvin (°K)
or	1.8	°F
Degree Fahrenheit (°F)	0.55556	°C
Temperature		
Degrees Fahrenheit (°F)	[(°F – 32) × 5/9]	°C
Degrees Celsius (°C)	[(°C × 9/5 + 32]	°F
Degrees Kelvin (K)	add 273.15 to °C	K
Energy		
Joule (J)	1.0	N.m
British thermal unit (BTU)	1.055	J
or	0.252	Cal
or	778.0	ft.-lb.
Calory (international)	4.1868	J
Kilowatt hour (kWh)	3.6	MJ (megaJoule)
Watt second (Ws)	1.0	J
Power		
Watt (W)	1.0	J/s
BTU/hr	0.2931	W
Horsepower (hp)	746.0	W
Heat flow (conductivity)		
BTU /(ft.^{2} × hour × °F)/inch thickness	0.1442	W/(m × °C)/cm
Viscosity		
Dynamic:		
Poise (P)	1.0	N.s/m^2
or	1.0	dyn.s/cm^2
Kinematic:		
Stokes (St)	0.1	cm^2/s
ft.^{2}/s	0.0929	m^2/s

[a] The use of the term liter (L) is discouraged by SI standards. Volumes should be expressed in mm^3, cm^3, or m^3.

[b] "Relative density" replaces terms "specific gravity" or "specific density".

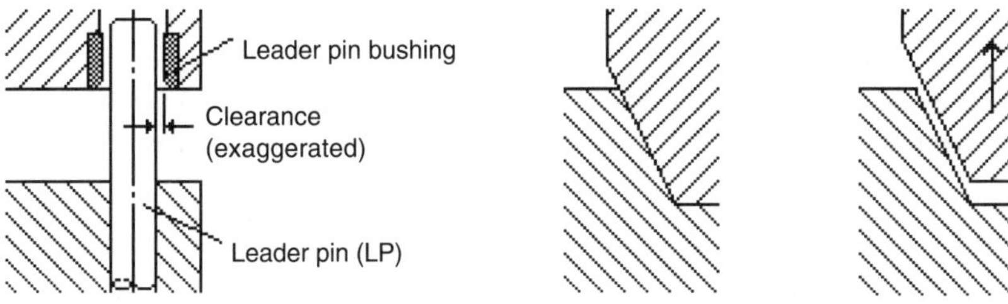

Figure 22.11 Leader pins and bushings require clearance (left), taper fits allow mating elements to seat only at the moment of closing or opening (right).

With taper fits, at the moment of closing or opening, the mating elements seat on, or lift off, in relation to each other without sliding relative to each other, as do the leader pins in their bushings. The difference between these two methods is shown in Fig. 22.11. However, after contact has been made, when closing, the mating taper faces slide over each other for the distance of the preload while the female taper expands slightly. Similarly, when opening, they again slide over each other until the preload distance is used up. Therefore, over time, there will be a certain, even if small, wear; but as a result of the preload, the wear of the taper surfaces will not affect alignment until the amount of preload has worn off.

As long as the preload is greater than zero (positive preload) or exactly zero, the proper alignment is maintained. However, once there is the slightest clearance between the mating tapers (no preload), the mold halves (or the mold parts) lose their alignment, and the tapers must be repaired to again provide preload. Tapers with heavy preloads will wear more rapidly than those with lesser preloads, since the abrasive effect increases with the force at which the surfaces contact each other.

22.14.2 Purpose of Preloads in Tapers

The following paragraphs refer to Fig. 22.12 as they explain in simple terms the reason for preloads:

View 1. No preload (negative preload). There is even some clearance between mating tapers, and no alignment is provided at all by this set of tapers, since the male taper can slide sideways by the amount C. This taper fit is useless.

View 2. Ideal condition (zero preload). The tapers meet but do not expand the female taper. However, this "ideal condition" is impossible to achieve because of manufacturing

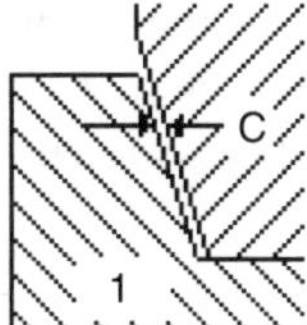

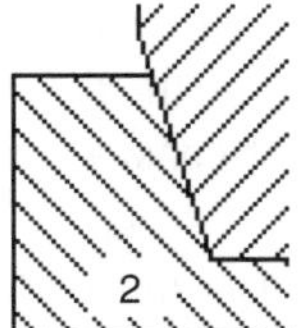

Figure 22.12 Preload in tapers: 1. No preload (negative preload), 2. ideal (zero preload), and 3. positive preload.

variations within the limits of the allowable and practical tolerances required to produce the mold parts.

View 3. Preload X (positive preload). This is the only practical method for providing proper alignment. Alignment is ensured as long as the wear is not creating a condition such as in view 1.

Some basic areas where mating tapers are used are listed below:

1. To provide alignment between mold plates instead of, or in addition to, leader pins, with taper pins. Where leader pins are also used (e.g., as protection for the cores), the clearance in the LP bushings must be large enough that the taper pins will not "fight" the leader pins. If, in multicavity molds, pairs of taper pins are used to align individual stacks, either cavity or core must be mounted so that it can be aligned by the taper pins without undue force.
2. To provide alignment between cavities and cores in individual mold stacks, with cavity lock or core lock. One of the matching stack assemblies, usually the core, must be mounted to its plate so that it can move sideways to allow the taper lock to move the stack into alignment.
3. To provide seating for stripper rings on their cores. The stripper ring must be allowed to float with regard to the core so as to find its seat without undue forces (and excessive wear) on the taper faces of the seat. Preload is the only practical way to prevent flashing between core and stripper ring. With circular shapes, this is fairly easy to produce. Oval or oblong stripper rings are much more difficult to produce, but with today's technology (e.g., using wire EDM), it is not impossible, but it is always more expensive than conical tapers.
4. Taper fits are used for other areas as well, such as the accurate alignment of gate inserts in gate pads, or the alignment of gate pads in cavities. In this latter application, a preloaded taper is especially important to prevent flashing after the gate pad and the cavity separate a short distance S at each cycle, as shown in Fig. 22.13 for moving cavities.

There are other applications for preloaded tapers, but the principle is always the same.

Taper fits and preloads can be fairly easily repaired (reset) if they are worn and have lost their preload. Such repair usually requires grinding on one of the faces which are at right angles to the taper axis, and shimming or regrinding a face on the mating part, to reestablish the original preload.

In the example in Fig. 22.14, the (floating) core is aligned to the cavity via the lock ring taper (a). The stripper ring is also floating, and seats on the core taper (c). The surfaces (b) and (d) are fully loaded when the mold is clamped up, and both the lock ring and the stripper ring are stretched to ensure good fit.

In some designs, the core and the locking ring are made from one piece of steel. However, this is expensive, difficult to make, and nearly impossible to adjust in case the taper fit is lost.

In Fig. 22.14, the locking ring is a separate part with a cylindrical fit with the core. This is only acceptable if the core is pressed into the ring. Modern designs have the ring seated on the core, with a very small taper (about 3° per side). This permits relatively easy repairs to the core assembly.

In the arrangement in Fig. 22.15, the stripper ring has two tapers: (a) seats in the cavity, and (c) on the core. Both core and stripper must be able to float sideways to permit alignment. This design is quite common where the stripper ring must split to release the product (e.g., a thread, as with preforms) but must be held with preload during injection.

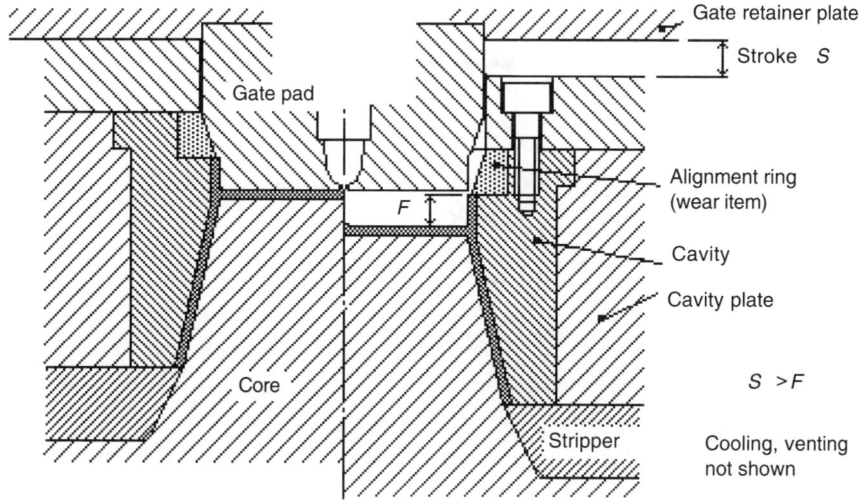

Figure 22.13 Taper in stripper and in moving cavity.

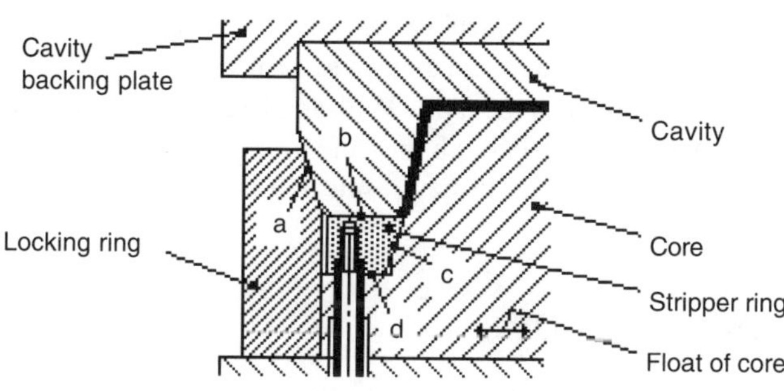

Figure 22.14 Core lock and stripper ring tapers.

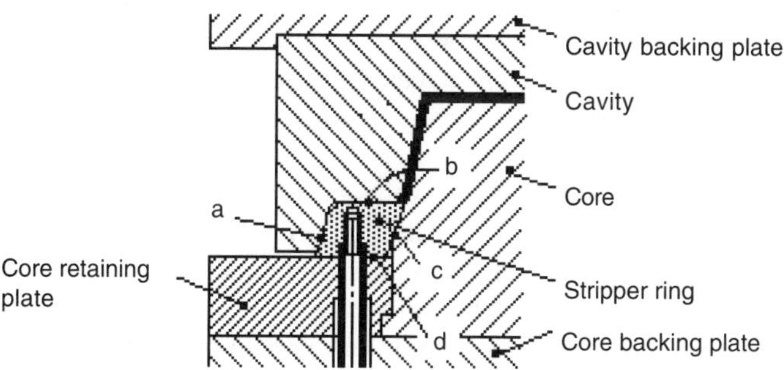

Figure 22.15 Cavity lock and stripper ring taper.

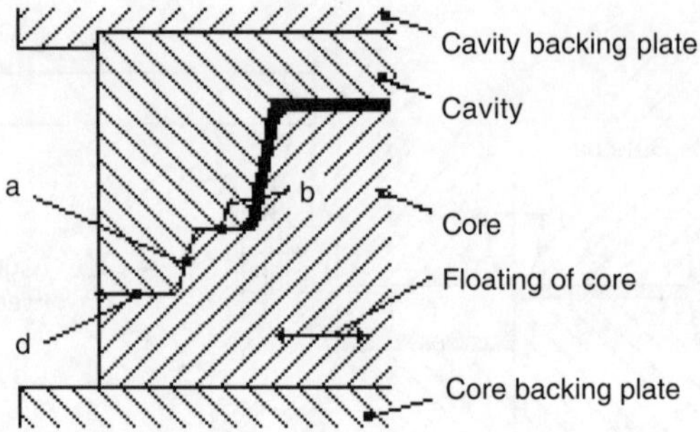

Figure 22.16 Cavity lock and air ejection.

The principle for cavity lock and air ejection, shown in Fig. 22.16, is the same as that for cavity lock and stripper ring. Taper (a) ensures perfect alignment as long as the core is allowed to float sideways.

Even though Figs. 22.15 and 16 show cavity lock, as a rule, core lock is preferable in molds with free fall ejection for this reason: When the product is ejected, it "flies" freely toward the cavity and has a greater tendency to hang up on the lip of the taper (a) in the cavity than the same product would in the mold with the core lock.

To prevent such hanging up, which will cause production stoppages, it may be necessary to increase the mold stroke more than would otherwise be necessary. This will affect the cycle time, and thereby the performance of the mold. With air ejection, it may be possible to control the air pressure such that there is not enough pressure to shoot the product into the other side, but this may lead to marginal tuning; it may result in a failure to eject at all and a stop in production.

22.14.3 Wedges and Shut-Offs

Anything said in this chapter regarding conical tapers applies also to wedges used for alignment. Wedges are in fact four short segments of the circumference of an infinitely large taper. Where wedges are used in shut-offs to produce openings in a product, preload must be provided to avoid flashing.

Alignment of two matching parts with tapers works only if there is some preload; to create accurate alignment, the tapered faces must contact before the full (clamping) force is reached. As the full clamp is applied, the part with the female taper is slightly expanded by the wedging action of the male taper. *A taper without preload is useless for alignment or to prevent flashing.*

Preload, even though it is a force, is often expressed as the distance the engaging tapered part travels from making first contact with its mating taper until the parts are fully seated under the preload force. Preload is usually indicated either with dimension lines (Fig. 22.17A) or with a note and a leader pointing to the taper (Fig. 22.17B).

It is important to understand that a portion of the clamping force is used to expand the female tapers; this amount is, therefore, not available to hold the mold shut when injecting. If there is no preload, the mold or its stack parts will not align properly. If there is too much

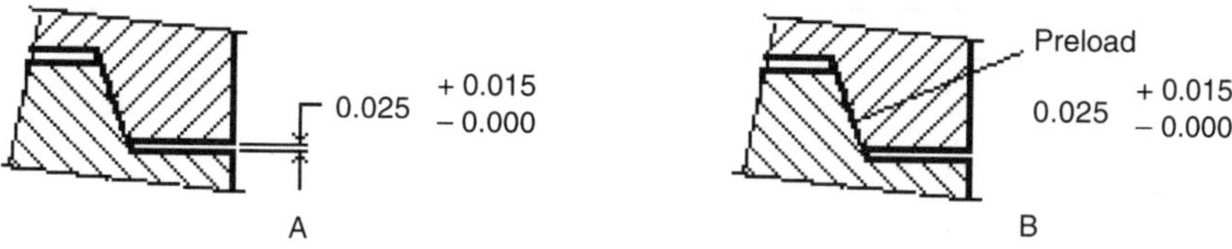

Figure 22.17 Preload indication on a part drawing: A. dimension lines, or B. note and leader pointing to taper.

preload, the mold will flash because too much of the available clamp force is used up in the tapers. This condition is more serious with multicavity molds using many stripper rings and/ or locking rings between cavities and cores.

Especially for molds with 16 or more cavities, the manufacturing tolerances for the tapered components must be selected so that:

- With all dimensions at a minimum, there is never less than zero preload, and
- with all dimensions at maximum, the force required to seat all tapers must not affect the clamping capability of the machine and risk flashing in the mold.

The resistance to expanding the tapers increases with the increase in the taper (wedge) angle; therefore, the preload distance must be reduced as the taper angle increases.

Taken from past experience, Fig. 22.18 on page 20 gives some guidelines for selecting preloads for certain mold combinations up to 16 cavities. The illustrations show suggested minimum/maximum values of tolerances for the preload. The tolerances for the matching components must be selected so that the final preload is within the suggested values.

22.15 How to Dimension Tapers

22.15.1 External Tapers

Three alternative methods for dimensioning tapers are shown in Fig. 22.19. If accuracy is not important, specify taper with the important diameter *d*, (in Fig. 22.19, the small end was selected "important"), the angle α, and the height of the tapered section *h* (Fig. 22.19A).

For high accuracy, use a ring gage. Specify (in addition to *d*, α, and *h*) the gage number and the dimension *DG* (with close tolerance) from the reference plane (Fig. 22.19B). For high accuracy if no ring gage is available, specify (in addition to *d*, α, and *h*) the size of micro rollers and dimensions *DB*1 & *DB*2 (with close tolerances).

The plane on which the micro rollers rest must be well defined; it may be a shoulder of the part, it may be a surface plate, or it may be built up with precision gage blocks (Jo-blocks) from the surface plate. The roller size must be selected so that the roller touches clearly both the plane and the taper. For dimensioning over rollers, see Section 22.3.1

22.15.2 Internal Tapers

Four alternative methods are shown in Fig. 22.20 for dimensioning internal tapers. If the dimensions do not have to be very accurate, specify a, *d*, and (if required) *h* (Fig. 22.20A). For

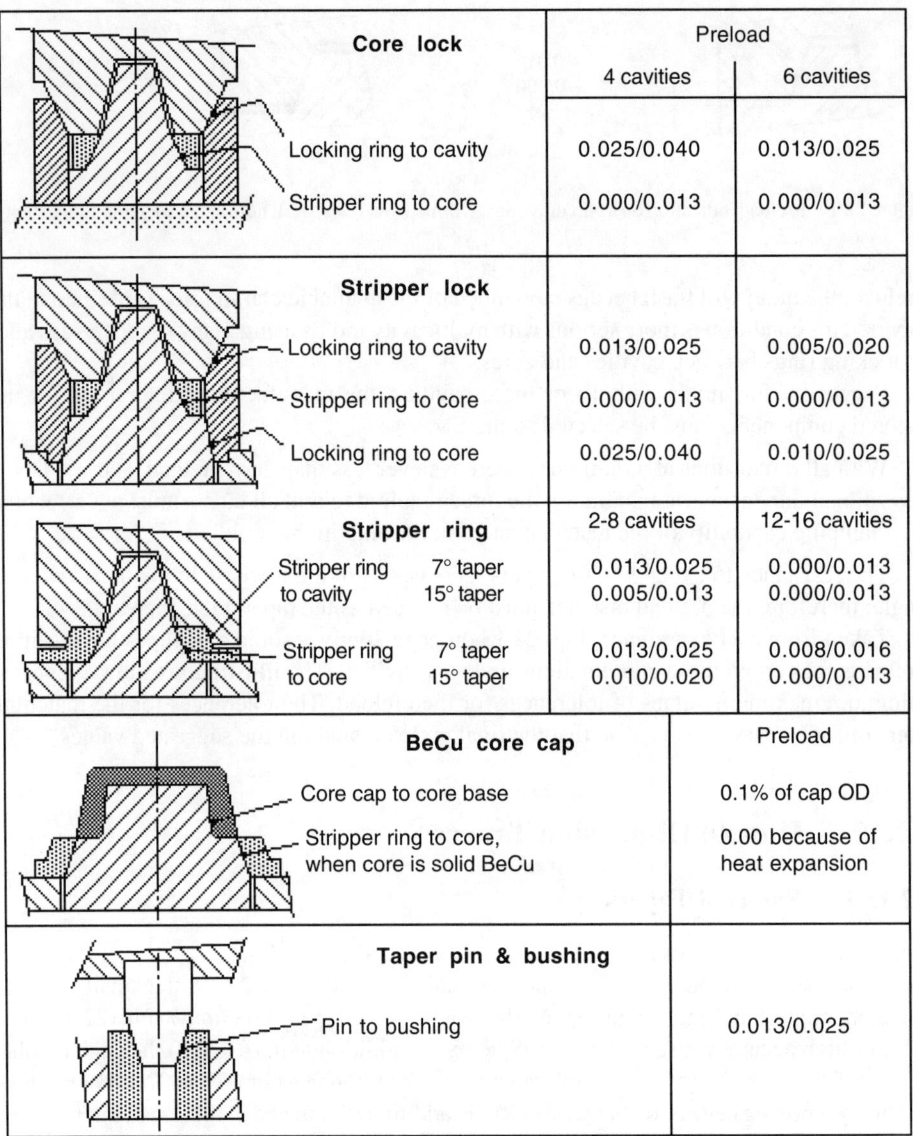

Figure 22.18 Guidelines for selection of preload for various mold combinations.

high accuracy, a use plug gage (Fig. 22.20B). Specify (in addition to *d*, a, and *h*) the gage number and the dimension *DG* (with close tolerance) from the reference plane.

If no plug gage is available, use micro balls (Figs. 22.20C and D). For large diameters, specify (in addition to *d*, α, and *h*) the size of micro balls and dimension *DB* (with close tolerance). If taper is longer, two balls must be specified, similar to alternative C for external taper. For small diameters, specify (in addition to *d*, α , and *h*) the size of micro ball and dimension *DH* (with close tolerance). For dimensioning over micro balls, see Section 22.3.1

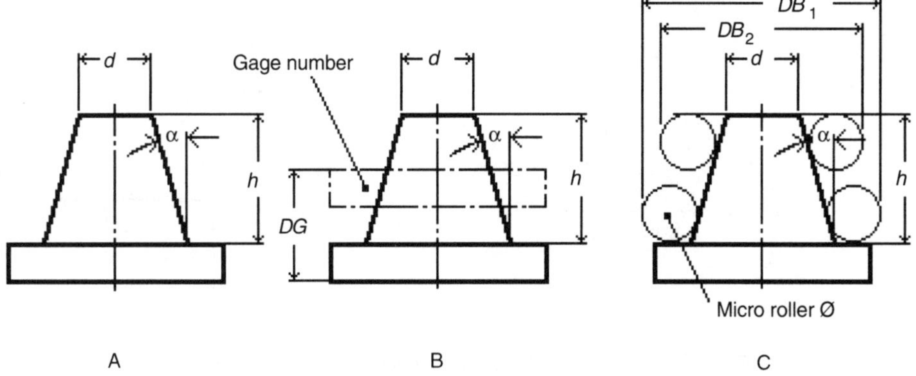

Figure 22.19 Dimensioning external tapers: A. simple dimensions when accuracy is unimportant, B. use of ring gage for high accuracy, or C. use of micro rollers also for high accuracy.

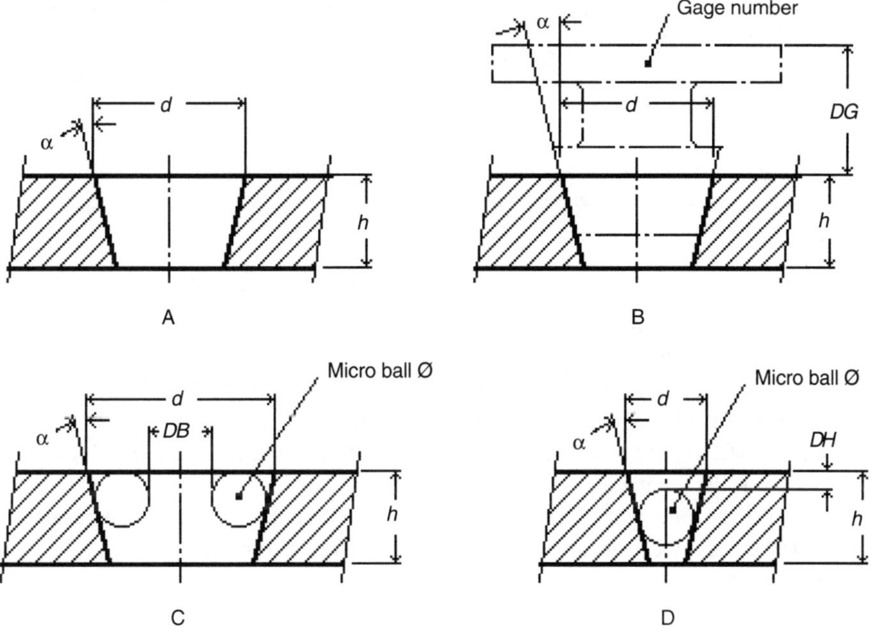

Figure 22.20 Dimensioning of internal tapers: A. simple but inaccurate, B. use of a plug gage for high accuracy. If no plug gage is available, then use methods: C. use of micro balls for high accuracy with large diameters, or D. use of micro balls for high accuracy with small diameters.

22.15.3 Angled Surfaces

Four examples are shown in Fig. 22.21 for accurate dimensioning of angled surfaces. For method A, angle α and either h or w is shown. Indicate diameter of micro roller and calculate dimension DR.

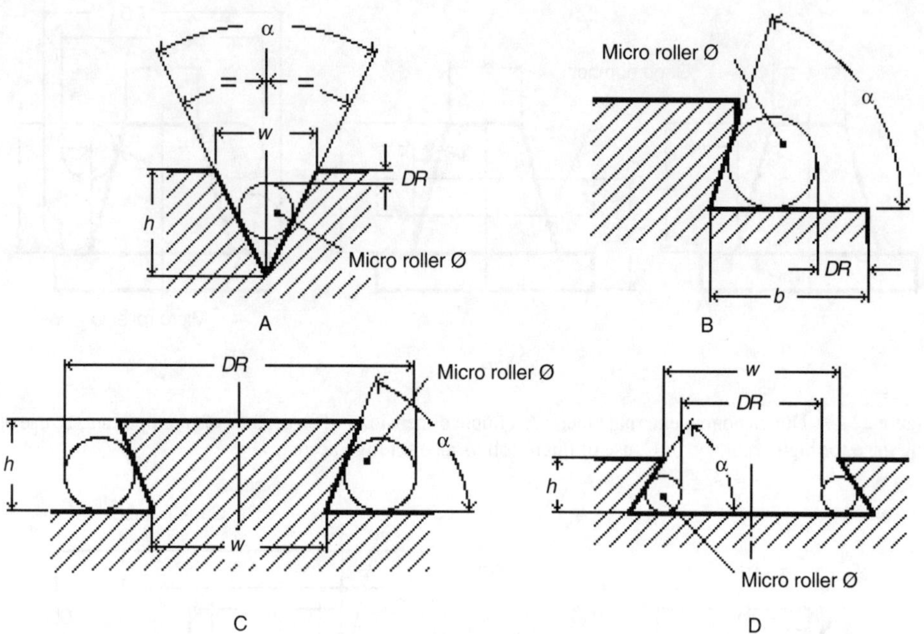

Figure 22.21 Dimensioning of angled surfaces using micro rollers to determine *DR*: A. V-shaped channel, B. angled surface, C. positive dovetail, and D. negative dovetail.

In method B, angle α and distance *b* from edge is shown. Indicate diameter of micro roller and calculate dimension *DR*.

For a positive dovetail angle α, width *w*, and height *h* is shown (Fig. 22.21C). Indicate diameter of micro rollers and calculate dimension *DR*.

In Fig. 22.21D, negative dovetail angle α, width *w* and height *h* is shown. Indicate diameter of micro rollers, and calculate dimension *DR*.

NOTE: Either the narrower (as shown) or the wider width of the dovetail can be specified, but not both, unless one is shown as reference. For dimensioning over rollers, see Section 22.3.1.

22.15.4 Construction (Checking) Balls

Where accurate dimensioning (and machining) of angled surfaces or bores is required, an auxiliary hole (tooling hole) and ball is introduced into the design. The hole is accurately dimensioned (*D* and *C*) from a reference edge (or hole) (*E*), with close tolerance. A specially constructed ball with stem *A* (specify diameter and checking ball number) is then inserted into the hole, and the angled surface (α_1) is then dimensioned (*F*) with close tolerance and measured from the ball, as shown in Fig. 22.22. An angled bore (α_2) would similarly show its centerline accurately dimensioned (*G*) from the ball. Specifications for some common standard checking balls are listed in Table 22.5.

Table 22.5 Specifications for Some Standard Checking Balls

I.T.I.	A	B	C	D
Catalog#			(minimum)	
448-8	12.700±0.005 (0.5000±.0002)	10.160±0.005 (0.4000±.0002)	17.0 (0.67)	6.345+0.005 (0.2498+0.0002)
448-12	19.050±0.013 (0.7500±.0005)	12.700±0.005 (0.5000±.0002)	23.0 (0.91)	9.521+0.005 (0.3748+0.0002)

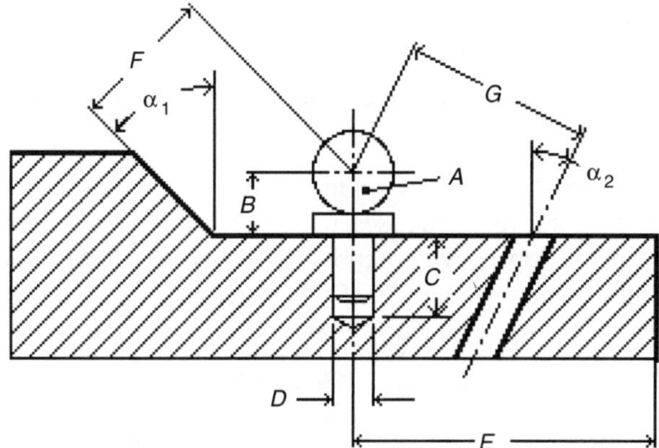

Figure 22.22 Dimensioning using a checking ball.

22.16 Fatigue in Stripper Rings or Any Preloaded Taper

22.16.1 Stripper Ring

Figure 22.23 (next page) shows how the geometry of the stripper ring affects the stresses in the ring. Since these parts are stressed at every cycle from Ø to full load, the stress ratio R = Ø.

Usually, the designer knows the clamping force F_c by knowing the type of machine used for the job; also, the designer knows that F_c must be greater than the injection force trying to open the mold.

$$F_c > n \times A_p \times p \tag{22.9}$$

where F_c is the clamping force, n is the number of cavities, A_p is the projected area of the product, and p is the injection pressure. This equation again highlights one of the major unknowns: What is the injection pressure p?

In Chapter 29, Mold Forces, Section 29.5, Injection Forces, further discussion of the difficulty of determining the actual pressure p, and the factors affecting it, is provided. At this point, however, simply continue and understand that a reasonable value of p for each mold must be assumed.

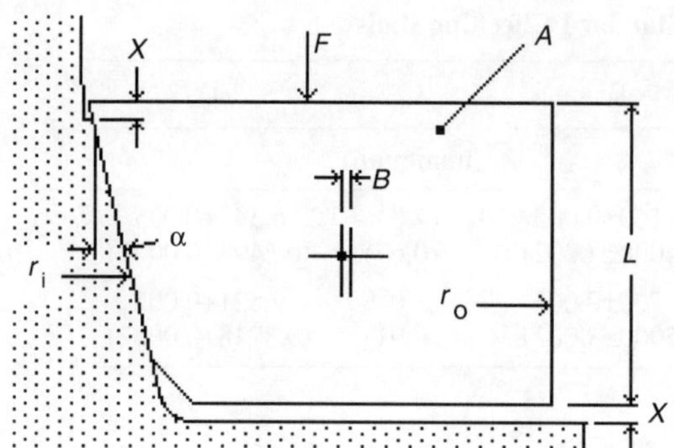

Figure 22.23 Stresses and forces acting on preloaded stripper ring.

As stated earlier, F_c must not only overcome the force caused by p but also must seat all tapers. We will call F the force required to seat one taper lock. If there is more than one cavity, and more than one taper lock per cavity or mold—and possibly some additional taper locks to align the mold plates—a number of forces F will be required, each with its own value. The sum of all the forces F (ΣF) must be added to the force generated by p.

$$F_c > (n \times A_p \times p) + \Sigma F \tag{22.10}$$

In general, for a thick-walled ring, F (the force to seat a taper) is directly related to the design variables a, r_o r_i and L. The designer must keep F (and ΣF) small enough so that these forces do not reduce the locking force F_c so much that the mold could be forced open during injection, resulting in flash. (In the equations below, E is the modulus of elasticity.)

$$F = \pi \, E \, L \, X \; \tan^2\alpha \left(1 - \frac{r_i^2}{r_o^2}\right) \tag{22.11}$$

or

$$X = \frac{F}{\pi \, E \, L \, X \; \tan^2\alpha \left(1 - \frac{r_i^2}{r_o^2}\right)} \tag{22.12}$$

22.16.2 Self-Locking of Tapers

If a is greater than the self-locking angle for *lubricated* steel on steel (coefficient of friction $\mu = \tan \alpha > 0.16$, or $\alpha > 9°$), the stretched ring will try to overcome the force F and assist to push the mold open when F_c ceases. If the angle is self-locking ($\alpha < 9°$), there will be a resistance (= F) to the closing, and there will be also a resistance to the opening. The smaller the locking angle α, the more force is required to open the taper lock.

NOTE: The coefficient of friction for *dry* steel on steel is $\mu = 0.8$. This corresponds to a taper angle $\alpha = 38°$. However, in molds, all tapers should be lubricated to reduce wear. A problem with dry tapers is that a higher clamp opening force is required to overcome self-locking. This force may not always be available and/or may create a noisy mold opening.

From experience, the taper angle α should be held within 5° and 15°, and preferably standardized to certain sizes to use standard measuring taper gages. When designing the ring shape, keep in mind that any (particularly threaded) holes act as stress risers, which will seriously affect the fatigue life of the ring.

22.16.3 Stresses

When F bears down on the ring to make it seat at the base, the ring will be stretched (radially) by the amount $B = X \tan \alpha$. The designer is interested in the hoop stress S_H (the stress caused by the stretching of the ring) when the force F is applied. The formula below is based on a given X dimension. More information will be provided about X later.

$$S_H = \frac{E\,X\,\tan\alpha}{2r_i}\left(1 + \frac{r_i^2}{r_o^2}\right) \quad (22.13)$$

22.16.3.1 Contact Stress S_C

Contact stress is the stress (pressure) created by the ring bearing down on the mating taper, or the pressure exerted by this taper on the ring. It may be important to know the contact stress if the seating area is small. If S_C is too high, the ring will hob into the taper seat. The following two formulae are based on the known force F:

$$S_C = \frac{F}{2\pi\, r_i L\ \tan\alpha}. \quad (22.14)$$

The designer can also determine the hoop stress S_H caused by contact stress S_C:

$$S_H = \frac{F}{2\pi\, r_i L\ \tan\alpha}\left(\frac{r_o^2 + r_i^2}{r_o^2 - r_i^2}\right). \quad (22.15)$$

Equations 22.9 and 22.10 can only be used if F is equal to, or smaller than F from equation 22.11.

The following formula for contact stress S_C is based on the known preload X:

$$S_C = \frac{E\,X\,\tan\alpha}{2r_i}\left(1 - \frac{r_i^2}{r_o^2}\right). \quad (22.16)$$

The effect of the angle α on the force F is shown in the chart below:

α	tan α	tan 2α	Compared with 5° angle
3°	0.0524	0.0027	⅓
5°	0.0875	0.0076	1
10°	0.1763	0.0311	4.1 times
15°	0.2679	0.0718	9.4 times

When selecting the material for the stripper ring, make sure that it is hard (and tough) enough not to break at the sharp corner at the parting line, and that it can withstand the wear and contact stress, but it must be soft enough to get into a hardness range with maximum fatigue life.

As stated earlier, tapers between 5° and 15° are suitable for most applications. The designer cannot do much about r_o or r_i, which depend on the product size.

Make sure, by proper selection of size and tolerances of the preload *X*, that there will always be some preload, without it becoming too large.

22.16.4 Effect on Preload by Grinding of the Taper

When grinding any surface, it is not practical to make cuts smaller than 0.0025 mm (0.0001 in.). When removing this (minimum) amount of a taper, it will change the dimension (preload) *X* by the amount D*X* (Fig. 22.24).

For example, for every 0.0025 mm (0.0001 in.) removed, ΔX equals:

α	sin α	ΔX (in.)	ΔX (mm)
3°	0.0523	0.0019	0.049
5°	0.0872	0.0011	0.029
10°	0.1736	0.0006	0.015
15°	0.2588	0.0004	0.010

It is, therefore, not practical to ask for tolerances that are smaller than the effect of grinding would permit. This can be shown best graphically, as in Fig. 22.25. If ΔX is equal to or smaller than the tolerance, and the part measures more than the maximum size (*D*), the resultant di-

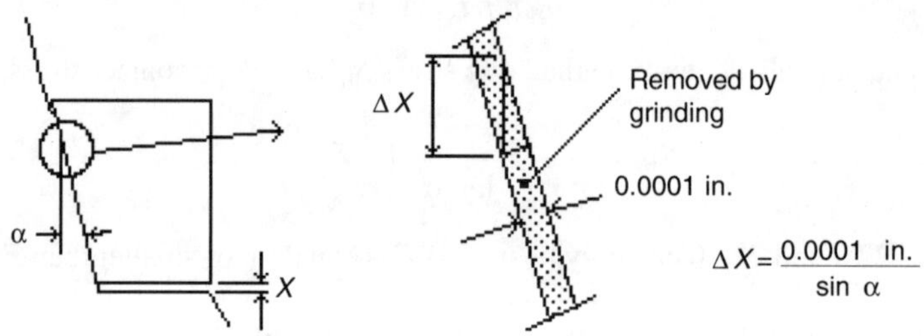

Figure 22.24 Grinding of the taper with ΔX = 0.0001 in. removed.

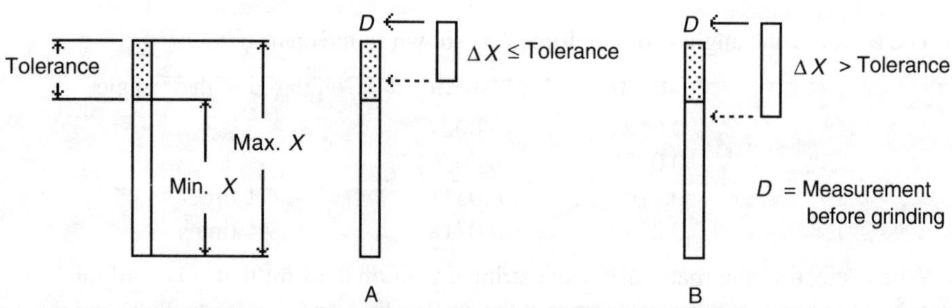

Figure 22.25 Dotted area shows minimum and maximum tolerance: A. grinding is less than or equal to tolerance, and B. grinding is greater than tolerance allows.

mension after grinding will be within tolerance. If ΔX is larger than the tolerance, it is possible that the part which is at first larger than the maximum D, will after grinding be smaller than the minimum D. *ΔX must be smaller than the tolerance for D.*

For this reason, preloads of rings with small taper angles are more difficult to produce. The foregoing also shows that in a multicavity mold, because of manufacturing variations during the final grinding steps, so-called "identical" parts will have different tapers, and thereby create different preloads.

22.17 Chamfers and Radii

22.17.1 General Information

Chamfers are generally specified on all outside edges for safer handling of the mold part; also, chamfers reduce the risk of stress cracking at the outer surfaces. Radii are generally specified on all inside corners to reduce the risk of cracking as a result of stress concentration.

The size of desired chamfers and radii can be stated with a general note on the drawing. Relative size of matching chamfer and radius is shown in Fig. 22.26. This ensures that the parts will seat properly on the flat surfaces, not on the radius.

All mold plates must be chamfered at least 1.5 mm × 45°.

All edges of counter bores and all mold part edges should be chamfered, unless a chamfer would not be be permissible because of the product design or the path of plastic flow, to facilitate assembly and to reduce the risk of personal injury (cut fingers) when handling the parts. Any edge which will be dragged on assembly over an O-ring or other soft seal must be indicated with a notation to "round-off and polish".

The *absence* of chamfers must be indicated with the notation "sharp" and a leader pointing to the affected edge. A corner must never be sharp. If a sharp corner is necessary for the product design, create it by using separate mold parts (inserts).

Some product designs show a sharp corner (e.g., at the base of a product) where a small radius could be perfectly acceptable for the function of the product. In such a case, the designer should request a change in design to permit a small radius; the benefit to the customer would be a longer mold life (less risk of cracking) or savings by avoiding the use of inserts (Fig. 22.27).

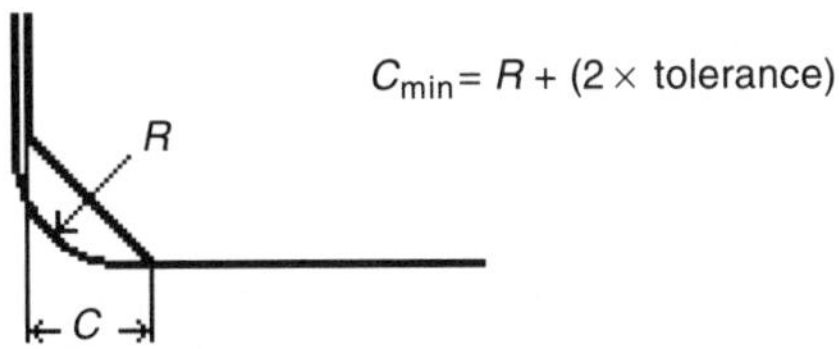

Figure 22.26 Relative size of matching chamfers and radii.

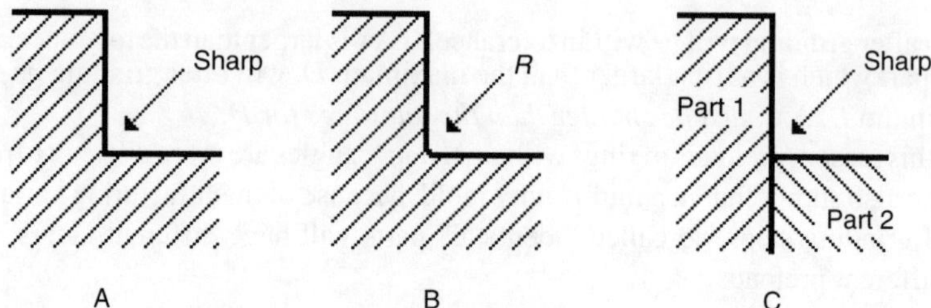

Figure 22.27 Corners of mold parts should not be sharp: A. desired sharp corner must either be B. given a suggested radius or C. created by using an insert.

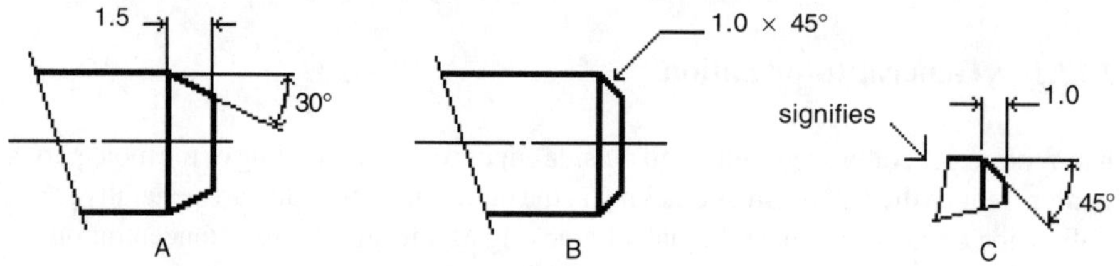

Figure 22.28 Specifying chamfer: A. show angle and length with dimension lines, B. with 45° chamfers, simply indicate length and angle with a leader to signify C.

22.17.2 Specifying Chamfers

The general method for specifying chamfers is to show its angle and length (Fig. 22.28A). A shortcut method that can be used only with 45° angles is illustrated in Fig. 22.28B and C. In this case, simply indicate length and angle, with a leader pointing to the chamfer.

22.17.3 Specifying Radii

The drawing board method for specifying radii is shown in Fig. 22.29. The preferred method is to show the center and dimension, within leader (Fig. 22.29A). If there is not enough space for the preferred method, show dimension at leader extension (Fig. 22.29B). If the center is outside the drawing area, shorten leader (Fig. 22.29C). For small radii only, no center is shown, but the letter R is added to the dimension notation (Fig. 22.29D).

The CAD/CAM method differs from the drawing board method by not showing the centers, and by showing Rad. (or Sph. R, for spherical radius), instead of R. The preferred method is shown in Fig. 22.30B, with a horizontal, short leader pointing to the dimension, regardless of the size of the radius.

22.17.4 Radii in Corners

Every inside corner, particularly in hardened steel, creates a severe hazard because of the stresses raised when forces are applied. Practically all mold material failures start from cracks originating in sharp corners.

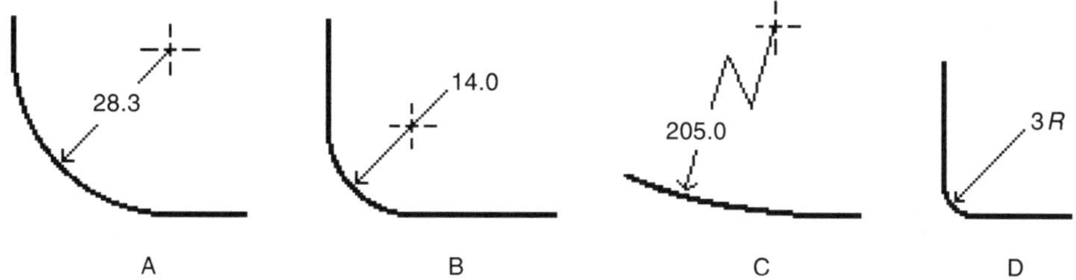

Figure 22.29 Specifying radii: A. preferred method, B. with space limitation note dimension at leader extension, C. if center is outside drawing area, shorten leader, and D. add letter R to dimension for small radii and show no center.

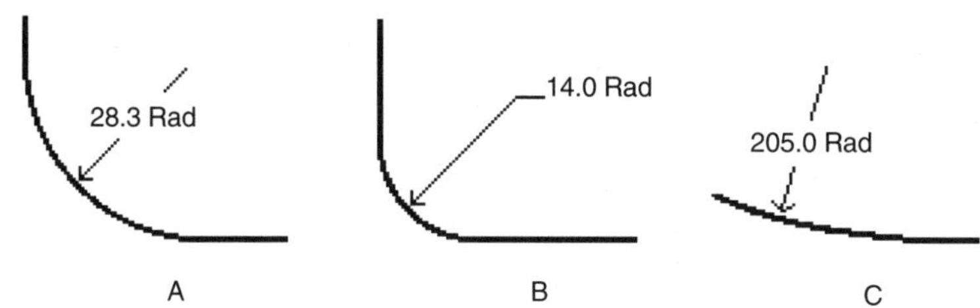

Figure 22.30 Specifying radii with CAD/CAM: A. small radius, B. horizontal, short leader at dimension is preferred for all sizes, C. large radius.

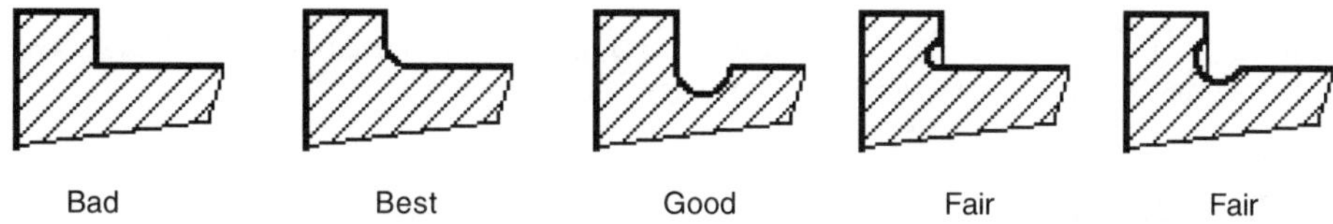

Figure 22.31 Examples of radii in corners.

Grinding creates very fine scratches. In hardened steel, where severe loads are expected, such corners must, therefore, not only be specified "ground" but also "polish" to remove any scratches created in grinding.

If it is not possible to have a radius in a corner, as suggested in Fig. 22.31 by the drawing labeled "Best", the part must be designed with a "hidden" or recessed radius. Typical high-risk areas where radii must be provided are:

- Any change in a diameter of a mold part,
- inside snap ring grooves,
- inside O-ring grooves,
- roots of cut threads,
- inside counter bores at seat of screw head, and
- heels.

(See also Chapter 18, Metal Fatigue.)

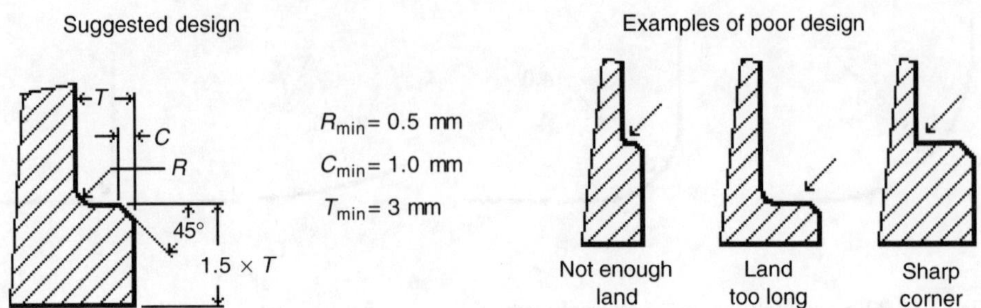

Figure 22.32 Suggested design for heels (left) and examples of poorly designed heels (right).

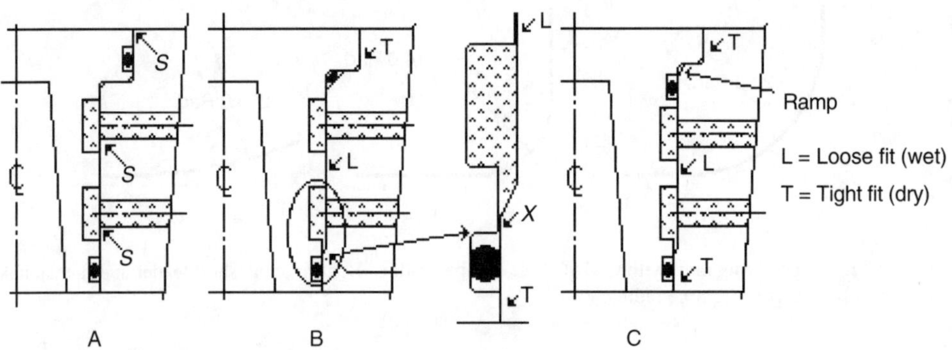

Figure 22.33 O-ring installation: A. Wrong—never drag O-rings over interrupted surfaces, B. Correct—ramp at lower O-ring protects it, and C. Correct—two ramps protect O-rings.

22.17.5 Heels

"Heeling" inserts into plates or into other inserts is a very good method of assembly. If there is not enough space for screws, it is the only practical and satisfactory method of retaining the part. Figure 22.32 shows a recommended proportion for heels and three examples of poorly designed heels.

22.18 O-Ring Seal Installation

(For O-ring groove sizes, see O-ring manufacturer's standards.)

When assembling any mold part, an O-ring must never be dragged over an interrupted surface, as illustrated in Fig. 22.33 where two cooling lines enter the bore for the cavity insert. The sharp edges of the bore (S) will cut the O-ring and make it useless (cause leaking). The correct method is to create a ramp, as indicated in Fig. 22.33 center and right, by stepping the diameters.

The amount of step *R* must be more than the amount the O-ring projects from the groove before it is compressed. This amount is less for smaller O-rings, more for larger ones.

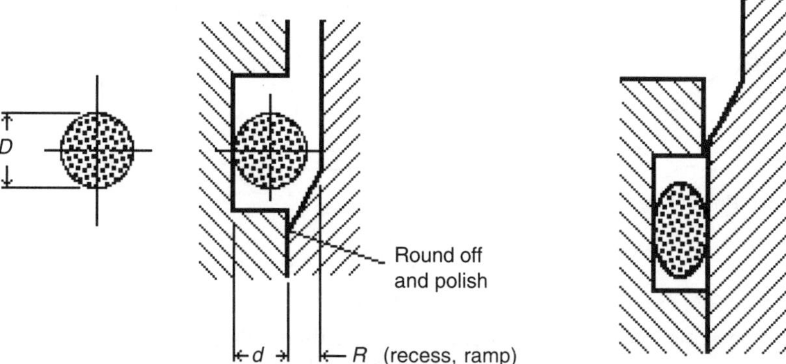

Figure 22.34 Recess or ramp *R* for an O-ring.

To calculate *R*, the minimum amount required is the maximum actual O-ring cross sectional size *D* (*not* the nominal size) less the minimum depth of the groove *d*, then rounded to the next one- or two-tenths of a millimeter (Fig. 22.34). For example, for a nominal ⅛ size O-ring, $D = 3.53 \pm 0.10$ mm, therefore $D_{max} = 3.63$. If $d_{min} = 2.87$ mm, then

$$D_{max} - d_{min} = 0.76 \text{ mm.}$$

Rounding to the next one-tenth means that the *R* selected = 0.8 mm.

22.19 Drilling in Molds

Two types of drills are used for most operations: twist drills and gun drills. (Occasionally, carbide drills are used for very hard materials.)

22.19.1 Twist Drills

Because of the web, the center of the drill has the form of a chisel edge and does not cut as the lips do, but is forced into the material to be drilled (Fig. 22.35). With larger drill sizes, it is necessary to predrill with a smaller size drill to overcome the resistance of the chisel edge in the material.

22.19.1.1 Accuracy of Hole Location and Diameter

There are two cutting edges (lips), and the chips from both edges move out of the hole through the two twisted channels, or "flutes". If there is the slightest difference in length of lips (caused by poor sharpening), one lip will cut more than the other, and the drill will wander off its intended path and increase the size of the hole beyond the drill diameter; it may even fail, causing seizure and breakage of the drill. This is recognized in American drill standards, in two ways:

1. The drill size diameter is always undersize (i.e., the diameter tolerance is minus (–), according to drill size, from –0.0005 to –0.002 in.

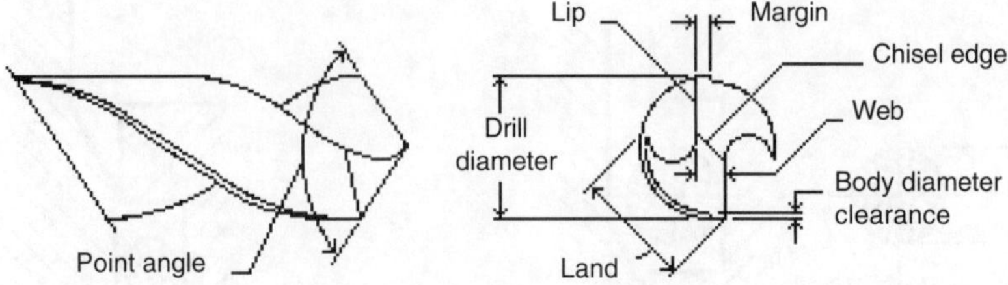

Figure 22.35 Side (left) and tip (right) cross sections of a twist drill.

2. The permissible oversize (U.S. Standard) of a drilled hole is specified as:
   Maximum oversize = 0.005 in. + .005 D
   Minimum oversize = 0.001 in. + .003 D

In regular drilling operations, the drill is held in a chuck, distant from the cutting tip, so that the drill has the opportunity to wander off its intended location. Once in the hole, it is guided by the hole walls and the narrow "margins" at the outer end of the cutting lips (Fig. 22.35).

This is usually not a problem with short, stubby drills, but it is a serious problem with long, slender drills, when the free length is more than 6 times the drill diameter ($L > 6D$).

22.19.1.2 Cooling of the Cutting Edges

The energy from the drive is used to cut chips as the drill advances through the material. This energy is transformed into heat, which must be removed to prevent overheating and annealing of the cutting edges of the drill. Depending on the material cut (hardness, composition, heat conductivity), the appropriate drill speeds and feeds must first be determined.

With vertical drill presses and milling machines, the drill enters from the top of the work piece; the coolant flows into the hole by gravity, from the outside into the drilled hole. Because of the rotation of the drill, the flutes act as a pump which removes the chips, it also drives back much of the coolant that enters the hole. At best, the cooling is marginal. (Newer drilling machines provide for "through the spindle and drill" cooling, to pump the coolant through the drill to the cutting edges. The coolant and the chips then flow back through the flutes, out into the open.)

22.19.1.3 Length of Drilled Holes

Because of the limited vertical motion available in drilling and milling machines, the length of holes is limited to the available strokes. Deep holes may, therefore, need several setups. With very deep holes, the twist drill has a tendency to wander off excessively from the desired (true) center. For example, in a hole 300 mm deep and 10 mm in diameter, even a properly sharpened drill could wander as far as 4 mm from its true center, or break. Smaller diameter high-speed steel (HSS) drills are more flexible than larger diameter HSS drills.

22.19.1.4 Guiding of the Drill Point

The drill is held in a chuck, but there is no additional support or guide for the drill to find its location in the work piece. To ensure that the drill will enter at the proper location, a short, stubby center drill is used to "predrill" the hole location. Production drilling using NC or CNC machines should always use center drilling prior to the regular drilling operation.

In manual mass production applications, drill jigs are often provided to support the drill very close to the start of the hole using hardened drill bushings. This is especially useful for long, slender drills ($L > 6D$), but it is costly for jobs where only one or very few work pieces are produced, as would be the case in many mold-making applications.

22.19.1.5 Finish of Bore

The finish of a hole drilled with a twist drill is usually rough and shows spiral drilling marks and possibly other scratches. If such marks are not permissible, the holes must be reamed or otherwise smoothed out, at additional expense. Drilling marks are potential stress risers, especially in very hard materials, and may lead to early failure of the part during heat treat or when cyclically stressed during molding.

22.19.1.6 Twist Drill Materials

Carbon steel is cheap and can be used for low productivity requirements. It will need frequent sharpening and/or replacing.

High speed steel (HSS) is more expensive, but it permits faster drill speeds and feeds because its annealing temperature is higher. The cutting edges stay sharp longer, and need less frequent sharpening. With very small holes, it may be cheaper to use the cheaper carbon steel drills and throw them away when they are dull, rather than sharpening them. With larger holes, and where high productivity is required, high speed steel drills are preferred.

Carbide drills are used primarily in cases where hardened steel (up to about 55–57 Rc, depending on the steel) must be drilled. They are sensitive to shocks and interrupted cuts, and break more easily than HSS drills. (Instead of drilling, the hole in the hardened work piece could be produced with EDM, which is suitable for any hardness of steel.)

22.19.1.7 Advantages and Disadvantages of Twist-Drilled Holes

In most cases, holes are short and enter from the flat side of a work piece, similar to milling. The drilling can, therefore, be combined with milling operations, reducing the number of setups. The drilling operation of short holes is cheap and fast, and the drills are relatively cheap and readily available.

In general, however, twist drills are not practical for deep holes in molds because of the shortcomings explained earlier, such as wandering of drill, poor cooling, etc. This applies particularly to cooling and air channels in plates, and to plastic channels in hot runner manifolds and sprue bars, where the deep holes enter from the edge of the flat work piece. It also applies to cooling and air channels in cavities or cores, where it is critical to maintain an accurate path of the drill, without wandering off.

22.19.1.8 Availability of Twist Drill Sizes

Virtually all twist drill sizes are readily available.

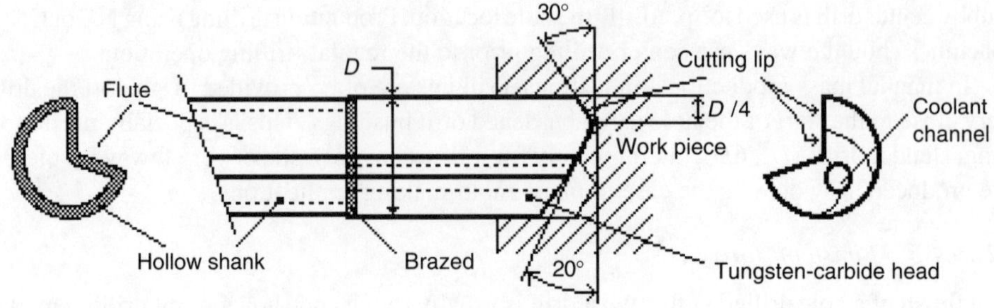

Figure 22.36 Deep hole drilling in cross section.

22.19.2 Deep Hole Drilling (Gun Drills)

This method requires a special "deep hole drilling" machine or a deep hole drilling adaptor to another machine, such as a milling machine. The drill operates in a horizontal plane. There are four essential differences from ordinary drilling or milling machines:

1. The stroke of the machine (depth of hole) can be considerably larger.
2. The drill is supported very close to the work piece, as with a drill jig.
3. The cutting edge of the drill is directly, pressure lubricated.
4. The drill works in one pass through solid material. It does not require predrilling.

There are two types of drills, featuring either internal or external chip removal. The external chip removal method is mostly used and is illustrated in Fig. 22.36.

22.19.2.1 Gun Drill Material

The head (tip), at the working end of the drill, is made from a tungsten-carbide alloy, which is much harder and longer lasting than high-speed steel. The head is brazed to a long steel tubing (shank), which is held at its other end in the machine chuck.

The tip is about 40 mm long when new; the cutting edge can be reground until the length of the remaining head is not long enough to act as a guide within the hole. The shorter the tip, the more the risk of wandering.

22.19.2.2 Cutting Edge of the Drill

The angles of the cutting lips depend on the material to be cut and are about 30° on the short lip and 20° on the long lip (Fig. 22.36). This is the most visible difference between twist drills and deep hole drills. There is no chisel edge, but the drill has a very sharp, well-defined V-shaped cutting edge, which spins around the drill center and describes a W-shaped groove, which keeps the drill on center, even when the cutting edge is far inside the work piece, away from the guide bushing.

As the drill progresses into the work piece, the ¾ cylindrical shape of the head and shank provides a better guide than the very narrow margin on the twist drills. This helps to prevent wandering of the drill.

22.19.2.3 Positioning the Drill

This is done using a method similar to a drill jig. A *drill bushing* belongs to and is stored with the drill; this bushing (*guide*) is fastened to the machine, at the tip of the *chip box*, which receives the returning coolant and the chips and locates the drill accurately. The face of the guide is held tight against the surface of the work piece (Fig. 22.37) so that the coolant (with the chips) returning through the *flute* can pass into the chip box without leaking into the open.

The designers must know the layout of the chip box sizes. This helps in the layout of cooling, air, and melt flow channels, as it shows where the gun drilling head might interfere with the part to be drilled.

There is another bushing, or *whip guide*, mounted in a ball bearing to support the long shank somewhere between the drive and the cutting edge. This guide prevents vibrations of the drill which may also affect the work piece.

Normally, a standard guide bushing is sufficient for location of the drill on the work piece. However, if the front of the chip box where the standard guide bushing is mounted interferes with the work piece, an extension must be provided (Fig. 22.38).

If the hole is at an angle (A) to the surface (Fig. 22.39) where the drill enters, but within the horizontal plane of the drill, the front of the drill bushing must be ground to this angle to ensure that the guide bushing seats tightly on the work piece surface. Angle A should be not more than 25°.

Note that the use of special drill bushings can be avoided by providing (milling) a spot-faced surface (Fig. 22.40) at the desired angle so that a standard drill guide can be used. The diameter of the spot face should be kept to a minimum.

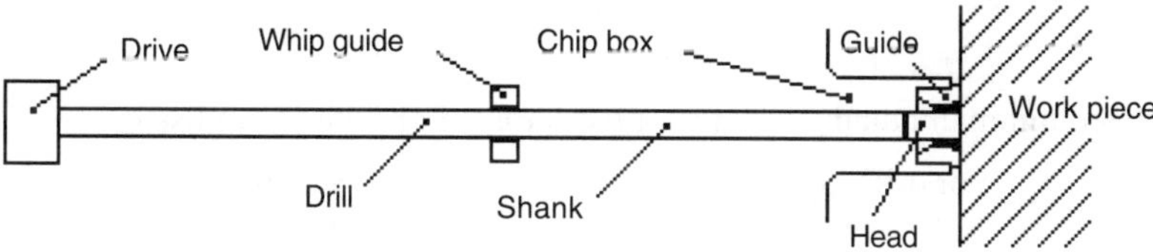

Fig. 22.37 Positioning of the gun drill on the surface of the work piece.

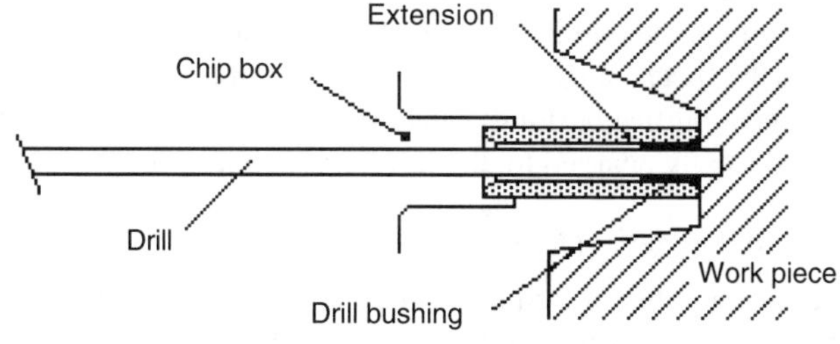

Figure 22.38 Extended guide bushing prevents interference of chip box with the work piece.

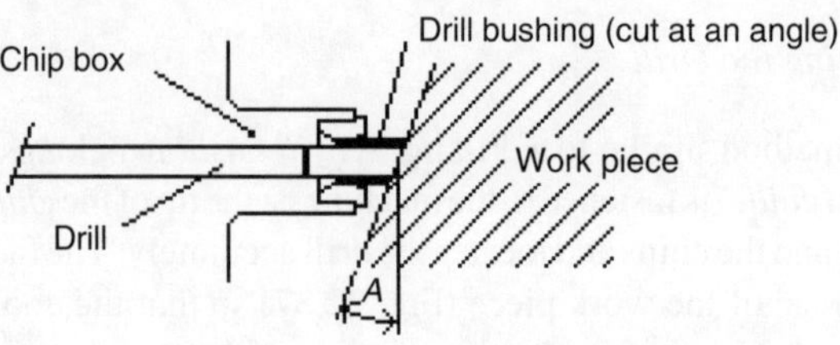

Figure 22.39 Drill bushing cut at angle *A* to match the angle of the work piece surface.

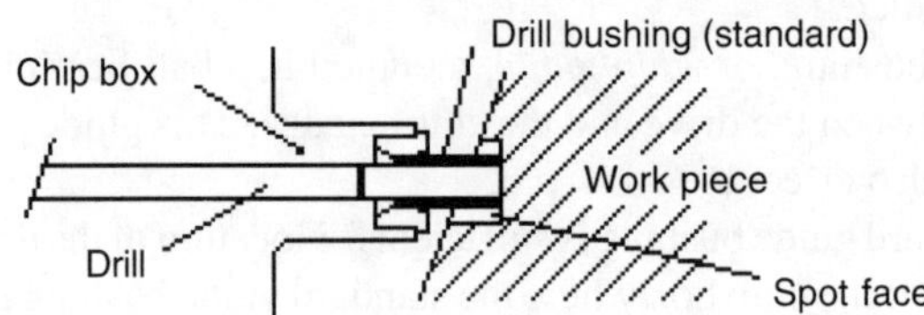

Figure 22.40 Spot face milled on work piece surface allows use of standard drill bushing and guide.

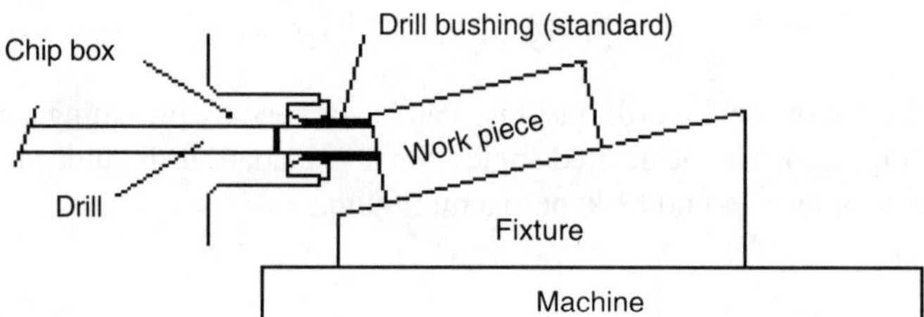

Figure 22.41 Specially designed fixture built at required angle from horizontal plane, and modified bushing at angle of work piece.

Suggested minimum diameters of spot faces for drill bushings are listed:

For drills up to:	Spot face diameter:
10.0 mm	14.3 mm
18.0 mm	22.0 mm
19.1 mm	26.0 mm

If the hole enters a work piece at a compound angle, the piece must be mounted on a specially designed and built fixture at the required angle which deviates from the horizontal plane of the drill, in addition to the special angle required at the face of the bushing (Fig. 22.41). Such fixtures, extensions, and modified bushings can be expensive. They are considered "tooling" that belongs to, and is usually charged to, the job for which they are required.

22.19.2.4 Diameter and Length of Gun Drills

The mold maker keeps a number of frequently used sizes of drills of various diameters and lengths in stock. The drills are fairly expensive, and the designer should select an available size, if at all possible.

Special diameters are available but are expensive and may entail a long delivery period from the vendor. If special drills are required, the production department must be advised immediately to enable them to avoid lengthy delays later on. The designer should keep a list of stock diameters and lengths of drills.

22.19.2.5 Cooling of Cutting Edge

The drill has a hole along its entire length through which coolant is pumped at high pressure (between 300 and 900 psi). The coolant exits right behind the cutting edge. It cools and lubricates the drill for minimum friction within the hole. It also washes the chips out through the open sector of the the drill (about 100° of the area of the drill). The chips are then separated from the coolant, which is then filtered and pumped again through the work piece.

Standard tolerances for diameters are +0.000, –0.01 mm (+0.000,–0.0004 in.). The head is ground with a back taper of ≈ 0.01–0.02 mm/mm (0.0004–0.0007 in./in.) to reduce rubbing in the hole. Maximum permissible oversize of hole is 0.1 mm (0.004 in.). Maximum allowable deviation from the straight is, depending on depth of hole, 2 mm (0.08 in.).

22.19.2.6 Hard Spots in Material

Poor material quality, with nonhomogeneous grains dispersed throughout the steel, may produce hard spots. This affects the progress of any drill into the steel. This is less serious with deep hole drills than with twist drills, but it can still be enough to force the drill to wander off more than is permissible.

22.19.2.7 Effects of Wandering of Drill

When drilling two deep holes from opposing sides to create an extra long hole, they may meet only partially. To prevent a flow restriction, the hole depths should be specified so that the holes overlap at least 10 mm at the meeting point.

If the deep hole is too close to a surface, there may not be enough metal around it to act as an evenly distributed heat sink, and the coolant is insufficient to remove all the heat generated by drilling. The material will anneal at the side of drilling closer to the surface and cause the drill to wander off in the direction of the surface. Setup of drilling may prevent this from happening by providing an extra heat sink if the surface is flat (e.g., by placing a suitable piece of steel on this surface).

A deep hole may also wander off enough and break through into the open, or may weaken the material behind a molding surface or supporting surface. Experience has shown that it is safer (and cheaper in the long run) to use plates of a better grade steel, such as P20 or stainless steels, rather than the cheaper AISI 4140 to avoid catastrophic drilling errors caused by hard spots and resulting wandering drills.

22.19.2.8 Finish of Deep Hole Drilled Channels

The finish inside the deep hole is usually satisfactory for coolant and air, and no reaming is required. For plastics flow channels, as in hot runner manifolds, it may be required to subject the work piece to a liquid honing or to a burnishing process to remove, or at least to smooth out, any roughness inside the bore where the plastic could hang up.

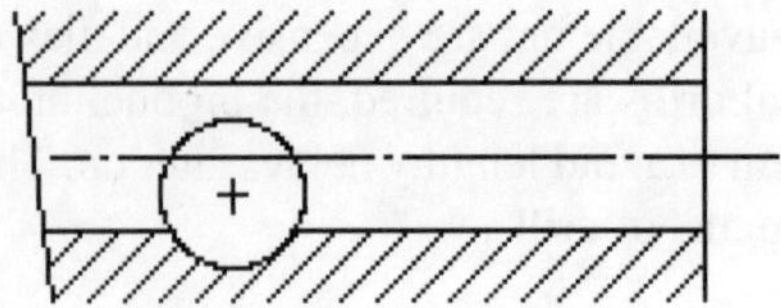

Figure 22.42 Intersecting channels with offset centerlines cause wandering off of drill.

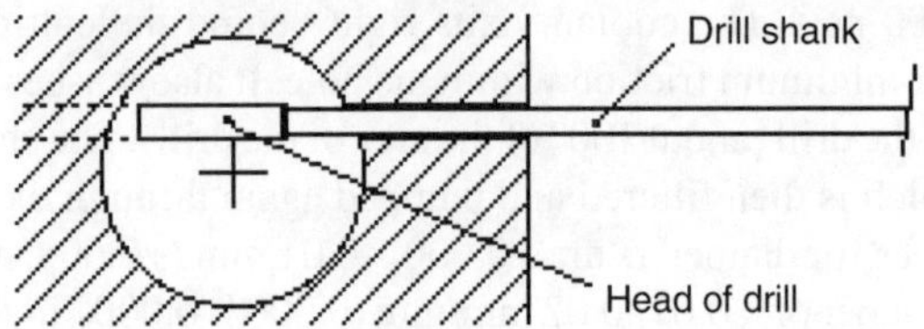

Figure 22.43 Small drill loses guidance as it passes through larger hole.

22.19.2.9 Suggestions for the Design of Deep Holes

Avoid interrupted cuts. Because the tungsten carbide head is fairly brittle, a gun drill should always cut into solid material to avoid breaking of the cutting edge.

This is not always possible because the channels used in molds do frequently intersect (water, air, plastic channels). However, the designer and the machine operator can take certain measures to minimize the risk of damage to the drill. Today, each sharpening costs about $100, (the equivalent of 1–2 hours), plus the time lost because of the interruption of work.

As for the size of the hole, the larger the drill, the easier it is to drill, within reason. Particularly for air lines, if the final hole is small (e.g., 4 mm diameter), the approach holes should be 8 or 10 mm if possible, rather than drilling a very long hole with a 4 mm drill.

Avoid offsetting of centerlines of channels. Where the holes meet, there is an interrupted cut; also, the cutting edge finds more resistance away from the centerline of the already existent hole and makes the drill wander off in the direction of least resistance (Fig. 22.42). (In a way, this is similar to the drill finding a "soft spot" in the material.)

For intersecting channels having a large difference in diameter, the small hole must always be drilled first, before drilling the larger hole. Otherwise, the small drill would lose its guidance as it passes through the large hole, as shown (exaggerated) in Fig. 22.43.

Finally, always locate the entrance of a deep hole so that it can be drilled with standard bushings (Fig. 22.44).

22.19.3 Drilling of Cooling and Air Channels in Molds

Cooling channels are for any fluid circulated within a mold with the purpose of maintaining the mold at a controlled temperature. It may be cold or hot water, oil, air, or a refrigerant circulated through the mold. Air channels are for compressed air, vacuum, and venting.

The following are some guidelines regarding designing coolant lines. However, there are no hard and fast rules, and there is no substitute for common sense.

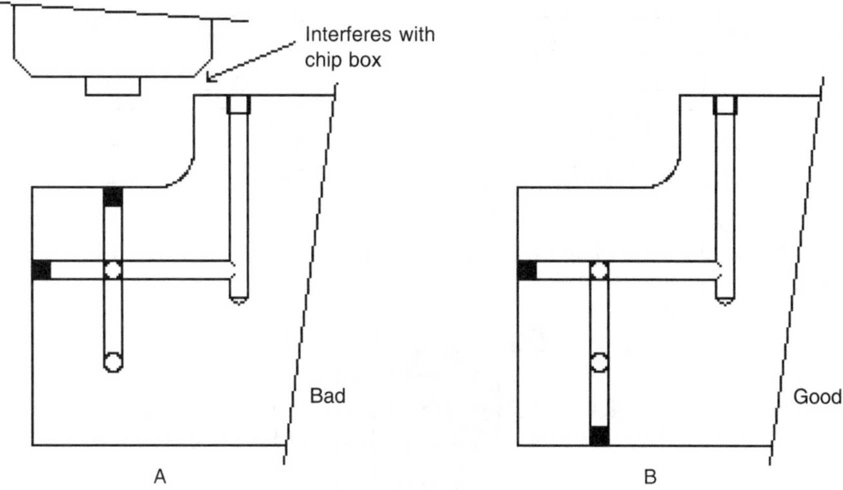

Figure 22.44 Entrances to deep holes should not prevent use of standard bushings: A. poor placement of hole causes interference with drill, and B. good placement of hole.

- For short holes, twist drills are usually satisfactory. The amount the drill wanders off its intended path is usually acceptable within the general tolerances.
- For long (deep) holes, or where the positional accuracy of a short hole is important, deep hole (gun) drills should be used.

22.19.3.1 Cross Drilling

There are many (usually inexpensive) molds which are not cross drilled. The connection for water (or air) circuits is done outside of the mold plate, cavity, or core using flexible hoses or rigid tubing. This is sometimes unavoidable (e.g., when connecting moving cooled mold parts).

It is the mark of a well-designed mold that cooling (and air) circuits are self-contained within the mold to reduce leakage possibilities and to improve accessibility to the mold. Typical patterns of cross drilling, and their sizing, are shown in Chapter 13, Mold Cooling. The designer is concerned about directing the flow of the coolant or bringing in compressed air (or vacuum) from the outside to a spot within the mold where it is required. Occasionally, vents also end in cross drilled channels to evacuate the cavity space.

22.19.3.2 Baffles for Cooling Flow

Unless the drillings can be arranged so that there is a continuous path for the fluid (coolant or air), baffles must be positioned in places to direct the flow. Illustrations A and B in Fig. 22.45 show how the same flow pattern can be achieved, with or without a baffle. In Fig. 22.45A, the part is drilled from three sides. In Fig. 22.45B, it is drilled from two sides at the expense of an added baffle and some additional drilling length. In general, B is less expensive to produce. This is a very simple example, but the principle must be considered every time a part is to be cross drilled.

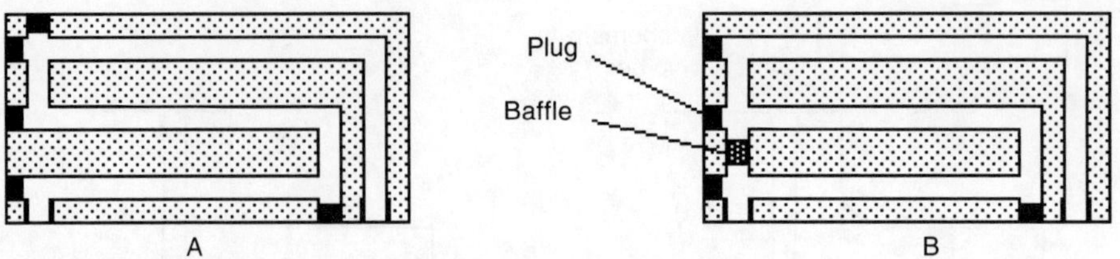

Figure 22.45 Cross drilling produces the same coolant path from A. drilling on three sides or B. drilling on two sides and use of a baffle.

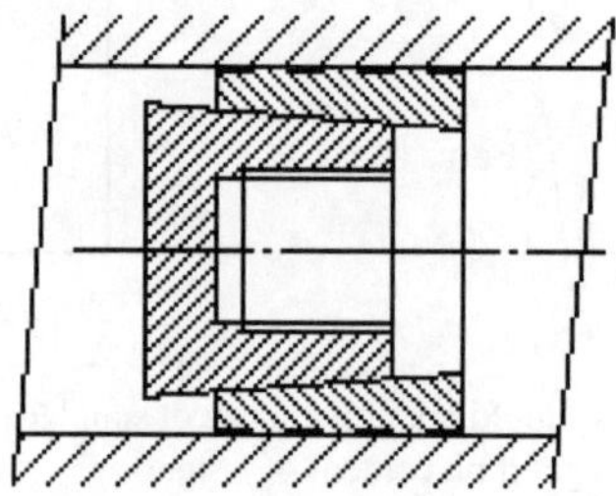

Figure 22.46 Plug baffle inserted in drilled hole seals it to prevent flow in undesired direction.

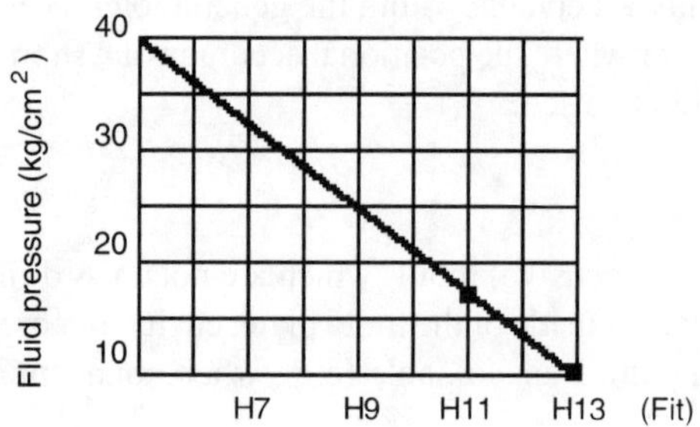

Figure 22.47 Fluid pressure varies with the fit of the hole.

22.19.3.2.1 Types of Baffles There are several types of baffles used in the industry. There is generally no problem with using any of them. A good method is the Hasco®, or similar plug, as illustrated in Fig. 22.46.

The plug is inserted into the hole using a special tool, which is calibrated for the insertion depth. At the required depth, the tool, which is threaded into the plug, pulls on the inner cone while holding the outer ring, thereby expanding the plug to make it seal the hole.

The advantage of these plugs is that they do not require soldering or other methods of securing baffles onto a baffle rod. Also, they are relatively easy to remove.

The fluid pressure that can be sustained by these baffles depends on the size (fit) of the hole, as shown in the graph (by Hasco) in Fig. 22.47. For example, with an H11 fit, the baffle will hold about 17 kg/cm^2; with an H13 fit, about 10 kg/cm^2. Since standard coolant pressures in molds are usually in the 4–6 kg/cm^2 range, either fit is satisfactory. It is suggested to use an H13 fit.

Since flow depends on pressure differences, and the pressure on both sides of the baffle are virtually the same, even a small leak between baffle and bore caused by scratches or tool marks from drilling will not affect the cooling pattern created by the baffle.

There are 5 standard (nominal) sizes of Hasco plugs: 6, 8, 10, 12, and 16 mm diameter. The corresponding bore tolerances are as follows:

Diameter (mm)	6	8	10	12	16
H13 (mm)	+0.180 –0.0	+0.220 –0.0	+0.220 –0.0	+0.270 –0.0	+0.270 –0.0

Compare this with tolerances of holes drilled with properly sharpened twist drills:

Maximum diameter	+0.157	+0.167	+0.177	+0.187	+0.207
Minimum diameter	+0.043	+0.049	+0.055	+0.061	+0.073

This shows that holes drilled with twist drills of the nominal sizes (6, 8, etc.) are acceptable, since even in the worst case (maximum bore size), the baffles will hold more than 10 kg/cm^2 pressure from one end of the baffle to the other.

Reaming of holes to closer tolerances is, therefore, not required. Tolerances of holes drilled with gun drills are smaller than those drilled with twist drills, and, therefore, are also acceptable. As a rule, these plugs will not be used as plugs to close the channels against the outside.

22.19.3.3 Plugging of Coolant Channels

Channels for *plastic* flow must be plugged so that there are no pockets where the plastic can remain stagnant and degrade when exposed to high temperatures for longer periods of time. For *cooling* lines, it is quite satisfactory to have locations where the coolant does not move. However, stagnant coolant does not provide any cooling. Only moving fluid removes heat!

The standard method of plugging coolant lines is to use standard NPT (National Pipe, Tapered) pipe plugs. Levlseal® (Unbrako SPS Corp.) pipe plugs for coolant lines are made from brass. As a rule, they are installed flush (level) with the surface of the part (Fig. 22.48A). Occasionally, they may be used deep inside a part (e.g., if the coolant channel is crossed near the surface by another bore, as illustrated in Fig. 22.48B.

Brass (rather than steel) plugs are used because they are softer than the material they thread into. They deform easier to fit the thread and will less overstress the mold part by its wedging action when tightening, especially if the wall surrounding the plug is thin. It is often a good

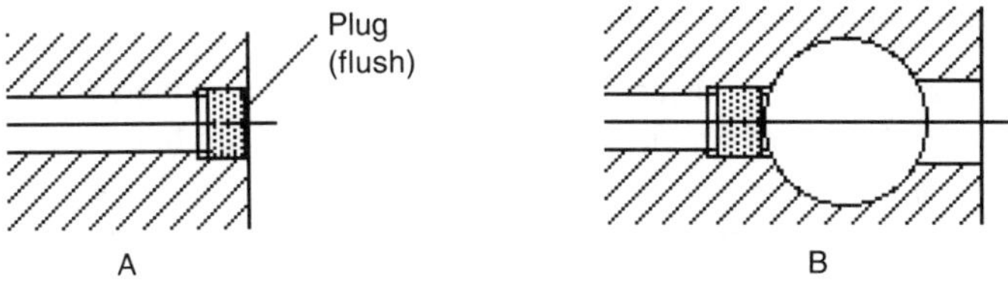

Figure 22.48 Plug installations in coolant channels: A. flush with part surface or B. inside part to plug entrance to cross channel.

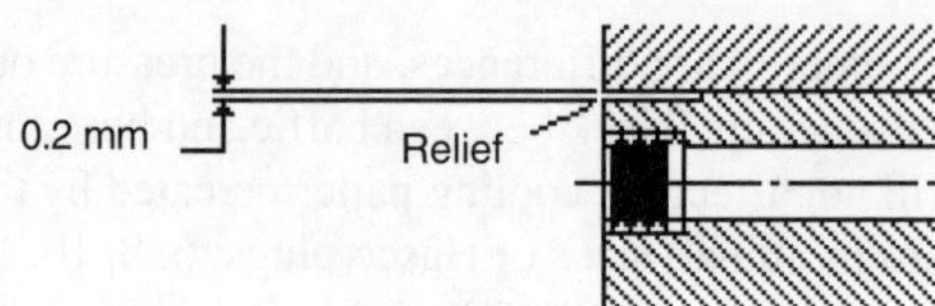

Figure 22.49 Relief machined on part surface where it adjoins another part near a plug.

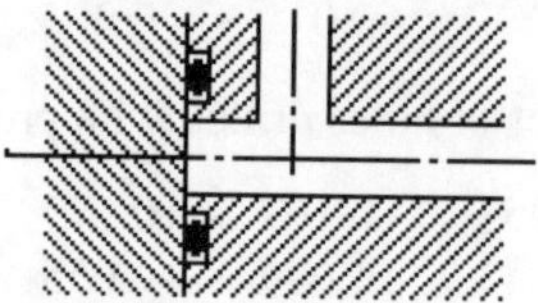

Figure 22.50 O-ring plugs are used where channel is plugged by another mold part or a plate.

precaution to create a relief on the outside of the part so that even if the part should be deformed, it will not affect the fit to an adjoining part (Fig. 22.49). Another reason for using brass plugs is that they do not rust and can be fairly easily removed, even after a long time.

SAE straight O-ring plugs are rarely used in molds. Occasionally, however, the channel is not plugged by a pipe plug but by another (mold) part. In this case, an O-ring must be provided, with the O-ring groove preferably in the part with the bore (as shown in Fig. 22.50) to ensure concentricity of the O-ring with the bore. However, the O-ring could also be in the matching part. The "plugging" mold part could also be an auxiliary plate screwed on the outside of the plate where the bore ends.

22.19.3.4 Distance of Waterlines from Holes and Surfaces

The longer the length and the smaller diameter of the drilled hole, the more the drill may wander off its intended path. It is recommended to stay away at least 5 mm from other bores or any surface at the far end of the drilled hole. At the start of the hole, this distance may be reduced to 4 mm. This applies also to long holes produced with deep hole drills, even though they stay much better in line than do twist drills.

Recommended minimum distances of *X* as shown in Fig. 22.51:

- From the edge of the steel (A): 5 mm.
- From intersecting holes (B): near the start, 3 mm; otherwise 5 mm.
- From the OD of a thread (C): 4 mm.

Stresses generated in the cross section *X* can be very high where there is a definite notching effect present caused by poor bore finish, and in a thread. This condition is worse in hardened steels. Also, some steels are very sensitive to the presence of corroding fluids, such as untreated cooling water, and are more likely to develop stress cracks. If in doubt, consult the steel supplier about risks from corrosion.

The size of the cooling channel (bore) is usually selected to be suitable for tapping the standard pipe thread sizes (see Table 22.6). If the cooling channel is smaller than the size for

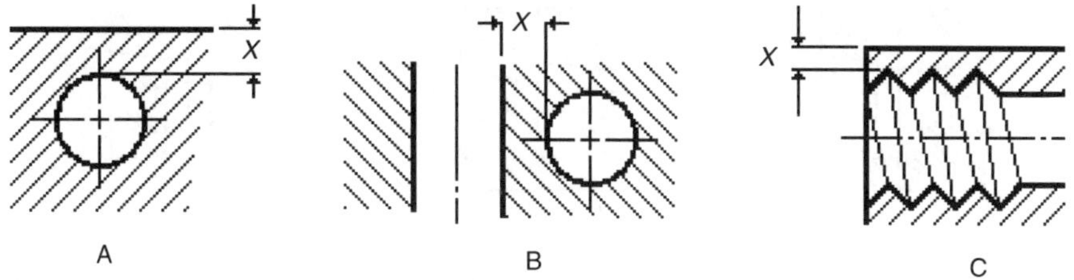

Figure 22.51 Recommended distances between drill hole and other holes or walls: A. hole and steel edge, B. intersecting holes, and C. threads and wall.

Table 22.6 Specifications for Standard NPT, BSP, and German Pipe Thread Sizes

Nominal pipe size BSP (or G)	Standard bore diameter (inch)	(mm)	Maximum OD of pipe thread (mm)	Cross sectional area (mm^2)	Minimum depth of tapping (mm)
1/16 - 28	.261 (G)	6.63	7.72	35	7.4
1/8 - 28	.3438	8.73	9.73	60	7.4
1/4 - 19	.453	11.51	13.16	104	11.0
3/8 - 19	.591	15.00	16.66	174	11.4
1/2 - 14	.750	19.00	20.96	283	15.0
3/4 - 14	.969	24.61	26.44	476	16.3

a standard plug, it must be drilled out with a standard bore size for a minimum depth suitable for the tap to achieve a full thread.

For *Standard NPT* tap drill sizes and cross sectional area of bores, follow recommended standards found in machinery hand books, etc. The NPT standard thread taper is 1:16, corresponding to a 1°47' angle with the centerline of the bore. For good quality threads, it is preferable to ream the entry of the bore with this angle, particularly with sizes of 1 in NPT and larger. This taper is desirable to reduce stresses in hardened steels and to provide clean threads. It is good practice to specify the taper on the drawing with a note: "Taper ream before tapping." This is also important for tapered threads used in conjunction with high-pressure hydraulics.

British standard pipe (BSP) threads differ significantly from NPT in that in BSP, the tapped bore is straight and the pipe thread is tapered. The number of threads per inch is different, and the threads are not interchangeable with NPT threads. The thread angle is 55°; taper is 1:16, similar to NPT. A sample designation: BSP ¼"-19.

German standard (VSM 51100) for pipe threads (*Rohrgewinde*) is interchangeable with BSP, but not with NPT. A sample designation: G ¼" VSM 51100.

22.19.3.5 Dimensioning of Drilled Bores

The end of a hole drilled with a twist drill is a flat "V"; the end of a hole produced by a deep hole drill is a flat "W". The end of a hole drilled with a slot drill is flat (Fig. 22.52).

The depth of the hole D is usually indicated from the surface of the mold part to the point of the drill. The mold designer is only concerned that the intersecting bores are fully open to

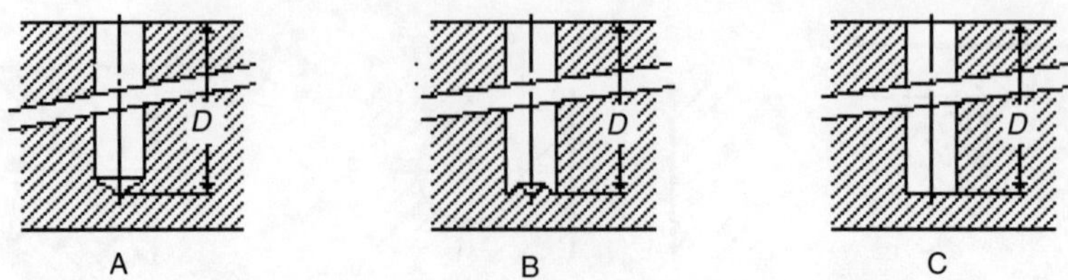

Figure 22.52 Ending of hole, depending on the type of drill used: A. twist drill V-shaped end, B. gun drill W-shaped end, and C. slot drill flat end.

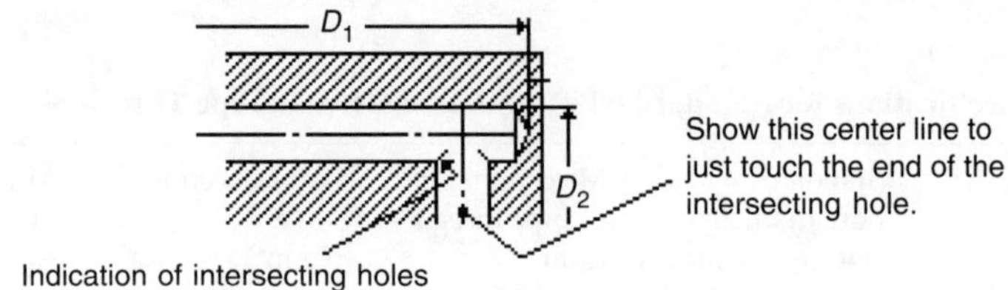

Figure 22.53 Indicating intersecting drill depths on the drawing.

free flow of coolant or air. There is no need to extend the drill depth beyond the intersecting bore.

The methods used to indicate drill depth are shown in Fig. 22.52. Occasionally, it may be necessary to show the actual drill point, if the end of the drill is near another feature in the mold. Conversely, it may be necessary to show that a flat bottom is required.

To indicate drill depths at intersections of bores, draw them as shown in Fig. 22.53. Where a smaller hole enters a larger hole, show the depth to the end of the intersecting hole.

Part II Mold Performance

23 Mold Performance and Mold Life

The purpose of any investment is to make a profit (basic economics.) The decision to invest in anything—whether in real estate, securities, machines, or equipment—is always based on the anticipation of good performance and healthy returns. The same holds true for buying, making, and using injection molds.

Performance of a mold is not only a measure of its productivity; it is also a measure of comparison with other molds when considering the reliability of operation, the quality of products made in the mold (accuracy, uniformity, finish, etc.), how the mold cost affects the cost of the product, and, finally, the life expectancy of the mold. The following chapters highlight and explain in simple language some of the basic relationships affecting the performance of injection molds.

For the newcomer to this field of technology, we recommend the reading of *Understanding Injection Molding* by H. Rees [1]. It will familiarize the reader with some of the common terms used but not explained in this book and will also assist in understanding some of the points made which refer to the mold design.

Chapters 24 through 28 explain the relationship between the various process settings of an injection molding machine and the influence of product design, mold design, machine characteristics, plastic used, etc., and how they affect the quality of the molded products and the productivity of the mold. Chapters 29 through 32 discuss forces in injection molds and the various influences which affect the useful life of a mold; this can be an important factor in a mold's overall performance.

This information is intended to be not only a reference for anyone who is connected with designing and operating molds (mold designers, estimators and molding technicians) but also for anyone charged with decision making for specifying and procuring injection molds and molding machines. It will also greatly assist product designers, plant (operating) engineers, as well as sales and service personnel, in understanding the various terms and relationships encountered in operating injection molds and machines. These chapters also highlight how changes in conditions affect the products and the productivity of the mold, and all factors which affect the physical life of a mold.

The graphs shown are greatly simplified and are used only to show certain relationships and their effect on the mold and the molding operations. Actual graphs, as may be required for accurate calculations or for more in-depth study of the subjects, are available from the design and research specialists in the various fields, such as those for stress (finite elements) analysis, cooling, melt flow, etc., and from the plastic materials suppliers.

Frequently asked questions, such as "How long will this mold last?" or, similarly, "What is the life of a mold?" will be answered. As will be shown, there are many factors to consider before answering such questions.

Theoretically, a properly designed, well-built and well-maintained mold should last forever and usually will outlive the need for the product made by the mold. However, some molds are required only for low production; these include experimental molds or molds for making a limited number of pre-production samples. In this category are also production molds where the total lifetime requirements are small, maybe in the order of one million or fewer pieces in the case of a small product, or even much fewer with some larger products.

If the requirements are small, many of the (more or less expensive) quality features of the mold may then be omitted to keep the mold cost low, even if as a result the unit cost of the product becomes higher than if it were molded in a production mold designed for very high output. Never forget that the overall cost of investment in the mold *plus* the cost of the total production for the molded product must be a minimum; the goal is always the *lowest unit cost*.

The optimal (minimum) cost of the product can be calculated by (roughly) estimating the cost of several, different molds for the same product by assuming fewer or more cavities, simpler or more intensive cooling, different runners and gating systems, etc., and then estimating and comparing the expected productivity and the resulting piece cost for each of these molds.

There are many molds built for consumer goods with a long product life; millions of products are required every year, for years to come. But as was discussed in Chapter 18, Metal Fatigue, the critical limit for the mold life is not the total number of pieces produced by the mold but the number of *cycles* the mold must run to produce the requirements.

The following examples may appear obvious but must be well understood when talking about mold life:

Example: To produce a (life) total of ten million pieces of a product in a mold with one cavity only, ten million cycles are required, and the fatigue life of the mold steels becomes an important consideration. Another product could be molded in a 32-cavity mold. To produce the ten million pieces required would require only $10{,}000{,}000 \div 32 = 312{,}500$ cycles. In this case, the mold parts could be stressed higher (as shown in the fatigue curves) and could be built lighter and thereby at lower cost.

Useful mold life. When speaking of mold life, reference is made to the *usable* life of a mold during which it can be *safely and efficiently* run. It is expected that a properly designed mold will resist deterioration as a result of normal wear during its operation, provided it:

- Is used in the molding machine specified for the particular mold;
- uses clean, filtered, and treated cooling water;
- uses the plastic originally intended to be molded; and
- is handled with proper care in the machine during operation, maintenance, and storage.

Catastrophic failure. Any mold can be severely damaged in a moment of carelessness. This obviously reduces the mold life or at least requires extensive repairs. Typical cases are:

- Damage caused by incorrect setup of the molding machine;
- damage to the molding surface, or breaking of a delicate core pin with a hard tool when removing a stuck product or part of a product;
- gates damaged during unblocking, which may be necessary during start-up;
- sloppy mounting of the mold in the machine, which can cause the mold to move or fall; or
- lack of alignment of the mold in a poorly maintained machine.

Wear and corrosion. Conditions causing wear and corrosion are always present. If the designer knows that the mold will be used with abrasive or corrosive plastics, or that the mold will operate in a very humid atmosphere, he or she can specify the proper materials and finishes to resist wear and corrosion. The mold should be designed so that common wear points such as gates can be easily replaced with simple gate inserts. Similarly, very delicate core pins, which are easily damaged or are expected to wear, should also be easily accessible for replacement. Some mold life problems in this category are:

- The molder switches plastics, and the mold, which was originally intended for non-abrasive or noncorrosive plastic, is now used with the "wrong" plastic.
- High humidity attacks molding surfaces, requiring frequent polishing. Eventually, the product will lose its dimensional accuracy.
- High humidity affects the mold plates and destroys parallelism.
- Untreated coolant attacks the cooling channels in the mold and reduces the efficiency of the mold cooling.

It is hoped that this book will contribute to the better understanding of the considerations that are important when designing, building, buying, and using injection molds. It should help to avoid costly mistakes in the planning of manufacture of any product, whether the production requirements are low or high.

References

1. Rees, H. (1994). Understanding Injection Molding, Hanser Publishers, Munich, and Hanser/Gardner Publishers, Cincinnati, OH.

24 Frequently Asked Questions About Mold Performance

24.1 What is the Life Expectancy of a Mold?

24.1.1 Product Obsolescence

The life of most molds is determined by the usefulness of the product. These molds become useless when the products for which they were made become obsolete.

24.1.2 Mold Obsolescence

A mold may become obsolete as a result of poor production planning. The output of a mold for a new product may have been poorly planned, and after a while, the production figures may exceed the forecast requirements. It then becomes necessary to increase the number of cavities—or otherwise increase the productivity—by improving the method of molding (better mold design, added features such as automation), rather than building an additional mold similar to the first one.

Molds may also become obsolete as a result of improved technology. Such improvements in mold and molding technology affect the mold life. Lower productivity molds, rather than left in continued use, are often scrapped in favor of higher productivity molds.

PET bottle preform molding started in 8-cavity molds with slow cycles. Improvements in the hot runner technology, with better mold-making methods, better molding machines, and with automation, have enabled the molder to reduce the cycle times considerably and, at the same time, to gradually increase the number of cavities to 12, 16, 24, 32, 48, 64, 72, and now to 96 cavities.

Example: At 35-second cycles and 8 cavities, the productivity of a preform mold is calculated at

$$\frac{3600 \text{ sec/hour}}{35 \text{ sec/shot}} = 103 \text{ shots/hour} \times 8 \text{ cavities} = 824 \text{ preforms/hour.}$$

At 22 seconds and 48 cavities, the productivity is

$$\frac{3600 \text{ sec/hour}}{22 \text{ sec/shot}} \times 48 \text{ cavities} = 7{,}855 \text{ preforms/hour.}$$

The productivity for the 48-cavity mold has, therefore, increased 9.5 times as compared to the older 8-cavity mold operation. More importantly, only one machine is required, rather than the more than nine machines that would be required with the old technology, using 8-cavity molds. Of course, to achieve this increase in productivity, a larger machine with a much larger plasticizing capacity to accommodate the larger mold is required. Provisions on the molding machine for a take-off mechanism which provides a significant portion of the necessary cooling time is also required.

The life of the "old", fewer-cavity mold has, therefore, ended, even though it is still in perfect operating condition.

24.2 What Affects Physical Mold Life?

24.2.1 Mechanical Components

In general, the physical mold life for most products is practically unlimited, provided the mold is properly designed and engineered, and not damaged by neglect, poor operation and maintenance, or by accidents. Major points that affect the mold life adversely are listed below. (It is interesting to observe that most of these points are routinely taken care of when engineering and building any kind of machinery, but it took many years before the mold making industry abandoned its "going-by-feel" approach and gradually started to adopt accepted, conventional engineering methods and calculations when building molds.)

- Inadequate physical strength of mold parts to withstand the loads encountered and caused by clamping forces, injection pressures, ejection forces, and forces created by differential heat expansion.
- Improper mold materials selection.
- Improper allowance for fatigue in mold steels is a frequent cause for breakage. (This is frequently overlooked but is a major source of mold breakdowns. Every mold part is cyclically loaded and unloaded when clamping and unclamping, and every cavity is cyclically stretched with every injection.)
- Improper heat treat specifications.
- Lack of lubrication of sliding faces.
- Incorrect machine setup.
- Poor maintenance practices and rough handling of the mold, especially if the molder does not adhere to the suggested start-up and maintenance procedures (i.e., when unblocking gates, which may be necessary if they freeze during start-up, when using contaminated materials, or when opening the hot runner portion of the mold to check the nozzles and nozzle heaters).

Other factors affecting the mold life:

- The use of dirty and contaminated plastics.
- Abrasive, erosive or corrosive fillers such as glass and some other additives.
- Corrosive plastics, which may attack (corrode) poorly suited mold materials.
- High humidity.
- Dirty and corrosive cooling water.

24.2.2 Electrical Components

Many of the electrical components of a mold may also affect mold life. Several major points are listed below:

- At rated voltages, most electrical (power) components such as mold heaters have a limited life, usually of 5,000–10,000 hours. Operating at lower voltages can increase the life of heaters; at 90% of rated voltage, the heater life can be almost unlimited. Operating at voltages higher than rated voltage, by as little as 10%, can dramatically reduce the heater life expectancy.
- Frequent ON–OFF switching at full voltage reduces the heater life.
- Poor contact of heaters with the heated mold part (lack of "heat sink") can drastically reduce the heater life.
- Hot, leaking plastic in contact with live power (at connectors, heater coils, etc.) will soon carbonize, causing short circuits and heater failure at these points.
- Coolant leaking onto electrical wiring, contacts, and connectors causes short circuits and possible damage to control equipment.
- All electrical components are sensitive to rough handling and accidental physical damage.
- Mechanical damage may be caused by poor location of cables, connectors, etc., where they are exposed to damage in operation—typically, by cables rubbing against moving parts or as a result of being pinched between moving mold and/or machine parts.
- Most electronic devices are sensitive to sudden voltage surges and to short circuits in the system.
- All electrical and electronic components are sensitive to high ambient temperatures, usually any temperature above 50 °C, unless such components are specially designed for hot applications or have adequate ventilation and/or cooling provided. If these components run at excessive temperatures, they may fail much sooner than at their normally expected life, and/or they may operate erratically and unreliably. Air conditioned plant operation is often suggested to ensure proper operation of the machines.

24.3 What is the Difference Between Production, Productivity, and Efficiency?

The terms production, productivity, and efficiency are often misunderstood or misinterpreted. The following sections will explain the proper meaning of each term as related to molding and mold making.

24.3.1 Production

Production is the total output of a molding machine, a group of machines, or the whole molding plant during a given time (hour, day, week, etc.), measured either in the number of pieces molded or in the amount of plastic "converted" (i.e., in tonnes per day, per week, etc.).

One can have very high production even though the plant is run very inefficiently, provided there are enough machines, molds, and manpower to give the required output.

24.3.2 Productivity

Productivity is the output of a machine, group of machines, or the whole molding plant per floor area and number of operators required. As a rule, the better the molding machines, the molds, and the methods used to operate them and the plant as a whole, and the fewer operators required, the higher is the productivity.

Example: A machine with a single-cavity mold and an operator can produce a certain article at a rate of 120 units/hour. Therefore, 4 machines and 4 operators will be necessary to meet the production requirements of 480 units/hour.

Assume the same machine can accommodate a 4-cavity mold for the same article and run fully automatically (i.e., no operator is needed). Output again will be 480 units/hour, but the productivity (per floor area is) 4 times that of the other setup.

Similar reasoning applies to the mechanization (automation) of other fields, such as materials handling (supplying the machines with plastic) and product handling (removing and storing the molded products).

24.3.3 Efficiency

There are several methods used to define efficiency of a molding plant. A good definition in the form of a formula is as follows:

$$\text{Efficiency (\%)} = \frac{\text{Number of good pieces actually produced}}{\text{Total number of pieces possible to produce in allocated time}} \times 100 \qquad (24.1)$$

where allocated time is the time a machine is available and designated for a specific job.

Efficiency can similarly be defined as:

$$\frac{\text{Number of hours producing good pieces}}{\text{Total number of hours machine is reserved for mold}} \times 100. \qquad (24.2)$$

An example will best explain the meaning of the above formulas. Suppose that for a specific production requirement, the time allocated for a mold is 40 hours. The installation time is assumed to be 6 hours. Installation time (in this context) includes:

- Mounting the mold in the machine,
- making all connections,
- heating up the mold,
- running it until it can run "on-cycle" fully automatically, and
- disconnecting and removing the mold, to get the machine ready for the next mold.

Assuming that all pieces produced are good (acceptable), the efficiency can be calculated as:

$$E = \frac{40\text{ hours} - 6\text{ hours}}{40\text{ hours}} \times 100 = 85\%.$$

Assuming further that only 90% of the products are good, the total efficiency E_T of this operation becomes

$$E_T = 0.85 \times 0.90 \times 100 = 76.5\ \%.$$

By halving the installation time (to 3 hours), the efficiency will be quite improved:

$$E = \frac{40 \text{ hours} - 3 \text{ hours}}{40 \text{ hours}} \times 100 = 92.5\%.$$

If the production run is longer, the effect of the installation time becomes less significant. If, for example, the run is planned for 240 hours (2 weeks of 5 days at 24 hours/day), the efficiency will be

$$E = \frac{240 \text{ hours} - 6 \text{ hours}}{40 \text{ hours}} \times 100 = 97.5\%$$

for a 6-hour installation and

$$E = \frac{240 \text{ hours} - 3 \text{ hours}}{40 \text{ hours}} \times 100 = 98.75\%$$

for a 3-hour installation.

This shows that the "quick mold changer" methods have significant advantages only if the production runs are short and the time for installation can be significantly reduced. Note that some quick mold changers today permit changes in less than 0.5 hour (time lost) between two producing molds. For high efficiency and productivity, it is much more important that the mold produces *good* (acceptable) products, without interruptions caused by poor housekeeping or breakdowns because of poor equipment such as molds, machines, robots, etc.

In some molds, the start-up time (to get the mold "on cycle") is long and can be a significant portion of the installation time, as defined above. The start-up time is usually directly related to the complexity of the product, the mold design, and the design of the hot runner system. Good record keeping of earlier setups will greatly reduce the time for starting up and tuning the machine.

24.4 How Does Performance Affect the Cost of a Molded Product?

A number of factors affect the cost of a product, most of them relating to the engineering side of this question. The cost of a product per unit (or per 1,000 units) produced can be broken down in the following elements:

1. Plastic cost,
2. power, water, and compressed air costs,
3. direct labor cost required to make this product,
4. mold cost,
5. machine cost,

6. maintenance cost, and
7. fixed plant costs (overhead, administration).

A complete explanation of the effects of many of the above factors from an engineering standpoint is provided in Chapter 3, Mold Requirements, Section 3.8, Reasonable Mold Cost. Any relevance of these factors from a performance standpoint is discussed below.

24.4.1 Plastic Cost

In most products, this cost is the major component of the total cost. In some products, it may reach as high as 90% of the total cost. It is, therefore, necessary to try to reduce the mass of the product as much as possible.

Plastic mass reduction is usually achieved by relying on proper design and engineering to arrive at a product with the least mass that will still satisfy the demands required of the product (lightweighting). A further advantage of such light weighting will be a reduction in the required injection and cooling times, with a corresponding savings in cycle time.

However, poor performance of the mold and/or machine and poorly controlled quality of the plastic will result in scrap (unusable products) and add to the cost of the plastic. Scrapping or reprocessing represents an added cost to the material cost.

With large production runs, under good operating conditions, the amount of scrap (scrap rate) should be less than 1%. With short production runs (frequent start-ups and adjustments), the scrap rate may be as high as 5, and even higher. Obviously, high scrap rate will increase the cost of the product.

24.4.2 Power, Water, and Compressed Air Costs

With efficient molding machines and with well-designed molds, these (operating) costs are directly related to the amount of plastic "converted" (processed). Power is used to operate the machine, the robot, and, where required, to dry (and/or heat) the plastic outside the machine and heat the plastic in the extruder.

Power is also required to provide the necessary circulation of the cooling water and to remove the heat from the cooling water. Therefore, the more plastic used, the more power and water will be required. Here, too, lightweighting will have beneficial effects on these costs.

Power is also used to provide any compressed air (and/or vacuum) used in the mold, usually for product removal and handling. Compressed air can be very expensive if used in large volumes, and the need for its use should be carefully evaluated. In many cases, the above costs are included in the plant overhead.

The cost of plastic and power represent the main difference between costs of operating hot runner and cold runner systems of injection molding.

The following are true for cold runner systems:

1. The plastic used for the runners can often (but not always) be reused. But, even if all of it can be reused, there is always at least some loss of material as a result of handling, and contamination during transporting and grinding of the runners.
2. There will be more power required in the processing of plastic than for hot runners. First, power is required to plasticize not only the amount of plastic required for the products but

also the plastic required in the runners. The requirement for this extra amount of plastic may also affect the selection of the molding machine, in that one with a larger plasticizing and shot capacity may be selected. With some smaller products in multicavity molds, the mass of the runner can be as large as, or is often even larger than, the total mass of the products molded each shot.

3. The additional heat energy (in the runners) must then be removed by cooling and adds to the cooling capacity required for the mold. Additional energy is also required to transport, grind and mix the reground plastic with virgin material.

In a hot runner system, since there are no runners to be reprocessed, there are none of the losses or additional power requirements as described for cold runners above in 1, 2, and 3. It can be stated, therefore, that hot runner molds are more economical to operate than cold runner molds. On the other hand, cold runner molds are usually much less expensive to build and are frequently used if the total requirement of the product is small, and the extra cost of the hot runner mold cannot be justified.

There are, occasionally, technical reasons to build a cold runner mold, usually for very small products, which, with today's technology, cannot be hot runnered satisfactorily. Such molds may use a combination of hot and cold runners, where each of several groups of cavities is fed from a small cold runner; they, in turn, are supplied from a larger, hot runner system.

24.4.3 Direct Labor Costs Required to Make a Product

There are several areas related to performance to consider in determining the cost of direct labor required to make a specific product:

1. *Handling of the raw material from the warehouse to the machine hopper.* Today, in many mold plants, virgin plastic is handled automatically by piping the plastic from silos to the machine where it may be, if necessary, mixed with color additives and/or reground material, which is usually handled manually using trucks, etc. This cost is usually included in overhead and is not directly added to the product cost.
2. *Cost of handling scrap, such as transporting, sorting, storing, etc.* This is also considered overhead and not directly added to the product cost.
3. *Actual work to operate the machine (to mold).* This refers to the method where an operator is used full time to remove the products from the machine (considered to be "semi-automatic" operation). Today, most molds can be designed so that operators are not required at the machines. However, with poorly designed and/or built molds, operators may be required to remove the products from the mold, to remove flash or runners from the products, or to place the products on shrinking fixtures to prevent the shape of the piece from deforming after ejection.
4. *Insert molding.* In many molds requiring inserts, such inserts are placed manually into the mold while it is open, thus requiring an operator even if the mold could otherwise run automatically. As a rule, the production quantities of such products are rather small.

   It can be stated that where large production quantities are required, every type of insert which can be inserted manually can also be inserted mechanically, automatically, using robots of some sort. Such equipment may be costly but pays for itself readily not only because of the elimination of manual labor but also because of increased productivity and better quality of the product, since the operation is not dependent on the skills of an operator. There is also the safety factor to consider: it is better to have nobody near any operating power press such as an injection molding machine.

5. *Transporting the products.* This includes the transporting of the molded pieces to any "postmolding" operation or moving them directly to storage or shipping. With large volumes produced, many of these operations can be fully or partly automated, and the amount of direct labor can be greatly reduced. With smaller production runs, or where products are molded in a wide variety of shapes and sizes, this is not always easy and will still require much labor. To go into more details of this area of production is beyond the scope of this book and is mentioned here only to make the reader aware that these costs can be substantial.
6. *Postmolding operations.* This includes costs for such operations as stacking, sorting, labeling, assembling, decorating, or packing a product. There are no definitive solutions to eliminate or reduce the high labor content of these operations. As a rule, provided the quantities are large, any one of these operations can be automated, thereby removing most or all direct labor costs. But today, even with smaller quantities, in many instances most of these operations can be mechanized economically, thereby using only a minimum amount of direct labor.
7. *Set-up* includes the cost of installing the mold, adjusting it for optimum operation, and removing it from the machine at the completion of the production run. This cost can be substantial if the mold changes are frequent (a few days or weeks only), but can be negligible if the mold stays in the machine for a considerable length of time. Typically, to establish the best operating condition of a mold, the best part of a shift could be required, since (after the extruder is adjusted with the first setting) it is necessary to run at least 10 shots before the products can be checked for quality and size.

   After each change in settings of temperatures and pressures of the injection unit, or of the cycle time elements, it will take some time to stabilize the system before the next samples can be run and checked again. This adjustment process continues until optimum settings are found. Note that with multicavity molds, products from every cavity should be checked after every setting. This highlights the advantages of testing every mold at the mold maker, before shipping, to establish the best molding conditions for the plastic specified and to record these settings for future use by the customer. This method of testing not only serves to prove that the mold runs as planned and that the products are okay but also saves the customer considerable time in starting a new mold and in the costs of the time disruption, plastic, and the power used for testing.
8. *Calculation of efficiency* (shown earlier). It is very important for a molder who changes molds frequently to reduce the set-up time. This can be achieved by standardizing the mold mounting methods and by simplifying the method of connecting services (power, water, air) to the mold as much as possible. The ultimate plan is to provide for automatic mold changing and mounting. Changeover times of a medium-size mold (≈500 kg) have been reduced from 4 hours to 0.5 hour. With hot runner molds, and using all quick change advantages, an important time-saving feature is to heat up the mold prior to installation, rather than to wait until it is installed.

   Set-up time is often considered overhead and is not directly attributed to the cost of the product. However, with frequent, short runs, it should be part of the product cost.

24.4.4 Mold Cost

There are several ways a mold cost per unit (or per 1,000 units) can be looked at. It depends greatly on the total volume of products required per year, or during the lifetime of the mold.

Typically, a mold for a product required for only one model year of an automobile or for a promotional item required for only one sales campaign will be written off for this limited number of products.

Example 1: A mold for an automotive part costs $60,000. The total requirements are estimated at 50,000 units. The mold cost per unit is:

$$\frac{\$60{,}000}{50{,}000 \text{ units}} = \$1.20/\text{unit}.$$

In such a case, the mold cost per unit amounts to a relatively high portion of the piece cost.

There are also molds for technical articles, of which a very limited number of pieces are expected to be produced per year, even though they could be required for many years. In this case, the mold cost per unit may be based on the total expected number of pieces to be molded for seven or even ten years.

Example 2: A mold for a precision test bar costs $35,000. There are only 1,000 test bars required per year, but it is expected that the same design will be used for many years to come. By depreciating the mold over 7 years, the final cost is:

$$\frac{\$35{,}000}{7{,}000 \text{ units}} = \$5.00/\text{unit}.$$

(In this case, the mold cost/unit is considerably higher than the cost of plastic used.)

On the other hand, if the product is "timeless", or in other words, it will be required without changes for many years, the mold could be written off in two or even more years, and the cost per unit (or 1,000 units) is then based on the expected production in two or more years:

Example 3: A mold for a 6-cavity cassette box costs $200,000. At an estimated cycle of 8 seconds, the mold can produce:

$$\frac{3{,}600 \text{ sec/hour}}{8 \text{ sec/shot}} \times 6 \text{ cavities/shot} = 2{,}700 \text{ units/hour},$$

or 14,580,000 units/year based on 5,400 hours/year operation.* In two years (selected write-off period), the mold will produce 29,160,000 units. Dividing the mold cost by this figure determines the final unit cost:

* The 5,400 hours/year operation figure used above is arrived at as follows:
Possible hours: 24 hours/day × 7 days/week × 52 weeks/year = 8,736 hours/year. Practical hours (for a conservative machine hour rate): 24 hours/day × 5 days/week × 50 weeks/year = 6,000 hours/year. Assuming a machine utilization of 90%, the total is 5,400 hours/year.

$$\frac{\$200{,}000}{29{,}160{,}000 \text{ units}} = \$0.0069/\text{unit, or } \$6.98 \text{ per } 1{,}000 \text{ units.}$$

The mold cost is, therefore, an almost negligible portion of the cost of the product. A higher quality mold will probably cost more, but it may produce more; however, such increase in cost will only negligibly affect the cost of the product, while it could considerably increase the productivity of the machine.

24.4.5 Machine (Hour) Cost

A machine usually costs much more than a mold, but the machine purchase price, including costs for installation and the cost of the money (interest), can be spread over a much longer period, often seven or ten years, depending on accounting practices. An exception to this would be if a (standard or special) machine (for molding, product handling, etc.) is required for one specific job only, in which case the machine is really not more than an extension of the mold and would be written off in the same time period as the mold.

As a rule, the molder will establish a machine hour rate that is then charged for the time a mold runs a specific job. Basically, the machine hour rate is established by dividing the cost of the machine (per year) by the number of hours the machine will be in operation. This assumes that the machine will be busy for all or most of the time during the year.

Example 1: A machine costs \$500,000. By adding costs for transport, installation, and the cost of money, assume the machine costs a total of \$700,000. By using seven years for write-off, the annual cost of the machine is then

$$\frac{\$700{,}000}{7 \text{ years}} = \$100{,}000/\text{year.}$$

Dividing this by 5,400 hours (see earlier explanation), the hourly rate for this machine.is

$$\frac{\$100{,}000/\text{year}}{5{,}400 \text{ hours/year}} = \$18.52/\text{hour.}$$

Note that this is very conservatively calculated, and that there is another theoretical 3,336 hours available on this machine (8,736 possible hours – 5,400 practical hours = 3,336 remaining hours). If the machine is planned to be used more than 5,400 hours, the machine hour rate will be less than the \$18.52. The machine hour cost is then apportioned to the cost of the product by dividing the machine hour rate by the number of pieces produced per hour. Note that this cost represents the actual cost of the machine and does not include any overhead or profit.

Example 2: Assume the above-mentioned machine can be used to produce a large pail weighing 1,000 g. Using a single-cavity mold, the cycle could be 25 seconds. The machine hour cost for this mold will be, therefore:

$$\frac{3{,}600 \text{ sec/hour}}{25 \text{ sec/shot}} \times 1 \text{ shot (cavity)} = 144 \text{ pails/hour.}$$

$$\frac{\$18.52/\text{hour}}{144\ \text{pails/hour}} = \$0.13\ \text{machine hour cost/unit.}$$

If the plastic required for the pail costs \$1.50/unit, the machine hour cost is about 9% of the cost of plastic. (Note that the machine needs to plasticize at least 144 kg/hour.)

A better designed mold may be able to produce this same pail in 20 seconds, or produce 180 pails/hour. The machine hour cost for this mold is now less:

$$\frac{\$18.52/\text{hour}}{180\ \text{pails/hour}} = \$0.10/\text{unit,}$$

or 7% of the cost of plastic; it will also give a 25% increase in productivity of the machine. Note: The machine now needs to plasticize at least 180 kg/hour.

Example 3: The same machine could also be used for a mold for an 8-cavity container, weighing 40 g, running at an 8-second cycle.

$$\frac{3{,}600\ \text{sec/hour}}{8\ \text{sec/shot}} \times 8\ \text{shots (cavities)} = 3{,}600\ \text{units (containers)/hour.}$$

$$\frac{\$18.52/\text{hour}}{3{,}600\ \text{units/hour}} \times 1{,}000\ \text{units} = \$5.14\ \text{per}\ 1{,}000\ \text{containers.}$$

The plastic for 1,000 units costs 40 g/unit × 1,000 units × \$1.50/kg = \$60.00; the machine hour cost is about 8.6% of the cost of the plastic. (Note: The machine needs to plasticize at least 144 kg/hour.)

Example 4: The same machine could be considered to run the same product in a 2 × 8 stack mold (two faces of 8 cavities each), at the same cycle as the single-face mold:

$$\frac{3{,}600\ \text{sec/hour}}{8\ \text{sec/shot}} \times 16\ \text{shots (cavities)} = 7{,}200\ \text{units (containers)/hour.}$$

$$\frac{\$18.52/\text{hour}}{7{,}200\ \text{units/hour}} \times 1{,}000\ \text{units} = \$2.57\ \text{for}\ 1{,}000\ \text{containers.}$$

The plastic for 1,000 units costs the same as before (\$60.00), but the machine hour cost is only 4.3% of the cost of the plastic, a significant reduction in the cost of the product. However, the machine needs now to plasticize at least 288 kg/hour, with a corresponding increase in the injection capacity. If this is beyond the capability of this machine, using this mold will require that the cycle time be increased, with the result that fewer pieces will be produced per hour, and the advantage of using a stack mold will partly be lost. As an alternative, a larger machine, with a corresponding, higher machine hour cost, will be required to run the stack mold.

Similar calculations can be made for any combination of molds, products, and cycle times and must be done in every case in which a decision is made regarding a new mold.

24.4.6 Maintenance Cost

For molds, annual maintenance costs can be conservatively estimated at 10–20% of the purchase price of the mold, depending on the complexity of the product. This may appear high for the first year or two, but will be realistic after a few years of operation of the mold, when a complete refurbishing, involving dismantling, cleaning, checking, and replacing of worn and damaged parts may become necessary, particularly if the plastics processed are abrasive or corrosive.

For machines, periodic oil changes, electrical and hydraulic maintenance, and normal wear and tear can be assumed to cost annually about 10% of the purchase price of the machine. Usually, maintenance costs for both mold and machine are part of the overhead of running the plant.

24.4.7 Fixed Plant Costs

Fixed plant costs usually include all other costs required to run a molding plant, and they are included in the plant overhead. Typical cost elements to be considered are listed below:

- Cost of building and land;
- building maintenance;
- costs of installations for services;
- power, light, and water, including cost of distribution;
- cost of hardware for transportation and handling;
- labor costs not attributable (chargeable) to a specific product cost; and
- taxes, administration, and others.

25 Cycle Time

25.1 Effect of Product Design

25.1.1 Wall Thickness *t*

Plastics are poor heat conductors, about 20–30 times worse than steel. The thicker the plastic, the longer will the heat travel from the hot plastic to the cold cavity walls. The values are different for various plastics, but in all cases the relationship between the wall thickness *t* and the cooling time follows a curve similar to that as shown in Fig. 25.1.

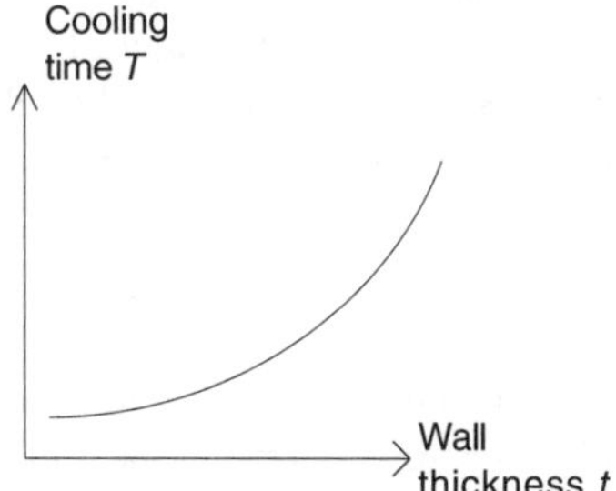

Figure 25.1 Curve shows thicker walls require longer cooling time.

25.1.2 Uniformity of Walls

Even if most of the walls are thin, the governing factor in the cycle time is the heaviest thickness that must be cooled before the product can be ejected.

25.1.3 Feasibility of Good Cooling Provided to All Product Areas

If the product design is such that it is impossible to provide equally good cooling to all areas, the cycle time will increase. In some cases, it could be very costly to provide such "good" cooling, and it is important to consider the economics—whether the added cost for better cooling is economically viable.

The following sections describe areas where a product designer familiar with mold design, or at least in cooperation with the mold designer, can save much in cycle time and thereby reduce the cost of the product.

25.2 Effect of Mold Design

The features of a mold can be broken down into several elements. All of these elements can affect the molding cycle:

- Heat exchange,
- plastic distribution,
- venting, and
- ejection.

25.2.1 Heat Exchange

The mold is a heat exchanger. Most of the heat in the injected plastic must be removed within the mold before the product can be ejected. The cooling layout within the mold must be able to remove the heat as fast as possible.

Regarding mold materials selection, different materials have different heat conductivity. In some cases, even the difference between various mold steels can be significant. The use of beryllium-copper or other high heat conducting materials can make a big improvement.

25.2.2 Plastic Distribution

The mold distributes the plastic from the machine nozzle to the cavities. The strength of mold is important in that higher filling speeds require higher injection pressures, which in turn require that the mold be solidly built.

The shape of runners and gates also affects distribution. The less resistance against plastic flow, the faster the cavities can be filled. This must be balanced against other considerations, such as gate appearance, inventory in hot runner manifolds, etc. (If resistance to plastic flow within the cavity space is substantial, the effect of runners and gates may be less significant.)

25.2.3 Venting

The mold must provide venting to permit all air to escape rapidly from the cavity space, ahead of the inrushing plastic. Otherwise, the pressure build-up of the trapped air will slow down the injection speed. The difference in filling time between good and poor venting in high-speed, thin-wall molds, could be 0.2–0.3 seconds, and could make a 10% difference in cycle time. Also, the trapped air could heat so much that the plastic might degrade or cause burn marks, usually at the rim of the product or in corners at unvented pockets or ribs.

25.2.4 Ejection

The mold must provide for the ejection of the molded product. The ejection time depends on the ejection method used. For fastest cycling, the ejection should take place during the opening stroke of the machine, so that no mold open time is required. This is not feasible with manual operation or with robots, where any removal operation between the mold halves must take place while the mold is in the mold open position. Note: "Swing chutes" (HUSKY patent) are a kind of robot; however, the motion to remove the products takes place entirely during the dry cycle.

25.3 Effect of Machine Size Selection

The dry cycle time of the machine selected for a job will affect the overall cycle and can be significant with products that can be molded at rapid cycles. As a rule, a larger machine will have a slower dry cycle, because heavier masses must be accelerated and decelerated. With fast-cycling molds, a reduction in dry cycle of even a few tenths of a second can lead to a considerable increase in productivity.

25.4 Effect of Molding Material (Plastic)

25.4.1 Viscosity

The higher the viscosity of the injected plastic, the slower will be the injection speed, and the longer the cycle time. Decreasing the viscosity by increasing the stock temperature is usually counterproductive, because the molding cycle will then require more cooling time.

25.4.2 Crystallinity

Crystalline plastics take longer to heat to the required stock temperature than amorphous plastics because of the additional heat required to change the crystalline structure of the molecules. The same additional heat must be removed during cooling, which prolongs the cooling time. This is best shown in the two graphs in Fig. 25.2.

These graphs are oversimplified and are used here only to illustrate the difference between the two types of plastics. Also, many plastics are a mixture of amorphous and crystalline

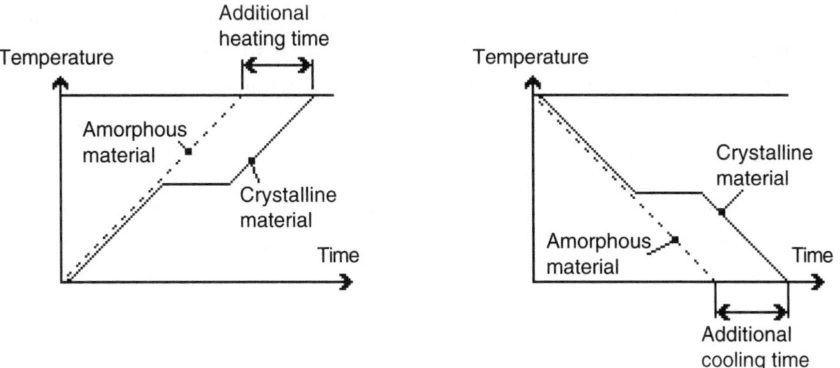

Figure 25.2 Graphs show A. temperature rise in amorphous and crystalline plastics, and B. cooling (temperature decrease) in amorphous and crystalline plastics.

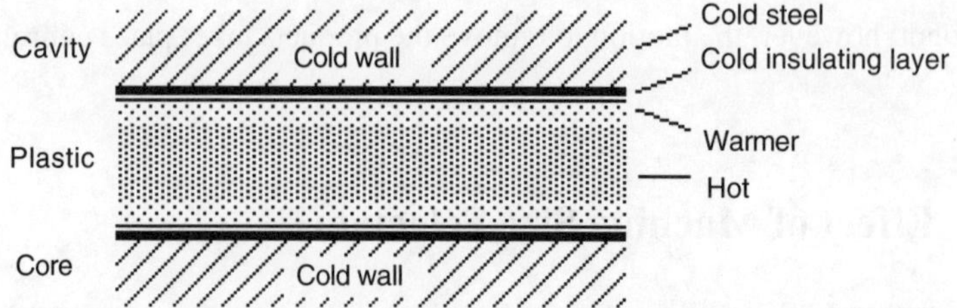

Figure 25.3 Schematic of plastic layer between cavity and core shows relative temperatures.

plastics, making the conditions still more complicated. Actual curves for various materials are available from the various materials suppliers.

As shown on the left graph in Fig. 25.2, for amorphous materials (PS, PVC, SAN, PC, etc.) the rise in temperature is more or less linear, over time. For crystalline plastics (PE, PP, etc.), the temperature first increases more or less linearly over time, then reaches a plateau during which "latent" heat is added, but the temperature of the plastic does not rise, and finally, when all crystals are molten, the temperature rises again, similar to amorphous materials. The right graph in Fig. 25.2 shows that a similar time delay (additional *cooling* time) takes place as the crystalline plastic cools down.

25.4.3 Stock Temperature

The higher the stock temperature, the longer it will take to cool down to the required ejection temperature. The machine should always be adjusted for the lowest temperature at which it will run reliably.

25.4.4 Heat Conductivity

In general, all plastics are good heat insulators. Also, without giving any specific values, it can be stated that amorphous materials are better heat conductors than crystalline materials.

The lower the conductivity, the longer it takes for the heat to travel from the hot plastic to the cold mold steel. This is illustrated in Fig. 25.3. Obviously, the thicker the plastic, the thicker will be the insulating layer as the plastic cools down in the mold, and the longer it will take before the product can be ejected without distortion or damage.

25.4.5 Sources of Plastic

It should be noted that the same plastic supplied by a manufacturer at different times (from different "batches") can, and often does, behave differently than earlier batches. This can be simply a case of accumulation of tolerances in the manufacture of a plastic.

Similarly, and sometimes more aggravating, a plastic purchased under the exact, identical specification supplied from one manufacturer will behave differently than a plastic from another manufacturer, and may require different pressures, time settings, and/or temperatures. The reasons for this may be the use of different types of additives, different shear viscosity, different molecular weight distribution (MWD), etc.

25.5 Effect of Product Temperature at Ejection

The plastic within the cavity space must be stiff enough so that the product will not be distorted or damaged during ejection. The higher the temperature of the product at ejection, the shorter will be the cycle time, provided all other conditions are equal.

Since the plastic continues to shrink as it cools outside the cavity space, not being restricted by the core, the final product dimensions depend on the temperature at ejection. The hotter the plastic at ejection, the smaller will be the product when it is finally cold. If the plastic is allowed to stay longer in the mold, the dimensions will be closer to the size of the cavity and core (steel) dimensions.

25.5.1 Stresses in the Product

Cooling the product completely within the mold may induce stresses in the plastic which could cause failure of the product (cracks, breakage). Ejection at higher temperatures helps in reducing such stresses.

25.5.2 Stripping of Undercuts or Threads

Stripping of products with internal undercuts is always preferred to using molds with the much more complicated collapsible cores, or with any unscrewing methods. However, there are limits as to how much the plastic will permit itself to be stretched during stripping off the core without cracking or breaking.

The permissible stretch depends greatly on the type of plastic (stiff, elastic), on the shape and depth of the undercuts, and on the relation of this depth to the overall length of the material stretched. For example, with the same depth of undercut, a small overcap will be more difficult to strip than a larger one.

Plastic, even a type that is stiff or brittle when cold, will permit quite a bit of stretching when still hot during ejection, as long as it is not too hot and does not deform permanently. In critical cases, it may be necessary to eject at higher temperatures and to use warmer coolant to ensure that the products will eject reliably. In such cases, it would be impossible to strip good products from a cold mold.

Note: This also has an effect on the machine selection. Higher stripping forces may require ejection forces higher than those normally used. It may also require the addition of a hydraulic ejection system within the mold to provide the necessary forces.

25.6 Effect of Injection (Filling) Speed

Faster filling speeds reduce the cycle time. Limits to the injection speed depend on:

- Machine capability,
- type of plastic,
- product design, and
- mold design.

25.6.1 Machine Capability

Injection speed is proportional to the hydraulic pressure in the injection cylinder and to the capability of the injection system to supply enough high-pressure hydraulic oil to the injection cylinder to maintain the forward motion during injection. In many machines, the hydraulic pump(s) alone may not have enough capacity to maintain the required forward motion of the ram screw (or injection piston), and the injection speed is limited by the output (volume and pressure) of the hydraulic pump supplying the injection unit.

To remedy this, either larger pump(s) and motor(s) will be required, or accumulators must be added to store high-pressure oil during the time the machine does not require much, or any, pressure oil. This stored high-pressure oil is then added during injection to the high-pressure oil coming from the pump.

NOTE: Hydraulic valves and piping must also be adequately dimensioned so that they will not offer too much resistance to the oil flow during injection.

25.6.2 Type of Plastic

Some plastics will lose some essential properties if injected too fast, beyond their permissible shear rate. In general, this does not apply to most (low cost) commodity plastics such as PP, PE, and PS.

25.6.3 Product Design

Long flow distances, frequent changes in direction of flow, and restrictions in narrow spaces will slow down the filling speed.

25.6.4 Mold Design

Most importantly, shape and size of runners and gates will affect the filling speed. Figure 25.4 shows a schematic cross section of a gate. *L* is the length of the narrowest section of the gate. The shorter the land *L*, the less resistance to flow and the faster the filling speed.

D is the gate diameter. The larger the gate, the less resistance there is to filling. However, there are practical limits to the size of the gate. Larger gates take longer to freeze off and are often not acceptable for aesthetic reasons. In operation, large gates may drool, especially in molds with longer cycles, unless valve gates, which are shut except during injection, are used.

Multiple gating into one cavity applies mainly to very large products, such as large pails, large automotive panels, etc., with relatively thin walls. The cavity space is filled from several

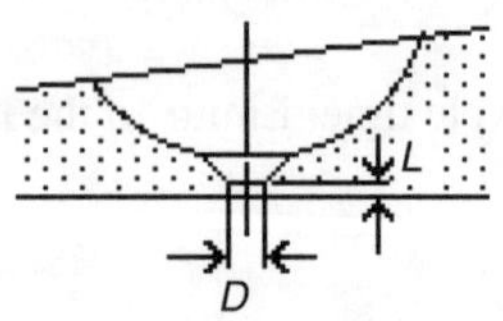

Figure 25.4 Cross section of gate.

gates, (2–6 or more) so that the flow length within the cooled cavity space from each gate is greatly reduced as compared to the filling length when injecting from one (usually centrally located) gate.

Example: A large pail of $D = 300$ mm (diameter) and $H = 400$ mm (height) has a flow length of

$$L = \frac{D}{2} + H = 550 \text{ mm},$$

and a wall thickness of $t = 1.8$ mm is desired. The L/t ratio is, therefore, 305. This is very high and may require the maximum injection pressure of a machine, or it may be impossible to mold at all.

By selecting 3 gates, evenly distributed 100 mm from the center, the flow length L can be reduced to

$$L = (150 \text{ mm} - 100 \text{ mm}) + 400 \text{ mm} = 450 \text{ mm}.$$

L/t now equals 250. This ratio is quite acceptable, and the machine will be able to produce this container.

In the case of multiple gating, the plastic is distributed for longer distances in the low resistance runners, rather than in the cold, narrow cavity space, where the resistance to the flow is very high. This applies not only to hot runners but to any runner system.

Venting is of special importance with multiple gating. Vents (vent pins, etc.) must be placed at all points where the plastic fronts coming from different gates meet, to prevent air entrapment and/or bad welding of the plastic.

Advantages of multiple gating are:

- Smaller gates, with better gate vestiges;
- less injection pressure required;
- less clamp pressure required, because of the lower injection pressure;
- possibility of using smaller wall thicknesses and lower product mass; and
- faster cycle time, due to higher injection speeds.

Obviously, multiple gating will cost more than single gating, and the additional cost must be considered before deciding on this method of injection.

With regard to the effect of multiple gating on product quality, from studies conducted on 6-gallon pails, it appears that, while all the other points as described above are favorable to multiple gating, there have been no significant benefits (but neither disadvantages) to the required quality of pails, such as resistance to top load and impact strength.

Venting is another part of mold design that may affect injection speed. During injection, as the plastic rushes in to fill the cavity space, the air within this space must be given the opportunity to escape, to make room for the plastic. The trapped air will be compressed by the incoming plastic and not only slow down the filling speed but also cause the air to increase its temperature to such an extent that the plastic will burn at its leading edge. Well-vented molds, therefore, will not only improve the filling speed but also produce better products.

25.7 Injection Capacity

25.7.1 Plasticizing Capacity

Every extruder is rated for the amount (Q) per hour of plastic it can "plasticize", that is, bring the cold plastic pellets to the required injection ("stock" or "melt") temperature. Normally, the extruders are rated for a Q in kg/hour of PS. Allowing for the different specific density, this corresponds to a Q of approximately 10% less for PP or PE. This rating applies to the output of a *continuously running extruder screw* .

25.7.1.1 RS Machines

In RS machines, the screw must stop during the injection and hold time. (The thrust bearings of extruders are not strong enough to support the full injection force while turning.) Also, the screw cannot plasticize against any high (back) pressures as would be present at even low injection pressures. Therefore, only a certain percentage of the rated Q is possible to achieve.

Example: Running a mold at a possible cycle of 12 seconds, the injection and injection hold time may be about 3.5 seconds. The actual time available for the extruder to "recover" (prepare plastic for the next cycle) is, therefore, 8.5 seconds, or 71% of the 12-second cycle.

Assume the extruder in this example is rated at 110 kg/hour. PS, which has a specific gravity of 1.05 kg/L (L = liter, or 1,000 cm^3).

$$\frac{110\text{ kg/hour}}{1.05\text{ kg/L}} = 104.762\text{ L/hour.}$$

At the required injection temperature, the plastic volume is greater than when cold by the shrink factor of 0.004 (0.004 cm^3/cm^3). The actual (hot) plasticizing capacity is

$$104.762\text{ L/hour} \times 1.004 = 105.181\text{ L/hour.}$$

If the same mold and machine should be used to make this product from PE instead of PS, the actual plasticizing capacity must be reduced by a factor which not only represents the difference in specific gravity but also accounts for the increase in volume of plastic at the injection temperature. If the specific gravity of PE is 0.9 kg/L, the difference in specific gravity between PS and PE is 1.05 kg/L – 0.9 kg/L = 0.15 kg/L. The cold (rated) capacity is, therefore,

$$\frac{110\text{ kg/hour}}{1.15\text{ kg/L}} = 95.65\text{ L/hour.}$$

With a heat expansion (or shrinkage) of 0.02 for PE between cold and injection temperature, the actual rated injection volume will be

$$95.65\text{ L/hour} \times 1.02 = 97.56\text{ L/hour.}$$

For the preceding example, since only 71% of the time will be available for plasticizing, the maximum plasticizing capacity for PE will be only 69.27 L/hour, or about 40% less than the "stated rating" for PS of the extruder.

The above example assumes that the extruder screw will plasticize PS and PE equally well. This is normally not the case; a screw designed to plasticize PS is less effective when plasticizing PE. In general, screw designs can and should be optimized for the material processed.

For long production runs, the "universal screw" supplied with a molding machine can be a liability; in other words, it is not very efficient for any plastic. It is suggested that screws specially designed for the plastic that will be molded should always be used.

Example : A mold has a shot size of 1,000 g and could cycle at 30 seconds, so that 120 pieces could be molded per hour, which would require 120 kg/hour plasticizing capacity. Assuming that the cycle could be adjusted so that 80% of the time is being used to recover, the extruder would need a rated Q of

$$\frac{120 \text{ kg/hour}}{0.80} = 150 \text{ kg/hour}.$$

If the extruder in the machine selected for this job has a rated capacity of only 120 kg/hour, it will not be possible to run this mold at its best possible cycle. By calculating 80% of 120 kg, only a practical supply of plastic of 96 kg/hour is obtained, and the molding cycle must be slowed down to 96 shots/hour rather than the planned 120 shots/hour.

25.7.1.2 P Machines

In P machines, it is possible to plasticize 100% of the time, since the thrust bearing of the extruder is loaded with a relatively low force (the back pressure required for proper plasticizing) and can continue to turn even while transferring plastic from the extruder to the shooting pot. This capability for full use of the extruder makes it possible to use a smaller extruder for a P machine than that required for an equivalent (clamp) size RS machine.

25.7.1.3 Selection of Extruder Size

In general, the plastic required by a mold should not be more than about 80% of the plasticizing capacity of the extruder. On the other hand, the plastic required by a mold should not be less than about 20% of the plasticizing capacity of the extruder. However, using low plasticizing capacities is not recommended because, while it is possible to run the extruder at such low throughput rates, the low efficiency of an electrical motor, idling much of the time at a low power factor, can result in penalties from the power companies or will require costly electrical apparatus to regulate the power factor.

25.7.2 Shot Size and Number of Cavities

In general, in the same machine, the larger the shot size, the slower the injection, since more plastic must flow through the same passages between the extruder and the mold. Also, as the shot volume increases, it takes longer for the extruder (RS) or the shooting pot (P) to push the plastic forward toward the machine nozzle.

A stack mold can effectively double the number of cavities possible to mold in any size of machine of suitable shut height, when considering just the clamp size. However, the demand for twice the plastic processed in the extruder and the need for twice the volume to be injected will require careful evaluation of the capacity of the injection unit to ensure the full advantage of using a stack mold in such a machine.

25.7.2.1 *Number of Cavities*

For cold runner systems, the more cavities there are in a mold, the longer will be the flow path of the plastic to these cavities, and the longer it will take to fill the cavities, since the runner system must also be filled. With small shot volumes per cavity, this difference will be less significant than with large volumes.

For hot runner systems, the runner system from machine nozzle to the gates are already filled (but not pressurized), and there is no appreciable difference in filling time per cavity, regardless of the number of cavities. However, the plastic entering the cavities must be replaced in the runner system, under pressure, from the machine nozzle. Also, the larger the hot runner system, the more volume of plastic must be compressed, which requires some additional volume of plastic, about 1–2% of the total volume in the runner system.

Example: Assume a mold has two cavities, each with a volume of 50 cm³. The volume coming into the hot runner from the machine nozzle will be

$$2 \times 50\ \text{cm}^3 = 100\ \text{cm}^3.$$

It will take a certain time t_1 to "push" 100 cm³ into the hot runner. To mold the same product in a six-cavity mold, the volume to replenish is

$$6 \times 50\ \text{cm}^3 = 300\ \text{cm}^3$$

in the hot runner. The time t_2 to push 300 cm³ into the hot runner will be longer than t_1 for 100 cm³. With small volumes, the difference between t_2 and t_1 may be small, but with larger volumes it could be significant and affect the expected cycle time.

25.7.2.2 *Pressure Drop*

Pressure drop in the runner system is greater as the runners increase in length, and with higher pressure drop, the injection speed will decrease (Fig. 25.5). The runner sizes could be adjusted (made larger) to compensate for the increased length, and thereby to reduce the pressure drop. However, this will increase the inventory in the hot runner system, slow down start-up, and may cause degradation in a heat-sensitive plastic if the plastic resides inside the system for more than one or two shots.

Minimum injection pressure p is a function of shear heating and shear stress. A minimum Δp will be found which balances these two factors.

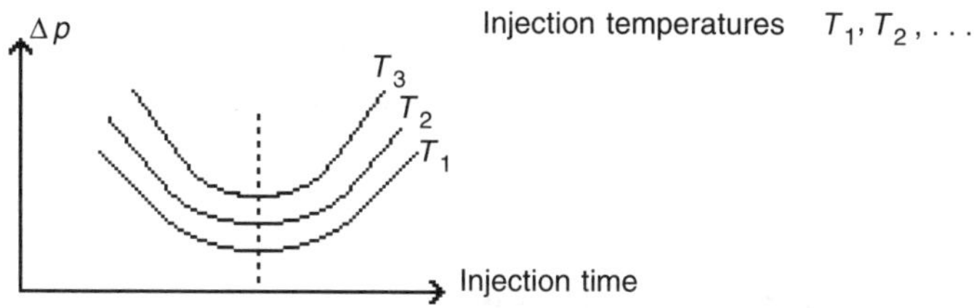

Figure 25.5 Optimum pressure drop as a function of injection time and temperature.

25.7.2.3 *Shot Capacity*

Shot capacity of the machine, whether RS or P, must be greater than the required shot volume of the mold. Similarly to *plasticizing capacity*, it is best if the *shot volume* required for a mold is between 20 and 90% of the shot capacity of the machine.

25.7.2.4 *Flow Molding*

In exceptional cases, where the shot volume required is greater than the shot capacity of the injection unit, the machine could be cycled so that the extruder feeds the plastic into the mold for a certain length of the molding cycle before the actual injection with high pressure is started. This method, called flow molding, is much slower than when the complete shot is ready in the (suitably larger) injection system before being injected.

25.8 Ejection Timing

Cycle time is also dependent on the timing of the start of ejection and the time required from the moment the products are pushed off the core (or out of the cavity) until they have cleared the molding area so that the mold can start re-closing. Several factors are important in determining the ejection timing.

Ejection should start as early as possible after the clamp starts opening, but not so soon that the products are pushed (or blown) into the other mold half and thereby fail to clear the mold.

If possible, ejection should be complete, that is, the products must have cleared the molding area, by the time the mold arrives in the mold open position. Air jets or air curtains to speed up the free fall motion of the products are often desirable, provided the products will not be damaged by high falling speeds.

With positive product removal systems, which are mechanically interlocked with the clamp motion (such as the HUSKY patented swing chutes), it is possible to use a portion of the mold opening and closing cycle for the mechanical take-off elements holding the products, to clear the molding area ahead of the closing mold. In all other cases, however, mold open time must be added (and the cycle lengthened) to ensure that the molding area is clear before the mold can re-close.

With product removal by robot, for safety reasons, the mold must be open before the robot is allowed to enter the molding area, and the mold must not close before the robot holding the products securely is safely out of the molding area.

25.9 Length of Clamp Stroke

As a rule, the longer the clamp stroke, the longer is the cycle time. A mold should be set up such that it uses the minimum clamp stroke possible, while still ensuring safe ejection of the products.

A problem occasionally encountered is that, while a minimum stroke would be possible for the operation of the mold, it would make it difficult for servicing the mold if easy access to the cavities and/or cores is required, for example, to remove partially molded pieces during start-up or for opening a frozen gate.

A shorter clamp stroke in combination with a positive take-off, such as a swing chute, can be faster than the operation of the same mold without take-off, where the mold needs a longer clamp stroke or even requires mold open time to ensure enough time for the products to fall free of the molding area.

Robots usually require longer clamp strokes to make room for the length of the end-of-arm tooling head, which must be added to the length required for the product itself.

25.10 Mold Cooling

The coolant must remove from the mold most of the heat energy added while the plastic passed through the extruder. The remaining heat dissipates into the plant atmosphere. Faster heat removal usually means faster cycles.

"Coolant" may be ordinary water, with certain additives to inhibit rust or chemical deposits, and/or additives to lower the freezing temperature of the coolant. Depending on these additives, the capability of the coolant to absorb heat will be less than with pure, clean water.

For many commodity plastics, cooling water supply is in the order of 4–6 °C, at pressures of about 4–5 kg/cm^2, similar to the line pressures of a city water supply. Colder water may require the addition of antifreeze.

Excessively cold water may cause condensation on the molding surfaces, especially if the mold stays open for more than a few seconds, and on the other mold surfaces. Water droplets condensed on the cold surfaces may cause surface defects on the products. In such cases, it is better to operate the mold with cooling water temperatures a few degrees above the dew point of the surrounding atmosphere. This condition is quite common in humid environments, especially in summer.

Coolant temperatures can be in the range from below freezing to near the boiling point of water, depending on the type of plastic being processed and the most suitable conditions for fast molding. Water at temperatures above 95 °C is not safe to use because of the possible danger if pressurized, overheated water should leak and flash into scalding steam.

Some "engineering type" plastics require molding at elevated mold temperatures, such as polycarbonate, SAN, and others. Refer to data sheets by the resin manufacturers to check this information.

Coolant can also be hot oil at temperatures above the boiling point of water. For example, such a coolant is used in injection blow molds, where temperatures of 120 °C may be required in the cores to keep the plastic from cooling before blowing takes place.

It should be well understood that "cooling" does not mean necessarily to cool the product down to room temperature, or close to it, but to keep the mold at a specified temperature, which should vary only minimally from cycle to cycle.

25.10.1 Cooling Efficiency

"Cooling efficiency" is a measure of the ability of the cooling medium to carry the heat away from the plastic. The cooling efficiency depends on many factors, as discussed in the following paragraphs.

25.10.1.1 Temperature Differential (ΔT) Between Plastic and Coolant

The greater the temperature differential ΔT between the hot plastic and the cold water, the faster will the heat be removed. Some plastics (ABS, PC, etc.) need to be molded at temperatures of about 50–80 °C. The operation of molds will, therefore, need heating units rather than chillers to supply a steady coolant circulation at this temperature. Since the temperature difference is not as large as for other materials, the cycle times will be longer.

Note: Cooling tower water is usually in the temperature range of 15 to 20 °C. In some hot countries, cooling towers will deliver water at much higher temperatures, up to 35–40 °C.

25.10.1.2 Distance Heat Must Travel from Hot Plastic to Coolant Channels

This heat travel distance is a function of the mold design. In good designs, the cooling channels are evenly distributed and close to the molding surfaces, and areas that cannot be directly cooled will be cooled by using indirectly cooled high heat conducting inserts, or heat pipes.

In some critical areas, even where direct cooling with channels is possible, mold materials such as beryllium-copper are used to speed up the heat transfer rate.

25.10.1.3 Heat Conductivity of Mold Materials

Usually, molds are made from hardened high carbon steels, which, amongst the steels, have a fairly poor heat conductivity. Mild, low carbon steels are better heat conductors but are rarely used because of their lack of strength and wear resistance.

Unfortunately, there is little that can be done about this poor heat conductivity. Selecting beryllium-copper is one solution for better cooling transfer, but the material is very expensive, not as strong as steel, and is hard to get in larger pieces without the risk of inclusions and voids, which often cannot be seen before much work has gone into the piece. Also, beryllium-copper requires special precautions in the tool room, such as exhausting the poisonous fumes created during machining.

Aluminum has excellent conductivity but is too soft for injection molds. It is mostly used for blow mold cavities, which do not see high pressures but do require good heat transfer. Areas subject to high wear in these molds are made from hardened steel inserts.

25.10.1.4 Dirt in Coolant, and Corrosion

Any dirt in the coolant will eventually deposit on the walls of the cooling channels in the mold; this creates a heat barrier that slows down the heat transfer from steel to the coolant and results in longer cycle times.

The same applies to the corrosive action of the cooling water on mold steel. Additives to the cooling water, and filtering, therefore, are highly recommended to keep these influences to a minimum.

25.10.1.5 Specific Heat of the Coolant

Specific heat is the quantity of heat required to raise the temperature of a unit mass of homogeneous material by one degree, in a specified way. Specific heat can also be expressed as the ratio of the quantity of heat required to raise the temperature of a unit mass of homogeneous material by one degree to the quantity of heat required to raise the temperature of an equal mass of a reference material, usually water, by one degree. The following chart (Table 25.1) shows values of specific heat for pure water at the temperature range encountered in molding.

Colder water has a slightly higher specific heat. Therefore, the same volume of cold water flowing through the mold will remove more heat than hotter water, all other conditions being equal. The importance of using a chiller is, therefore, not only the larger temperature differential between plastic and coolant but also the increase in the specific heat, which will make the cooling process somewhat more efficient. On the other hand, additives to the cooling water will reduce the specific heat values of the water and, therefore, reduce the cooling efficiency.

It is beyond the scope of this book to go more deeply into this field.

25.10.1.6 Flow Characteristic of Coolant Inside Channels

The concern is whether the flow is laminar or turbulent. For cooling purposes, laminar flow as in a slow-moving coolant, is not much more useful than stagnant coolant (in pockets,

Table 25.1 Specific Heat Values for Pure Water at Molding Temperatures

Water temperature		Specific heat	
°C	°F	kJ/kg-°C	BTU/lb-°F
0	32	4.225	1.009
10	50	4.195	1.002
21.1	70	4.179	0.998
37.8	100	4.174	0.977
54.4	130	4.179	0.998

Table 25.2 Viscosity and Density Values for Pure Water at Molding Temperatures

Temperature T (°C)	Density ρ (kg/m³)	Absolute viscosity μ (kg/m-s)	Kinematic viscosity ν (m²/s)
5	999.5	1.55×10^{-3}	1.5508×10^{-6}
10	999.2	1.31×10^{-3}	1.3110×10^{-6}
20	997.6	1.01×10^{-3}	1.0124×10^{-6}
40	992.0	6.60×10^{-4}	0.6653×10^{-6}

without any circulation), and does little or nothing to remove heat. The designer must always try to achieve turbulent flow.

Flow characteristics are expressed with a dimensionless coefficient called the "Reynolds number" (Re). As a rule, turbulent flow is present when the Re is greater than 4,000. The Re can be calculated as follows:

$$\mathrm{Re} = \frac{V \times D}{\nu}, \tag{25.1}$$

where V is velocity (m/sec), D is diameter (m), and ν is the kinematic viscosity (m²/sec), determined by the formula

$$\nu = \frac{\mu}{\rho}, \tag{25.2}$$

where μ is absolute viscosity (kg/m-sec) and ρ is density (kg/m³).

For values of ν for pure water at various temperatures, see Table 25.2. Note that the viscosity of water varies with temperature, as is shown for temperatures encountered in mold cooling.

Note that with "hotter" cooling water, the kinematic viscosity decreases and Re becomes larger, therefore increasing the cooling efficiency, all other conditions being equal. This could be significant in a condition where Re is at the border between linear and turbulent flow, at about Re = 4,000. However, since good mold cooling systems are designed to have flows with an Re well above 10,000 (10,000 < Re < 20,000), an increase of 10–15% in Re is practically insignificant.

The channels must be dimensioned so that the flow characteristic (Re) in all channels is about equal if uniform cooling is expected. Large channels with little volume moving through the channels create laminar flow conditions and provide less cooling than smaller channels where the coolant moves rapidly through the channel.

25.10.1.7 Volume of Coolant per Unit of Time

The total volume of coolant per minute passing through the mold is probably the most important factor affecting the cooling. The total amount of coolant required in a mold to cool a given mass of plastic in an assumed cycle time can be calculated as explained in Chapter 13, Mold Cooling. Here, only the various factors affecting the amount of coolant required by and available to a mold are listed.

The plant main supply lines (piping and valving) must be sufficiently large to be able to supply all molds in production with the required flow rate (volume per minute). The branch lines and valves between the main line and the mold in a machine must also be large enough to handle the flow without excessive restrictions. There is no point in having a mold with a well designed (and often expensive to achieve) cooling layout in a machine that has insufficient coolant volume supply.

The line pressure must be adequate to ensure that the required coolant will be "driven" through the mold. As a rule, the cooling line pressures are 4–5 kg/cm^2 (57–71 psi). If the line pressure is too low, or if there is not enough flow volume available from the supply system, the coolant will take the path of least resistance within the mold and flow only through some channels, bypassing the others. In those channels, there will be no flow and, therefore, no cooling. The mold will not be cooled properly, and will run slower than planned.

Some systems may require higher pressures, which could be provided by installing secondary pumps to provide the necessary flow speeds within the cooling channels. This is only practical, however, if the main lines can supply the required volume per minute. The flow volume is directly related to the pressure differential between the supply line pressure and the return line pressure, and to the resistance to flow within the supply lines, connecting hoses, and the cooling channels inside the mold.

The return line pressure at the machine is usually slightly above atmospheric pressure, to ensure that the coolant will return to the central water supply reservoir. It is good practice to make the size of the return line larger than the supply line, to reduce the return line pressure drop to a minimum.

Pressure drops occur as a result of changes in direction of flow in the cooling channels; they are usually calculated as an "equivalent flow length" for each turn or reversal. Such equivalent flow lengths have been established experimentally and are usually shown in tables or graphs in hydraulic engineering handbooks. Factors affecting the flow (pressure drop) within the mold are :

$$\Delta\rho = \rho f\left(\frac{L}{D}\right)\left(\frac{V^2}{2}\right), \tag{25.3}$$

$$V = \frac{Q}{A}, \tag{25.4}$$

$$\text{and } f = \frac{0.3164}{\text{Re}^{.25}}, \tag{25.5}$$

where A is the cross sectional area of the cooling channels (m^2), D is the channel diameter (m), L is the length of the cooling channels (m), r is the density of the coolant (kg/m^3), Re is the Reynolds number (10,000 < Re < 20,000), V is the velocity of the coolant (m/sec), Q is the volume of coolant (m^3/sec), and f is a friction factor.

Roughness and corrosion of the cooling wall also affect the pressure drop and are expressed in experimental figures, which are also shown in tables or graphs as a factor affecting the Reynolds number. Computer programs are available to make all these calculations, which are necessary to assist the mold designer in arriving at the best (most economical) cooling channel layouts.

26 Wall Thickness

The amount (mass) of plastic used constitutes a major portion of the cost of a product. The mass of plastic may represent as much as 90% of the total product cost in a fully automated process in which labor is mostly eliminated.

The cost of the equipment, such as mold, molding machine, robot, and other handling and postmolding operations, is usually insignificant when compared to the cost of the plastic per unit, provided the quantities are large. It is, therefore, most important to reduce the mass of the product as much as possible if major cost savings are to be achieved.

It is also important to understand that if there is a strength problem in the product, increasing the wall thickness does not necessarily increase the strength. A properly designed product, with even wall thicknesses and adequate ribbing, is usually stronger and stiffer than a product with thicker and/or uneven walls.

There are other advantages in "thin-walling" a plastic product. Savings as a result of faster molding cycles can be significant, which means not only faster production but also better use of the manufacturing equipment.

26.1 Injection Pressure

During injection, the hot plastic is pushed by the injection pressure and flows between the cooled walls of the cavity and the core to fill this "cavity space". In general, the filling speed is proportional to the injection pressure p after the gate (inside the cavity space); the higher the pressure, the higher is the filling speed.

Note that in a thin-walled product, the speed after the gate is most important to ensure that the cavity is filled before the plastic freezes in the narrow passages between the cooled mold walls.

26.1.1 Injection Pressure Behind the Machine Nozzle

The injection pressure behind the machine nozzle (at the tip of the extruder screw) can be fairly easily calculated for RS machines:

$$p = \frac{F}{\text{extruder screw} A}, \tag{26.1}$$

or for P machines:

$$p = \frac{F}{\text{injection piston } A}, \tag{26.2}$$

where F is the force of the hydraulic injection piston (kg, lb., or N), A is the cross sectional area (cm^2, in.^{2}, or m^2), and p is the injection pressure behind the nozzle (kg/cm^2, psi, or Pa).

If the area of the hydraulic injection piston of the injection unit is known, the injection pressure at the machine nozzle can be easily calculated as follows:

$$p = \frac{A_p}{A} \times \text{hydraulic injection pressure.} \qquad (26.3)$$

where A_p is the cross sectional area of the hydraulic injection piston and A is the cross sectional area of either the piston or the extruder screw. The hydraulic injection pressure can be read on the pressure gage of the injection unit.

26.1.2 Actual Pressure Inside the Cavity Space

The actual pressure after the gate, inside the cavity space, can only be estimated or measured by placing pressure transducers inside the cavity space or by using flow modeling techniques. The actual pressure depends on the following factors:

- Resistance in the (cold or hot) runners and in the gates, which are both related to the mold design;
- temperature of the plastic, which is related to the setup or operating conditions of the machine (temperatures of the plastic and the coolant);
- viscosity of the plastic, or the molecular weight, since the viscosity is a function of temperature and shear rate;
- distance of the plastic from the gates, within the cavity space. Friction losses slow down the plastic as it flows through the narrow spaces between the cavity and core walls; and
- the flow rate of the resin, that is, the injected mass of the product per time required for injection.

Forces opposing the flow (and the filling speed) in the cavity space are resistance caused by friction on the mold surface, the effect of cooling, and the wall thickness of the product. Resistance from friction on the mold surface can be especially important in very thin-walled products, such as disposable drinking cups. A very high polish alone may not be sufficient, and flash chrome plating or other special surface treatments may be required to make the surface more slippery and to facilitate the filling of the cavity space. For example, with disposable PS drinking cups, flash chrome plating of the surfaces after polishing has resulted in two to three more shots per minute than was possible before plating.

On the other hand, a certain roughness of the surface may be necessary to facilitate ejection of the product, for example, when molding PE. As happens often in the molding business, what is good for one material may not be suitable for another.

The effect of cooling will also affect flow within the cavity space. As the plastic flows away from the gate, it is in contact with the cold molding surfaces. This makes the plastic freeze and will, therefore, reduce the gap available for the hot plastic still coming from the gates, as shown in the schematic in Fig. 26.1. Note that near the gate, the plastic remains hot as the high velocity of the hot plastic wipes the walls, and no frozen layer accumulates. This is the reason why the gate area must be especially well cooled if fast cycles are to be achieved.

If the wall thickness t is relatively large, the effect of the build-up of solidified layer on the walls will not be as significant as with thin-walled products. Products with small wall

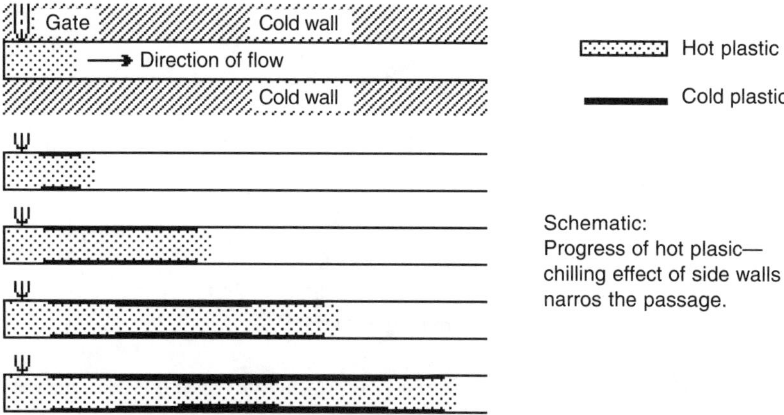

Figure 26.1 Schematic showing progress of hot plastic and the chilling effect of side walls, which narrows the passage.

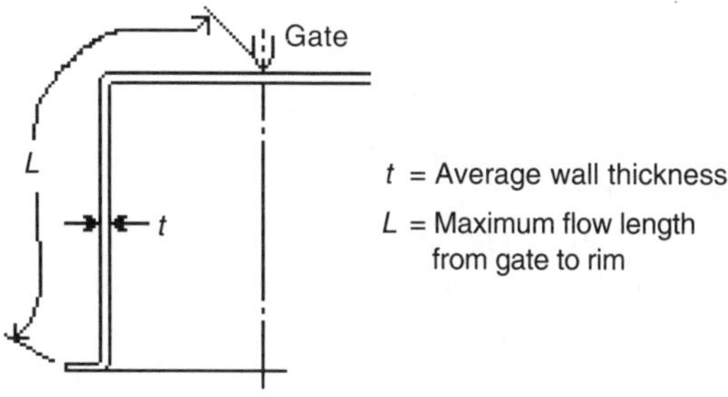

Figure 26.2 Flow length and wall thickness in a mold.

thickness *t* will require higher flow speeds for filling than those with large *t* for proper filling and shorter cycles.

26.2 Flow Length

An important criterion to consider when thinking about the moldability of a product is the L/t ratio; this is the ratio of the flow length L from the gate to the farthest point the plastic must flow to fill the cavity space, to the average wall thickness t of the product. This relationship is illustrated in Fig. 26.2.

From experience, the L/t ratio can be classified into four groups:

1. $L/t < 100$ Heavy-walled products, easy to mold.
2. $100 < L/t < 200$ Most products fall into this category, relatively easy to mold.
3. $200 < L/t < 300$ Difficult to mold, needs special considerations.
4. $300 < L/t$ Very difficult to mold, may need special equipment.

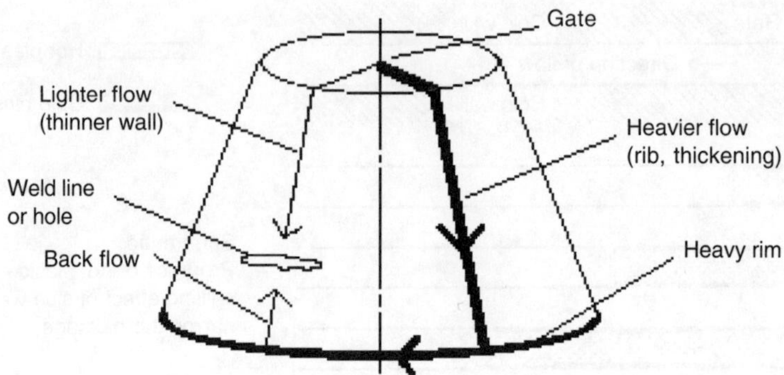

Figure 26.3 Schematic of a pail shows backflow from heavy rim toward lighter flow, creating weld or hole.

The influence of the L/t ratio is not linear. For example, for a 20-liter pail with a wall of 1.8 mm, a ratio of 250 may be quite easy to fill, while a similar ratio in a ½-liter container with a wall of 0.5 mm could be very difficult to fill.

The L/t ratio achievable is also dependent on the type of resin injected. A high viscosity resin such as polysulfone, PC, acrylic, PET, etc., has a higher resistance to flow because of its chemical nature (cross linking, higher molecular weight) and, thus, requires more pressure to fill the cavity space with sufficient speed.

To facilitate filling, it is preferable that the wall thickness near the gate is larger and gradually diminishes toward the rim. It is always more difficult to fill a heavier section of the product from a thinner section. Unfortunately, inexperienced product designers often ask for thick rims in an otherwise thin-walled product. This causes filling and packing problems, and such visible defects as voids and sinks in the rim; it may then require different molding parameters, using higher melt and/or mold temperatures, which all contribute to longer cycle times. It may also require higher clamping forces and a change to a larger machine than would otherwise be necessary.

A heavy rim may also cause *back flow* , in case the rim is filled faster from a heavier wall section (or a rib) before the rest of the product wall is filled. In this case, some of the plastic will flow from the rim back toward the plastic that approaches (slower) through the thinner wall sections and trap the air between the two approaching plastic streams (Fig. 26.3). This can result in visible weld lines, burn marks, and/or holes in the wall, and produce rejects.

26.3 Clamp Force

The thinner the walls, the more clamp force is required to hold the mold shut during injection. This is because of the high pressure required to fill the cavity space rapidly, before the plastic freezes. (An exception to this is during starve feeding.)

The theory is quite simple: To hold the mold closed during injection, the force required can be calculated by multiplying the total projected area of the product(s) by the *actual* injection pressure inside the cavity.

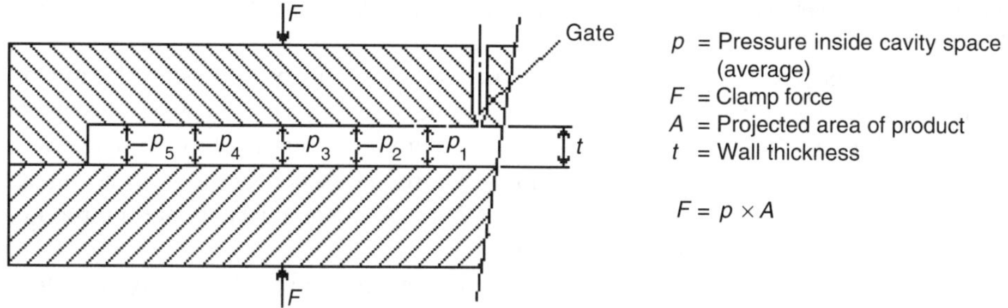

Figure 26.4 Actual pressure inside a cavity space decreases with increased distance from the gate. The average of these pressures is used to calculate the clamp force.

The total projected area is easy to calculate; refer to Chapter 22, Rules and Calculations for Designers, Section 22.1, Projected Area A_p. Note that with 2-plate molds, the area of the runners must be included.

The actual injection pressure is very difficult to establish, except by actual measurements in the cavity space. The pressure in the cold runners is usually higher, and with large runners may approach the injection pressure at the machine nozzle, which can be calculated as shown above (Section 26.1.1).

The pressures inside the cavity get lower as they are farther away from the gate, or, as shown in Fig. 26.4, $p_5 < p_4 < p_3 < p_2 < p_1$. For calculation of the required clamp force, the designer must work with an assumed, average pressure *p*.

The average cavity pressure *p* to be used for calculating has been established through experience. See Sectopm 29.5.1 for approximate values of *p* in tons per square inch.

Note that thin-walled products may need even higher pressures for proper filling. Also, if there is fine resolution required on the molding surfaces, as for example with audio or video grooves in molded record disks, the higher figures should be selected to make sure that the clamp will be sufficiently strong to hold the mold shut without flashing.

There are no hard rules about the pressure inside the cavities, and the only way to determine the actual pressures in the cavities is to measure them by using pressure transducers in various locations. A computer program covering all the possible variables affecting the pressure could also be used.

26.4 Melt Index

Before looking at the relationship between wall thickness and melt index, consider the term "melt index" (MI). The relationship between molecular weight (M_w) and MI can be expressed in an empirical formula as follows:

$$\log \mu = A + B \times M_w^{0.5} \tag{26.4}$$

where μ is the viscosity, M_w is the molecular weight, and *A* and *B* are constants depending on the nature and the temperature of the resin. This discussion will not go into more details but shows only that there is a definite relationship, which can be calculated. For more details, see Rubin [1].

In general, the physical properties of a polymer *increase* with the increasing M_w However, above a certain M_w, specific for each polymer, the rate of improvement in physical properties rapidly diminishes. On the other hand, the processability (viscosity) of the resin *decreases* exponentially with increasing M_w, and a point will be reached where the polymer becomes so viscous that it cannot be processed without degradation. A commercial polymer compromises between physical properties and processability.

Because the M_w of a resin is rather difficult and slow to determine, it is convenient to use the fairly easily determined MI as a measure for the M_w. The MI indicates the viscosity of a polymer at low (Newtonian) shear rates, which is in effect an indication (or measure) of the M_w.

The MI is measured with an extrusion rheometer, as shown in the schematic in Fig. 26.5, which depicts one typical setup of this test. After the temperature of the rheometer is stabilized at the specified value, the weight is applied to the plunger. The MI number is the number of grams of weight extruded in ten minutes.

MI is commonly used for polyolefins (PP and PE), and occasionally for PS, ABS, acrylics, nylon, and acetal. An MI number for one type of plastic, tested under one set of specifications, cannot be used to evaluate the moldability of another type of plastic with a similar MI number but tested under different specifications.

When using the MI as characteristic, one must always compare the MI numbers for similar type resins. In the USA, the American Society for Testing Materials (ASTM) specifies the values (temperatures, weights, orifice, and land sizes) to be used for various resins.

Designers are usually interested in the ease with which plastics can be injected and fill the cavities (moldability). Table 26.1 shows approximately the degrees of moldability for three different commodity plastics.

The following highlights explain relationships between MI and other characteristics:

- *M_w*: High M_w resins entangle more than lower molecular weight resins. Therefore, high M_w resins are more resistant to flow; in a test rheometer, they extrude less material (mass) under equal conditions, that is, they have a lower MI.

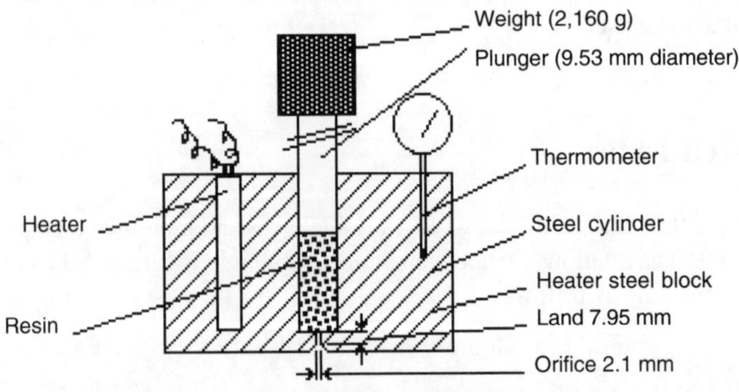

Figure 26.5 Extrusion rheometer used to measure melt index (MI) of a polymer resin.

Table 26.1 Comparison of Moldability of Three Resins

MI number :	High range	Medium range	Low range
Moldability :	Easy	Medium	Difficult
PE	< 50	25	< 7
PP	< 30	15	< 5
PS	< 5	2.5	< 1.0

- *Density*: As the M_w increases (MI decreases), the resins become less dense. It can be shown, for example, that for PPs of similar crystallinity, under identical molding conditions, that as the M_w increases (and MI decreases), the product will yield lower crystallinity. This lower crystallinity explains some of the property changes caused by the different molecular weights.
- *Notched izod impact strength*: As the M_w increases (MI decreases), the impact strength increases. This is probably a result of the lesser crystalline structure and the increased strength of molecular entanglements.
- *Tensile (yield) strength*: As the M_w increases (MI decreases), the yield strength increases. This increase reflects the longer molecules and the increased degree of entanglements.
- Stiffness: As the M_w increases (MI decreases), the stiffness decreases. This is probably the net result of the effects of crystallinity and chain lengths.
- *Percentage elongation to failure*: As the M_w increases (MI decreases), the percentage elongation to failure increases sharply. The longer molecules are more entwining and more resistant to stretch.
- *Brittleness temperatures and impact strength*: As the M_w increases (MI decreases), the brittleness temperature decreases. The impact strength increases as the M_w increases (MI decreases).
- *Environmental stress cracking*: As the M_w increases (MI decreases), the resistance to environmental stress cracking increases. Probably, the greater entanglement of higher M_w molecules leaves less room for the "invading" molecules.
- *Shrinkage and warpage*: As the M_w decreases (MI increases), fewer internal stresses are induced during molding, so that less shrinkage and warpage will be caused.
- *Creep*: Creep is the permanent deformation resulting from a prolonged application of stress below the elastic limit. (At room temperature, creep is also called "cold flow".) As the M_w increases (MI decreases), creep decreases.

The effect of M_w on polymer properties can also be shown in a simple chart (Table 26.2). As the M_w increases (MI decreases), properties change as shown in the table.

If a cavity cannot be filled with a low MI material, there are several options available:

1. Increase the melt and/or mold temperature to reduce the viscosity of the plastic. This may make it possible to fill the cavity but will lengthen the molding cycle significantly. Also, it may adversely affect the physical properties of the cooled resin (and the molded product).

Table 26.2 Effects of M_w Increase on Polymer Properties

Property	Increase	Decrease
Crystallinity		X
Density		X
Notched izod impact strength	X	
Tensile strength	X	
Stiffness		X
Percentage elongation to failure	X	
Brittleness temperature		X
Resistance to environmental stress cracking	X	
Shrinkage and warpage	X	
Creep		X

2. Change to a material with a higher MI. It will be possible to mold the product, but the physical properties will be lower. This may be unacceptable. Also, the cost of high MI material is higher than that of material with low MI.
3. Increase the wall thickness (reduce the *L/t* ratio). This will make it possible to fill, but will increase the weight and the cooling time. For both of these reasons, the cost of the product increases.
4. Decrease the pressure drop in the hot runner system by increasing the melt channel sizes. However, this will also increase the residence time of the resin in the hot runner components and may cause increased degradation. Also, a larger volume of resin in the hot runner system will increase the start-up time from cold, and will increase the time required for color changes. Another disadvantage of increasing the channels is the reduction of effective wiping action as the plastic flows near the walls of the channels; this also results in longer time required for color changes.
5. Select a more powerful injection unit that can push the stiffer material into the cavity and thereby insure proper filling.

The latter choice is the preferred option, since it allows one to use the stiffest, lowest cost material, at the lowest possible stock temperature and shortest cycle. This will result in the product with the lowest cost.

26.5 Shrinkage

Shrinkage depends on two factors: heat expansion and compression during injection (See Chapter 8, Shrinkage). The effects of these two influences are described below:

1. Heat expansion: Hot plastic has a larger volume than cold plastic. The amount of expansion is directly proportional to the temperature of the plastic. The amount of heat expansion (percent of linear expansion) is different for various plastics but can also vary for the same type of plastic made by different suppliers.
2. Compression during injection: Plastic under pressure has a smaller volume than when not compressed. The amount of reduction of volume is directly proportional to the pressure.

The following relationships between heat and pressure, and shrinkage, apply:

a. Injecting hot, at low pressure, will result in much shrinkage.
b. Injecting hot, at high pressure, will result in less shrinkage.
c. Injecting cooler, at high pressure, will result in the least shrinkage.

26.5.1 Wall Thickness and Shrinkage

In general, wall thickness of a product should decrease as the distance from the gate increases. However, this is rarely possible to achieve, except in some containers.

- Thick walls will fill easily, with little pressure, but will shrink more. To make sure that the product is well filled, without sink marks or voids, the pressure must be kept on the plastic after injection, using hold pressure, to make up for the volume that has shrunk as the plastic cools down within the cavity space. This requires that the gates are large enough and not yet frozen so that the additional plastic can enter the cavity space.
- Average walls: This is a “grey area”. Such walls behave somewhere between thin- and thick-walled products and will usually require hold pressure.
- Thin walls require much higher pressure to fill the cavity space at high speed and will, therefore, not shrink as much as heavy walls. Since thin walls will cool rapidly, and the gate also freezes rapidly, there is often no point in using injection hold pressure, except to prevent sinks in and around the gate area, which remains hot longer after the initial injection.
- Thin walls with heavy bosses and/or other local, heavy cross sections: Usually poorly cooled, heavy bosses or other heavy sections will shrink more because the hold pressure (if used at all) will be ineffective after the thin walls freeze and block the flow to these heavy sections. This can be often seen by the shrink marks on surfaces behind bosses and ribs. To remedy, heavy bosses, and heavy sections in general, should be cored out.
- Thin walls and heavy rims. The pressure of the plastic diminishes as the distance from the gate increases. Therefore, there will be less pressure to fill the rim and to maintain the pressure on the rim (if hold pressure is used). This results in the rim shrinking more (in percent) than the areas of the product leading to the rim. A cored out rim will shrink less.

Nonuniform shrinkage will cause the product to warp during cooling after ejection, particularly with rectangular products, such as trays, or with technical products that have uneven shapes and wall thicknesses. The cooling cycle may have to be lengthened (more cost) to ensure that the products will cool enough so that they will not warp after ejection.

In some cases, such as snap-on lids and over-caps, shrinkage in the rim may be desired to create a “toe-in” effect, which helps the holding action on the matching container. By deliberately increasing the thickness of the rim, the amount of toe-in can be increased. Note that the toe-in effect can also be increased by higher stock temperature and/or faster cycles, and decreased by lower stock temperature and/or longer cooling cycles.

The ideal condition is to provide the right steel dimensions so that the lowest stock temperature can be used with the shortest cycle time. Usually, this can be determined only by experimenting with different setups using one cavity of a multicavity mold before completing all the cavities of the mold. From the recorded shrinkages, temperatures, and times, the best dimensions of the cavity and the core can then be calculated. This may be time consuming and, often, the experimental cavity and/or core must be discarded after these experiments, but it will eventually result in obtaining the best performance of the mold.

A single-cavity experimental mold may not give the same results as when experimenting with one cavity in an otherwise completed multicavity mold; this is because the flow length in the runners and the cooling layout in an experimental mold is usually different; also, the machine used for testing it is probably much smaller than the planned production machine.

Similar considerations apply to steel sizes supplied from an earlier mold for the same product, which may have had fewer cavities, a different cavity layout, a different cooling arrangement, or may have been run in a machine with a different injection system. Such "supplied steel dimensions" are seldom the optimum steel sizes for a new mold for a different machine. On the other hand, for small dimensional differences of the product, the steel dimensions of a successful earlier mold can be extrapolated for a new mold, provided all other conditions are the same and the same mold layout is used.

26.6 Cooling Time

In Section 1.3.4.2. and 26.2, the difference between thin and heavy walls was defined by using the L/t ratio. Here in this section, the focus is on the effect of the actual thickness of the product. The following considerations are based on the assumption that the mold is well cooled.

26.6.1 Thin Walls

A thin wall will be arbitrarily defined as less than 1 mm thick ($t < 1$ mm). The heat is rapidly carried away on both sides of such a thin wall by the cold cavity and core walls. Since the wall is thin, the core of the hot plastic will also be rapidly cooled, and the insulating effect of the plastic has not much influence. Both cavity and core require the best possible cooling layout for efficient operation. Cycle time is short (usually less than 10 seconds).

26.6.2 Heavy Walls

A heavy wall will be arbitrarily defined as more than 2 mm thick ($t > 2$ mm). As long as the pressure is kept on the plastic, the heat will move toward the cavity and core walls and dissipate, but because of the insulating properties of the plastic, the heat of the plastic between the cooled boundary layers will take much longer to be absorbed by the cooling water flowing through the mold.

After the pressure on the plastic is gone (hold pressure off, frozen gate, or valve gate closed), no more plastic enters the cavity space, and as it cools, the plastic begins to shrink. This shrinkage has the effect that the plastic loses contact with the surrounding cavity wall, while shrinking on to the core, thus maintaining good contact with the core touching the inside the product. This reduces the cooling efficiency of the cavity wall, and less heat will be absorbed by the cavity than by the core cooling system. Cycle time is long (often more than 20 seconds).

For thick-walled products, core cooling is usually more important than cavity cooling. Fortunately, cores are usually easier to cool than the cavities, where the areas around the gate often make a good cooling layout difficult. For the same reason, inside center gating is particularly difficult if fast cycling is required, because it is often virtually impossible to provide efficient cooling to the inside of a core while also providing space for the (cold or hot) runner leading to the gate.

Note that in molds for perfectly flat and symmetrical products (e.g., audio records), both cavity and core sides must be cooled equally well. Uneven cooling will also produce warping when the products are ejected, unless they are kept very long in the mold to cool completely.

26.7 Core Shift

Core shift can be defined as the spatial deviation of the position of the core during molding from its original position in the mold before the plastic was injected into the cavity space. Core shift is experienced most often in molds for thin-walled containers, although it may also occur in molds for other shapes. It is a frequent problem with long, slender, and not necessarily thin-walled products, such as vials, test tubes, pen barrels, etc.

During injection, the plastic will always take the path of least resistance. As the plastic enters the cavity space, it will find less resistance in any wider space and flow in that direction before it starts flowing into any narrower spaces. This will rapidly cause a build-up of more pressure on the side of the heavier wall than on the opposing, narrower side. A pressure differential is thereby created which will tend to deflect (bend) the core toward the thinner side, and make any original difference in the wall thickness even greater than it was before injection.

Core shift can be caused by several factors. Only those factors affecting core shift in container molding will be considered here.

26.7.1 Inaccuracies in Machining

If the cavity and core are not machined “perfectly” concentric with each other, and if they are not properly aligned when the mold arrives in the clamped-up position, there will be an offset position at the start, with the resulting different wall thicknesses on opposite sides of the product.

26.7.2 Multiple Gates

Multiple gates into the bottom of a container used to improve the filling speed, as for large pails, will have to be “perfectly” symmetrical. The gate sizes must be “identical” to avoid preferential filling from one gate.

For multiple gates near the open end of a deep container or any long, slender product, it is important that at least two gates are used, located at 180° opposite each other, to direct the flow from both sides of the core simultaneously so as to create pressure on the core from opposing sides; this will practically eliminate core shift or deflection of the core. This method has long been used and is still used in cold runner molds for such products as pen barrels, etc. Occasionally, three gates, at 120° around the open end, or a diaphragm gate over the whole open end, is used.

Multiple hot runner gating near the open end of a large, thin-wall container has been developed by Husky (patented).

There are also mechanical means used to prevent core deflection by supporting the core at the closed end of vials, etc., or by having the core enter into the cavity in the case of tubings, syringes, etc.

26.7.3 Deflection

Deflection may result and depends on the length of the core in relation to its base diameter. The core can be considered a cantilevered beam. Under perfect conditions, if there is no pressure differential, there are no forces to deflect the beam. However, because of manufacturing tolerances, there will always be an inaccuracy in the alignment between cavity and core, and, therefore, every core will see some side load caused by the pressure differential created by the incoming plastic stream.

The greater the ratio of unsupported length (L) over base diameter (D), the more the core will tend to deflect. It usually becomes critical for cores with $L > D$. It should be noted that the forces F are greater if the product is cylindrical than if it is conical, since the total projected area at right angles to the centerline is greater (Fig. 26.6).

26.7.4 Core Support

The core support in the core base at diameter D may affect core shift. If the core is part of a much wider core base, the support is as stiff as it can be made. However, if there is a stripper ring, the core is supported solidly only much farther down as L increases by the thickness of the stripper ring (Fig. 26.7).

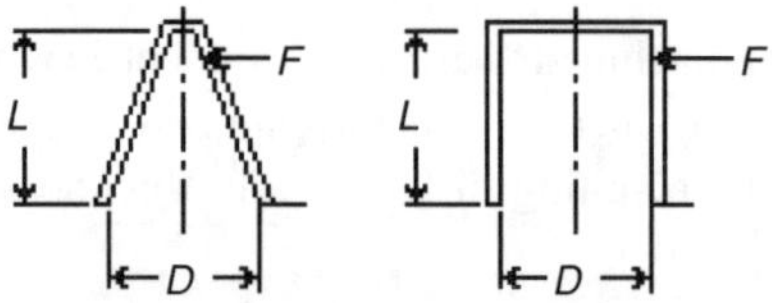

Figure 26.6 Deflection forces are greater for cylindrical products than for conical products.

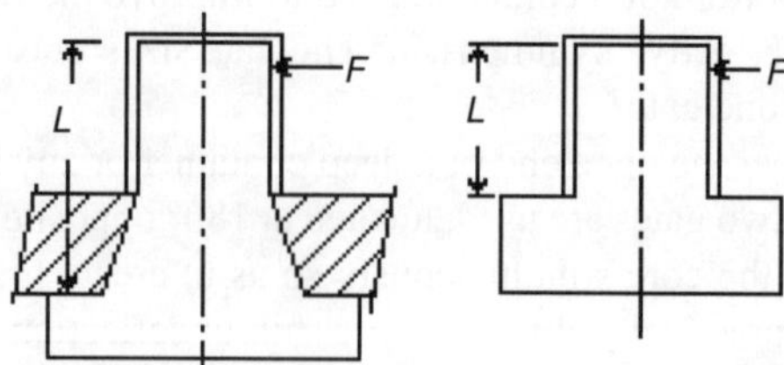

Figure 26.7 Varying lengths L between A. core with stripper ring and B. core with no stripper required show greater deflection when stripper ring increases L.

A well-fitting stripper ring with proper preload will help to support the core but will add new variables as a result of the additional tolerances of stripper ring taper and tapered seat on the core. Note that any taper fit has value for alignment or support only if it is preloaded—that is, under tension when seated firmly on the opposing taper. In multicavity molds, the preload per taper will be quite small, but it must be specified if the mold is to function properly.

Figure 26.7 also shows that for air ejection (provided the plastic permits air ejection without the need for strippers), the core length is reduced considerably. The *L*/*D* ratio is thereby reduced, the core is much stiffer, and there is much less core deflection and core shift.

26.7.5 Quality of Alignment Features

Dependence on only leader pins will provide very poor alignment between cavity and core, and is not practical for thin walled products. Any mold for thin-walled products must depend on tapered fits between cavities and cores to provide the required concentricity between them. However, because there are at least two pairs of tapers, and with the possible need for stripper rings, three or more pairs of matching tapers per cavity stack must be manufactured to ensure the alignment of just one cavity with one core, it is practically impossible to achieve 100% perfect alignment all the time.

Usually, the product designer will allow a certain amount of variation in wall thickness, which can often be achieved by carefully tolerancing and machining the tapers required for alignment. From experiments and from experience with many thin-walled molds of all sizes and shapes, it was found that, often, a deviation of only 0.0025 mm before pressure being applied results in a deviation in wall thickness of 0.025 mm, or a ratio of about 1:10. In some critical applications, it is then necessary to mechanically shift the core by about one-tenth of the difference of the measured wall thickness deviation after the first samples are produced. Needless to say, this is a very costly alteration to the mold, particularly if there is more than one cavity involved.

26.7.6 Wedge Lock

The “wedge lock” method of mold alignment makes it relatively easy to shift the core in relation to the cavity and to achieve such small changes. Even though the mold cost becomes higher than with circular tapers, it is the preferred method of molding such items as large pails.

26.7.7 Other Factors Affecting Core Shift

Many other factors in the mold and machine design may affect core shift:

1. Platen deflection of the machine, and/or weak mold plates: If the platens or plates are not strong enough, they will deflect and cause the centerline of the cavity and core to move out of parallelism and alignment.
2. *L*/*t* ratio: The greater the *L*/*t* ratio, the higher the injection pressures required; this can affect the core shift. The effect of the core shift is more significant with thin-walled products.
3. Melt index (MI): The lower the MI, the higher the injection pressures required to fill the mold, and the greater the effect on core shift.

References

1. Rubin, I. I. (1972). *Injection Technology, Theory and Practice,* John Wiley and Sons, New York.

27 Product Size and Shape

The following sections provide information about factors affecting product size and shape which holds true for all plastics but is of particular importance for products made from materials with high shrinkage factors, such as PE, PP, and some technical plastics.

27.1 Shrinkage Allowance

It is practically impossible to predict any shrinkage allowance. Shrinkage depends on many factors, such as those listed below:

- Material specifications: Even though identical specifications may appear on data sheets for the "identical" plastic made by various manufacturers, the shrinkage can vary from different sources, and even from the same source, it can vary from batch to batch.
- The adding of reused plastic to virgin material will affect shrinkage. The higher the percentage of regrinds, the greater the uncertainty of the effect.
- The addition of fillers (glass, talcum, colorant, etc.) affects the shrinkage. Usually, the higher the percentage of fillers, the smaller is the shrinkage.
- Product temperature at ejection: The higher the temperature of the product leaving the mold, the more it will shrink after ejection.
- Injection pressure and hold pressure: The higher these pressures, the less shrinkage will take place, since they affect the pressure in the cavity space.
- Wall thickness and resistance to flow in the cavity space: The heavier the wall, the more shrinkage is to be expected.
- Product shape: Frequent changes in the direction of flow in the cavity, especially at sharp corners, resist the flow, reduce the pressure, and thereby increase the shrinkage. This can be significant if heavy sections in a product are filled with plastic through restrictions such as very thin walls.

Many of the above conditions are often beyond the control of the mold designer or the mold maker, and are related to the product design or governed by economic requirements, such as the desire to use cheaper plastic and/or to use a relatively high percentage of reground material. However, the mold designer should be aware of any possible future plans the molder may have to use the same mold with other raw materials, which may negatively affect the expected productivity of a mold and the accuracy of the molded products. The mold designer should cooperate with the product designer and the molder and discuss the reasons behind any request for changing or eliminating any design features that could affect the desired size, and possibly the shape, of the products.

27.2 Molding Material

In Section 27.1, the influence of the quality of molding material on shrinkage was explained. The other effect related to the molding material is the ease of flow (or pressure required) to fill the cavity space:

- The harder the material flows, the more pressure will be required to fill the cavity space; the more pressure is on the plastic, the less it will shrink.
- Conversely, the easier the plastic flows, the less pressure will be required, and the more the plastic will shrink.

27.3 Operating Conditions

Once the best compromise has been reached with regard to shape and material, it is up to the molder to follow the results of the mold test run and the operating guidelines to achieve equally good performance comprising dimensional accuracy of the products and high productivity. It may still be necessary to readjust the molding conditions (temperatures, pressures, and times) to ensure the accuracy of the products when changing to molding materials different from the test run, or to compensate for a different plant environment, coolant supply, etc., which could affect the mold's performance.

By changing molding conditions, it is often possible to improve the fit of matching products, such as lids and tubs. However, such changes can slow down the productivity, and it is usually better to change the mold dimensions at a one-time cost, rather than live with a less productive setup. The economics of such a change can be easily calculated by comparing the additional machine time cost (as a result of the longer cycle) with the cost of the alteration required to achieve the faster molding cycle.

27.4 Accuracy of Mold

The accuracy required, or the tolerances specified for most plastic components, are usually not very demanding. However, as was listed above, there are many influences affecting the size of the molded product.

In practice, the mold designer must carefully scrutinize every dimension shown on a product drawing to make sure that it is really possible to mold the product to the specified tolerances. This must be done before starting to design a mold, and better yet, before accepting an order for a mold.

It is well within the capability of the average mold maker to machine steel or metal parts to very close tolerances. (In some cases, special machine tools or machining methods may be required.) Also, for very close tolerances, the temperatures of the work piece on the machine tool could affect the steel sizes, especially on larger parts where heat expansion can be significant.

The real difficulties start when wall thicknesses of the product are very small (< 1 mm), and the mold parts which define these walls need to be worked to very close tolerances to ensure—especially in multicavity molds—that the wall thicknesses from cavity to cavity are virtually the same. It is important to understand that even though the mold parts (steel) can be made very accurately, such accuracy does not guarantee that the product will be as expected.

The mold cost increases exponentially as the tolerances become smaller, requiring higher qualified machining personnel, better and more accurate machine tools, and more attention to mold operation during the manufacturing. The required parts inspection must be more thorough, and the rate of rejection of mold parts that are outside the close, permissible dimensional limits will increase.

27.5 Mold Cooling

Poorly cooled areas of the mold will leave the plastic hotter than well-cooled areas. After ejection (while the plastic is still warm), the hot areas will continue to shrink; such shrinkage will often distort the product. This is significant with high shrinkage materials, such as PP and PE, and will be especially damaging to flat surfaces in a product if it is surrounded by a rim, which cools slower than the flat surface. Typical examples include any flat tray or cover, whether round or rectangular.

Special attention is required by the mold designer to ensure that every area of the product can be cooled as good as possible, and that no areas are left uncooled. Most of the difficulties in providing adequate cooling in a mold are found in corners, in very thin core sections, and in small and/or delicate pins and complicated inserts that are necessary because of the product design. Even though such pins and inserts are usually poorly cooled, for good mold life they are often preferred instead of machining slender projections from the solid; this ensures that they will be strong enough but also that they can be easily replaced when worn or damaged.

28 Crystallinity

As has been stated earlier, plastics are divided into amorphous and crystalline materials. It is not practical to list all known plastics; the following lists show only some of the more common plastics processed by injection molding.

A few typical amorphous resins are:

- Polystyrene (PS),
- acrylics, and
- polyvinyl chloride (PVC).

A few typical crystalline resins are:

- Acetal,
- polyethylene (PE),
- nylon (polyamide),
- polypropylene (PP),
- polyethylene terephthalate (PET), and
- polysulfone

Note that for reasons of more efficient manufacturing of the resins, all crystalline plastics contain a mixture of crystallinity in a wide range (distribution) of molecular weights.

Some plastics are a mixture of amorphous and crystalline plastics, developed to enhance the required physical and chemical properties. A typical example of such a plastic is acrylonitrile-butadiene-styrene (ABS). Note that, above the melting point (melt temperature) all plastics are amorphous.

28.1 Properties of Injection Molding Grade Plastics

Table 28.1 shows some limited, approximate, but still interesting data for 39 selected, frequently used plastics. The column labelled "Amorphous/crystalline" describes the general structure of the resin. Although, strictly speaking, resins with crystalline structures should be referred to as "semi-crystalline" because they contain amorphous regions as well, the terms "crystalline" and "semi-crystalline" are both used to describe relative differences in crystallinity. For example, a resin labelled "semi-crystalline" would contain less crystalline structure than a "crystalline" resin. Note that for some resins, specific values of crystallinity are listed (in percent form). Because processing conditions can have a significant effect on the crystalline content of a resin, any specific value that is listed in Table 28.1 should be considered approximate, at best.

The column labelled "Shear rate or temp. sensitive" indicates whether the viscosity of a resin is sensitive to changes in shear rate or temperature. A resin that is shear- or temperature-

Table 28.1 Properties Chart for Injection Molding Grade Plastics

ASTM Acronym	Full chemical name	D792 Solid density (g/cm3)	Melt density (g/cm3)	D1003 Transp. (TP) Transl. (TL) Opaque (O)	Amorphous Crystalline	D648 Shrinkage (%)	Drying required (hours @ °C)	Melt temp. (°C)	Mold temp. (°C)	Injection speed	Shear rate or temp. sensitive	Resid. time sensitive	Max L/t ratio (1mm thick)	No flow temp. (°C)	Tonnage required (ton/in.2)	Typical inj. press. (MPa)
ASA	Acrylonitrile-styrene-acrylate	1.07		O, TP	Amorphous	.4 to .6		230 to 260	40 to 90							
ABS	Acrolonitrile-butadiene-styrene	1 to 1.2	0.9	O, TP	Amorphous	.5 to .6	2-3@88-77	195 to 240	38 to 93	slow, even	No	Yes	30-150:1	135 to 150	4 to 6	120-140
CA	Cellulose acetate	1.3		TP		.3 to .7	2-3@60-70	180 to 230	50 to 80	various		Yes		130 to 170		69 to 180
CAB	Cellulose acetate butyrate	1.2		TP		.3 to 7	2-3@60-70	180 to 230	50 to 80	various		Yes		130 to 170		69 to 180
CAP	Cellulose propionate	1.2		TP		.3 to .7	2-3@60-70	180 to 230	50 to 80	various		Yes		130 to 170		69 to 180
EVA	Ethylene-vinyl-acetate	.94 avg.	0.8	TL, TP				200 to 210	20 to 60					100		
HDPE	High density polyethylene	.93 to .97	0.81	O, TP	Crystalline	1.2 to 2.2	none	200 to 280	10 to 70	fast	No	No	250:1	120 to 130	2 to 2.5	100 to 200
HIPS	High impact polystyrene	1 to 1.1	0.9	O, TP	Amorphous	.4 to .7	none	180 to 280	10 to 85	fast	No	No	200-250:1	130	2 to 4	100 to 200
Ionomer	Ionomer	.94 to .95		TP	Semi-cryst.		none (8 @60)	210 to 260	5 to 50	slow, mod.	Yes	No		100	2 to 3	70 to 110
LCP	Liquid crystal polymer	1.5 to 1.7		TP, TL	Liq. crystal	.2 to .8	4-8@150	400 to 430	240 to 280	medium	Yes	No	200-300:1	370	2 to 3	80 to 120
LDPE	Low density polyethylene	.91 to .93		TP, TL	Crystalline	1.5 to 3	none	170 to 240	10 to 50	fast	No	No	275:1	100 to 110	1.5 to 2	100 to 150
LLDPE	Linear low density polyethylene	.90 to .92		TP, TL	Crystalline	1.5 to 5	none	170 to 200	10 to 50	fast	No	No	250-300:1	90	1.5 to 2	100 to 150
MDPE	Medium density polyethylene	.93 to .95		O, TP	Crystalline	1.2 to 2.2	none	190 to 260	10 to 70	fast	No	No	250:1	110	2 to 2.5	100 to 150
PA66	Polyamide (Nylon 6/6)	1.1 to 1.4f	1.0 to 1.2	TL	Crystalline	1 to 2.2	3-4@71-60	270 to 320	20 to 100	fast	Yes	Yes	140-340:1	240 to 260	4 to 5	100 to 150
PA6	Polyamide (Nylon 6)	1.1 to 1.4	1.2 to 1.3	TL	Crystalline	.8 to 2.1	3-4@71-60	260 to 310	20 to 100	fast	Yes	Yes	140-340:1	190 to 200	3 to 5	90 to 150
PAEK	Polyaryletherketone	1.3 to 1.5f		O	Semi-cryst.	.1 to .6	2-4@175-150	370 to 400	160 to 200	slow	No	No	170-200:1	370	3–4 (6f)	160 to 200
PAI	Polyamide-imide	1.4 to 1.6		O	Amorphous	.1 to .2	5@175	305 to 370	205 to 200	fast	Yes	Mild	140-340:1		5 to 6	160 to 200
PAR	Polyarylate	1.2	0.77	O	Amor./cryst	.6 to .9	3-4@175-150	260 to 380	65 to 150	moderate	No	Yes	30-100:1		3–5 (4-6f)	138 to 200
PAS	Polyaryl-sulfone	1.36 avg.		O	Amorphous	0.6	3-4@175-150	340 to 370	120 to 155	moderate	No	No	140-170:1			138 to 200
PBT	Polybutylene-terephthalate	1.3 to 1.6f	1.1 to 1.2	TL	Semi-cryst.	1.5 to 2(.5f)	3@150	240 to 270	50 to 100	mod., fast	Yes	Yes	160-200:1	220 to 250	3 to 5	80 to 120
PC	Polycarbonate	1.2 to 1.5f	1.1	TP	Amorphous	.4f to .7	3-4@120	270 to 325	80 to 110	fast	No	Yes	30-100:1	195	3–5 (6f)	138 to 200
PEI	Polyether-imide	1.3 to 1.5f		O	Amorphous	.5 to .7(.2f)	4-6@150-130	340 to 425	65 to 175	med. fast	No	Yes		230		100 to 160
PEK	Polyetherketone	1.3 to 1.5f		O	Semi-cryst.	.1 to .6	2-4@175-150	370 to 400	160 to 220	slow	No	No	170-200:1	370	3–4 (6f)	160 to 200
PEEK	Polyether-ether-ketone	1.3 to 1.4f		O	30% cryst.	.1 to 1.4	2-4@175-150	370 to 400	160 to 220	fast	No	No	200:1	370	2–4 (6f)	160 to 200
PES	Polyether-sulfones	1.2 to 1.6f		TP	Amorphous	.3f to .6	3-4@150-135	340 to 380	140 to 160	fast	Yes	Yes	60-120:1	300	5 to 10	160 to 200
PET	Polyether-terephthalate	1.4 to 1.7f	1.2	O, TP	Cryst./amor	.2 to 2	2-4@180-160	260 to 300	7 to 80	slow, even	Yes	Yes	80-200:1	240 to 250	2 to 6	70 to 160
PETG	PET (copolymer)	1.2 to 1.3		O, TP	Amorphous		4@66	190 to 275	20 to 30	slow–fast	Yes	Yes	80-200:1			80 to 100
PMMA	Polymethyl-methacrylate (acrylic)	1.1 to 1.2	1.0 to 1.1	O	Amorphous	.4 to .8	2-3@77-93	200-260	38 to 60	various	Yes	No	130-150:1	160 to 170	2.5 to 3	100 to 200
Polyester	Thermoplastic polyester	1.3		O	Amorphous	1.5 to 1.8	2-4@120-77	230 to 260	40 to 100	slow, even	Yes	Yes	80-200:1	70 to 80	3 to 5	80 to 100
POM	Polyoxymethylene (acetal)	1.4 to 1.6f	1.2	O	80% cryst.	.8f to 2	none	180 to 230	80 to 100	med-fast	Yes	Yes	100-200:1	160 to 170	3.5 to 5	100 to 170
PP	Polypropylene	.90 to .92	0.77	TP, TL	Semi-cryst.	1 to 2.5	none	230 to 275	15 to 65	fast	No	No	200-300:1	170 to 180	2 to 3	100 to 130
PPO	Polyphenylene-oxide	1.1 to 1.2f	.96 to 1.0	O	Amorphous	.2f to .7	2-4@115-107	250 to 315	82 to 110	fast	Yes	Yes		150 to 200	2.5 to 5f	120 to 180
PPS	Polyphenylene-sulfide	1.3 to 1.9f	1.5 avg.	O	65% cryst.	.1 to .5	3-4@150	300 to 360	40 to 150	slow	Yes	No	150:1	260 to 280	2 to 3	50 to 140
PS	Polystyrene	1.0 to 1.1	0.9	O, TP	Amorphous	.4 to .7	none	180 to 280	10 to 85	fast	No	No	200-250:1	130 to 160	2 to 4	100 to 200
PSU	Polysulfone	1.2 to 1.6f		O, TP	Amorphous	0.7	5@120	310 to 400	100 to 170	slow	No	No				
PVC	Polyvinyl-chloride	1.2 to 1.4	1.2 to 1.3	TP	Amorphous	.2 to .5	none (2@60)	180 to 204	20 to 40	slow-mod.	Yes	Yes	100:1	120		70 to 140
SAN	Styrene-acrylonitrile	1.1 to 1.3	0.9	O, TP	Amorphous	.3 to .7	2-3@80-70	220 to 270	5 to 60		Yes	Yes		130 to 170		35 to 140
TPUR	Thermoplastic polyurethane	1.2 to 1.3		O, TL	Amorphous	.8 to 2	2-3@110-104	190 to 220	30 to 65		Yes			120	.5 to 2	70 to 140
UHMWPE	Ultra high molecular weight PE	.93 to .94	0.9	O, TP	Crystalline	1.2 to 2.2	none	200 to 280	10 to 70	fast	No	No	150:1	120	3 to 4	130 to 200

Notes: f = filled property
MPa x 145 = psi
°F = (°C - 32)/1.8

sensitive will thin considerably at higher shear rates or temperatures. Although the true meaning of the terms "shear-sensitive" and "temperature-sensitive" refers to the relationships between shear rate and viscosity or temperature and viscosity, the meaning of these terms is not very well understood and is often assumed to describe a resin's thermal stability.

For complete, accurate data, always refer to an up-to-date data sheet issued by the manufacturer of a selected plastic.

28.2 Effect of Cooling on Crystalline Materials

In the molten stage, all plastics are amorphous. As they are cooled down (within the cavity space, after injection), they gradually crystallize back to their original structure. However, crystallization requires time for the crystals to grow. If the cooling is very efficient, the heat is removed so fast that the resin has not enough time to crystallize, and the material, even though it is "crystalline", will remain in an amorphous state after being cooled. In other areas of the mold not too close to the cooling, the plastic has enough time to crystallize before cooling and will again become crystalline in structure. Areas "in between" may result in smaller size crystals than areas farther removed from the immediate effect of cooling.

When filling a cavity, the resin closest to the mold walls will cool down rapidly and form an amorphous layer (Fig. 28.1). This layer is not oriented, since the plastic stopped flowing immediately after cooling down so rapidly.

Somewhat closer toward the center (between the walls), the plastic is now heavily oriented because of the fast flow during filling. However, this outside layer cooled down quickly and retained the heavy orientation. Only small crystals are formed in this layer.

More toward the center, the plastic also flows fast but is cooled less. After the mold is filled, some of the orientation is lost because of the still higher temperature. The crystals in this layer are larger.

The center layer is cooled last and maintains heat long enough to eliminate much of the orientation imparted by the filling of the cavity space. This center layer is characterized by the largest crystals, since it has the longest time to allow the resin to crystallize.

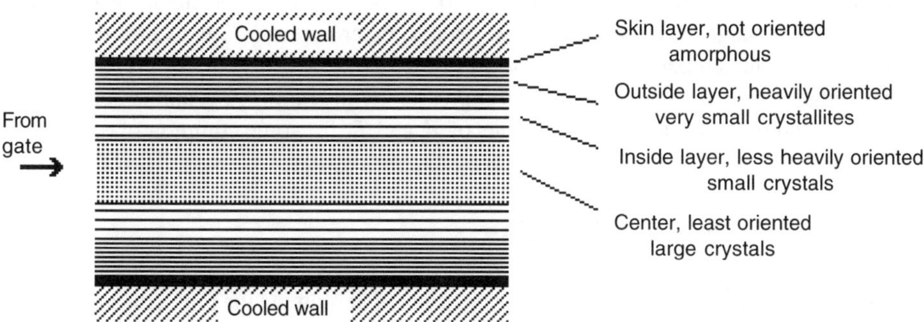

Figure 28.1 Variations in crystallization through cross section of product wall in a cooled mold.

Typically in large pails, it is the size of the crystals in the center layer which give the pail the top load strength. The thicker the pail, the larger the crystals, and the larger the strength to support the top load.

In other products where top load is not of such great importance, but where toughness and impact strength is required, the amorphous and the small crystalline layers are more important, and to increase these properties, little or nothing is gained by increasing the wall thickness. In such cases, uniform wall thickness with well-placed ribbing for reinforcement will produce a lighter, stronger product.

Actual photographs of the structure of the cross section of a wall of a PE pail have been made which prove these statements [1].

28.3 Practical Effect of Distribution Between Amorphous and Crystalline Layers

Amorphous structures (but also small crystallinity) improve impact strength. Crystalline structures (especially large crystals) improve compressive strength. Typically, in an industrial quality large pail, both these characteristics are required, and by adjusting the molding process (filling speed and cooling rate), the quality (crystallinity) of the final product can be adjusted to generate an acceptable compromise of both top load and impact strength.

28.4 Melting Point (T_m) and Glass Transition Point (T_g)

In crystalline plastics, the change from solid to liquid is abrupt and easily discernible. Above the melting point, crystals cannot exist. (For more details, the reader is referred to Rubin [2].)

In contrast, with amorphous (noncrystalline) plastics, the material softens over a wide range of temperatures, and there is no dramatic change in flow properties at any specific point. However, when plotting certain properties such as specific volume or heat capacity against temperature, there is a point of an abrupt change in the slope of the curve. This point is called the *glass transition point* (T_g).

Below the T_g, the polymer is stiff and behaves like a solid. It is relatively brittle, with little elasticity. Above the T_g, the plastic will behave like a viscous liquid. Additional heat will bring it to its melting range. A common method used to determine the T_g is to plot the specific volume against the temperature, as shown in Fig. 28.2.

The relationship between T_g and T_m is expressed by the ratio T_g/T_m, and is generally between 0.50 and 0.75. Table 28.2 lists values of T_g and T_m for some common plastics.

28.5 PET Preform Molding

PET is a crystalline material. However, for the process of blowing the preforms into finished bottles, the preforms must be amorphous. This is achieved by very intensive, rapid cooling of the plastic in the mold.

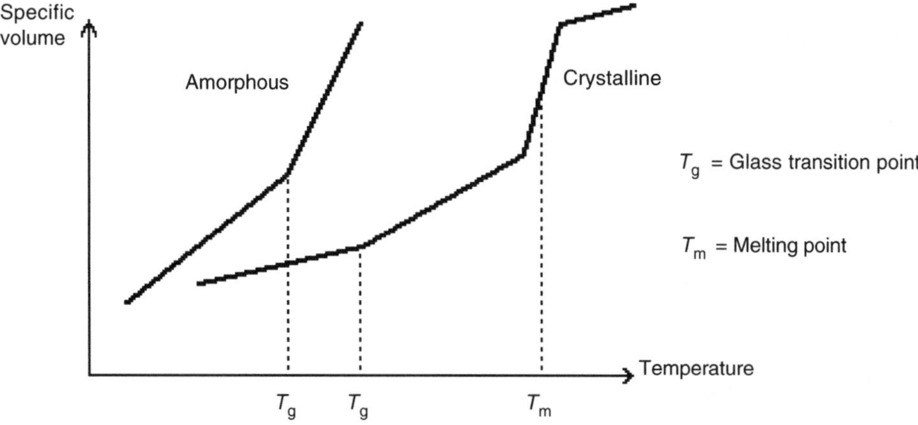

Figure 28.2 Specific volume vs. increasing temperature of amorphous and crystalline plastic showing glass transition and melting points.

Table 28.2 Glass Transition and Melting Points for Common Injection Molding Plastics

Material	Glass transition point (T_g) °C	Melting point (T_m) °C
Acetal	-85	190
Nylon 6	50	225
PC	152	225
PE	-125, -20	110–141
PP	-5, 45	172–176
PS (GP)[a]	81-100	235
PVC	70-80	200
PET	73	260

[a] GP = general purpose.

It also requires that the plastic be perfectly dry, to a degree not normally required for other resins. PET must be dried to less than 0.005% humidity before entering the extruder. This is achieved in about 3–4 hours, at a dew point of -40 °C (-40 °F) at a drying temperature of 182 °C (360 °F).

28.6 Drying of Plastic

Some (commodity) plastics such as PP, PE and PS do not require drying. Most other plastics need to be dried before processing to avoid bubbles and blisters in the melt which would impair the quality of the product. Approximate values for drying plastics are shown in Table 28.1.

References

1. Catoen, B. (1992). *Effects of injection time, melt temperature and gate configuration on the performance of industrial containers.* Research study, Mold Division, Husky Injection Molding Systems, Buffalo, NY.
2. Rubin, I. I. (1992). *Theory and Practice of Injection Molding*, John Wiley and Sons, New York.

29 Mold Forces

Before considering loads, preloads, and the effect of fatigue, one must define the basic forces that are common in every mold. (Forces within the molding machine are not discussed here.)

29.1 Clamping Forces

All clamp forces are cyclical, that is, they occur during every molding cycle, with values rising from zero to its maximum value and back to zero. Clamp forces act both in line with the motion of the clamp and at right angles to the motion of the clamp.

The clamp force acting in line with the motion of the clamp is the clamp force (compression) required to hold the mold shut so that plastic cannot escape during injection. The clamp force is supplied by the clamping mechanism of the molding machine, either by direct hydraulic action, or by toggle mechanisms. Occasionally, clamps and wedges in or on the mold are used instead of, or in addition to, the clamping action of the machine.

The clamping force is resisted by the sum of the parting line areas of all cavities and the areas of all stop blocks that may have been added in cases where there was insufficient parting line area because of space restrictions in the mold layout (Fig. 29.1).

There are two forces that act at right angles to the clamp forces: forces in side cores and forces caused by alignment elements. Forces in side cores include forces 1) to hold side cores in place against a shut-off face, 2) to counteract the injection pressure, or 3) a combination of both 1 and 2. These forces are generated either by the clamp forces, with the use of wedges as shown in Fig. 29.2, or with hydraulic actuators, as shown in Fig. 29.3.

In either case, the shut-off forces f_s in the wedge or in the screws are backed up by the mold, and create compressive, tensile, or bending forces, or a combination of them. In both cases,

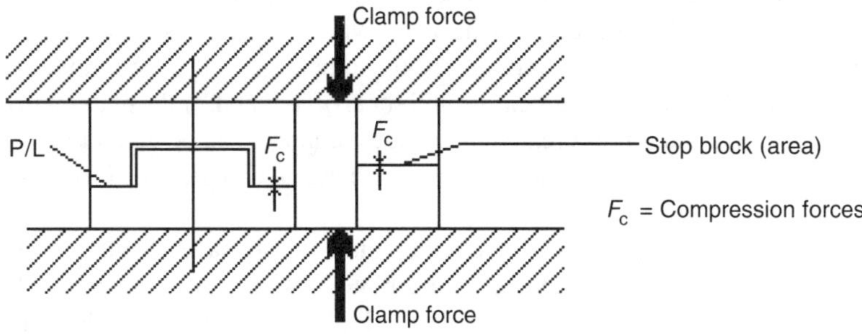

Figure 29.1 Clamp force (compression) acting in line with the motion of the clamp to hold the mold closed at the parting line.

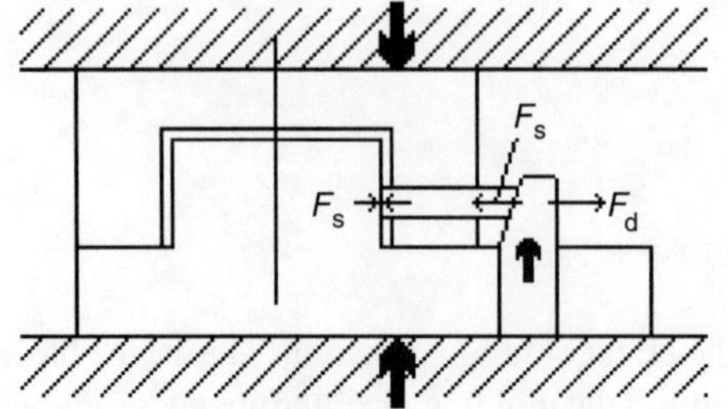

Figure 29.2 Shut-off force in side core and deflection force on wedge created by clamping.

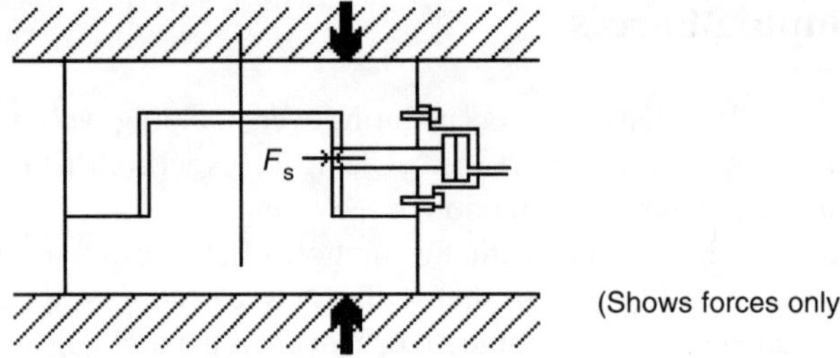

Figure 29.3 Shut-off force in side core from hydraulic cylinder during clamping.

the shut-off force will tend to deflect (bend) the core, which in thin cores may have a significant effect on the core.

Forces caused by alignment elements result from such mold elements as taper locks, taper pins, and wedges. Taper locks and taper pins, when perfectly aligned, are stretched slightly until seated. If improperly aligned (Fig. 29.4), the tapered faces will force the plates or components into proper location. Note that uneven temperatures of plates will create such misalignment of components.

With floating mold members, except for a relatively small force required to move the misaligned items into their proper location, there are no undesirable side forces acting on the mold. With mold members located on plates by spigots or dowels, forces caused by improper alignment can stretch, compress, and/or bend the affected plates and will rapidly wear the tapers and damage the mold. The forces in the tapers will "fight" the resistance against moving because of these dowels or spigots and, therefore, something will have to give—either the tapers will wear rapidly or break, the locators will shear off or bend, or the plates will otherwise be damaged. Note that the forces created by wedges or tapers can be greater than the forces of the mounting screws holding the mold plates together, and will permanently reposition any plates.

Wedges used to align mold halves, or cavities and cores, behave similarly to round taper locks as far as forces induced in the opposing plates or parts are concerned (Fig. 29.5). However, all wedges must be backed up by the mold (plates) against the wedge forces. These forces induce tensile, shear, bending, and compressive stresses in the mold plates. The amount of these forces depends also on the amount of preload selected for the action of the wedges.

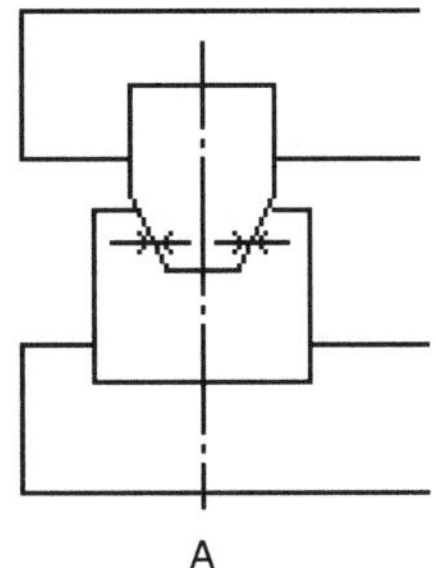

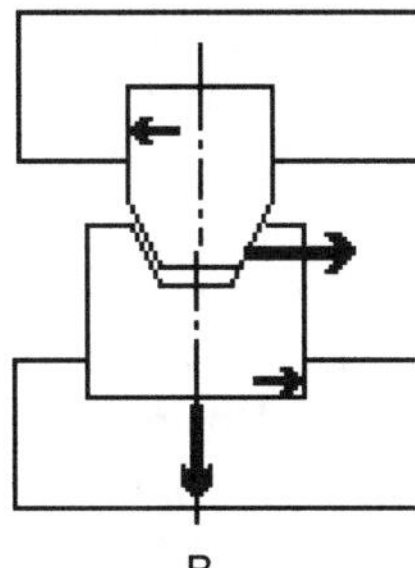

Figure 29.4 Alignment in tapered molds: A. good alignment, B. poor alignment results in stretching and bending of plates and wear of tapers.

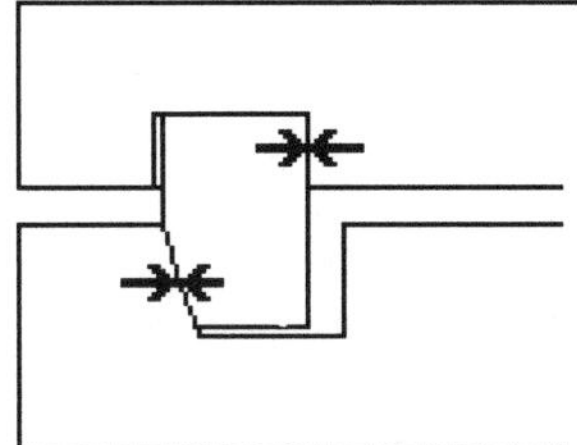

Figure 29.5 Wedge forces create opposing forces in plates.

29.2 Screw Forces

Mold plates are usually held together by screws. Provided the screws are properly designed and preloaded, the forces in the screws are *static*, not cyclical. However, the plates are loaded cyclically in compression.

29.3 Mold Opening Forces

Mold opening (tensile) forces appear when opening the mold after completion of the injection cycle. Depending on the shape of the molded product and its wall thickness, these forces can be small or considerable. Occasionally, such forces can even stall the molding machine and prevent the mold from opening.

Typically, a mold for a near-cylindrical, thin-walled container may be overpacked, either because of poor adjustment of the machine or weak construction of the cavity and/or core; the cavity expands (and/or the core is compressed) as a result of the injection forces and then returns (elastically) to its original shape after the injection pressure has ended. The thin plastic wall, which has not shrunk enough after cooling to fit in the existing space between the core and cavity walls, is now trapped and severely resists the opening of the mold. In extraordinary cases, such as when molding bobbins (for the textile industry) with heavy undercuts in both the cavity and core, hydraulic cylinders inside (or beside) the mold can then assist the machine clamp in opening the mold.

The opening force subjects the core itself to a cyclical, tensile stress. However, all screws in the mold holding the plates together and holding the cores are preloaded and are, therefore, only loaded statically at their preload. They are, therefore, not affected by the opening forces as long as the sum of all preloads of screws per plate or per core is greater than the opening force seen by these elements.

To repeat, the opening force is cyclical (i.e., it occurs during every molding cycle), and ranges between zero and a force that is usually much smaller than the compressive force during injection. The opening force can be significant in some cases, such as the bobbins mentioned above, or where there is a risk of overpacking the mold.

The meaning and significance of "static" and "cyclical" loads is explained in Chapter 18, Metal Fatigue.

29.4 Ejection Forces

Ejection forces may create tension, compression, bending, or torsion. These forces appear where stripper plates, ejector plates, or unscrewing methods are used. Ejection forces are cyclical, since they occur every molding cycle.

29.4.1 Tension

The ejection forces F_e depend on the type and temperature of plastic and, to a large extent, on the shape and surface finish of the product. Deep draw products, with little or no draft angles and/or undercuts, offer more resistance to the ejection forces than flat products or products with large draft angles and no undercuts. Note that in some cases, a warm plastic will eject easily, while the fully cooled product cannot be ejected or requires great force.

Ejecting the product from the core subjects the core to a tensile stress. In systems using two-stage ejection, tension occurs in members connecting the two ejector plates, or the stripper and ejector plate.

29.4.2 Compression

Ejector pins or sleeves are compressed between the ejector plate and the product that they eject (Fig. 29.6). Also, there are compressive forces between the stripper ring (or stripper bar) where it touches the product and the support of the stripper ring (or bar).

Note that during each injection, the ejector pins are also compressed; therefore, these elements are cyclically compressed *twice* every molding cycle, but not necessarily with the same force.

Example: A ¼" diameter ejector pin is 16" long. The injection pressure inside the cavity is assumed to be 5,000 psi. The modulus of elasticity for steel is 30,000,000 psi. The area of the pin is

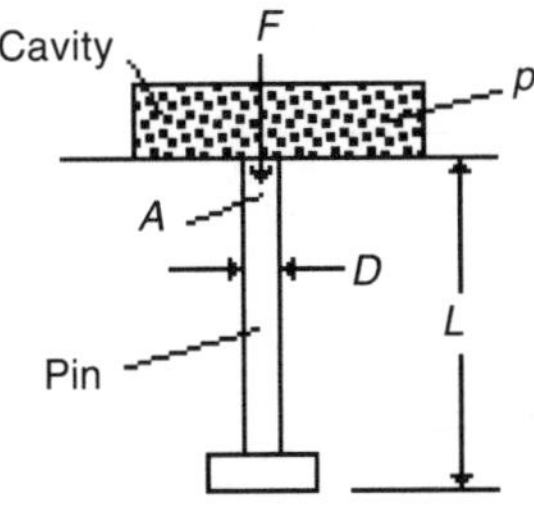

Figure 29.6 Compression force on ejector pin.

$$A = \frac{D^2\pi}{4}. \tag{29.1}$$

The force on the pin is

$$F = A \times p. \tag{29.2}$$

The formula for compressive strain is

$$f = \frac{F \times L}{A \times E}. \text{ Substituting, } f = \frac{A \times p \times L}{A \times E} = \frac{p \times L}{E}, \tag{29.3}$$

so for this example, the compressive force on the ejector pin is

$$f = \frac{5{,}000 \text{ psi} \times 16 \text{ in.}}{30{,}000{,}000 \text{ psi}} = 0.00267 \text{ in.}$$

The amount of compression can be significant, especially with very long ejector pins. In the example shown, the compression of the pin is almost 0.003 in. Note that the pin diameter D (or cross section A) is irrelevant. A thicker or a thinner ejector pin (or a core) will compress the same amount when subjected to the same pressure p.

Note also that to support the injection forces acting on the ejector pins, the ejector plate must be well supported, preferably right under or near the location of the ejector pins (see Fig. 29.7). Dimension d shown there should be as small as possible to minimize the bending forces in the ejector plate, but for practical reasons, the number of stop pins is usually much smaller than the number of ejector pins.

29.4.3 Deflection

Deflection, or bending, of an ejector plate is a result of two forces:

1. The forces described above that support the injection pressure, and
2. the forces required to eject the products.

Both may be present in molds.

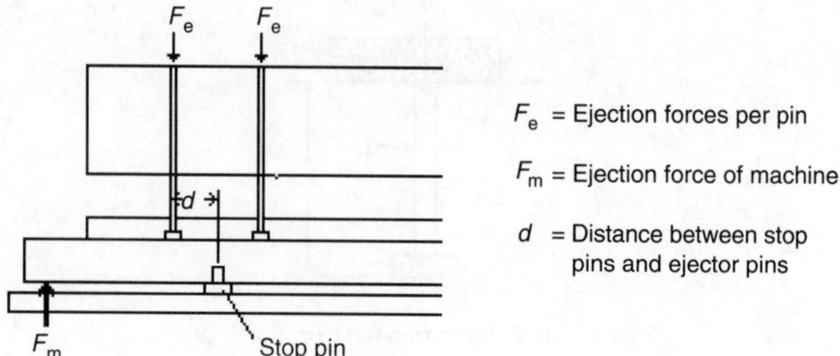

Figure 29.7 Ejection forces on ejector pins, with minimum distance between stop pins and ejector pins. The machine ejector in this mold is poorly located.

29.4.3.1 *Ejector Plate Deflection*

Consider an ejector plate system such as that shown in Fig. 29.7. The figure shows the machine ejector in a very poor location, too far out, thereby creating a large bending moment during ejection. The designer must always select locations of machine ejectors so that the average distances to the ejector pins are at a minimum. If there are no suitable locations, the ejector plate must be thick enough not to bend more than suggested below. This can be calculated by assuming the worst possible case (i.e., that the maximum ejection force available from the machine will be required to eject), and assuming the case of a straight beam loaded with a number of loads and supported in two places. Note from the figure that the stop pins assist only against the injection forces.

Shown below is a simplified method of calculating the strength (against deflection) of an ejector plate. Note that any thickness will deflect to some extent. The designer must select a reasonable thickness, and in general should keep deflection less than 0.005" at the worst point.

Two assumptions are necessary to perform this calculation:

1. Assume that the ejector plate is a bar, 1 inch wide. If it is wider, divide the total load by the width. For example, if the load is 10 tons, and the width is 10 inches, there will be (simplified) a load W of one ton to be worked with.
2. Assume that all ejector pins (and/or sleeves) load the plate as a continuous, uniform load within the distance from the first to the last ejector pin.

Figure 29.8 shows the calculations for deflection at the center of an ejector plate in a mold in which the machine ejectors are located far apart on the outside of the load. The formulas to the right of the figure show how the moment of inertia is determined, as well.

If the machine ejectors are within the load, the deflection under load is much smaller (Fig. 29.9). These calculations provide a fairly accurate and usable indication of whether the design thickness for the plate is satisfactory.

29.4.3.2 *Stripper Plate Deflection*

Since the stripper rings (or plate) seat on the core taper and are, therefore, supported by the core (and the core backing plate), the injection forces acting on the stripper rings (or plate) do not cause deflection of the stripper plate.

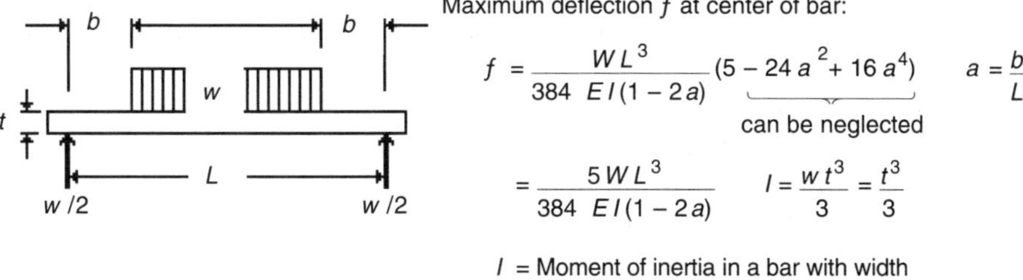

Figure 29.8 Maximum deflection f at the center of the ejector plate (bar).

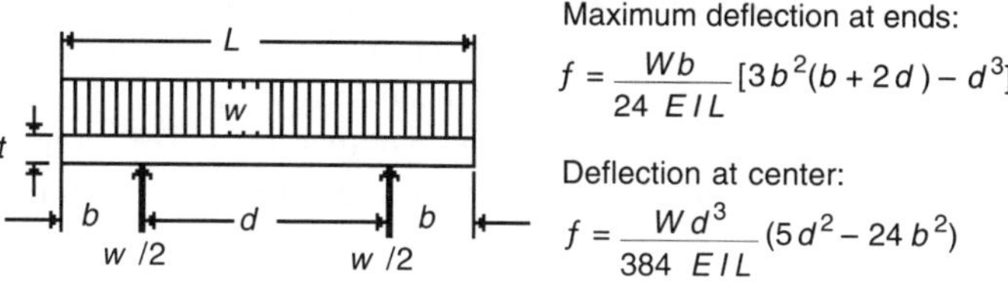

Figure 29.9 Deflection in a the plae where the machine ejectors are within the load.

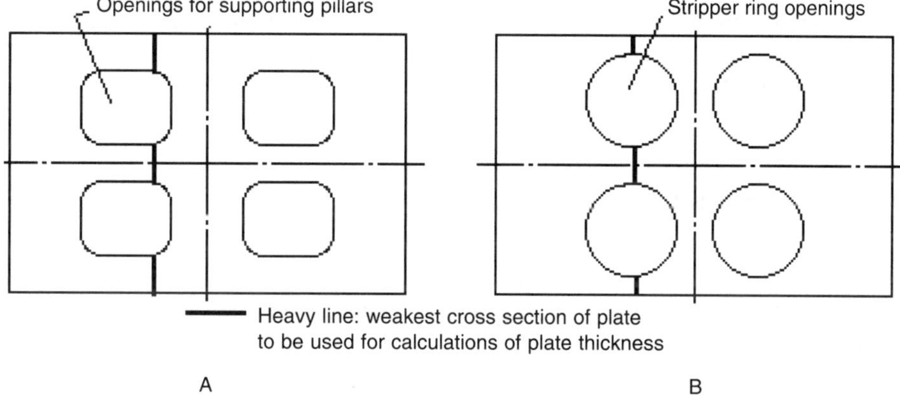

Figure 29.10 Openings in plates must be accounted for in calculations of thickness: A. typical ejector plate and B. typical stripper plate.

Calculation of stripper plate thickness is essentially the same as that for ejector plates. Care must be taken in both these cases to include the reduction of plate width as a result of the cutouts in the ejector plate (usually where support pillars are located) and in the stripper plate where the actual stripping action takes place (i.e., the usually circular openings for the stripper rings) (Fig. 29.10).

It is up to the mold designer to select the number and distribution of ejector pins and, also, the location and number of machine ejectors used, to minimize the distances between machine ejectors and ejector pins or stripper rings. This will reduce the deflection of the ejector or stripper plate. If large distances are unavoidable, the plates must be thicker to keep deflection to a minimum.

To repeat, it is usually impossible to predict the ejection forces, but the designer can calculate "from the other end" by designing all components assuming that the full available machine ejector force will be used. Except for extraordinary cases, such as the textile bobbin mentioned earlier, this method will usually give a safe basis for determining ejector or stripper plate thicknesses.

29.4.4 Torsion

Torsional forces appear in rotating cores during unscrewing, and in accessories such as the unscrewing head shafts. Torsional forces in the mold cores are usually small. However, it may be of historical interest that there is at least one application, in a Dow Chemicals patented method, where the core is rotated during injection. In this case, the torque requirement is extremely high and must be carefully considered in the design of the core and all rotating parts which impart the rotation to the cores. (The purpose of this rotation of about one turn was to biaxially orient the plastic in the product (e.g., a round container, or a preform) to increase its strength.)

Coil springs are torsion springs. Since the strength of steel in torsion is about one-half that in tension or compression, any application of these springs in molds must be carefully calculated to stress them well below their fatigue strength (see Chapter 18, Metal Fatigue).

For long life, springs should be loaded only to about 20% of its maximum stroke to solid. (Follow spring manufacturer's recommendations). It is usually impractical to use springs for long strokes; air actuators should be used instead.

29.5 Injection Forces

Injection forces appear in all molds, whether cold or hot runner. The injection forces are cyclical (i.e., they occur during every molding cycle, ranging from zero to a maximum value and back to zero).

29.5.1 Hydraulic Forces

During the filling phase, the pressure in the runner system can be quite high, but there is no significant pressure within the cavity space. As soon as the cavity space is filled, the injection pressure in the whole system rises, and, for a short time, the plastic within the cavity will act like a hydraulic fluid and push in all directions.

It is important to understand that plastics do not behave exactly like water or oil, and that there is a significant difference in actual pressures along the path of the plastic from the injection end of the machine to the end of the flow path within the cavity. The hydraulic pressure at the injection end can be easily expressed by a simple formula. Since the force on the plunger or screw is

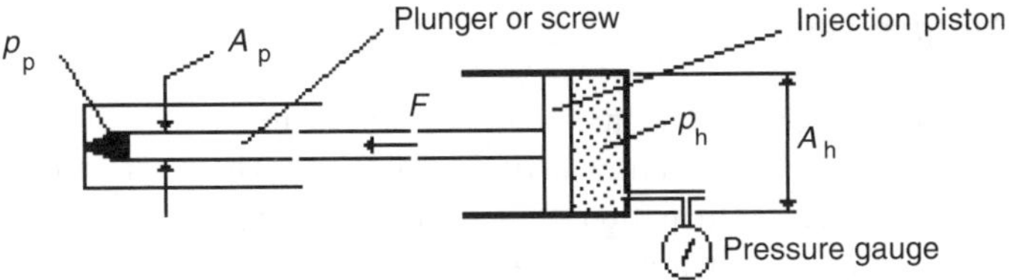

Figure 29.11 Hydraulic pressure on injection unit.

$$F = p_h \times A_h \,, \tag{29.4}$$

where p_h = hydraulic oil pressure and A_h = area of hydraulic piston, the pressure on the plastic can be determined by the following substitution (Fig. 29.11):

$$p_p = \frac{F}{A_p}\,, \text{ or } p_p = \frac{p_h \times A_h}{A_p}\,, \tag{29.5}$$

where p_p = pressure in plastic and A_p = area of plunger or screw. The pressure p_p of the plastic is proportional to the ratio of cross sectional areas of the hydraulic piston and the screw (or plunger), multiplied by the hydraulic (oil) pressure of the injection unit. With many machines, the ratio between A_h and A_p is exactly 10, so that any reading of the oil pressure multiplied by 10 gives the plastic pressure at the injection end, before the machine nozzle. Usually, there is a chart attached to the injection cylinder to enable the operator to easily convert the pressure p_h indicated on the gage into the plastic pressure p_p.

The pressure at the injection unit may be quite high; it is often 20,000 psi, and for the high injection speeds needed for thin-walled products, it may be as high as 29,000 psi. However, there are significant pressure drops between the original high pressure on the plastic before the machine nozzle, and the final pressure inside the cavity. The pressure drops occur in the machine nozzle, in the runner system, and in the gates, and there is also a significant pressure drop inside the cavity from the gate toward the end of the flow path in the cavity, especially with thin-walled products.

Without sophisticated computer simulation of the plastic flow within the hot runner system and the cavity, it is practically impossible to accurately predict the plastic pressure inside the cavity space, but it may be as low as 25% of the original pressure at the injection end. There are certain conventions used by designers to estimate the pressure in the cavities and to determine the clamp force required to hold the mold shut during injection.

A rule of thumb for pressures inside the cavity space is:

- 5 tons/in.^{2} for thin-walled products and
- 2–3 tons/in.^{2} for technical products.

More often, however, experience obtained from specific jobs will be used.

Molds equipped with pressure transducers along the path of the plastic give a good reading of the actual pressures in selected locations, but this is usually done only as a research project and is not practical for the general requirements of a mold designer or mold maker, unless the results (data) can be extrapolated from such experimental molds to the mold to be designed.

29.5.2 Compression

After the cavities are filled, the actual injection pressures in the cavity will compress the steel and other materials used in cavities, cores, and any mold parts. These parts are "squeezed" and push in the direction of the clamp motion between their molding surfaces and the supporting mold plates. These forces are then backed by the machine platens. Note that these are the same forces which tend to open the mold at the parting line, against the clamping force of the machine.

After the cavities are filled, the core is surrounded by plastic under pressure and will be compressed in the direction at right angles to its long axis. If cooling lines are drilled too close to the surface, there may not be enough steel to resist the pressure on the core, and the core will eventually fail.

Similarly, the cavity or gate inserts may collapse under hydraulic pressure if cooling lines are too close to the surface. This is of special concern where such inserts are made from beryllium-copper or aluminum, or from other materials that are weaker than steel.

Note: Blow mold cavities are usually made from aluminum; however, the forces generated by the blow pressures are much lower than the forces in an injection cavity, and aluminum, being a much better heat conductor than steel, is by far the better choice of material.

29.5.3 Tension

Consider the cavity to be a pressure vessel. After the cavities are filled, the pressure inside it will tend to expand the cavity and create tensile stresses (actually, combined tension and compression) inside the cavity walls. There are two areas of concern:

1. The cavity is often crisscrossed with cooling channels and has screw- and other holes, which, proportional to their size and number, reduce the effective steel cross section required to resist the load coming from the hydraulic pressure of the plastic. Also, these holes are considerable stress risers. The smaller the hole sizes, the worse are the stress risers that weaken the cavity walls disproportional to their size (see Chapter 18, Metal Fatigue).
2. Where there is interlocking between cavity and core, with the female taper section located in the cavity block, the expanding cavity will also expand this taper so much that the tapers will lose some or all of their contact force with the mating tapers. This can result in unacceptable misalignment and core shift.

The experienced designer usually has a "feel" for a suitable wall thickness of a mold part and will design proper layouts and sizes for distances of cooling channels and screw holes from the molding surface. Usually, being a little "too heavy" is no real problem in a mold, as long as it will be strong enough for its expected mold life.

There are some crude approximation calculations available to the designer for calculating whether the planned mold part will withstand these tensile forces. More recently, Finite Elements Analysis (usually quite expensive) can determine with greater accuracy whether a mold part will last its expected lifetime. But both these methods for calculating the wall thickness depend principally on the accuracy of the assumption of the injection pressure within the mold.

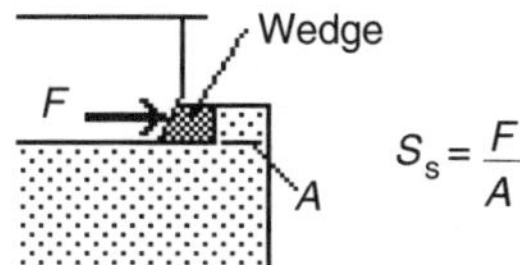

Figure 29.12 Shear stress S_S is created on area A of mold plate by alignment wedge.

29.5.4 Deflection

Deflection usually happens with long, unsupported, slender cores, but even "stubby" cores, such as those for 7-oz. thin-wall tumblers, may deflect significantly if the plastic flow from the gate is not perfectly centered. If the flow favors one side and fills there earlier, by the time the pressure inside the cavity rises there is not enough balancing force on the opposite side of the core, and the core will deflect toward the side with the lesser plastic flow.

29.6 Shear Forces

Shear forces may appear in many locations in the mold. Shear forces are often created by misalignment of mold parts as a result of poor workmanship or by uneven heat expansion of components.

Shear appears, for example, where alignment wedges are backed up by the mold plates or other mold parts (Fig. 29.12). In this case, the shear stresses are cyclical, that is, they occur at every molding cycle, and can cause the weak area to break when its fatigue limit (number of cycles to fail) is reached. Shear may appear in dowels (or spigots) if the position of the dowel holes in the mating parts are not perfectly matched.

29.7 Forces Caused by Unequal Mold Cooling

29.7.1 Heat Expansion

Unequal temperatures between mold plates will create some elongation of the warmer plate in relation to the colder plate. In case of uneven heat expansion between two parts, one part increases in size relative to the other. An example of heat expansion and shear caused by uneven mold cooling follows.

Example: Two plates are located by a pair of dowels at a distance L of 20 in. (Fig. 29.13). During operation, because of uneven cooling, there is a temperature difference of 25 °F between the two plates.

The heat expansion of steel is 0.000006 in./°F-in. The elongation f of the warmer plate is, therefore $f = .000006$ in./°F in. × 25 °F × 20 in. = 0.003 in. This means that there is a force generated in this plate which is equivalent to a force that would stretch the plate by $f = 0.003$ in.

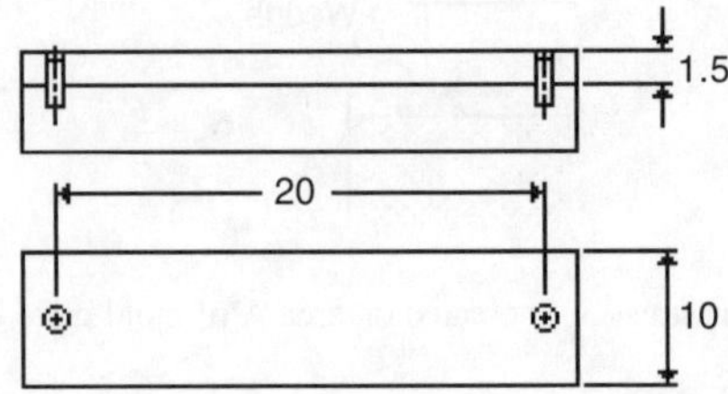

Figure 29.13 Cross section (top) and top view (bottom) of two plates located by a pair of dowels.

If a cross section A of the plate (between the two dowels) is assumed of 1.5 in. × 10 in., or $A = 15$ in.^{2}, the force F required to stretch such a plate by 0.003 in. equals

$$F = \frac{f \times A \ \times E}{L}, \text{ or} \tag{29.6}$$

$$F = \frac{0.003 \text{ in.} \times 15 \text{ in}^2 \times 30{,}000{,}000 \text{ psi}}{20 \text{ in.}} = 67{,}500 \text{ lb., or } 33.75 \text{ tons.}$$

This is a considerable shear force to be taken by the dowels. A small dowel could be sheared right off by this force. A larger dowel may become deformed and/or may damage the precisely machined dowel holes.

These forces can be considered static, rather than cyclical, since they happen only whenever the mold is started up, and possibly not more than a few hundred times during its life.

The designer must therefore be aware that:

- uneven heat expansion can introduce considerable forces into a mold,
- dowels must be of adequate size to be able to withstand such forces should they arise, and
- a better-suited method of alignment may be required than that of using two dowels.

As a rule, when using two dowels for proper alignment. It is good design practice to make the distance between the two dowels as large as possible. However, whenever there is a possibility of a temperature differential between the mating parts, other methods of alignment must be considered to allow "floating" of aligned parts to avoid these unwanted high forces in the mold. For example, a hot runner manifold is dowelled only in the center and kept from rotating with sliding faces distant from the center, while it is free to expand in all directions.

Common causes for unequal plate temperatures are poor mold design, such as poor location or lack of cooling channels, wrong installation of the mold, or other factors affecting the cooling system, such as plugged lines, internal corrosion, etc. Of particular importance is the cooling arrangement near inherently hot areas, such as the plates surrounding the hot runner manifold, and any areas in injection-blow molds where very hot fluids heat a portion of the cores while the rest of the cores and the cavities are cooled.

Unequal plate temperatures will result in misalignment of leader pins and bushings or locating tapers that were well aligned while the mold was cold. When clamping up, sideways (tensile and compressive) forces are exerted which will stress these parts in a direction and to an amount for which they were not intended. These forces also cause rapid wear and breakage of these parts. These forces are cyclical (i.e., they occur during every molding cycle, ranging from zero to a maximum value).

Floating cores and/or floating stripper rings are desirable because they prevent misalignment of the cavity stack. However, leader pins and bushings are very sensitive to plate expansion, unless they have large enough clearances and are only used for guiding the mold halves during servicing.

29.7.2 Hot Runner Components

Some hot runner systems depend on the heat expansion for proper operation. In such systems, the manifold is suspended freely between the surrounding plates by steel insulators. As the manifold heats up, it expands in all directions. In its long axis, there are no restraining supports, and the expansion takes place without creating any forces within the mold. The increase in thickness, however, is significant, and, similar to the heat expansion of the insulators, it must be taken into account when calculating the space into which these components are mounted. In all such designs, it is important to adhere to the manufacturer's specifications, which often include very tight tolerances.

29.8 Injection Unit Force

In some molding operations, during every cycle, the injection unit (machine nozzle) moves toward the sprue bushing and is held there during injection with the force of usually one or two hydraulic cylinders, which are part of the injection system of the molding machine. The force is cyclical, that is, it occurs during every molding cycle, ranging from zero to its maximum value, which depends on the molding machine and its adjustment. It is usually between 5 and 20 tons.

In other molding operations, the machine nozzle is held permanently against the sprue bushing. In this case, the force from the injection unit on the spruc is static rather than cyclical. It is obvious that the seat between nozzle and bushing must be strong enough to resist this (compressive) force. But, in addition, the designer must assess where the force exerted by the machine nozzle will be supported. This depends essentially on the type and the design of the mold.

In a hot runner mold, the sprue sits on the hot runner manifold, and the force is added to the force of the insulators supporting the manifold in the cavity plate. In single-face molds, this force is taken up by the screws holding the mold half to the stationary platen. In stack molds, the manifold is located in the plates between the cavity plates, and the force of the machine nozzle will push on the center portion of the mold. This force must, therefore, be added to the clamp force required to hold the mold closed during injection.

29.9 Effect of Surface Definition

The forces that hold a mold closed against the injection pressure are also closely related to the requirement for proper surface definition. *Surface definition* (or "definition") is the accuracy

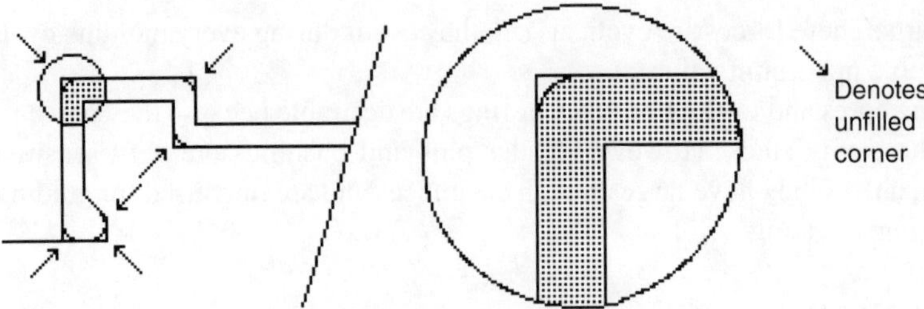

Figure 29.14 Coffee can lid with unfilled corners.

of reproduction of the mold surface on the surface of the molded product. Two extreme, but typical examples will explain this concept:

1. A coffee can lid molded in PE is shown in Fig. 29.14. The cavity and core have sharp corners, but the plastic will not fill these corners unless forced into them by high injection pressure. In a product such as an inexpensive lid, it may be of little or no importance to have the exact shape of the corners reproduced on the product. The mold can be filled with little pressure, and even be operated with starve feeding, that is, the amount of plastic injected is exactly the right amount to fill the cavity, and no significant injection pressure will be seen within the cavity space (and by the clamp).
2. Disks for audio or video recordings. The surface of a record consists of fine grooves, which are "embossed" with the signals that the pickup will read when playing the recording. The original master recording disk is nickel-plated until the deposited metal is thick enough to be self-supporting (about 0.015–0.020 in. thick,) and can be removed from the master without deformation. The plated deposit gives, in negative, a perfect reproduction of the original grooves with all the signals. This "stamper" is then used in the mold as a cavity insert. The microscopic variations of the surface contain all data required for replaying the record, and it is, therefore, of great importance that, during injection, the grooves with all peaks and valleys on their surface are exactly reproduced in the plastic.

Note that there is even a large difference in the quality of the reproduction required for analog audio and digital video signals. The digital signals for audio (in compact disks) and in video disks consist of dots and spaces at the bottom of the grooves, which are much smaller than the grooves and the surface variations in the sides of the analog audio records. Very high injection pressures are, therefore, required to reproduce these engravings, especially for digital recordings.

How this can affect the clamping force of the molding machine can be shown in an actual instance, which happened in the early 1960s, when most mold design was still done without calculations or without much knowledge of the injection molding technology which is available today. At that time, a mold maker built many 4-cavity molds for "401" coffee can lids for a 100-ton clamp, which ran very successfully. With a diameter of 4.1", the projected area per cavity is:

$$A = D^2 \frac{\pi}{4}, \text{ or } (4.1)^2 \frac{\pi}{4} = 13.20 \text{ in.}^2 \text{ per cavity, or } 52.8 \text{ in.}^2 \text{ for the mold.}$$

For a 100-ton clamp, this means less than 2 tons/in.^{2} (100 tons ÷ 52.8 in.^{2}) were available to hold the clamp shut during injection. However, since definition was not a problem, the mold could be injected using the starve-feeding method, and there was no flashing.

When a customer inquired about building a mold for a (rather crude-quality) voice record for a children's toy, on a 3.5-in. diameter disk, the mold maker was confident that the same mold design as for the above lids could be used. After completing the mold as planned, the mold flashed badly, and the disks were unusable because of lack of groove definition.

Obviously, there was not enough pressure to properly fill the grooves. After blocking one cavity, the records were acceptable. However, obviously, the mold was not as productive as planned. The actual molding conditions were:

$$A = (3.5)^2\frac{\pi}{4} = 9.62 \text{ in.}^2 \times 3 \text{ cavities} = 28.86 \text{ in.}^2 \text{ for the mold.}$$

In the 100-ton clamp, this provided 3.46 tons/in.^{2}, which is considerably more than that required for the 4-cavity lid mold.

This example is given to emphasize the importance of accurately forecasting the injection pressure required for a job, and therefore to more accurately estimate the forces acting within the mold which in turn affect the fatigue life of the various mold components. The designer must also remember what influence the melt temperature and the injection speeds have on filling, and how they are related to the actual injection pressures within the cavity space.

Because of the slow injection speeds in many molding machines available in earlier periods, the 12-inch "long-playing" (LP) records could, for a long time, be produced only by the "injection–compression" process, in which (in a vertical clamp) a measured hot "blob" of plastic (usually PVC) is placed automatically between the open mold halves, which then close and press the plastic into the grooves of the stampers.

With the slow demise of the LPs, injection molding these disks was never seriously done, although HUSKY completed a very successful project not only to mold analog disks such as LPs, but also 12-inch digital video disks of the kind used today for the Karaoke entertainment industry. Today, all compact disks (CDs), made from PC, are injection molded in single-cavity molds.

30 Preload

In every machine assembly, two (or more) parts are held together by the force of screws, clamps, power cylinders, etc., depending on the need of the application. In molds, plates or other mold parts are held together with screws or wedges, and their accurate relative position is often achieved with taper fits under pressure. When molding, the mold halves are held together or "clamped up" during the mold closed times with a force generated by the machine clamping mechanism.

30.1 What is Preload?

The following three figures (Figs. 30.1–3) illustrate the simple explanation of the principle of preload:

First, assume a solid frame and a weightless loose bar touching the frame. A spring under the bar has the exact same uncompressed length as the height L_0 of the bar from the bottom of the frame (Fig. 30.1).

If a force F is now applied on top of the bar, it will compress the spring and create a gap between the frame and the bar. The spring will resist the force F and be compressed to a length L until the force of the compressed spring F_s equals F (Fig. 30.2).

Now assume a spring that is longer than the original spring ($L_1 > L_0$). To place the spring into the frame, it must be first compressed, or "preloaded", from L_1 to L_0. The preload in the spring is proportional to the length of the preload, L_p. If a force F is now applied, the bar will

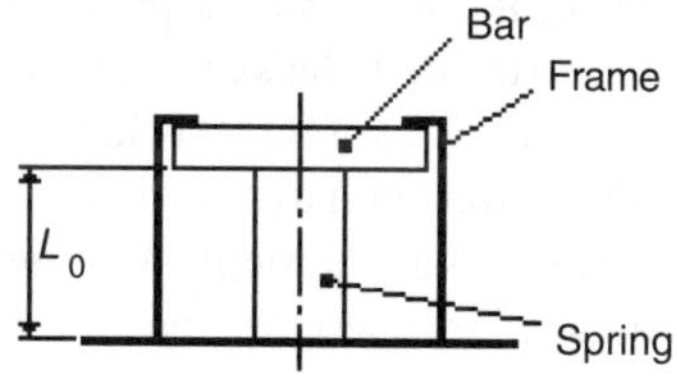

Figure 30.1 Solid frame with "weightless" bar over uncompressed spring at height L_0.

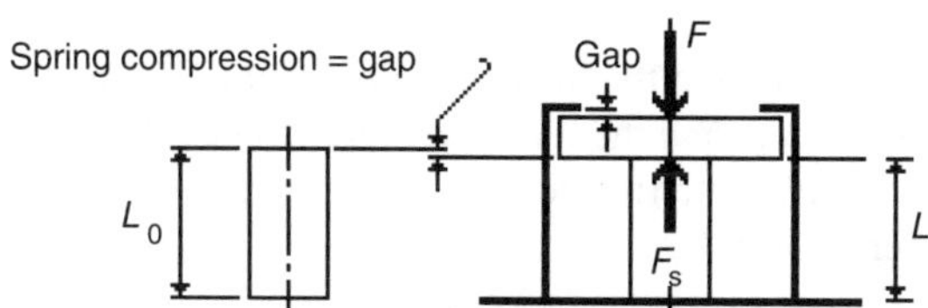

Figure 30.2 Force applied on top of bar compresses spring to height L until $F = F_s$.

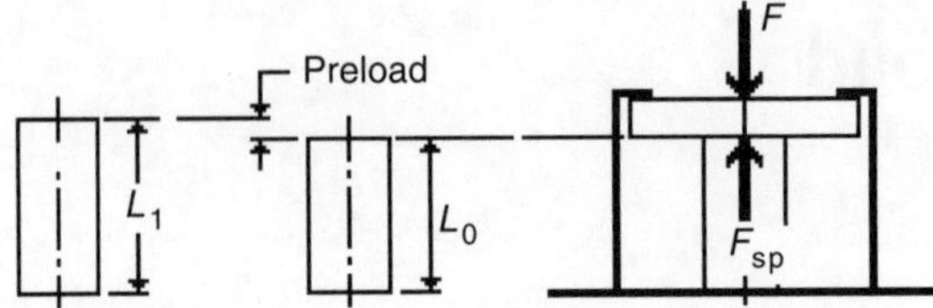

Figure 30.3 Preload on the spring holds the bar in place against the frame until $F > F_{sp}$.

be held in place against the frame (Fig. 30.3) until the force of the preload in the spring, F_{sp}, has been reached. If F becomes greater than F_{sp}, the gap will reappear.

As long as the spring force (preload) is greater than the force F, the bar will remain touching the frame. If F becomes greater than F_s, a gap will again appear. In general, every joint must be preloaded to function as required.

Preload is the minimum force required to hold (clamp) two pieces together against an expected load, F. Typical areas where preloads are required will be presented in the following sections.

30.2 Clamp Force

Clamps must be preloaded to counteract injection pressures inside the mold. When the mold closes without any preload (tonnage) on the clamp, the cavity and core halves will "just touch" on the parting line of the mold, but not exert any force against each other. If the plastic is injected at this point, the forces created by the hydraulic pressure of the plastic acting against the molding surfaces will open the mold. The gap created will let the plastic escape, and the mold will flash along the parting line. Similarly, if the selected clamp force (tonnage) is insufficient, the parting plane will separate as soon as the injection forces become greater than the clamp force.

As in the example with the spring, the mold must be preloaded, that is, the clamp must exert (before injecting) a pressure between the mold halves which is at least equal to, but preferably *somewhat greater than*, the expected forces of the plastic which tend to open the mold. The force (preload) is generated in the power train as shown in Fig. 30.4.

NOTE: Fig. 30.4 shows a hydraulic clamp system. With mechanical (toggle) clamps, the force F is created by compressing the links between the moving platen and the clamp frame.

The following are some important statements regarding the clamp. It is very important to understand them and their relationship to the fatigue life of both the mold and machine parts.

1. The force F is equal to the sum of forces f_t in the tie bars. There are usually four (4) tie bars, so $F = 4f_t$.
2. The force F is equal to the sum of the forces f in the parting line (really, a parting plane).
3. F must be *somewhat greater than* the forces tending to open the mold during injection.
4. The forces f_t (in each tie bar) stretch the tie bars. The amount of tie bar stretch f is expressed in the following general formula:

$$f = \frac{F \times L}{A \times E}, \tag{30.1}$$

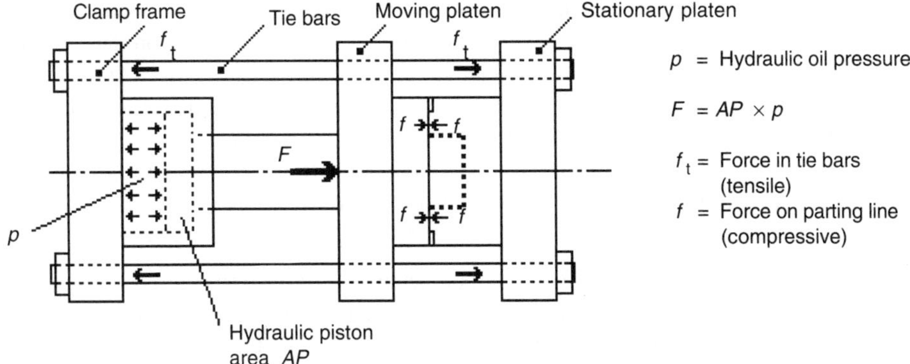

Figure 30.4 Preload force F on parting line in a hydraulic clamp system.

where f = elongation or compression of tie bar, F = tensile force on tie bar, L = distance of bar stretched, A = cross sectional area of bar, and E = modulus of elasticity.

The tensile strength of the steel (or its hardness) does not appear in the above formula, and, therefore, has no influence on the amount of stretch. However, as was shown in Chapter 18, Metal Fatigue, the harder the steel, the lower is the fatigue life of the part. On the other hand, a hard surface on the tie bars is desirable to guard against wear and abuse by objects hitting the tie bars, usually during mold handling.

5. The size (diameter) of tie bars in better machines is selected so that they will be stressed to about 9,000–10,000 psi or less, at the rated, full force of the clamp. For this reason, in better quality machines the tie bars are larger, and in cheaper machines they may be smaller, and may be stressed up to 15,000 psi. Since the tie bars are cyclically loaded, the heavier tie bars are less stressed and ensure longer life. (Tie bar breakage is common in cheaper machines.)
6. It is important that all tie bars are installed so that they stretch equally, within very close variations, not more than 0.001–002 in. for most machines. If they stretch unequally, the stress in the "shorter" tie bars will be greater than planned or permissible, and they will fail sooner; also, the mold will be loaded unevenly on the parting plane.

   Uneven stretch leads to loss of parallelism of machine platens and can result in various mold problems, such as flashing, parting plane wear, core shift, etc. Recognizing this stretch is important for the mold designer and mold service technician, because malfunction of a mold (flashing, uneven product walls, etc.) can often be traced back to poor setting of the tie bar stretch, and not to any trouble within the mold.

In point 3 above, "somewhat greater than" is shown in italic type. This is an important point to consider. A mold is usually designed for a specific machine size and tonnage that is large enough to allow the mold to run without flashing. The clamping force should only be somewhat greater than required to protect the parting plane from excessive compressive load.

To calculate the area of the parting plane, the designer must be aware that this plane is not loaded by the *static* force of the clamp (adjusted or rated tonnage of the machine), but by a *cyclical* force (from zero to full load), which happens at the rate of once every molding cycle.

The loads *f* (Fig. 30.4) must be less than the fatigue strength of either of the two steels touching on the mating faces. This is fairly easy to achieve if there is enough space on the surface of the mold; it means that mating faces between the cavities, cores, and stripper rings (if used) must have enough supporting surface. If there is not enough space, additional support blocks between the cavity and core halves of the mold are required to add area so that the parting planes will not be crushed. Typically, a mold that is overstressed in the parting plane will perform well until it reaches its critical number of cycles, where fatigue will begin, usually first visible as pitting of the parting line surface and disappearance of the vents.

Example: A mold is designed for a particular machine with a 150-ton clamp, and the critical areas were calculated on this basis. In production, this machine is for some reason not available, and is substituted by a 200-ton machine. If the setup technician uses the full tonnage, rather than the design tonnage, the critical areas will be overstressed by 33% (Fig. 30.5). This may be above the permissible fatigue strength, and, although the mold will not fail immediately, it will fail after a number of cycles corresponding to the now different stress levels.

This becomes clear if one looks again at an illustration similar to Fig. 18.8 in Chapter 18, Metal Fatigue. At the design tonnage, the mold should last "forever". At the 33% higher level, the curves intersect, and the maximum expected number of cycles can be read at X to be about 200,000 cycles before the mold will fail.

30.3 Forces on Side Cores

A *side core* is a core entering the mold from the side, usually, but not necessarily, at right angles to the centerline of the mold, to produce a hole or opening in the side of the product which would prevent the ejection of the product after molding.

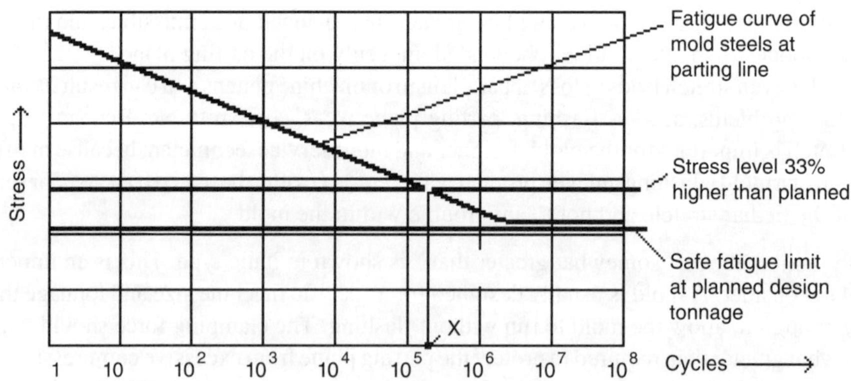

Figure 30.5 Fatigue curve of mold steels at the parting line for planned 150-ton clamp and actual 200-ton clamp shows higher stress on mold P/L.

Every side core requires two distinct forces: those for moving the side cores in and out of the mold, and those holding the side core against the plastic pressure during injection.

30.3.1 Forces Moving Side Cores In and Out

The forces that move side cores in and out of the mold can be generated with:

1. Springs, usually in the "Out" motion, but unless the springs are very long and easily accessible, springs should be avoided if long mold life is desired. (The "In" motion can be with angle pins, cams, or wedges.)
2. cams or angle pins for "In" and/or "Out" motion, and
3. hydraulic actuators for "In" and "Out" motion.

The forces required to move the side core are relatively small, just enough to overcome friction and any resistance by the plastic itself against the withdrawal of the side core. They are not intended to counteract the injection forces.

30.3.2 Forces Holding Side Cores Against Plastic Pressure During Injection

The forces holding the side cores against the plastic pressure of injection are major forces, which are generated with wedges or hydraulic actuators (See also Chapter 29, Mold Forces, Figs. 29.2 and 29.3).

30.3.2.1 Wedges

With wedges, the force F is generated by the clamp force of the machine. A portion of the clamp force is applied between the (angled) rear face of the side core and a wedge mounted on the mold half opposite the mold half where the side core is located. The force can be calculated from geometric relationships between wedge angle and location of mating wedge surfaces at the moment when the mold is fully clamped up.

The advantage of wedges is that they are simple. However, small spatial variations in positions and lengths can result in large variations in force, and manufacturing tolerances must be kept very small. Otherwise, there must be some possibility of solid and permanent adjustment, by shimming etc., to arrive at the proper force F. The disadvantage is that the force F can easily be too small, resulting in flashing, or too large, resulting in excessive stresses in the wedges and their supports, which then results in fatigue problems and shortened mold life.

Usually, the side cores are relatively small, and the portion of clamp force required to activate the wedges is negligible compared to the total clamp force. With some large wedges, such as those required for the four side cores of large beverage bottle crates, the wedges can use a considerable portion of the clamp force, which will affect the selection of machine clamp size for such a mold.

30.3.2.2 Hydraulic Actuators

The use of hydraulic actuators is only practical where the forces are relatively small. For safety reasons, the hydraulic pressure of actuators is usually limited to 500 psi, and for the often quite high forces, a hydraulic actuator could become too large to be practical.

The advantage is that the force F can be easily regulated with pressure controls, and overstressing can be avoided. The disadvantages are the space occupied by the actuator and its unwieldy and/or often difficult installation in or on the mold. In addition, hydraulic actuators need valving, piping, and controls, which complicates the installation and increases the services required.

For small side cores, hydraulic actuators can sometimes be used to move *and* to hold the side cores, and may have advantages over other systems.

30.3.3 Preload on a Side Core

The forces in the wedges or the actuator must be sufficiently large so that they provide a preload against the (assumed) plastic pressure p acting on the side core in the direction of its motion, from the inside of the mold. There are several possibilities to consider:

1. The side core does not shut off against the core (Fig. 30.6). The full injection pressure p acts on the projected area A of the side core. The preload must be greater than $A \times p$. Since the side core does not rest against the core, its stroke must be limited by a fixed stop.

   The side core and the stop are subjected to cyclical loads once every machine cycle. The stop surface must be large enough to provide a safe stress level for these two parts. Ideally, the preload is kept only somewhat larger than the expected (assumed) force F. This will ensure that during injection the side core remains in position, but at the same time, F_p (or F_s) is small enough to avoid overstressing the stop area.

   The preload determines the required size of the stop area. If the stresses are too high, the stop surface will begin to deteriorate once the number of cycles to fatigue failure is reached. This can be seen by a collapse of the stop, permitting the side core to enter too far into the mold, thus creating wrong depth dimensions where the side core enters into the product.
2. The side core shuts off against the core. Theoretically, there should not be any force to be backed up. However, even the smallest gap caused by dirt or poor seating against the core will let the plastic enter the contact area and thereby create a condition similar to that above, where the side core enters only partly into the cavity space. The designer should still use a preload F_p greater than $F = A \times p$ to back up the side core.

   NOTE 1: Supporting the preload by resting the side core on the core (instead of using an external stop) can be quite acceptable if the contact areas are large and the stresses on the surfaces are below the fatigue level.

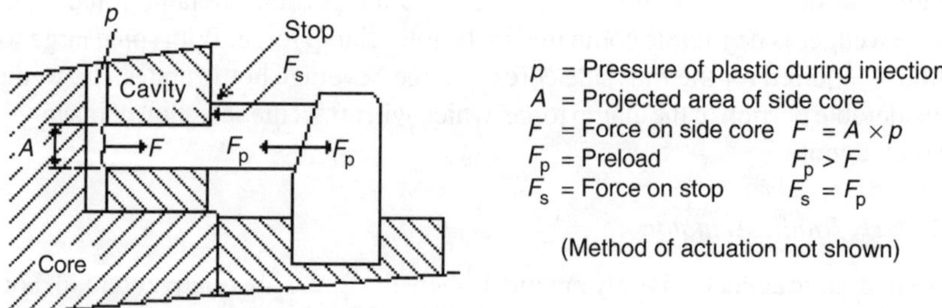

Figure 30.6 Preload Fp must be greater than the force of the injection pressure acting on the side core.

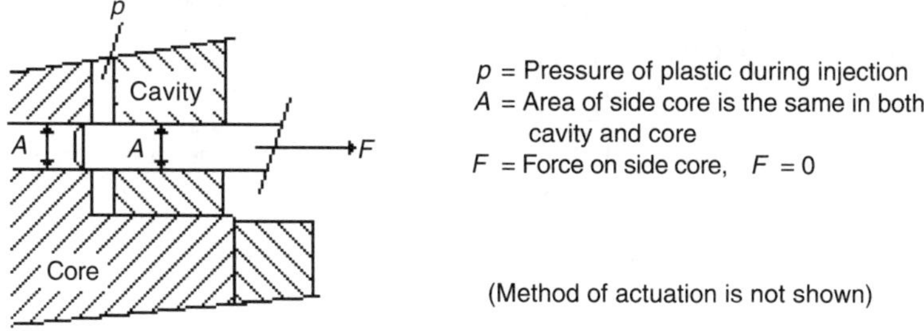

Figure 30.7 The side core enters the core, with no force on the side core. (The only forces required are for moving the side core in and out against friction created by the mold and the plastic.)

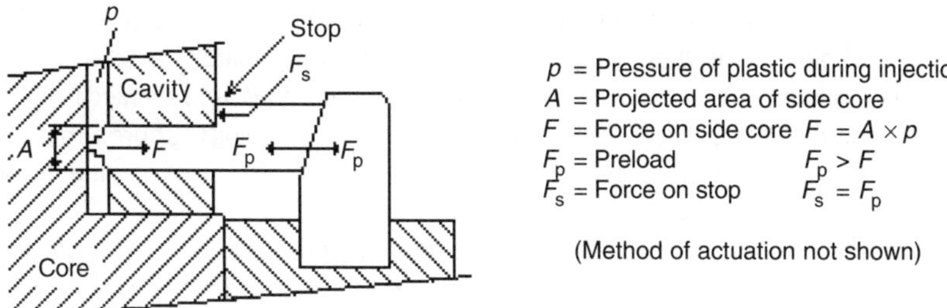

Figure 30.8 Portion of the side core shuts off against the core.

NOTE 2: If the side cores are small, stopping them on the core could create unacceptably high loads in the contact areas and would soon damage both the side cores and the core. In this case, an external stop *must* be used, similar to that in Fig. 30.6 above. Under full clamp, the side core should "just touch" the core, thereby ensuring a flash-free molding. This requires precise mold-making tolerances.

In either case of the notes above (1 or 2), there is always the risk of flashing over the opening to be produced by the shut-off face.

3. The side core enters the core. This method is used occasionally to ensure that flashing does not occur over a (usually round) hole (Fig. 30.7).

   NOTE: This method requires close tolerances to ensure alignment of holes in cavity and core.

4. A combination of 1 and 2 above. In this case, the force on the total projected area should be backed up, even if some portion touches the core and theoretically sees no force (Fig. 30.8).

30.4 Mold Forces at Right Angles to Clamp Force

In most molds, forces at right angles to the clamp force (i.e., forces parallel to the machine platens) are contained (and are balanced) within the cavities. The cyclical injection forces

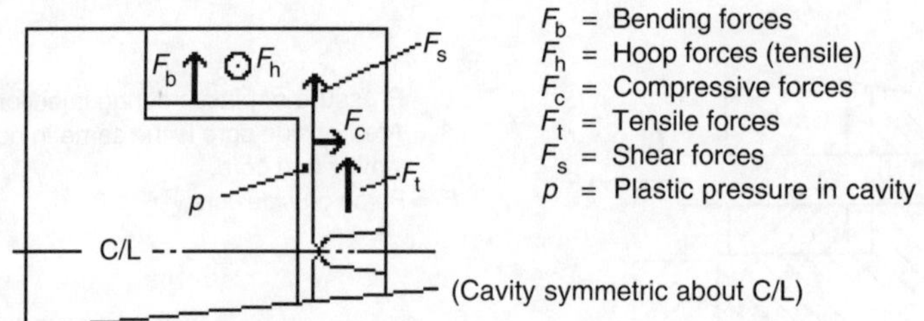

Figure 30.9 Mold forces acting at right angles to the clamp force F_c.

stress the cavity walls only, as shown in Fig. 30.9. (Note that the core is also subjected to most of these forces, but they seldom affect the core as they affect the cavity.)

The actual stresses consist of a combination of tensile, hoop, compressive, bending, and shear stresses. Analysis of these stresses can be very complicated or even impossible, especially when other factors such as the effect of drilling, notches, and surface finishes must be taken into account. For practical purposes and in most cases, some approximation in the calculations and the experience of the designer will be sufficient to arrive at the proper wall thickness required to ensure long mold life.

The designer must *never* forget that *the most important input into any calculation is the injection pressure p of the plastic within the cavity space*, which is in many cases only a rough estimate, also frequently based only on past experience. (This is where computerized mold filling analysis can be of invaluable help to determine mathematically the expected plastic pressures within the cavity space.) All mold calculations depend, therefore, on the accuracy of the assumed pressures within the mold.

The reason why the strength of mold parts is calculated is to optimize the design, that is, to arrive at the smallest size of components while not sacrificing mold life. Here are some considerations:

- Mold materials (steels, etc.) are expensive, but they usually amount only to about 10–15% of the total mold cost. However, breakdowns caused by "under designing" can easily and very quickly wipe out any "savings" in materials.

  The effect of "over designing", however, can also be very costly:

- The larger the cavities, the larger the mold. It will then require a greater and more expensive machine to operate.
- The effect on cycle time and power cost is significant. A heavy mold means that heavy masses must be accelerated each cycle, which requires more power; also, to keep the acceleration forces within acceptable limits, the open and close speeds may have to be slower than with a lighter mold.

30.5 Inserted Cavities

Where the cavities are inserted in a "frame" or "shoe", the cavity walls are backed up by the surrounding steel, and they can be thinner than when free standing, or "modular". In this case, the shoe takes up a portion of the forces pressing from the inside of the cavity.

Fatigue plays an important role in a mold of this construction, and stresses in both the cavity walls and the shoe must be considered and should be calculated to avoid unpleasant surprises when the mold fails after a short period of operation. There are three possibilities:

1. The cavities are pressed into the shoe; both are prestressed. If the preload is greater than the force expected from the injection pressure, cavities and shoe will see only a static load (the preload). This condition is ideal for long mold life, with the least material required for the construction; but it requires accurate forecasting of the injection pressure and very accurate machining of the (press) fitting diameters.
2. The cavities are located in the mold shoe with "transition fit" (i.e., there is accurate location but no distinct preload). At every injection, the cavity will see the full plastic pressure that will expand the cavity against the surrounding shoe. The cavity is stressed cyclically, and so is the shoe. Since the shoe is usually made from milder steels (25-30 Rc), the effect of fatigue is not as significant as for the hardened mold steels of the cavity.
3. The cavities are pressed into the shoe, but are not preloaded enough for the full expected forces during the injection cycle. In this case, the fatigue condition is somewhat better, as was explained in Chapter 18, Metal Fatigue, Section 18.4, with the notion of "stress ratio *R*" in fluctuating loads. Since the cavities take some of the load, the cyclical loads in the shoe will be less than in 2 above, where all of the cyclical load was taken up by the shoe.

30.6 One-Piece (Modular) Cavities

The relationship between product size and depth and wall thickness of the cavity are very important. The deeper the cavity, the thicker must the walls be to ensure that the stresses are below the fatigue limit (Fig. 30.10). This may seam obvious but is frequently missed in mold designs.

30.7 Composite Cavities

In some molds, cavities cannot be made from one piece, either because of availability of material or because of the method of construction and/or size of the cavity. In some molds,

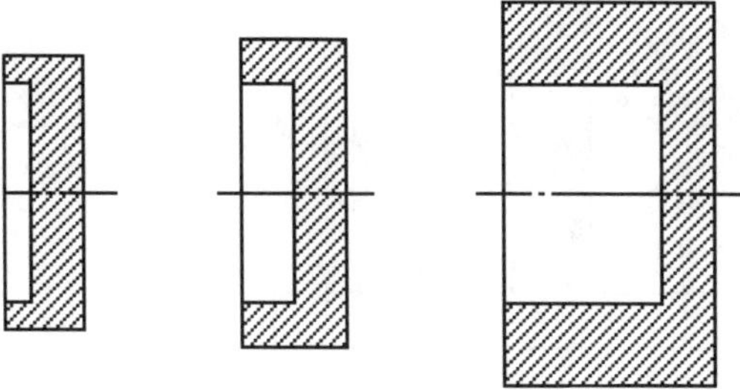

Figure 30.10 Increasing cavity depth (left to right) requires increasing product wall thickness to keep stress below the fatigue limit.

large parts of the cavities move at an angle to the clamping motion of the mold, similar to side cores. Both these cases will be explained using typical examples.

Example: A fairly large two-cavity mold for a waste basket is made by turning the round cavity portions from two separate rectangular steel blocks (Fig. 30.11).

With a separating insert between the two blocks, the three pieces are then joined by heavy bars (width *b* and height *W*) on the outer faces (Fig. 30.12).

The forces required to keep the mold closed against the injection pressure must be calculated. To get a better feel for the forces and pressures involved, the following actual dimensions and values of the mold are selected:

1. To determine the force *F*, assume the following:

   Injection pressure in the cavity space p = 5,000 psi (from experience)
   Height of product H = 8 in.
   Mean diameter D = 7 in.

   Therefore, the force is

   $F = H \times D \times p$, or 8 in. × 7 in. × 5,000 psi = 280,000 lb. or 140 tons.

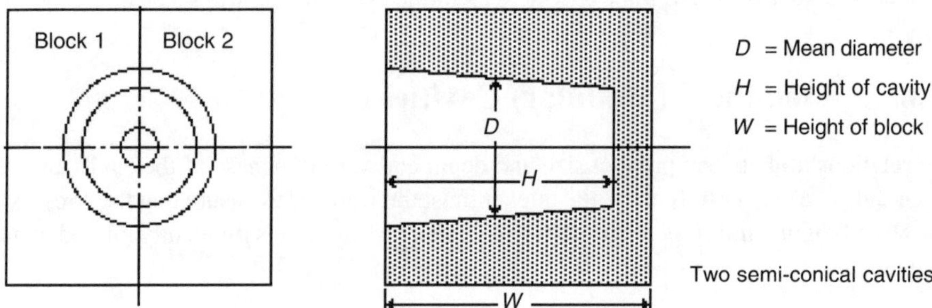

Figure 30.11 Schematic of top view (left) and cross section (right) of two rectangular blocks acting as two semi-conical cavities for production of a waste basket.

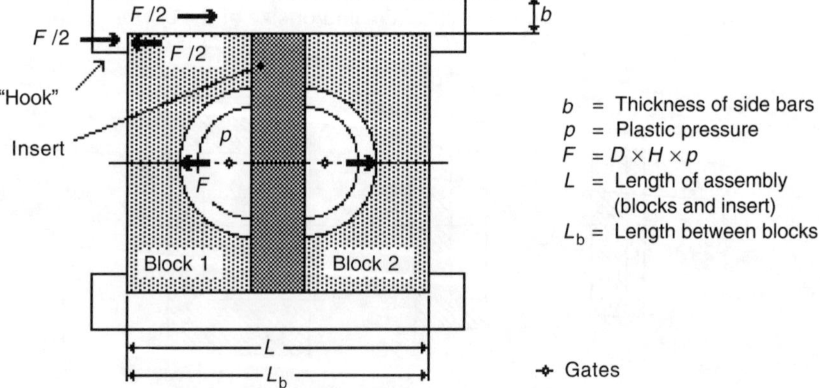

Figure 30.12 Two block cavities separated by an insert and joined by heavy side bars.

One-half of this force F (or 140,000 lb.) is taken up by each side bar.

2. The stressed length of the bar is L_b (= L) Each block is 7 in. wide, and the insert is 3 in. wide. Therefore, the length L_b (or L) is 7 in. + 7 in. + 3 in.= 17 in.
3. The cross sectional area of the bar is $A = W \times b$. $W = H$ + height of the cavity bottom, selected to be 2 in., so W = 7 in. + 2 in. = 9 in. Assuming the thickness of bar b = 2 in., the cross section A of the bar is, therefore $W \times b$ = 9 in. × 2 in. = 18 in.^{2}.
4. To calculate the stresses in the bar, select a steel bar of SAE 4140, preheated to 30 Rc, about 150,000 psi UTS, or a yield strength of 120,000 psi. The bar is stressed by the force $F/2$ = 140,000 lb. The tensile stress in the bar is:

$$S = \frac{F}{A}, \text{ or } \frac{140{,}000 \text{ lb.}}{18 \text{ in.}^2} = 7{,}777 \text{ psi.}$$

This seems reasonable, as it represents a safety factor (fs) of about 15 over the yield strength of this steel, which is also well below the expected fatigue strength for this steel.

5. Now check the elongation of the bar when subjected to the full (estimated) plastic pressure within the cavity. For steel, E = 30,000,000 psi (approximately). By using the above calculated values,

$$f = \frac{F \times L}{A \times E}, \text{ or } f = \frac{140{,}000 \text{ lb.} \times 17 \text{ in.}}{30{,}000{,}000 \text{ psi} \times 18 \text{ in.}^2} = 0.0044 \text{ in.}$$

If the side bars would just fit perfectly without any preload over the cavity assembly, as shown (Fig. 30.12), they would stretch cyclically (once during each injection) by 0.0044 in., enough to make the mold flash between the cavity blocks and the insert. This is, of course, unacceptable.
It is obvious now that the side bars must be preloaded, that is, the length L_b between the "hooks" of the bars before assembling must be slightly less than the length L of the assembly. This can be done by "shrinking" the side bars over the assembly as follows:

$$L_b \text{ (should)} = L - (f + \text{about } 25\% \text{ of } f). \tag{30.2}$$

In this case, L_b should be 17.000 in. – (0.0044 in. + 0.0016) = 16.994 in.

The designer does not want to preload too much, because the preload will increase the stresses in the bar. In this case, there is ample safety factor. The assembly is then rather simple: The bars are heated just enough that they expand more than the required 0.006 in. and are then easily pushed over the blocks. As they cool down, they shrink onto the blocks.

Note 1: The sketch did not show any screws that hold the side bars to the cavity blocks and inserts. The reduction in cross section of the bars as a result of the screw holes was omitted here but must be taken into account when calculating the size of bars.

Note 2: The area of the faces between the hooks of the bars and the outside of the cavities are loaded (statically) by compression, which also must be calculated to be within permissible limits.

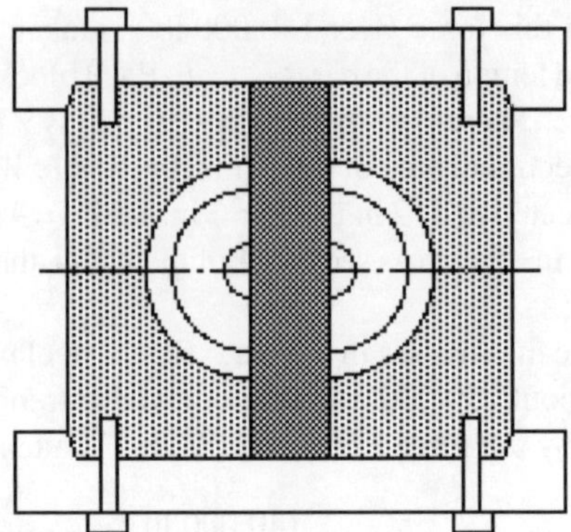

Figure 30.13 Hooks on the side bars are designed with wedges to create the preload.

NOTE 3: The fact that the bars are preloaded above the expected load (force F) makes their load not cyclical but static, which means a longer life for the bars. Under these conditions, a smaller dimension b could be possible, but would increase the stresses in the bars.

NOTE 4: The hooks also see a shear stress, and also a bending stress caused by the moment between the forces $F/2$ in the cavity block and the side bar. Both these stresses are relatively small and could be negligible, but should be checked to ensure that the hook will not shear off.

NOTE 5: Instead of shrinking the side bars, the hooks could also be designed with wedges, as shown in Fig. 30.13, so that when tightening the screws that hold the side bars to the cavity blocks, the wedges would create the required preload, similar to the one calculated above. This method requires great accuracy in manufacture to ensure proper preload. The side bars should rest against the cavity blocks without any clearance when the required tension is reached in the bars.

30.8 Side Split Cavities

In side split cavities, one or more large portions of the cavities move at usually right angles to the clamping motion of the mold. This can be best explained with typical examples, such as:

- Mugs with handles, which split into two halves with a parting line right through the handles (See Section 30.8.1);
- beverage bottle crates, which usually have four full side portions on the cavity, which must move away from the core to release the undercuts in the side walls of the products (see Section 30.8.2); and
- large pails with circumferential reinforcement ribs, where the cylindrical portion of the cavity is in one piece but the ribbed section is usually produced by four moving segments. The forces in such side splits are quite similar to the forces found in side cores (see Section 30.8.3).

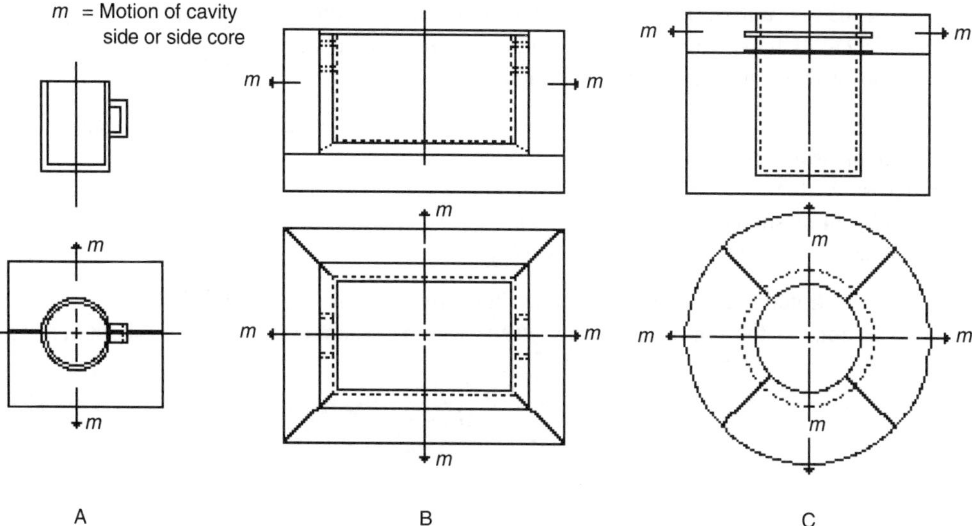

Figure 30.14 Three examples of products molded with side split cavities: A. mug with handle, B. crate, and C. pail.

The number of possible variations is too large to show, or even to attempt to list. The illustrations in Fig. 30.14 show schematically the three examples described above. The heavy lines indicate the parting lines (parting planes) where the moving sections meet.

30.8.1 Mug with Handle

Figure 30.15 is a schematic only, and is used to explain some forces and problems with the design of a mug with a handle. It is only practical if H is not as shown but actually much

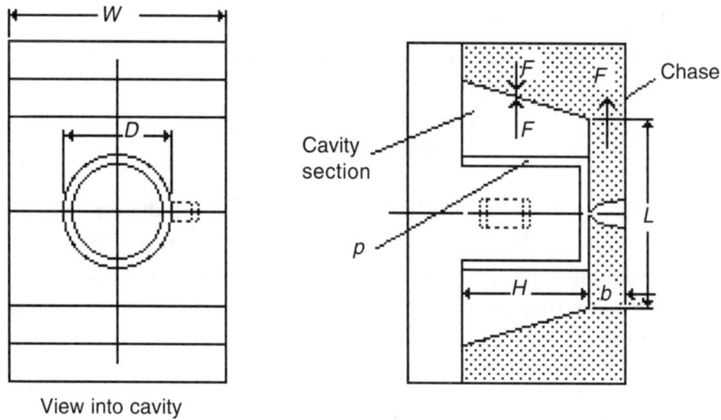

Figure 30.15 Cross sectional views into cavity of mold for a mug with a handle.

smaller than D. With a H/D ratio as shown in Fig. 30.15, there must be additional wedges to lock the cavity halves from the open end, as will be shown later.

The variables in Fig. 30.15 are defined as follows:

Projected area: $A = H \times D$ (area of handle is small, in this case, and is neglected)
Plastic pressure (estimated): p
Force on wedges: $F = A \times p$
Modulus of elasticity: E = 30,000,000 psi
Cross section near gate: $G = W \times b$

As illustrated, the whole force F (distributed over the tapered surface) is taken up by the cross section G of the mold near the gate, over the length L. The stress S in this area G is, therefore,

$$S = \frac{F}{G}. \tag{30.3}$$

The elongation f caused by the force F is

$$f = \frac{F \times L}{G \times E}. \tag{30.4}$$

Example: Assume some dimensions for this layout: H = 4 in., D = 5 in., L = 9 in., b = 1.5 in., W = 11 in., and p is approximately 5,000 psi. The steel (chase) is P20, prehardened to 30 Rc, with a yield strength of 120,000 psi. Entering these values in the formulas,

$F = A \times p = H \times D \times p$, or 4 in. × 5 in. × 5,000 psi = 100,000 lb. or 50 tons.

$$S = \frac{F}{G} = \frac{F}{W \times b}, \text{ or } \frac{100{,}000 \text{ lb.}}{11 \text{ in.} \times 1.5 \text{ in.}} = 6{,}060 \text{ psi.}$$

This may appear to be quite safe, as it is only about 1/20 of the yield strength. However, do not forget to subtract the areas of any openings (screw and dowel holes, cooling channels, gate, etc.) which can reduce the cross section considerably, and thereby the safety margin.

Now calculate the elongation in the narrow section of the chase:

$$f = \frac{F \times L}{G \times E}, \text{ or } \frac{100{,}000 \text{ lb.} \times 9 \text{ in.}}{30{,}000{,}000 \text{ psi} \times 11 \text{ in.} \times 1.5 \text{ in.}} = 0.0018 \text{ in.}$$

This means that when injecting, the plastic forces acting on the wedges and the chase will stretch the narrow section of the chase by about 0.0018 in. Considering that the actual cross section G is smaller because of the holes, etc., as mentioned above, the stretch will be greater, and the mold will definitely flash at the parting line between the cavity sections.

The designer must, therefore, preload the wedge portions of the chase by an amount greater than the above calculated amount f. A reasonable practice is to preload it to between 125 and 150% of f.

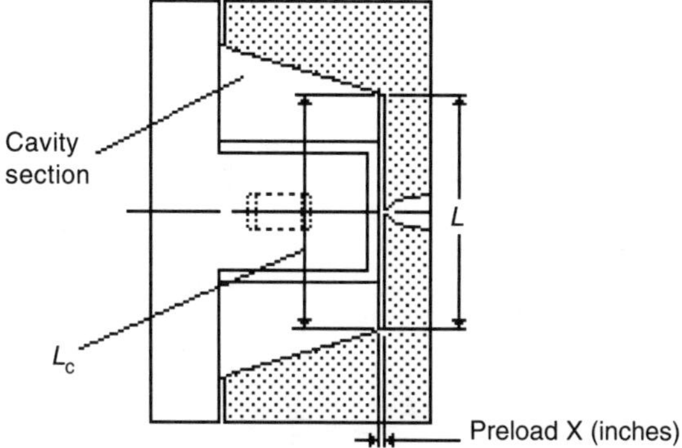

Figure 30.16 Side cross section of cavity for mug with a handle, showing preload *X*.

30.8.1.1 Determining Preload

Look again at part of the sketch shown in Fig. 30.15 (detailed in Fig. 30.16). First, make the dimension *L* somewhat smaller than the corresponding dimension L_c taken over the top face of the cavity sections. If the mold is now closed but no clamp force applied, the cavity sections will not go all the way to the bottom of the chase, but will stay a distance *X* away from the bottom. This distance *X* is commonly called preload, even though it is a length, not a force.

By applying the full clamp force, the wedges of the cavity section slide over the tapered surfaces of the chase until the cavity sections seat firmly against the bottom of the chase. The resulting wedge action pushes the tapers on the chase outward, and creates the required preload force in the narrow section of the chase. Now, during injection, the cavity sections will not move (outward) unless the forces generated by the plastic pressure exceed the calculated preload.

The required preload (dimension *X*) is fairly easy to calculate, but the accuracy in manufacture is critical because even small variations can affect the amount of preload considerably. Note an important difference between this application and the one with the composite cavity shown in Fig. 30.11. In that application, the preload was generated by heat shrinking or fixed wedges, and the prestressed members where subject to a *static* load. In the case of moving cavity sections, the preload is *cyclical* , that is, it occurs at every cycle, and can significantly affect the mold life.

There must be an ample safety factor, and the stress level should be well below 10% of yield strength of the material. Note also that in such cases, it is better to stay away from very hard mold steels, which are much more sensitive to cyclical loads.

If high wear resistance is required (e.g., for the surfaces where the wedges meet and slide against each other), special bronze or hardened steel wear inserts should be used with a milder steel chase, such as P20, rather than making the whole chase from a hardened mold steel. It is easier and less expensive to replace inserts if they wear; also, inserts can be shimmed to eliminate the need for close tolerancing in manufacturing to arrive at the correct preload.

To repeat: it is vital that the proper preload is found *before* the mold goes into operation. If the preload is insufficient, the mold will flash. With excessive preload, the stress level in the steel may be raised so high that it can substantially reduce the mold life.

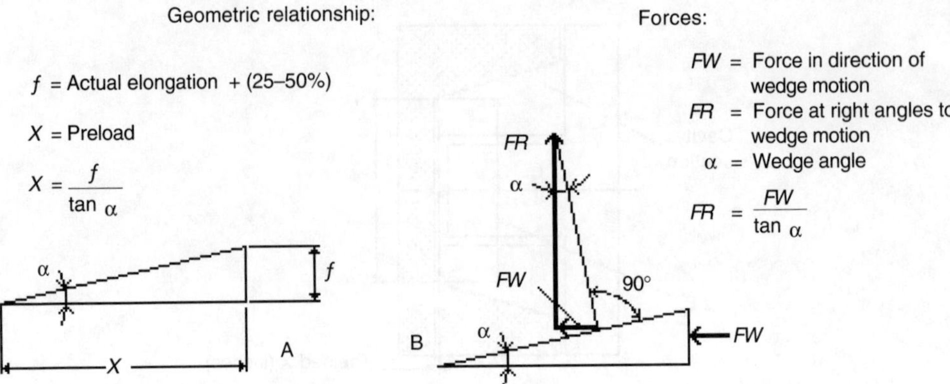

Figure 30.17 Theory of wedges (distances and forces): A. Preload in wedge, and B. wedge force.

30.8.1.2 *Calculating Preload in Wedges*

General relationships regarding forces in wedge motion are shown in Fig. 30.17. Referring to this figure and going back to the example of the mug, if an actual preload of 0.0027 in. (0.0018 in. stretch (*f*)+ 50% safety factor) is desired and the wedge angle is 15°, then from Fig. 30.17A,

$$X = \frac{f}{\tan\alpha}, \text{ or } \frac{0.0027\text{ in.}}{\tan 15°} = \frac{0.0027\text{ in.}}{0.27} = 0.01\text{ in.}$$

Also from Fig. 30.17B, one can calculate how much of the (available) clamping force will be diverted into the preload. The force in the narrow section of the chase was 100,000 lb. By adding 50% to properly preload, a force F_R of 150,000 lb. is required. The wedge force F_W to be supplied by the clamp is, therefore,

$$F_W = F_R \tan\alpha, \text{ or } 150{,}000\text{ lb.} \times 0.27 = 40{,}500\text{ lb. (20.25 tons).}$$

This is a significant amount and must be carefully considered when selecting the clamp size required for the mold.

30.8.1.3 *Shims*

When shimming a mold part, it is better practice to use "heavy" shims, rather than thin, commercially available "shim stock", which come in a limited range of thicknesses.

The disadvantages of using standard shim stock are:

- Any special thickness must be made up from two or more standard size thicknesses of shim stock.
- Cutting shim stock can be time consuming and difficult. Bent edges can affect the actual thickness of the resulting shim pack.
- When dismantling a mold, thin shims can be easily damaged or lost.
- Failure to replace them where they were before dismantling the mold will result in incorrect mold (preload) dimensions and flashing.

Advantages of using standard shim stock are:

- Low cost and
- easy availability.

A *heavy shim* is a piece of (usually hardened) steel of substantial thickness, at least about 1 mm (0.040 in.) thick, ground to a calculated dimension. Advantages of using heavy shims are:

- The shim is shown on the assembly drawing, and carries a mold part number.
- It can be safely handled without becoming damaged.
- Overlooking the shim during assembly is virtually impossible because the absence of it is clearly evident.

Disadvantages of using heavy shims are:

- The added cost of additional mold parts, and
- changes to the size can be made only by grinding (reducing the thickness). To increase the thickness, the shim could be plated heavily and then reground, but it is better to replace the shim with a new one with the proper dimension. In either case, the new dimension(s) must be recorded on the drawings for the mold.

30.8.1.4 Stresses and Deflection in Side Cavities

Look again at a sketch (Fig. 30.18) that is similar to Fig. 30.15. As already pointed out, the design is only practical if H is much smaller than D.

In a condition similar to the illustration in Fig. 30.18, there is a pronounced bending moment $M = m \times F$ which adds to the stresses in the chase, particularly in the corners where the wedges meet the bottom of the chase. The dimension m should be kept small to ensure that the stresses caused by the moment will not bring the total of all stresses beyond permissible limits.

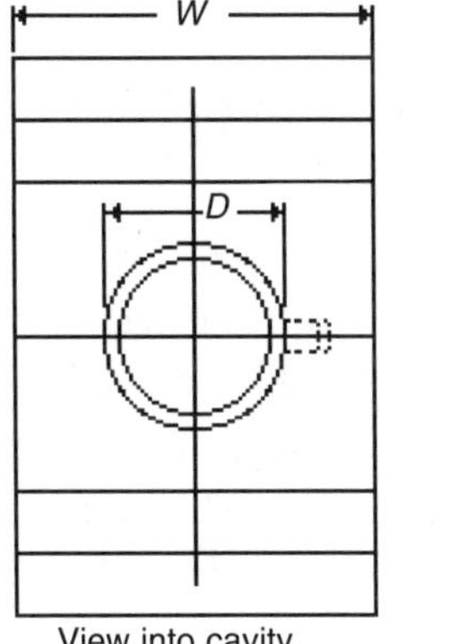

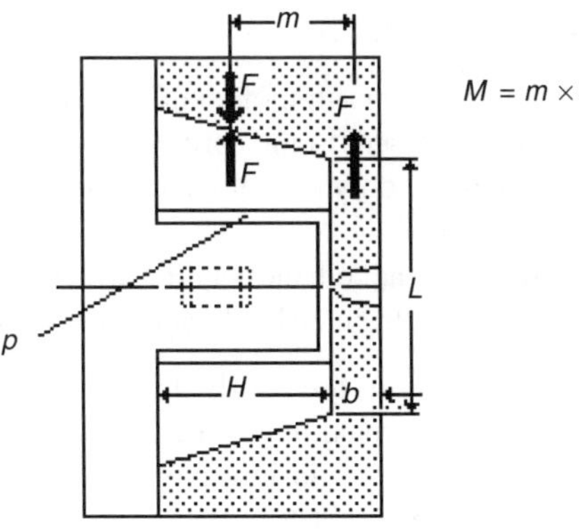

Figure 30.18 Cross sectional views of a cavity for a mug with a handle, showing dimension *m* for calculating the bending moment of the chase.

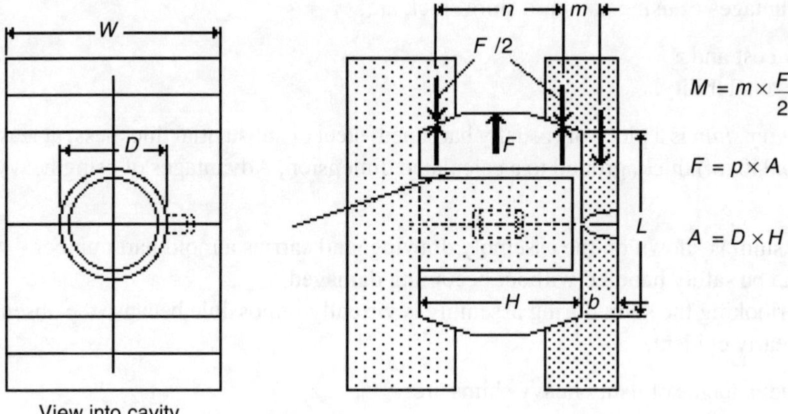

Figure 30.19 Cross sectional views into cavity of mug with handle with wedges on the core side similar to those on the cavity side.

With an H/D ratio as shown or even greater, there must be additional wedges to lock the cavity halves from the open end. Products such as water pitchers, or any deep draw technical products, fall into this category. The design must be modified to eliminate all or most of the bending stresses in the chase.

This is usually done by providing wedges on the core side similar to, but not necessarily the same as, those on the cavity side, as shown in the simplified sketch in Fig. 30.19. There must be some provision (which is not shown) to guide and move the cavity slides. This can be done either within the cavity chase with angle pins or hydraulically, or with a system of mechanical links. The slides must move 1) away from the cavity chase to unlock the wedges by following the motion of the core, and 2) sideways, away from the centerline of the mold, to release the undercuts (in this case, handles).

Provided the calculations were correct and properly executed, various differences can be seen from the first design with wedges on the cavity side only (remember, all stresses caused by the force F are cyclical):

1. The forces seen by each mold half are only one-half what they were before.
2. The bending moment in the corners is reduced for two reasons:
   a. the forces are only half of what they were before, and
   b. m can be small.
3. A new problem is added: the bending effect of the "beam" formed by the cavity half loaded with an evenly distributed load p and supported by the wedges, at distance n (Fig. 30.20A).

If the cavity sections are poorly designed, too thin, and have many channels within, the bending stress and the maximum deflection can be significant and must be carefully checked. If excessive, it could cause flashing along the parting lines where the sections meet, especially halfway between the height H of the side cores.

Calculating the deflection of a beam as illustrated in Fig. 30.20 can be quite a challenge, but since the thinnest cross section is t, to save time and, if anything, to be conservative, one can assume that the cross section of the beam is $q \times t$, and ignore the stiffening effect (the

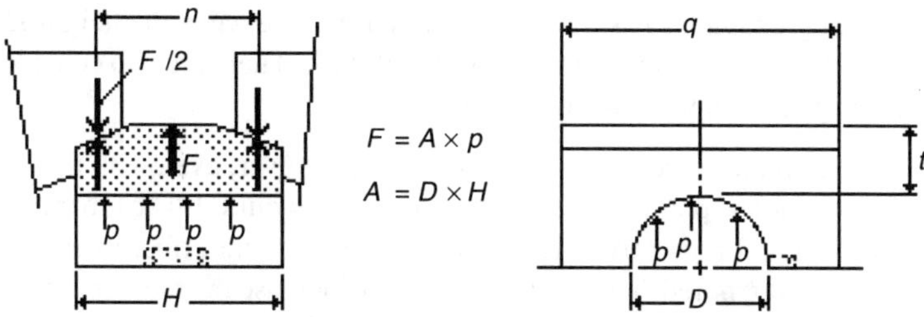

Figure 30.20 Deflection stress on beam created by wedges and loaded cavity half (left) and cross section of the beam (right).

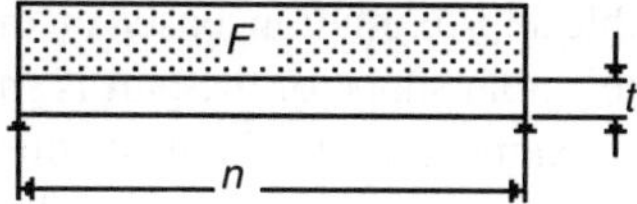

Figure 30.21 Maximum deflection on a rectangular, uniformly loaded beam (simplified from Fig. 30.20).

moment of inertia) of the curved sides. This approach will in part offset the weakening effect of the cooling lines, etc., in the beam. Of course, in critical cases, the proper, complete calculations may be required to find the exact maximum deflection of the beam to ensure that the parting line will not flash.

Figure 30.20 can be simplified, as shown in Fig. 30.21. The general deflection formula for a uniformly loaded beam, using the same letters as in the figures, is as follows:

$$f = \frac{5}{384} \cdot \frac{(F \times n^3)}{E \times I}. \tag{30.5}$$

For a rectangular section, the formulas for the moment of inertia, I, and the section modulus, Z, are:

$$I = \frac{q \times t^3}{12} \text{ and} \tag{30.6}$$

$$Z = \frac{q \times t^2}{6}. \tag{30.7}$$

Example: A pitcher is 10 in. tall with a 6 in. diameter, $n = 8$ in., $q = 12$ in., $t = 3$ in., $p = 5{,}000$ psi, and $E = 30{,}000{,}000$ psi.

$$F = 10 \text{ in.} \times 6 \text{ in.} \times 5{,}000 \text{ psi} = 300{,}000 \text{ lb.}$$

$$I = \frac{12 \text{ in.} \times (3 \text{ in.})^3}{12} = 27 \text{ in.}^4.$$

$$f = \frac{5}{384} \times \frac{300{,}000 \text{ lb.} \times (8 \text{ in.})^3}{30{,}000{,}000 \text{ psi} \times 27 \text{ in.}^4} = 0.0025 \text{ in.}$$

In the preceding example, f is the deflection halfway between the height H of the side cores. This deflection is very high, and the mold will flash. There are two considerations that may help the designer deal with this problem:

1. The pressure could have been assumed too high. However, even with half that pressure (at which it may not be possible to fill the mold), there would still be too much deflection.
2. The designer should now seriously consider the beneficial effect of the curved sides, which were at first ignored. This will significantly increase the moment of inertia (I), but to be fair, the designer must then also consider what effect the drillings in the side cavities have, and deduct the corresponding areas. This could be quite time consuming but may be well worthwhile, to keep the mold as small as possible.

While an accurate calculation of the moment of inertia, I, can be quite cumbersome, the designer should consider a possible approximation, based on the assumptions outlined below and illustrated in Fig. 30.22. The actual shape of the part is shown to the left of Fig. 30.22.

Assumption 1, as used in the example above, is most conservative, with $I = 27$ in.^{4}, and can lead to unnecessarily heavy constructions. Assumption 2 is really misleading and should never be used, since it omits the reducing effect of the cavity itself, as well as the effect of all cross drillings. For similar dimensions as in the example, I would be 216 in.^{4}, or 8 times greater than I from assumption 1.

Assumption 3 considers the cross section as an arch, fitting into the actual shape. The stiffening effect of the two corners, which were not included in the calculation, is estimated to make up for the losses as a result of cross drillings, etc. In this example, $I = 103$ in.^{4}, or about 3.8 times greater than the I from assumption 1. Substituting the value $I = 103$ in.^{4} in the previous calculation for maximum deflection, the actual deflection would be about 0.0006 in. This is still too much, and some redesign will be required.

Note that each mold half will deflect by this amount; therefore, the gap through which the plastic could flash at the center of the span will be twice the amount of the deflection, in this case, 0.0012 in. The designer should target the maximum total deflection to be less than the venting gap of the mold for the material planned to be injected. This is usually ≈0.0002–0.0004 in. There is nothing wrong with any gap anywhere, in any mold, even if it is caused by deflection, as long as only air can escape and the plastic is restrained from passing through. There can never be too much venting!

Figure 30.23 shows stresses in the sliding cavity of the above example. Maximum stress will occur at the center of the cavity, where the maximum deflection occurs at F.

The general formula for stress in a uniformly loaded beam, supported on both ends, is:

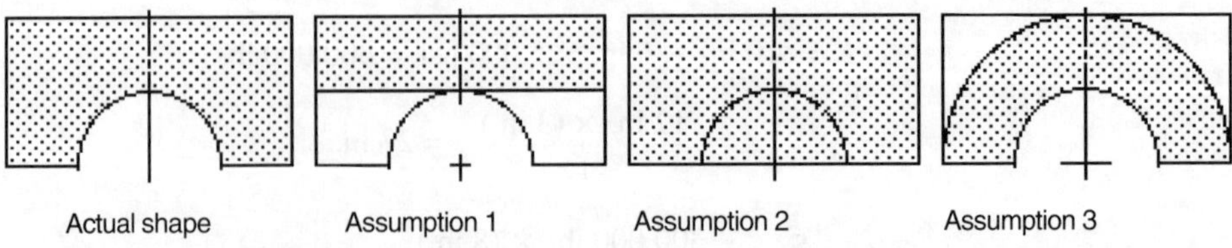

Figure 30.22 Calculation of the moment of inertia for actual shape using assumptions 1, 2, and 3.

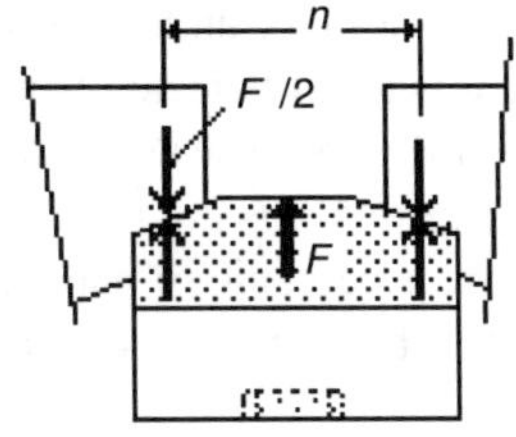

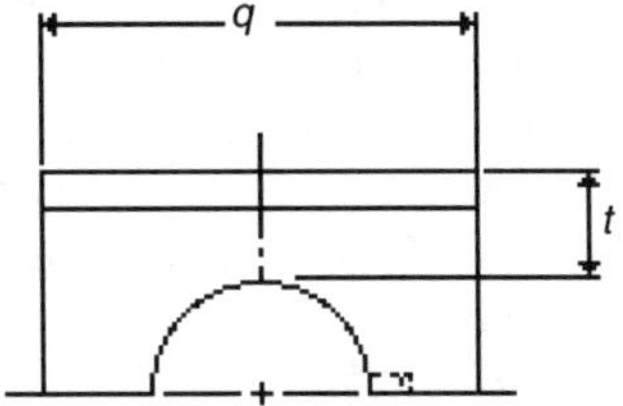

Figure 30.23 Deflection stresses in sliding cavity are maximum in center of cavity.

$$S = \frac{W \times L}{8Z}. \tag{30.8}$$

In this example, the load, W, is equal to F. Also $L = n$, in the example, and Z = section modulus.

Here, the designer again faces the same dilemma as with the moment of inertia. Unless there is an accurate calculation of Z, which could be quite cumbersome, the designer can approximate the cross section of the sliding core in a manner similar to the approximation for the moment of inertia I.

Going back to the dimensions used in the example, note the difference in Z between assumptions 1 and 3, above. In assumption 1, $Z = 18$ in.^{3}, and in assumption 3, $Z = 34.7$ in.^{3}, almost twice as much. Again, it can be stated that if S is acceptable with assumption 1 (the simplest one), there is no worry about excessive bending stresses in the sliding cavity. Using approximation 1 in the example,

$$S = \frac{F \times n}{8Z} = \frac{300{,}000 \text{ lb.} \times 8 \text{ in.}}{8 \times 18 \text{ in.}^3} = 16{,}666 \text{ psi.}$$

This may be too high for long mold life, since it is close to 10% of the yield of the steel selected. Using approximation 3, and taking advantage of the "reserve strength" in the sides of the cavity block, Z is about double, and the stresses become about one-half of the above, or about 9,000 psi, which should be acceptable. This method of calculation is quite simple, and in general yields good results.

One must never forget that in mold design, there are other factors besides the consideration of fatigue strength which affect the selection of steel sizes. For example, availability of steel, space in the molding machine, machine tool capacity, and other factors may limit the designer's choices. But this must not be used as an excuse to build a mold which is inherently weak and will not last.

30.8.2 Crate Mold

In the mold for a mug with a handle, the design was split in one plane only and had to consider only simple, opposing forces. In the case of a crate mold, where usually all four sides of the crate move away from the center, conditions are somewhat more complicated, but the principle is still the same. The arrows in Fig. 30.24 indicate the motion of the side cores.

To hold the side cores in place during injection, some method is needed to apply preload to each side core, as shown in Fig. 30.25. Not shown are the methods of moving and guiding the side cores, which must move in step with the retracting core until the undercuts in the sides are freed and the product is able to follow the core before it is finally ejected from it.

Preload is similarly applied as described in Section 30.8.1. It should be noted that the long taper on the side core may be necessary to guide the side core, but it is not necessarily used to take up the force F_2 over its full length. In fact, it is quite acceptable to make the acting portion of the wedge for the preload similar in length to the taper where F_1 is contained, so that in fact $F_1 = F_2 = 0.5\ F$.

This means that the bottom of the cavity and the bottom of the core share equally in providing the preload necessary to prevent the side cores from moving away from the core and letting the mold flash. Both these "bottoms" must be analyzed to make sure that the stresses are well below the fatigue limit of the steels selected. The forces F, F_1, and F_2 are cyclical, since they occur at every molding cycle.

When the mold is closed but not yet clamped up, the side cores will be away from their seated positions by the amounts of their preloads X_1 and X_2 shown in Fig. 30.25. The diagonal ends of the side cores will be, therefore, somewhat distant from each other (see "Gap" in

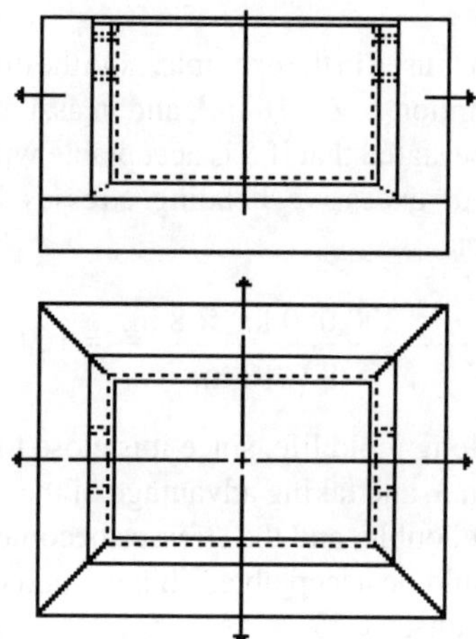

Figure 30.24 Motion of four side cores away from the center.

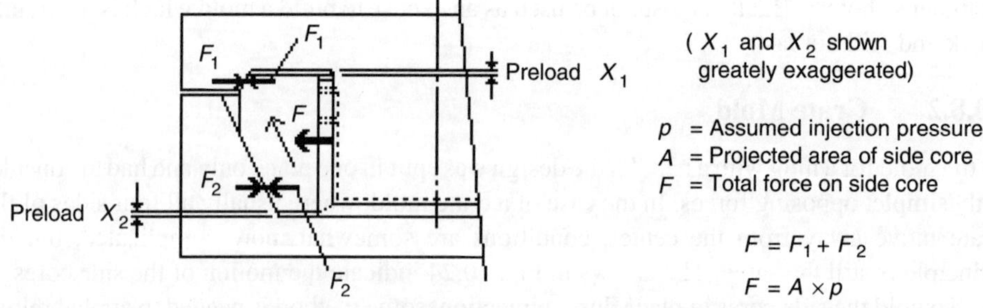

Figure 30.25 Preload on side cores to hold them in place during injection.

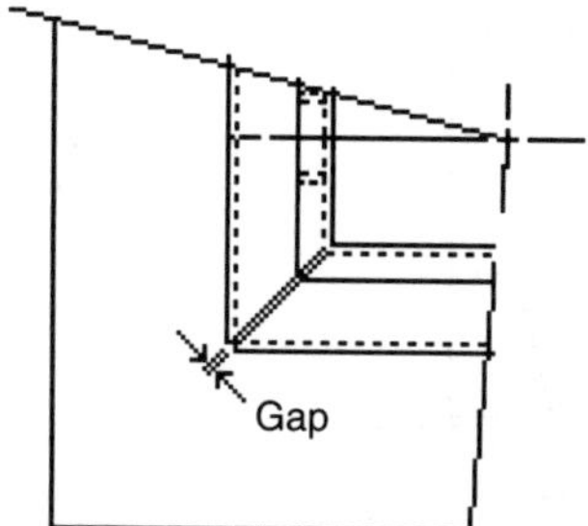

Figure 30.26 The gap is open when the mold is closed but not yet clamped.

Fig. 30.26) and come together (to just touch) only after the mold is clamped up. There is no need for preloading the diagonal parting planes. If the side cores are properly preloaded to prevent any outward motion, the diagonal parting planes will not separate.

The calculations for the proper preload and the resulting stress levels are similar to the ones shown for the mug with a handle and will not be repeated here. The suggested taper angles are preferably between 10° and 15°. A taper angle of less than 10° is rarely used; a 7° taper angle is the minimum recommended, because it approaches the friction angle for lubricated steel on steel. The advantage of using a smaller taper angle is that less force is required and, therefore, a smaller percentage of the available clamp force will be diverted to prevent the side cores from moving away from the core, during injection.

30.8.3 Pail Mold

Figure 30.27 is only a schematic; there are many different pail designs and ways of designing such molds. The design variations depend mainly on considerations such as shape of the product, *L*/*t* ratio, melt index, total number of the product to be molded over the life of the mold, mold cost, machine size, etc. But the basic principles are the same in all cases. There is always:

1. A motion of the segments in the direction parallel to the core, while moving radially outward (away from the center of the pail) to release the undercuts on the cavity side;
2. a preload to prevent the segments from moving during injection (force F_s) (With the mold closed but not yet clamped up, there will be a space *X* separating the faces at the parting line, and a gap on the radial faces of the segments.); and
3. a preload to ensure that the alignment between cavity and core is maintained during injection (force F_a. The alignment is achieved either with a circular taper fit or with four straight wedges, but should never depend on leader pins.

Leader pins (not shown in Fig. 30.27) are used in addition to tapers or wedges to ensure that the mold halves will be *approximately* aligned before the tapers or wedges can engage. They also protect the cores during servicing of the mold. The fit in their bushings must be loose enough to ensure that they will not interfere with the final alignment by the tapers.

The preloads F_a and F_s are necessary to operate the mold successfully (i.e., without flashing and without misalignment); they must be calculated using the methods and assump-

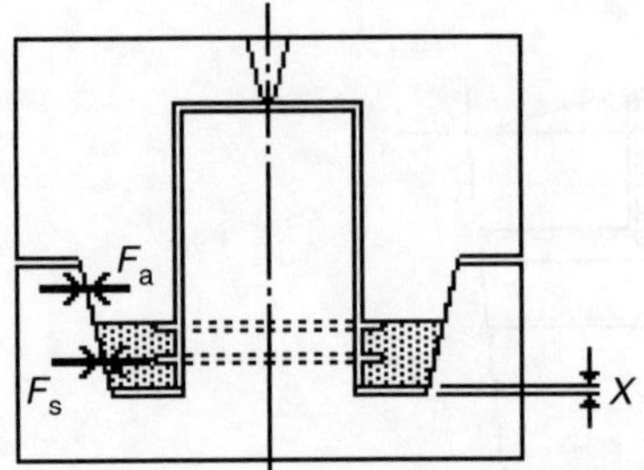

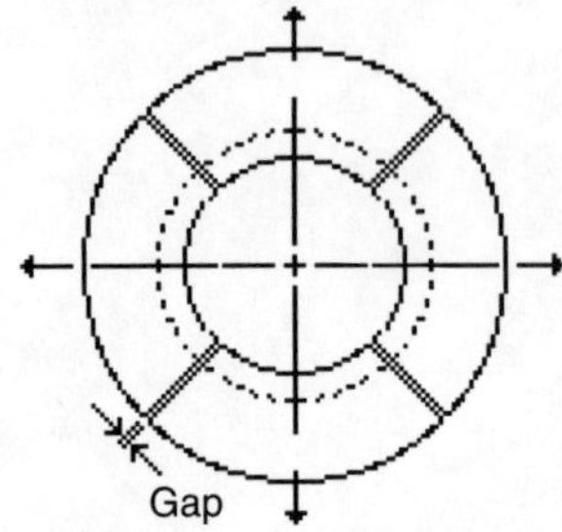

Figure 30.27 Schematics of a mold for a large pail, showing preload *X* and preload to ensure alignment between cavity and core, and to prevent movement during injection.

tions outlined earlier in this chapter. Of course, it is also necessary that the calculated dimensions are executed properly during the manufacture of the mold.

Very close tolerancing will be necessary in all cases. The use of four separate wedges instead of circular tapers adds to the number of components in the mold (and to its cost), but the wedges can be more easily adjusted than taper fits if this becomes necessary, and they can be easily replaced in case of severe wear.

The alignment taper force F_a and the forces F_s are cyclical (i.e., they appear during each cycle, and their value ranges from a minimum of zero to a maximum F_a and F_s). In Fig. 30.27, where the core is shown in a one-piece construction, they add to the stresses seen by the outside portion of the core (alignment ring). They are resisted by the hoop strength of this ring and, to some extent, by the bottom of the core block.

In Fig. 30.28, the alignment is reversed, so that the force F_a will tend to open the cavity block but will assist in counteracting the forces F_s acting on the core block. Comparing the forces in Fig. 30.28 to those in Fig. 30.27:

- In the cavity, in both cases, there are hoop stresses as a result of the injection pressure.
- In Fig. 30.27, the forces F_a, because of the preload of the circular locating taper, act to restrain the cavity from expanding. In Fig. 30.28, these forces F_a add to the hoop stresses in the cavity.
- In the core (or rather in the locating ring), there are, in both cases, hoop stresses as a result of the preload on the segments; in Fig. 30.27, the preload force from the alignment tapers adds to the hoop stresses in the locating ring; in Fig. 30.28, the preload from the locating tapers in the cavity works against the expansion of the locating ring and reduces the stresses in the ring.

In this design all above stresses are cyclical.

For practical reasons, in many molds, the locating ring and the core are made from two separate pieces. The ring is either shrunk onto the core or mounted with a small (self-locking) taper on the core. In both these designs, the ring is preloaded, either from the shrink fit or from the taper (Fig. 30.29). If the preload is selected to be greater than the forces from the locating tapers or the wedges, the stresses in the ring are static, which is the preferred condition for long mold life. (Not shown in Fig. 30.29 are the screws used to hold both the core and the locating ring of either design to the core backing plate.)

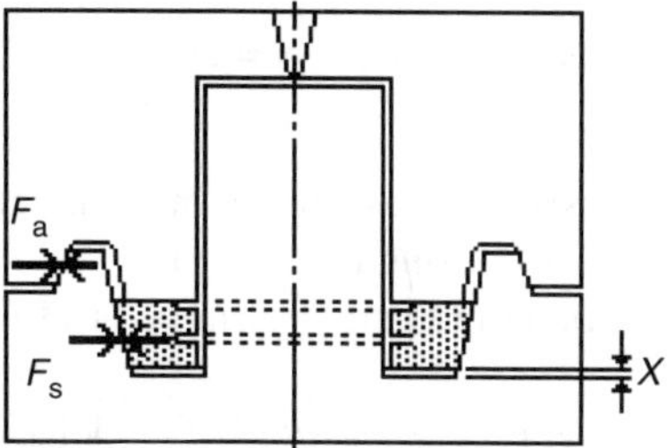

Figure 30.28 Alignment (locating ring) is reversed from that in Fig. 30.27, and the core locks into the cavity.

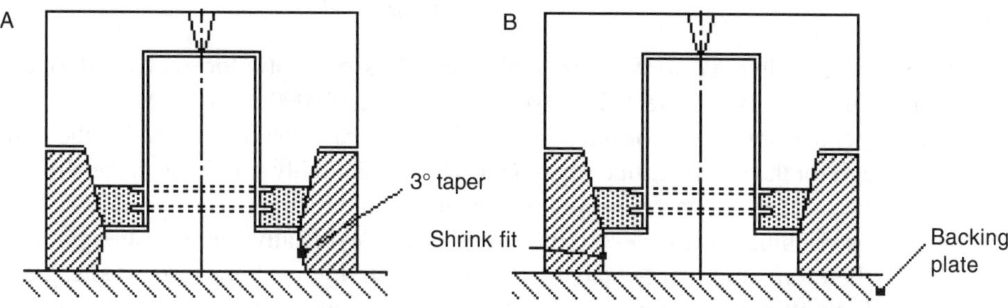

Figure 30.29 Locating ring is preloaded by A. taper fit or B. shrink fit.

30.8.3.1 Hoop Stress and Cavity Expansion

Hoop stress and cavity expansion apply primarily to thick-walled, tubular cavities and to heavy rings, such as the locating ring (Fig. 30.30). The formulae are approximations but are usually suitable for mold design. Allowance must be made for cross drillings, holes, etc., which will affect the effective wall thickness.

$$\text{Stress at inside of hoop: } S = p\,\frac{R^2 + r^2}{R^2 - r^2}, \text{ and} \tag{30.9}$$

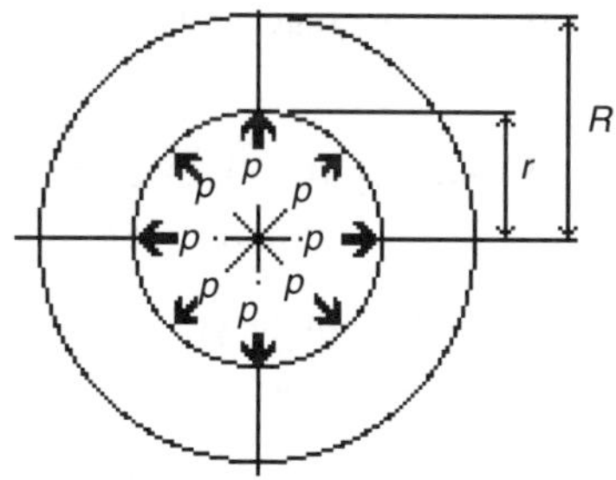

Figure 30.30 Hoop stresses acting on a thick-walled ring.

$$\text{Stress at outside of hoop: } S = 2p\,\frac{r^2}{R^2 - r^2}, \tag{30.10}$$

where S = hoop stresses in kg/cm² (psi), p = pressure on ring in kg/cm² (psi), R = outer radius of ring in cm (in.), and r = inner radius of ring in cm (in.).

The expanion of the inner and outer radii are determined as follows:

$$\text{Increase in } r\text{: } f = p\,\frac{r}{E}\left(\frac{R^2 + r^2}{R^2 - r^2} - V\right) \text{ and} \tag{30.11}$$

$$\text{Increase in } R\text{: } f = p\,\frac{R}{E}\,\frac{2r^2}{R^2 - r^2}, \tag{30.12}$$

where f = expansion in cm (in.), V:= Poisson's ratio (Poisson's ratio for steel $V = 0.3$), and E = modulus of elasticity (for steel, 2,100,000 kg/cm² (30,000,000 psi)).

The cavity expansion can be considerable, and S must be calculated to ensure that the walls are thick enough for the job. Note that stress S must be considerably smaller than the ultimate stress S_u for the cavity material, to prevent the cavity from bursting.

Since the stretching (loading) of the cavities occurs cyclically, fatigue strength is in volved. It is, therefore, important *not* to use the ultimate strength S_u but a suggested value of about 10% of the ultimate strength S_u, or better and more conservatively, 10% of the yield strength S_y of the cavity material.

Note that in Figs. 30.27 and 30.28, the cavity is shown with a solid bottom. This was done only to simplify the illustrations. In reality, most cavities of this type of mold have large gate inserts or even a moving bottom so that the cavity is really a ring. If there is a solid bottom, or only a small bore to accommodate the gate, the bottom will reinforce the ring and assist in resisting the plastic pressure. For practical purposes, however, this "reinforcement" can usually be ignored.

Example: A pail has a diameter of 10 in. and a height of 15 in. First, assume that the cavity will require a 5-in. thick wall, to allow for the losses in strength because of the water channels drilled in the walls. Note that the height is immaterial, since the strength of a "slice" of 1 in. is considered at right angles to the pail axis through the ring.

To use equation 30.9, $r = 5$ in., $R = 5$ in. + 5 in. = 10 in., and the cavity pressure is assumed at $p = 5{,}000$ psi, average, and the steel has a yield strength of 120,000 psi.

$$S = 5{,}000 \text{ psi}\left(\frac{100 \text{ in.}^2 + 25 \text{ in.}^2}{100 \text{ in.}^2 - 25 \text{ in.}^2}\right) = 8{,}333 \text{ psi}.$$

S is, therefore, well below 10% of the yield strength, since 8,333 psi < 12,000 psi.

The increase of the inner radius r, using equation 30.11, is

$$f = 5{,}000 \text{ psi} \times \frac{5 \text{ in.}}{30{,}000{,}000 \text{ psi}} \times \left(\frac{100 \text{ in.}^2 + 25 \text{ in.}^2}{100 \text{ in.}^2 - 25 \text{ in.}^2} - 0.3\right) = 0.0011 \text{ in.}$$

NOTE: The inner radius increases by 0.0011 in., or, the wall thickness at the moment of maximum injection pressure will be more than 0.001 in. larger. If there were no shrinkage and no draft angle, this could be enough to lock the product between the cavity and the core, and prevent the mold from opening.

The increase in the outer radius R is

$$f = 5{,}000 \text{ psi} \times \frac{10 \text{ in.}}{30{,}000{,}000 \text{ psi}} \times \left(\frac{2 \times 25 \text{ in.}^2}{100 \text{ in.}^2 - 25 \text{ in.}^2} \right) = 0.0011 \text{ in.}$$

In this case, the outer radius R will expand by the same amount as the inner radius, r.

This knowledge is important to understand how much more (or less) force will be exerted on the locating taper or wedges used to align the cavity and core. In Fig. 30.27, where the cavity locks into the core, the additional expansion will spread the core taper and may loosen the fit of the side core sections. In Fig. 30.28, where the core locks into the cavity, the spreading of the cavity will reduce and could even eliminate the preload planned for the taper lock. The result could be loss of alignment between cavity and core.

Note that the average radius of the tapers is less than R but larger than r; but a spread of even one-half of the above calculated expansion (0.0011 in. ÷ 2= 0.00055 in.) could have a significant effect on the mold. Note also, that the assumed 5,000 psi will be higher nearer the gate and lower near the open end of the cavity. In any case, 5,000 psi is just an assumption and could be much more as the plastic wall thickness is decreased. It is, therefore, very important to try to determine, by experimentation and instrumentation of the mold, the actual pressures inside the cavity under different molding conditions and with different plastics and melt indices.

30.9 Preload in Screws

Essentially, every screw consists of a head and a threaded shank. When the screw is turned into the receiving (female) thread in the work piece (or in the nut), there is at first no resistance except friction, until the joint touches and the head comes to rest on the work piece. Further turning (torquing) of the screw will begin to stretch the screw, as a result of the wedge action exerted by the threads. The screw is thus preloaded.

All screws used for fastening have a lead- or wedge angle smaller than the friction angle between steel and steel, even when lubricated; thus, the screw is inherently self-locking (i.e., it will, under normal circumstances, not unscrew after tightening).

For in-depth discussion of preload on screws, the reader is referred to Chapter 19, Mold Screws. Section 19.5 covers the holding action and preload of screws, including detailed calculations. Section 19.7 describes the effects of cyclical loads on screws.

30.9.1 Shoulder Screws

A shoulder screw is illustrated in Fig. 30.31. With shoulder screws (commonly known as stripper bolts), the effective length L is almost Ø, and, therefore, they can never be tightened properly. Their use should be avoided in molds.

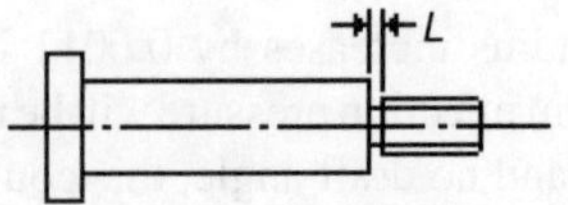

Figure 30.31 Shoulder screw.

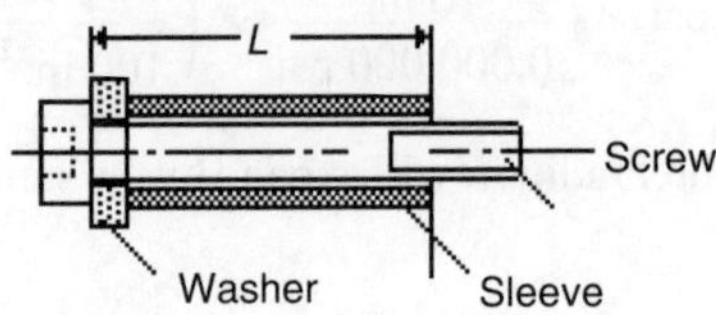

Figure 30.32 Sleeve, washer and screw combination instead of shoulder screws.

Instead, a sleeve, washer, and screw combination should be used, as shown in Fig. 30.32. *L* is now substantial, and the screw can be tightened effectively.

Heavy vibrations have a similar effect as frequent shocks and may cause the screws to loosen, particularly if the screws are short and it is not possible to stretch them sufficiently. If vibration or shock is present, screws should be prevented from turning (loosening) by such methods as shown in Section 19.12.

In molds, the plate sizes and cross sections A_p are usually large and in proportion to the clamp size and force; the effect of plate compression can, therefore, usually be ignored. However, with the full clamp force acting on smaller units (e.g., cavity modules), the compression f_p may become significant. Note that plate compression f_p must be calculated to include cored-out areas and steps to arrive at a relevant value.

30.9.2 Mold Mounting Screws

Screws to hold the mold in the machine serve two purposes: 1) to hold the mold on the platens so that the mold halves will not fall down, and 2) to resist the opening forces to the mold.

The mold opening force available on a molding machine is usually about 10% of the clamping force, and is indicated in the specifications for every machine. Thus, mold mounting screws or other methods of holding the mold on the platens are subject to a cyclical load.

However, the mounting screws are preloaded to a value much higher than the expected maximum mold opening force of each screw, and see only the static force applied to them by torquing during mold installation. It is very important that all screws are torqued equally, and above the expected opening force per screw. If not, a less tightened screw will be stretched cyclically by the force exceeding the (insufficient) preload. Screws should be tightened to approximately 60–75% of their yield strength.

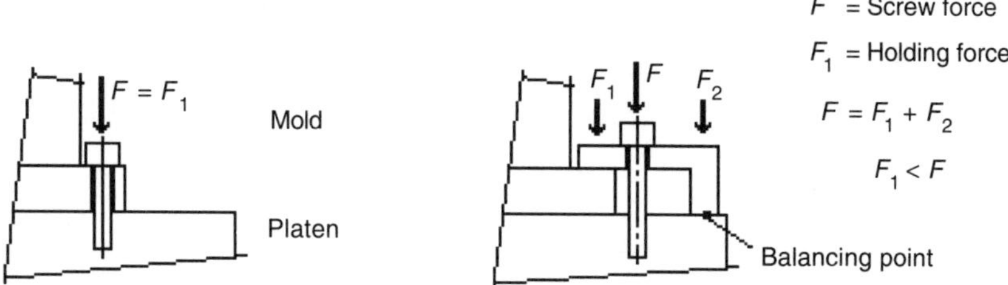

Figure 30.33 Screws and clamps holding the mold to a platen.

30.9.3 Mold Clamps

When using mold clamps, only a portion of the screw force is applied to holding the mold, while the remaining portion is used up by the balancing point of the clamp (Fig. 30.33). Even if the screw holding the clamp is very close to the mold, only 60–80% of the holding force is available. As a rule, clamps (indirect mounting) should use larger screws than would be necessary if the mold is held to the platen directly.

31 Wear and Lubrication

Both wear and lubrication are important considerations for the life of a mold. This chapter will focus on mold materials and surface finish selections, and how the mold design and the upkeep of the mold on the shop floor will affect the life of the mold.

When speaking of mold wear, there are three distinctly different areas to consider:

1. Wear caused by the flow of plastic through the runner system, including the gates, and over the molding surfaces during injection;
2. wear caused by chemical reactions; and
3. wear as a result of mold parts sliding over other mold parts during cycling of the mold.

One additional type of wear should also be mentioned, but it will not be further discussed here, and that is wear as a result of abuse caused by rough and/or improper handling of the mold, or by poorly trained operating and maintenance personnel.

Physical damage to a mold occurs frequently during start-up, shutting down, servicing and storing of the mold. Also, there may be damage caused to the mold shoe, the molding surfaces, heaters and wiring, or any mechanisms within or attached to the mold if it is badly reassembled after cleaning or other maintenance procedures.

To reduce damage to the mold from these causes, and to ensure long life of the mold, some mold makers issue a "mold manual" with each mold, including clear instruction on mold operation and maintenance. The manual may include a comprehensive troubleshooting guide, which is intended to provide the mold technician with helpful information on how to correct troubles during molding without doing damage to the mold.

31.1 Wear Caused by Resin Action

Many basic commodity plastics, such as PP, PE, and PS, and some engineering plastics, such as PC, PSU, and acrylic, are neither abrasive nor corrosive. However, the addition of colorants, fillers, and reinforcement may change this characteristic and, according to the type of additive, may make the plastic abrasive and/or corrosive, in which case it may act to wear down the molding surfaces and wear out the gates.

31.1.1 Abrasion

Abrasion is actually wearing down of the mold and its components by friction of the plastic passing through it. The speed of the hot melt passing through the gate is extremely high. Because of the fast injection speed required to fill the mold of a thin-walled product, the velocity of the flow through the gate may approach the speed of sound! At such high speeds, combined with the high temperatures generated when the hot melt is forced through the small opening, even a very low-viscosity and nonabrasive resin will gradually wear down the gate.

There are two options to consider when deciding on the material for a gate:

1. Make the mold part in which the gate is located from such a material that the heat surrounding the gate (and created there) will be easily dissipated into the coolant. The advantage gained thereby is a better (smaller and smoother) gate vestige and faster molding cycles. A beryllium-copper (BeCu) alloy is frequently used for the purpose of good heat conductivity; however, this material is not very resistant to abrasion, and the gate will increase in size much faster than in steel, and eventually cause the product to have a poor gate vestige, which may not be acceptable. If BeCu is selected, it is important that the gate is located in a (preferably simple and, therefore, low-cost) gate insert, which can be readily replaced during routine maintenance.
2. Make the gate from a very hard and abrasion-resistant tool steel, or even a tungsten-carbide steel alloy. There are, however, disadvantages of such steel gates:
   a. the heat conductivity of such materials is much lower than that of Be-Cu, and the gate area will be hotter, thereby resulting in a longer cycle time; and
   b. the manufacturing cost is higher, especially for tungsten-carbide steel parts.

Preferably, even a steel gate should be placed in a gate insert. Even a steel gate will not last forever, because of wear and, possibly, physical damage caused by cleaning during start-up. However, where it is not acceptable to have a (circular) witness line around the gate where the gate insert fits into the cavity, the whole cavity must be made from hardened steel. This is more expensive if it should be necessary to replace the gate when it is worn or damaged.

With some steels, such as H13, the gate area can be welded and remachined when worn or otherwise damaged, but this requires special skills and equipment. For other materials, welding is usually done only as a "quick fix" before replacing the cavity.

If the plastic is very abrasive, there is little option. The designer will usually have to select a very hard, steel gate insert to avoid frequent gate replacements, even if it means slower cycles and less productivity as compared to a mold with a better heat-conducting material.

31.1.2 Corrosion

Corrosion is the gradual wearing away of the mold and/or its components by chemical action, in this discussion, of the resin. A few plastics (typically, PET and PVC) are corrosive; their chemical composition is such that mold materials will be attacked when in contact with such plastics, usually by corrosive fumes released from the hot plastic when injecting.

Usually, the mold maker selects the proper mold material and/or special surface coating to protect against corrosion. Since, in many cases, the corrosive action is generated by gases of hydrochloric acid (HCl), a good solution is to make the mold entirely from high chrome (stainless) steel or, in a mold made from ordinary mold steels, to chrome plate at least all surfaces expected to be exposed to the corrosive fumes. However, over time, even chromium is attacked by HCl, and it is important that special shut-down procedures are established and followed, such as careful internal purging with a neutral resin and cleaning and protecting (spraying with a mold protection liquid) all molding and external surfaces before storing.

HCl vapors can enter into the tiniest cracks of the molding surface. In the case of plated molding surfaces, the vapors attack the chrome plating base (usually consisting of a coat of copper plating) through a virtually invisible crack and will eventually cause the chromium to lift and flake off. (A test to see if the plating is worn off is quite simple: apply a solution of copper sulfate to the suspected area. If the area turns to a copper color, the plating is gone.)

The HCl vapors also eat their way into the base steel, even if it is a chromium steel. Given enough time, such corrosion will cause a crack (stress raiser), which will eventually cause the part to fail, particularly if it is severely stressed.

Vents and vent channels, where the corrosive fumes pass with high velocity and at high temperature over the surface of the mold, are particularly at risk. If not carefully monitored for erosion and properly maintained, they may be a source of trouble, such as unacceptable flashing.

31.2 Corrosion Caused by Coolant

Coolants are usually water with some additives, mainly to remove harmful constituents of the raw water supply, and inhibit rust. Also, antifreeze solutions are added where the chiller used could be operated near or below freezing temperatures. Sooner or later, however, even with rust inhibitors, water attacks practically every mold steel, except some stainless steels.

The harmful effect of corrosion is twofold:

1. The corroded surfaces are a heat barrier, and the cooling efficiency of the mold gradually decreases, with the effect that the mold will have to be cycled slower than with the originally clean cooling surfaces.
2. Corrosion will start eating into the steel, particularly from surface scratches, which are difficult to avoid when drilling the cooling channels. (Note that gun drilling leaves a smoother hole surface than drilling with ordinary twist drills.)

It is quite common for a stressed mold part to break where the cooling channel is located near the molding surface; cutting the failed part open usually reveals a small crack that started from a surface scratch and then corroded toward the molding surface. When the area that is (cyclically) stressed by the molding process reaches its fatigue number of cycles, the part just cracks, and the first indication that something is amiss is the coolant leaking into the cavity space.

Electroplating (as used for outside (visible) surfaces) is not practical inside the cooling channels. Electroless nickel plating is quite successful for penetrating the drilled holes, but since it does not reach remote parts of the cooling system equally well, its efficiency as rust preventive is limited. (To deposit the nickel, the fluid nickel solution must be in constant motion.)

The best solution would appear to be to make the mold from stainless steels (SS). The problem with this is that there are other considerations as well, such as availability and cost (higher than steel); usually, a compromise will have to be chosen, particularly for the "stack" parts, that is, all mold parts which come into direct contact with the plastic during molding. Some mold makers now make all mold plates from SS, and many or all of the stack parts also from SS where possible.

31.3 Corrosion Caused by Atmosphere

In many localities, the ambient humidity can reach quite high values, and if the mold is cooled below the dew point, the humidity in the air will condense on the mold surfaces. This, in turn, will corrode the mold over time.

If molds plates are made from SS or from plated machinery steels, there is no problem. Some molders paint the molds from the outside to protect them against rust. This is not very good protection, and it does nothing to protect the rest of the mold parts. In this case, with high ambient humidity, it is important that all parts are frequently cleaned and well lubricated to protect them from early wear by corrosion.

31.4 Wear of Tapers

As previously discussed, tapers must be preloaded to be effective. This means that after the matching tapers first come into contact with each other, they slide with ever increasing pressure over their surfaces for the short preload distance, until they come to a stop when all parts are finally seated.

To slide two metals over each other under pressure is inherently undesirable, and should be avoided unless the contacting surfaces are well lubricated with a lubricant able to support the high pressures on the surface without being squeezed out. Lubrication will be discussed later.

(When two pieces of the same metal contact each other under great pressure, the surfaces will merge with each other. *Pressure welding* is used in some instances in the manufacturing of steel products by pressing two pieces together with extremely high impact force.)

In any mechanical device (machine, mold, etc.), any poorly lubricated (or nonlubricated) sliding of identical metals over each other under pressure will sooner or later result in a similar welding of the surfaces. This “seizing” can be seen where small pieces of one part rip out and weld to the other part. If the sliding speed is high, as is common in rotating shafts, friction between the damaged surfaces will cause these parts to heat up excessively and result in catastrophic failure.

With rotating shafts and high surface speeds, even when different metals are used, such as steel shafts and bronze bearings, seizing will occur unless the surfaces are well lubricated; that is, a film of lubricant is maintained at all times between these surfaces. In molds, the speeds are low but the specific pressures are high, and seizing will also occur.

In molds where good lubrication is not acceptable or feasible, the materials sliding over each other must be made from materials with distinctly different crystalline structure. Where the surface pressures are sufficiently low, certain plastics which do not require lubrication (Delrin®, etc.) may be used.

A common misconception is the believe that different hardness is the equivalent of different crystalline structure. There is no advantage gained as protection against seizing by using the same steel, but at different hardness, for a matching taper fit. Even if in practice matching tapers have been occasionally made from the same steel (e.g., H13), one hardened to 51 Rc and the other to 46 Rc, and they appear not to have seized, the reason is probably that either there was no preload, the tapers where greased, or the mold has not been through enough cycles to show damage.

As a general rule in mold design: Do *not* use the same steel for two parts sliding over each other under load. Use combinations such as (AISI) H13 with A2, H13 with S7, S7 with A2, and so on.

The only exception to the above rule is when a surface of a steel can be nitrided. Not every steel will accept nitriding. A typical good nitriding steel is H13. Nitriding not only creates an extraordinary hardness at the surface (as high as 70 Rc for a depth of only a few thousands of an inch) but actually penetrates the crystalline lattice on the surface of the steel and makes it difficult for a similar crystalline structure to penetrate and cause damage.

Ejector pins and sleeves, which are mass produced by mold supply houses, are often made from H13 and nitrided (and in addition, covered with a solid lubricant). This way, they can slide even for very long periods in an H13 core block without seizing. There is also very little (if any) side pressure on them against the walls of the holes where they slide, which is helpful for long life. Even so, ejector pins and sleeves do need replacement from time to time as they wear out; also, the bores in which they slide wear out with time. There are standard, "oversize" pins available just for this reason to facilitate replacement.

31.5 Lubrication in Molds

As explained above, metal parts sliding over each other require lubrication for long mold life. Health regulations require that molds which produce articles used in the food and health industry must not have lubricants any place where they could come into contact with the molded products in the mold, during ejection, or when being conveyed away from the mold.

These regulations preclude the use of lubricants in many locations in a mold where they would be required for long life, and the designer must know whether lubrication is acceptable for a mold before starting to design. Lubrication requirements can affect the selcction of the mold materials and may require special shielding to separate necessary lubricated areas from the products.

For molds not subject to such stringent regulations, certain areas are lubricated, either by hand during mold stoppage for periodic routine maintenance, or automatically while the mold is running, using pressure lubrication.

31.5.1 Leader Pins, Ball Bearings, and Bushings

Leader pins are usually greased by hand. Lubricant is wiped over the surface while the mold is open. Theoretically, they could also be pressure lubricated, but this could be quite complicated and not worthwhile. Today, for precision alignment, molds rarely depend on leader pins and bushings.

In some high-quality molds, where the alignment of the mold halves depends on leader pins, ball bushings that require only occasional oiling, or which are permanently oiled, are used instead of steel or bronze bushings, which require fairly frequent lubrication. However, such ball bushings are more sensitive to misalignment and can be destroyed rapidly if the leader pins enter with excessive force. Today, many molds do not depend on leader pins for alignment but on taper locks, and the leader pins are only used as a safety feature to protect the cores during maintenance, etc.

Sealed, lubricated ball bushings are used in some molds to guide the connecting rods between the ejector plate and the stripper rings. This ensures that the stripper ring tapers stay in position relative to the core taper and minimize the wear between these tapers that could be caused when gravity pulls the heavy stripper rings out of alignment.

In some cases, plastic bushings are also used. These bushings are usually based on the low coefficient of friction of plastics such as Teflon®, Delrin®, etc., and are used often in combination with metals such as lead or bronze. They do not require any lubrication, but they will wear, too, if used above their specified permissible loads and speeds.

Sealed, lubricated ball bearings are used where linkages in molds activate the ejection mechanism and/or separate plates during ejection.

31.5.2 Pressure or Central Lubrication

Some molds with sliding surfaces which require lubrication are built with high-pressure grease or oil lines (metal or plastic tubing, or drilled channels) leading to these surfaces. These lines come from a central distribution point from which the service technician can lubricate all points during periodic (daily, weekly, monthly) maintenance shutdowns. In some molds, the lubricated points can be connected to the central lubrication network of the molding machine.

It is easy to understand that neglecting to lubricate any area in a mold with oil or grease of the right kind will sooner or later cause damage to the unlubricated surfaces. Such a result could require extensive repairs to the mold; in other words, it will reduce the life of the mold.

31.6 Special Surface Treatments to Reduce Wear

31.6.1 Hard Chrome Plating

Hard chrome is occasionally used to increase the hardness of a through-hardened steel and to give the surface a different crystalline pattern, which will be less likely to seize with another metal. In a very thin coating ("flash chrome") of less than 0.025 mm (0.001 in.), it may be suitable for some jobs; however, in general, it is not recommended for this purpose.

A very thin flash chrome coat is often used in certain molds for thin-walled products to improve the flow of the resin over the surface during injection. From actual experience with high-speed, thin-walled, disposable products molding, the filing speed of some products has been increased by 0.2 seconds and resulted in an increase in production of 10% and more.

Hard chrome plating is commonly used in molds that process corrosive plastics. Hard chrome is very sensitive to shock; any heavily plated coat will soon crack and flake off, and the remaining very hard and rough chrome surface will damage a matching surface. Except in the case of an emergency (e.g., if necessary to continue production while waiting for a replacement part), hard chrome plating to repair wear surfaces is not recommended.

31.6.2 Nitriding (Hard or Soft)

Nitriding is a (heat) treatment of a previously hardened steel. The steel is subjected to a temperature of about 540 °C (1,000 °F). Not every steel is suitable for nitriding, and only such steels can be used which will not reduce their hardness at the nitriding temperature.

The skin is very hard and about the same thickness as flash chrome plating, but while the chrome plating adds thickness to the steel, the nitriding causes carbides to enter the steel matrix, and the part does not measurably increase in size.

31.6.3 Other Coatings

There are many other coatings for prevention of wear, all with various claims by their manufacturers. A few of these, which have been used in molds with varying results, will be named. The main problem is that they do not last as long as would be desirable, and the parts must be re-treated from time to time.

When considering the following coating products, designers and molders must differentiate between coatings applied to the molding surfaces to improve the plastic flow within the cavity space for easier filling or better release of the product, and coatings applied to lubricate mating sliding surfaces.

- Poly-Ond (nickel, phosphorus) is baked onto the mold part at about 400 °C.
- Ovonic(47) (Diamond Black™), an organic substance, is applied at about 120 °C.
- Ionotriding is nitriding at a lower temperature of about 350 °C.
- The Lindure™ process is soft nitriding at about 590 °C.
- Titanium nitrides are applied via a chemical or physical deposition process that takes place at temperatures between 700 and 1,065 °C.

Each of the above processes have their advantages and disadvantages, which must be carefully investigated to determine whether they are suitable for the material and the shape of the part. More information about these processes should be requested from the appropriate suppliers.

32 Hobbing-In of Mold Parts

32.1 Hobbing

Hobbing is a manufacturing process for producing a number of identical (usually small) cavities. A "hob" is a male punch that has the same shape as the outside of the molded product (the cavity shape); its dimensions are slightly increased to allow for the shrinkage of the plastic during molding.

In a hobbing press, the hob, which is made from a very tough steel hardened to over 62 Rc and highly polished, is pushed with forces of thousands of tons into an annealed, low carbon (mold quality) steel blank. This blank, into which the exact shape of the cavity has now been embossed, must be machined to its final size to fit the mold shoe and then case-hardened to create a hard surface to be suitable for molding. Polishing of the molding surfaces is usually not required after hobbing.

Using this method, large numbers (30 or more) of practically identical cavities can be produced from one hob. This process is of particular interest when making cavities with intricate shapes, such as longitudinal ribs, fancy engravings, etc., which are difficult (or expensive) to machine inside small cavities. Typical applications of this method are cavities for closures, for tooth paste tubes, bottles caps, electrical connectors, and other products requiring many identical cavities.

The main advantage of this process is the relatively low cost of making a large number of identical cavities. It has been used for many years and is still in use by some mold makers.

The main disadvantage is that the mold steel is soft, except for a relatively thin carburized skin (about 0.030–0.045 in. thick), which can be hardened to about 56–58 Rc. The problem is that this hard skin is not properly backed up by the underlying matrix, and under the repeated molding cycles, the steel surface will collapse when its fatigue limit is reached, not only at the parting line faces but, over time, even at the molding surfaces.

Another disadvantage is that the cavities can warp during heat treatment after hobbing, with the result that the cavities are not as "identical" as might be expected; however, such small variations are usually acceptable with many products.

Today, most such cavities are made using *electric discharge machining* (EDM). The required copper or carbon electrodes are similar in shape to the above-described hob but are much easier to make. Only one, or one small set of, electrodes can be used for a large number of cavities.

One advantage is that the cavities can be made from through-hard steel, and they usually are finish cut after the steel is hardened so that there will be no distortion of the cavity shape. The other advantage is that the molding surface is well supported by the underlying steel and not as easily damaged by the cyclical loading of the molding process. A disadvantage with this method is that the cavity finish depends on the quality of the EDM process and usually requires polishing of the molding surfaces.

In the foregoing, *desired* hobbing was discussed as a method of making molds. But in the operation of molds, there is occasionally the case where a steel surface is subjected to excessive pressures, thus creating *undesired* hobbing-in into the mold surfaces. *This chapter covers the undesired hobbing action in mold steels.*

32.2 Typical Areas Subject to Undesired Hobbing-In

Areas that are subject to hobbing-in can usually be recognized as depressed dimples in the mold surface which can be deep enough to affect the product size and/or the appearance of the molded surface. It can also be the cause of flashing at the parting line or at shut-offs within the cavity space. Typical areas include:

1. Parting line areas,
2. molding surfaces,
3. shut-offs between pins or inserts and the opposite wall or matching inserts,
4. support areas under plates or under inserts, and
5. taper seats.

32.3 Control of Hobbing

There are two methods used to control the effect of and to avoid undesired hobbing:

1. Proper selection of steel (yield strength) and
2. adequate size of the contact area.

32.3.1 Steel Selection

For most steels used in molds, the compressive strength is about the same as the tensile strength. The hardness of steel, indicated by its Rockwell (Rc) number, is about directly proportional to its tensile strength. The ratio between "hardness" (in Rc) and "tensile strength" (in psi) is about 1 to 5,000. In other words, a steel hardened to 30 Rc has a tensile strength of about 150,000 psi.

As mentioned above, a hard surface supported by a relatively soft base will collapse easier; for this reason, carburizing steels such as P5 and P6 are rarely used in molds today. For the same reason, a surface covered with (hard-) chrome or treated with a soft or hard nitriding process will not prevent the steel from hobbing-in (collapsing) when overloaded. Such surface treatments, however, may be quite useful for other reasons, such as reducing the coefficient of friction or protecting the steel from corrosion.

32.3.2 Size of the Contact Area

32.3.2.1 Venting

Venting can affect the size of the contact area. Every mold requires good venting at the parting line, so the mold faces do not touch at the parting line along the vents, vent grooves, and vent

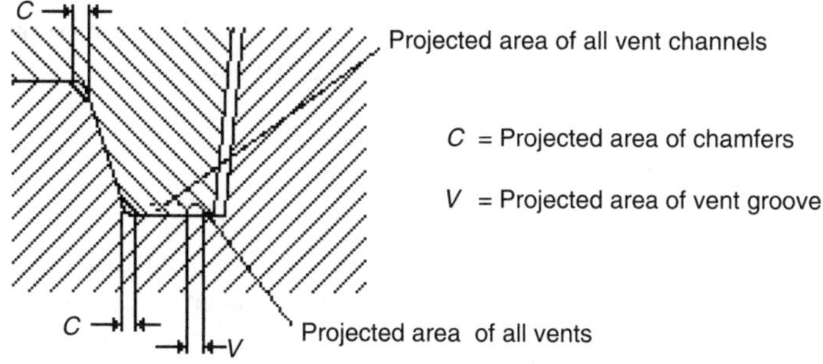

Figure 32.1 Schematic of parting line shows projected area of vent grooves, vent channels, and chamfers.

channels. These portions of the parting line will not assist in supporting the clamping force, and, therefore, the sum of the projected areas of the vents, vent grooves, and channels must be considered (i.e., subtracted) when calculating the total supporting area between cavity and core halves. Also, any loss of contact (of the supporting) areas, such as in the corners where a radius or chamfer meet, must be considered (Fig. 32.1). The sum of all noncontacting areas can be considerable in relation to the supporting areas—ignoring them may result in a major mold failure.

Since the vents (vent gaps) are always very small (sufficient to let the air pass through but not so large that the plastic can flash through), even a small collapse of the mold steel because of insufficient supporting area will eventually diminish the vent gap, with resulting malfunction of the mold and the need to repair it. Such (sometimes costly) repairs consist of regrinding of the vents and may require reseating of some tapers, thus affecting the life of the mold.

32.3.2.2 Dirt on the Parting Line

Small particles left on the parting line after flashing and improper cleaning are frequently responsible for local hobbing-in, because with repeated cycling, the dirt acts like the ball end of a ball-peen hammer repeatedly hitting the surface.

The best way to eliminate flashing is to provide sufficient clamping force and to design the mold so that the clamping force is well supported when the mold is clamped up. A serious problem arises when the mold is used in a machine with a clamp force larger than that originally intended for the mold. If the full available clamp force is then used, the mold steels can be stressed above their fatigue strength, and the mold life will be reduced, even though it was properly designed for the smaller clamp size.

32.3.2.3 Parting Line Stops

If, for reasons of available space, it is not possible to provide a sufficiently large parting line area, stops must be provided with sufficient area to make up for the lack of parting line area. The combined areas of parting line plus stops must be large enough that the permissible (fatigue) stresses are not exceeded.

These parting line stops must be designed so that, upon clamping, they contact each other at the same moment as the parting line. The material for the stops must be strong enough, and the area large enough, that the elastic limit of the stops is not exceeded, causing them to deform (collapse) permanently and, therefore, lose their effectiveness. The stops should be located symmetrically about the center of the mold. These are cyclical loads.

32.4 Taper Locks

As explained earlier, a taper lock is only effective for alignment if it is preloaded. It is important that the amount of preload is kept to a minimum to keep both the axial and the radial forces small, thereby increasing the life of the mating parts.

As the mold arrives in the "almost closed" position, just before the parting line (and any stops) begins to support the whole clamping force, the various taper lock elements (stripper rings, taper pins, wedges, etc.) are, for a short moment, subjected to a portion of the full clamping force. The actual amount of clamping force present at this moment depends on the preload and the size of the machine. To achieve full clamping force, the tie bars must be stretched according to the machine size, about 0.5–2.5 mm (0.020–100 in.), or more.

At the first contact of the mold halves with the tapers, the process of stretching the tie bars begins. This force on the tapers increases until the parting line touches.

For practical purposes, it is suggested to assume a load of 10–20% of the full available clamp force, as acting in the axial direction of the tapers. This force will drive them into the supporting plates; the design of these tapers and supports is, therefore, subject to the same considerations as when dimensioning the parting line area. If the backing of these elements is insufficient, and if the fatigue limit of the steel is surpassed, the element will gradually hob into the supporting surface, change the desired interlock dimensions, and thereby lose its usefulness for mold alignment.

32.5 Support of Mold Plates

As long as the plates are solid and facing each other or the machine platen, there is usually no problem with hobbing-in. However, a plate may be supported by parallels, by a "frame" created from a plate by removal of part of the supporting area (as in the case of hot runner plates or ejector boxes), or by pillars (either separate mold parts or cut out from the solid plate); in these cases, the total sum of the actual contacting areas must be considered by calculating the specific pressures acting on the steel. These pressures must be less than the permissible fatigue stresses.

Plate and support areas located over slots and other large openings in the platens must be considered (deducted) when calculating the total supporting area; small mounting and ejector holes in the platens can usually be ignored, unless there is a large number of these holes in the supporting area of the platen.

Example 1: A stack of plates has an area of 20×30 in., or a total area of 600 in.^{2}; the maximum clamp force is 600 tons. The specific load is

$$\frac{600 \text{ tons}}{600 \text{ in.}^2} = 1 \text{ ton/in.}^2\text{, or 2,000 psi.}$$

which is very low, and therefore presents no problem for strength.

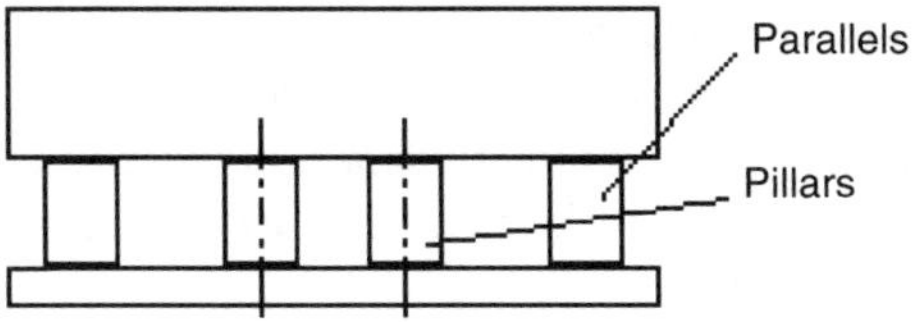

Figure 32.2 Mold stack supported by parallels and pillars. (Heavy lines indicate areas under compressive forces.)

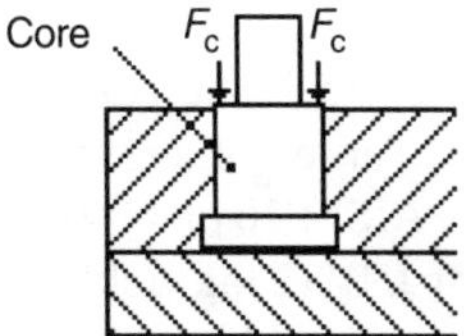

Figure 32.3 Area under compression at the base of the core on the plate (heavy line).

Example 2: Here, the stack is only supported by two parallels of 2×30 in. each, and by eight round, 2-in. diameter pillars (Fig. 32.2). The total supporting area is

$$2 \times 60 \text{ in.}^2 + 8(\pi\ 1^2) = 145.12 \text{ in.}^2,$$

and with the same maximum clamp force as in Example 1 above, the specified load is

$$\frac{600 \text{ tons}}{145.12 \text{ in.}^2} = 4.13 \text{ ton/ in.}^2, \text{ or } 8{,}269 \text{ psi.}$$

Even though this value is higher than that normally suggested (10% of yield), it has been explained (in Chapter 18, Metal Fatigue) that the fatigue limit of mild steel is (relatively) higher than that of P20 or tool steels, and will be acceptable for this application.

32.6 Support of Cavities, Cores and Other Inserts on Mold Plates

Even though, in the above examples, the selection of mild steel appears to be acceptable to support the mold plate, it may be unacceptable when considering the hobbing-in effect on the plate caused by the cavities, cores, or any other elements in the mold which are cyclically loaded during the operation of the mold.

For calculations, the designer must consider the sum of all areas of the bases of cavities, cores (Fig. 32.3), and other elements, such as locating taper pins, that are seated on a mold plate and loaded by the clamping force. It must be assumed that all these elements are seated equally; that is, they all become loaded at the same moment that the mold becomes clamped up. If not, the part loaded first will be pushed by the total available clamping force into the supporting plate and cause a permanent depression in the plate.

Dividing the maximum clamp force by the total contact areas of these elements yields the specific load on the supporting plate caused by these elements. The calculated value must still be below the permissible stress for mild steel. If not, a stronger grade of steel must be selected (e.g., AISI P20), or the seating areas of the elements must be increased to ensure that they will not hob into the supporting plate.

32.7 Influence of Elevated Temperatures

The yield strength of all mold materials (steels, etc.) diminishes as the working environment temperature increases. This is of particular importance for hot runner components. The yield strength of a machinery steel, such as AISI P20 or 4140, at 200 °C is only slightly less than its strength at room temperature; but at 450 °C, the yield strength is only about one-half of the strength at room temperature. For this reason, high-heat (called "hot work") die steels such as AISI H13 must be selected where elevated temperatures are present and the loads are high, because their yield strength does not diminish significantly until the steel is heated to temperatures above about 480 °C. In all such critical areas, the designer must ascertain proper data from suppliers data sheets.

Glossary of Molding Terminology

Time Elements in a Molding Cycle

Cycle time: The time in seconds (s) from the start of one shot to the start of the next shot. Cycle time includes the mold open time. Usual method of measuring is to start timing the stop watch when the mold reaches the mold closed position.

Mold open time (MO time): The time in seconds (s) from the moment the mold arrives in the fully open position to the moment when it starts closing for the next cycle. Ideally, a mold should run without MO time, but MO time may be necessary for ejection, or mechanical product removal.

Mold closed time (MC time): The time in seconds (s) from the moment the mold is fully clamped up, to the moment when it starts opening, for ejection of the product. MC time includes injection time, hold time and cooling time (see passive cycle, below).

Dry cycle (time): In general, it is the time in seconds (s) from the moment the mold starts closing from its fully open position, to the moment it arrives again in the fully open position after having closed completely. Times for injection, hold and cooling are not included (MC time).

Dry cycle includes the mold closing time, the time to clamp up after mold protection signals that it is OK to clamp up fully, the unclamping time (mold break time), and the mold opening time.

Dry cycle is an important criterion of the clamp performance of a molding machine. The shorter the dry cycle, the less time is wasted.

Besides the design of the clamping mechanism, dry cycle time depends very much on the distance the platen has to move. The shorter the stroke, the shorter is the dry cycle.

Passive cycle: In hydro-mechanical machines such as Husky machines using shutters, we have another consideration: When the machine is ready to open up, after the cooling cycle, the clamp force is first relaxed, which takes a fraction of a second, then, the shutters are withdrawn, which takes another fraction of a second. The total of these time elements, before the clamp starts actually opening, is called the passive cycle.

The mold remains closed until this moment, and the actual cooling time (see below) includes the time for the passive cycle. The set cooling time is then actually less than the required cooling time, by the length of the passive cycle.

On straight hydraulic machines, and on toggle machines, there is no passive cycle. As soon as the cooling time is ended, the mold starts to open up.

Injection time: The time in seconds (s) from the moment the valve admitting high pressure to the hydraulic cylinder of the injection piston (P machine) or the ram screw (RS machine) is energized, to the moment when the high pressure oil supply is shut off.

Injection hold time (usually simply called "hold" time): The time in seconds (s) from the moment the valving admitting lower pressure to the hydraulic cylinder of the injection piston (P machine) or the ram screw (RS machine) is energized, to the moment when they are shutting down the low pressure oil supply.

Hold time is required to push additional plastic into the cavity space (after the mold has been filled during the injection time) to make up for the volume lost by shrinkage of the plastic as it cools down inside the mold.

With thin walled products, there is usually little need for injection hold time, except possibly to improve the gate vestige.

Profiled injection: With machines controlled by microprocessors, injection and hold pressures can be selected to change relative to injection time and/or stroke. The pressures or velocities can be increased or decreased in steps, or at a variable rate, to "profile" the injection rate.

With older machines, injection and hold times are distinct times which are individually adjusted, for each pressure and time.

Cooling time: This is commonly identified only as the time in seconds (s) from the moment Injection Time (or, if used, Injection Hold Time) ends, until the moment the mold starts to open. (Actually, "cooling time" is the same time as the "cycle time", since the coolant flows through the mold without any interruption, regardless of whether the mold is moving, or stopped open or closed.)

Ejection time: (In seconds) The time required to push the products from the cores (or, occasionally, out of the cavities) after the mold starts opening, and to clear the molding area so that the mold can re-close without interfering with the removed products or any mechanical removal devices.

Ejection time is within the cycle time. Long ejection time can affect the cycle time seriously because it may require slowing down the opening speed of the clamp (the dry cycle), or it may require the addition of mold open time, or the lengthening of any mold open time already in the cycle.

Ideally, the ejection time should be within the mold opening time, and be completed by the time the mold starts re-closing, without requiring any mold open time.

Occasionally, it is possible to extend (overlap) the ejection time somewhat into the mold closing time, provided the products and any mechanical product removal system are out of the way before the closing mold would collide with them. If it is unsafe to do so, ejection time must take place and be completed during the mold open time.

Mold Operating Terminology

Shot weight: Total mass (weight) of all products molded during one molding cycle. (In cold runner molds, the shot weight is the sum of the mass of the products and the mass of the runners.) As a general rule, shot weight should not be less than 20% and not more than 80% of the shot capacity of the machine selected for a job.

Shot capacity: Used for defining the maximum amount of plastic that can be injected from a shooting pot (P machine) or from the ram screw (RS machine). Shot capacity is usually given as the mass in ounces (oz) or in grams (g), of polystyrene (PS), with a specific gravity of 1.05. When molding other materials, with a different specific gravity, the indicated mass must then be divided by 1.05, to get the corresponding shot volume, and then be multiplied by the value of the specific gravity for such material.

Cushion: When adjusting the length of the injection stroke of an RS machine, the stroke should be about 3/8 in. (or 10 mm) more than the actual shot volume required to fill the mold. This means that when the mold is filled after the end of the injection stroke, there will be still a small amount of plastic left in the barrel, a "cushion", at the point of the extruder screw. With amorphous plastics, (Acrylics, PC, SAN, and PS) there is less or no need for cushion since they shrink little, compared to crystalline plastics.

The reasons for this cushion are:

1. To have a reservoir for hold pressure, to make up for shrinkage.
2. It is difficult to control the stopping of the screw exactly in the right position as it retracts during plasticizing when preparing the shot for the next injection.
3. To ensure that even in case of some plastic leaking back through the check valve at the tip of the screw during injection, there will always be enough plastic to fill the mold. In a P injection system, the injection stroke of the shooting pot (which determines the shot volume) is better controlled, also, there is no check valve where the plastic could leak and cause inadequate ("short") shot volumes.

When molding crystalline materials and/or thick walled products, even a P injection system requires a cushion to have a material reserve in the shooting pot to be used for filling during hold pressure.For thin walled products, there is often no cushion required.

Starve feeding: A method of adjusting the injection system so that the shot volume is exactly the same as the volume of the cavity space. This method is only practical with a P machine where the shot volume can be exactly set for the volume required to fill the mold. The advantage of this method can be significant in thin wall molding where there is no need for hold pressure. (As the plastic is injected, it rapidly fills the cavity and almost immediately freezes the gate and the walls, so that hold pressure would not push any more plastic in the cavity space.)

As long as the cavity is not filled, the plastic exerts relatively little pressure on the mold walls. By the time the mold is full, there is no more plastic added to that already in the cavity space, so that there will be no significant force trying to open the mold against the clamping force.

The advantage of this method is that a (thin walled) product with a large projected area can be molded in a relatively small tonnage clamp. Another advantage is that molds with long, slender cores can be filled with minimal core deflection since the pressures inside the cavity space never reach the high pressures associated with ordinary molding.

Note that starve molding requires much skill in setup and operation and is not normally used by molders.

Projected area (A): Total area of product(s) and runners seen in the direction of the clamping force *F*. In hot runner molds, only the projected area of the cavities must be considered. In cold runner 2-plate molds, the projected area of the runners must be added to the area of the products, when determining the necessary clamping force. In 3-plate molds, the runners are in a different plane than the cavities. The clamping force required in this case corresponds to the larger of either the projected areas of the runner system, or the projected area of the cavities, but not the sum of these areas.

The forces due to the projected area at right (and other) angles to the clamping force is usually taken up by the strength of the mold walls. With moving side cores, the injection force on the projected area of the plastic pressing on the side cores must be resisted by wedges which lock the side cores in position when the mold is clamped up. The force of the actuator of the side core is rarely strong enough by itself to counteract the injection force.

Clamping force (F): The force that holds the mold together ("clamped up") so that the injection pressure of the plastic inside the cavity space, will not open the mold, and cause it to flash at the parting line. The required clamping force *F* can be determined by multiplying the total projected area of all cavities (plus any additional area due to cold runners—see above—with the assumed injection pressure p of the plastic, inside the cavity space, i.e., the pressure of the plastic after it has passed through the gate.

Clamp force, tie bar stretch and preload: To ensure that the mold will not open during injection, it must be held closed with a force, or "preload", greater than the necessary clamping force (*F*), see above. This preload is generated by the clamping mechanism (straight hydraulic, toggles, or hydro-mechanical) of the molding machine. When the clamp closes, the mold halves touch at the parting line. At this moment, there is still no force on the mold halves, and the tie bars are in their original, unstretched length.

As the clamping mechanism starts to exert its full force, the mold halves which had only touched are now being compressed, and force the stationary platen away from the clamping mechanism. Counteracting this motion are the tie bars which connect the stationary platen with the machine clamping mechanism. They will therefore be stretched, and provide the necessary preload to keep the mold closed during injection.

Tie bars and tie bar stretch: The tie bar stretch is proportional to the clamp force. Knowing the tie bar stretch and the tie bar size (diameter and length), the actual clamp force (preload) can be easily calculated. The tie bar size (diameter) determines the maximum possible clamp force of a machine. The specific tie bar stretch should be well below the yield point of the selected steel (less than 10%) to ensure long life (fatigue life) for the tie bars.

Excessive clamp force (over-clamping) should be avoided if:

- The compressive forces on the parting surfaces of the mold may be too high, and could damage these surfaces.
- If a small mold is used on a large machine (not generally recommended), the mounting surfaces of the machine platens can be damaged by excessive specific pressures of the metal to metal contact.
- Unnecessarily high tie bar stretch reduces the fatigue life of the tie bars.
- Excessive stretch constitutes a waste of clamping energy.

"Heavy wall" and "thin wall" products:Such definitions are arbitrary and are based on experience. A common method is uses the *L/t* ratio, which applies mainly to containers, with a more or less uniform wall thickness.

Operating window: Operating window is a term describing the permissible range of operating settings (or adjustments) of the molding machine and mold, such as temperatures (melt and cooling), pressures and cycle time elements, without significantly affecting the moldability, the sizes and the quality of the product. The operating window is influenced primarily by the mold design, the design of the hot runner system including the selection of gate size and location.

The larger the operating window, the easier it is for the molder to get the mold to produce satisfactory products. It avoids costly, lengthy or frequent fine tuning of the operating settings which are necessary with a narrow operating window. It also makes the molding operation less dependent on a stable electric power supply (voltage) and on stable cooling water temperatures.

A large operating window gives the molder more freedom in using a wider range of quality of plastic, including the use of regrinds.

Also, a large operating window can make it easier for the molder to select from a wider range of machines where such mold could be run, since there would be fewer restrictive conditions (pressures, clamp force, size and number of controls, etc.) which force the molder to use only a certain type or size machine.

Index